PROCESS AND DEVICE MODELING FOR INTEGRATED CIRCUIT DESIGN

NATO ADVANCED STUDY INSTITUTES SERIES

Proceedings of the Advanced Study Institute Programme, which aims at the dissemination of advanced knowledge and the formation of contacts among scientists from different countries.

The series is published by an international board of publishers in conjunction with NATO Scientific Affairs Division.

A	Life Sciences	Plenum Publishing Corporation
B	Physics	London and New York
C	Mathematical and Physical Sciences	D. Reidel Publishing Company Dordrecht and Boston
D	Behavioural and Social Sciences	Sijthoff International Publishing Company Leyden, The Neth. and Reading, Mass., USA
E	Applied Science	Noordhoff International Publishing Leyden, The Neth. and Reading, Mass., USA

Series E: Applied Science — No. 21

PROCESS AND DEVICE MODELING FOR INTEGRATED CIRCUIT DESIGN

edited by

FERNAND VAN DE WIELE
Professor of electrical engineering
Université Catholique de Louvain, Belgium

WALTER L. ENGL
Professor of electrical engineering
Technische Hochschule, Aachen, Germany

PAUL G. JESPERS
Professor of electrical engineering
Université Catholique de Louvain, Belgium

NOORDHOFF — LEYDEN — 1977

Proceedings of the NATO Advanced Study Institute
on Process and device modeling for integrated circuit design
Louvain-la-Neuve, Belgium
July 19-29, 1977

ISBN-13: 978-94-011-7585-2   e-ISBN-13: 978-94-011-7583-8
DOI: 10.1007/978-94-011-7583-8

© 1977 Noordhoff International Publishing.
Softcover reprint of the hardcover 1st edition 1977
All rights reserved. No part of this publication may be reproduced, stored in a retrieval system, or transmitted, in any form or by any means, electronic, mechanical, photocopying, recording, or otherwise without the prior permission of the copyright owner.

PREFACE

An Advanced Study Institute on process and device modeling for integrated circuit design was held in Louvain-la-Neuve, Belgium on July 19-29, 1977 under the auspices of the Scientific Affairs Division of NATO. The Institute was organized by a scientific organizing committee consisting of Professor F. Van de Wiele of the Université Catholique de Louvain, Professor W.L. Engl of the Technische Hochschule Aachen and Professor P. Jespers of the Université Catholique de Louvain. This book represents the contributions of the lecturers at the Institute and the chapters present a concise treatment of a very timely subject, namely, process and device modeling for integrated circuit design. The organization of the book parallels the program at the Institute with an introduction comprised of a review of modeling and basic semiconductor physics. This is followed by the chapters devoted to basic technologies, modeling of bipolar and MOS devices. The last chapter of the book presents the specific topic of process modeling.

The subject matter of this book is suitable for a wide range of interests from the advanced student, through the practising physicist and engineer, to the research worker. Although a novice may find some difficulty with the mathematical development, he can acquire a perspective into the field of process and device modeling for integrated circuit design with this book. Likewise, portions of this book may be used as a textbook since the chapters are intructional and self-contained.

The editors would like to express their appreciation to Monseigneur Ed. Massaux, Rector of the Université Catholique de Louvain, for the facilities and accomodations which made the Institute a success, and to Dr. T. Kester of the Scientific Affairs Division of NATO for his encouragement and support. The personnel of the Microelectronics Laboratory of the Université Catholique de Louvain are to be commended for their help during the Institute, in particular Prof. E. Demoulin, Dr. M. Declercq and A. Fontaine.

TABLE OF CONTENTS

Preface.	v
Section I. Introduction.	1
W.L. Engl, O. Manck, A.W. Wieder	
Device modeling	3
D.L. Scharfetter, J.G. Ruch	
Semiconductor physics and characterization of bipolar transistors	19
Section II Basic technologies and measurements	31
H.S. Rupprecht	
Diffusion phenomena in silicon	33
J.D. Meindl, R.W. Dutton, K.C. Saraswat, J.D. Plummer, T.I. Kamins, B.E. Deal	
Silicon epitaxy and oxidation	57
H.S. Rupprecht	
Ion implantation	115
D. Widmann	
Pattern generation for integrated circuit fabrication	141
W.H. Schroen	
Test structures and diagnostic techniques	173
H.S. Rupprecht	
Defect characterization	213
H.S. Rupprecht	
Measurement techniques	231
D. Widmann	
Fundamental limits in integrated circuits	261

Section III Modeling of bipolar devices   281

H.C. de Graaff
    Review of models for bipolar transistors   283
P.G.A. Jespers
    Measurements for bipolar devices   307
D.L. Scharfetter
    Bipolar transistor model for IC design   365
W.L. Engl, O. Manck, A.W. Wieder
    Modeling of bipolar devices   377
H.C. de Graaff
    High current density effects in the collector of
    bipolar transistors   419
H.C. de Graaff
    Emitter effects in bipolar transistors   443
R.W. Dutton, D.A. Divekar
    Bipolar models for statistical IC design   461
F.M. Klaassen
    Survey of I^2L modeling   519

Section IV Modeling of MOS devices   539

F.M. Klaassen
    Review of physical models for MOS transistors   541
F.M. Klaassen
    Characterization and measurements of MOST devices   573
G. Merckel, J. de Pontcharra
    Surface characterization. C-V technique   589
G. Merckel
    Surface characterization. Weak inversion   609
E. Demoulin, F. van de Wiele
    Ion implanted MOS transistors   617
G. Merckel
    Ion implanted MOS transistors. Depletion mode devices   677
W.H. Schroen
    Physical MOS models   689
G. Merckel
    Short channels. Scaled down MOSFET's   705
G. Merckel
    SOS MOSFET's   725
F.M. Klaassen
    A MOST model for CAD with automated parameter
    determination   739
G. Merckel
    CAD models of MOSFET's   751

Section V Process modeling 765

W.H. Schroen
  Process modeling 767
H.S. Rupprecht
  Process modeling 795
D.L. Scharfetter, J.G. Ruch
  Process oriented IC design 807
W.H. Schroen
  Modeling of I^2L and process selection 813
D.A. Antoniadis, R.W. Dutton
  Simulation of integrated circuits fabrication processes 837

Participants 865

Lecturers 867

Scientific organizing committee 867

Section I

INTRODUCTION

DEVICE MODELING

W.L. Engl, O. Manck, A.W. Wieder

Institut für Theoretische Elektrotechnik,
Technische Hochschule Aachen, Germany

1. INTRODUCTION

In order to fabricate a semiconductor device, the technologist needs to know a set of technological parameters, like temperatures, flow rates, implantation energies, diffusion- and implantation times and others. The device can then be characterized by a set of physical parameters like geometries, impurity profiles, mobilities, lifetimes and so on. Finally, the device will be applied by circuit designers, who are only interested in its electrical performance data, like current gains, switching times, cut-off frequencies, and others.

The relation between technological processing parameters and resulting device parameters is the topic of a relatively new research field, which we call "PROCESS MODELING" and to which the second part of this summer course is dedicated. Device modeling on the other hand - well established due to large research efforts, going on for many years - aims at relating physical device parameters to device terminal characteristics. According to the dual aspect of circuit and device simulation, there is interest in device modeling by the circuit designer who wants his device being characterized as simple and as accurate as possible in order to predict the performance of a circuit which uses this particular device. And on the other hand, the device designer seeks information about which physical parameters he should aim at to give the device the requested electrical characteristics. Historically, the development led very early to models describing terminal behavior on the basis of a few measured constants, which suffice in many instants up to now, whereas tracing terminal characteristics back to physical parameters was not equally successful. In recent years, great progress was achieved in understanding the phenomena occuring within a device by applying numerical methods to device modeling. Key words of device modeling are given in figure 1.

2. MODELS FOR CIRCUIT SIMULATION

Circuit simulation requires a model of a device which allows for calculating the terminal currents $\underline{I}(t)$ and terminal voltages $\underline{V}(t)$ as n-port quantities, the device being considered as a n-port. Several of these, together with passive components form a circuit, which is to be analyzed. The accuracy of circuit simulation depends strongly on the level of complexity of the device model. A hierarchy of models with different complexity is desired in order the simulation to be cost effective. For the purpose of circuit simulation the relation of the n-port parameters of the device to device physics and technology is of little significance.

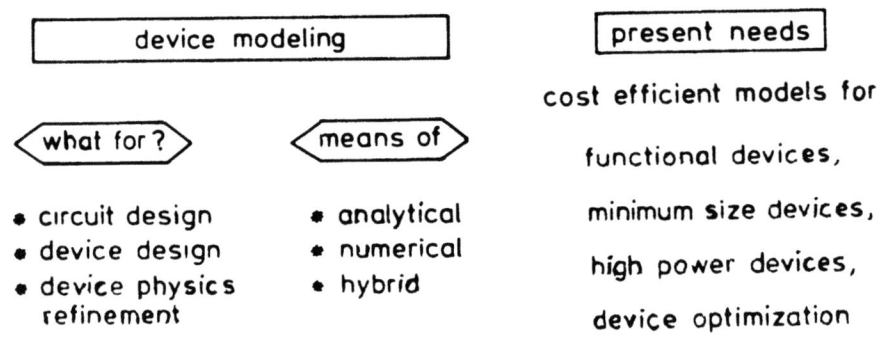

Fig. 1. Key words of device modeling [1,2]

Representative for bipolar devices, the development of transistor models will be reviewed. The basic idea of the regional approach [3] leads to an analytical solution of the field equations of the device. From this solution result models which have been expressed in terms of network element like the Ebers-Moll model (figure 2) [4,5], or they aimed to maintain close contact to physics, like the Beaufoy-Sparkes model [6]. Finally both aspects were tried to be combined in the Linvill lumped model [7] with its network like elements. In their simplest form these models are equivalent as has been shown by Hamilton et al. [8]. Based on the extrinsic Ebers-Moll model further refinements have been introduced. Current dependent current gain has been modelled by a parallel input diode [9]; current output conductance has been taken into account by introducing the Early voltage [10]. These and further refinements lead to the desired model hierarchy [1]. The price for improved accuracy is an increase in the computational time.

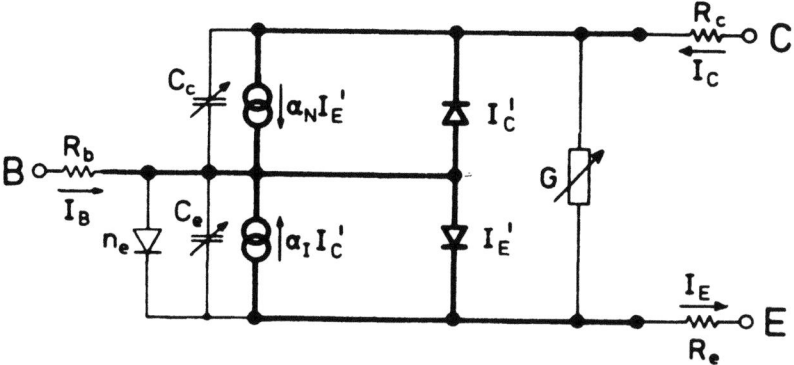

Fig. 2. Ebers-Moll-Model (EMM) for a NPN-transistor.

A parallel line of development is based on Gummel's integral charge relation [11], which relates electrical terminal characteristics to the base charge. In this approach most of the secondary effects have been regarded for and partially traced back to physical data. Only the base current is described by phenomenological fitting techniques in the Gummel-Poon model [12]. Hence this model is highly accurate over a large range of operation. Extensive measurements necessary to characterize a device by Gummel-Poon parameters and considerable computer time needed for the solution of the implicit equations counteract the improved accuracy. Thus simplified versions of the Gummel-Poon model have been derived, which finally lead back again to the basic Ebers-Moll model.

Modern network analysis programs offer an option of different models taken from these hierarchies and allow for user defined models. Simulation of steady state circuit behavior can be performed to any meaningful degree of accuracy. Transient behavior is modelled by phase shifting of controlled sources and by delay networks.

Modeling a field effect transistor leads to reduced device equations, since minority carriers often can be neglected. With Shockley's gradual channel approximation [13], a one-dimensional analytical integration yields the channel current as a function of source, drain and gate voltage. Frohman-Bentchkowsky and Vadasz [14,15] achieved improvements by taking surface and velocity saturation effects into account. For long channel devices their results are in good agreement with experiments. Circuit models developed on the same basis by Shichman and Hodges [16] have been implemented in most network analysis programs. They are simple enough, such that a relatively large number of devices can be handled by a computer. Simulating the transient behavior of long channel devices intrinsic transit times due to channel formation [17] come into play in addition to the usual RC-time constants.

Short channel devices are approximated for circuit analyse by one-dimensional models which are adapted to experimental results by physical and empirical curve fitting techniques. This is mostly being done by introducing an effective channel length, by modeling the field dependent mobility in the drain region and by taken the influence of substrate voltage on threshold voltage into account. Based on firmer physics ground are the models by El Mansy and al. [18], Merckel and al. [19] and the extension by Höfflinger and al. [20] which give very detailed results, but due to its complexity may rather be categorized as device oriented models than circuit designer oriented models.

The crucial disadvantage of models refined by curve fitting techniques is not the elaborate parameter evaluation [21] but is founded in the modeling procedure itself. The device to be simulated has to be built first and then measured. Hence the designer is very restricted in changing geometrical dimensions and technological parameters in order not to falsely predict the terminal behavior. For each new device new models have to be developed. In the process of IC-design, where circuit and device design cannot always be clearly separated, simulation of devices which have not yet been implemented is of keen interest. Hence modeling should start from geometrical, physical and technological data and allow to describe the field of the internal device quantities by solving the basic field equations. Thus useful information on internal device mechanisms is gained for the device designer. Integrating the fields over the terminals yield simultaneously the terminal characteristics for the circuit designer. For the design of functional logic, where device design and circuit design merge such a procedure seems even to be mandatory.

3. EXACT SOLUTIONS

For bipolar devices almost all model parameters except the base charge in the Gummel-Poon model are without direct physical significance. In trying to achieve analytical solutions approximamations are necessary, e.g. with respect to doping profiles, Fermi statistics, transport and recombination models [1]. These approximation yield an oversimplified picture, which is not always satisfactory. For FET's the situation is more favorable, because analytical solutions can be extended to most of the technologically relevant parameters [22,23]. Generally however special treatment is necessary for each different device and even for different ranges of operation of the same device. A universal description with analytical methods alone is not possible.

This unsatisfactory situation has first been overcome by Gummel who introduced numerical methods for solving the basic device equations [24]. Neglecting for a moment the finite computer speed we conclude that complete numerical solutions avoid all the disadvantages just mentioned. There exists a direct coupling between physical device parameters and electrical device performance. Knowing physical data such as doping profiles, mobilities, recombination rates and bandgap narrowing mechanisms, device performance can be predicted before fabrication is carried out, which is important for new devices. One can also check beforehand whether a technological improvement effort is justified by a corresponding performance gain. Finally, discrepancies between measurement and simulation allow even to detect validity limits of the device equations the simulation is based upon. In this idealized case all requirements of device modeling can be fully met.

Bipolar as well as field effect transistors are three-dimensional devices. In many cases however one dimension is large compared to the others and hence two-dimensional computations are sufficient. Many solutions for various devices MOS as well as bipolar have been published in the meantime [1]. These calculations, although carried out for only a few operating points, require much effort with regard to program structure and consume a lot of computer time. Hence systematic parameter variations are not feasible and a restriction to informative examples and a few variations is imperative. On the other hand, these model calculations allow to look into a working device and watch the different internal mechanisms. Contrary to analytical or partially numerical procedures one can exactly check which effects are dominant where and which can be neglected, respectively. From the understanding of the internal mechanisms one gains valuable hints for technological improvements and limits. Furthermore, one could

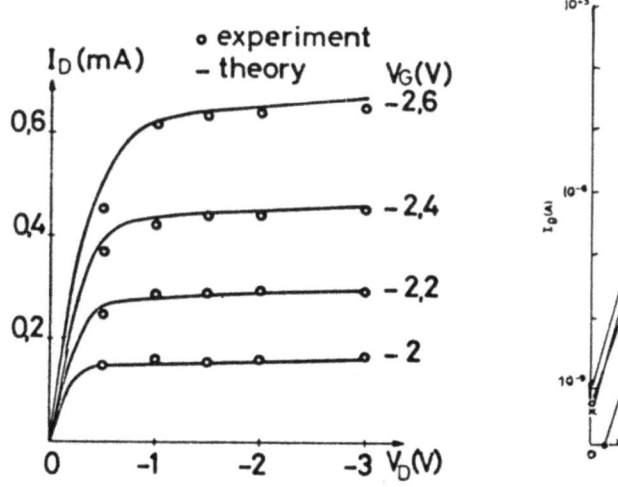
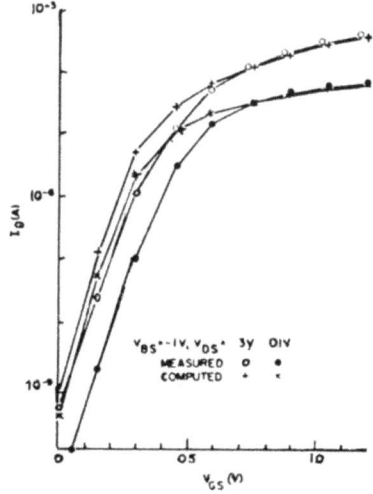

Fig. 3. Terminal characteristic of a MOSFET. Comparison of theory with experiments. a) linear plot of the saturation region [19], b) logarithmic plot of saturation and subthreshold region [25].

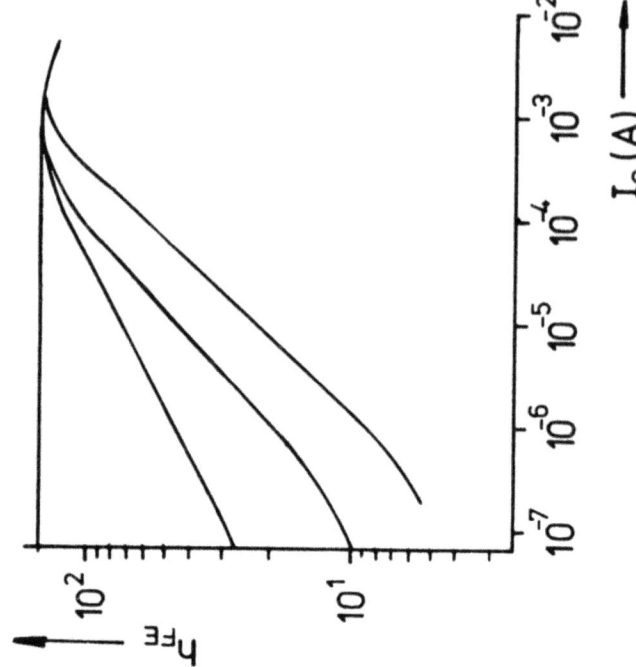

Fig. 4. Collector current of a vertical NPN-transistor. Comparison of two-dimensional calculations with experiments.

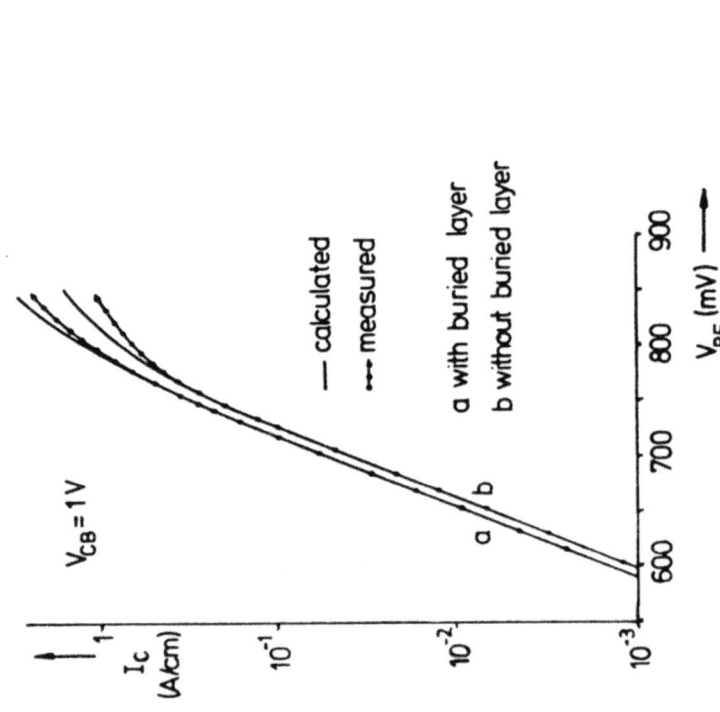

Fig. 5. Current gain of a vertical NPN-transistor versus collector current.

find starting points for approximate analytical solutions and
justify the approximations. This will be demonstrated later in
the paper on numerically based analytical modeling for well-known
transistor effects.

Numerical device modeling is so highly accurate that the
consequences of the underlying physical models can be detected
in the device characteristics. Hence the question naturally arises
to which extent the classical device equations themselves are a
good model of the physical reality and to which extent the measured
results can be matched simply by parameter variation, respectively.

One can say in general, that transport phenomena are better
modelled than recombination phenomena which still need some clarification. Therefore field effect devices are very well modelled,
as the results of Gaensslen [23], Borel [22], Hachtel [25] and
El Mansy [18] have proven. A comparism of theory with experiments
is given in the figures 3a and b. Only mobility and the onset of
velocity saturation have to be matched from experimental data.
Good agreement is obtained in the saturation region. Terminal
characteristics in the subthreshold region are given with still
sufficient accuracy.

For bipolar devices more parameters have to be matched from
measurements which opens different possibilities. The total doping
profile as well as the electrically active profile must be known
accurately for the whole device. Some regions cannot be resolved
by direct measurements, and others only with small accuracy, although various measurement techniques are available. In addition
lifetime in small regions of non uniformly doped material is not
accessible for direct measurement but can be of dominant influence
for the base current. The situation is similar for the effect of
bandgap narrowing in heavily doped silicon which still needs some
clarification especially in heavily doped emitters. Despite of
this, agreement between measured and calculated device characteristic is quite good for bipolar devices, too, as is demonstrated
by the data for the collector current (figure 4). The difference
in current for large bias will be referred to later. Base current
and current gain is simulated with still sufficient accuracy after
matching the effects of bandgap narrowing as well as bulk and surface recombination parameters (figure 5).

Exact numerical modeling is a very valuable tool to trace
back physical parameters from electrical terminal measurements by
reversely looking into the solutions. Hence this tool should contribute in the future to finding out in which instances the models
of solid state physics which we apply, need to be refined. For this
reason, we shall discuss the basic device equations.

The basis for device modeling are the continuity equations for
electrons and holes

$$\frac{\partial n}{\partial t} + \text{div } j_n = - R \qquad (1)$$

$$\frac{\partial p}{\partial t} + \text{div } j_p = -R, \qquad (2)$$

where R is the recombination rate. The current densities of electrons and holes are proportional to the gradients of the corresponding quasi-Fermi potentials $\phi_{n,p}$

$$j_n = -\mu_n \, n \, \text{grad } \phi_n \qquad (3)$$

$$j_p = -\mu_p \, p \, \text{grad } \phi_p, \qquad (4)$$

where $\mu_{n,p}$ are the carrier mobilities. Carrier concentration, electrostatic potential and quasi-Fermi potentials are related by an integral relation, where $D_{c,v}$ denotes the effective density of states in the conductivity and valence band.

$$n = \int_{E-E_c}^{\infty} \frac{D_c(E)\,dE}{1 + \exp((E + q\phi_n)/kT)} \qquad (5)$$

$$p = \int_{-\infty}^{E-E_v} \frac{D_v(E)\,dE}{1 + \exp(-(E + q\phi_p)/kT)}. \qquad (6)$$

The potential ψ can be calculated by Poisson's equation

$$\nabla^2 \psi = -\frac{q}{\varepsilon}(p - n + N_D^+ - N_A^-), \qquad (7)$$

where N denotes the net impurity concentration.

If the quasi-Fermi potententials $\phi_{n,p}$ are at most a few thermal voltages apart from the band edges the integral relations can be simplified and yield the familiar Boltzmann approximation. Bandgap narrowing caused by heavy doping effects [1,26-29] can be taken into account by a doping dependant intrinsic concentration.

$$n = n_{in}(N) \exp((\psi - \phi_n)/V_T) \qquad (8)$$

$$p = n_{ip}(N) \exp((\phi_p - \psi)/V_T) \qquad (9)$$

$$n_{ie}^2 = n_{in} \cdot n_{ip} \qquad (10)$$

This formulation also allows for the consideration of Fermistatistics and only implies the quasi neutrality condition to hold, which is generally true for heavily doped regions. Physical parameters contained in these equations are doping, mobility, recombination rate and bandgap shrinkage.

There are great difficulties to accurately measure the number of active and inactive dopants within a device but results of process modeling look very promising to overcome this restriction [30].

For the mobilities the dependance on impurity concentration due to electron-ion scattering [31] and the dependance on mobile carrier concentration due to electron-hole scattering [32] has to be taken into account. The latter especially for high injection conditions in lowly doped regions. The difference in collector current between simulation and experiment we have noticed before stems from the influence of electron-hole scattering. This effect is even more pronounced in lateral PNP's and thyristors [33]. Furthermore the dependance on electric field strength caused by velocity saturation effects has to be taken into account [31] for MOS devices and especially for transient calculations of bipolar devices. In MOS work finally the tensor property of the mobility has to be considered [34]. Although there are some questions about miniority carrier mobility [30], in general, the dependance of mobility on different parameters is numerically known with sufficient accuracy to be introduced into the calculation a priori.

With the recombination rate the situation is quite different. SRH trap recombination mechanisms are dominant in silicon and have to be applied in the bulk as well as at the surface. The parameters involved are lifetimes and recombination velocities. They are dependant on technology and are not easily accessible for direct measurement at actual device structures. In addition they are often not known with sufficient accuracy and therefore have rather the character of fitting parameters than of a priori parameters. Numerical calculations itself however suggest new measurement techniques, which have to be developed to collect the relevant parameters. First steps in this direction look quite promising [35]. Furthermore Auger recombination processes have proven to be important for devices with either heavily doped regions or very high injection conditions to happen.

Quite similar is the situation with the parameters describing the effects of heavy doping. The shrinkage of the bandgap depends on the species of the dopant, on the impurity concentration and likely on the mobile carrier concentration itself. These effects have a strong impact on device performance in steady state and its temperature dependance as well as on transient performance. However the parameters describing these effects are not known with sufficient accuracy, so that their quantitative influence is still under investigation. Although experimental data show large discrepancies recent measurements by Slotboom and al. (figure 6) lead to parameters which in turn result in good simulations as far as boron doping is concerned. But still no final conclusion seems to be possible at present especially for the most interesting regions of heavily doped emitters and subcollectors.

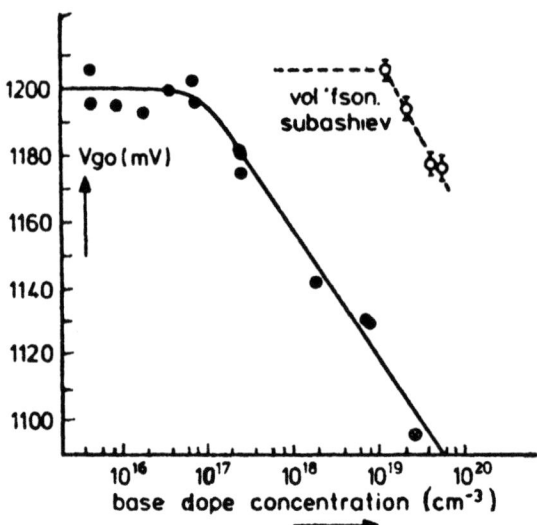

Fig. 6. The measured extrapolated bandgap narrowing as a function of impurity concentration. After J.W. Slotboom [20].

Degeneration effects in field effect devices are rather caused by strong inversion in the channel region than by doping. At sufficient high gate voltage a vertically deeper channel is formed with less surface concentration. Taking these effects into account the model of El Mansy et al. [18] characterizes device behavior over a wide dynamic range including weak channel and high level operation, using only a constant effective channel mobility.

In summary, refinement of the basic equations themselves and the functional dependence of the "constants" therein is an important goal for future device modeling. The advent of exact numerical solutions has created a sensitive tool for checking out different possibilities and they allow for an improved understanding of device performance. Finally they can lead to new measurement techniques.

4. ANALYTICAL, NUMERICAL AND HYBRID MODELING APPROACHES

Earlier it was stated, that exact numerical simulation leaves nothing to desire, if computation time is of no importance. In reality, this is not so. The computational effort imposes a rather severe restriction on the modelist, which allows to solve a problem only for a rather limited number of parameter values. Hence the ideal situation of device optimization by parameter variation is far from being true for complex two-dimensional calculations and the problem arises how to reduce the necessary effort, without loosing the correlation between physical and electrical parameters. Only if this problem can be solved, device simulation on a computer will become a useful enginering tool for the device designer.

Three approaches are possible. Firstly the problem is still tackled numerically but the equations can be simplified either by dropping some, or be reducing the dimension. Examples are some papers on two-dimensional FET's. There, the minority carrier current can be neglected in many cases and the system of 3 equations reduces to 2 equations. This system no longer contains the substrate current generated in steady state in the space charge regions or caused by charge pumping during switching transients. With this simplification the computational effort is reduced almost by a factor or two.

For a bipolar transistor, neglecting one of the transport equations is not possible, but as we have learned from the two-dimensional simulation, one-dimensional steady state solutions are quite appropriate and allow to study base push out, punch through and the influence of doping profiles and bandgap narrowing and volume recombination. Only lateral effects like current crowding, sidewall injection, current spreading, lateral discharge and surface effects require two-dimensional treatment. For one-dimensional calculations the gain in computation time is about a factor of 50 compared with typical two-dimensional calculations.

Secondly we can start with an analytical treatment of the problem. As we know already complete solutions are not achievable by this approach and one must restrict oneself to a limited range either with respect to operating points, space or time or all together. Simplifying assumptions which must be made in order to allow for solvability are either postulated or may be derived from exact numerical solutions. An example for bipolar devices is the relation for the transfer current flowing into the collector, which Gummel found by studying his numerical solutions. It is defined here for electrons threspassing a p-region,

$$J_{cc} = qn_i^2 \frac{e^{\frac{V_1}{V_T}} - e^{\frac{V_2}{V_T}}}{\int_C \frac{p}{D_n} \cdot \frac{n_i^2}{n_{ie}^2} ds}$$

where V are the voltages at the junctions and C means the path of integration following current flux lines. It holds within any region where the quasi-Fermi potential of majority carriers may assumed to be constant and the influence of recombination on minority carrier flow may be neglected. It is useful also for three-dimensional considerations, when applied to a current path assumed by intuition. Numerous applications could be given.

Modeling FET's all analytical descriptions are ruled by the gradual channel approximation, which assumes that space charge in

the channel is only a source for vertical electrical fields. With
respect to terminal characteristics this approximation is so suc-
cesful, that it is underlying all analytical investigation. A
complete description of nearly all additional terms necessary to
take care of secondary effects is given by Merckel, Borel and
Cupcea [19] and by El Mansy et al. [18], rsp. Their model and an
extension to the subthreshold region are an example how information gained by two-dimensional simulation can be used for generating simplified yet accurate models. The same holds for the JFET
model by Chiu and Gosh [36], which is based on Kennedy and
O'Briens's [37] numerical results.

Thirdly, there remains the hybrid approach which is a combination of analytical and numerical treatment of the modeling problem. With simplifying assumptions and, or analytical partial solutions simpler equations are derived from the basic ones. These
equations are in turn solved numerically. So far the hybrid approach
was applied to FET's most successfully De la Moneda [38] reduces
the problem such, that only the Poisson equation remains in two-
dimensional form, whereas the transport equations must only be
solved numerically in one dimension. For the JFET Bergmann [39]
yields only a single equation with one dimension in space for the
potential by assuming displacement current flowing only in vertical and carrier current only in lateral direction. This equation
is solved numerically for the transient behavior and the agreement
between the calculated and measured curves is good.

CCD simulation is often based on a one-dimensional equation
for the potential which was derived from the transport equation
[40,41]. This allows to describe the charge transfer only, but
this is sufficient for many applications.

For bipolar transistors, Hachtel [42] reduces the system of
3 equations to a system of two for the one-dimensional problem
without loss of generality. Other simplified but efficient models
for bipolar devices will be derived later in this text exploiting
that in certain regions the quasi-Fermi potential for majority
carriers is constant, quasi neutrality holds, and recombination
can be neglected. In the most simplified case, only the equation
for the minority carriers is to be solved [43].

5. CONCLUSIONS

In conclusion, the modern modelist must dig into basic solid state
physics, in order to know how to refine the device equations. For
predicting terminal characteristics the computational effort of
exact numerical calculations especially if carried out in more than
one dimension or in the transient mode of operation is rather high.
However, the hybrid approach allows to generate a hierarchy of models
with different trade offs with respect to accuracy and computational
effort. From this hierarchy the user may select the most appropriate
model for his application. The modelist's dream are predictions of

device behavior, for devices which have not yet been built. In order to make this dream come true new measurements have to be devised which allow to evaluate the parameters for the basic equations more precisely. The final goal is to replace fitting parameters by a priori measured parameters which may farther in the future be the result of process modeling.

REFERENCES

1. Engl, W.L., Manck, O., Wieder, A.W., Device Modeling, (Solid State Devices 1975, Journal de Physique, Paris) p.1. This lecture is an updated version of 1.
2. Agajanian, A.H., A Bibliography on Semiconductor Device Modeling, Solid-State Electronics. 18 (1975) 917.
3. Shockley, W., The Theory of p-n Junction in Semiconducters Junction Transistors, Bell Syst. Techn. J., 28 (1949) 435.
4. Ebers, J.J. and Moll, J.L., Large Signal Behavior of Junction Transistors, Proc. IRE 42 (1954) 1761.
5. Moll, J.L., Large-Signal Transient Response of a Junction Transistor, Proc. IRE 42 (1954) 1773.
6. Beaufoy, R. and Sparkes, J.J., The Junction Transistor as a Charge-Controlled Device, Automatic Telephone a. Electric Com. J. 4 (London) 1957, p. 310.
7. Linvill, J.G., Lumped Models of Transistors and Diodes, Proc. IRE 46 (1958) 1141.
8. Hamilton, D.J., Linholm, F.A., Marshak, A.H., Principles and Application of Semiconductor Device Modeling, (Holt, Rinehart and Winston, Inc., New York) 1971.
9. Johnson, E.D., Kleiner, C.T., McMurray, L.R., Steel, E., Vassolla, F.A., Transient Radiation Analysis by Computer

Program (TRAC)1(Users Guide, Harry Diamond Labs, Autonetics Div., North American Rockwell, Anaheim, Cal.) 1968.
10 Early, J.M., *Effects of Space-Charge Layer Widening in Junction Transistors*, Proc. IRE 40 (1952) 1401.
11 Gummel, H.K., *A Charge Control Relation for Bipolar Transistors*, Bell Syst. Tech. J. 49 (1970) 115.
12 Gummel, H.K., Poon, H.C., *An Integral Charge Control Model of Bipolar Transistors*, Bell Syst. Tech. J. 49 (1970) 827.
13 Shockley, W., *A Unipolar Field-Effect Transistor*, Proc. IRE 40 (1952) 1365.
14 Frohman-Bentchkowsky, D., Vadasz, L, *Computer-Aided Design and Characterization of Digital MOS Integrated Circuits*, IEEE J. Solid State Circuits SC-4 (1969) 57.
15 Frohman-Bentchkowsky, D., Grove, A.S., *Conductance of MOS Transistors in Saturation*, IEEE Trans. Electron Devices ED-16 (1969) 108.
16 Shichman, H., Hodges, D.A., *Modeling and Simulation of Insulated-Gate Field-Effect Transistor Switching Circuits*, IEEE J. Solid State Circuits SC-3 (1968) 285.
17 Goser, K., *Channel Formation in an IGFET and its Equivalent Circuit for CAD*, IEEE Int. Solid State Conf., Philadelphia (1970) Dig. Tech. Pap. p. 98.
18 El Mansy, Y.A., Boothroyd, A.R., *A New Approach to the Theory and Modeling of Insulated-Gate Field-Effect Transistors*, IEEE Electron Devices, ED-24, No. 3, pp. 241-253, 1977.
----, *A Simple Two-Dimensional Model for IGFET Operation in the Saturation Region*, IEEE Electron. Devices, ED-24, No. 3, pp. 254-262, 1977.
19 Merckel, G., Borel, J., Cupcea, N.Z., *An Accurate Large-Signal MOS Transistor Model for Use in Computer-Aided Design*, IEEE Trans. Electron Devices ED-19 (1972) 681.
20 Schemmert, W., Gabler, L., Hoefflinger, B., *Sub-Threshold and Active Region Characterization of Ion-Implanted Buried-Channel MOS Transistors*, IEEE Int. Electron Devices Meeting, Washington D.C. (1974) Dig. Tech. Pap. 546.
21 Rohrer, R., Fan, S.P., Claudio, L., *Automated Bipolar Junction Transistor DC Model Parameter Determination*, IEEE J. Solid State Circuits SC-6 (1971) 260.
22 Borel, J., Merckel, G., Monnier, J., Saintot, P., Vandorpe, I., Stern, M., Maffei, M., *A Connection between Technology and Models using a Computer Analysis*, IEEE Int. Solid State Circuits Conf., Philadelphia (1973) Dig. Tech. Pap. 102.
23 Gaensslen, F.H., Le Blanc, A., *Simulation of Micron MOS Device Characteristics*, Eur. Solid-State Devices Conf., Munich (1973).
24 Gummel, H.K., *A Self-Consistent Iterative Scheme for One-Dimensional Steady State Transistor Calculations*, IEEE Trans. Electron Devices ED-11 (1964) 455.
25 Hachtel, G.D., Mack, M.H., *A Graphical Study of the Current Distribution in Short-Channel IGFETS*, IEEE Int. Solid-State Circuits Conf., Philadelphia (1973) Dig. Tech. Pap. p. 110.

26 Van Overstraeten, R.J., De Man, H.J., Mertens, R.P., Transport Equations in Heavily Doped Silicon, IEEE Trans. Electron Devices ED-20 (1973) 290.
27 Mock, M.S., Transport Equations in Heavily Doped Silicon, and the Current Gain of a Bipolar Transistor, Solid State Electron. 16 (1973) 1251.
28 Slotboom, J.W., The pn-product in Silicon, Solid State Electronics, 20 (1977) 279.
29 Jacobini, C., et al., A Review of some Charge Transport Properties of Silicon, Solid State Electron. 20 (1977) 77.
30 NATO Advanced Study Institute on Process and Device Modeling for IC Design, Louvain la Neuve, Belgium, July 1977.
31 Caughey, D.M., Thomas, R.E., Carrier Mobilities in Silicon Empirically Related to Doping and Field, Proc. IEEE 55 (1967) 2192.
32 Dannhäuser, F., Die Abhängigkeit der Trägerbeweglichkeiten in Silizium von der Konzentration der freien Ladungsträger, I, Solid State Electron. 15 (1972) 1371.
33 Anheier, W., Engl, W.L., Manck, O., Wieder, A.W., Rigorous Numerical Analysis of a Planar Thyristor, Int. Electron. Devices Meeting, Washington D.C. (1975) Dig. Techn. Pap. 363.
34 Gautier, J., Borel, J., Merckel, G., Short Channel MOSFET Modeling, Eur. Solid State Devices Conf., Nottingham (1974) Dig. Tech. Pap.
35 Flocke, H., Manck, O., Engl, W.L., Determination of Surface Parameters with Field-Induced Junctions, to be presented at the ESSDERC 1977 in Brighton, England.
36 Chiu, T.L., Gosh, H.N., Characteristics of the Junction-Gate-Field-Effect Transistor with Short Channel Length, Solid State Electron, 14 (1971) 1307.
37 Kennedy, D.P., O'Brien, R.R., Computer Aided Two-Dimensional Analysis of the Junction Field-Effect Transistor, IBM J. Res. Dev. 14 (1970) 95.
38 De la Moneda, Threshold Voltage from Numerical Solution of the Two-Dimensional MOS Transistor, IEEE Trans. Circuit Theor. CT-20 (1973) 666.
39 Bergmann, G., Die Dynamik des Sperrschichtfeldeffekttransisters, Thesis, Universität Stuttgart, Germany, 1975.
40 McKenna, J., Schryer, N.L., Analysis of Field-Aided, Charge-Coupled Device Transfer, Bell Syst. Techn. J. 54 (1975) 667.
41 Mohsen, A.M., McGill, T.C., Mead, C.A., Charge Transfer in Overlapping Gate Charge-Coupled Device, IEEE J. Solid State Circuits SC-8 (1973) 191.
42 Hachtel, G.D., Joy, R.C., Cooley, J.W., A New Efficient One-Dimensional Analysis Program for Junction Device Modeling, Proc. IEEE 60 (1972) 86.
43 Dunkley, J.L., Kang, S.D., Nygaard, P.A. Modular Bipolar Analysis, IEEE Int. Electron Dev. Meeting, Washington D.C. (1976) Dig. Tech. Pap. 312.

Semiconductor Physics and Characterization of Bipolar Transistors

D. L. Scharfetter
J. G. Ruch

Bell Laboratories
Murray Hill, New Jersey 07974

Introduction

This paper reviews briefly the fundamental equations which are used by semiconductor device analysts to predict bipolar transistor behavior.[1,2,3] The computer program which obtains the numerical solution of these equations is referred to as a transistor simulator. These simulators, computer programs which calculate device characteristics as determined by materials properties and impurity profiles, can be employed directly in the performance evaluation of circuit designs. To be practical, two-dimensional effects must be included while simulation costs must not be prohibitive. The program TRANSIM[4] (TRANsistor SIMulator), which will also be described in this paper, employs a highly efficient algorithm specialized for bipolar transistor simulation, which reduces simulation costs for one-dimensional analysis by an order of magnitude over previously published approaches.[3] Two-dimensional effects are approximated without the complexity of a full two-dimensional analysis.

Charge Transport Equations[2,3]

The distribution and motion of carriers within a one-dimensional semiconductor device structure can be obtained by solving three basic equations: (1) the continuity equation for holes, (2) the continuity equation for electrons, and (3) Poisson's Equation:

$$\frac{\partial p}{\partial t} = g - \frac{1}{q} \frac{\partial J_p}{\partial x} , \quad (1)$$

$$\frac{\partial n}{\partial t} = g + \frac{1}{q} \frac{\partial J_n}{dx} , \quad (2)$$

and

$$\frac{\partial E}{\partial x} = \frac{q}{\epsilon}(p-n+N_D-N_A) \tag{3}$$

where

$$J_p = q\mu_p pE - kT\mu_p \frac{\partial p}{\partial x} \tag{4}$$

$$J_n = q\mu_n nE + kT\mu_n \frac{\partial n}{\partial x} \tag{5}$$

Boundary conditions are imposed at the contacts by introducing the appropriate restrictions in Eqs. (1) through (5). For example, current boundary conditions for a P-N device are introduced by requiring that (5) is equal to the terminal current density at the N contact and that (4) is equal to the terminal current density at the P contact. Voltage boundary conditions are imposed by requiring that the integral of E(x,t) over the interval between two contacts equals the voltage. In addition, the electric field at the metallic contacts is assumed to be zero. Initial values for the hole and electron densities are given by the quiescent zero bias solution, i.e., the solution when

$$J_n = J_p = 0 \quad \text{and} \quad \frac{\partial p}{\partial t} - \frac{\partial n}{\partial t} = 0. \tag{6}$$

In general, the functions g, μ_p, μ_n vary at each point in the device according to the value of n, p and E at that point.

Carrier Generation-Recombination

The carrier generation term g is composed of two components: (1) carrier generation and recombination through defects, and (2) impact or avalanche ionization.

Hole electron generation and recombination through defects are represented by a Shockley-Read-Hall (SRH) single level model[17] which characterizes defects with neutral and single charge states. The generation-recombination rate through a single level center is given by

$$g_d = \frac{pn - n_i^2}{\tau_{po}(n+n_1) + \tau_{no}(P+P_1)} \tag{7}$$

The impact ionization rates are strongly dependent upon the electric field intensity and the hole and electron current densities. These generation terms are given by

$$g_I = \frac{1}{q}\left[\alpha_n(E)|J_n| + \alpha_p(E)|J_p|\right] \tag{8}$$

where $\alpha_n(E)$ and $\alpha_p(E)$ are given by the relations,[3]

$$\alpha_n(E) = 2.25 \times 10^7 \exp(-3.2 \times 10^6/E) \tag{9}$$

$$\alpha_n(E) = 3.80 \times 10^6 \exp(-1.75 \times 10^6/E) \tag{10}$$

3. Mobility Expressions

It is necessary to include in the analysis the field (E) and the ionized impurities density (N_D).

The theoretical mobility is approximately by the following expression[3]

$$(\mu_0/\mu)^2 = 1 + \left[\frac{N_D}{N_D/S+N}\right] + \left[\frac{(E/A)^2}{E/A+F}\right] + (E/B)^2$$

	μ_0	N	S	A	F	B
HOLES	480	4×10^{16}	81	6.1×10^3	1.6	2.5×10^4
ELECTRONS	1400	3×10^{16}	350	3.5×10^3	8.8	7.4×10^3

Solution Procedures

Because of the nonlinearities in the equations describing the hole, electron and field distributions, obtaining a transient or even a steady state dc solution poses a very difficult numerical problem. The structure to be analyzed is first subdivided into a number of small cells. The equations are then normalized to reduce redundant coefficient calculations and standard difference approximations are used to approximate the spatial derivatives in Poisson's Equation and the continuity equations for each increment:

$$\frac{dp}{dt}(N) = g(N) - \left[J_p(M) - J_p(M-1)\right]/\Delta x \tag{13}$$

$$\frac{dn}{dt}(N) = g(N) - \left[J_n(M) - J_n(M-1)\right]/\Delta x \tag{14}$$

$$\frac{E(M) - E(M-1)}{\Delta x} = \frac{q}{\epsilon}\left[p(N) - n(N) + N_D(N) - N_A(N)\right] \tag{15}$$

where the Mth mesh point is located midway between the major mesh points N + 1 and N.[3] It is customary to next employ the standard difference approximations in the current density expressions (4) and (5) and substitute these results in (13) and (14). However, it can be shown that this procedure leads to numerical instability whenever the voltage change between mesh points exceeds $2kT/q$.[3] Rather, (4) and (5) are treated as differential equations in p and n with J_n, J_p, μ_p, μ_n and E assumed constant between mesh points. The solutions of these differential equations then relates J_n and J_p to the other variables:[3]

$$J_p(M) = E(M)\left[\frac{P(N)\mu_p(M)}{\left[1.0 - exp(-E(M)\Delta x)\right]} + \frac{p(N+1)\mu_p(M)}{\left[1.0 - exp(E(M)\Delta x)\right]}\right] \tag{16}$$

$$J_n(M) = E(M)\frac{n(N+1)\mu_n(M)}{\left[1.0 - exp(-E(M)\Delta x)\right]} + \frac{n(N)\mu_n(M)}{\left[1.0 - exp(E(M)\Delta x)\right]} \tag{17}$$

These equations provide numerically stable estimates of the current density under all conditions; if the intermesh point voltage is small, these equations approach the standard difference relations; whereas when the voltage change is large, they approach the drift current density at either mesh point N or N + 1.

A Transistor Simulator[4]

The expanding use of computer simulation methods for IC analysis and design has created a need for numerous and accurate transistor models parameters. The conventional approach, namely extraction by curve fitting, can be laborious and may not be suitable while a new technology is evolving. Transistor simulation programs, including full two-dimensional analysis capability (see references) are expensive to run. These programs accept as input structure dimensions, impurity profiles, and materials parameters. This section describes a transistor simulator TRANSIM,[4] developed to allow rapid and inexpensive characterization of bipolar devices. The computer program TRANSIM (TRANsistor SIMulator) employs a highly efficient algorithm specialized for one-dimensional bipolar transistor analysis. Two-dimensional effects are included by a combination of several one-dimensional calculations. Device characteristics are obtained by combining the electrical characteristics of the various elements; they include characteristics and effects associated with a) the parasitic collector diode due to unequal collector and emitter area; b) the parasitic emitter diode due to the emitter side-wall effect and c) emitter crowding. Provision can be easily made for the automatic generation of model parameters suitable for use in circuit simulation.

Mathematical Model and Solution

The mathematical model, although simpler than the one described above which uses the general transport/continuity/space charge equations has, in general the same accuracy and detail in its solutions. This is achieved by utilizing certain characteristics of the solutions common to most bipolar transistor structures. Briefly, these are that: a) the recombination current is small relative to the electron current (NPN structure) and can be obtained by self-consistent perturbation methods from a recombinationless solution and b) the charge-control relation developed by Gummel[5] is exact for a recombinationless model and can be utilized to facilitate convergence. As a result, from the approximation a), the hole quasi-Fermi level through the structure is set constant before the recombination perturbation is applied and the mathematical model reduces to a system of two coupled nonlinear equations:

1)   1) The electron transport equation:

$$J_n = - q\mu_n(N_t, E, T) n \nabla \delta_n \qquad (18)$$

2)   2) Poisson's equation:

$$-\nabla^2 \phi = \frac{q}{\epsilon} (p - n + N_D - N_A) \qquad (19)$$

where: p, n, N_o, N_A have the usual meaning and ϕ_n is the quasi-Fermi level for the electrons and μ_n is the (impurity (N_t), electric field (E), and temperature (T) dependent) mobility.

Numerically, these equations reduce to a set of nonlinear algebraic equations which are solved by a Newton iteration. An important aspect of our formulation is that the coefficient matrix has a four-diagonal sparsity structure, as compared to a seven-diagonal one for a full one-dimensional analysis. Since most of the computing time is spent solving the linear system of equations, for each iteration step, the smaller coefficient matrix makes TRANSIM most efficient. A factor of about 10 reduction in cost compared to a full one-dimensional analysis program[3] has been observed.

The recombination current is obtained by perturbing the zero-order solutions (described above) for the holes, electrons and potential distribution. To obtain the true solution for holes (NPN structure) with finite recombination, we calculate the deviation of the hole quasi-Fermi level in the emitter region from its assumed constant value V_{BE} through the device. This requires solving an equation of the form:

$$\frac{d}{dx}(e^{\delta}\phi p) = \frac{1}{D_p p_O} \int_O^X p_O \frac{e^{\delta}\phi p}{\tau_p} dx$$

where

$\delta\phi_p$ is the deviation of the hole quasi-Fermi level
p_O is the zero-order hole concentration
D_p is the normalized diffusion constant.

Because p << n in the emitter region, the Shockley-Read-Hall model for recombination, used to derive the above expression could be greatly simplified. In the base region, the full Shockley-Read-Hall expression is used since electron and hole concentrations can be comparable at high injection. The maximum deviation in the calculation of the recombination current, using this approach, from the self-consistent one-dimensional analysis is less than 0.5% for devices having current gain larger than 50. The weakness of this approach is that the base current cannot be specified as an input. Only the two pairs; (V_{CE}, V_{BE}) and (V_{CE}, I_c) can be specified as an input; the pair (V_{BE}, I_c), because of the construction of the initial guess solution is not suitable.

The convergence to the solution is facilitated by using a trial solution, constructed according to a set of approximate, basically analytical solutions. These solutions are determined by the driving conditions: when the pair (V_{BE}, I_c) is specified, the base-collector junction is depleted to satisfy V_{CE} and the electron distribution is obtained by a) filling the depletion region with enough electrons so that, when moving with the saturation velocity, they create a current I_c and b) distributing the electrons through the base so that their gradient generates a diffusion current I_c and finally c) neutralizing the net impurity concentration in the collector and emitter with electrons. Consequently, TRANSIM converges rapidly to any given bias state automatically without the user having to resort to time evolution tricks necessary with other programs.[3]

Model for Two-Dimensional Structures

Quasi-two-dimensional models have been developed in the past[2,6,7] to take into account emitter crowding. They consist of a number of one-dimensional transistor models coupled with lateral base resistances. The basic assumption is that apart from the base current, the transistor behavior is one-dimensional. In fact, other two-dimensional effects occur in a structure as shown in Fig. 2. For high current densities, when base widening occurs, current spreading takes place and emitter side-wall injection as well as base-collector parasitic diode effects also drastically influence transistor characteristics. Collector current spreading is not directly accounted for in TRANSIM. Device characteristics effected by spreading ar e b and f_T fall-off at high currents, and would require a full two-dimensional calculation to be predicted from first principles.

The Program TRANSIM dissects the transistor structure into basically three types of elements: a) an ideal transistor, b) a diode connected between base and emitter representing the emitter side-wall effect and c) a diode representing the collector-base parasitic diode.

In the calculation of the ideal part of the transistor, the structure is divided into a series of one-dimensional transistors interconnected by a conductivity-modulated resistance. Similarly, the side wall is divided into segments; for each segment, the program calculates the doping profile and derives the electrical characteristic. The axial component of the injected current is assumed to be collected by the collector while the lateral component contributes to the base current. In inverse operation the base-collector parasitic diode is taken into account. The program combines the electrical contribution of the various elements to obtain the electrical characteristic in forward and inverse operation for the equivalent structure.

Input/Output

The inputs to the program are the doping profiles through the active and inactive part of the device (Sections A-B and C-D in Fig. 2) as well as the base and emitter area and perimeter. The temperature and a recombination parameter (which is the only adjustable parameter) also have to be specified.

Various output options are desirable. For a given bias condition, the simulator predicts the currents and/or voltages as well as the spatial distribution of holes, electrons, quasi-Fermi level, and recombination rates. On the other hand, provisions can be made for the automatic generation of models parameters suitable for use in circuit simulators. In this case it is desirable to have this program generate internally various characteristics, such as $I_c - V_{BE}$, $I_B - V_{BE}$, $\tau - V_{BE}$, and the junction capacitance-voltage characteristics. Parameter extraction is performed by curve fitting the appropriate characteristics.

As an illustrative example shown, in Fig. 3 is a comparison between simulation and experimental measurement of the gain-current characteristic. The two curves correspond to two different epitaxial layer thicknesses. Similar agreement is obtained when comparing transistors having different emitter stripe aspect ratios.

References

1. H. K. Gummel, "A Self-Consistent Interactive Scheme for One-Dimensional Steady-State Transistor Calculations," IEEE Trans. Electron Devices, Vol. ED-11, pp. 455-465, 1964.
2. C. W. Gwyn, D. L. Scharfetter and J. L. Wirth, "The Analysis of Radiation Effects in Semiconductor Junction Devices," IEEE Trans. on Nuclear Science, Vol. NS-14, pp. 153-169, 1967.
3. D. L. Scharfetter and H. K. Gummel, "Large-Signal Analysis of a Silicon Read Diode Oscillator," IEEE Trans. Electron Devices, Vol. ED-16, pp. 64-77, 1969.
4. J. G. Ruch, D. L. Scharfetter, "Characterization of Bipolar Devices," 1973 Internal Electron Devices Meeting, Washington, D.C., December, 1973.
5. H. K. Gummel, "A Charge Control Relation for Bipolar Transistors," BSTJ, vol. 49, pp. 115-120, Jan., 1970.
6. V. A. Dhaka, "Distributed Model for High-Frequency Bipolar Transistors," Internation Elec. Devices Meeting, Washington, D.C., Paper 5.1, Oct., 1965.
7. H. N. Gosh, P.H. de la Moneda, and N. R. Dono, "Computer-Aided Transistor Design Characterization and Optimization," Solid State-Elec., Vol. 10, pp. 705-726, 1967.
8. A. DeMari, "An Accurate Numerical Steady-State One-Dimensional Solution of the P-N Junction," Solid-State Electronics, Vol. 11, pp. 33-58, 1968.
9. B. V. Gokhale, "Numerical Solutions for a One-Dimensional Silicon N-P-N Transistor," IEEE Trans. Electron Devices, Vol. ED-17, pp. 594-602, 1970.
10. G. D. Hachtel, R. C. Joy and J. W. Cooley, "A New Efficient One-Dimensional Analysis Program for Junction Device Modeling," Proc. IEEE, Vol. 60, pp. 86-98, 1972.
11. O. G. Petersen, R. A. Rikoski and W. W. Cowles, "Numerical Method for the Solution of the Transient Behavior of Bipolar Semiconductor Devices," Solid-State Electronics, Vol. 16, pp. 239-251, 1973.
12. J. W. Slotboom, "Interactive Scheme for 1- and 2-Dimensional D.C. Transistor Simulation," Electronics Letters, Vol. 5 pp. 47-50, 1971.
13. M. S. Mock, "A Two-Dimensional Mathematical Model of the Insulated-Gate Field Effect Transistor," Solid-State Electronics, Vol. 16, pp. 601-609, 1973.
14. J. W. Slotboom, "Computer-Aided Two-Dimensional Analysis of Bipolar Transistors," IEEE Trans. Electron Devices, Vol. ED-20, pp. 669-679, 1973.
15. H. H. Heimeier, "A Two-Dimensional Numerical Analysis of a Silicon N-P-N Transistor," IEEE Transactions on Electron Devices, Vol. ED-20, pp. 708-714, 1973.
16. O. Manck, H. H. Heimeier, and W. I. Engl, "High Injection in a Two-Dimensional Transistor," IEEE Trans. on Electron Devices, Vol. ED-21, pp. 403-409, 1974.

17. O. Manck and W. L. Engl, "Two-Dimensional Computer Simulation for Switching a Bipolar Transistor Out of Saturation," IEEE Transactions on Electron Devices, Vol. ED-22, pp. 339-347, 1975.
18. W. Shockley and W. T. Read, Jr., "Statistics of the Recombination of Holes and Electrons," Physical Review, Vol. 87, pp. 835-842, 1952; R. N. Hall "Electron-Hole Recombination in Germanium," Physical Review, Vol. 87, p. 387, 1952.

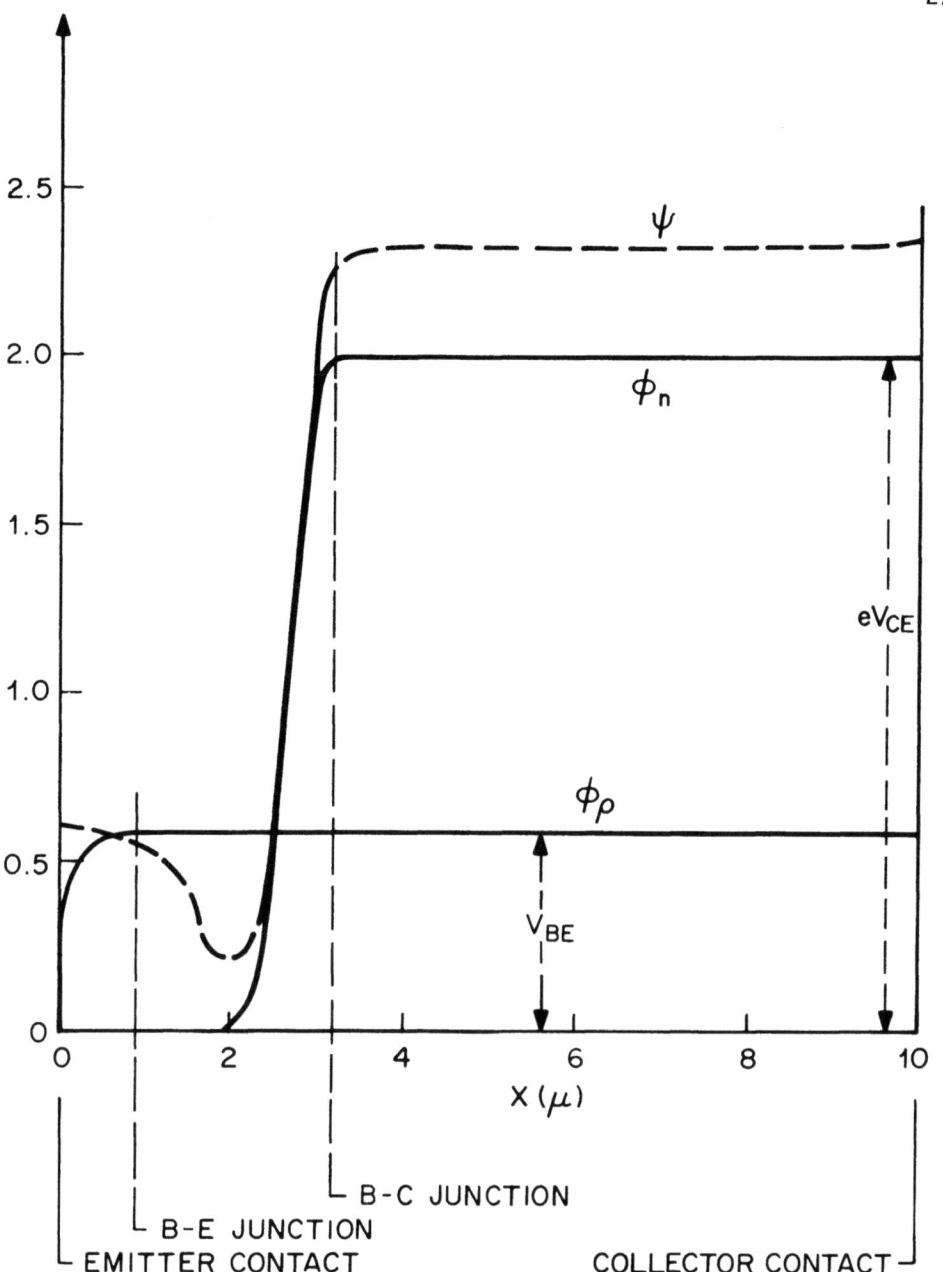

FIG. 1 QUASI-FERMI LEVELS AND ELECTROSTATIC POTENTIAL FOR AN NPN TRANSISTOR

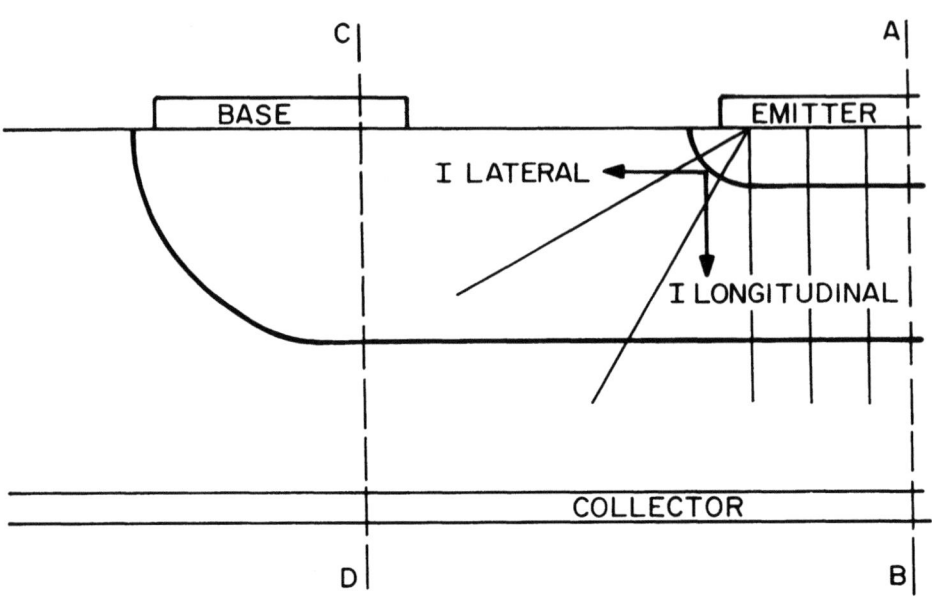
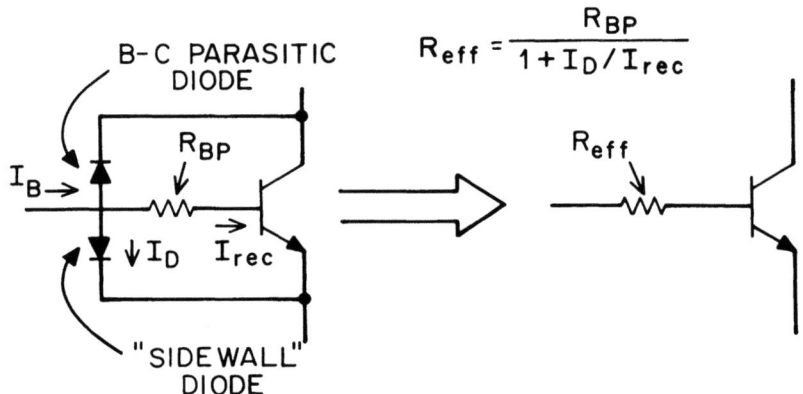

FIG. 2 GEOMETRICAL STRUCTURE FOR AN NPN TRANSISTOR

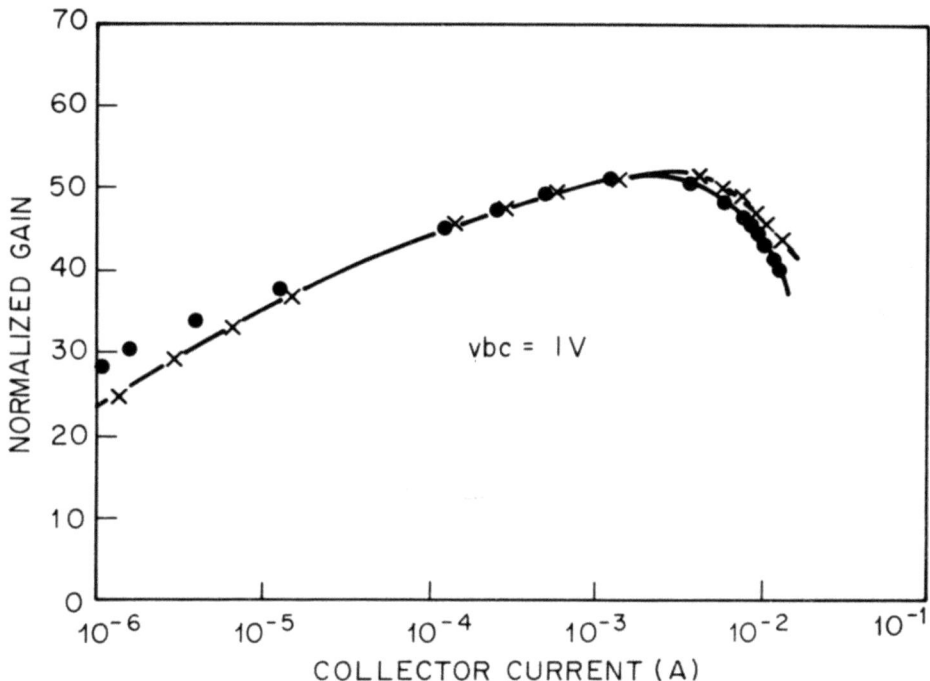

FIG. 3 CURRENT GAIN vs. COLLECTOR CURRENT, EXPERIMENT AND MODEL (SOLID)

Section II

BASIC TECHNOLOGIES AND MEASUREMENTS

DIFFUSION PHENOMENA IN SILICON

H. S. Rupprecht

IBM System Products Division-East Fishkill
Hopewell Junction, New York 12533 USA

ABSTRACT. Onsager's treatment of irreversible processes leads
to a generalized, phenomenological theory of diffusion, which
can be used to describe complex, interactive diffusion systems.
This concept will be applied to a substitutional diffusion
model, as it is valid for impurities such as As, P, and B, and
a correlation will be established between phenomenological and
atomistic parameters. Finally, recent results illustrating
means of modifying the temperature dependence of thermal
diffusion by radiation enhancement are discussed.

1. INTRODUCTION

The progress in the area of silicon device technology has been
aided by several favorable circumstances. One is the discovery
that natural oxides grown on silicon surfaces can be used as
effective masks in thermal diffusions. This holds true
particularly for those doping elements in groups III and V of
the periodic chart that also have the highest solubility
in silicon, such as As, P, and B.

The masking behavior combined with advanced lithographic
techniques has permitted selective doping and, hence, opened
the way for large-scale integration and all its technological
ramifications.

It is basically for these reasons that diffusion has developed
into one of the cornerstones of silicon technology. Only in
recent years has another technology emerged in this area, which
provides even better process control than standard diffusion,

namely, ion implantation. This technology will be discussed in a later lecture.

In this lecture we shall familiarize ourselves with the fundamental physical phenomena of diffusion pertinent to silicon technology. We shall find, perhaps to our surprise, that, despite the importance of this technology, there are still many areas in which detailed knowledge is lacking.

Diffusion phenomena are frequently described by means of Fick's laws. The relation given in Fick's first law represents the most idealized case, namely, strong dilution of the solute (doping atom) in the solvent (host crystal) without electrical and chemical interference.

Fick's first law essentially relates the flux J_A of dopant atoms A to the concentration gradient dN_A/dx in the following linear way:

$$J_A = -D_A \frac{dN_A}{dx} . \qquad (1)$$

In the case of multicomponent systems, however, it is convenient to resort to a formalism based on Onsager's treatment of irreversible processes, which will be discussed in detail in Section 2 of this lecture. By choosing only the first diagonal matrix element, we find

$$J_A = L_{AA} X_A . \qquad (2)$$

X_A is related to the chemical potential of the solute A as follows:

$$X_A = - \frac{d\mu_A}{dx} \qquad (3)$$

and

$$\frac{d\mu_A}{dx} = \frac{kT}{N_A} (1 + \frac{d \ln \gamma}{d \ln N_A}) \frac{dN_A}{dx} , \qquad (4)$$

γ being the activity coefficient.

Equations (3) and (4) permit us to rewrite Eq. (2) in the following way after setting $\gamma=1$:

$$J_A = -L_{AA} \frac{kT}{N_A} \frac{dN_A}{dx} . \qquad (5)$$

Comparing Eqs. (1) and (5) leaves us with the relation

$$L_{AA} \frac{kT}{N_A} = D_A. \qquad (6)$$

We can now write Eq. (2) in a form that essentially represents a generalization of Fick's first law.

$$J_A = -D_A \cdot (1 + \frac{d \ln \gamma}{d \ln N_A}) \frac{dN_A}{dx}. \qquad (7)$$

Equation (7) is no longer restricted to dilute systems or electrical and chemical noninterference.

The following sections of this lecture will give a rigorous treatment of (a) multicomponent systems, (b) the thermodynamics of a vacancy-impurity-semiconductor system, and (c) the establishment of a correlation between the Onsager phenomenological coefficients and the diffusion coefficients based upon random-walk theory.

2. FUNDAMENTAL DIFFUSION EQUATIONS FOR MULTICOMPONENT SYSTEMS

In this section we shall introduce the fundamental diffusion equations governing the phenomena in multicomponent systems. The treatment will follow closely those described by de Groot and Mazer [1] and later on extended by S. M. Hu [2] to a substitutional model suited to deal with the diffusion of elements such as arsenic, phosphorus, and boron into silicon.

The main underlying concept is based upon Onsager's studies of irreversible processes. One of the salient features of Onsager's treatment is that the rate of entropy production σ can be described by

$$\sigma = \sum_{i=1}^{n} J_i X_i, \qquad (8)$$

where J_i and X_i are any of the Cartesian components of the independent fluxes and thermodynamic forces.

Under those conditions the currents can be represented as a linear function of the thermodynamic forces:

$$J_i = \sum_{k}^{n} L_{ik} X_{xk}. \qquad (9)$$

For the phenomenological coefficients L_{ik} the following relations hold true according to Onsager:

$$L_{ik} = L_{ki} \quad \text{(Onsager's Reciprocity Relations)}. \tag{9a}$$

If one applies Eq. (9) to a substitutional diffusion system involving the host crystal (silicon) and various diffusants, one is faced with the fact that the fluxes and the thermodynamic forces are interrelated.

For the currents one finds

$$\sum_{i=1}^{n} J_i = 0. \tag{10}$$

This is a consequence of the conservation of lattice sites.

The thermodynamic forces are interrelated by means of the Gibbs-Duhem relation:

$$\sum_{j=1}^{n} N_j \frac{\partial \mu_j}{\partial x} = 0, \tag{11}$$

where $-\partial \mu_j/\partial x = X_j$ and N_j is the concentration of the component j of the system.

Under those circumstances the rate of entropy generation can be expressed after elimination of J_n by

$$\sigma = \sum_{i=1}^{n} J_i X_i = \sum_{i=1}^{n-1} J_i X_i - X_n \sum_{i=1}^{n-1} J_i \tag{12}$$

$$= \sum_{i=1}^{n-1} J_i (X_i - X_n).$$

If one eliminates X_n in Eq. (12) using Eq. (11), $X_n = -\sum_{j=1}^{n-1} N_j/N_n X_j$, one obtains for the entropy production σ

$$\sigma = \sum_{i=1}^{n-1} J_i (X_i - \sum_{j=1}^{n-1} \frac{N_j}{N_n} X_j). \tag{13}$$

The phenomenological equations can therefore be written as

$$J_i = \sum_{k=1}^{n-1} l_{ik} \{X_k + \sum_{j=1}^{n-1} \frac{N_j}{N_n} X_j\} \quad i=1,2\ldots n-1 \tag{14}$$

or, after interchanging indices,

$$J_i = \sum_{k=1}^{n-1} [l_{ik} + \frac{N_k}{N_n} \sum_{j=1}^{n-1} l_{ij}]X_k \quad i=1,2\ldots n-1 \tag{15}$$

$$J_n = -\sum_{i,k=1}^{n-1} [l_{ik} + \frac{N_k}{N_n} \sum_{j=1}^{n-1} l_{ik}]X_k. \tag{16}$$

The new phenomenological coefficients l_{ik} now follow again the Onsager reciprocity relations

$$l_{ik} = l_{ki}. \tag{17}$$

The relation between the coefficients L_{ik} defined by Eq. (9) and l_{ik} of Eq. (14) can be found in the following way. X_n in Eq. (9) can be eliminated by the use of Eq. (11), resulting in

$$J_i = \sum_{k=1}^{n-1} [L_{ik} - \frac{N_k}{N_n} L_{in}]X_k \quad i=1,2\ldots n-1. \tag{18}$$

By comparing the coefficients in Eqs. (15) and (18), one obtains the desired relationship between matrix elements L_{ik} and l_{ik}:

$$L_{ik} - \frac{N_k}{N_n} L_{in} = l_{ik} + \frac{N_k}{N_n} \sum_{j=1}^{n-1} l_{ik} \quad i,k=1,2,\ldots n-1 \tag{19}$$

$$L_{nk} - \frac{N_k}{N_n} L_{nn} = -\sum_{j=1}^{n-1} \{l_{in} + \frac{N_k}{N_n} \sum_{i=1}^{n-1} l_{ij}\} \quad k=1,2 \ldots n-1 \tag{20}$$

From Eqs. (19) and (20) follows

$$\sum_{i=1}^{n} (L_{ik} - \frac{N_k}{N_n} L_{in}) = 0 \quad k=1,2\ldots n-1. \tag{21}$$

3. APPLICATION TO A 3-COMPONENT SYSTEM

To get a better understanding of the formalism elaborated in the previous sections, we shall now discuss the case of a dopant (subscript A) diffusion in a silicon lattice (subscript B) via a vacancy (subscript V) mechanism. The objective is to determine the correlation between the phenomenological coefficients.

It will be further demonstrated that the off-diagonal matrix elements in L_{ik} are expressible as functions of the diagonal matrix elements.

From Eq. (19) one finds

$$L_{AA} - \frac{N_A}{N_B} L_{AB} = 1_{AA} + \frac{N_A}{N_B} (1_{AA} + 1_{AV}) \tag{22a}$$

$$L_{AV} - \frac{N_V}{N_B} L_{AB} = 1_{AB} + \frac{N_V}{N_B} (1_{AA} + 1_{AV}) \tag{22b}$$

$$L_{VA} - \frac{N_A}{N_B} L_{VB} = 1_{VA} + \frac{N_V}{N_B} (1_{VA} + 1_{VV}) \tag{22c}$$

$$L_{VV} - \frac{N_V}{N_B} L_{VB} = 1_{VV} + \frac{N_V}{N_B} (1_{VA} + 1_{VV}) \tag{22d}$$

A set of symmetric solutions for Eq. (21) is given by the following relations:

$$L_{AA} = 1_{AA} \tag{23a}$$

$$-L_{AB} = 1_{AA} + 1_{AV} \tag{23b}$$

$$L_{AV} = 1_{AV} \tag{23c}$$

$$L_{VA} = L_{AV} = 1_{AV} \tag{23d}$$

$$-L_{VB} = 1_{VA} + 1_{VV} \tag{23e}$$

$$L_{VV} = 1_{VV} \tag{23f}$$

An additional set of relations is obtained from Eq. (21) by rewriting them in the following form:

$$L_{AA} - \frac{N_A}{N_B} L_{AB} + L_{VA} - \frac{N_A}{N_B} L_{VB} + L_{BA}$$
$$- \frac{N_A}{N_B} L_{BB} = 0 \quad (24)$$

$$L_{AV} - \frac{N_V}{N_B} L_{AB} + L_{VV} - \frac{N_V}{N_B} L_{VB} + L_{BV}$$
$$- \frac{N_V}{N_B} \cdot L_{BB} \quad 0$$

Combining Eq. (24) with Eq. (23) results in

$$L_{BA} = L_{AB} \quad (25a)$$

$$L_{BV} = L_{VB} \quad (25b)$$

It can be further demonstrated, by use of Eqs. (23), (24), and (25), that the off-diagonal matrix elements are given by

$$L_{AV} = \frac{1}{2} (L_{BB} - L_{AA} - L_{VV}) = L_{VA} \quad (26a)$$

$$L_{AB} = \frac{1}{2} (L_{VV} - L_{AA} - L_{VV}) = L_{BA} \quad (26b)$$

$$L_{VB} = \frac{1}{2} (L_{AA} - L_{BB} - L_{VV}) = L_{BV}. \quad (26c)$$

As an example of the importance of the cross terms, we consider the case of self-diffusion. For the phenomenological coefficients, we obtain from Eqs. (26a)--(26c), by setting all those coefficients equal to zero that contain a subscript A,

$$L_{BB} = L_{VV} = -L_{BV} = -L_{VB}. \quad (27)$$

Using Eq. (18) we find for the fluxes

$$J_V = -L_{VV} \frac{\partial \mu_V}{\partial x} - L_{VB} \frac{\partial \mu_B}{\partial x} \qquad (28a)$$

or with Eq. (27)

$$J_V = L_{VV} \left(1 + \frac{N_V}{N_B}\right) \frac{\partial \mu_V}{\partial x} \qquad (28b)$$

$$J_B = -L_{VB} \frac{\partial \mu_V}{\partial x} - L_{BB} \frac{\partial \mu_B}{\partial x} \qquad (28c)$$

or with Eq. (27)

$$J_B = -L_{BB} \left(\frac{N_B}{N_V} + 1\right) \frac{\partial \mu_B}{\partial x} \qquad (28d)$$

As $N_B \gg N_V$ is always fulfilled, we see immediately that the condition $J_V = -J_B$ necessary for reasons of lattice site conservation can be maintained only by means of the cross term

$$L_{VB} \frac{\partial \mu_V}{\partial x} = L_{BB} \frac{N_B}{N_V} \frac{\partial \mu_B}{\partial x},$$

which actually controls J_B. (The diagonal term is negligible.)

Furthermore, one finds using Eqs. (2) and (27)

$$L_{BB} = \frac{D_B N_B}{KT} = L_{VV} = \frac{D_V N_V}{KT} \qquad (29)$$

or

$$D_B = D_V \frac{N_V}{N_B}. \qquad (30)$$

The self-diffusion coefficient is smaller than the vacancy diffusion coefficient by a factor N_V/N_B.

So far we have treated the diffusion phenomena in a multicomponent system in a very general way. The only specification that we have imposed so far is that the diffusion take place via a vacancy model.

In the following sections, we shall apply an increasingly detailed atomistic model to gain further insight into the diffusion mechanism of impurity atoms in silicon.

4. CHEMICAL POTENTIALS

4.1 Electrically Neutral System

For reasons of simplicity we shall restrict ourselves again to a 3-component system, using the same symbols as in Section 3 (B for semiconductor host atoms, A for dopant atoms, and V for lattice site vacancies). We further assume $N_B > N_A \gg N_V$, a condition frequently fulfilled as in the case of an emitter diffusion (As or P) in Si.

The Gibbs free energy of such a neutral system is then given by

$$G_1 = G_1^o + N_A U_A + N_V U_V - N_{AV} E_\alpha - kT\ln\Omega, \qquad (31)$$

where

G^o is the free energy of the host lattice (undoped)
U_A is the free energy associated with the exchange of a host atom with an impurity atom
U_V is the free energy of vacancy formation
E_α is the binding energy of a vacancy to an impurity atom.
N_{AV} is the number of vacancy impurity pairs.

$\Delta S_M = k\ln\Omega$ is the entropy of mixing as defined by

$$\Delta S_M = k\ln\Omega = k\ln\frac{N_S!}{N_A!(N_S-N_A)!} + k\ln\frac{N_S!}{N_V!(N_S-N_A-N_V)!} \qquad (32)$$

where

$$N_S = N_A + N_B + N_V.$$

In the further treatment, we just consider the first term on the right-hand side of Eq. (28). (The second term can be treated in an analogous way.)

By applying Sterling's relation,

$$\ln W! \simeq W\ln W - W, \qquad (33)$$

to Eq. (32) one obtains

$$\Delta S_M = k[N_S \ln N_S - N_A \ln N_A - N_S \ln(N_S - N_A) + N_A \ln(N_S - N_A)] \qquad (34)$$

Hence,

$$\mu_A = \left.\frac{\partial G_1}{\partial N_A}\right|_{N_S, N_V = \text{const.}} = U_A + kT\ln\left(\frac{N_A}{N_S - N_A}\right) \qquad (35)$$

or, by writing μ in the form of $\mu = U_A + kT \ln(\gamma \frac{N_A}{N_S})$, one finds

$$\gamma_A = \frac{N_S}{N_S - N_A} = \frac{1}{1 - \frac{N_A}{N_S}} . \qquad (36)$$

Similarly, one gets:

$$\mu_V = \frac{\partial G_1}{\partial N_V}\bigg|_{N_S, N_A} = U_V + kT \ln \frac{N_V}{N_S - N_V} \qquad (37)$$

and, for $N_S \gg N_V$,

$$\mu_V = U_V + kT \ln \frac{N_V}{N_S} \qquad (38)$$

and, hence, as the equilibrium condition,

$$N_V^o = N_S \exp - \frac{U_V}{kT} . \qquad (39)$$

Equation (38) requires some modification if we want it actually to comply with the conditions outlined in Eq. (31). In deriving the Gibbs free energy, we have allowed for pair formation between dopant atoms and vacancies accompanied by a reduction of the free energy. This effect has to be taken into account when deriving the chemical potential of a vacancy.

We can go about it in the following manner. Each dopant atom has Z next-nearest lattice sites, which can be occupied by a vacancy. The configurational probability for a vacancy dopant pair is therefore the product of the probability that a particular lattice site is occupied by a vacancy $p_V = (1/N_S - N_A) N_V$ and the coordination number Z. Hence,

$$N_{AV} = N_A \cdot \frac{N_V \cdot Z}{N_S - N_A} . \qquad (40)$$

If an attractive force exists between a dopant atom and a vacancy, the coordination number increases virtually from Z to

$$Z^* = Z \exp - E_\alpha / kT. \qquad (41)$$

The statistical effect is equivalent to increasing the total number of lattice sites to

$$N_S^* = N_S + Z^* - Z. \tag{42}$$

Therefore Eq. (40) goes over into

$$N_{AV} = \frac{N_A N_V}{N_S} \frac{Z^*}{1+(Z^*-Z-1)\frac{N_A}{N_S}} \tag{43}$$

A similar correction now has to be made for Eq. (37):

$$\mu_V = \frac{\partial G_1}{\partial N_V}\bigg|_{N_S, N_A, N_{AV}=\text{const}} = U_V + kT \ln \frac{N_V}{N_S + (Z^*-Z-1)N_A}$$

$$\mu_V = kT \ln \frac{N_V}{N_V^0} - kT \ln(1+\frac{N_A}{N_S}(Z^*-Z-1)), \tag{44}$$

and the activity coefficient is given by

$$\gamma_V = \frac{1}{1+\frac{N_A}{N_S}(Z^*-Z-1)} . \tag{45}$$

4.2 Electrically Active System

Diffusion systems in which the involved species could be ionized had been first discussed by Longini and Green. Hu extended their approach to the formation of vacancy-impurity pairs. Mathematically this can be accomplished by adding another term, G_2, to Eq. (31) of the form

$$G_2 = n_C E_C + n_V E_V - p_A E_A - p_p E_p - n_\beta E_\beta - kT \ln\Omega_e, \tag{46}$$

where E_C, E_p, E_V, and E_A are the energy levels connected with the conduction band edge, valence band edge, vacancies (acceptors), dopants (donors or acceptors), and their respective densities N_C, N_p, N_V, and N_A. The charges pertaining to those levels are denoted by n_C, p_p, n_V, and p_A. The impurity-vacancy pair can assume the following charge states:

$$A^0V^0; \ A^0V^-; \ A^+V^0; \ A^+V^-.$$

For the respective concentration the following definitions are used:

$$N_{A^+V^-} = n_\beta$$

$$N_{A^0V^-} = n_{AV} - n_\beta$$ (47a)

$$N_{A^+V^0} = P_{AV} - n_\beta$$

$$N_{A^0V^0} = N_{AV} - n_{AV} - P_{AV} + n_\beta$$

E_β is the coulombic interactive energy related to the A^+V^- pair.

Ω_e is the configurational term representing the charge permutations.

The mathematical treatment of this problem is fairly involved and will not be repeated here. For a detailed discussion the reader is referred to Ref. 2.

According to S. M. Hu, one obtains the following relations for the total system (electrically active and inactive) by using the following definitions:

$$\beta = \exp(E_\beta/kT)$$

$$= g_A \exp[(E_A-E_F)/kT] \quad (47b)$$

$$\xi = g_V \exp[(E_F-E_V)/kT]$$

with g_A, g_V being the degeneracy factors of donor and vacancy states.

$$\mu_V = \frac{\partial(G_1+G_2)}{\partial N_V} \quad (48)$$

$$= kT \ln \frac{N_V}{N_V^*} - kT \ln \frac{1+\xi}{1+\xi^*}$$

$$- kT \ln[1+ \frac{N_A}{N_S} [\frac{Z^*}{1+\xi}(1+ \frac{1+\beta\zeta}{1+\zeta}\xi) - Z + 1)].$$

The asterisks denote intrinsic conditions, e.g., N_V^*, ξ^*.

For the chemical potential of the dopant species, one obtains

$$\mu_A = kT \ln \frac{N_A}{N_S} - kT \ln(\frac{1-\zeta}{1+\zeta} *) - kT \ln (1- \frac{N_A}{N_S}). \qquad (49)$$

Figure 1 depicts the dependence of the donor activity coefficient calculated by S. M. Hu [2] for constant-energy band structure and degeneracy factors of unity. If one compares those results with the ones obtained by evaluating experimental diffusion profiles, such as for As, one finds that the activity coefficient has a maximum in the neighborhood of $N_A \sim 10^{20}$ cm^{-3} and then decreases with increasing concentration.

This fact has been explained by invoking arsenic cluster formation, which will be treated in the next section.

At this point it seems appropriate to make some comments as to the validity of the treatment outlined in this section.

One basic assumption in Eq. (46) is that the free-carrier concentrations can be expressed by an effective state density at the respective band edge. It is known, however, that the band edges are markedly influenced at high doping concentrations owing to band tailing [3,4].

It is the lack of detailed, quantitative knowledge of heavily doped semicondcutors (and silicon in particular)| that still poses a great dilemma for any quantitative evaluation.

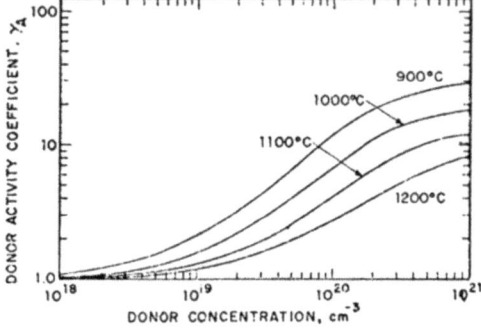

Fig. 1. Donor activity coefficient in Si calculated using Fermi statistics, on the assumption of constant donor level and constant energy-band structure. (Source: Ref. 2)

5. EFFECT OF CLUSTER FORMATION AND IMPURITY (DONOR ACCEPTOR) PAIRING

At higher doping concentrations, cluster formation between dopant atoms of identical species can become important in diffusion processes. This fact had been recognized first by S. M. Hu [5]. He had proposed, originally, the formation of dimers. Mantano measurements and vapor pressure data on As, however, could be fitted much better by the assumption of tetramers [Hu, 1969-see also Refs. 6 and 7].

These clusters are presumed to be immobile and thus to reduce the number of atoms participating in the diffusion. If these centers are ionized, they still impact the position of the Fermi level.

The concentration of clusters involving atoms as a function of the total concentration can be derived readily from the following consideration of the chemical reaction:

$$mA \leftrightarrow A_m \tag{50}$$

or

$$N_{A_m} = K_C N_A^m, \tag{51}$$

with K_C being the reaction constant for clustering.

If we denote the total concentration of dopant A (as measured by radiotracer techniques, for instance) by N_A^T, we have

$$N_A^T = N_A + mN_{A_m} \tag{52a}$$

or, with Eq. (51),

$$N_A^T = N_A + m K_C N_A^m. \tag{52b}$$

Therefore, the concentration of the dopant atoms, which actively participate in the random-walk process, will be smaller by the factor

$$\frac{N_A}{N_A^T} = \frac{1}{1+m \cdot K_C \cdot N_A^{m-1}}. \tag{53}$$

Schwenker et al. [8] have concluded from changes of the electrical conductivity as a function of annealing that the clusters in arsenic-doped silicon exist in the form of dimers (m=2) for their specific experimental conditions.

In general, however, one has to assume that all possibilities coexist and that the prevalence of one is a function of temperature and of total concentration.

In an analogous way, one can calculate the contributions from donor-acceptor pairing.

If one assumes

$$A^+ + D^- \leftrightarrow (AD) \tag{54a}$$

$$K_p N_A \cdot N_D = N_{AD} \tag{54b}$$

with K_p being the equilibrium constant for pairing. N_A^T then becomes

$$N_A^T = N_A + N_{AD} = N_A(1 + K_p N_D) \tag{55}$$

and the fractional factor is given by

$$\frac{N_A}{N_A^T} = \frac{1}{1 + K_p N_D}. \tag{56}$$

6. DETERMINATION OF PHENOMENOLOGICAL COEFFICIENTS

A general relation between the phenomenological coefficients and the diffusion coefficients was given in Eq. (6). Further insight into the particular diffusion mechanism can be obtained by specifying the atomistic model and applying random walk theory.

From random walk theory one obtains the following relation:

$$D = \frac{<R^2> \theta}{2 \cdot 3\tau}, \tag{57}$$

where $<R^2>$ is the mean square displacement of the diffusion species, τ the time during which the displacement takes place, and θ the fraction of particles that simultaneously and independently participate in the random walk process.

In the case of vacancies, Eq. (57) can be written as

$$D_v = <|\sum_i^{n_v} r_i|^2>/6<\sum_i^{n_v} \frac{1}{w_i}> , \qquad (58)$$

with r_i the ith jump vector of n_v jumps and $1/w_i$ the dwell time between jumps. S. M. Hu has worked out the relation governing the random walk of a vacancy that interacts energetically with a dopant atom. For this purpose he has devised a four-frequency model as shown in Fig. 2.

The assumptions made in his calculations are that the saddle point lowering is given by $Q = 1/2\ E_b$, with E_b being the binding energy between the vacancy and the dopant.

Under those conditions one obtains:

$$D_A = \frac{1}{6} r^2 w_A f_A \cdot N_{AV}/ZN_A \qquad (59)$$

with

$$f_A = <|\sum_i^{n_v} rj|^2>/n_A r^2 .$$

The correlation factor, f_A, takes into account that sequential jumps of the dopant A are interdependent because of the attractive force that the dopant exerts upon the vacancy.

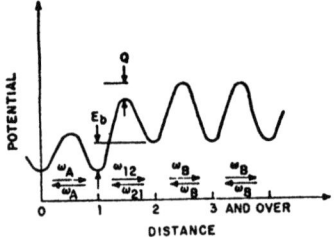

Fig. 2. Schematic diagram of vacancy potential as a function of distance from the impurity atom at the origin, and the four distinguishable vacancy jump frequencies; 1, 2, and 3 denote the coordination order of the sites. (Source: Ref. 2)

That is, in a tight binding case, the two species (A and V) would tumble along together, with the vacancy acting like a satellite.

For the tight binding case the expressions obtained for D_V and D_B become much simpler than otherwise and are as follows:

$$D_V = \frac{1}{6} r^2 [w_B - (N_{AV}/ZN_V) \; Zw_B - w_A f_A)] \tag{60}$$

and

$$D_B = \frac{1}{6} r^2 \frac{N_V}{N_B} [w_B - (N_{AV}/ZN_V) \cdot (Zw_B - 2w_A f_A)]. \tag{61}$$

7. COMPARISON OF THEORY WITH EXPERIMENT

In 1968, S. M. Hu [9] had predicted a dip in the base (boron) profile at the metallurgical junction with an emitter, when formed by arsenic diffusion. This prediction was essentially based upon his theoretical treatment of interactive diffusion systems.

The reason for the "dip" is to be found in the strong electric field at the leading edge of the diffusing arsenic (Fig. 3).

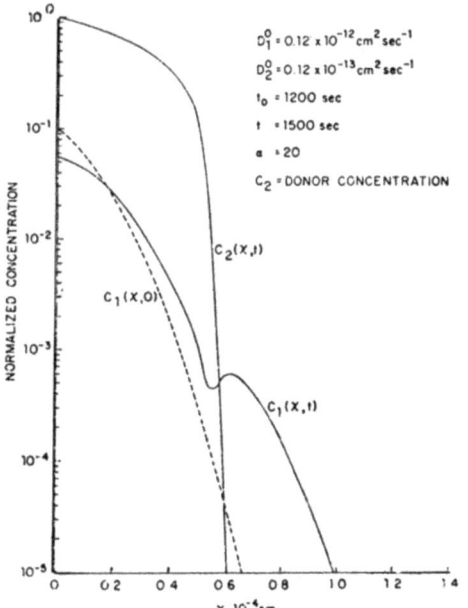

Fig. 3. Combined effect of internal field and equilibrium vacancy concentration on the base and the emitter profiles in sequential diffusion process. (Source: Ref. 9)

Unfortunately, no adequate analytical technique existed at that time to verify the boron profile.

In 1972, J. Ziegler et al [10] came up with a novel profiling concept for boron combining nuclear reaction $^{10}B(n,\alpha)^{Li}$ and backscattering techniques, which for the first time verified experimentally Hu's theory. Figures 4 and 5 show the experimental boron profiles and, superimposed, the arsenic distribution, as

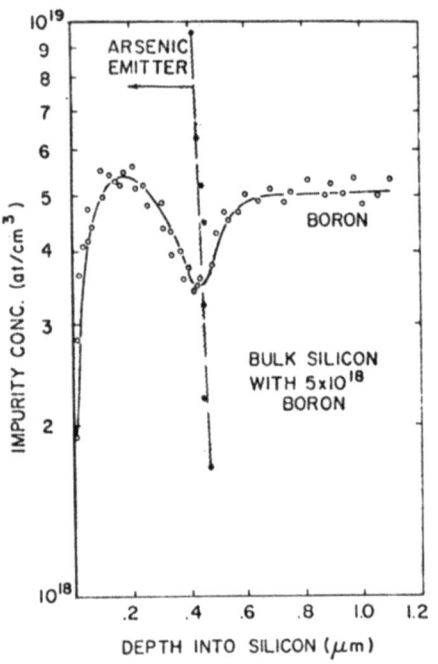

Fig. 4 Shown are the final distributions of impurities for a high-concentration arsenic diffusion into a silicon sample grown with 5×10^{18} atoms/cm^3 of boron impurities. The shaded area indicates probable experimental error. The accumulation of excess boron in the arsenic layer is incomplete because of outdiffusion from the wafer. (Source: Ref. 10)

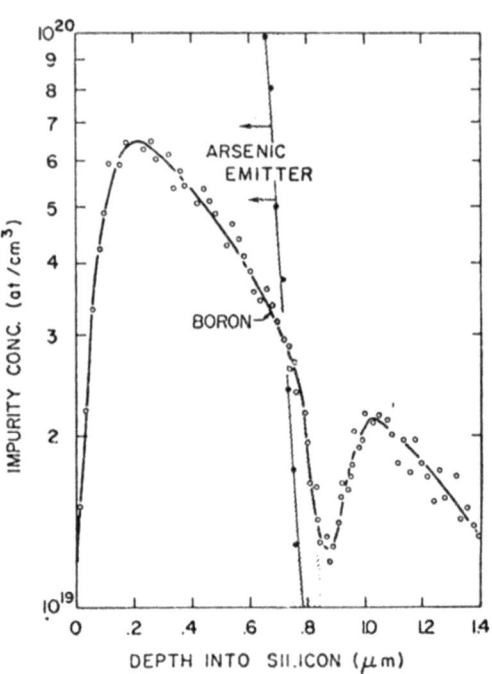

Fig. 5. Shown are the final distributions of impurities for a transistor structure. The arsenic profile comes from three independent measurements: nuclear back-scattering, neutron activation, and junction staining. The back-scatter arsenic profile is shown with dots. (Source: Ref. 10)

found separately from neutron activation analysis.* This phenomenon has since been confirmed by several groups using SIMS techniques. Jones et al.[11], for example, found a similar behavior for Ga.

8. RADIATION-ENHANCED DIFFUSION

Diffusion is a thermally stimulated phenomenon, which, as such, is controlled by an Arrhenius type dependence of the diffusion coefficients upon temperature:

$$D_i = D_{oi} \exp(-Q_i/kT).$$

Let us consider, in the following, a substitutional diffusion, which takes place in a diamond lattice. We further allow for chemical and electrical interaction between vacancies and dopants. The potential energy of a vacancy as a function of its position relative to the dopant atom can then be described in the schematic diagram given in Fig. 6.

---

*Ziegler et al. later reported a possible error in the depth scale of the As profile: the junction should be at a greater depth (9-10 μm).

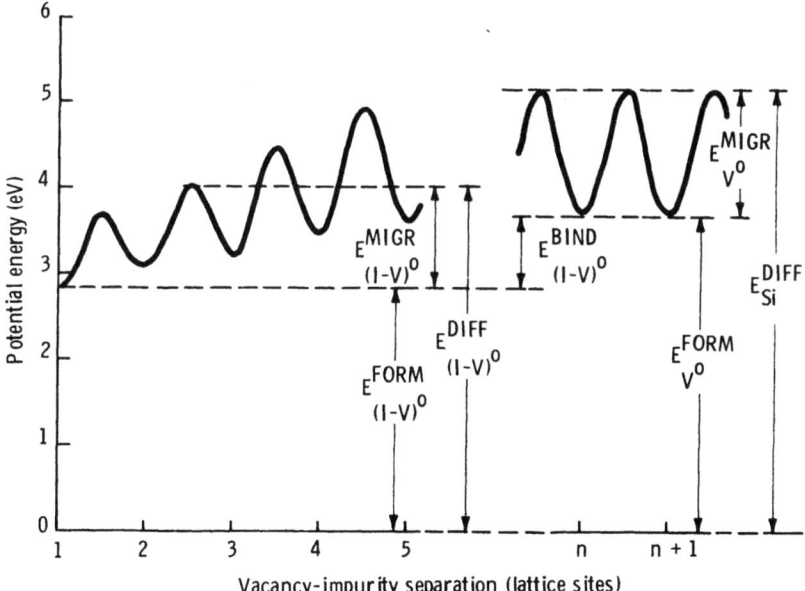

Fig. 6. Potential energy of a vacancy in instrinsic silicon (Fermi level fixed near center of bandgap) as a function of its position relative to a substitutional Group V impurity atom. (Source: Ref. 14)

In deriving this model, one assumes that, in the diamond lattice, a vacancy must first jump to a 3rd coordination site in order to move from a 1st coordination site to another 1st coordination site.

With the definitions taken from Fig. 6, one finds the following relations:

$$Q_B = E_{V^o}^{Form} + E_{V^o}^{Migr} \quad \text{self-diffusion} \tag{62a}$$

$$Q_A = E_{(AV)^o}^{Form} + E_{(AV)^o}^{Migr} \quad \text{dopant diffusion} \tag{62b}$$

or

$$Q_A = E_{V^o}^{Form} - E_{(AV)^o}^{Bind} + E_{(AV)^o}^{Migr} \tag{62c}$$

$E_{V^o}^{Form}$ corresponds to U_v in Eq. (31), and $E_{(AV)^o}^{Bind}$ is identical with $E_b = E_\alpha + E_\beta$, which have been defined in Eqs. (31) and (46).

The temperature dependence for self-diffusion and impurity diffusion can be markedly reduced under conditions known as radiation-enhanced diffusion [12,13]. In this case additional vacancies are generated by displacement reactions, a non-thermal process. As a consequence the activation energy Q_A or Q_B no longer depends upon the energy of vacancy formation $E_{V^o}^{Form}$. Furthermore the diffusion coefficients for a given temperature can be several orders of magnitude higher because of the additional vacancy concentration.

At temperatures above 600°C, proton and He$^+$ bombardment, as an example, generates point defects, presumably Frenkel pairs.

Figure 7 depicts experimental results obtained by B. Masters [14] under intrinsic conditions, $N_A < n_i$, for B, P, and As. For comparison, the diffusion coefficients for the thermally activated diffusions are also given (straight lines at the left).

Masters interprets his data by means of a theory by Dienes and Damas [15]. The steady-state conditions for Frenkel pairs (interstitials and vacancies) are given according to them by

$$\frac{d N_v}{d t} = k_1 \phi - k_2 w_i (N_v + N_v^*) N_i - k_3 w_v : N_v = 0 \tag{63a}$$

$$\frac{d N_i}{d t} = k_1 \phi - k_2 w_i (N_v + N_v^*) N_i - k_3 \cdot w_i N_i = 0 \tag{63b}$$

where $k_1 \phi$ represents the production rate of Frenkel pairs during bombardment with a flux ϕ.

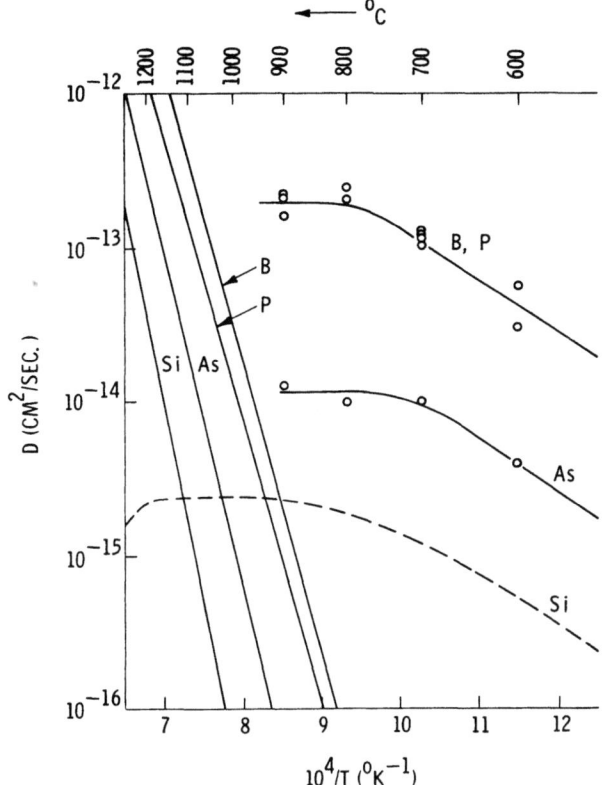

Fig. 7. Temperature dependence of enhanced diffusion at $\phi = 5 \times 10^{11}$ protons/cm^2--sec. The lower broken curve is the silicon-enhanced self-diffusivity. The straight lines at the left represent the thermally induced diffusivities of silicon, arsenic, boron, and phosphorus. (Source: Ref. 14)

N_i is the density of interstitials, w_i and w_v the interstitials and vacancy jump frequencies, and N_v^* the thermally generated vacancy concentration.

The second term describes the recombination of the Frenkel pairs; the last terms describe the recombination of vacancies and of interstitials, respectively, on sinks in the bulk of the crystal or on the surface.

From (63a) and (63b) one obtains:

$$N_i w_i = N_v w_B$$

and

$$N_v = \frac{1}{2} \{[(N_v^* + k_3/k_2)^2 + 4k_1\phi/k_2 w_v]^{1/2} - N_v^* - k_3/k_2\}$$
$$N_i = \frac{1}{2}\frac{w_v}{w_i} [(N_v^* + k_3/k_2)^2 + 4k_1\phi/k_2 w_v]^{1/2} - N_v^* - k_3/k_2\} \quad (64)$$

The enhanced self-diffusion coefficient D_B^{Enh} is then in the first approximation given by

$$D_B^{Enh} = \frac{r^2}{6} (w_v \frac{N_v}{N_B} + w_i \frac{N_i}{N_B}) \quad (65)$$

$$D_B^{Enh} = \frac{r^2}{6} \frac{w_v}{N_B} \{[N_v^* + k_3/k_2)^2 + 4k_1\phi/k_2 w_v]^{1/2} - N_v^* - k_3/k_2\}$$

From Eq. (65) one can make the following predictions:

a. At low temperatures and high fluxes, the flux-related term in [ ] dominates and the enhanced diffusion coefficient depends on the flux by a square root law.

$$D_B^{Enh} \sim \frac{r^2 w_v}{6 N_B} (4k_1\phi/k_2 w_v)^{1/2} \quad (66)$$

b. On the other hand, at very low fluxes, we expect

$$D_B^{Enh} \sim \frac{r^2}{6} \frac{w_v}{N_B} \cdot \frac{2 k_1 \phi}{k_2 w_v} \cdot \frac{1}{N_v^* + k_3/k_2} ;$$

that is, D_B depends linearly on the flux.

These predictions agree well with Masters' experimental results obtained on impurity diffusion (Fig. 8).

At extrinsic conditions $N_A \gg n_i$ the situation becomes much more complex. The diffusion length of the point defects, estimated by Masters [14] from some of his results to be in the order of microns for the intrinsic case, is markedly reduced. This leads to profiles first reported by Ryssel et al. [16] in Fig. 9.

9. SUMMARY

We have intentionally refrained in this presentation from any discussion of specific mathematical solutions of the general differential equations, dealing, for instance, with moving

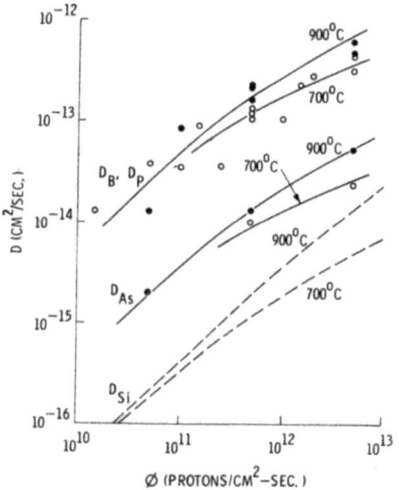

Fig. 8. Flux dependence of enhanced diffusion at 700°C (open circles) and 900°C (closed circles). The broken curves represent enhanced silicon self-diffusivity. (Source: Ref. 14)

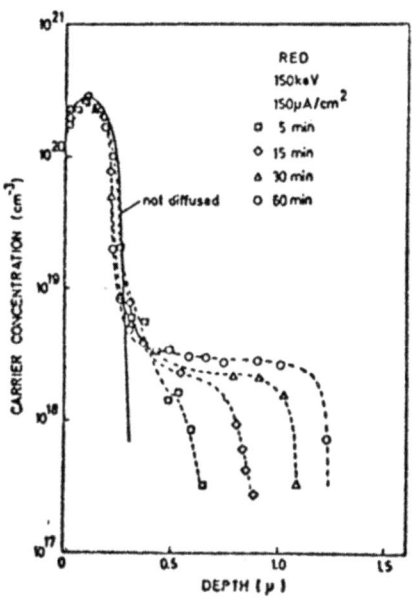

Fig. 9. RED Profiles of As implanted layers. (Source: Ref. 14)

boundaries or chemical reactions at the diffusion edge. The interested reader is referred to the standard literature, such as Mathematics of Diffusion, by Crank, or Heat Conductions in Solids, by Carlslaw and Jaeger.

Rather, it was our objective to give a brief introduction to the general theory of diffusion and the underlying physical concepts. A formalism has been described which permits the treatment of complex, interactive systems as they occur in semiconductor processing. The application to process modeling will be the subject of a later presentation.

ACKNOWLEDGMENTS

I should like to thank my colleagues S. M. Hu, B. J. Masters, and R. O. Schwenker for many helpful discussions and for their critique of the manuscript. Dr. Masters also kindly provided some of his recent results.

REFERENCES

1. S. R. deGroot and P. Mazur, Nonequilibrium Thermodynamics (North-Holland Publishing Co., Amsterdam, 1962), pp. 64-69).
2. S. M. Hu, Phys. Rev. 180, 773 (1963).
3. E. O. Kane, Phys. Rev. 131, 79 (1963).
4. V. L. Bonch-Bruyevich, The Electronic Theory of Heavily Doped Semiconductors (Elsevier Publishing Co., Inc., New York, 1966).
5. S. M. Hu, Private communication.
6. T. L. Chiu and H. M. Ghosh, IBM J. Res. Dev. 15, 472 (1971).
7. S. M. Hu, in Atomic Diffusion in Semiconductors (Plenum Press, London and New York, 1973), pp. 306-310.
8. R. O. Schwenker, E. S. Pan, and R. F. Lever, J. Appl. Phys. 42, 3195 (1971).
9. S. M. Hu and S. Schmidt, J. Appl. Phys. 39, 4272 (1968).
10. J. F. Ziegler, G. W. Cole, and J. E. E. Baglin, Appl. Phys. Lett. 21, 177 (1972).
11. C. L. Jones and A. P. W. Willoughby, Appl. Phys. Lett. 25, 114 (1974).
12. J. C. Pfister and P. Baruch, Proc. Intern. Conf. on Crystal Lattice Defects, J. Phys. Soc. Jap. 18, Supplement III, 251 (1963).
13. W. K. Hofker et al., Appl. Phys. 2, 265 (1973) by Springer Verlag (1973).
14. B. J. Masters and E. F. Gorey, Bull. Am. Phys. Soc. 22, 328 (1977).
15. G. J. Dienes and A. C. Damask, J. Appl. Phys. 29, 1713 (1958).
16. H. Ryssel et al., in Ion Implantation in Semiconductors (Plenum Press, New York and London, 1975), p. 169.

SILICON EPITAXY AND OXIDATION*

J.D. Meindl, R.W. Dutton, K.C. Saraswat, J.D. Plummer
Stanford University Integrated Circuits Laboratory
Stanford, California 94305

T.I. Kamins
Hewlett-Packard IC Laboratories
Palo Alto, California

B.E. Deal
Fairchild Semiconductor, Research & Development
Palo Alto, California

ABSTRACT. The first-order process models for silicon epitaxy and oxidation are described. Epitaxial dopant inclusion, autodoping, and transient effects are discussed, and experimental results are presented. Silicon orientation, surface doping, and ambient effects are considered for silicon-oxidation rates.

PREFACE. The work described herein is a program report on two subgroup activities at Stanford University in IC process and device modeling. This work is sponsored through DARPA. Ion implantation is directed by Dr. J.F. Gibbons, and further details can be found in Stanford Technical Reports prepared under a DARPA contract. The technical contributions of T.J. Rodgers, R. Reif, D. Hess and C. Ho are gratefully acknowledged.

1. INTRODUCTION

Epitaxy and oxidation are two critical steps in IC technology. Process models date back to the mid-1960's. Epitaxy often controls bipolar circuit performance and yields. Oxide growth is essential in all IC technologies, and segregation frequently dominates fabrication limits and the process control of surface effects. The following sections will review the basic models and introduce considerations for more recent experimental work.

---

* This work has been supported by DARPA contract DAA-B07-75-C-1344.

2. SILICON EPITAXY

2.1 Introduction

There is an extensive literature on the chemistry and reactor dynamics of silicon epitaxy [1] and polycrystalline deposition [2]. The recent advent of low-pressure CVD systems will undoubtedly add substantially to this body of knowledge [3].

The focus of this discussion is on process modeling for silicon epitaxy in a horizontal reactor [4]. This emphasis is based on the experience at Stanford University and, in large part, on that of technologists using commercial equipment. Figure 1 is a schematic of a typical reactor. Table 1 describes the various source gases and their functions, and Table 2 summarizes the key features of several silicon gaseous sources. Table 3 presents a typical reactor sequence used during epitaxial silicon growth. The following sections will elaborate on the fundamental models that describe the kinetics of silicon growth, dopant inclusion, and transient system response.

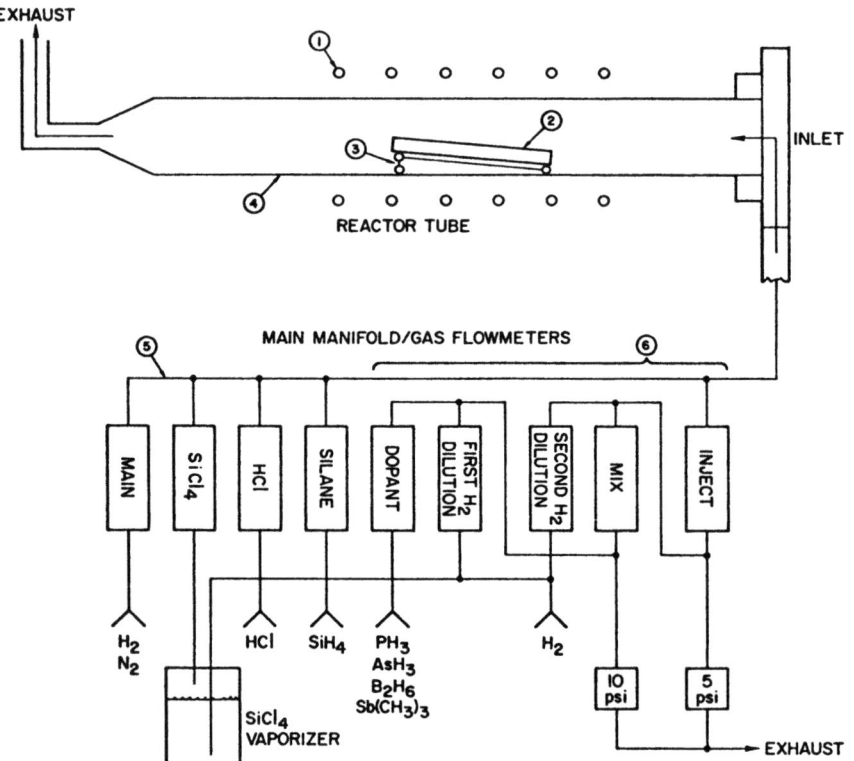

Fig. 1. Schematic of the horizontal epitaxial reactor.

Table 1

GASES USED IN EPITAXIAL DEPOSITION

Gas	Function	Comments
N_2	main flow	Purges out explosive/poisonous gases prior to opening the reactor tube to air
H_2	main flow	Most common ambient for growth or epitaxial layers
$SiCl_4$	Si source	Common liquid Si source Vaporized in an H_2 bubbler Corrosive vapor
SiH_4	Si source	Common gaseous Si source Pyrophoric gas
HCL	Si etchant	Most common Si etchant used for substrate preparation Corrosive poison gas
PH_3	Si dopant	Most common phosphorous source for doping epitaxial silicon Flammable poison gas
AsH_3	Si dopant	As PH_3
$Sb(CH_3)_3$	Si dopant	A liquid antimony source used as a vapor at a concentration of a few hundred ppm in H_2 Used because SbH_3 is unstable Poisonous vapor
B_2H_6	Si dopant	As PH_3

Table 2

SOURCES OF SILICON IN EPITAXY *

Source	Typical Conditions		Comments	Reference
	Temperature (°C)	Rate (μ/min)		
SiH_4	1000 to 1050	0.2 to 1.0	Pure gaseous source Pyrophoric gas Low-temperature deposition Low autodoping Moderate growth rates Surface quality sensitive to O_2	1,3,4,5,8 9,10,12 13
$SiCl_4$	1150 to 1200	0.5 to 1.5	Corrosive liquid source High-temperature deposition Moderate autodoping, outdiffusion Moderate-high growth rates Most common source for linear bipolar integrated circuits Easy to obtain good crystal quality on thick layers	1,2,6,11 13,14 15
$SiHCl_3$	1150 to 1200	1.0 to 10	Corrosive liquid source High-temperature deposition Moderate autodoping Very high growth rates Most common source of poly-Si dielectric isolation Very high purity epitaxial layers used in high-voltage devices	1,7 20 16,17
SiH_2Cl_2	1050 to 1100	≥ 1.0	Gaseous source at 7 psi Properties: intermediate $SiCl_4$, SiH_4	1,37
$SiBr_4$	---	---	Rarely used source	18
$Si(CH_3)_4$	1150	0.4	Rarely used source	19

Table 3

TYPICAL EPITAXIAL GROWTH CYCLE

Step	Time	Temperature	Gas Concentration	Comment
N_2 purge	2'	R.T.	---	Purge out O_2
H_2 purge	2'	R.T.	---	Charge to H_2 ambient
Heat	2'	1200°C	---	In H_2 ambient
HCL etch	2'	1200°C	1%	Etch 0.26 μ of silicon
H_2 purge	2'	to 1050°C	---	To lower temperature and remove HCL
Growth	8'20"	1050°C	0.05% SiH_4 0.3 ppb PH_3	Growth of 5 μ, 1 Ω-cm P-doped Si layer
H_2 purge	1'	1050°C	---	Purge reactants prior to cooling
Cool	4'	R.T.	---	In H_2 ambient
N_2 purge	2'	R.T.	---	Before opening to air

2.2 Film growth

For the kinetics of film growth, the silicon source gas reacts and decomposes as described by a typical equation for silane,

$$SiH_4 \xrightarrow{1000°C} Si + 2H_2 \qquad (1)$$

The silicon growth rate is proportional to the partial pressure of silane,

$$V = K_I(SiH_4)_I \tag{2}$$

where V is the growth rate, K_I is the surface-reaction rate constant, and the subscript denotes the surface quantities. For halide-bearing source gases, there are reactive conditions whereby silicon etching occurs; however, for model simplicity, these source gases will not be considered because of the complexity of the multispecies reaction.

Equation (2) describes growth only under the conditions of surface-reaction control; that is, surface reaction is the limiting step. Figure 2 is a typical Arrhenius plot for silane growth vs inverse temperature with gas partial pressure as a parameter [4]. The silane concentrations are those of the turbulent gas stream.

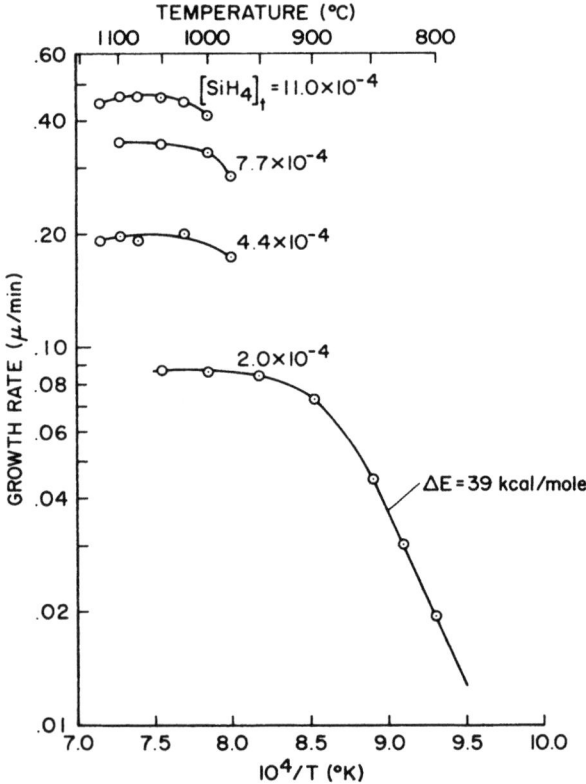

Fig. 2. Arrhenius plot of SiH_4 growth rate.

Figure 3 is a cross section of the reactor. A boundary layer [5] (often labeled a "stagnant" layer) exists between the turbulent gas and reactive wafer interface I; also shown are plots of reactant gas species under two limiting cases.

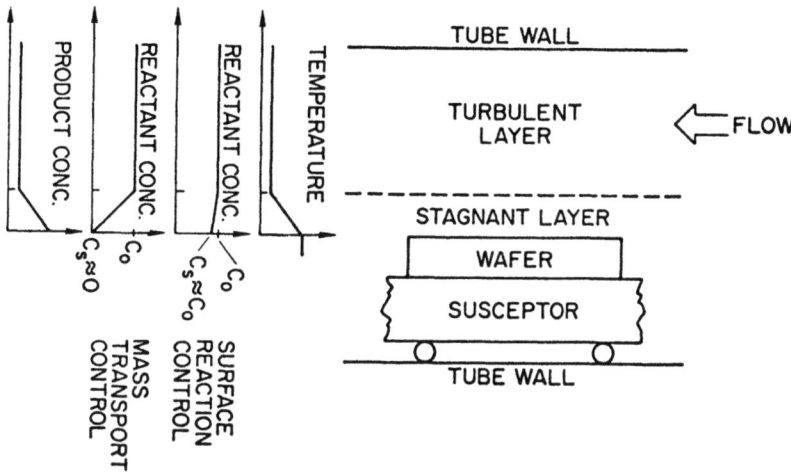

Fig. 3. Cross section of horizontal reactor with thermal and chemical profiles during growth.

Epitaxial growth proceeds by the following steps:
- (1) mass transfer of the reactant molecules (such as SiH_4) by diffusion from the turbulent-layer reservoir across the boundary layer to the silicon surface.
- (2) adsorption of reactant atoms on the surface
- (3) one or more chemical reactions at the surface
- (4) desorption of product molecules (such as H_2)
- (5) mass transfer of the product molecules by diffusion through the boundary layer, back to the turbulent layer
- (6) lattice arrangement of the adsorbed silicon atoms (may occur as part of 3)

Concentration gradients across the boundary layer are such that there is a diffusion flux of product molecules away from the surface. The turbulent layer is depicted as having no concentration or temperature gradient.

The overall deposition rate is determined by the slowest of processes (1 to 6 above). The expected temperature dependence of the growth rate varies markedly for different controlling mechanisms. If diffusion of either reactant or product across the stagnant layer (so-called mass-transport control) is the slowest part of the raction, the growth rate would not depend on the temperature to first order. On the other hand, if a surface chemical reaction is the slowest process (surface-reaction rate control), one would expect the deposition rate to have the same temperature dependence as the chemical reaction, which is an exponential function of inverse temperature. Furthermore, because the various silicon sources undergo different chemical reactions with different activation energies, one would expect them to have diverse growth-rate temperature dependences under surface-reaction rate control.

The curve shown in Figure 2 displays a low-temperature reaction realm in which surface-reaction rates dominate the growth rate and a high-temperature nonactivated reaction realm in which mass transport dominates the growth rate. One striking feature is that similar plots for $SiCl_4$, SiH_4, SiH_2Cl_2, and $SiHCl_3$ have the same activation energy; (roughly 37 kcal/mole or 1.6 eV/molecule). This common activation energy in the temperature-dependent region implies that the dominant surface "reaction" is step 6; (the lattice arrangement of the silicon atoms). Indeed, the activation energy for surface diffusion of adsorbed silicon atoms on silicon has been measured to be 36 ± 6 kcal/mole [6].

At higher temperatures, growth reaction is nonactivated, suggesting that the slowest process is mass transport of either products or reactants. The rate-limiting step is probably reactant transfer. This hypothesis is supported by two considerations. First, in the boundary layer under reactant mass-transport limited conditions, the growth rate is proportional to the diffusion coefficient of the reactant species in hydrogen. This coefficient, in turn, is proportional to the inverse square root of the molecular weight of the reactant molecule. Second, the large temperature gradient in the boundary layer tends to retard the gaseous diffusion of reactants toward the susceptor and to increase the diffusion of products away from the susceptor.

In summary, boundary-layer theory and empirical observations produce the following qualitative results for the epitaxial growth rate of common silicon sources.

(1) The growth rate indicates two distinct regions, a low-temperature region in which the growth rate fits an Arrhenius plot and a high-temperature region in which the growth rate does not depend on temperature.

(2) In the low-temperature region:
- (a) Activation energy is approximately 37 kcal/mole, a number consistent with the activation energy for surface diffusion of adsorbed silicon atoms.
- (b) Activation energy is independent of source type.
- (c) For low source concentrations, the growth rate is directly proportional to the input silicon source concentration.
- (d) The linear growth rate/source concentration relationship applies at high growth rates only for SiH_4. The choloride-containing sources do not obey this relationship because a reverse reaction, HCl etching, occurs [7].

(3) In the high-temperature region:
- (a) The temperature insensitive growth rate is controlled by mass transport of the silicon source from the turbulent layer to the silicon interface by diffusion.
- (b) For low source concentrations, the growth rate is directly proportional to the input silicon source concentration.
- (c) The growth rates for chloride-containing sources are not linear functions of source concentration at high concentrations because of HCl etching.

The changeover between the surface-controlled and mass-transport controlled realms of SiH_4 deposition can be described by the balancing of molecular fluxes because no reverse reactions complicate the SiH_4 problem. The growth rate V is directly proportional to the SiH_4 concentration at the silicon surface, as given by Eq. (2). The surface-reaction rate constant K has the characteristic activation energy of 37 kcal/mole and the units of the growth rate (cm/sec) because $[SiH_4]_I$ is a unitless ratio of gas flows. The flux of SiH_4 atoms across the boundary layer F_1 is proportional to the SiH_4 concentration gradient across the layer, which is the difference between SiH_4 concentration in the turbulent layer $[SiH_4]_T$ and that at the interface $[SiH_4]_I$,

$$F_I = h(SiH_4)\left[[SiH_4]_T - [SiH_4]_I\right] \qquad (3)$$

where $h(SiH_4)$ is the mass-transport coefficient for SiH_4 and is a temperature-insensitive constant that relates molecular flux to the concentration gradient; it is proportional to the SiH_4 diffusion coefficient in H_2 and is inversely proportional to the boundary-layer thickness. The flux of silicon atoms being incorporated into the lattice is

$$F_2 = N_{Si} V \tag{4}$$

in which N_{Si} is the atomic density of silicon, $N_{Si} = 5.0 \times 10^{22}\,\text{cm}^{-3}$. A simultaneous solution of Eqs. (2) through (4) obtains the interface silane concentration,

$$[SiH_4]_I = [SiH_4]_T \left[1 + \frac{N_{Si}}{h(SiH_4)} K_I \right]^{-1} \tag{5}$$

There are two limiting cases:

Case 1: <u>Surface-reaction control.</u>

At low temperatures, K_I is small enough so that $[SiH_4]_I \approx [SiH_4]_T$ (See Figure 3). In this case, from Eq. (2),

$$V = K_I [SiH_4]_T \tag{6}$$

Case 2: <u>Mass-transport control.</u>

At high temperatures, K_I becomes large, and $[SiH_4]_I \approx 0$; then from Eqs. (2) and (3),

$$V = \frac{h(SiH_4)}{N_{Si}} [SiH_4]_T = K_M [SiH_4]_T \tag{7}$$

The rate constant for mass-transport control is K_M (cm/sec).

Although Eqs. (6) and (7) predict a first-order growth-rate dependence on silane concentration, the surface-reaction rate constant K_I is exponentially dependent on temperature and the mass-transport rate constant K_M has only a linear dependence on temperature (essentially no dependence over the narrow 1000° to 1100°C range commonly used for SiH_4 epitaxy).

Experimental SiH_4 growth rate is plotted vs temperature in Figure 2. The surface-reaction region has an activation energy of $\Delta E = 39$ kcal/mole. Transfer to the mass-transport region is nearly complete at 1000°C for all growth rates from 0.1 to 0.5 µ/min. From 1000°C, the maximum variation observed is only 10%. Figure 4 is a graph of growth rate vs silane concentration for T = 1050°C. The slope of the line is the mass-transport rate constant value $K_M = 7.5 \times 10^{-4}$ cm/sec = 450 µ/min. These values for activation energy [6] and the rate constant [5] are consistent with the literature.

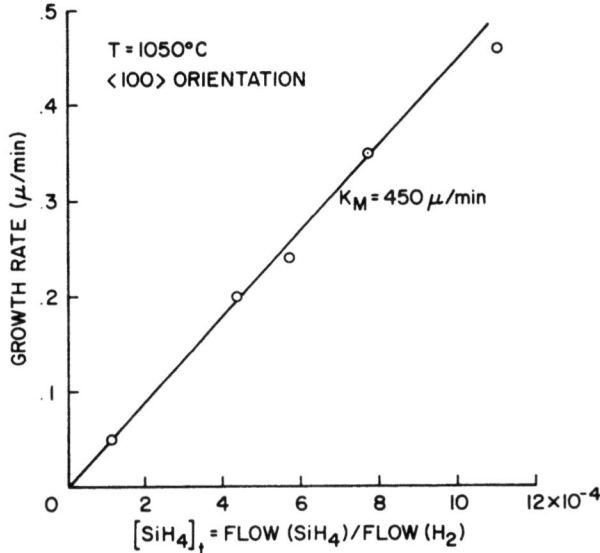

Fig. 4. SiH_4 growth rate vs SiH_4 concentration.

2.3 Doping of epitaxial layers

2.3.1 Injected species. Epitaxial layers are most commonly doped by the incorporation of a small amount of dopant hydrides (PH_3, AsH_3, B_2H_6). Phosphine, arsine, and diborane are gaseous sources that can be diluted to the 10^{-9} level required for lightly doped epitaxial layers. The doping of SiH_4 epitaxial layers with dopant hydride sources has been described theoretically [8,9]. The dopant is transported across the boundary layer by diffusion as described in the previous section for SiH_4. The analysis is complicated, however, by the fact that the incorporated dopant is nonuniformly distributed near the surface of the growing layer. The relevant dopant concentrations in the reactor and in the wafer are presented in Figure 5 for PH_3. The turbulent layer (input) dopant molar concentration $[PH_3]_T$ is the flow ratio PH_3/H_2. The concentration of phosphine near the surface of the growing epitaxial layer is $[PH_3]_I$. The concentration of incorporated phosphorous atoms near the surface of the layer $[P]_I$ is higher than the phosphorous concentration deeper in the bulk material $[P]_B$ [8,9]. All of the incorporated phosphorus is considered to be ionized. The solid-state phosphorous concentrations $[P]_I$ and $[P]_B$ have units of (atoms P/atoms Si) and must be multiplied by the atomic density of silicon ($N_{Si} = 5.0 \times 10^{22}$ atoms/cm^3) to obtain the conventional doping concentration.

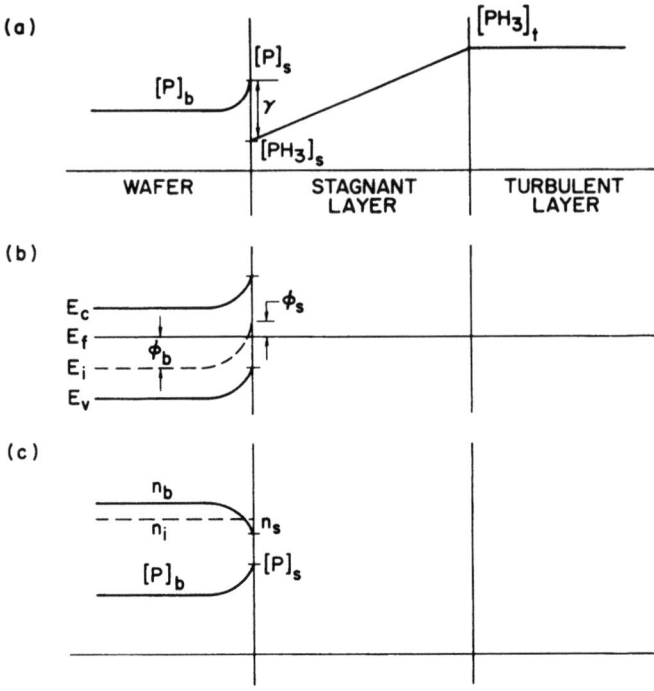

Fig. 5. Dopant concentrations in the reactor and in the wafer.
- (a) Dopant concentrations near an epitaxial interface
- (b) Band bending near a growing silicon surface
- (c) Dopant and free-electron concentrations near a growing silicon surface

The goal of this analysis is to predict the solid-state doping $[P]_B$ from the turbulent-layer phosphine concentration $[PH_3]_T$ as a function of temperature and growth rate. The effective segregation coefficient γ_{EFF} relates these two variables as

$$\frac{[P]_B}{[PH_3]_T} = \frac{1}{\gamma_{EFF}} \tag{8}$$

For doping concentrations lower than the concentration of free electrons in silicon at epitaxial temperatures (10^{19} cm^{-3}), the segregation coefficient is a function of only temperature and growth rate, but not doping level [8]; that is, for a set temperature and growth rate, solid-state doping is a linear function of gas-phase dopant concentration. With respect to the dopant concentrations defined in Figure 5a, Eq. (8) can be expressed as

$$\frac{[P]_B}{[PH_3]_T} = \frac{1}{\gamma_{EFF}} = \underbrace{\frac{[P]_B}{[P]_I}}_{(a)} \cdot \underbrace{\frac{[P]_I}{[PH_3]_I}}_{(b)} \cdot \underbrace{\frac{[PH_3]_I}{[PH_3]_T}}_{(c)} \qquad (9)$$

The overall partition coefficient γ_{EFF} can now be evaluated term by term. Ratio (a) on the RHS of Eq. (9) relates the surface and bulk solid-phase dopant levels that may be different because of band bending at the growing silicon surface. Ratio (b) relates the actual solid and gas doping concentrations at the interface. Ratio (c) is essentially a measure of the boundary layer doping gas gradient which is the driving force that provides the required flux of dopant to the surface.

Following the notation in Figure 5 and the derivations of Rodgers [4] and Bloem [8], the expression for an effective segregation coefficient is

$$\frac{[P]_B}{[PH_3]_T} = \frac{1}{\gamma_{EFF}} =$$

$$\left\{ \frac{\gamma}{\exp\frac{\phi_S - \phi_B}{kT} + \exp\frac{V_C}{V}\left[1 - \exp\frac{\phi_S - \phi_B}{kt}\right]} \right.$$

$$\left. + \frac{h(SiH_4)}{h(PH_3)} [SiH_4]_T \right\}^{-1} \qquad (10)$$

where ϕ_S is the surface potential pinned at one-third the gap above the valence band [8], ϕ_B is the bulk Fermi potential, and V_C is a critical growth rate at which trapping starts to occur. The parameter $h(PH_3)$ is the phosphine transport coefficient similar to $h(SiH_4)$, and the other parameters have their standard meanings. Figure 5 defines the equilibrium segregation term

$$\frac{1}{\gamma} = \frac{[P]_I}{[PH_3]_I} \qquad (11)$$

For low growth rates, the effective segregation coefficient is equal to the equilibrium segregation coefficient γ, modified to account for trapping. Reactant incorporation is surface-reaction limited. For high growth rates, the effective segregation coefficient is proportional to the input silane concentration $[SiH_4]_T$. Four limiting cases of Eq. (10) exist as a result of combining the two limiting conditions in the crystal (bulk-surface equilibrium and complete trapping) with the two limits for dopant transport to the crystal (surface reaction and mass transport). These cases are depicted schematically in Figure 6 for an arbitrary donor hydride DH_3.

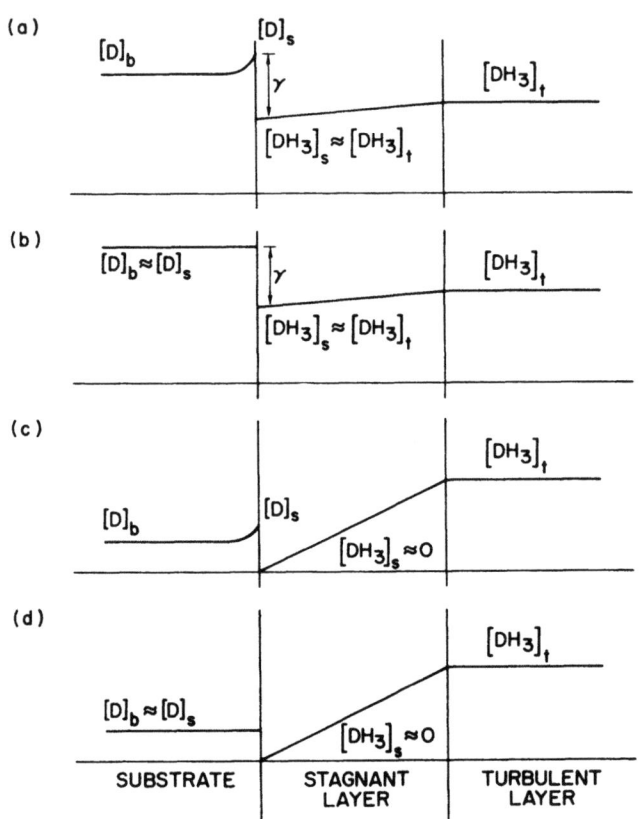

Fig. 6. Limiting-case dopant-concentration profiles for silicon epitaxial layers.

Case 1.

Surface-reaction limited dopant transport and bulk-surface equilibrium conditions would occur during growth at moderately high temperatures, low growth rates, and with a fast diffusing dopant (1150°C, 0.1 μ/min, PH_3) [9]. In this case, Eq. (10) can be approximated by

$$\frac{[P]_B}{[PH_3]_T} = \frac{1}{\gamma_{EFF}} = \frac{1}{\gamma/\exp\frac{\phi_S - \phi_B}{kT}} \qquad (12)$$

The effective segregation coefficient is the equilibrium segregation coefficient modified by band-bending effects.

Case 2.

Surface-reaction limited dopant transport and surface dopant trapping conditions would occur during growth at low temperatures, moderate growth rates, and with a slow diffusing dopant (1000°C, 0.5 μ/min, AsH_3). The effective segregation coefficient is then the equilibrium segregation coefficient,

$$\frac{[As]_B}{[AsH_3]_T} = \frac{1}{\gamma_{EFF}} = \frac{1}{\gamma} \qquad (13)$$

Case 3.

Mass-transport limited dopant delivery and bulk-surface equilibrium conditions would occur during growth at very high temperatures, moderately high growth rates, and with a fast diffusing dopant (1225°C, 2 μ/min, PH_3). The effective segregation coefficient is determined by mass transport of dopant across the boundary layer. Equation (10) can be approximated as

$$\frac{[P]_B}{[PH_3]_T} = \frac{1}{\gamma_{EFF}} = \frac{1}{[h(SiH_4)/h(PH_3)][SiH_4]_T} \qquad (14)$$

Case 4.

Mass-transport limited dopant delivery and surface dopant trapping conditions would occur during growth at relatively low temperatures, high growth rates, and with a slow diffusion dopant (1050°C, 4 μ/min, AsH_3). The effective segregation coefficient is given in Eq. (14).

Although all of the growth conditions depicted in Fig. 6 are possible, practical considerations limit SiH_4 epitaxial growth to the 1000° to 1050°C range and a 0.2 to 1.0 µ/min growth rate. At higher temperatures, the SiH_4 process has the following problems.

(1) At high growth rates, SiH_4 tends to decompose in the boundary layer, forming a "cloud" in the reactor that quickly coats the exit end of the reactor tube.

(2) Silicon deposition on the reactor tube increases for air-cooled tubes.

(3) Autodoping and outdiffusion begin to become problems.

At temperatures below 1000°C, crystal quality suffers; there is a maximum epitaxial growth rate for evey temperature below which polycrystalline deposition occurs. Growth rates below 0.2 µ/min are impractical for common 3 to 20 µ layers because long deposition times are not only wasteful, but lead to excessive outdiffusion. Typical silane-reaction conditions are intermediate, therefore, to the conditions depicted in Figure 6a and 6b. Some surface-dopant trapping is present; an increase in growth rate leads to an increase in solid doping level, all other variables held constant.

The effective segregation coefficient has been measured experimentally over a wide range of conditions. Bulk doping concentration vs monotomic gas-phase dopant concentration is presented in Figure 7 for a combination of different dopant sources, silicon sources, growth rates, and temperatures. The doping concentration is calculated as the effective flow ratio of dopant gas to main hydrogen flow. With reference to the reactor in Figure 1, this ratio is

$$[DX_3] = \frac{f(dopant)}{f(dopant) + f(1st\ H_2)} \cdot \frac{f(mix)}{f(mix) + f(2nd\ H_2)} \cdot \frac{f(inject)}{f(main)}$$

(15)

where f is the flow in liters per minute. The concentration in the right-hand scale is the normal concentration per cubic centimeter divided by N_{Si}. All the dopants have higher concentrations in the bulk than in the gas phase by a factor of 10 to 200 and have approximately the same effective partition coefficient (≈ 0.01) except for AsH_3/SiH_4 at 1050°C and $(CH_3)_3Sb/SiH_4$.

The fact that the effective partition coefficients are independent of growth rate, temperature, and even silicon source for PH_3 and B_2H_6 suggests a surface-reaction limited process with

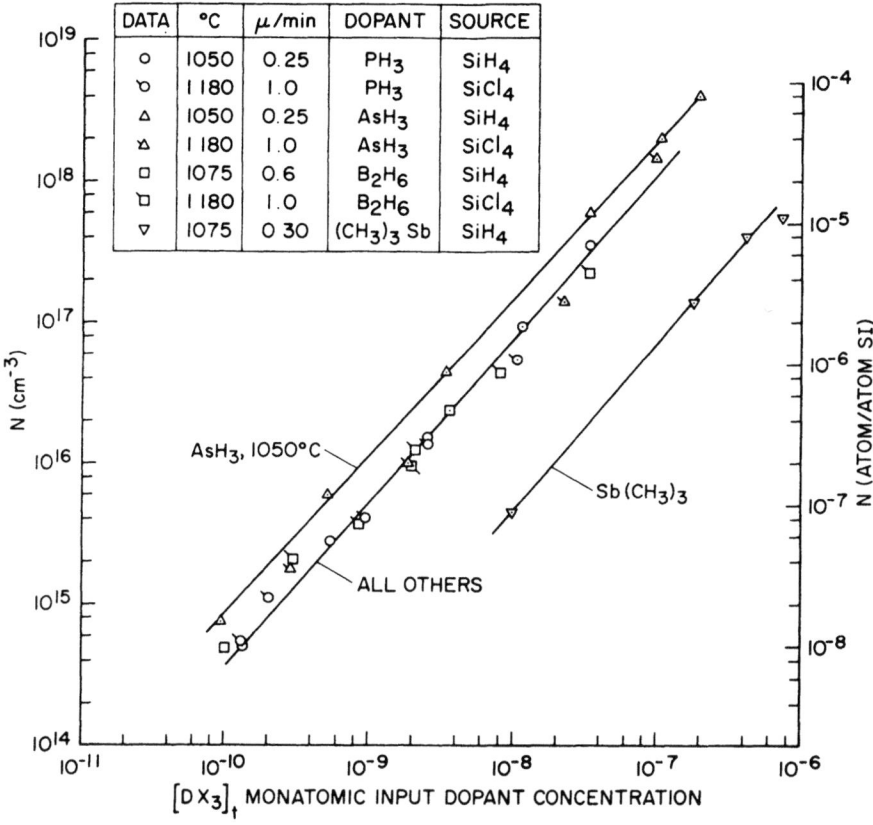

Fig. 7. Solid-state doping vs gas-phase dopant concentration for various dopant sources, silicon sources, and temperatures.

bulk-surface equilibrium (Figure 6(a)). Trapping is not present to any appreciable degree for PH_3/SiH_4 because of the low 0.25 μ/min growth rate. For $PH_3/SiCl_4$, B_2H_6/SiH_4, the high growth temperatures (1075° to 1180°C) establish surface-bulk equilibrium. There is some trapping in the 1050°C AsH_3/SiH_4 system relative to the 1180°C $AsH_3/SiCl_4$ system, as evidenced by the higher solid doping concentration for a given input arsine concentration.

The trapping phenomenon in AsH_3/SiH_4 epitaxial layers has been considered as a function of temperature and growth rate. According to Eq. (10), doping should decrease as temperature (V_c) increases in the presence of trapping. This trend is verified for three growth rates in Figure 8. All the curves begin to flatten out at 1025°C, indicating that trapping is nearly complete, $[As]_B \approx [As]_I$.

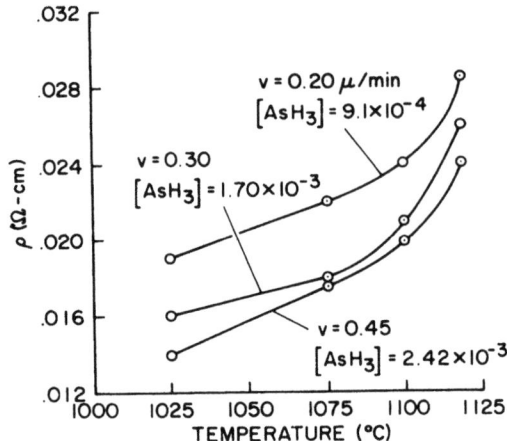

Fig. 8. Resistivity of arsine-doped silane-grown epitaxial layers as a function of temperature with growth rate and arsine concentrations as parameters.

2.3.2 Autodoping. One of the limiting and most important considerations in epitaxial deposition is the movement of substrate impurities into epitaxial layers. Two components are considered -- outdiffusion (movement caused by solid-state thermal diffusion in silicon) and autodoping (extra movement of substrate doping into an epitaxial layer, which cannot be explained by diffusion effects). The following dopant fluxes from a wafer into an epitaxial layer must be considered:

(1) dopant "evaporated" from the back of the wafer, which mixes into the turbulent layer

(2) dopant "evaporated" from the front of the wafer into the boundary layer

(3) dopant that outdiffuses from the substrate as a result of thermal diffusion alone.

For standard buried-collector bipolar processing, a moderately doped (10^{15} to 10^{16} cm^{-3}) epitaxial layer is grown over localized N^+ buried layers. The backside of the wafer is usually masked during the buried-layer diffusion to eliminate this autodoping source. The wafer is then subjected to very high temperature-time cycles to produce junction isolation; in this case, the sealed-wafer backside does not provide an autodoping source, and frontside autodoping is dominated by outdiffusion from the high-temperature isolation step. Consequently, for standard buried-collector processing, outdiffusion can explain dopant migration from the buried layer into the epitaxial layer [10].

For low-temperature silane epitaxial growth, especially when high-temperature processing steps do not follow, there is a definite movement of dopant into the growing layer from the substrate that cannot be explained by outdiffusion. Figure 9 is the profile of an undoped epitaxial layer grown with silane (1100°C, 0.4 μ/min) on an arsenic-doped substrate.* For the 10 min period at high temperature, the diffusion length for arsenic is $\sqrt{\overline{DE}} = 0.041$ μ. The profile solution for outdiffusion [10] is

$$N(epi) = \frac{1}{2} N(substrate) \; erfc \left(\frac{x}{2 \sqrt{Dt}}\right) \tag{16}$$

in which x is the distance into the epitaxial layer measured from the interface. The calculated outdiffusion curve is graphed in Figure 9.

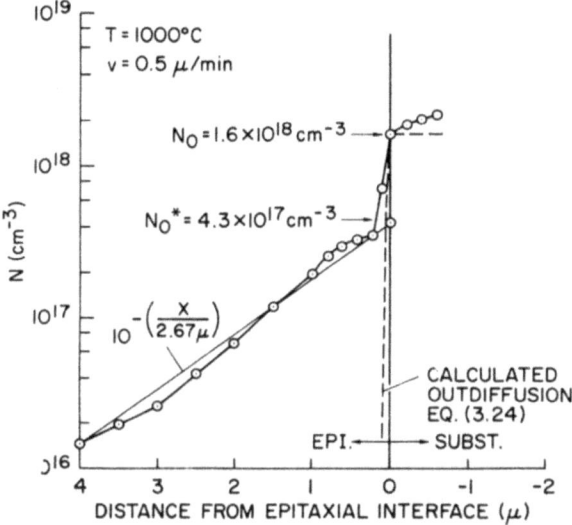

Fig. 9. Autodoping profile for an intrinsic silane-grown epitaxial layer on an arsenic-doped substrate.

In this case, the dopant movement (here called "autodoping") is dominant over the "outdiffusion" term. Because, for this experiment, the arsenic-doped "wafer" was actually an arsenic-doped epitaxial layer deposited on a lightly boron-doped wafer, the sole component of autodoping contributing to the measured profile

―――――――――――――――――――
*All profile measurements were made using the spreading-resistance probe technique.

was evaporation from the front of the wafer into the growing layer. A well-known feature of the frontside evaporation solution is that the evaporation rate is proportional to the surface concentration [11]. As the surface concentration, resulting from autodoping decreases, the evaporation rate decreases which, in turn, reduces concentration in the growing layer; the doping profile is a simple decaying exponential extrapolating back to an effective surface concentration N_o^* which is determined empirically for a given dopant, silicon source, and set of growth conditions. For arsenic-silane in Figure 9, the profile is close to exponential, dropping a factor of 10 every 2.67 µ.

This behavior has also been measured in layers grown at higher temperatures with $SiCl_4$. Figure 10 shows the measured autodoping tail of a boron-doped substrate into an undoped epitaxial layer grown with $SiCl_4$ at 1180°C and 1.0 µ/min. Here, the calculated outdiffusion is comparable to the autodoping. Outdiffusion is much more prominent than in the previous case because of the higher growth temperature and the higher diffusivity of boron relative to arsenic. Again, an exponential fit is within experimental error. The curve decays a factor of 10 every 0.86 µ. If a substantial subsequent heating step such as diffusion isolation had been done to the substrates used in the boron $SiCl_4$ experiments, the outdiffusion tail would have become dominant over autodoping.

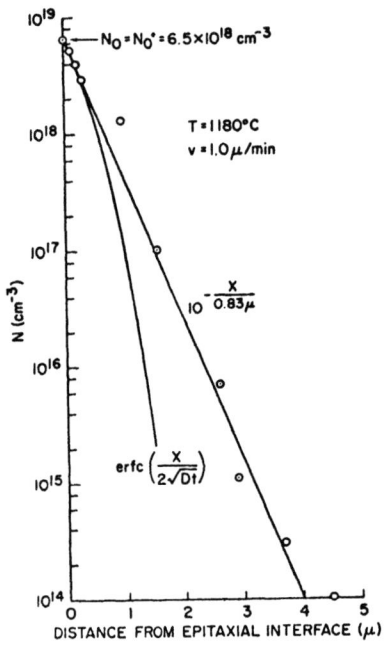

Fig. 10. Autodoping profile fo for an intrinsic silane-grown epitaxial layer on a boron-doped substrate.

One immediate conclusion that can be drawn from a comparison of Figures 9 and 10 is that the dopant species is a very important consideration. One would expect a priori the slow-diffusing dopant, As, grown at the lower temperature to have much less autodoping than the fast-diffusing dopant, B, grown at the higher temperature. Because the opposite is true, arsenic must be a poorer autodoping element than boron. One significant factor is vapor pressure. Arsenic "evaporates" more easily than boron. The exact nature of the autodoping differences between the elements have not been reported, but empirical data to be presented here demonstrate that the worst-to-best autodoping order of common silicon dopants is As >> B > P > Sb.

The second source of dopant for autodoping is evaporation from the backside of the wafer. The amount of autodoping is proportional to the evaporation rate. This rate and the autodoping profile depend to first order on wafer dopant concentration N_o. The net effect is that the wafer backside supplies dopant to the epitaxial layer at a rate initially set by its concentration, evaporation rate, diffusivity, and temperature. The amount of dopant supplied then decreases slowly in time because of dopant-depletion effects on the wafer backside. The time variation is usually weak so that the backside appears as a relatively constant autodoping source. The solution to this boundary condition is simple. For a given set of growth conditions, the autodoping profile is constant and dependent only on initial wafer-backside doping. In a study [12] of intrinsic epitaxial layers (SiH_4, 1050°C, 0.35 μ/min) grown on arsenic-doped wafers, the following empirical relation was derived:

$$N(\text{constant autodoping level}) = (2.5 \times 10^5) \, N(\text{backside}) \quad (17)$$

The constant autodoping levels were measured at 4 to 7μ from the original wafer surface, after the point at which the frontside evaporation and outdiffusion transients had died out.

With a general description of the three components of dopant movement into an epitaxial layer (outdiffusion and frontside and backside autodoping), a conceptual picture of autodoping can be drawn. Figure 11 shows that outdiffusion dominates nearest the interface, frontside evaporation effects become apparent as soon as the outdiffusion tail dies out, and backside evaporation dominates the autodoping profile farthest from the interface because the backside autodoping source remains relatively constant and the frontside source is continuously buried under growing silicon. Figure 11 is highly idealized; for example, the analysis does not consider lateral autodoping variations that include systematic worse autodoping at the wafer edges as a result of the backside source. Any of the three regions in Figure 11 may not be present

under certain conditions, (large outdiffusion tail may go completely
to the surface of a thin epitaxial layer).

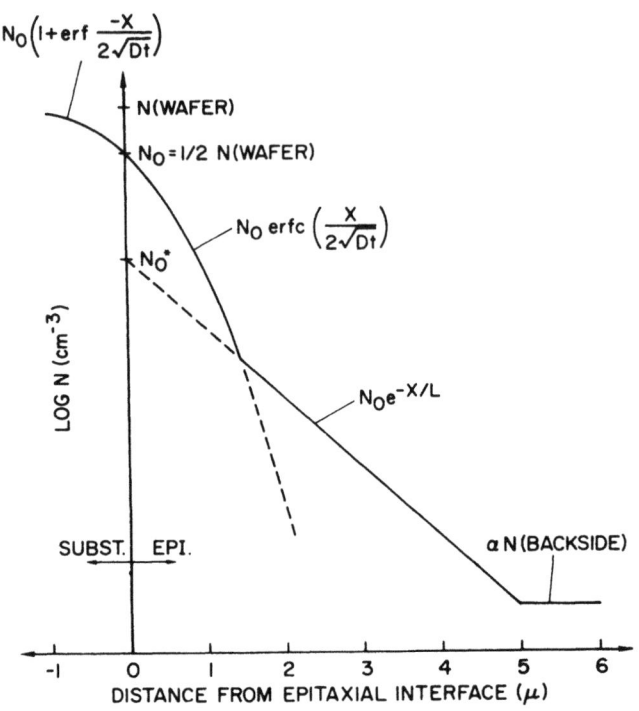

Fig.11. Conceptual diagram of autodoping and outdiffusion profiles.

2.3.3 <u>Time-dependent effects</u>. The above subsections have
defined injection and autodoping effects primarily for the steady-
state condition. Recent work has demonstrated practical means for
obtaining a predefined doping profile by utilizing transient doping
effects. A two-step process [13] using silane has achieved abrupt
epitaxial interfaces by allowing the boundary layer vapor pressure
P_g to decay with time after the growth of a first layer. The concentration in the first layer is reduced from the bulk value N_B to a
value set by the ratio of loss coefficients ξ_{BI} from bulk to the
first boundary layer, and ξ_{IE} from the second boundary layer to
the epitaxial layer,

$$N_B^* = \frac{\xi_{IE}}{\xi_{BI}} N_B \qquad (18)$$

Following this decrease in surface concentration, the vapor pressure decays to a second equilibrium value with an exponential time behavior. Subsequent epitaxial growth after the decay of P_g results in an abrupt doping profile.

A transfer-function characterization method has been reported that predicts profiles for time-varying injected dopant species [14]. Figure 12a is a block diagram of the epitaxial reactor as a system, the input is the gas flow as a function of time f(t), and the output is the doping density profile in the epitaxial layer N(x). All four blocks interact, With respect to the doping mechanism and despite such complexity, the reactor can be considered analogous to a linear system within limitations. The implication of this is that it is possible to obtain a transfer function of the reactor that relates the dopant gas-flow input f(t) to the doping profile output N(x) as shown in Figure 12b. After the transfer function has been determined, it is possible to calculate the dopant gas flow as a function of time required to achieve a desired dopant profile in an epitaxial film.

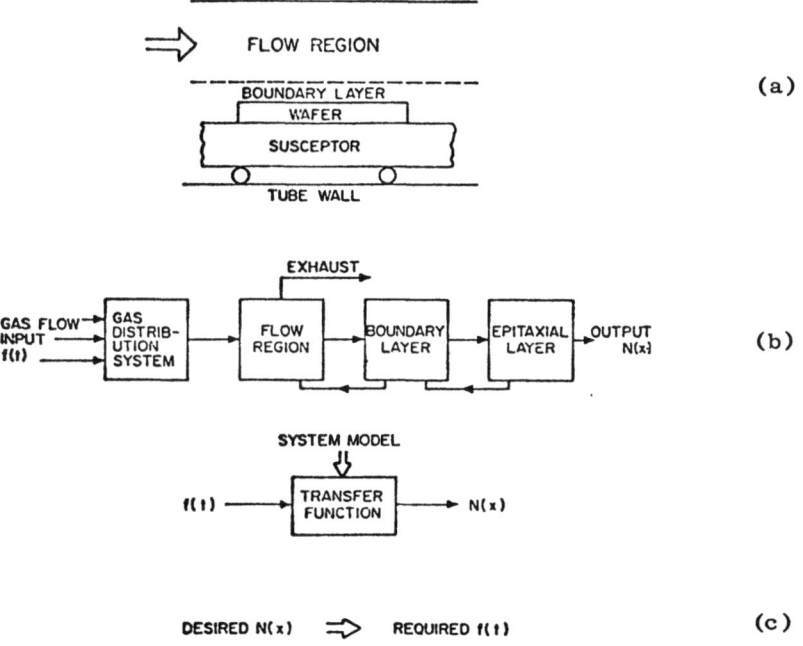

Fig. 12. Block diagram of reactor system.
  (a) Cross section of the reactor chamber
  (b) The reactor as a system
  (c) One transfer function relating the dopant gas-flow input f(t) to dopant concentration in the epitaxial film N(x)

2.3.3a Transient response of the dopant incorporation mechanism. A linear system can be characterized by obtaining its transient response. After this response is known, it is a relatively easy task to determine the transfer function.

Although there are definite limits to this approach, it has been used successfully in several chemical systems. For example, diffusion-rate processes in catalytic reactors have been characterized by "admittance functions" analogous to ac electrical parameters [15]. The time-varying output-gas concentrations were related to the time variations of the input gases. Further work has considered diffusion through a boundary layer, adsorption on a surface, and diffusion into a solid [16]. Several of the same kinetic processes are present in our study of dopants in epitaxial layers; however, additional chemical processes are involved because the output is not a time-varying gas concentration, but a variation of the solid dopant concentration as a function of position in the deposited epitaxial layer.

In addition to providing a method for fabricating a desired dopant profile, this investigation will produce basic information concerning the various mechanisms involved in the dopant-inclusion processes. As indicated by Kobayashi and Kobayashi [17], transient studies may reveal the importance of various mechanisms that may not be apparent in steady-state studies. In particular, consideration of limiting cases may allow separation of the several mechanisms involved; dopant gas flow without film growth, for example, would involve only a selected number of mechanisms involved in the total epitaxial doping process.

The reactor was first optimized for a nominal deposition of approximately 0.6 μ/min and uniform doping during the entire deposition process. Silane was the source of Si, and deposition was accomplished at 1070°C. Arsine was selected as the dopant gas, and flow settings were determined for the typical partial pressure of approximately 2×10^{-6} and 6×10^{-6} atm, corresponding to dopant concentrations of roughly 1×10^{15} and 3×10^{15} cm^{-3} in the epitaxial layer. The substrates were <100>-oriented silicon wafers with phosphorous dopant concentrations in the 10^{15} cm^{-3} range. The epitaxial layers were grown with a step-function change in the dopant gas flow during the continuous deposition; during the deposition, the dopant gas flow was changed from one of the above described to the other to simulate the step input. Both the increasing and decreasing steps were used in these experiments.

The dopant profiles in the epitaxial layers were obtained by capacitance-voltage measurements on deep-depletion MOS structures and on planar and mesa p-n junctions. The thickness and dopant

concentrations in the samples were chosen to be compatible with
the C-V technique. Spreading-resistance measurements and stacking-
fault thickness determinations were used to confirm the capacitance-
voltage measurements. Figures 13 and 14 are the resulting dopant
profiles. To ensure that the experimental profiles were not limited
by the resolution of the capacitance-voltage data-reduction technique,
a theoretical capacitance-voltage curve was generated by solving
Poisson's equation. The data-reduction program was then applied
to this curve. The results revealed that the experimental profiles
varied slowly compared to the resolution profile, indicating that
the major features observed were not artifacts of the analysis
technique.

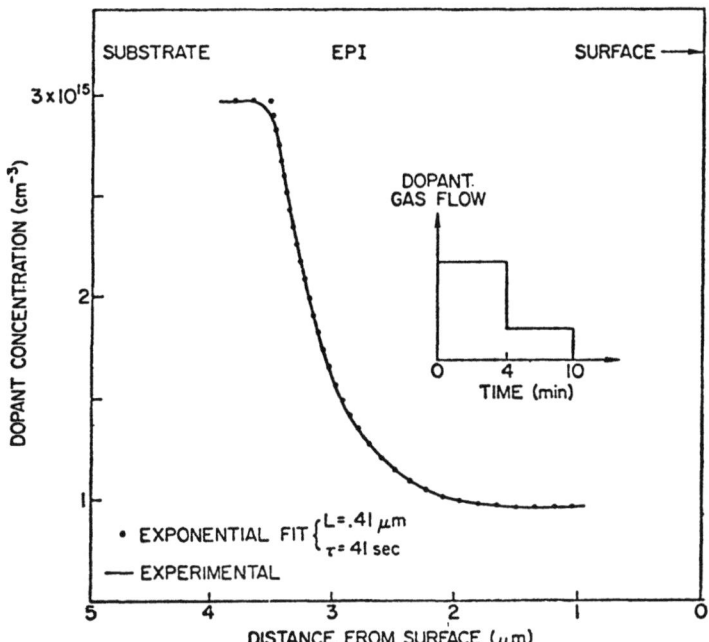

Fig. 13. Experimentally observed dopant concentration as a function
of distance from the surface of an epitaxial film for a decreasing
step change in the dopant gas flow. Also shown is an exponential
fit to the experimental curve.

As can be seen in Figures 13 and 14, the transition from one
dopant concentration to the other occurred in approximately 1.2 μ,
corresponding to roughly 2 min. The heat cycling during the fabrication of the MOS and p-n junction C-V samples produced a square-
root of Dt of only 0.07 μ because of the low diffusivity of the
arsenic dopant in the epitaxial layer. Spreading-resistance data

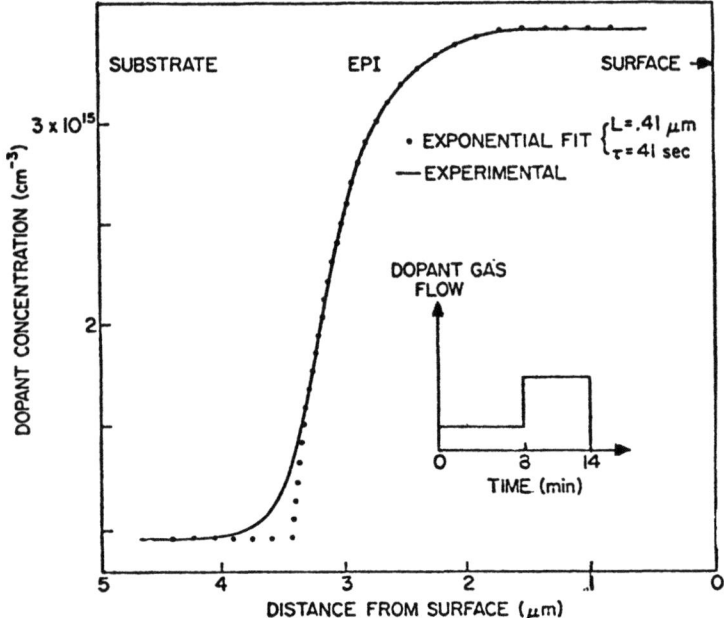

Fig. 14. Experimental dopant profile for an increasing step change in dopant gas flow and exponential fit to the experimental profile.

confirm that there is no significant deviation between samples measured after epitaxial deposition and after complete C-V sample fabrication. The increasing and decreasing steps reveal an abrupt onset of the transition region followed by a gradual decay. As seen in Figures 13 and 14, the decay curve can be fitted to an exponential curve with fairly good accuracy, and a decay length of 0.41 μ corresponds to a time constant of 41 sec. As a result, the transient response of the epitaxial factor can be approximated by a single exponential function, and the response is given by

$$H(t) = 1 - \exp\left(-\frac{t}{T_1}\right) \qquad (19)$$

where T_1 = 40 sec.

2.3.3b <u>System analysis.</u> After the transient response of a system has been characterized, it is possible to predict the output for a certain input. One of the simplest methods is to use the convolutional integral.

In a linear system, the output g(t) to an input f(t) can be calculated by convolving the input with the impulse response I(t),

$$g(t) = \int_0^\infty I(\tau) \, f(t - \tau) \, d\tau \tag{20}$$

The impulse response can be determined by taking the time derivative of the step response $h(t)$; $I(t) = \partial h(t)/\partial t$. Because $h(t) = 0$ for negative values of t, $I(\tau)$ is also zero for negative values of τ. The lower limit of integration can be simplified, therefore, to

$$g(t) = \int_0^t I(\tau) \, f(t - \tau) \, d\tau \tag{21}$$

The limits of integration can be further simplified, depending on the form of the input. For **example,** if $f(t) = 0$ for $t < 0$, then $f(t - \tau) = 0$ for $t < \tau$, and thus the upper limit of the integration becomes

$$g(t) = \int_0^t I(\tau) \, f(t - \tau) \, d\tau \tag{22}$$

Equations (21) and (22) can be evaluated analytically or numerically through numerical integration.

Similar calculations can be performed in the frequency domain via the Laplace or Fourier transform. The advantage of the Laplace transform is that it is possible to draw analogs to electrical circuits; the advantage of the Fourier transform is that numerical methods are very efficient. In both techniques, repeated transforms are necessary to convert the information from the time domain to the frequency domain and then back to the time domain.

In the epitaxial reactor, differentiating Eq. (19) obtains

$$I(t) = \frac{1}{T_1} \exp\left(\frac{-t}{T_1}\right) \tag{23}$$

Because the growth rate G is constant in time, the doping density $N(x)$ is

$$N(x) = G \, g(t) \tag{24}$$

By combining Eqs. (22), (23), and (24),

$$N(x) = \frac{G}{T_1} \int_0^\infty \exp\left(-\frac{\tau}{T_1}\right) f(t - \tau) \, d\tau \tag{25}$$

As noted earlier, the limits of integration can be simplified, depending on the type of f(t).

2.2.3c <u>Verification of the approach</u>. In a series of experiments, the dopant gas flow consisted of an increasing step followed by a decreasing step, approximating a pulse in the input. Four pulse widths were used (4.3, 3. 1.8, and 0.8 min). The resulting doping profiles in the substrates were measured using the C-V techniques, and Figure 15 plots the results. For pulsewidths of 4.3 and 3 min, the resulting profiles appear to be a simple superposition of profiles obtained in Figures 13 and 14 because the pulse widths are longer compared to the system time constant (41 sec). When using shorter pulses of 1.8 and 0.8 min, however, the higher limit of doping was never reached, thereby indicating the limits of the deposition system in responding to arbitrary changes in the rate of dopant gas flow.

To verify the validity of the transfer-function approach, doping profiles were calculated using Eq. (25) for the four pulse inputs. The results are plotted in Figure 15. There appears to be excellent agreement between theory and experiments except at the onset of the high-to-low region. Even for the shortest pulse, where the pulse width of 48 sec is comparable to the system time constant of 41 sec, the model works with good accuracy.

In another experiment, the shape of the dopant gas-flow input function was a ramp followed by a step. The results of measurements and theoretical calculations are plotted in Figure 16 which, again, displays excellent agreement except for one discrepancy -- the initial and final values of the dopant gas flow are the same but are different for doping profiles in the epitaxial layer. This can be attributed to some mechanical malfunction in one of the flowmeters.

2.4 Summary for epitaxy

This section has defined several key problem areas in modeling silicon epitaxy. The kinetics of film growth and dopant inclusion have been discussed. First order models have been presented. In the final subsection recent results in transfer function modeling work in progress has been discussed.

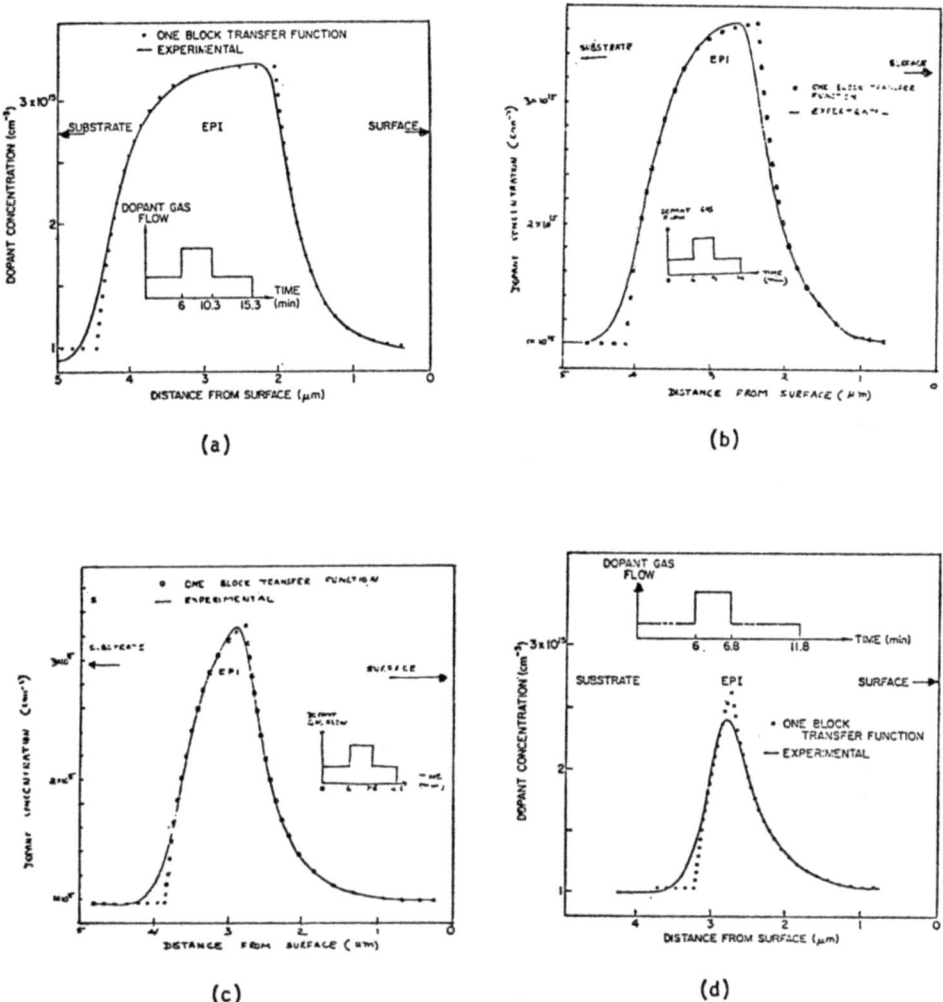

Fig. 15. Response of the reactor to pulse inputs in the dopant gas flow. The pulse width in time is
(a) 4.3 min (b) 3 min (c) 1.8 min, and (d) 0.8 min.

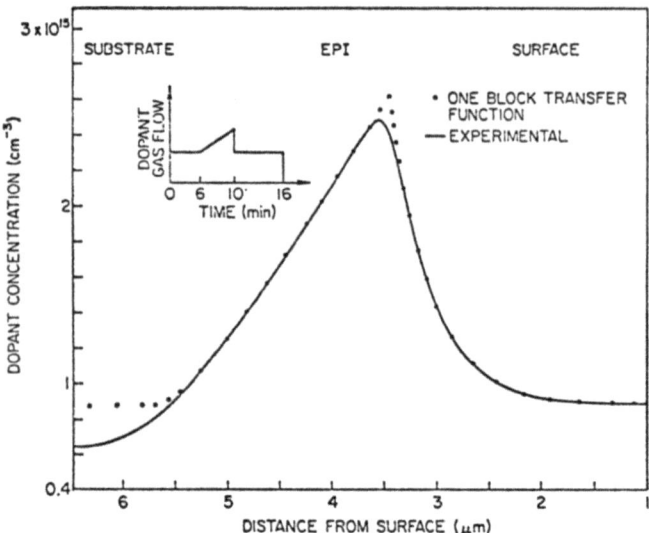

Fig. 16. Response of the reactor to a ramp followed by a decreasing step input in the dopant gas flow.

3. SILICON OXIDATION

3.1 Introduction

Modeling the kinetics of silicon oxidation is well known [18]. The first-order model is based on the diffusion and surface reaction of a single oxygen species at the silicon interface. Present trends in the processing technology, however, are quickly outstripping the utility of the simple model. For example, concentration-dependent effects can change oxide thickness by as much as 300% for otherwise identical conditions. Low-temperature processing makes this effect even more pronounced. Partial pressures of oxygen, HCl, and other ambients also change growth parameters. Two-dimensional local oxidation with inherent strain-dependent effects is now a common technology tool which requires more extensive understanding of physics beyond the SiO_2-Si interface [41]. In addition, there are electrical alterations of the SiO_2-Si interface that result from physical properties. Deal [42] has recently reviewed this subject and it is clear that further work in this area is required. The presence of oxide and surface states can substantially change MOS threshold leakage, surface mobility, and overall device stability.

The multistream models for oxidation kinetics and two-dimensional dependencies will not be discussed here. It is anticipated that model development in these areas will benefit directly from other numerical modeling work that couples and conserves diffusing particles The following discussion will be confined to problem formulations that have evolved as direct extensions of the first-order oxidation model [18]. This model will be reviewed and experimental results and interpretations will be considered for orientation, concentration, and ambient-dependent oxide growth conditions. In all cases, modifications amount to coefficient changes in the basic model.

3.2 The basic model

The first-order model solves for the oxide film growth by integrating a flux equation for oxygen as it crosses the silicon-oxide layer of thickness, x_o, and reacts at the silicon surface. Figure 17 is a schematic of the model, with the appropriate flux terms. According to Henry's Law, the equilibrium solid concentration is proportional to the bulk gas partial pressure, P_G,

$$C^* = H P_G \tag{26}$$

where C^* is the maximum silicon-oxide concentration for a given P_G and H is the Henry's Law coefficient. Because we are treating non-equilibrium, the solid value is less than C^*. The flux, F_1, is determined by the difference in solid concentrations,

$$F_1 = h(C^* - C_o) \tag{27}$$

Here, C_o is the silicon-oxide surface concentration and h is the mass-transfer coefficient. The value of h in the solid can be related to the gas-transfer coefficient by

$$h = \frac{h_G}{HkT} \tag{28}$$

where h_G is the mass-transfer coefficient in the gas. The second and third flux terms shown in Figure 17 represent diffusive flux (Fick's Law) and the surface reaction. The diffusive flux is determined by the concentration gradient and the effective diffusivity D_{eff},

$$F_2 = D_{eff} \frac{\partial C}{\partial x} \approx D_{eff}\left(\frac{C_o - C_i}{x_o}\right) \tag{29}$$

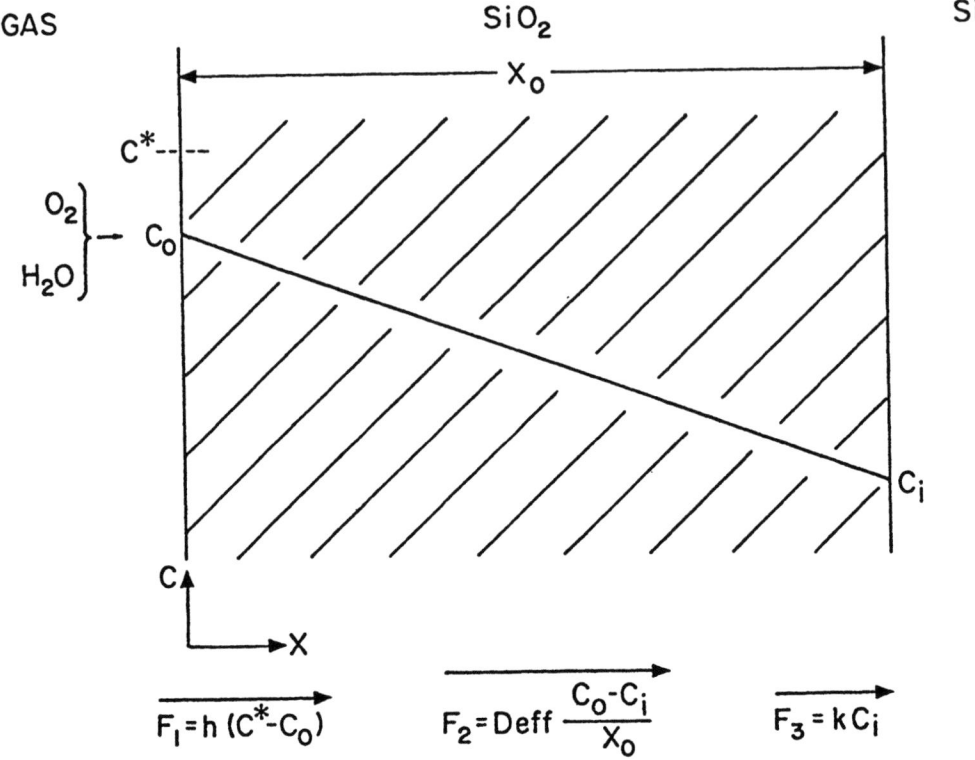

Fig. 17. Boundary and Flux conditions for the Gas-SiO$_2$-Si system.

In this case, C_i is the interface value of the oxidizing species in the silicon oxide. The interface reaction rate is

$$F_3 = k\, C_i \tag{30}$$

where k is the surface-reaction rate constant for oxidation.

Using the condition at steady state ($F_1 = F_2 = F_3$), one can solve for C_i and C_o in terms of C^*:

$$C_i = \frac{C^*}{1 + \frac{k}{h} + \frac{kx_o}{D_{eff}}} \tag{31}$$

and

$$C_o = \left(1 + \frac{kx_o}{D_{eff}}\right) C_i \tag{32}$$

Two limiting cases can be defined. Surface-reaction rate control occurs when D_{eff} becomes large (compared to the surface rate constant k) and

$$C_i \approx C_o \approx \frac{C^*}{1 + \frac{k}{h}} \tag{33}$$

This condition implies that surface-reaction kinetics is the slowest, and hence dominant, effect. Mass transport, or diffusion control, occurs when $k \gg D_{eff}/x_o$ and C_i approaches zero. In turn, $C_o \approx C^* = Hk\,T\,C_G$ where C_G is the gas-phase oxidation concentration. Under these conditions, the flux toward the surface becomes

$$F = \frac{D_{eff}}{x_o} C^* \tag{34}$$

and the dominant effect is transport across the oxide layer.

To describe the rate of oxide growth, the flux equation at the interface can be written as

$$N_1 \frac{dx_o}{dt} = F_3 = \frac{kC^*}{1 + \frac{k}{h} + \frac{kx_o}{D_{eff}}} \tag{35}$$

where N_1 is the number of oxidant molecules incorporated per unit volume ($2.2 \times 10^{22}/cm^3$ for dry O_2, and $4.4 \times 10^{22}/cm^3$ in a wet ambient). The following integral defines the relationship of x_o and t:

$$N_1 \int_{x_i}^{x_o} \left(1 + \frac{k}{h} + \frac{kx_o}{D_{eff}}\right) dx = kC^* \int_0^t d\tau \tag{36}$$

which results in

$$x_o^2 + Ax_o = B(t + \tau) \tag{37}$$

where

$$A \equiv 2 D_{eff} \left(\frac{1}{k} + \frac{1}{h}\right) \tag{38}$$

$$B \equiv \frac{2 D_{eff} C^*}{N_1} \tag{39}$$

and

$$\tau \equiv \frac{x_i^2 + Ax_i}{B} \tag{40}$$

The quadratic equation for x_o can be solved by the following limiting cases.

Case 1.

$$x_o \approx \frac{B}{A}(t + \tau) \qquad \text{for } t + \tau \ll \frac{A^2}{4B} \tag{41}$$

and B/A is the "linear" rate constant.

Case 2.

$$x_o^2 \approx B(t + \tau) \qquad \text{for } t + \tau \gg \frac{A^2}{4B} \tag{42}$$

and B is the "parabolic" rate constant.

Examination of the limiting forms indicates that the oxidation process in the parabolic domain is diffusion limited and, in the linear region, it is surface-reaction limited (generally, h >> k). Both regions should be directly dependent on the equilibrium concentration of oxidant in the oxide. Factors affecting the diffusion process should be most influential, therefore, over long oxidation times.

Experiments have indicated that, with an intially clean bare Si surface, $x_i = 0$ for wet O_2 oxidation, and an effective $X_i \simeq 200$ Å is found for dry O_2 oxidation (as a result of an initial phase of rapid oxidation by a different mechanism).

At the present time B, B/A, and τ are known only for (111)-oriented lightly doped conditions. Under these restrictions, these parameters can be expressed in the following form:

$$B = C_1 e^{-E_1/kT} \tag{43}$$

$$B/A = C_2 e^{-E_2/kT} \tag{44}$$

$$\tau = (X_i^2 + AX_i)/B \tag{45}$$

where

dry O_2: $C_1 = 7.72 \times 10^2 \, \mu^2/hr$
$C_2 = 6.23 \times 10^6 \, \mu/hr$
$E_1 = 28.5$ kcal/mole = 1.23 eV/molecule
$E_2 = 46.0$ kcal/mole = 2.0 eV/molecule

wet O_2: $C_1 = 2.14 \times 10^2 \, \mu^2/hr$
$C_2 = 8.95 \times 10^7 \, \mu/hr$
$E_1 = 16.3$ kcal/mole = 0.71 eV/molecule
$E_2 = 45.3$ kcal/mole = 1.97 eV/molecule

Orientation, the addition of a chlorine species, and heavy substrate doping have all been shown to affect B, B/A, or both rate constants. The following sections discuss these individually.

3.3 Silicon crystal orientation effects

A log-log plot of oxide thickness vs oxidation time for the thermal oxidation of silicon in dry oxygen at temperatures ranging from 700° to 1200°C is shown in Figure 18. As expected, the difference in oxidation rate between (100)- and (111)-oriented silicon decreases gradually from 700°C (where the differences are on the order of 40%) to 1200°C (where the differences are less than 2% for all points) with the (111) oxidizing faster than the (100). It should be noted that, in the region below 100 Å at 700°C, the oxide thicknesses obtained for (100) and (111) silicon are essentially the same. Only after the oxide thickness exceeds 100 Å do the normal orientation differences appear. This observation suggests that, in the initial oxide growth regime at 700°C, the silicon surface

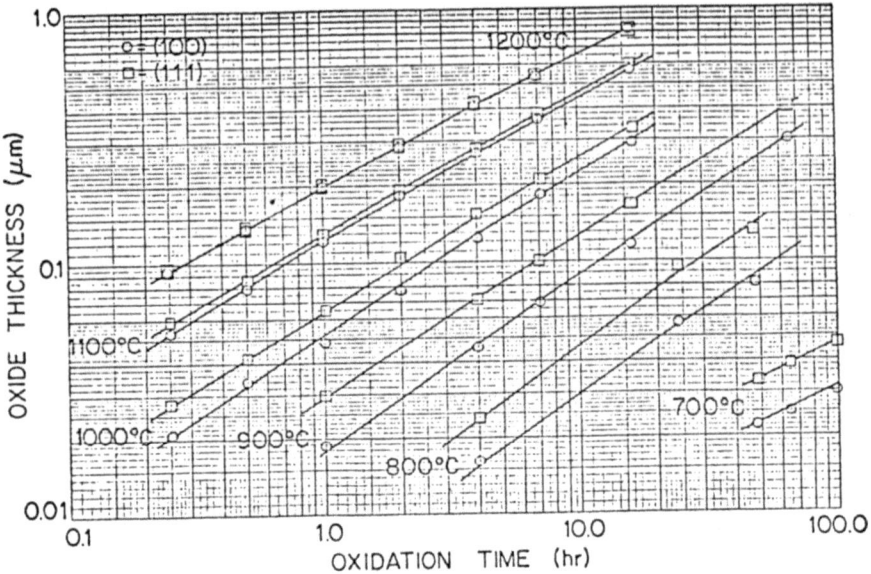

Fig. 18. Oxide thickness vs time for <111> and <100> silicon.

plays a minimal role. Such conclusions are consistent with previous data [19] obtained by Auger analysis of thin oxides grown between 200° and 800°C, which show that the hydration of the silicon surface plays a crucial role in silicon oxidation. The results can be of interest in the fabrication of MNOS devices.

The oxidation rate data were evaluated using the general relationship for the thermal oxidation of silicon given in Eq. (37). Because the parabolic rate constant B depends on the partial pressure of oxidant in the oxidizing ambient, the solubility of the oxidizing species in SiO_2, and the effective diffusion coefficient of the oxidizing species in SiO_2 [18], there should be no variation of B with silicon orientation. On the other hand, the linear rate constant B/A involves the reaction rate constants at the SiO_2 surface and at the silicon surface [18] so that any orientation dependence of the oxidation rate should appear in this term via the constant A.

Because the method of analysis utilized depends on the correction factor τ, which is determined by extrapolation and is thus prone to some error, oxidation times at least three times greater than the τ value for any set of oxidation conditions were generally used for the determination of B and B/A. In this way, any error in τ should have little effect on the calculated values of B and B/A. The results of least-squares analyses of oxidation-rate data for (100)- and (111)-oriented silicon wafers using Eq. (37) are given in Table 4. It should be mentioned that the oxide thickness (x_i) at zero oxidation time as determined by extrapolation of the x_o vs t curve to zero oxidation time, was found to be 160 ± 40 Å for all oxidation conditions. Table 4 demonstrates that, according to theory, there is essentially no difference in B for the two orientations, at least from 1200° to 900°C. Below 900°C, very long oxidation times (> 100 hr) become necessary to eliminate the effects of τ, and the scatter observed was considerable. At 700°C, a value for B could not be obtained because a plot of oxide thickness versus time yielded a straight line, indicating that the parabolic contribution to the overall oxidation rate was negligible. The linear rate constant B/A, shows the expected orientation dependence, with B/A for (111) always being equal to or greater than that for (100). It should be noted that the kinetic parameters for (111)-oriented silicon in Table 4 are in very close agreement with those obtained by Deal and Grove [18]. Such agreement is remarkable when one considers the different oxidation conditions (humidity, gas quality, flow rates, furnaces, etc.) prevailing in two laboratories carrying out similar experiments 11 years apart.

Activation energies were obtained by fitting the data in Table 4 to an Arrhenius equation in the form of

$$k = k_o e^{-E_a/kt} \tag{46}$$

where E_a is the activation energy, T is the temperature in °K, and k is a gas constant. Plots of B and B/A vs temperature are shown in Figures 31 and 32 in connection with the HCl data described later. This analysis obtained values of 1.3 eV (30 kcal/mole) and 1.2 eV (28 kcal/mole) for the activation energy of B for (100)- and (111)-oriented silicon, respectively, which agrees well with previously determined values for the dry thermal oxidation of (111)-oriented silicon [19, 20].

A similar analysis for the linear rate constant for the (111) orientation yielded a value of 2 eV (47 kcal/mole), which is in close agreement with the energy required to break a Si-Si bond [21]. Determination of the activation energy for B/A for (100)-oriented silicon, however, affords two choices. If the data from 900° to 1200°C are analyzed and the low-temperature points are neglected because of large scatter, a value of 2.5 eV (57 kcal/mole) is obtained -- an increase of ≈ 25% over that for (100). This

Table 4

RATE CONSTANTS FOR <111> AND <100> SILICON

Temperature (°C)	Orientation	τ (hr)	A (μm)	B (μm^2/hr)	B/A (μm/hr)
1200	(100)	0.03	0.0399	0.0453	1.14
	(111)	0.03	0.0404	0.0458	1.13
1100	(100)	0.09	0.101	0.0247	0.246
	(111)	0.09	0.0845	0.0244	0.289
1000	(100)	0.35	0.195	0.00913	0.0467
	(111)	0.35	0.120	0.00956	0.0797
900	(100)	3.2	0.429	0.00332	0.00775
	(111)	1.2	0.214	0.00381	0.0178
800	(100)	10.0	0.441	0.000755	0.00171
	(111)	5.0	0.354	0.00119	0.00335
700	(100)	--	--	--	0.000222
	(111)	--	--	--	0.000348

increase was predicted by Ligenza from consideration of steric hindrance in the (100) and (111) planes of silicon during thermal oxidation [22].

On the other hand, if the data from 700° to 1200°C are analyzed, the 900°, 1100°, and 1200°C points could be considered as scatter from a line drawn parallel to the one for (111) orientation, passing through the 700°, 800°, and 1000°C points, thereby resulting in the same activation energy as (111)-oriented silicon. Indeed, some faith could be put in such as assumption, because the 700°C data give B/A directly from an x_o vs t plot as indicated above. One could argue, therfore, that the higher temperature analyses were in error due to the inaccuracy in determining A from a plot of x_o vs $(t + \tau/x_o)$. The slopes of these lines (B) are quite large; therefore, a small change in the slope would shift the intercept (A), and thus B/A, considerably. At present, it is not known which of the above alternatives is the correct one. In any case, the preceding discussion indicates that there has been little or no work reported since 1961 concerning the differences between oxidation rates and corresponding activation energies of (100)- and (111)-oriented silicon in dry oxygen.

3.4 Impurity doping effects

The effects of impurity doping levels on thermal oxidation rates are intimately connected with a widely encountered phenomenon in semiconductor processing -- namely, impurity redistribution. Figure 19 illustrates how redistribution and thermal oxidation interact.

As a thermal oxide is grown over a doped silicon substrate, redistribution of the impurity results. In the case of phosphorus, arsenic, and antimony, the dopant atoms tend to pile up at the surface, resulting in a higher surface concentration than background concentration ($C_s \gg C_B$). In boron, the opposite effect takes place, resulting in surface depletion ($C_s \ll C_B$).

Very heavily doped substrates ($C_B \approx 10^{19}$), it has been observed [23] that for both phosphorus and boron, the oxidation rates can be substantially different (generally faster) than those observed on lightly or moderately doped substrates. With respect to Figure 19, the two parameters that have been correlated [23] with this increased oxidation are C_s (the dopant concentration in the silicon at the surface) and C_{ox} (the average impurity concentration in the oxide).

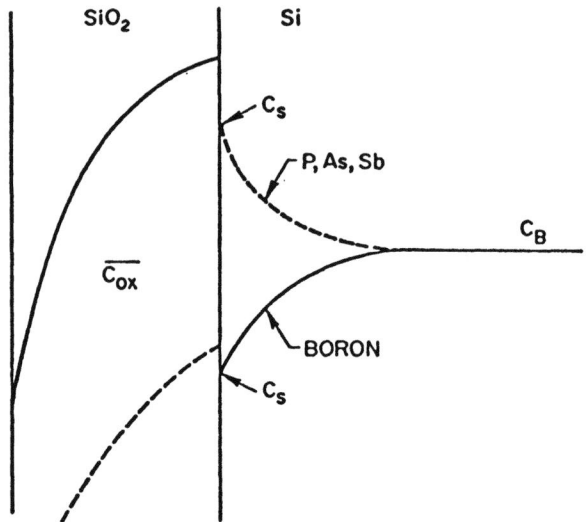

Fig. 19. Effect of redistribution on impurity doping profiles.

Intuitively, the effect of C_s is to reduce the amount of energy required to break Si bonds and thus to affect the surface reaction; it would thus be expected to influence the linear rate constant B/A and not B. The effect of C_{ox} is to change the diffusion constant for the O_2 or H_2O oxidizing species in the SiO_2, and it would thus be expected to affect the parabolic rate constant B.

To study these enhanced oxidation rates in more detail, a large number of heavily doped wafers were prepared as illustrated in Figure 20. Because of the difficulty in purchasing wafers with doping levels > $10^{19}/cm^3$, the samples were prepared by diffusion of phosphorus into standard lightly doped (111) substrates. The predeposition and drive-in schedules were chosen so that the diffused profiles were flat (within 10%) over the first 2 μ into the silicon. The drive-ins were done in N_2 to prevent oxidation and hence redistribution during the formation of heavily doped layers.

Five types of samples were prepared with surface doping levels between $5 \times 10^{19}/cm^3$ and solid solubility ($\approx 3 \times 10^{20}/cm^3$), as indicated in Figure 20. The resulting diffused profiles were measured using spreading resistance and anodic sectioning techniques. Agreement between the two measurement techniques was excellent.

It is important to note that the doping concentrations tabulated in Figure 20 represent <u>electrically active concentrations</u>, not chemical concentrations.

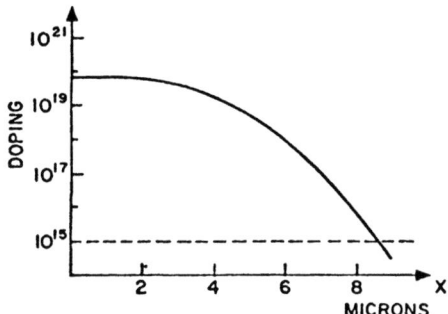

SAMPLE	C_s	
	SPREADING ρ	ANODIC SECTIONING
A	5×10^{19}	5.2×10^{19}
B	7.2×10^{19}	
C	1.8×10^{20}	
D	2.8×10^{20}	2.8×10^{20}
E	3.1×10^{20}	3.3×10^{20}

Fig. 20. Surface concentrations of diffused heavily doped samples.

Following preparation of the samples, a series of initial experiments were conducted to evaluate the magnitude of the enhanced oxidation effect over N^+ regions. A typical result is shown in Figure 21. The horizontal scale is again an electrically active surface concentration. The data indicate approximately a three-to-one range in oxide thickness for these wafers which were oxidized simultaneously.

The choice of a low oxidizing temperature tends to maximize the enhanced oxidation rate for the common N-type impurities because, as explained above, they tend to pile up at the silicon surface. As a result, the dominant effect is on the reaction occurring at the Si-SiO$_2$ interface, and hence the linear rate constant B/A which dominates the overall reaction at low temperatures. It should be noted that low-temperature oxidation is becoming increasingly important in the semiconductor industry because of the trend toward larger diameter silicon wafers.

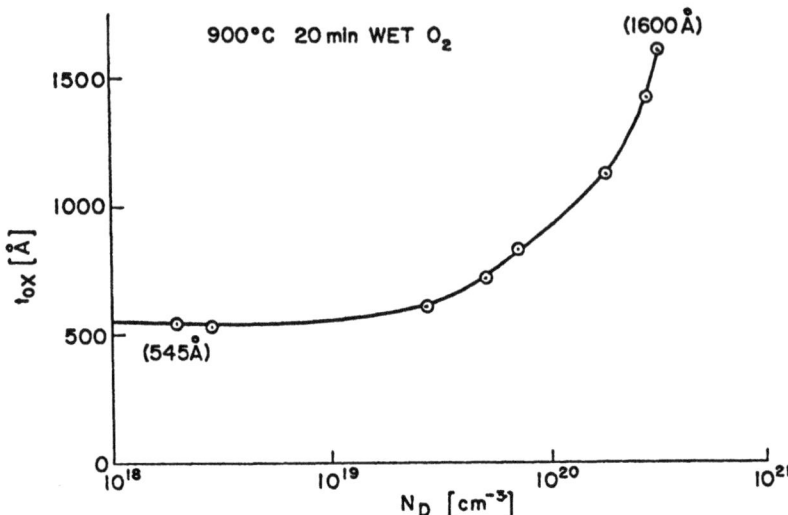

Fig. 21. Oxide thickness vs doping. Samples oxidized simultaneously.

Wet oxidation also tends to increase the effect observed in Figure 21 because there is more redistribution for wet oxides when N-type impurities are involved (Cs/Cb is larger for wet than dry oxides) [23,24].

Following these initial experiments, a more carefully designed series of oxidations were made to evaluate the enhanced oxidation over N^+ regions under a wide variety of temperatures and ambient conditions. Typical of the results obtained are the data shown in Figures 22 and 23. The bottom curve (A) in each case is lightly doped (1×10^{15}) material, and the results agree with previously published data [18]. The upper curve (F) corresponds to a substrate doped to approximately solid solubility.

The following observations can be made about these data, even in the form shown in these figures.

(1) The enhanced oxidation effect is more pronounced at

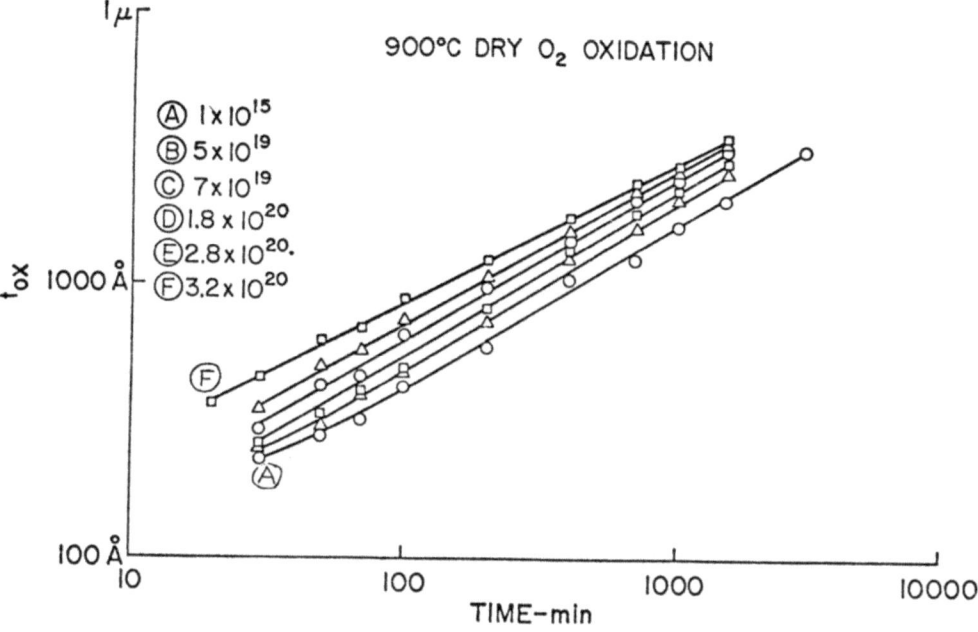

Fig. 22. Oxide thickness vs time for various doping levels at 900°C.

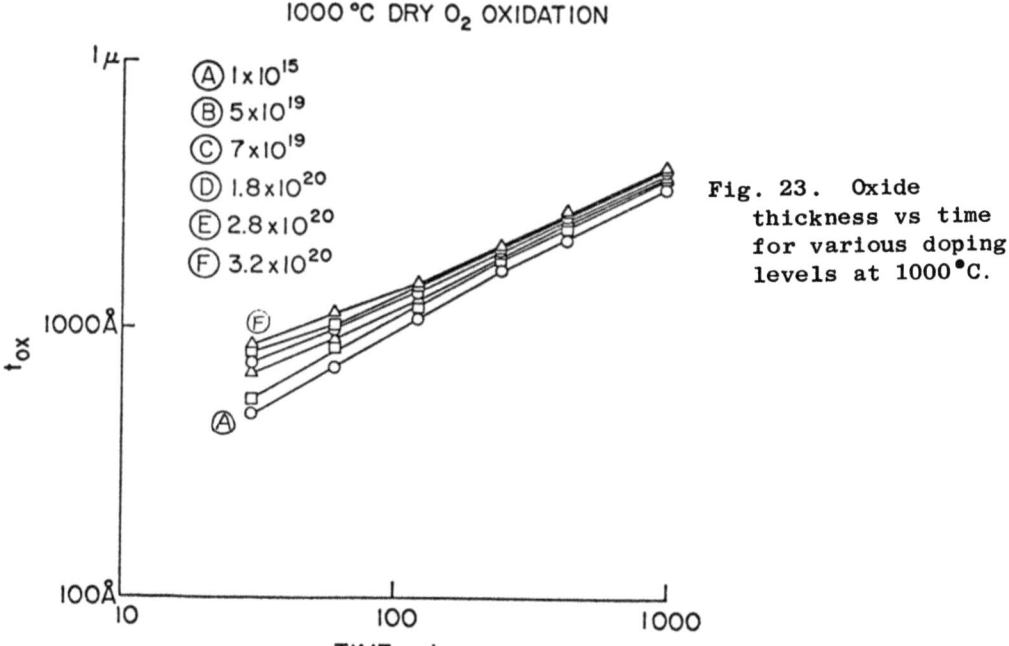

Fig. 23. Oxide thickness vs time for various doping levels at 1000°C.

lower temperatures (900°C) than at higher temperatures (1000°C).

(2) These effects are also more pronounced for shorter times and thinner oxides than they are for longer times and thicker oxides.

Both of these observations are in agreement with our previously stated expectations; that is, N-type dopants that pile up at the silicon surface and segregate into the Si should affect the reaction kinetics at the Si-SiO$_2$ interface far more than they affect the diffusion of the oxidizing species through the SiO$_2$. As a result, we would expect B/A to be affected and not B, in agreement with the observations stated above.

Based on the general oxidation relationship, we can consider from a theoretical point of view what the curves in Figures 22 and 23 should look like if the only effect of N$^+$ doping was on the surface-reaction kinetics. Figure 24 is a series of theoretical curves generated from this oxidation relationship. The various

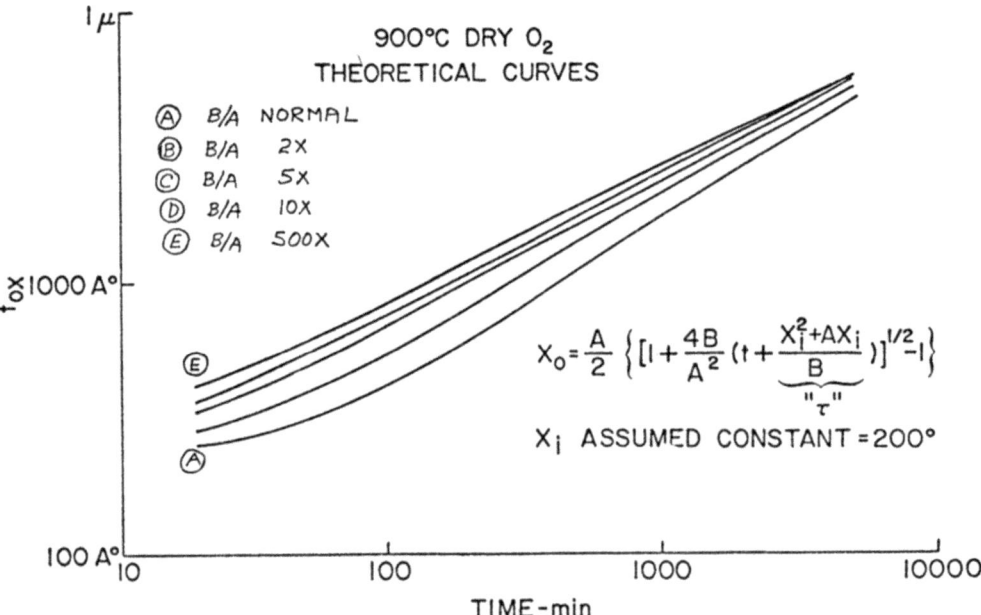

Fig. 24. Theoretical curves of oxide thickness vs time for various linear rate constants.

curves were generated for values of the linear rate constant B/A between the normal or lightly doped substrate value (2.8 x 10^{-4} μ/min at 900°C) and 500 times this value (corresponding to a much enhanced surface-reaction rate). The qualitative agreement between these theoretical curves and the experimental results in Figure 22 is apparent.

The analyses of these data in Figures 22 and 23 can be carried one step further by extracting linear and parabolic rate constants according to standard techniques [18]. The results for the 900°C oxidation are shown in Figure 25. The following conclusions can be drawn from these data.

(1) The parabolic rate constant is essentially unchanged by the N^+ substrate. This is in accordance with our expectations because most of the phosphorus segregates into the silicon.

(2) The linear rate constant is virtually unchanged for doping levels < 5 x 10^{19}/cm^3. Above this concentration B/A increases approximately one order of magnitude as the doping level is increased to the solid-solubility limit.

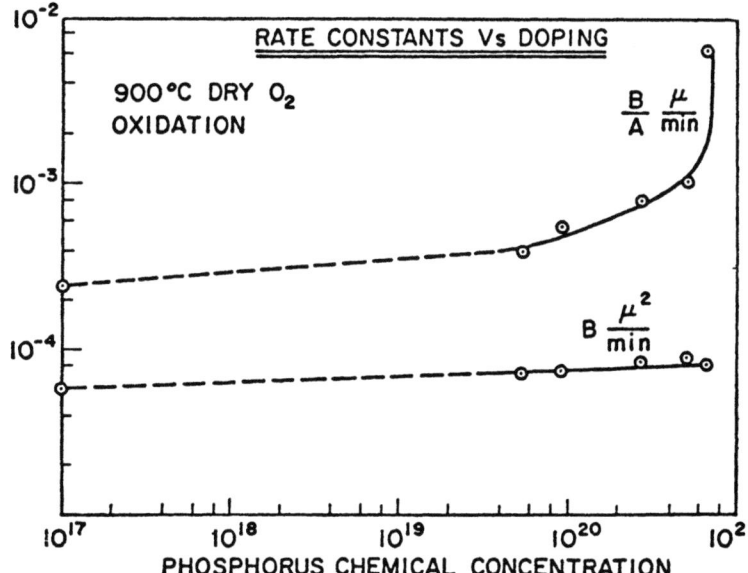

Fig. 25. Linear and parabolic rate constants vs chemical dopant concentrations at 900°C.

It should be noted that the horizontal scale in this figure has been corrected to a chemical phosphorus concentration in the substrate, rather than the electrically active concentration previously used. The chemical phosphorus concentration (and hence the percentage of the species electrically active in our diffused samples) was determined by sputtering Auger techniques by Johannesen et al. at Stanford University [25]. There is still some question concerning the absolute accuracy of the horizontal scale in Figure 25 because of the sensitivity limitations of the Auger technique. Independent measurements on these samples are currently being performed using SIMS techniques. These should resolve the question of chemical vs electrically active phosphorus concentrations in these samples.

The 1000°C extracted rate constants show a form very similar to Figure 25. Samples have also been oxidized at 1100° and 800°C, and the resulting oxide thicknesses are being measured. Completion of these data at the four temperatures will allow us to plot both B and B/A as a function of temperature and to extract activation energies for the rate constants as a function of substrate doping level.

The reported activation energy for lightly doped wafer for B/A, the linear rate constant, is approximately 46 kcal/mole [18]. This compares well to the energy required to break a Si-Si bond which is 42.2 kcal/mole [21]. The 900° and 1000°C data indicate that B/A becomes extremely large for the samples doped close to solid solubility. The activation energy for B/A in this case, therefore, approaches zero, thereby indicating that the reaction at the Si-SiO$_2$ interface is no longer a rate-limiting step at these doping levels; that is, it appears that the incorporation of dopant atoms in concentrations close to solid solubility so strains the lattice structure that bond breaking becomes extremely easy. Whether this simplified model is correct can only be verified by a better model of the molecular and atomic structure near the Si-SiO$_2$ interface. Such a model is necessary also for prediction of charge densities and is, therefore, a goal of our research program. We are also beginning to pursue it in the manner indicated at the beginning of this section.

3.5 Oxidation in HCl/O$_2$ mixtures

One of the most significant developments in the passivation of thermally grown silicon-dioxide films over the past several years has been the addition of a chlorine species during silicon oxidation. It has been demonstrated [26,27,28] that such additions result in improved threshold stability and increased dielectric strength. In addition, it has been observed that chlorine additions increase the rate of silicon oxidation [29,30,31], but extensive data for widely varying oxidation conditions have not been reported. In this investigation, therefore, the thermal-oxidation kinetics of silicon in O_2/HCl mixtures is characterized as a function of oxidation temperature, HCl concentration, and silicon orientation; the results are compared to those obtained from silicon oxidations performed in a dry-oxygen atmosphere.

3.5.1 Rate constants. Log-log plots of oxide thickness vs oxidation time for the thermal oxidation of (100)- and (111)- oriented silicon in O_2/HCl mixtures at temperatures of 900°, 1000°, and 1100°C are shown in Figures 26, 27, and 28, respectively. It is clear that the (111) silicon always oxidizes faster than does the (100). It can also be seen that the effect of the HCl addition is to increase the oxidation rate relative to oxidation in dry oxygen. In particular, a relatively large increase in the oxidation rate occurs with the initial 1% HCl addition. Subsequent additions have a somewhat smaller effect, but the overall rate is systematically increased. It can also be observed in these figures that the overall increase in oxide thickness for a particular oxidation time on increasing the HCl concentration from 0 to 10% is larger for (100)- than for (111)-oriented silicon. This effect may be related to the silicon-etching phenomenon. It should be noted that very few data

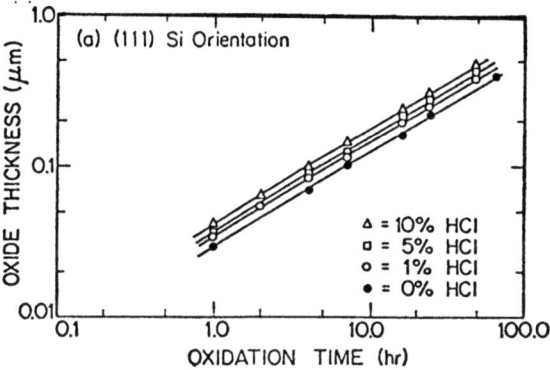

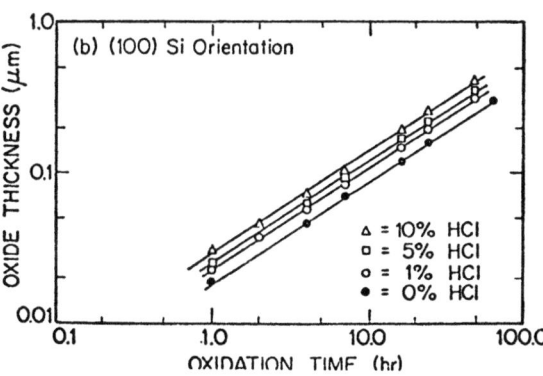

Fig. 26. Oxide thickness vs oxidation time for the oxidation of <100>- and <111>-oriented N-type silicon in various O_2HCl mixtures at 900°C.

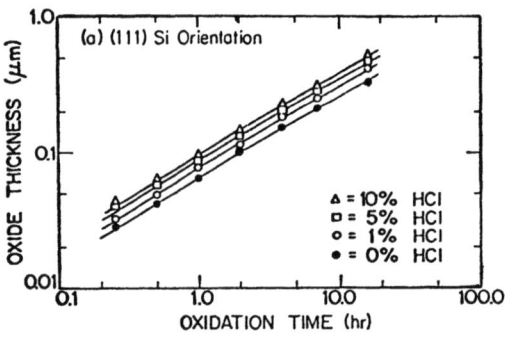
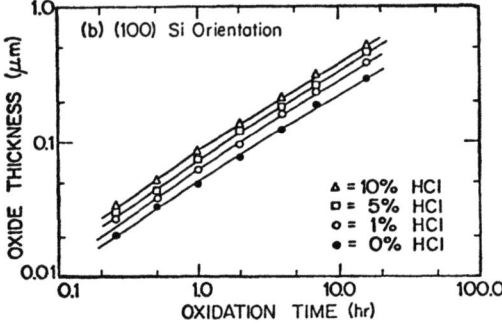

Fig. 27. Oxide thickness vs oxidation time for the oxidation of <111> and <100> oriented N-type silicon in various O_2/HCl mixtures at 1000°C.

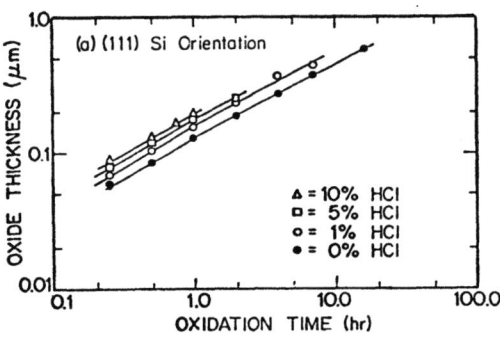
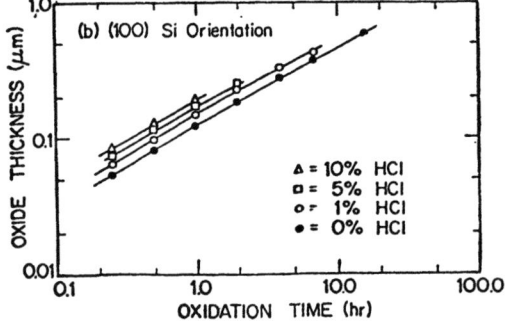

Fig. 28. Oxide thickness vs oxidation time for the oxidation of 111- and 100-oriented N-type silicon in various O_2HCl mixtures at 1100°C.

points are plotted for oxidations carried out at 1100°C with HCl concentrations greater than 1% because of silicon etching and oxide-bubble forming oxidation with O_2/HCl mixtures at 1100°C [30,31,32,33].

To separate the effects of HCl on the parabolic B and linear B/A rate constants and to obtain some indication of the role HCL plays in silicon oxidation, Eq. (37) was utilized. During the determination of τ, x_i was assumed to be essentially constant at 160 ± 40 Å for all oxidation conditions investigated. Although x_i should decrease with HCl addition due to water-vapor generation as per the reaction of O_2 and HCl

$$4 \text{ HCl} + O_2 \rightarrow 2 H_2O + 2 Cl_2 \tag{47}$$

the extrapolations performed to determine x_i and thus τ could not detect small changes in x_i. Nevertheless, because τ depends on x_i, B, and B/A, a decrease in τ was observed on HCl addition, probably in part as a result of water generation because x_i and, therefore, τ are zero for oxidation in a steam ambient[18].

The results of least-squares analyses on the data in Figures 26, 27, and 28 are summarized in Figures 29 and 30 which are semilog plots of the effect of HCl concentration on the parabolic and linear rate constants, respectively. It should be noted that the rate constants generated via Eq. (37) and plotted in Figures 29 and 30 are effective rate constants in that they represent the combined effects of oxygen, water, and chlorine species on B and B/A.

Figure 29 shows that essentially no orientation effect is observed in the parabolic rate constant B, in accordance with theory; also, increasing the HCl concentration above 1% results in a linear increase in B (the plots are put on a semilog scale for convenience). The increase in B going from 0 to 1% HCl in dry oxygen is not as consistent among the three temperatures, however. For 1000° and 1100°C, a large increase in B due to a 1% HCl addition occurs, whereas this trend is not observed at 900°C. These observations are in agreement with previous work that revealed a linear increase of B with a 0 to 9% HCl addition at 900°C.[31] and a large increase in B due to a 1% HCl addition at 1100°C [30]. Presumably, the large increase in B and the linear increase with subsequent HCl additions are the result of water generation as described in Eq. (47) because small amounts of water increase the thermal-oxidation rate of silicon in dry oxygen [34]. As pointed out by van der Meulen and Cahill [31], however, gas-phase thermodynamic equilibrium calculations indicate that the amount of water generated via Eq. (47) cannot account for the observed increase in the oxidation rate. This is consistent with the published results that a Cl_2 addition to a dry-oxygen atmosphere yields a considerably higher silicon-oxidation rate than

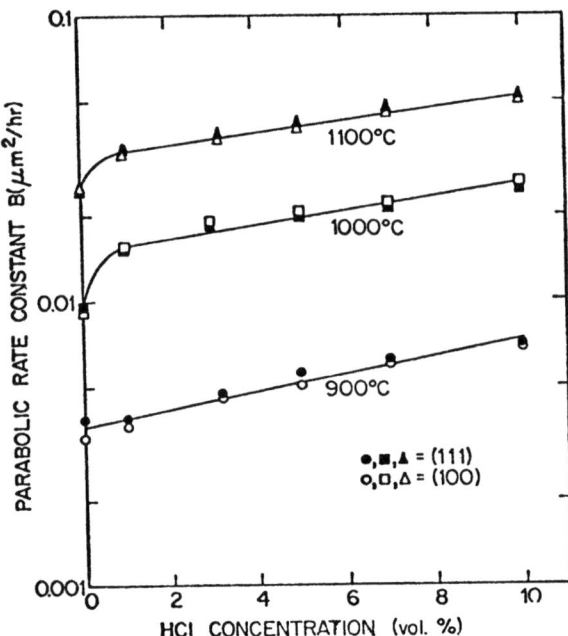

Fig. 29. Parabolic rate constant vs % HCl for <111> and <100> oriented N-type silicon at 900°, 1000°, and 1100°C.

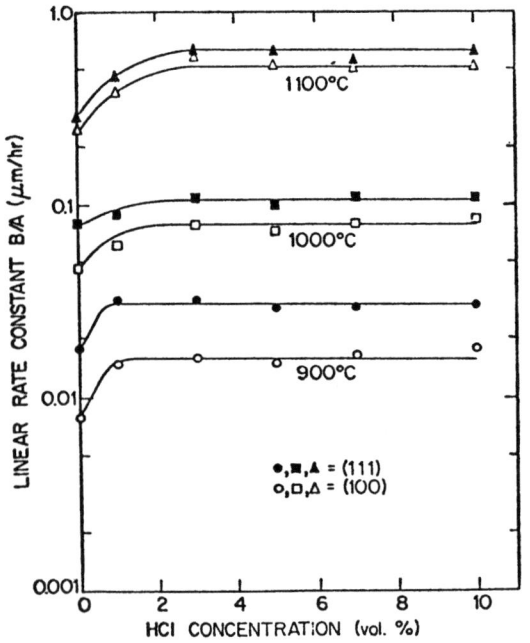

Fig. 30. Linear rate constant vs percent HCl for <111>- and <100>- oriented N-type silicon at 900°, 1000°, and 1100°C.

an equivalent HCl addition at 1150°C [34], suggesting that the chlorine species, rather than the water generated, may be primarily responsible for the increase in the oxidation rate at high temperatures. Because it is difficult to obtain Cl_2 which contains less than 3 ppm water, however, some effect, although small, would be anticipated from the water present.

In regard to the above argument concerning the amount of water generated via Eq. (47), thermodynamic calculations such as those in Ref. [31] give only the idealized gas-phase equilibrium conditions. The presence of a surface -- especially one such as SiO_2 which is useful for catalytic purposes -- could shift the equilibrium and thus alter the amount of water generated and incorporated into the oxide surface.

Some insight can be gained into the silicon-oxidation regime affected by water or chlorine [as generated via Eq. (47)] from consideration of the definitions of the parabolic (B) and the linear (B/A) rate constants. The definitions are as follows [18]:

$$B = 2 D_{eff} \frac{C^*}{N_1} \qquad (48)$$

where D_{eff} is the effective diffusion coefficient, C^* is the maximum concentration of oxidant in oxide, and N_i is the number of oxidant molecules incorporated into a unit volume of oxide,

$$B/A = \frac{C^*}{N_1 \left(\frac{1}{k} + \frac{1}{h} \right)} \qquad (49)$$

where k is the surface-reaction rate constant, and h is the gas-phase transport coefficient.

Inspection of these definitions and a comparison of Figures 29 and 30 reveal that the monatonic increase in B with HCl concentration is related to the effective diffusion coefficient D_{eff} because this term does not appear in the expression for B/A. It also appears that the rapid initial increase in B and B/A may be related to the solubility of the oxidizing species in the oxide film. Although the latter conclusion could be explained by water addition, this explanation would not account for the increase in D_{eff} because this parameter is known to decrease with water addition [18].

In light of the above arguments, the gradual increase of the parabolic rate constant with increasing HCl concentration could be partially related to a chlorine species because it has been established that a higher concentration of chlorine is incorporated into the silicon-dioxide film as the HCl concentration and/or the silicon-oxidation temperature are increased [35]. Consequently, increasing the chlorine concentration may cause the SiO_2 lattice to be strained (especially near the interface) thereby allowing diffusion of the oxidant to occur more easily and increasing the oxidation rate as observed in the oxidation of heavily boron-doped silicon [36]. Because the ionic radius of boron when substituted for silicon in a silicon lattice (which should be close to the radius of boron substituted for silicon in SiO_2) is 0.88 Å [37], and the ionic radius of chlorine is 1.8 Å [38], it would be expected that the SiO_2 lattice expansion would be even greater for chlorine as compared to boron. This speculation is also consistent with the previous observation that Cl_2 additions result in larger silicon-oxidation rates than do HCl additions at high temperatures because more chlorine is apparently incorporated into the SiO_2 with O_2/Cl_2 than with O_2/HCl mixtures [35].

Figure 30 reveals a strong orientation effect on the linear rate constant B/A, which apparently decreases with increasing temperature. This observation was discussed in the previous section on orientation effects. As generally found for B, an initial rapid increase in B/A can also be observed when increasing the HCl concentration in the oxygen atmosphere from 0 to 1%; however, a further increase resulted in essentially no change in B/A. Although an explanation for these observations is not apparent, it is conceivable that the breaking of Si-Si bonds could be promoted by etching the silicon substrate which is known to occur with the use of HCl [30,31,32,33] to increase the linear rate constant. Increased HCl concentrations, however, should then further increase B/A unless an equilibrium between oxidation and etching is established.

Additional investigation of this phenomenon is required if we are to understand the mechanism involved. Preliminary results obtained by Deal with O_2/Cl_2 oxidation atmospheres in the Fairchild Laboratory substantiate the above interpretation; relatively large increases in B/A caused by a Cl_2 addition at 1100°C were observed, with considerably smaller increases noted for B.

3.5.2 <u>Activation energies</u>. Figure 31 is an Arrhenius plot of the parabolic rate constant B for the thermal oxidation of silicon in O_2 and several O_2/HCl mixtures. Because essentially no orientation effect was observed for B, only the (111) orientation is plotted.

A least-squares analysis of the data for dry oxygen produced a value of 1.2 eV (28 kcal/mole), which agrees well with previously determined values for the thermal oxidation of silicon in dry

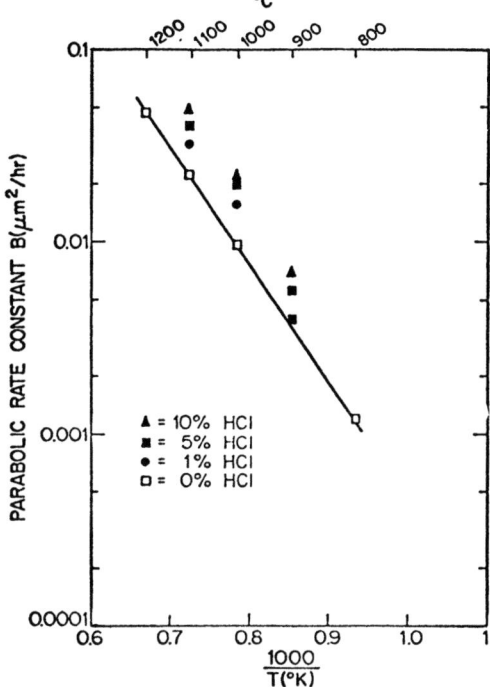

Fig. 31. Arrhenius plots of the parabolic rate constants for silicon oxidized in various O_2/HCl mixtures. Silicon orientation is <111>.

oxygen [18, 39] and for the diffusivity of oxygen through fixed silica [40]. It can be seen that the HCl addition to the dry-oxygen atmospheres resulted in a convex curvature (relative to dry-oxygen oxidation) of the Arrhenius plot.

Figure 32 is an Arrhenius plot of the linear rate constant B/A for the thermal oxidation of silicon in O_2 and several O_2/HCl mixtures. A least-squares analysis of the data for dry-oxygen (0% HCl oxidation of (111)-oriented silicon yielded a value of 2 eV (47 kcal/mole), which is in close agreement with the energy required to break a Si-Si bond [21]. Determination of the activation energy for (100)-oriented silicon, however, affords two choices as explained in the previous section on orientation effects.

Similar to the parabolic rate constant, the Arrhenius plot of the linear rate constant for HCl/O_2 mixtures (Figure 32) is curved, and the curvature is concave relative to dry-oxygen oxidation. Based on Eq. (47), it is not surprising that Arrhenius plots of the rate constants for oxides grown in HCl atmospheres should result in

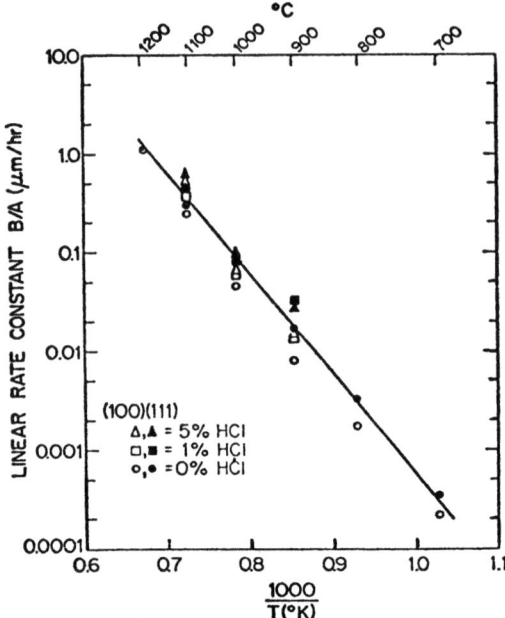

Fig. 32. Arrhenius plots of the linear rate constants for <111> and <100> oriented silicon oxidized in various O_2/HCl mixtures.

curved lines. It is apparent that more than one activation process is occurring (such as oxidation caused by oxygen and water vapor and several possible effects resulting from chlorine). As the oxidation temperature is changed, the relative importance of these activated processes may also change. It becomes necessary, therefore, to separate out the effects of water and chlorine on the silicon oxidation rate. Toward this end, experiments are being performed to investigate the oxidation rate of silicon in O_2/H_2O mixtures.

4. CONCLUSION

In the previous section the fundamentals of silicon oxidation and first order process models have been presented. The subsequent discussions illustrate changes in the model coefficients which can account for changes in oxidation rates based on orientation, ambient and surface doping effects.

In both the epitaxy and oxidation sections the discussion has focused on two areas. First, the basic process models have been defined. Second, recent advances have been summarized which extend the range of application of the basic models to include state-of-the-art processing techniques. This includes programmable doping of epitaxial layers and ambient and surface effects during oxidation.

REFERENCES

1. D.W. Shah, "Mechanisms in Vapour Epitaxy of Semiconductors," in Crystal Growth Theory and Techniques, Vol. 1, ed. by C.H.L. Goodman, Plenum Press, London, 1974.
2. T.I. Kamins, IEEE Trans. on Parts, Hybrids and Packaging (PHP-10), 4, Dec. 1976, p. 221-229.
3. M.L. Hammond and R.J. Gieske, "Hot Wall CVD -- Impact on the Industry," presented at Semicon, San Mateo, California, May 27, 1976.
4. T.J. Rodgers, "Advanced Integrated-Circuit Technology for Micropower IC's," SEL-75-034, Ph.D. Dissertation, Stanford University, Stanford, California, Aug. 1975.
5. F.C. Eversteyn et al., J. Electrochem Soc. (117), 7, Jul. 1970, pp. 925-931.
6. R.F.C. Farrow, J. Electrochem. Soc. (121), 7, Jul. 1974, pp. 899-907
7. H.C. Theuerer, J. Electrochem. Soc. (108), 7, Jul. 1961, pp. 649-653
8. J. Bloem, J. Crystal Growth, (13/14), 1972, pp. 302-305.
9. J. Bloem, "Trends in the Vapor Deposition of Silicon," in Semiconductor Silicon 1973, The Electrochemical Society, Inc., Princeton, New Jersey, 1973, pp. 213-226.
10. A.S. Grove et al., J. Appl. Phys., (36), 3, Mar. 1965, pp. 802-810.
11. C.O. Thomas et al., J. Electrochem. Soc. (109), 11, Nov. 1962, pp. 1055-1061.
12. G. Skelly and A.C. Adams, J. Electrochem. Soc. (120), 1, Jan. 1973, pp. 116-122.
13. T. Ishii, K. Takahashi, A. Kondo, and K. Shirahata, J. Electrochem. Soc. (122), 11, Nov. 1975, pp. 1523-1531.
14. T.I. Kamins, R. Reif, and K.C. Saraswat, "Transient Response of Dopant Incorporation into Silicon Epitaxial Films," Extended Abstract No. 230, p. 601, 150th Electrochemical Society Meeting, Oct. 17-22, 1976, Las Vegas, Nevada.
15. P.F. Deisler, Jr. and R.H. Wilhelm, Industrial and Engineering Chemistry, (45), 1953, p. 1219.
16. J.B. Rosen and W.E. Winsche, J. Chem Phys. (18), 1950, p. 1587.
17. H. Kobayashi and M. Kobayashi, Catalysis Reviews - Science and Engineering (10), 1974, p. 139.
18. B.E. Deal and A.S. Grove, J. Appl. Phys. (36), Dec. 1965, pp. 377-3778.
19. J. Ruzyllo, I. Shiota, N. Miyamoto, and J. Nishizawa, J. Electrochem. Soc, (123), 26, 1976.
20. A.G. Revesz and R.J. Evans, J. Phys. Chem. Solids. (30), 1969, p. 55
21. L. Pauling, The Nature of the Chemical Bond (3rd edition), Cornell University Press, Ithaca, New York, 1960.
22. J.R. Ligenza, J. Phys. Chem. (65), 1961, p. 2011.
23. B.E. Deal and M. Skalar, J. Electrochem. Soc. (112), Apr. 1965, pp. 430-435.
24. B.E. Deal, A.S. Grove, E.H. Snow, and C.T. Shah, J. Elec. Soc. (112) Mar. 1965, pp. 308-314.

25. J.S. Johannessen, W.E. Spicer, J.F. Gibbons, J.D. Plummer, and N.J. Taylor, "Observations of Phosphorus Pileup at the SiO_2-Si Interface," to be published.
26. R.J. Kriegler, Denki Kagaku (46), 1973, p. 446.
27. R.S. Ronen and P.H. Robinson, J. Electrochem. Soc. (119), 1972, p. 747.
28. C.M. Osburn, J. Electrochem. Soc. (121), 1974, p. 809.
29. R.J. Kriegler, J. Electrochem. Soc. (119), 1972, p. 388.
30. K. Hirabayashi and J. Iwamura, J. Electrochem. Soc. (120), 1973, p. 1595.
31. Y.J. van der Meulen and J.G. Cahill, J. Electron. Materials (3), 1974, p. 371.
32. R.J. Kriegler, Semiconductor Silicon, ed. by H.R. Huff and R.R. Burgess, The Electrochemical Society, Inc., Princeton, New Jersey, 1973, p. 363.
33. E.A. Irene, J. Electrochem. Soc. (121), 1974, p. 1613.
34. R.J. Kriegler, Y.G. Cheng, and D.R. Colton, J. Electrochem. Soc. (119), 1972, p. 388.
35. Y.J. van der Meulen, C.M. Osburn, and J.F. Ziegler, J. Electrochem. Soc. (122), 1975, p. 284.
36. B.E. Deal and M. Sklar, J. Electrochem. Soc. (112), 1965, p. 430.
37. Helmut F. Wolf, Semiconductors, John Wiley and Sons, Inc., New York, 1971, p. 199.
38. A.F. Wells, Structural Inorganic Chemistry, Oxford University Press, 1962, p. 68.
39. A.G. Renesz and R.J. Evans, J. Phys. Chem. Solids (30), 1969, p. 551.
40. F.J. Norton, "Nature" (171), 1961, p. 701.
41. T. Hirao, K. Kijima, and T. Nakano, Semiconductor Silicon, 1977, pp. 1005-1014.
42. B.E. Deal, Proc. of the Electrochem. Soc. Philadelphia, Pennsylvania, 9-13, May 1977.

ION IMPLANTATION

H. S. Rupprecht

IBM System Products Division--East Fishkill
Hopewell Junction, New York 12533   USA

ABSTRACT. Ion implantation has become widely accepted as a new
technology to introduce dopants into silicon, particularly for
applications that require a high degree of dose and profile
control. The present state of the art permits a fairly accurate
prediction of impurity profiles. The basic concepts of the
theory of atomic collisions and its implications for range-energy
relations will be discussed following the treatment by Lindhard,
Schiott, and Scharff (LSS theory) [1].

1. INTRODUCTION

The advance in solid-state technology is largely determined by
our ability to control and modify certain materials parameters in
a well-defined manner. Semiconductors are a typical example,
where minute traces of impurities in the order of parts per
million and less will influence the electrical properties
markedly.

The concept of introducing such dopants into semiconductors by
means of high energetic particles was discussed many years ago.
In 1954, W. Shockley [2] submitted a patent describing the
"Forming of Semiconductor Devices by Ionic Bombardment." As the
semiconductor technology matured, developing shallower and
shallower device structures, the control aspect became vitally
important. And it is precisely in the field of control that ion
implantation offers major advantages over standard diffusion
technologies, for example,

a. Accurate control of the total amount of impurity transferred to the wafer by measurement of the accumulated charge during implantations.

b. A high degree of areal uniformity across the wafer obtained by mechanical or electrical scanning (uniformity better than 1%).

c. Accurate control of the depth distribution by a well-defined accelerating potential, thus making possible a wide range of dopant profiles for device configurations.

In contrast to standard diffusion profiles, which are frequently close to error-function types (with the maximum concentration at the surface), the impurity profiles in implanted structures resemble more Gaussian or truncated-Gaussian type distributions with the maximum concentration at a mean projected range, R_p, and with a standard deviation, ΔR_p (Fig. 1).

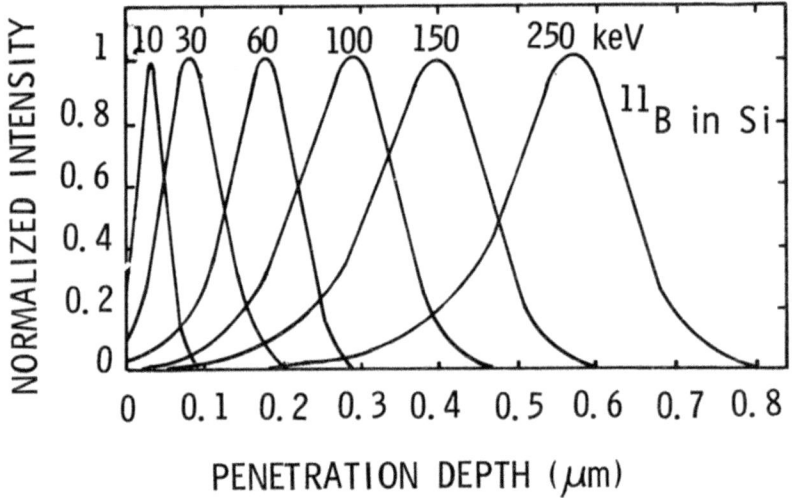

Fig. 1. Normalized range distributions of boron in amorphous silicon. Parameter boron energy. (Source: Ref. 3)

2. RANGE-ENERGY RELATIONS

It was the Danish school in particular that contributed to our present understanding of atomic collisions. The Lindhard, Scharff, and Schiott theory [1], known as the LSS theory, has become the basis for all theoretical investigations.

We are, today, in a position to make reasonable predictions of implanted profiles as a function of ion species, target material, implant energy, and dose.

2.1 Stopping Phenomena

Essential to an understanding of the range concepts are the stopping phenomena of heavy ions, that is, those phenomena that control the energy loss dE/dR of the projectile after entering the target.

Two effects are responsible for the energy loss -- electronic interactions and nuclear collisions:

$$\frac{dE}{dR} = \left(\frac{dE}{dR}\right)_e + \left(\frac{dE}{dR}\right)_n \qquad (1)$$

2.1.1 Electronic stopping

The exact treatment of the electronic interactions in atomic collisions is rather complex. For a more detailed review we refer the reader to Ref. [4].

Lindhard and Scharff [5] followed an approach by Fermi and Teller to treat the target as being composed of a Fermi gas of electrons into which the positively charged nuclei are imbedded. They derived the following expression for the electronic stopping:

$$-\left(\frac{dE}{dR}\right)_e = \xi_e 8\pi e^2 N\, a_o\, Z_1 Z_2\, (Z_1^{2/3} + Z_2^{2/3})^{-\frac{3}{2}}\, \frac{v}{v_o}, \qquad (2)$$

where

  N is number of scattering centers per volume
  Z_1 is atomic numbers
  v is velocity of projectile
  v_o is Bohr velocity (electron velocity in first hydrogen orbit)
  a_o is Bohr radius of hydrogen atom
  $1, 2$ is index of projectile, target
  ξ_e is a constant in the order of $Z_1^{1/6}$

Boundary condition $v < v_o Z_1^{2/3}$ must be fulfilled for Eq.(2) to be valid. Equation (2) is frequently written as

$$-\left(\frac{dE}{dR}\right)_e = N \cdot S_e ,$$

where
$$S_e = \xi_e \, 8\pi e^2 \, a_o \, Z_1 Z_2 (Z_1^{2/3} + Z_2^{2/3})^{-\frac{3}{2}} \frac{v}{v_o}$$

2.1.2 Nuclear stopping

Under certain conditions, as Bohr has pointed out [6], atomic collisions can be treated in a semiclassical way. The kinetic equations controlling the trajectories resemble very closely the one with which you might be familiar from the motion of planets. Of course, the reason is that one deals in both cases with central forces, attractive forces in the case of planetary motion and repulsive ones for atomic collisions.

Understandably, the solutions for the trajectories will differ: elliptical orbits for attractive forces and hyperbolic orbits for repulsive forces.

One conveniently introduces an impact parameter, p, defined by the distance of the target atom from the original (undeflected) trajectory of the projectile. (See Fig. 2.)

The energy, T, transferred during the collision from the projectile to the target atom is a function of this impact parameter and of the energy the projectile has prior to the collision.

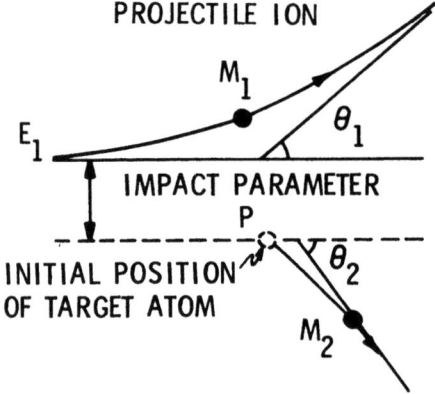

Fig. 2. A typical two-body scattering process with an impact parameter ρ. (Source: Ref. 4)

One finds the following relations:

$$T(E, p \to \infty) = 0$$

$$T(E, p = 0) = T_{max} = \frac{4M_1 M_2}{(M_1+M_2)^2} E. \qquad (3)$$

The probability that, during a collision, an energy transfer occurs, with T between T and T+dT, is proportional to an annulus of $2\pi p \, dp$.

One can, hence, define a differential scattering cross section for the collision, given by

$$d\sigma = 2\pi p \, dp. \qquad (4)$$

It is worth noting that expression (4) is not a true probability, as it is not normalized. This can be accomplished, however, by assuming that the target consists of randomly distributed atoms (amorphous body) of concentration N. We can now assign an average area to each target atom in a layer segment ΔR, given by

$$\pi b^2 = \frac{1}{N \Delta R}. \qquad (5)$$

With Eq. (5) we modify the differential scattering cross section as follows:

$$d\sigma^* = \frac{2\pi p \, dp}{\pi b^2} = N \Delta R \, 2\pi p \, dp = N \Delta R \, d\sigma. \qquad (6)$$

Definition (6) permits us to write the energy loss due to nuclear collisions, as the projectile passes through a film of thickness ΔR, in the following way:

$$-\left(\frac{\Delta E}{\Delta R}\right)_n = N \int_{T_{min}}^{T_{max}} T \, d\sigma \, (E,T) = N \cdot S_n, \qquad (7)$$

where

$$S_n = \int_{T_{min}}^{T_{max}} T \, d\sigma \, (E,T).$$

S_n is called nuclear stopping power.

2.1.3 Atomic potentials

In section 2.1.2, we described the interatomic potential strictly as a coulombic potential. Frequently one has to take into account, however, the screening by the orbital electrons. There have been various proposals to modify the atomic potential, known as Bohr-, Born-Mayer, and Thomas Fermi potentials.

The difficulty in using some of these potentials lies in the mathematical treatment of the differential cross section -- only for very specific types of potential functions is an analytical solution obtainable.

Lindhard et al. [1] suggest an interatomic potential of the following form (power law potential):

$$V(R) = C R^{-s} \quad \text{with } s>1 \tag{8}$$

and
$$C = Z_1 Z_2 e^2 \, a_s^{s-1} \, s^{-1}$$

$$a = 0.8853 \, a_o \, Z^{-1/3}$$

$$Z = (Z_1^{2/3} + Z_2^{2/3})^{3/2}$$

$$a_s \simeq a$$

This leads to an expression for the cross section

$$d\sigma_n = \frac{C_n}{T_m^{1-1/s}} \cdot \frac{dT}{T^{1+1/s}} \tag{9}$$

for $s \geq 1$

with $C_n = \dfrac{\pi}{s} \, (B^2 \, a_s^{2s-2} \, \dfrac{3s-1}{8s^2})^{\frac{1}{s}}$. $T_m = (1-\dfrac{1}{s}) \, Sn$

$$B = 2 \, Z_1 Z_2 e^2 \, (M_1+M_2)/M_1 M_2 v^2$$

2.1.4 LSS theory

With the definitions of the preceding sections, we can describe the concept of the LSS range theory.

We define a probability function P (E,R) dR in such a way that it measures the probability for a projectile of energy E to come to rest between R and R+dR, when R is measured from its momentary position.

The function will be normalized so that

$$\int_0^\infty P \, dR = 1 \qquad (10)$$

and

$$\langle R^m \rangle \int_0^\infty R^m P \, dR \qquad (11)$$

defines the mth moment of R.

We assume a particle of energy E enters the target at R=0. By the preceding definition, it will have a probability P (E,R) dR of coming to rest at R. By traversing a thin film of target material ΔR, two possibilities exist:

1. It can be scattered and lose energy by nuclear collisions and electronic interactions.

   Let us assume the nuclear event caused an energy loss T_n and the electronic contributions can be described by $\sum_i T_{ei}$. When leaving the infinitesimal layer ΔR the probability of coming to rest at the same distance R from the origin is now given by $P(E-T_n-\sum_i T_{ei}; R-\Delta R)$. On the other hand, the probability that an event will take place leading to the energy loss $T_n + \sum_1 T_{ei}$ is given by

   $$N\Delta R \, d\sigma_{n,e}(E,T),$$

   where $d\sigma_{n,e}$ is scattering cross section related to energy loss

   $$T = T_n + \sum_1 T_{ei}.$$

2. If no collision has occurred, the particle leaves the layer with its original energy E and the probability of reaching a total distance R is given by $P(E, R-\Delta R)$.

The chance that there is no collision is given simply by

$$1 - N\Delta R \, d\sigma_{n,e}(E,T).$$

This situation is depicted schematically in Fig. 3.

LSS THEORY

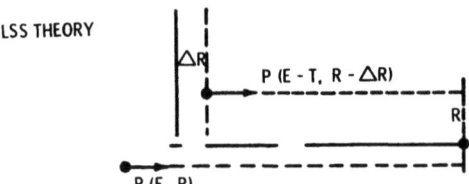

PROBABILITY THAT COLLISION OCCURS   PROBABILITY THAT NO ENERGY
LEADING TO ENERGY LOSS T   TRANSFER TAKES PLACE

N∆R dσ (E,T)   1−N∆R dσ (E,T)

$$P(E,R) = N\Delta R \int_0^{T_M} P(E-T, R-\Delta R)d\sigma + (1-N\Delta R \int_0^{T_M} d\sigma) P(E, R-\Delta R)$$

Fig. 3. Schematic diagram illustrating concept of probability function.

We can describe the function P (E,R) as the sum of two products, where we take all possible energy transfers into account by integration over all T's.

For simplicity's sake, we drop $\sum_i T_{ei}$. The mathematical treatment remains pretty much the same. We have only to keep in mind that $d\sigma_n$ actually, for the general case (including electronic interaction), has to be replaced by $d\sigma_{n,e}$.

We find for P(E,R)

$$P(E,R) = N\Delta R \int_0^{T_{max}} d\sigma_n \cdot P(R-\Delta R, E-T_n) \qquad (12)$$

$$+ (1-N\Delta R \int_0^{T_{max}} d\sigma_n) \cdot P(R-\Delta R, E)$$

and in the limit of ∆R→0

$$\frac{\partial P(E,R)}{\partial R} = N \int_0^{T_{max}} \{P(E-T,R) - P(E,R)\} d\sigma_n. \qquad (13)$$

By multiplying Eq. (13) by R^m and integrating over R by parts, we obtain

$$\int_0^\infty R^m \frac{\partial P}{\partial R} dR = R^m P \Big|_0^\infty - m\int R^{m-1} P \, dR \qquad (14)$$

$$= N\iint P(R,E-T_n)R^m \, d\sigma_n \, dR - N\iint P(R,E)R^m d\sigma_n dR$$

as $R^m P\big|_0^\infty = 0$ for both boundaries

and by use of Eq. (11), Eq. (14) becomes

$$m<R^{m-1}(E)> = N\int [<R^m(E)> - <R^m(E-T_n)>] d\sigma_n . \tag{15}$$

Considering the first moment m=1, one obtains from Eq. (15)

$$1 = N\int (<R(E)> - <R(E-T_n)>) d\sigma_n . \tag{16}$$

Expanding Eq. (16) in Taylor series with respect to T_n, under the assumption $T_n \ll E$, one gets

$$1 = N\int \left(\frac{\partial <R(E)>}{\partial E} \cdot T_n - \frac{1}{2} \frac{\partial^2 <R(E)>}{\partial E^2} T_n^2 + \ldots \right) d\sigma_n . \tag{17}$$

Using only the first-order term of the expansion, together with Eq. (17), gives

$$\frac{d<R(E)>}{\partial E} = \frac{1}{N \cdot \int T_n d\sigma} = \frac{1}{N \, S_n} \tag{18}$$

and, after integration,

$$<R(E)> = \int_0^E \frac{dE'}{N \cdot S_n} . \tag{19}$$

Equation (19) can easily be expanded to include the electronic stopping effects, which we had dropped in our derivation from Eq. (12). This is accomplished by substituting $S = S_n + S_e$; S_e is defined in Eq. (12).

By including second-order terms, one obtains from Eq. (17)

$$<R(E)> = \int_0^E \frac{1}{N \, S(E')} \cdot \left\{ 1 + \frac{\Omega^2(E')}{2} \frac{d}{dE'} \left(\frac{1}{S(E')} \right) \right\} dE' \tag{20}$$

where

$\Omega^2 = \int T^2 \, d\sigma_{n,e}$ which is related to straggling.

Here we have used the approximate solution from Eq. (19) to substitute for $d^2<R(E)>/dE^2$ in Eq. (17).

Equation (20) expresses the total range in parameters characteristic for the atomic collision process.

For the range straggling, one finds from

$$<\Delta R^2> = <R^2> - <R>^2$$

as a first-order approximation

$$\langle \Delta R^2 \rangle = \int_0^E \frac{\Omega^2}{N^2 S^3} \, dE'. \tag{21}$$

Equations (20) and (21) describe, in essence, the range-energy relations, which have become known as the LSS theory [1]. They relate to the total range of the projectile. The quantity of interest for most applications, however, is the projected range. That is the projection of the total range on the original ion-beam direction.

The mathematical expressions are then much more complicated; for a detailed discussion, we refer the reader to a publication by Furukawa, Matsumura, and Ishiwara [6].

The refinements of the original LSS theory are mostly in the mathematical approach, that is, in the particular choice of orthogonal functions used for the expansion in Eq. (16). Winterbon [8], Sanders [9], and Gibbons [10], for instance, chose Legendre polynomials. The exact impurity distribution can then be derived from a mathematical formalism, which relates the moments $\langle R^m \rangle$ of the distributions to the distribution itself.

Winterbon [8] gives examples in which the distributions deviate markedly from standard Gaussian profiles, depending on the mass ratios between the projectile and the target atoms and on the particular atomic potential ($m = \frac{1}{S}$) (Fig. 4).

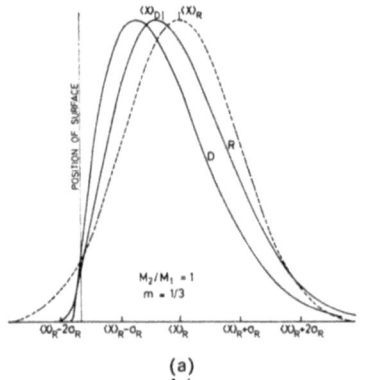
(a)
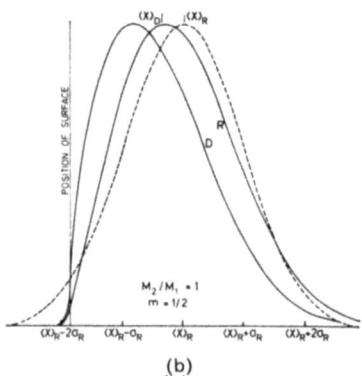
(b)

Fig. 4. Theoretical calculations of range (R) and damage (D) distributions. The dashed curve represents a Gaussian distribution for comparison. (Source: Ref. 8)

Gibbons, Johnson, and Mylroie [10] describe a procedure that permits the synthesis of the distribution function by use of Edgeworth's approximation, and higher moments are tabulated in their Projected Range Statistics [9].

Unfortunately higher-moment calculations are time-consuming and extremely difficult to perform.

2.1.5 Lateral scattering

In most of the applications, ion implantation is used to dope the silicon selectively by means of beam stopping masks. The masking material can be, for example, a photoresist film (specially prepared to avoid outgassing), a dielectric film (SiO_2/Si_3N_4), or a metal layer. The question therefore arises as to what extent the impurity distribution spreads laterally beyond the window in the mask.

Furukawa and coworkers [7] have studied this phenomenon and tabulated values of the lateral straggling for various ions and energies. They based their calculations on the LSS theory and determined the quantities R_p and $R_\perp$ as defined in Fig. 5. For this purpose one has to rewrite the integro-differential equation (16) in the following way:

For m=1,

$$1 = N\int(<R_p(E)> - <R_p(E-T)> \cos\phi)d\sigma, \qquad (22)$$

and for m=2,

$$2<R_p(E)> = N\int(<R_c^2(E)> - <R_c^2(E-T)>d\sigma \qquad (23)$$

$$2<R_p(E)> = N\int(<R_r^2(E)> - (1 - \frac{3}{2}\sin^2\phi) \cdot <R_r^2(E-T)>)d\sigma,$$

with ϕ the laboratory scattering angle,

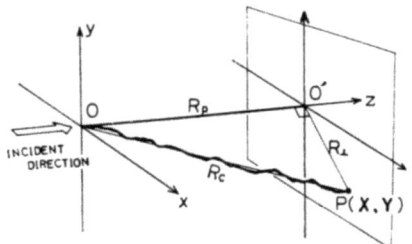

Fig. 5. Definition of range R_p, R_c and $R_\perp$. (Source: Ref. 7)

$$\langle R_c^2 \rangle = \langle R_\perp^2 \rangle + \langle R_p^2 \rangle$$

$$\langle R_r^2 \rangle = \langle R_p^2 \rangle - \langle R_\perp^2 \rangle / 2 .$$

The lateral standard deviations are defined by

$$\langle \Delta X \rangle = \langle \Delta Y \rangle = \sqrt{\langle R_\perp^2 \rangle / 2} .$$

With the assumption that the spatial distribution of the incident ion is Gaussian, the distribution probability function can be written as follows:

$$f(x,y,z) = \frac{1}{(2\pi)^{3/2} \langle \Delta R_p \rangle \langle \Delta X \rangle \langle \Delta Y \rangle}$$

$$\cdot \exp\left(-\frac{x^2}{2\langle \Delta X \rangle^2} - \frac{y^2}{2\langle \Delta Y \rangle^2} - \frac{(z-\langle R_p \rangle)^2}{2\langle \Delta R_p \rangle^2}\right) . \qquad (24)$$

The quantities $\langle \Delta X \rangle$, $\langle \Delta Y \rangle$, $\langle \Delta R_p \rangle$, and $\langle R_p \rangle$ are obtained from the solutions of Eqs. (22) and (23) by use of the given definitions.

Figure 6 schematically describes the energy dependence of the spatial straggling for P and B ions and for other heavy ions in silicon.

By means of Eq. (24), we can now calculate the impurity distribution obtained by implanting through an indefinitely long slit of width 2a.

We find, with ϕ_o, the ion dose per area:

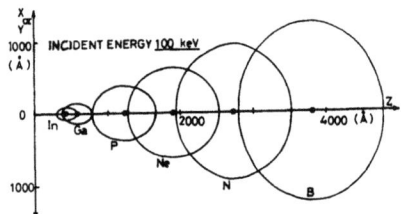

Fig. 6. Spatial distributions of 100 keV B^+, N^+, Ne^+, P^+, Ga^+ and In^- incident to Si. Contours show the positions where the ion concentration becomes $1/\sqrt{e}$ of the maximum value. (Source: Ref. 7)

$$F(x,y,z) = \int_{-a}^{+a} \int_{-\infty}^{+\infty} f(x-\xi, y-\eta, z)\phi_o \, d\xi d\eta \qquad (25)$$

$$= \frac{\phi_o}{\sqrt{2\pi}<R_p>^2} \exp\left(-\frac{(z-<R_p>)^2}{2<\Delta R_p>^2}\right)$$

$$\cdot \frac{1}{\sqrt{\pi}} \, \text{erfc}\left(\frac{x-a}{\sqrt{2}<\Delta x>}\right).$$

Figure 7 gives an example of a 70-keV B implantation into silicon.

2.1.6 Recoil collisions

Recoil collisions are inherent in nuclear stopping phenomena, as we have seen in the previous sections. The energy imparted by the projectile upon the target atoms in those collisions will produce a cascade of secondary projectiles, consisting predominantly of target atoms. We shall now consider the situation in which the target is a composite medium, a situation that arises quite generally in practical applications. A masking film, possibly dielectric or photoresist material, is deposited on the silicon surface to permit selective doping. Collisions in this film can lead to knock-on phenomena which inject undesirable impurity atoms into the silicon substrate and cause additional changes in the chemical and electrical behavior of the silicon.

Such knock-on phenomena were first experimentally investigated by Moline and coworkers [11] using ^{18}O from SiO_2 films. By means of the ^{18}O (p,α) ^{15}N nuclear reaction they have measured the total recoil yield per incident projectile. We shall give a brief description of the model and then describe an extension to Moline's model as suggested by W. K. Chu [12], which gives a good approximation of the distribution of knock-on impurities in silicon (Fig. 8).

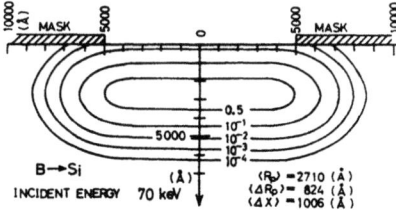

Fig. 7. Contours of equi-ion-concentration of x-z plane, in case of 70 keV B⁻ incident to Si through a 1-μm slit of the mask. If the width of the slit is increased, the flat parts of the contours will be expanded. (Source: Ref. 7)

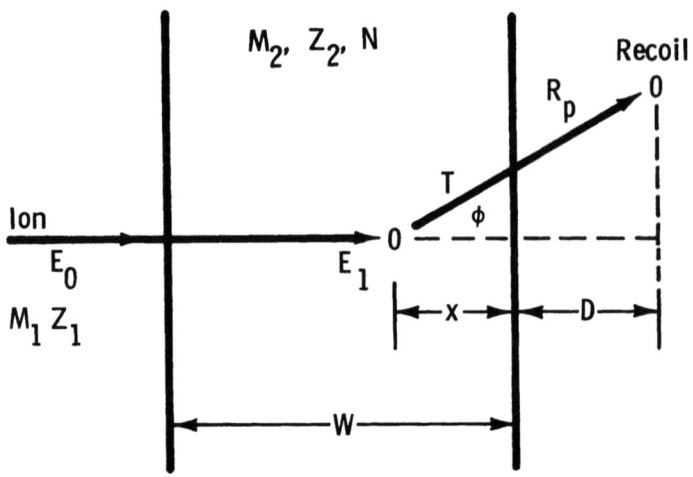

Fig. 8. Schematic diagram of recoil collision. (Source: Ref. 11)

Let us assume that a collison occurs at a distance x away from the film-silicon interface. For a recoil atom with energy T and a projected range $R_p(T)$, the following relation has to hold in order to reach the silicon substrate:

$$R_p(T) \cdot \cos \phi \geq x . \tag{26}$$

The minimum escape energy is therefore implicitly defined by

$$R_p(T_{min}) \cdot \cos \phi = x . \tag{27}$$

If we denote with n the total recoil factor per projectile (total number of recoil atoms per projectile), then this factor is made up from contributions,

$$dn^* = Ndx \cdot d\sigma (T), \tag{28}$$

where dn^*/dx is the differential factor, which relates to recoil events in an infinitesimal layer, x and x+dx, and in an energy transfer, T.

The differential factor $dn/dx = \int_{T_{min}}^{T_{max}} dn^*/dx \, dT$ includes all possible energy transfer out of film dx leading to an escape into the silicon.

$$\frac{dn}{dx} = N \int_{T_{min}}^{T_{max}} d\sigma(T) \qquad (29)$$

Under the assumption of a square law potential (s=2 in Eq. (8)) and $T_m > T$, we obtain

$$d\sigma = \frac{0.33 \pi a^2 T_{max}^{1/2} \, dT}{2 \varepsilon_1 T^{3/2}} \qquad (30)$$

Integration of Eq. (29) and use of Eq. (30) give

$$\frac{dn}{dx} = \frac{0.33 \pi a^2 N}{\varepsilon_1} \left[\left(\frac{T_{max}}{T}\right)^{1/2} - 1 \right] \qquad (31)$$

The total recoil fraction n is obtained by integration of Eq. (31) with respect to x:

$$n = \int_o^W \frac{dn}{dx} \, dx = 0.33 \pi a^2 \frac{E_1}{\varepsilon_1} N \int_o^W \frac{\left(\frac{T_{max}}{T_{min}}\right)^{1/2} - 1}{E_1} \, dx \qquad (32)$$

with $E_1 = E_o - (W-x) S_1$,

where S_1 is stopping power for projectile in surface film W

a is defined in Eq. (8)

ε_1 is dimensionless parameter (see Ref. 2)

E_1 is in LSS energy units

The minimum escape energy, T_{min}, can be expressed in the following way with S_2, the stopping power for recoils in the surface film:

$$T_{min} \cos \phi = S_2 x \quad . \qquad (33)$$

The total recoil fraction, n, is therefore given by

$$n = \frac{\pi a^2}{2} N \frac{E_1}{\varepsilon_1} \left(\frac{\gamma}{S_2}\right)^{1/3} \left(\frac{W}{E_o}\right)^{2/3}, \qquad (34)$$

where $\gamma = T_{max}/E_1$.

Moline et al. [10] have made, in deriving Eq. (34) from Eq. (32), two simplifying assumptions:

1. $E_1 = E_o$ (negligible energy loss in the thin film or $\Sigma T_i \ll E_o$).

2. S_2 is constant over film thickness.

To include the effect of cascades, they multiplied Eq. (34) by an empirical factor $(1+(W/W_o)^{2/3})$ in which

W_o is an empirical parameter.

In order to obtain the distribution of recoil atoms inside the silicon substrate, W. K. Chu [12] defines a recoil factor n*(D), which represents the contribution of recoil atoms beyond a distance D inside the silicon to the total factor n.

For a recoil atom to come to rest beyond a distance D the total range $R_p(T)$ has to be

$$R_p(T) \geq \frac{x-D}{\cos\phi} \qquad (35)$$

in analogy to Eq. (26). One finds, therefore, n*(D) from

$$n^*(D) = \int_0^{W+D} \frac{dn}{dx}\, dx, \qquad (36)$$

where the integration boundaries have been changed according to Eq. (33) from $x_i \to x_i + D$.

If we assume that the stopping power S in the film is the same as in silicon (a very good assumption for SiO_2), then we find the distribution of the recoil atoms from

$$\frac{dn^*(D)}{dD} = 0.33\pi a^2 \frac{E}{\epsilon} \frac{n}{E_o} \left[\left(\frac{\gamma E_o}{S(W+D)}\right)^{1/3} - \left(\frac{\gamma E_o}{SD}\right)^{1/3} \right] \left(1+\frac{W}{W_o}\right)^{2/3}.$$

The data presented in Fig. 9 are based upon Eq. (37), and the cutoff depth is defined as $D=R_p(T_m)$, the maximum range of a recoil collision right at the silicon-film interface.

Very shallow impurity distributions are obtainable, as we have seen, from recoil collisions. One application which requires such distribution in surface films has been described by Shannon

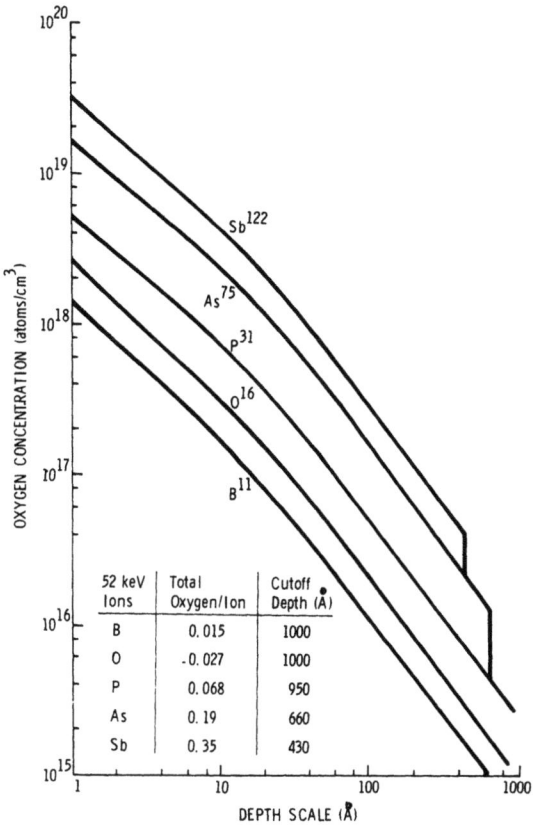

Fig. 9. Oxygen profile obtained from 50-keV recoil collision of 10^{13} ions/cm² traversing an oxide layer of 10 Å. (Source: Ref. 12)

[13]. It permits the control of barrier heights in Schottky devices because of changes of the internal field. The required distributions are less than 150 Å deep.

3. RADIATION DAMAGE

Radiation damage in the target is an unavoidable consequence of the stopping processes, which accompany the penetration of energetic particles. As discussed previously, the interactions may be of an electronic nature, leading possibly to polarization phenomena, or they may induce atomic displacements, cascading effects, etc., which eventually can destroy any lattice structure, leading to amorphous bodies.

3.1 Effect on Electrical Activity

This disorder effect manifests itself in a reduced electrical activity of the implanted impurities. This effect is illustrated very distinctly by the example given in Fig. 10, which describes the distribution of the free carriers due to a 240-keV boron implant into a <111> silicon wafer misoriented by $3°$. The data are taken from Ref. 14. Only after annealing temperatures of $900°C$ had been applied was the full electrical activity obtained.

This recovery behavior has been studied by many workers in the field [15, 16]. The typical differences between n dopants (P, As, Sb) and B are described in Figs. 11 (Crowder-Morehead) and 12 (Seidel-MacRae).

For the n dopants one finds a steady increase in the free-carrier concentration with increasing temperature for isochronal annealing. Boron, however, exhibits a marked region of retrograde annealing behavior. In this temperature region, interstitial silicon is competing with substitutional boron for the substitutional lattice sites displacing essentially the

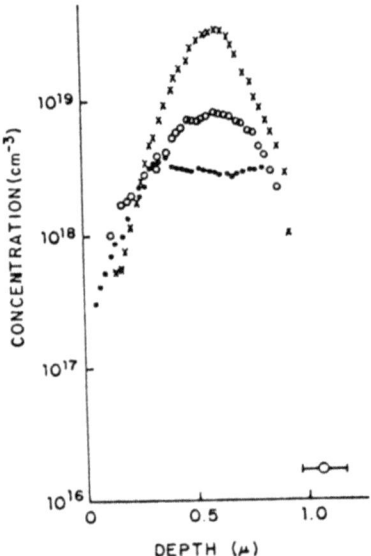

Fig. 10. Room-temperature implantation profiles of boron, 240 keV, into (111) silicon misoriented by $3°$, with total dose of 10^{15} cm^{-2}. Post-implant anneal temperatures were $700°C$ (O), $800°C$ (O), and $900°C$ (X), 1/2-hr duration. Stained junction depth is indicated by 1-0-1. (Source: Ref. 14)

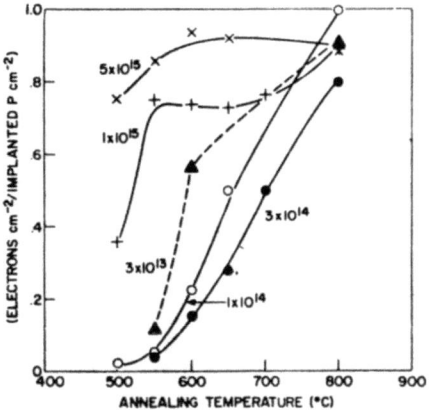

Fig. 11. The ratio of electrons/cm^2 (obtained from sheet Hall coefficients) to implanted atoms/cm^2 for P implantations into Si as a function of the isochronal annealing temperature. The total number of P ions/cm^2 implanted is indicated for each curve. The implantation energy was 280 keV. The duration of annealing at each temperature was 30 min. (Source: Ref. 15)

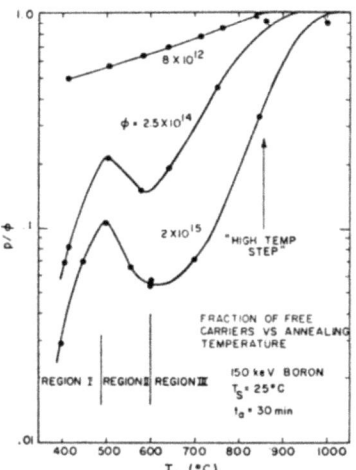

Fig. 12. Isochronal data in reduced form. The ratio of the free carrier content from Hall measurements to the dose is plotted against the annealing temperature for three doses. The maximum free carrier content is nearly equal to the measure dose pm $\simeq \phi$. (Source: Ref. 16)

boron. At higher temperatures, thermally generated vacancies accommodate interstitial boron in substitutional sites, thus increasing the electrical activity.

In general, it was observed that the formation of a continuous amorphous layer during the implant aided the annealing process. Crowder [17] gives the implant conditions listed in Table I for various III and V elements, which lead to the formation of an amorphous layer.

A question essential to many device applications in the area of charge storage concerns the leakage currents encountered in ion-implanted structures. Fang and Michel [18] have investigated this problem and found that, for anneal temperatures around $900°$ - $930°C$, the leakage currents for diffused and implanted junctions are identical. Figure 13 shows a typical example obtained on gated diode structures.

Table I. Ion dose required to produce a continuous amorphous silicon layer

Ion	D_0, cm^{-2}*	D_{300}, cm^{-2}*	D_{300} (measured), cm^{-2}
B	8×10^{14}	4×10^{17}	$>2 \times 10^{10}$
P	2×10^{14}	8×10^{14}	$>5 \times 10^{14}$
As	1×10^{14}	2×10^{14}	2×10^{14}
Sb	6×10^{13}	1×10^{14}	1×10^{14}
Bi	4×10^{13}	6×10^{13}	5×10^{13}

SOURCE: Ref. 17.

*D_0 (cm^{-2}) Low Temperature ⎫
$D_{300°K}$ (cm^{-2}) Room Temperature ⎬ Implant

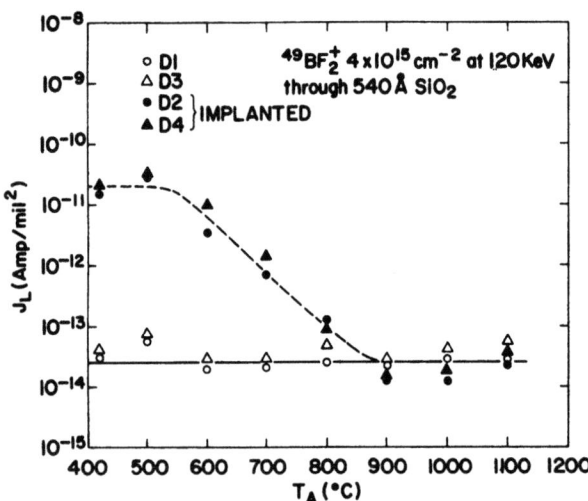

Fig. 13. Areal leakage density as a function of annealing temperature for boron-implanted junctions compared with the diffused junctions. Annealing time is 30 min for each data point. The gate plate was biased at 0 V relative to the substrate and V_{sub} = −5 V. (Source: Ref. 18)

3.2 Gettering Effects

Crystalline defects have been known, for many years, to be precipitation centers for impurities and, particularly, for heavy metal contaminants. B. Masters [19] was the first to use the decoration technique to determine the radiation-damage distribution due to ion implantation. Figure 14 depicts the P distribution obtained from a 300-keV implant and the damage profile decorated with radioactive copper.

Various workers in the field have since investigated the gettering efficiency of various species. T. Seidel [20] et al. compared the gettering of gold due to ion damage with the gettering due to P diffusions and found Ar to be more effective for annealing temperatures up to $1000°C$ (Ar 200 keV, dose $10^{16} cm^{-2}$).

M. Poponiak et al. [21] applied Ar backside gettering to device wafers at either prebase or pre-emitter processing steps and used the resulting pipe densities as a criterion for the gettering efficiency.

4. DEVICE APPLICATIONS

One can state several reasons for the relatively slow acceptance of ion implantation in the early years of silicon technology. One

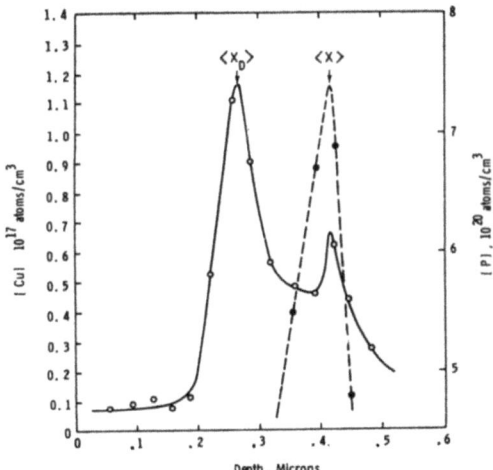

Fig. 14. Copper-decorated $^{31}P^+$ implant at room temperature, 1×10^{16} ions/cm^2 at 300 keV. Annealed at 30 min at $600°C$ and anodically sectioned. Comparison of radiocopper o and radiophosphorous o profiles (Source: Ref. 19)

obvious reason is that the penetration depth, R_p, of the standard dopants, such as As, P, and B, has been limited to less than 0.5 µm for technically practical energies, that is, E<400keV.

It is therefore understandable that the first commercial applications were directed towards the advancement of FET devices.

4.1 High-Performance FET Devices

Substantial improvements have been obtained in the speed performance of FET circuits [22]. The application of ion implantation enables one to fabricate n-channel enhancement and depletion devices on the same chip. It further makes it possible to use relatively high resistivity substrate material and, in conjunction with self-aligned gate concepts, to minimize parasitic capacitances. Figure 15 depicts schematically an 11-stage ring oscillator consisting of n-channel enhancement drivers and depletion loads. Another noteworthy design feature of those devices concerns the relative insensitivity of the depletion load device to substrate bias. The precise control of the dopant profile, which is needed to accomplish this objective, can be obtained only from ion implantation.

On circuits with 1-µm gate widths, turn-on delays as low as 150 psec were measured.

4.2 All-Ion-Implanted Bipolar Structures

In this section we shall describe an epiless bipolar structure, in which the subcollector region is formed by implantation of high-energy phosphorous [23]. Figure 16 depicts the device cross

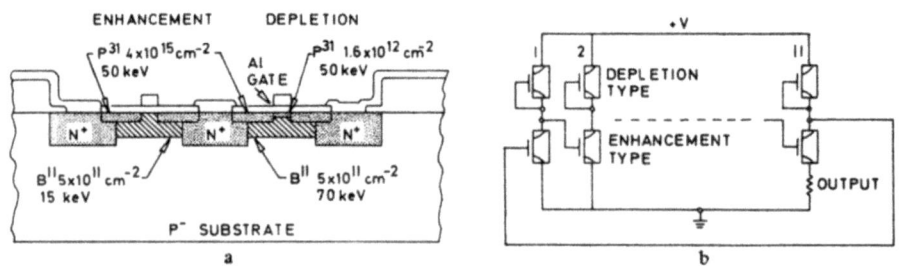

Fig. 15. (a) Device embodiment of ion-implanted enhancement/depletion self-aligned MOSFET's and (b) schematic diagram of an 11-stage oscillator using enhancement/depletion inverter circuits. (Source: Ref. 22)

section as well as the impurity distribution after the drive-in cycle of the implanted emitter. This drive-in cycle also serves to anneal the radiation damage due to the subcollector, to subcollector reachthrough, and to base implants.

The electrical results of those devices indicate that the residual damage in the surface region, where emitter and base junctions are formed, is low, yielding high-quality p-n junction and excellent transfer characteristics in the base region. Typical curve trace data are shown in Fig. 17.

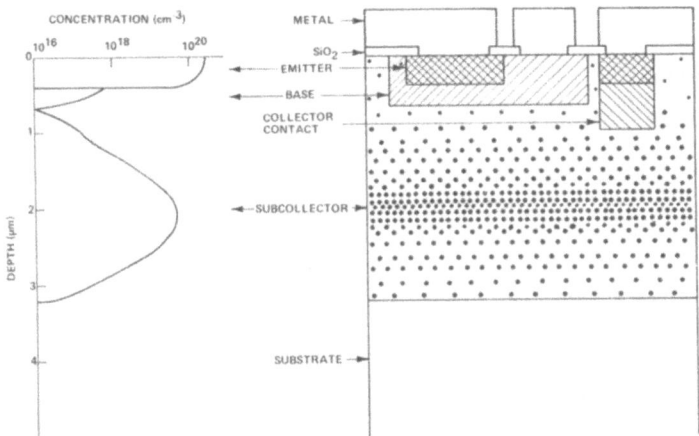

Fig. 16. Schematic of an epiless bipolar structure.
(Source: Ref. 23)

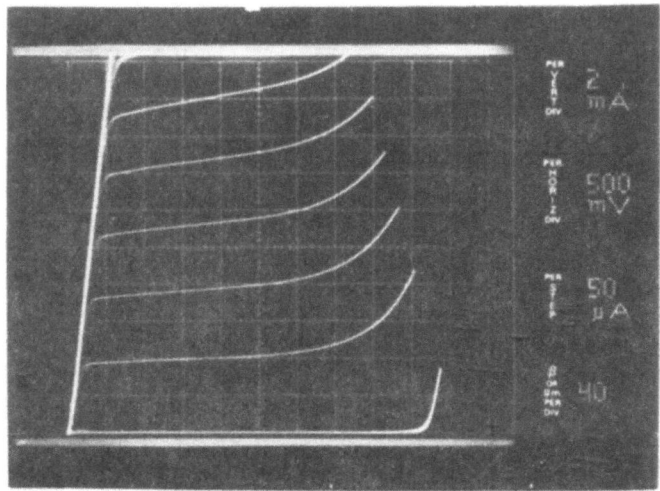

Fig. 17. Common emitter transfer characteristics.

ACKNOWLEDGMENT

The many helpful discussions with B. Masters and W. K. Chu, as well as their valuable critique of the manuscript, are greatly appreciated. I should also like to thank A. Michel and W. K. Chu for making available to me, prior to publication, some of the experimental results on bipolar device development (A. Michel) and the data on knock-on distributions (W. K. Chu).

REFERENCES

1. J. Lindhard, M. Scharff, and H. E. Schiott, K. Dan. Vidensk. Selsk. Mat.-Fys. Medd. 33, 14 (1963).

2. U. S. Patent 2,787,564.

3. K. Wittmack, J. Maul, and F. Schulz, in Ion Implantation in Semiconductors and other Materials (Plenum Press, New York - London, 1973), p. 119.

4. G. Dearnaley, J. H. Freeman, R. S. Nelson, and J. Stephen, Ion Implantation (North Holland Publishing Co., Amsterdam, London, 1973), p. 11.

5. J. Lindhard and M. Scharff, K. Dan. Vidensk. Selsk. Mat.-Fys. Medd. 27, 15 (1953).

   _____, Phys. Rev. 124, 128 (1961).

6. N. Bohr, K. Dan. Vidensk. Selsk. Mat.-Fys. Medd. 18, 8 (1948).

7. S. Furukawa, H. Matsumura, and H. Ishiwara, Jap. J. Applied Physics 11, 134 (1972).

8. K. B. Winterbon, Radiation Effects 13, 215 (1972).

9. J. Sanders, Can. J. Phys. 46, 455 (1968).

10. W. S. Johnson, J. Gibbons, and S. W. Mylroie, Projected Range Statistics (Halstead/Wiley, New York, 1975), 2nd ed.

11. R. A. Moline, G. W. Reutlinger, and J. C. North, Proc. 5th International Conference on Atomic Collisions in Solids (Plenum Press).

12. W. K. Chu, private communication.

13. J. M. Shannon, Solid-State Electronics 19, 537 (1976).

14. J. M. Fairfield and B. L. Crowder, J. Electrochem. Soc. 117, 671 (1970).

15. B. L. Crowder and F. F. Morehead, Appl. Phys. Lett. 14 (10), 313 (1969).

16. T. E. Seidel and A. V. MacRae, Radiation Effects 7, 1 (1971).

17. B. L. Crowder, J. Electrochem. Soc. 118, 943 (1971).

18. A. Michel, F. F. Fang, and E. Pan, J. Appl. Phys. 45, 2991 (1974).

19. B. Masters, J. W. Fairfield, and B. L. Crowder, in Ion Implantation ed. F. Eisen and L. T. Chadderton (Gordon and Breach, New York, 1971), p. 81.

20. T. E. Seidel, R. L. Meek, and A. G. Cullis, in Lattice Defects in Semiconductors (Institute of Physics, London and Bristol, 1974), p. 494.

21. M. Poponiak, T. Nagasaki, and T. H. Yeh, Abstract 185, p.392, The Electrochemical Society Extended Abstracts, Fall Meeting, Dallas, Texas, Oct. 5-10, 1975.

22. F. F. Fang and H. S. Rupprecht, IEEE J. Solid State Circuits, SC-10 (4), 205 (1975).

23. A. Michel, to be published.

PATTERN GENERATION FOR INTEGRATED CIRCUIT FABRICATION

D. Widmann

Research Laboratories, Siemens AG,
D-8000 München 70, F.R. Germany

ABSTRACT. The various techniques for generating patterns in resist films are outlined. Resolution, overlay precision, defect density, and cost are discussed as essential criteria for characterizing these techniques. In the second part pattern definition techniques, including etching, structured deposition, local oxidation, and indirect methods are treated.

1. INTRODUCTION

Current semiconductor technologies need thin film patterns which act as interconnection or masking patterns. Such patterns are commonly realized in two basic steps. In the first step the desired pattern is generated in a resist film deposited on the wafer, while in the second step the pattern is transferred to an insulator or metal film by a pattern definition technique, such as etching. Following the same sequence, a survey of pattern generation in resist films (Section 2) will be succeeded by a treatment of pattern definition (Section 3).

2. PATTERN GENERATION IN RESIST FILMS

The basic steps for generating a pattern in a resist film are shown schematically in Fig. 1. Exposure can either be performed by UV light or by electrons or by x-rays. The resist is called a positive resist if the exposed regions are resolved in the developer and a negative resist if they are not resolved in the developer.

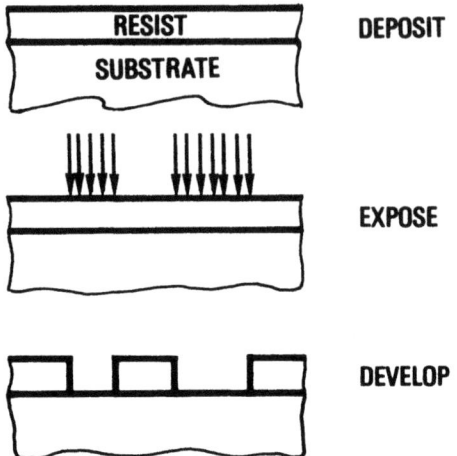

Fig. 1. Basic process steps for pattern generation in positive resist films. Each of the steps shown may be followed by baking.

2.1. Properties of a positive photoresist (AZ 1350 H)

For simplicity only a positive photoresist, namely AZ 1350 H[*], will be considered. Comprehensive reports on negative photoresists are to be found in /1/ and /2/.
 Positive photoresists contain three components, viz. a resin, a sensitizer and a solvent /3/. The viscosity of the resist can be controlled via the solvent percentage. The general chemical composition of the sensitizer of positive photoresists and its decomposition during exposure is as follows:

[*]Trademark of Shipley Company, Newton, Mass., USA

The result of exposure is an acid component which can be resolved in an alkaline developer. The development rate r_{dev} is a function of the product of exposure intensity I_{exp} by exposure time t_{exp}. For AZ 1350 H films this function is strongly non-linear (Fig. 2) if the baking temperature after photoresist film deposition is relatively low, e.g. about 65°. A non-linear behaviour is desirable because in this case variations in photoresist thickness have no serious influence on linewidth*. This is shown schematically in Fig. 3.

AZ 1350 H photoresist is photosensitive at violet and ultraviolet wavelengths (Fig. 4), but rather insensitive for instance to yellow light ($\lambda \approx 600$nm). This light can thus be used for the inspection of wafers coated with unexposed resist films, in order, for instance, to align the wafers with masks.

The photosensitivity of AZ 1350 H decreases with increasing baking temperature. Baking the deposited photoresist film above about 130° makes it completely insensitive.

The resolution capability of AZ 1350 H is extremely high, being better than 0.1 µm (see Section 2.5). The resolution limit has not as yet been determined, but is believed to be given by the length of the sensitizer molecules (≈ 80nm).

After development the photoresist pattern acts as a mask against etching, ion implantation, or other treatments. There are however certain limitations with regard to the stability of AZ 1350 H patterns:

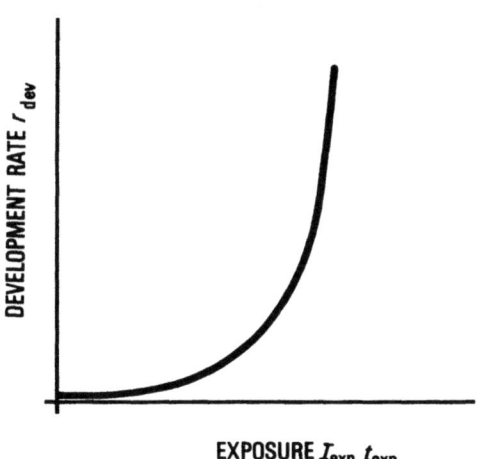

Fig. 2. Development rate (arbitrary units) versus exposure $I_{exp} \cdot t_{exp}$ (arbitrary units) for AZ 1350 H photoresist.

---
* Interference effects are not considered. For this problem, see Fig. 11.

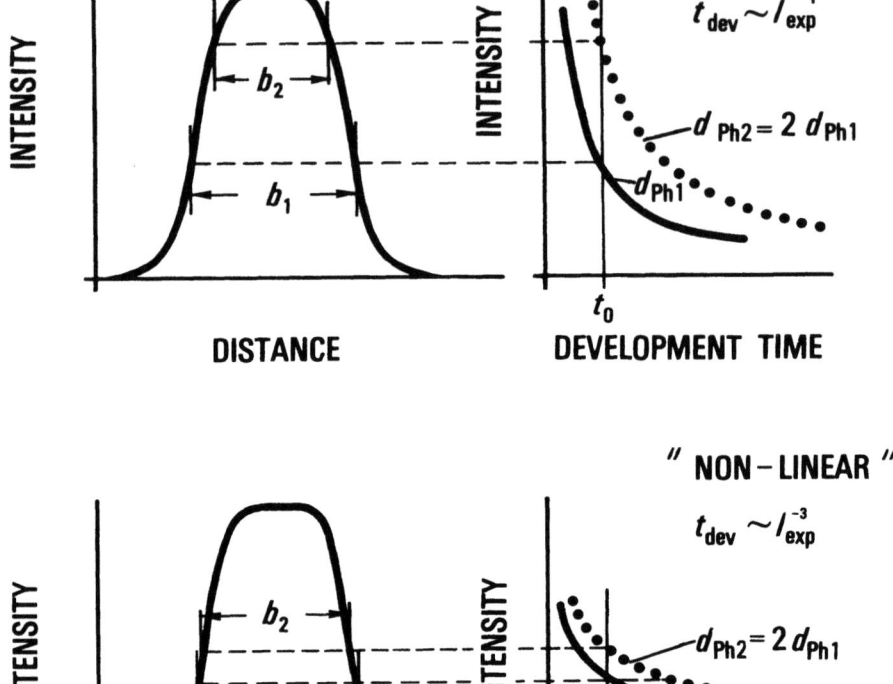

Fig. 3 Schematic representation of the influence of photoresist
thickness variations on linewidth. Thicknesses ranging
from d_{Ph1} to $2d_{Ph1}$ are assumed (see Fig. 5). If the photoresist acts in a "linear" manner (see above) large variations in linewidth from b_1 to b_2 result, while "non-linear" photoresists (lower part) result in only moderate variations
in linewidth, i.e. b_1-b_2 is small. A development time $t_{dev}=t_o$
is assumed.

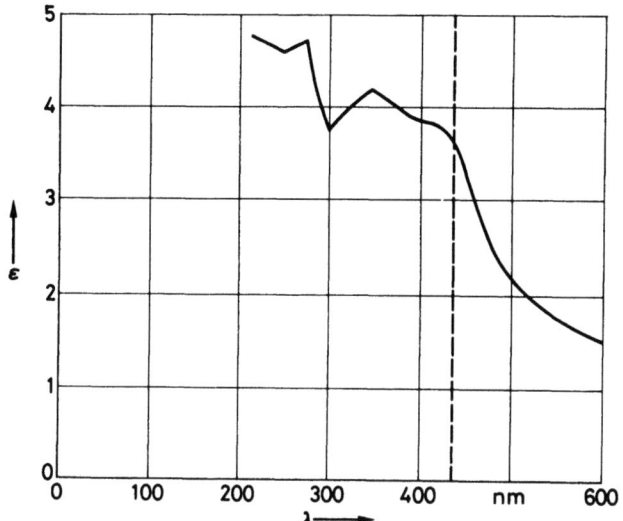

Fig. 4 Logarithm of the spectral sensitivity ε (arbitrary units) of AZ 1350 H photoresist

- Temperatures exceeding $120°$ lead to a rounding of the photoresist edges and to linewidth changes.
- AZ 1350 H does not withstand strong alkaline etchants, oxidizing acids (H_2SO_4, HNO_3) or organic solvents.
- High-dose implantations ($>10^{15} cm^{-2}$) may degrade the photoresist pattern. Adding benzoyl-peroxide, for example, to the photoresist will result in patterns with greater resistance to ion implantation /4/.

2.2. Photoresist film deposition and removal

Coating of wafers with a thin photoresist film is usually performed by spinning at spin speeds of 5000 to 10 000 rpm. Typical film thicknesses range from 1.0 um to 1.7 µm. On flat substrates the uniformity is excellent (about $\pm 3\%$) except near the substrate edges, where the film thickness may increase considerably.

If the surface of the wafers is not plane, i.e. if it contains steps large thickness variations will result in the vicinity of the steps (Fig. 5). As a rule of thumb, the required photoresist thickness should be larger than the difference between the highest and the lowest level of the surface topography in order to assure sufficient edge coverage over the whole wafer.

AZ 1350 H films are known to adhere well to metal surfaces such as Al or Cr if the photoresist is deposited immediately

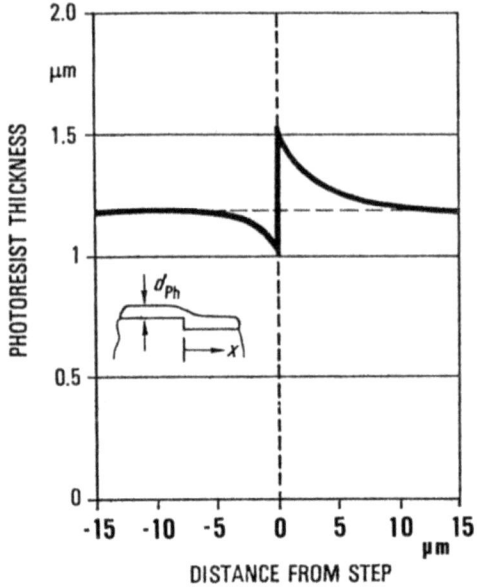

Fig. 5 Measured thickness of an AZ 1350 H film in the vicinity of a step on the wafer surface. The step height is 0.5 μm.

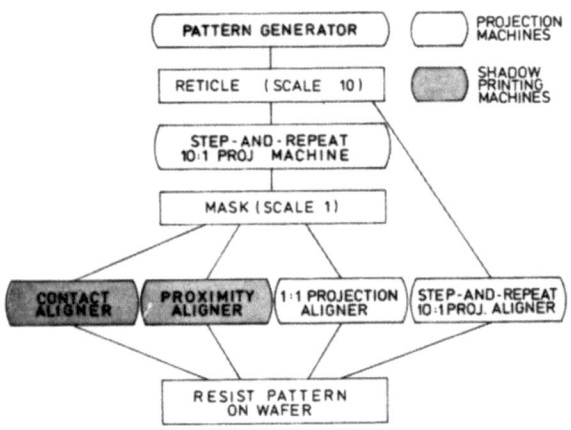

Fig. 6 Schematic illustration of the various ways for generating photoresist patterns on wafers.

after the evaporation or sputtering of the metal films. Adhesion problems may arise on SiO_2 surfaces, especially if the oxide is heavily doped with phosphorus (phosphorous silicate glass or PSG). Hexamethyldisilazane (HMDS) /5/ is a possible adhesion promoter (primer) recommended in these cases.

After pattern definition the AZ 1350 H photoresist pattern is stripped by dipping the wafers in an acetone or dimethylformamide bath. The stripping should be followed by an oxidizing treatment, such as a hot (100°C) mixture of sulfuric acid (90%) and nitric acid (10%) in order to remove any organic residues.

Another stripping method is plasma oxidation of the photoresist. Using this method care must be taken to prevent the penetration of the sodium (AZ 1350 H contains a few ppm of sodium) into silicon dioxide films.

2.3. Photoresist exposure techniques

Pattern generation in photoresist films on wafers can be performed by several methods shown schematically in Fig. 6. These are either projection or shadow printing methods. Projection printing can be performed with a 1:1 scale or with a 5:1 or 10:1 reduction. The image fields of currently available projection systems are limited. While with 1:1 projection image fields 3" in diameter are possible, 10:1 projection is today limited to a diameter of 14mm. With shadow printing methods only 1:1 replications are possible. Shadow printing covers both contact printing (no gap between mask and wafer) and proximity printing (small gap of several μm between mask and wafer to avoid mechanical damage).

The reticles and masks used for the optical pattern generation techniques (Fig. 6) are either emulsion or hard-surface masks. Emulsion masks are cheaper, however they have two important drawbacks. They are easily damaged mechanically, and the smallest geometries realizable under production conditions are lines and spaces of about 6 μm in width. The pattern-forming material of currently used hard-surface masks is chromium, "black chromium", iron oxide or silicon. The latter two materials are opaque in the ultraviolet wavelength range, but transmissive in the visible wavelength range.

2.4. Characterization of photoresist exposure techniques

The principal features characterizing photoresist exposure techniques are the following:
- Resolution
- Linewidth tolerance
- Overlay precision
- Defect density
- Exposure cost per wafer

In Table 1 a rough comparison is shown between the various

optical exposure techniques on the basis of the criteria listed above.

The following section treats the resolution capability of optical exposure systems and the overlay precision of optically exposed patterns in more detail.

2.5. Resolution of optical exposure systems

The resolution of optical shadow printing systems is limited by the diffraction of light at the mask windows. This is shown schematically in Fig. 7 for a mask pattern consisting of a periodic

	Contact printing	Proximity printing 4)	1:1 projec.	10:1 projec. 5)
Resolution 1) – plane wafer surface 2) – 15µm waviness – 15µm waviness and 1µm profile	1 µm 1 µm 1.5µm	4µm 5µm 6µm	4µm 5µm 6µm	1.2µm 1.5µm 2 µm
Linewidth tolerance 1) – plane wafer surface 2) – 15µm waviness – 15µm waviness and 1µm profile	± 0.2µm ± 0.2µm ± 0.5µm	± 0.4µm ± 0.5µm ± 0.6µm	± 0.4µm ± 0.5µm ± 0.6µm	± 0.2µm ± 0.3µm ± 0.4µm
Overlay error 1) 3)	± 1.5µm	± 1 µm	± 1 µm	± 0.5µm
Defect density (defects per cm^2)	1	0.1	0.01	0.01
Exposure cost ($ per 3" wafer)	1	1	2.5	10

Table 1. Comparison of optical contact printing, proximity printing and projection exposure on 3" wafers

1) The tabulated values can be obtained without sophisticated conditions. The limit values are better by a factor of about 2.
2) The mask surface is assumed to be completely plane
3) The effects of temperature variations and in-plane wafer distortions are not considered (see Section 2.6.)
4) Nominal gap between mask and wafer 15 µm
5) Step-and-repeat machine with automatic focusing

grid of opaque lines and transparent spaces with equal linewidths
b. The period of the grid is 2b; 1/2b is the spatial frequency
of the grid. As shown in Fig. 7 the distribution of the light
intensity incident on the photoresist surface differs from the
"ideal" distribution, which would be expected

for parallel light without diffraction. Fresnel's diffraction
theory /6/ predicts a square root relationship between the minimum transferable period $2b_{min}$ of a periodic grid and the product of the wavelength λ by the gap s between mask and wafer,

$$2b_{min} = k\sqrt{\lambda s} \qquad (1)$$

For intimate contact the effective "gap" is about half the photoresist thickness. The proportionality factor k can be determined
experimentally /7/. If small tolerances in resist, exposure and
development parameters are allowed, k is about 3 if collimated
light and non-reflective masks are assumed. This is, however,
a rather rough estimate. For instance, complicated diffraction
patterns appear near the corners of the windows.

Relation (1) shows that reducing the wavelength λ and/or
the gap leads to better resolution. Both approaches have been
investigated. Lin /8/ used a wavelength of 200nm instead of the
currently used wavelengths near 400nm. Rottmann /9/ proposed
to flatten wafers with the aid of a special wafer chuck, so that

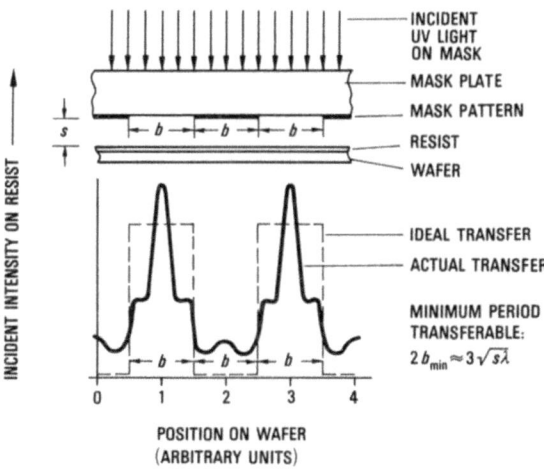

Fig. 7. Intensity transfer by shadow printing

smaller gaps between mask and wafer can be maintained. Another approach for reducing diffraction effects is already realized in most commercial proximity printers. This uses multiple coherent light sources generated from a single light source by a special mirror or prism configuration /10/.

The resolution of optical projection printing systems is limited by diffraction at apertures, by non-ideal lenses or mirrors, and by defocusing. If a periodic grid is again assumed for the mask pattern, the intensity transfer by the projection system can be characterized by the modulation M,

$$M = \frac{I_{max} - I_{min}}{I_{max} + I_{min}} \qquad (2)$$

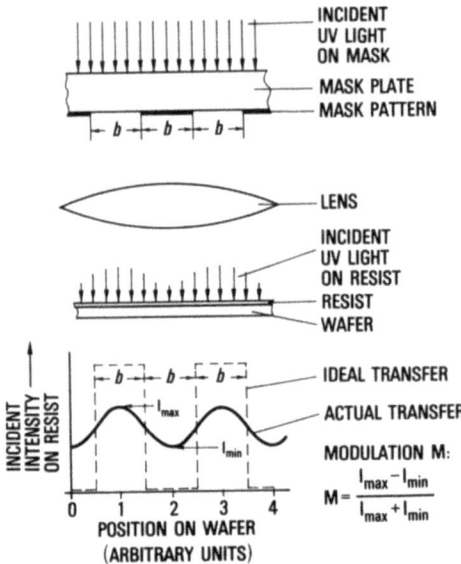

Fig. 8. Intensity transfer by projection printing

The meaning of I_{max} and I_{min} is illustrated in Fig. 8. The modulation transfer function (MTF) representing the dependence of the modulation on the spatial frequency of the periodic grid, is a figure of merit for the projection system. In Fig. 9 the MTF curves are shown for 1x and 10x reduction lenses available on the market. It can be seen that, for a given spatial frequency, projection with 10x reduction provides larger modulation values than 1:1 projection because of the larger numerical aperture of the 10x reduction lens. The image field diameter of the 10x reduction

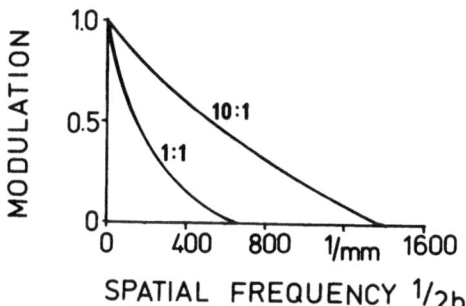

Fig. 9. Modulation transfer functions (MTF) for two different lenses. The 1:1 lens has a numerical aperture of 0.125, its image field is 75mm in diameter. The 10:1 lens has a numerical aperture of 0.33, its image field is 14mm in diameter. The wavelength λ is 436nm.

lens is however five times smaller than that of the 1x reduction lens.

In order to obtain reproducible results even in the case of small variations in the exposure parameters, a modulation of more than about 0.6 is required for wafer exposure. This means that with the 1:1 lens lines of 4 μm in width and with the 10:1 lens 1.2 μm lines can be achieved. 1:1 projection systems using mirrors as imaging elements /11/ instead of lenses seem to exhibit a somewhat higher resolution capability than 1:1 lens systems.

If the MTF curve for a given projection system is known, the intensity distribution over the photoresist surface can be calculated for any geometry on the mask. Fig. 10 shows the intensity distribution calculated by Dill /12/ for a single 1-μm line imaged by a lens projection system. The width of the transition region between the bright line and the dark environment is about 0.4 μm in this case.

Fig. 11 shows the calculated intensity distribution within the resist film during exposure of the single 1-μm line (due to Walker /13/). The interesting feature on this diagram is the intensity modulation in the normal direction. This modulation is caused by the interference of incident and reflected waves

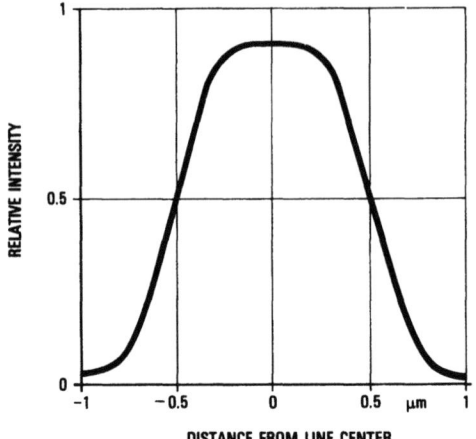

Fig. 10 Calculated intensity distribution for a single 1-µm line imaged by a high-resolution lens (after Dill /12/).

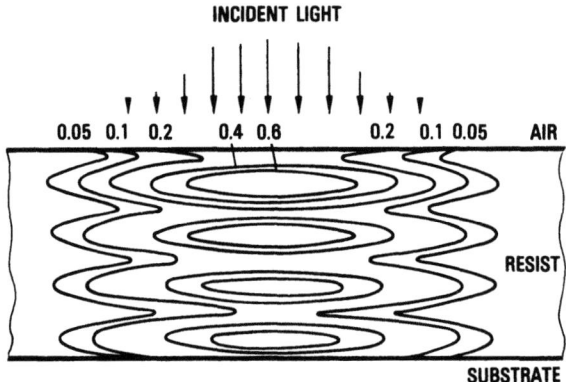

Fig. 11 Calculated intensity distribution inside the photoresist film during exposure of a nominal 1-µm line (after Walker /13/).

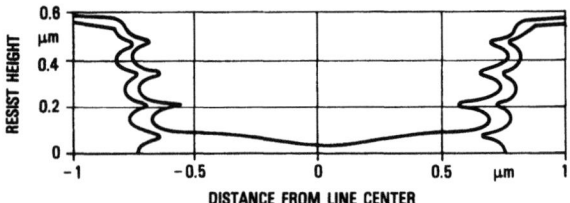

Fig. 12 Calculated photoresist profiles corresponding to two different development times (after Dill /12/). Positive photoresist AZ 1350 J is assumed.

/14/. The interference effect, which may be responsible for relatively large linewidth tolerances /13/, is less pronounced with mirror projection systems because there is in that case no requirement for a single exposure wavelength.

Developing the photoresist film exposed as in Fig. 11 leads to the photoresist profiles shown in Fig. 12 for two different development times (profiles calculated by Dill /12/). The linewidth increases with increasing development time, and the staircase-like edge contours (the step height is 0.13 µm) are the result of the interference effects demonstrated in Fig. 10. Real photoresist profiles are very similar to those calculated /12/. The fact that the small details along the photoresist sidewalls can be resolved indicates a resolution capability of the photoresist which is better than 0.1 µm.

2.6. Overlay precision of optically exposed patterns

The misregistration between two edges belonging to different levels of an integrated circuit process is defined as the difference between the actual distance W_A of the two edges and the designed distance W_D (Fig. 13). This misregistration $W_A - W_D$ is the result of the centerline-to-centerline offset ΔW_{CC} (see Fig. 13) and the linewidth deviations ΔW_1 and ΔW_2 of the geometries in levels 1 and 2,

$$W_A - W_D = \Delta W_{CC} + \frac{1}{2}(\Delta W_1 - \Delta W_2) \quad (3)$$

If the deviations ΔW_{CC}, ΔW_1 and ΔW_2 are measured on many wafers at many different points on the wafers (ΔW_{CC} can be measured by the use of verniers with an accuracy better than ± 0.1 µm), a Gaussian distribution of the deviations can be assumed. The standard deviation of $W_A - W_D$ is defined as the edge-to-edge overlay error ε_{EE}. If ε_{CC}, ε_1 and ε_2 are the standard deviations of ΔW_{CC}, ΔW_1 and ΔW_2, respectively, ε_{EE} is calculated as follows,

$$\varepsilon_{EE} = \pm \sqrt{\varepsilon_{CC}^2 + \frac{1}{4}\varepsilon_1^2 + \frac{1}{4}\varepsilon_2^2} \quad (4)$$

The linewidth tolerances ε_1 and ε_2 are caused by variations of the photolithographic processes on masks and wafers (see Table 1) including etching. Linewidth variations of photoresist stripes near surface steps /15/ may also contribute to ε_1 and ε_2.

The following factors contribute to the error ε_{CC} /16/:
- Positional errors on reticles and masks
- Thermal expansion of masks and wafers
- Process-induced in-plane wafer distortions
- Differential bending of masks and wafers
- Misalignment

These factors are of different importance for the different exposure techniques. Using contact printing state-of-the-art design rules assume $\varepsilon_{EE} \approx \pm 2$µm. The smallest error can be obtained with the 10:1 reduction projection (ε_{EE} better than ± 0.5 µm) because most of the error factors listed above are compensated by the chip-by-chip realignment.

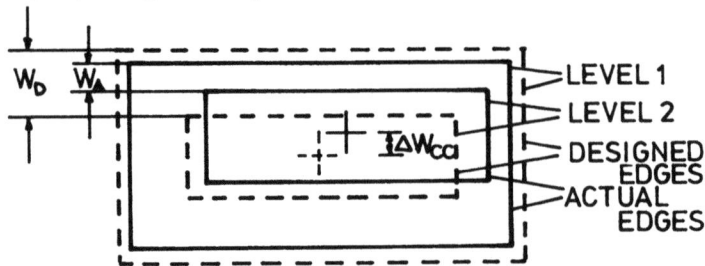

Fig. 13 Definition of misregistration. The distance of the actual positions of the upper horizontal edges is W_A, while W_D is the distance of the designed positions. $W_A - W_D$ is the misregistration of the two edges indicated. ΔW_{cc} is the misregistration of the centers of the two rectangles.

2.7. Electron beam exposure

Although full-image electron projection techniques using special masking methods have been developed, these techniques have not gained major importance so far. The following considerations will therefore be limited to scanning electron beam exposure. This technique uses an electron beam which is controllably deflected in order to expose the desired areas in the resist film. The main advantages of scanning electron beam exposure are low defect densities (no mechanical contact), flexibility, high-resolution capability (large depth of focus), and high overlay precision. Fig. 14 shows a schematic view of a scanning electron beam exposure system.

Round beams with typical current densities of $20A/cm^2$ and diameters ranging from 0.1 to 2 um have been used for exposure depending on the smallest element to be exposed. Another approach is the "shaped beam" /17/ with which shorter exposure times can be obtained.

There are two basic modes of electron beam exposure, the raster scanning mode and the vector scanning mode. In the raster scanning mode the electron beam is scanned row by row over the entire pattern area as in a television system, turning off the beam in the areas not to be exposed. In the vector scanning mode only those areas are scanned which are to be exposed.

Unfortunately, the area over which the electron beam can be scanned is limited due to deflection aberrations. The largest fields exposed so far are about 5mm x 5mm. If an entire wafer or mask area is to be exposed, step- and repeat techniques can be used. Another approach, the EBES system (electron beam exposure system) was developed at Bell Labs /18/. In this system the beam is only scanned in the x-direction while the table with the substrate is moved continuously in the y-direction.

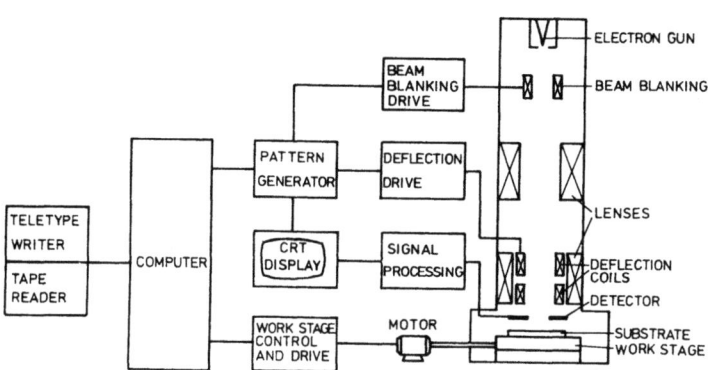

Fig. 14 Schematic diagram of the basic set-up of a scanning electron beam exposure system.

The maximum beam scanning speed provided by a deflection system can be used successfully only if the electron resist is sensitive enough. Assume an electron beam 0.2 µm in diameter and a beam current of 5×10^{-9} A. At a beam speed of 20 cm per second a resist sensitivity of 2×10^{-5} As/cm^2 is required. This requirement is met by PMMA resist (polymethylmetacrylate). The time for exposing an area of 1 cm^2 is 40 minutes under the above conditions. Shorter exposure times can be obtained by increasing the beam current, the beam cross-section area, and the scanning speed. Using a step-and-repeat exposure system, Chang /19/ has reported an exposure time of 5 minutes per 3" wafer with 2.5 µm minimum linewidth and with 20% of the wafer area requiring exposure.

Resolution of electron beam exposure is limited by electrons scattered in the resist and by electrons backscattered from the substrate. Fig. 15 shows a calculation of Kyser /20/ illustrating the broadening of the transition region between the exposed and the non-exposed region in the resist. An important result of electron scattering is the so-called proximity effect which means that the effective exposure intensity is larger for closely spaced lines than for single exposed lines. This proximity effect may be partially compensated by proper control of the beam scanning speed. However, the lateral spread of the dissipated energy at the line edges remains. This means that the resist parameters and the development conditions must be carefully controlled in order to arrive at the desired linewidths. Submicron lines with close tolerances are therefore not easily generated on profiled surfaces because of large resist thickness variations at the profile steps. On the other hand, the waviness of wafers is not very critical due to the large depth of focus of electrons beams.

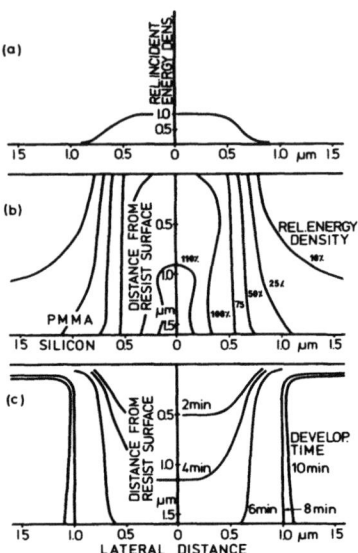

Fig. 15 Calculation of pattern formation by electron beam exposure.
a) Assumed distribution of the energy or charge density in the incident electron beam.
b) Calculated distribution of dissipated energy in 1.6 µm film of PMMA on silicon. The electron energy is 25 keV and the beam width is 1.2 µm.
c) Calculated PMMA contours after different development times (after Kyser and Viswanathan /20/).

With electron beam exposure systems high alignment accuracies better than ± 0.1 µm are possible, since steps on the wafer surface can be accurately detected by secondary or backscattered electron signals /21/. Overlay errors at any position of a chip or a wafer may, however, exceed the potential alignment error due to drifting of the beam which is caused by electronic instabilities or electrostatic charging fields along the beam path. An overlay precision of ± 0.5 µm has been reported /19/ using a step-and-repeat exposure system in which each 5mm x 5mm chip was realigned before exposure of the chip. Smaller chips would allow smaller overlay errors.

Electron beam exposure is compatible with semiconductor device fabrication, since the radiation damage can be completely annealed out at temperatures above 400°C.

The main drawback of electron beam exposure is cost. For

instance, the price for the EBES machine is about 1.25 million dollars, and exposure times are large compared to those obtained with optical exposure techniques. Nevertheless, electron beam exposure has proven advantageous, for instance, for mask pattern generation. Low-cost wafer exposure with electron beams seems also to be feasible, since further improvements in the exposure system and in electron-sensitive resists can be expected.

2.8. X-ray exposure

Since x-rays cannot be easily collimated or deflected, projection of scanning methods are not possible, so that x-ray exposure is limited to shadow printing.

A schematic diagram of an x-ray proximity exposure system is shown in Fig. 16 (after Smith /22/). This figure also illustrates the resolution limiting effect due to the finite size d of the source. If D is the source-to mask distance and s the width of the gap between mask and wafer, the width δ of the transition region between exposed and non-exposed regions is

$$\delta = s\frac{d}{D} \qquad (5)$$

For conventional x-ray sources d is in the order of 2 mm, while s should be about 50 μm in order to prevent any contact between mask and wafer.

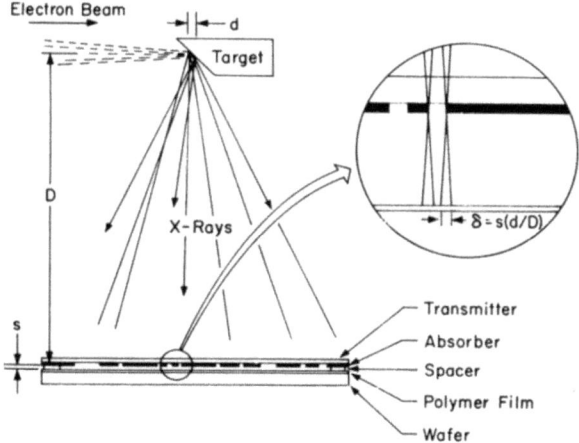

Fig. 16 Schematic diagram of an x-ray exposure system. The inset illustrates the resolution limiting effect due to the finite size d of the source (after Smith /22/).

For the reproduction of 1 μm lines and spaces δ should be smaller than about 0.5 μm. Using the above relation, D calculates at 200 mm under the above conditions.

Submicron lines can be replicated by increasing the source-to-mask distance D. In fact, 0.1 μm geometries have been demonstrated by Feder et al. /23/. They point out that the physical resolution limit of x-ray exposure is due to scattered secondary electrons whose range is of the order of 40 nm.

Three important drawbacks of x-ray exposure should be mentioned:
- Exposure times are relatively long (several hours under the above conditions using PMMA as a resist). This problem may be diminished by x-ray sources of higher intensity (e.g. synchrotron radiation) and more sensitive resists.
- The fabrication of high-contrast masks with submicron geometries is difficult /22/.
- Using point sources the replicated image is distorted as a result of the divergent x-rays.

Despite these drawbacks x-ray lithography offers the possibility of realizing submicron resist patterns even on profiled surfaces.

3. PATTERN DEFINITION TECHNIQUES

The generation of a resist pattern on a semiconductor wafer is generally followed by a pattern definition process, i.e. the resist pattern is transferred to an insulator (SiO_2, Si_3N_4) or conductor (Al, doped polycrystalline Si) pattern.

Several pattern definition techniques are available:
- Wet-chemical etching
- Plasma etching
- Ion etching
- Local oxidation
- Anodization
- Lift-off patterning
- Selective plating

The first three of these techniques are etching (or subtractive) processes while the other four techniques are build-up (or additive) processes.

3.1. Characterization of pattern definition techniques

The most important features characterizing pattern definition techniques are the following:
- Change of linewidths (definition see Fig. 17)
- Edge contour of etched grooves
- Etch rate selectivity
- Contamination or damage of semiconductor devices
- Pattern definition cost per wafer

Several factors contribute to the cost of pattern definition, such as equipment cost, safety precaution cost, etch rate, uni-

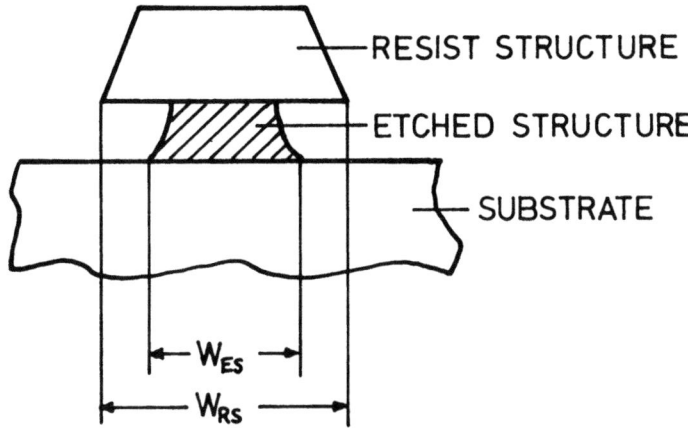

Fig. 17. Definition of the linewidth change caused by a pattern definition process. W_{RS} and W_{ES} are the linewidths of the resist structure and the etched structure, respectively, measured at the bottom of the structures. $W_{ES}-W_{RS}$ is defined as the linewidth change.

formity of etch rate, number of wafers per batch, and yield. This latter factor is especially important, since in many cases the wafers cannot be re-worked after unsuccessful pattern definition.

In the following sections the different pattern definition techniques including indirect methods (section 3.9) are discussed with regard to the criteria listed above.

3.2. Wet-chemical etching

The isotropic nature of the etch attack of wet etchants determines the shape of the etched grooves. In Fig. 18 the evolution of the contours of wet-chemically etched structures is illustrated for a pure SiO_2 film and a PSG/SiO_2 double layer (PSG is the abbreviation of phosphorous silicate glass). The generally used etchant for these films is a mixture of 7 parts of water and 1 part of a 40% solution of NH_4F in water.

Assuming good resist adhesion, in the case of pure SiO_2 cylindrical sidewall contours with a vertical upper region are formed /24/. If the etching process is stopped directly after the film has been completely etched through (case c in Fig. 18) no dimensional change, measured at the bottom of the film, will be found. A relatively fast widening of the etched line occurs however if etching is continued (case d in Fig. 18). Prolonging the etching time by only 10% results in a widening which about corresponds to the film thickness. In practice, the etching time is usually chosen at least 10% longer than the predetermined exact etching time, in order to assure that the film is etched completely through, irrespective of small variations of bath temperature, etchant composition, film thickness or other parameters. Dimensional changes in the order of the film thickness or even more must therefore be taken into account, if wet-chemical etching is used for pattern definition in amorphous films. Polycristalline films, such as polysilicon and aluminum behave similarly. The etch rate is however generally less uniform due to the grain structure of these films.

Fig. 18 shows on the right the contours obtained when a double layer of phosphorous silicate glass (PSG) and silicon dioxide is etched in buffered hydrofluoric acid. Due to the higher etch rate of PSG the etched lines are widened more rapidly at their upper portions than in the case of the pure SiO_2 film. The sidewall of the PSG film is nearly vertical, while tapered SiO_2 edges with a small sope angle are formed. Edges

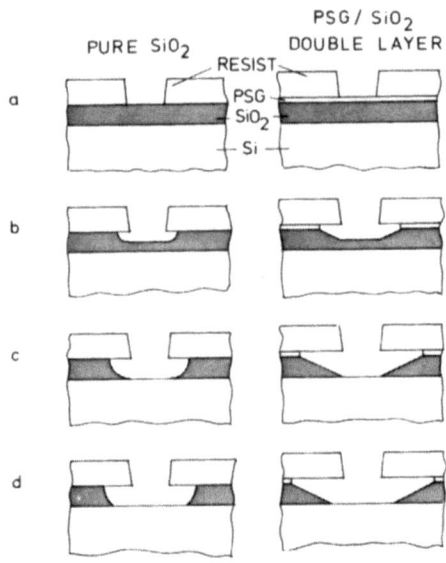

Fig. 18. Schematic cross-sections of wet-chemically etched films after different etching times (from (a) to (d)). Left: Pure SiO_2 film. Right: Double layer of phosphorous silicate glass (PSG) and SiO_2. In both cases buffered hydrofluoric acid is assumed for the etchant.

of this kind may occupy a considerable area disallowing high element densities.

Even smaller slope angles than those shown for the PSG/SiO_2 case in Fig. 18 may appear if adhesion of the resist to the substrate is reduced. For pattern definition of micron dimensions excellent resist adherence is thus of major importance in obtaining reproducible results.

Controlled tapering of etched structures consisting of a homogeneous material such as SiO_2 or polycrystalline silicon has been achieved by ion bombardment of the film prior to resist coating /25, 26/. The ions which are implanted near the film surface enhance the etch rate of the film. The implanted portion of the film is thus playing a similar role as the PSG layer in the case of the PSG/SiO_2 double layer.

A special application of wet-chemical etching in semiconductor technology is the etching of single crystal silicon with alkaline etchants, such as hot potassium hydroxide. As the etch rate in <100> directions is about 60 times faster than in <111> directions, V-shaped grooves with minimum undercutting under the mask edges are formed, if wafers with a (100) surface are etched and if the pattern edges are running along <110> directions (Fig. 19).

Some problems of wet-chemical etching in semiconductor technology should be mentioned:
- Uncomplete wetting of small details to be etched may lead to uncontrolled etch rates. The use of a proper wetting agent is therefore essential for most wet-chemical etching processes.
- Etching of aluminum interconnections running over steep steps may be accompanied by tunnel formation or even interruption along the steps.
- After etching of SiO_2 films with buffered hydrofluoric acid the underlying silicon surface may be contaminated by heavy metal atoms which may reduce the minority carrier lifetime.
- The formation of electrochemical potentials may cause locally enhanced or reduced etch rates.
- High cost may be required for safety precautions, e.g. for removing corrosive vapors while maintaining class 1 filtered-air environment.

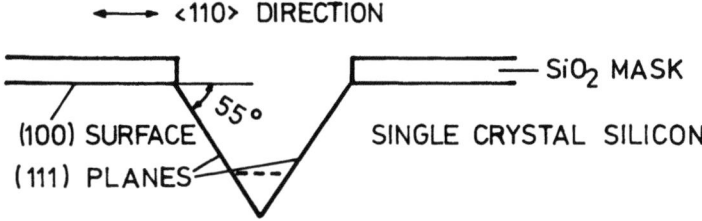

Fig. 19 Preferential etching of single crystal silicon in an alkaline etchant.

3.3. Plasma etching

The plasma in a plasma etcher is generated by an RF electric field (13,56 MHz). The plasma contains activated particles at a pressure in the 1 Torr pressure range. If these particles form stable volatile compounds with the film to be etched, etching takes place. For instance, using CF_4 as reactive gas, silicon is etched by the chemical reaction

$$Si + 4F^* \rightarrow SiF_4 \qquad (6)$$

Typical etch rates are 0.1 µm per minute.

A schematic view of a simple plasma etcher is shown in Fig. 20. RF power can be coupled to the plasma capacitively (as shown in Fig. 20, or inductively. Several improvements over this simple apparatus have been proposed:
- The wafers may be placed backside down on a termperature-controlled platform in order to maintain constant temperature during the entire etching cycle. The platform may also be used as one side of the capacitor across which the RF power is applied /27/.
- A perforated aluminum tube surrounding the wafers (a "etch tunnel"*) may be introduced into the plasma chamber in order to keep the RF field away from the wafers thus reducing radiation damage of the wafers during etching.
- The plasma discharge chamber may be separated from the reaction chamber /28/. The result is similar to that obtained with the etch tunnel.

*) Trademark of International Plasma Corporation

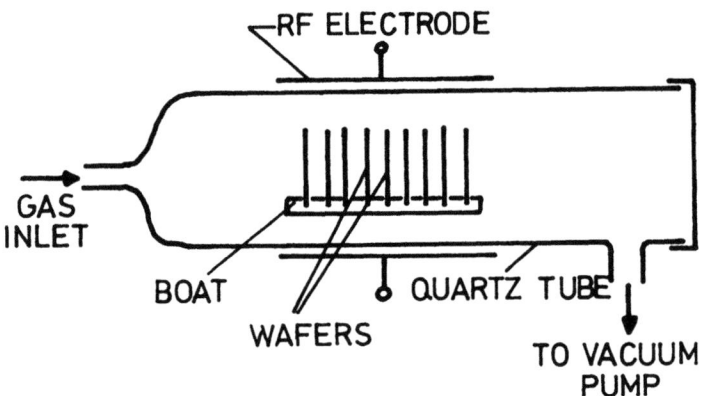

Fig. 20. Schematic view of a simple plasma etcher

The etch attack in the plasma etchers discussed above is substantially isotropic. Linewidth change and edge contours are thus similar to those of wet-chemically etched structures.

The radiation damage caused by plasma etching using an etch tunnel is readily annealed out during the conventional high-temperature processes used in semiconductor fabrication.

Plasma etching has actually replaced wet etching in many applications as it does not exhibit most of the drawbacks of wet etching (see section 3.2.). Table 2 gives a survey over the standard reactive gases used for plasma etching in present day semiconductor technology. Also indicated in Table 2 is the etch rate selectivity which may be a problem, e.g. in the case of SiO_2 films on silicon.

Film to be etched	Gas	Si	SiO_2	Si_3N_4	Al	AZ1350H
Si	CF_4		30		>100	>10
	CF_4+5%O_2		15	7	>100	15
SiO_2	CF_4+5%O_2	0.07		0.5	>100	1
Si_3N_4	CF_4+5%O_2	0.15	2		>100	2
Cr	He+18%Cl_2+2%O_2					
Mo, W	CF_4					

Table 2. Reactive gases used in standard plasma etching processes for different materials. Also indicated is the etch rate selectivity.

Attempts for plasma etching of Al films have been made using BCl_3 as reactive gas /29/, however this has not yet become a standard method.

3.4. Ion etching

Ion etching means etching with accelerated ions. Typical ion energies and ion current densities are 1keV and 0.5 mA/cm^2, respectively. The ions may either be inert (e.g. Ar ions), or they may chemically react with the material to be etched (reactive ion etching /30/). If ion etching is performed in a sputtering apparatus with the wafers placed on the cathode, the term "sputter etching" is used. In another ion etching approach the wafers are separated from the plasma discharge chamber /31/. This kind of ion etching which shall be treated in more detail is called "ion-beam etching".

A schematic diagram of the time evolution of the shape of the etched grooves /32/ during ion-beam etching using Ar ions is shown in Fig. 21. Compared to wet-chemical etching the following principal differences can be stated:

- No undercutting under the resist edges takes place.
- The masking resist pattern is also etched.
- The edges are faceted due to the enhanced ion etching rate of planes exhibiting an angle of about 60° with respect to the surface.
- There is no etch stop at the interface between the film to be etched and the substrate. Etch rates do not differ by more than a factor of 2 for Si, SiO_2 and AZ 1350 H films.
- The absolute etch rates are relatively small (in the order of 20nm per minute for Si, SiO_2 and AZ 1350 H). Using a reactive gas, the etch rate of the desired film may be increased.
- Etched geometries without dimensional change compared to the original resist linewidths may be realized if the resist sidewalls are steep and if the film thickness to be etched is relatively small (smaller than about 0.4 µm for SiO_2 films). Using a masking pattern consisting, for instance, of titanium instead of photoresist, thick layers of several microns can be etched without dimensional change compared to the mask linewidths /33/.

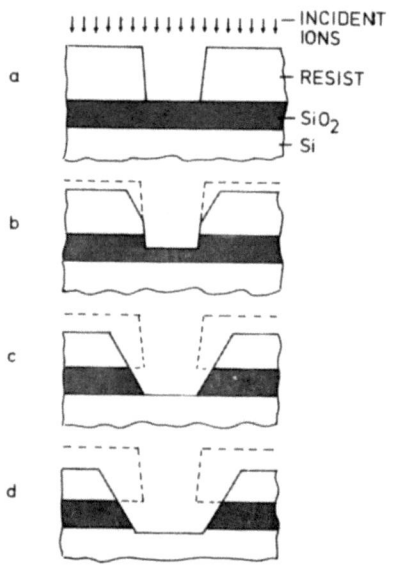

Fig. 21. Schematic diagram of the time evolution of the contours of grooves etched in a silicon dioxide film by ion-beam etching with a collimated argon ion beam. The final slope angle of the geometries is about 60°.

- Severe radiation damage, which is not annealable even at elevated temperatures of about 1000°C, may be found, if a silicon film is ion-etched down to the underlying silicon substrate /34/. To avoid the radiation damage, the ion etching process should be stopped before the SiO_2 film is etched through, or the damaged layer near the silicon surface (about 10 nm thick) should be chemically etched.

3.5. Local oxidation

The term "local oxidation" refers to the selective thermal oxidation of silicon. The principal steps of the local oxidation process are shown schematically in Fig. 22. The nitride pattern acts as a barrier against thermal oxidation of the underlying silicon, so that the oxide grows in the nitride-free regions only.

Etching is not fully avoided in this process. However, the undercutting problem is reduced since only a relatively thin nitride film (about 100 nm) rather than a thick oxide layer which may be 1 µm or thicker must be etched.

Thermal oxidation is governed by a diffusion meachanism. Thus, at the silicon nitride edges, the oxide grows laterally at about the same amount as vertically. This amount corresponds

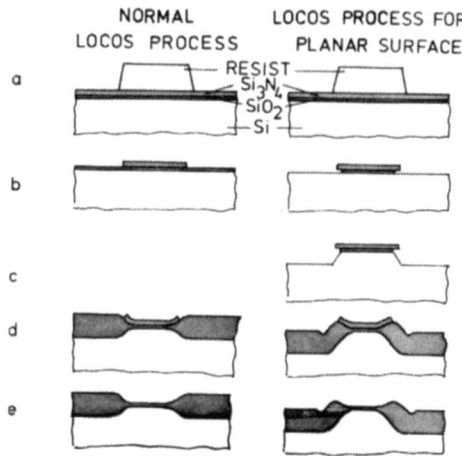

Fig. 22. Schematic illustration of the process steps for local oxidation. Left: Normal local oxidation process. Right: Local oxidation process leading to a nearly planar surface.
(a) After thermal oxidation (thin film), Si_3N_4 deposition and resist pattern formation.
(b) After Si_3N_4 (and SiO_2) etching and resist removal.
(c) After preferential silicon etching.
(d) After selective thermal oxidation.
(e) After Si_3N_4 removal.

to the thickness of silicon consumed during thermal oxidation, and is about half the thickness of the grown oxide. Consequently, the oxide rises about half its thickness over the original silicon surface, and the broadening of the oxide regions, compared to the linewidths of the windows in the silicon nitride, corresponds to about the oxide thickness. It is important to note that this linewidth change is well controlled by the oxidation mechanism, which is an advantage over most etching processes. Another important advantage of local oxidation is the possibility of realizing self-aligned doping of thick oxide regions /35/ (field oxide regions in MOS integrated circuits and isolation regions in bipolar integrated circuits).

As shown schematically in the right part of Fig. 22 the local oxidation process can be varied in such a way that a nearly planar surface is formed. This is accomplished by etching the silicon prior to thermal oxidation. The etch depth should correspond to half the desired oxide thickness. Small deviations from full planarity appear at the silicon nitride edges.

The benefits of this structure are apparent. Fine patterns in resist films are more easily realized if the substrate surface is plane and metallization failures often observed at steep steps are avoided. Moreover, the formation of raised silicon islands surrounded by silicon dioxide allows closely spaced transistors /36/ and makes plane pn-junctions possible. As a result, the breakdown voltage of the pn-junctions is increased and the pn-capacitance is reduced.

Some general problems associated with local oxidation should be mentioned. If the silicon nitride is deposited directly on the silicon surface, dislocations may be generated in the silicon. A silicon dioxide layer beneath the silicon nitride layer prevents the dislocations. However, this oxide layer acts as a "channel" for the oxgene during thermal oxidation leading to an additional lateral oxide growth, called "bird's beak" formation. In order to minimize this effect the oxide layer should be as thin as possible.

Another problem with locally oxidized structures is boron depletion at the sidewalls of the silicon islands due to the tendency of boron to segregate to the oxide. If this effect is not compensated by proper boron doping, it may lead to enhanced leakage currents.

Attempts have been made to avoid deposition and etching of the silicon nitride film used as the oxidation-preventing mask during thermal oxidation. For instance, Kooi /37/ has shown that a thin SiO_2 film treated in ammonia atmosphere at elevated temperatures withstands further film growth for some time during subsequent thermal oxidation.

3.6. Anodic oxidation

This technique is similar to the local oxidation technique. The difference is that the substrates must be immersed in an electrolyte and that electric fields rather than high temperatures are the driving force for the growth of the oxide. Good results have been obtained with selectively anodized aluminum films for metal interconnection patterns /38/. However, silicon dioxide films grown by anodic oxidation have not gained major importance.

Another technique similar to anodic oxidation may be mentioned here. Single crystal silicon can be converted into "porous" silicon by anodic treatment in hydrofluoric acid /39/. The oxidation rate of porous silicon is much higher than that of single crystal silicon. Thick oxide layers can thus be formed at relatively low temperatures. The diffusion of impurities such as boron, phosphorus and arsenic, is also enhanced in porous silicon.

3.7. Lift-off patterning

The lift-off technique is primarily used for the patterning of metal films which are deposited by evaporation. The principal steps of this technique /40/ are schematically illustrated in Fig. 23. No linewidth changes are expected compared to the pho-

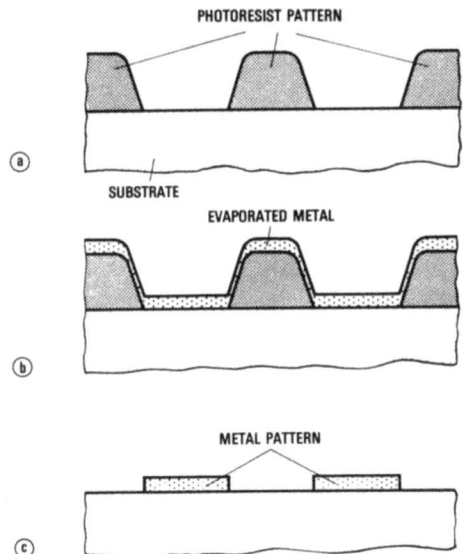

Fig. 23. Schematical illustration of the principal steps of the lift-off technique described by Stelter /40/.
(a) After resist pattern formation.
(b) After metal evaporation.
(c) After metal lift-off.

toresist pattern. There are, however, severe limitations for
the functioning of the lift-off mechanism. The allowable maximum metal film thickness is dependent on the slope angle of the
photoresist sidewalls /41/. Moreover, the metal should be brittle
so that it cracks easily at the photoresist edges if ultrasonic
agitation is applied. These drawbacks are overcome if overhanging
resist edges /42/ or resist sidewalls exhibiting a negative slope
angle /43/ can be realized. In these cases complete separation
of the metal portions deposited on the substrate and those deposited on top of the resist pattern is achieved.

The lift-off approach which uses overhanging photoresist
edges is treated in more detail. Fig. 24 shows schematically
the process steps. The calculation of the metal film growth in
the vicinity of the resist edges (Fig. 25) demonstrates that
for the evaporation conditions indicated the metal portions on

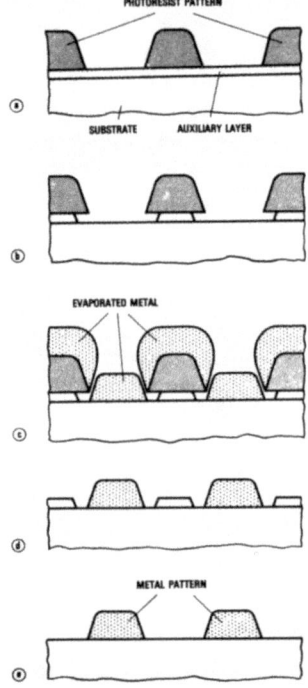

Fig. 24. Process steps for the lift-off technique using overhanging photoresist edges /42/.
  (a) After resist pattern formation on auxiliary layer
      (e.g. chromium)
  (b) After etching of the free portions of the auxiliary
      layer
  (c) After metal evaporation
  (d) After metal lift-off
  (e) After removal of the remaining portions of the auxiliary layer

the substrates are indeed separated from the metal portions on
the resist. The resist solvent can thus penetrate through the
microcrack and resolves the resist so that the metal portion
on top of the photoresist is lifted off. The calculated cross-section of the metal film in Fig. 25 indicates that the bottom edge
of the metal geometry is slightly apart from the photoresist
edge. This displacement corresponds to the auxiliary layer thickness which may be about 0.2 μm for a 1-μm thick metal film.

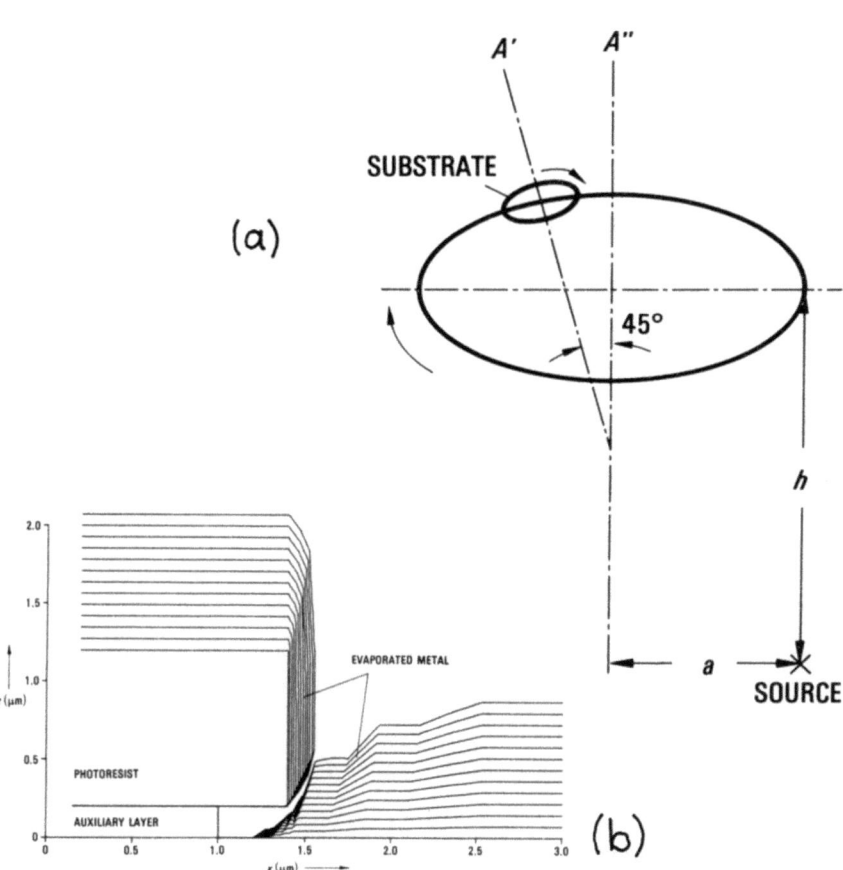

Fig. 25. (a) Assumed evaporation geometry for the lift-off metallization shown in Fig. 24.
(b) Calculated metal film growth in the vicinity of a resist edge assuming the evporation geometry shown in (a).

3.8. Selective plating

The selective plating technique is the only pattern definition technique which ensures exact replication of the resist pattern shape at least for thicknesses smaller than the resist thickness. Its application is however limited to those metals which can easily be deposited by plating, e.g. gold, silver, copper and nickel. The resist pattern must be generated on a metal film in order to connect the areas to be plated to the cathode contact. The metal film should also provide good adhesion to the substrate and good uniformity of the initial growth of the metal to be plated. Bi-metal layers rather than a single metal film may be necessary to meet these requirements. For instance, the beam-lead technique /44/ uses a Ti/Pt or a Ti/Pd double-layer over which a gold pattern is plated.

3.9. Indirect pattern generation

For high packing density of the circuit elements of integrated circuits high-resolution techniques with high overlay precision are required. If, however, only parts of an integrated circuit, such as the channel length of MOS transistors, shall exhibit small dimensions and/or high overlay precision, this can be obtained by indirect methods using conventional pattern generation techniques. The following list gives some examples of these indirect methodes:
- Small channel lengths of MOS transistors by use of a double-diffusion technique (DMOS transistors /45/)
- Small channel lengths of MOS transistors by use of V-groove etching (VMOS transistors /46/).
- Small contact holes generated by use of a crossed-stripe technique /47/.
- Small gaps between CCD electrodes by multilevel metallization /48/.
- Small gaps between CCD electrodes /49/ or small emitter widths /50/ by use of a lift-off technique similar to that described in Section 3.7.
- Small polysilicon electrode widths for CCDs by use of the enhanced etch rate of p-doped polysilicon in an acid etchant /51/.
- Self-aligning of the gate electrode of an MOS transistor with respect to the source and drain regions by use of a polysilicon gate (silicon gate technique).
- Self-aligning of a polysilicon gate electrode of an MOS transistor with respect to the field oxide edges by use of silicon nitride /52/.
- Self-aligned boron doping of the silicon substrate in the field oxide areas of MOS integrated circuits /35/ by use of silicon nitride (see Section 3.5.).

- Self-aligned emitter, base and collector regions of bipolar transistors with respect to the isolation regions by use of the local oxidation technique /53/ (see section 3.5.).
- Semi-self-aligned contact holes to doped regions by use of contact hole doping.

REFERENCES

1. K.G. Clark, Electronic Components, 621, (July 1974)
2. D.J. Sykes, Solid-State Technology, 53, (August 1973)
3. D.J. Elliott, Microelectronics, Vol. 5 No. 3, 3, (1974)
4. G.T. Elie and H.C. Kluge, IBM Techn. Discl. Bull., Vol. 17 No. 1, 85, (1974)
5. R.H. Collins et al., United States Patent No. 3, 549, 368, (Dec. 1970)
6. M. Born and H. Wolf, The Principles of Optics, 2nd Edition, Pergamon Press Oxford (1965)
7. Y. Wada and K. Uehara, Jap. J. Appl. Phys. $\underline{13}$, 2014, (1974)
8. B.J. Lin, J. Vac. Sci. Technol., $\underline{12}$, 1317, (1975)
9. H.R. Rottmann, Proc. KODAK Microel. Seminar, 79, (1974)
10. K.G. Clark et al., Solide State Technology, 79, (April 1976)
11. D.A. Markle, Solid State Technology, 50, (June 1974)
12. F.H. Dill, A.R. Neureuther, J.A. Tuttle and E.J. Walker, IEEE Trans. El. Dev., $\underline{ED-22}$, 456, (1975)
13. E.J. Walker, IEEE Trans. El. Dev., $\underline{ED-22}$, 464, (1975)
14. D.W. Widmann, Appl. Opt., $\underline{14}$, 931, (1975)
15. D.W. Widmann and H. Binder, IEEE Trans. El. Dev., $\underline{ED-22}$, 467, (1975)
16. J.A. Underhill, Proceedings of the Society of Photo-Optical Instrumentation Engineers, $\underline{80}$, 85, (June 1976)
17. H.C. Pfeiffer, J. Vac. Sci. Technol., $\underline{12}$, 1170, (1975)
18. D.R. Herriott et al., IEEE Trans. El. Dev., $\underline{ED-22}$, 385, (1975)
19. T.H.P. Chang et al., Proc. 7th Int. Conf. on Electron and Ion Beam Science and Technology, Washington, D.C., (May 1976)
20. D.F. Kyser and N.S. Viswanathan, J. Vac. Sci. Technol. $\underline{12}$, 1305, (1975)
21. H. Friedrich, H.U. Zeitler and H. Bierhenke, Proc. 7th Int. Conf. on Electron and Ion Beam Science and Technology, Washington, D.C., (May 1976)
22. D.L. Spears and H.I. Smith, Solid State Technology, 21, (July 1972)
23. R. Feder et al., J. Vac. Sci. Technol., $\underline{12}$, 1305, (1975)
24. T. Yanagawa and J. Takekoshi, IEEE Trans. El. Dev., $\underline{ED-17}$, 964, (1970)
25. R.A. Moline et al., IEEE Trans. El. Dev., $\underline{ED-20}$, 840, (1973)
26. G. Bell and J. Hoepfner, in "Etching for Pattern Definition", edited by M.J. Rand and H.G. Hughes (Electrochemical Society, Softbound Symposium Series, Princeton, 1976)
27. A.R. Reinberg, in "Etching for Pattern Definition", edited

by M.J. Rand and H.G. Huges (Electrochemical Society, Softbound Symposium Series, Princeton, 1976)
28. Y. Horiike and S. Masahiro, Proceedings 7th Conference on Solid State Devices, Tokyo, 1975
29. R.G. Poulson et al., International Electron Devices Meeting Technical Digest, Washington, (Dec. 1976)
30. J.A. Bondur, J. Vac. Sci. Technol., 13, 1023, (1976)
31. M. Cantagrel and M. Marchal, J. Mat. Sci., 8, 1711, (1973)
32. L. Mader and J. Hoepfner, J. Electrochem. Soc., 123, 1893, (1976)
33. H. Dimigen and H. Lüthje, Thin Solid Films, 27, 155, (1975)
34. H.R. Deppe et al., Solid State Electronics, 20, 51, (1977)
35. E. Kooi et al., Philips Res. Repts., 26, 166, (1971)
36. J.W. Evans et al., IEEE J. Solid-State Circ., SC-8, 373, (1973)
37. E. Kooi et al., Electrochemical Soc. Fall Meeting, Dallas, (1975)
38. H. Tsunemitsu and H. Shiba, NEC Res. Develop., 25, 74, (1972)
39. Watanabe et al., J. Electrochem. Soc., 122, 1351, (1975)
40. M.K.J. Stelter, SPC and Solid-State Technol. 9, 60, (1966)
41. H.I. Smith et al., J. Electrochem. Soc., 118, 821, (1971)
42. D.W. Widmann, IEEE J. Solid-State Circ., SC-11, 466, (1976)
43. M. Hatzakis, J. Electrochem. Soc., 116, 1033, (1969)
44. M.P. Lepselter, Bell Syst. Techn. J., 45, 232, (1966)
45. Y. Tarui et al., Proceedings 1st Conference on Solid-State Devices, Tokyo, 1969
46. T.J. Rodgers and J.D. Meindl, IEEE J. Solid-State Circ., SC-9, 239, (1974)
47. I.E. Magdo and S. Magdo, International Electron Devices Meeting Technical Digest, Washington, 276, (1974)
48. A. Mohsen and T. Retajczyk, IEEE J. Solid-State Circuits, SC-11, 180, (1976)
49. V.A. Browne and K.D. Perkins, IEEE J. Solid-State Circ., SC-11, 203, (1976)
50. A. Hayasaka et al., International Electron Devices Meeting Technical Digest, Washington, (1976)
51. K.H. Nicholas et al., Proceedings European Solid-State Device and Research Conference, Nottingham, U.K., 132, (1974)
52. V.L. Rideout and V.J. Silvestri, International Electron Devices Meeting Technical Digest, Washington, 593, (Dec. 1976)
53. R. Edwards et al., in "Semiconductor Silicon", edited by H.R. Huff and R.R. Burgess, The Electrochem. Soc., (1973)

TEST STRUCTURES AND DIAGNOSTIC TECHNIQUES

Walter H. Schroen

Texas Instruments Incorporated, Dallas, Texas, U.S.A.

1. INTRODUCTION

Many process and device parameters can most conveniently be measured using test structures. Figure 1 describes the principle of the plug bar concept. An ideal plug bar has four sets of test structures: The first set delivers physical parameters, the

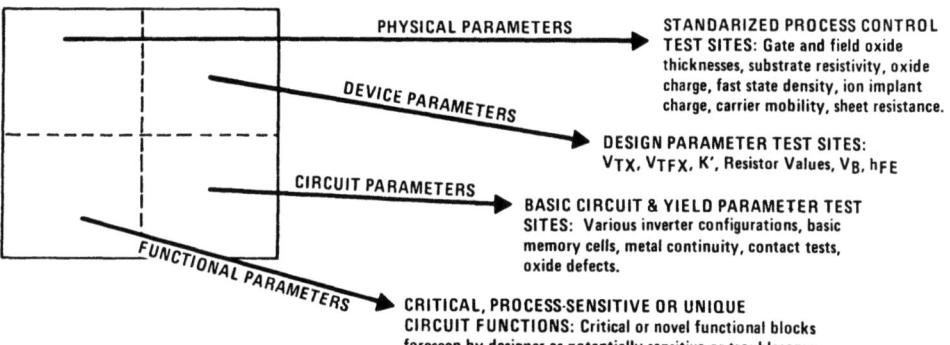

- PLUG BAR DESIGN BECOMES PART OF CIRCUIT DESIGN RESPONSIBILITY; STANDARDIZED TEST SITES & CRITICAL FUNCTION BLOCKS ARE LAID OUT TO FIT PROBE LOCATIONS FOR THE CIRCUIT, PERMITTING TESTING AT MULTIPROBE.
- BASIC TEST TRANSISTORS REMAIN ON ALL BARS TO PERMIT PARAMETER SLICE MAPPING AT MULTIPROBE.

Fig. 1. Plug bar concept.

second set measures device parameters, the third set evaluates circuit parameters, and finally the fourth set characterizes functional parameters. Figure 1 gives examples of parameters which each test structure set should deliver.

Standardized test site and critical function blocks are laid out so that they fit probe locations for the circuit, permitting testing at multiprobe. Basic test devices remain on all bars to permit parameter slice mapping at multiprobe. The mathematical detail of 33 test structures was calculated by Buehler[1], primarily for the evaluation of resistivity and dopant density. The test pattern includes MOS capacitors, pn junctions, bipolar and MOS transistors, diffused and metal resistors, alignment marker, and etch control and resolution structures. The pattern requires four mask levels for its fabrication.

Another test structure with well-established theory is the tetrode transistor[2]. The tetrode transistor has been used to separate the leakage current components originating at the surface and in the bulk[2,3]. A particularly powerful process control tool is the combination of a simple field plate device and the tetrode transistor[3] (Fig. 2). With the help of the tetrode transistor, bipolar device stability can be predicted through process monitoring. The following description is based on Aiken's theory[2].

FIELD-PLATE DEVICE		TETRODE TRANSISTOR	
APPLICATIONS	LIMITATIONS	APPLICATIONS	LIMITATIONS
1. INDICATES IF SURFACE IS DEPLETED OR INVERTED	1. BASE SURFACE CONCENTRATION MUST BE $< 8 \times 10^{18} cm^{-3}$	1. QUANTITATIVE MEASURE OF PCT. SURFACE AND BULK	1. DIFFICULT TO ANALYZE AT HIGH CURRENT DENSITIES
2. DETECTS MOBILE IONIC CHARGE IN OXIDE	2. QUALITATIVE MEASURE OF SURFACE AND BULK COMPONENTS ONLY	2. USEFUL FOR ANY BASE SURFACE CONCENTRATION	2. DISCLOSES NO INFORMATION ON SURFACE DEPLETION OR INVERSION
3. INDICATES WHICH COMPONENT IS CHANGING DURING LIFE TESTS, MAGIC BAKE, ETC.		3. INDICATES WHICH COMPONENT IS CHANGING DURING LIFE TESTS, MAGIC BAKES, ETC.	

Fig. 2. Separation of surface and bulk components.

2. THE BIPOLAR TETRODE TRANSISTOR AS A PROCESS CONTROL TOOL

2.1 Principal and application of tetrode transistor

The degree of process control of bipolar device parameters is determined by the ability to minimize contamination levels and variations in the dopant impurity profiles. The fluctuations in the net emitter-base doping profile, the surface recombination velocity and the material lifetime affect nearly every dc and ac terminal characteristic of bipolar transistors. One of the most important parameters, the current gain, is influenced by all of the above factors (see Fig. 3) and is one of the most difficult to analyze in order to establish process controls. The bipolar

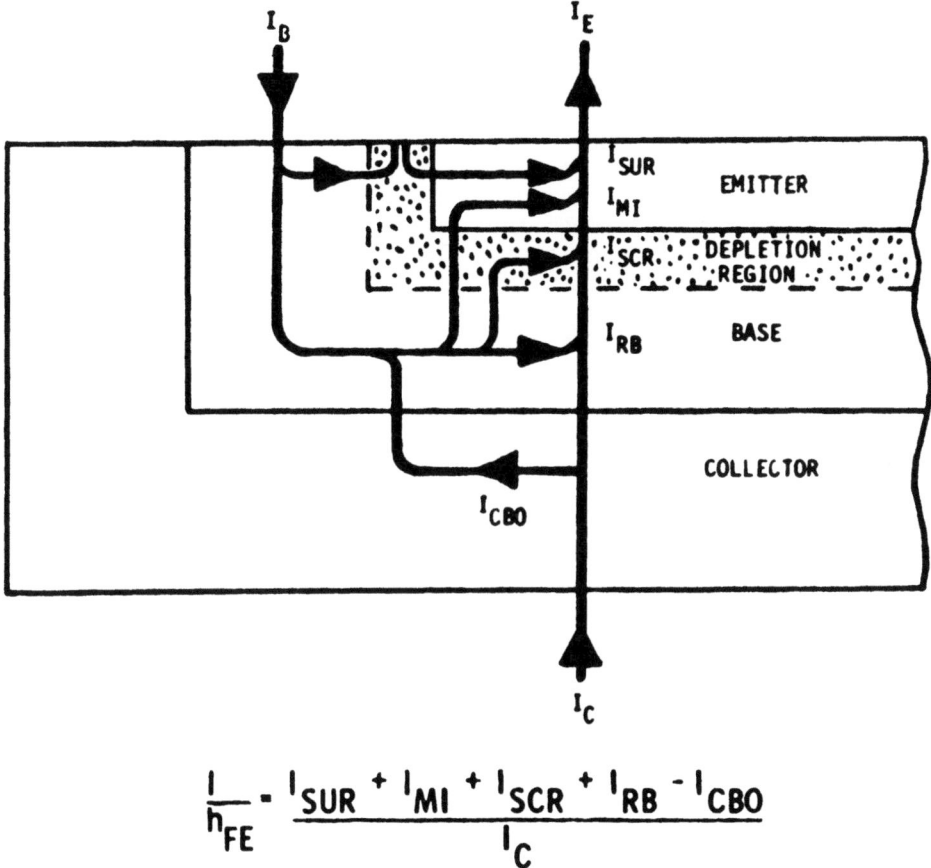

$$\frac{1}{h_{FE}} = \frac{I_{SUR} + I_{MI} + I_{SCR} + I_{RB} - I_{CBO}}{I_C}$$

Fig. 3. Components of h_{FE} and leakage current.

tetrode device is the only means of separating surface and bulk effects plus detecting profile variations in state-of-the-art shallow devices. Field plate devices[4] (Fig. 2) have been utilized successfully to separate surface and bulk generation and recombination currents but cannot be used on shallow, heavily doped devices due to the inability to vary the surface potential over a sufficient range. The bipolar tetrode does not suffer from this limitation plus, as mentioned, it can also detect variations in the base doping under the emitter.

An approximate theory[5,6] for the separation of surface and bulk components of current gain using a tetrode device has been available for the case where the current gain is either entirely bulk or entirely surface controlled. Aiken's[2] work extended the theory in two respects. First, transistors of arbitrary dimensions can be analyzed allowing the utilization of practical and actual emitter geometries in the test device. Second, combinations of surface and bulk components, as normally observed in most devices, are included in the analysis.

The bipolar tetrode device has two independent base contacts separated by a standard emitter (see Fig. 4). The surface and bulk contributions to the current gain are determined by applying a transverse potential between the two base contacts and intentionally debiasing the transistor. The relative importance of surface and bulk recombination currents is established by noting the variation in current gain with transverse bias and comparing this with the analytical model. Physically, the variation in current gain with transverse bias is due to the forcing of the

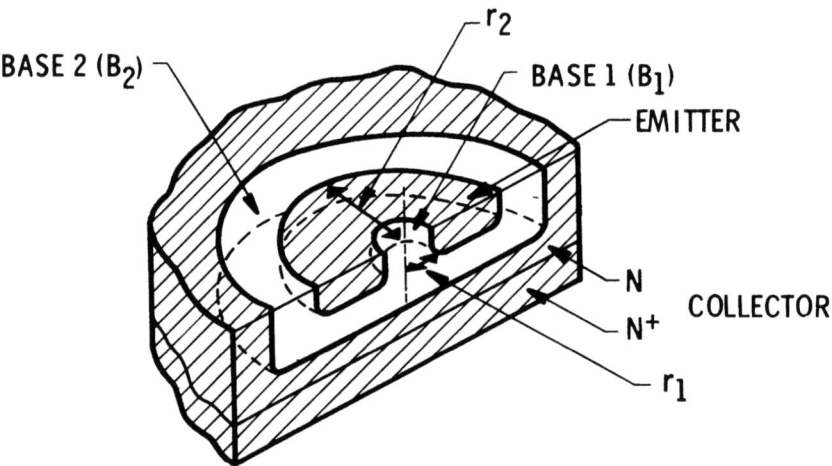

Fig. 4. Layout of experimental tetrode transistor.

injected emitter current from the bulk to the surface which causes a corresponding change in the base current resulting from the increased importance of the surface recombination velocity. The base doping under the emitter is determined simply by measuring the resistance between the two base contacts.

Experimentally, the tetrode can be utilized to determine whether deep levels in the band-gap, which control the transistor gain at low currents, are in the bulk or at the surface. Furthermore, both the origin of the controlling component of gain at intermediate currents plus the cause of magnitude variations in gain can be established. Thus, the importance of profile variations as well as the spatial locations of key recombination centers (a necessity in establishing the source of contamination and/or defects) can be identified. These data can in turn be used as process control indicators.

2.2 Bipolar tetrode model*

The current gain, h_{FE}, is defined as the ratio of the collector current I_C to the base current I_B, $h_{FE}=I_C/I_B$ (see Fig. 3). Both currents, as a function of the base-emitter voltage V_{BE}, follow analogous relationships:

$$I_B = I_{BO} \exp(qV_{BE}/nkT) \qquad (1)$$

$$I_C = I_{CO} \exp(qV_{BE}/mkT) \qquad (2)$$

where I_{BO} and I_{CO} are constants, n=1 for large I_B and n>1 for small I_B, m=1 for all I_C. Combining Equations (1) and (2) renders the dependence of h_{FE} on V_{BE}:

$$1/h_{FE} = I_B/I_C = I_{BO}/I_{CO}) \exp\left[(qV_{BE}/kT)[(1-n)/(n)]\right] \qquad (3)$$

Plotting $\ln(1/h_{FE})$ as a function of $\ln I_C$ for a typical bipolar device results in Fig. 5. The current-dependent contribution to h_{FE} is indicated by the straight line at lower I_C; the slope of this line is $(1-n)/n$, as can be verified by taking the derivative of $\ln(1/h_{FE})$ with respect to $\ln I_C$. The current independent component of $1/h_{FE}$ indicated in Figure 1 is due to a second component of base current having n=1.0 [therefore the slope, $(1-n)/n$, is zero].

The value of h_{FE} is determined by the various recombination components comprising I_B (see Fig. 3). Portions of the emitter current recombine in the emitter-base depletion region, in the neutral base, along the surface periphery, and in the emitter itself. The low current component of I_B results from recombination

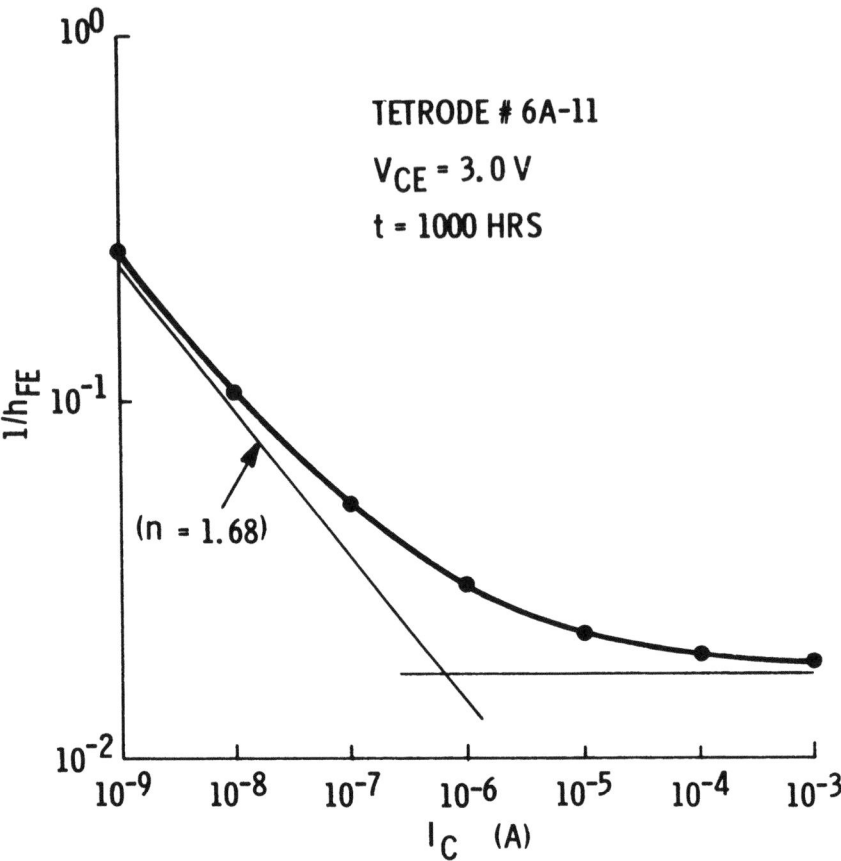

Fig. 5. Plot of experimental data relating $1/h_{FE}$ to I_C resulting in determination of two components having values of n equal to 1.68 and 1.0.

in the emitter-base transition region and is caused either by surface recombination or by bulk recombination. When deep levels are present near the surface or throughout the bulk, both the surface and bulk recombination currents can be approximated by Equation (1). Since deep levels result in surface and bulk currents of the same form, the origin of the space charge recombination current cannot be determined by measurements made on a single-base transistor.

The geometrical structure of Figure 4 allows a transverse voltage, V_T, to be applied between the two base contacts B_1 and B_2. If the collector current is held constant, such a voltage increases the injected current density on one side of the emitter while lowering the density on the other side. The result of spatially varying the injected current density is a variation in the relative importance of the surface and bulk recombination currents since forcing the injecting current density to the surface increases the importance of the surface recombination. Therefore the variation of I_B or h_{FE} at fixed I_C with V_T is an indication of the relative importance of surface and bulk recombination currents at that value of I_C. Experimentally, I_C is held fixed by adjusting V_{BE}.

The mathematical relationship between I_B and V_T for transistors of arbitrary dimensions will now be derived.[B] The annual geometry of Figure 4 is used in this derivation although the derivation for linear geometries follows the same procedure with a change in coordinate systems. The following assumptions are made in the analysis:

1. The base transport factor is approximately unity. This is invalid only when the high current h_{FE} approaches unity.

2. The transverse voltage V_T applied between the base contacts appears wholly across the width of the emitter.

3. The transverse voltage produced by the base current is small compared with V_T. This condition is satisfied by using appropriately low collector currents.

4. The base is homogeneous in the transverse direction.

5. The collector current is to be held constant.

Under these conditions:

$$I_C = \int_{r_1}^{r_2} 2\pi j_D e^{\beta V(r)} r \, dr \qquad (4)$$

where $\beta = q/kT$, j_D is a proportionality constant in current density, and r is the radius. As explained by Coppen and Matzen[5], there are two cases depending on the polarity of V_T.

Case 1: $V_{B2} > V_{B1}$
    If V_1 is the potential between the emitter and base contact B_1, the potential variation as a function of radius for $V_{B2} > V_{B1}$ is:

$$V(r) = V_1 + \left(\frac{b}{\beta}\right) \ln\left(\frac{r}{r_2}\right) \tag{5}$$

where

$$b = \beta V_T / \ln(r_2/r_1) \tag{6}$$

Also at $V_T = 0$:

$$I_C = \pi j_D \, e^{\beta V_{10}} \left(r_2^{\,2} - r_1^{\,1}\right) \tag{7}$$

where V_{10} is the value of V_1 at $V_T = 0$. Combining Equations (4), (5), and (7) results in

$$e^{\beta(V_1 - V_{10})} = \frac{1}{2} \left[\frac{r_2^{\,2} - r_1^{\,2}}{r_2^{\,(b+2)} - r_1^{\,(b+2)}}\right] (b+2) r_2^{\,b} \tag{8}$$

If the h_{FE} is surface controlled, the base currents at $V_T = 0$ and $V_T \neq 0$ are respectively:

$$I_{BO} = 2\pi j_x (r_1 + r_2) e^{\beta V_{10}/n} \tag{9}$$

$$I_B = 2\pi j_x \left(r_1 \, e^{\beta(V_1 - V_T)/n} + r_2 \, e^{\beta V_1/n}\right) \tag{10}$$

where j_x is the surface current density proportionality constant. Combining Equations (8), (9), and (10) results in the variation of the normalized base current, I_B/I_{BO}, with V_T for a fixed collector current and a surface controlled h_{FE}:

$$\frac{I_B}{I_{BO}} = \left(\frac{r_2}{r_1 + r_2}\right) \left[\left(1 + \frac{r_1}{r_2}\right)^{\left(\frac{b}{n}+1\right)}\right] X \tag{11}$$

$$X \left[\left(\frac{r_2^2 - r_1^2}{r_2^{(b+2)} - r_1^{(b+2)}} \right) \left(\frac{b+2}{2} \right) r_2^b \right]^{\frac{1}{n}}$$

If the h_{FE} is bulk controlled, Equations (9) and (10) are replaced by:

$$I_{BO} = \pi j_x \, e^{\beta V_{10}/n} \left(r_2^2 - r_1^2 \right) \qquad (12)$$

$$I_B = \int_{r_1}^{r_2} 2\pi j_x r \, e^{\beta V(r)/n} \, dr \qquad (13)$$

Integrating Equation (13) and combining with Equations (12) and (8) leads to the variation in I_B/I_{BO} with V_T for a bulk controlled h_{FE}:

$$\frac{I_B}{I_{BO}} = 2 \left[\frac{r_2^{\left(\frac{b}{n}+2\right)} - r_1^{\left(\frac{b}{n}+2\right)}}{\left(\frac{b}{n}+2\right) r_2^{b/n} \left(r_2^2 - r_1^2 \right)} \right] X$$

$$X \left[\left(\frac{r_2^2 - r_1^2}{r_2^{(b+2)} - r_1^{(b+2)}} \right) \left(\frac{b+2}{2} \right) r_2^b \right]^{1/n} \qquad (14)$$

Equations (11) and (14) represent the extreme cases of 100% surface and bulk controlled h_{FE} respectively when $V_{B2} > V_{B1}$. The equations are valid for any value of n with the remainder of the quantities as defined in Figure 4 and Equation (6). Since base recombination components are independent of one another, Equations (11) and (14) may be superimposed to cover the general case of partial bulk and partial surface recombination currents.

Case 2: $V_{B1} > V_{B2}$

The derivation for this condition follows from the previous analysis:

$$V(r) - V_1 = \left(\frac{b}{\beta}\right) \ln\left(\frac{r}{r_2}\right) \tag{15}$$

Combining Equations (4), (7), and (15) leads to

$$e^{\beta(V_1 - V_{10})} = \left(\frac{2-b}{2r_1^b}\right) \left[\frac{\left(r_2^2 - r_1^2\right)}{r_2^{(2-b)} - r_1^{(2-b)}}\right] \tag{16}$$

If the h_{FE} is completely surface controlled, the base current is:

$$I_B = 2\pi j_x \left(r_1 e^{\beta V_1/n} + r_2 e^{\beta(V_1 - V_T)/n}\right) \tag{17}$$

Combining Equations (9), (16), and (17) results in the variation of the normalized base current with V_T at a fixed I_C for the surface controlled situation:

$$\frac{I_B}{I_{BO}} = \left(\frac{r_1}{r_1 + r_2}\right) \left[1 + \left(\frac{r_2}{r_1}\right)^{\left(1 - \frac{b}{n}\right)}\right] \times$$

$$\left[\frac{(2-b)\left(r_2^2 - r_1^2\right)}{2r_1^b \left(r_2^{(2-b)} - r_1^{(2-b)}\right)}\right]^{1/n} \tag{18}$$

If the h_{FE} is bulk controlled, combining Equations (12), (13), and (16) leads to the solution for this case:

$$\frac{I_B}{I_{BO}} = 2\left[\frac{r_2^{\left(2-\frac{b}{n}\right)} - r_1^{\left(2-\frac{b}{n}\right)}}{r_1\left(\frac{b}{n}\right)\left(r_2^2 - r_1^2\right)\left(2-\frac{b}{n}\right)}\right] \times$$

$$\times \left[\frac{(2-b)\left(r_2^2 - r_1^2\right)}{2r_1^b\left(r_2^{(2-b)} - r_1^{(2-b)}\right)}\right]^{1/n} \tag{19}$$

As in Case 1, Equations (18) and (19) may be superimposed to cover the general case of partial bulk and surface components. The difference between Cases 1 and 2 is due to the asymmetry of the annular test structure. In the limiting case of $r_2 > r_1$ and $b > 2n$, the solution Equations (11), (14), (18), and (19) reduce to those of Coppen and Matzen[5].

2.3 Experimental results.*

An example of utilizing the tetrode to determine the controlling factors of the h_{FE} as a function of I_C is shown in Figures 5, 6, and 7. The plot of $1/h_{FE}$ versus I_C in Figure 5 indicates two components of base current having values of n equal to 1.68 and 1.00. The predicted tetrode response for the negative V_T variation is illustrated in Figure 6 for the case of n=1.68. Matching the experimental results at I_C=10nA discloses that the base current is 85% surface controlled as shown in Figure 7. The remaining 15% is due to the component of I_B having n=1.00. This component becomes dominant at high currents.

Variations in the magnitude of h_{FE} from device can be due to either variations in the emitter-base profile (the net doping in the base under the emitter) or variations in the surface charge, surface recombination velocity, or bulk lifetime. If the control of the current gain magnitude is determined only by the variation in the net base doping under the emitter, then the h_{FE} varies linearly with the base resistance under the emitter[7].

2.4 Relationship of h_{FE} to transverse base resistance.**

For bipolar devices not limited by the base transport factor (base width < 10μm), the current gain can be defined as:

$$h_{FE} = [I_{co} \exp(q\, V_{BE}/kT)] / I_{B1} \exp(q\, V_{BE}/nkT) \quad (20)$$
$$+ I_{B2} \exp(q\, V_{BE}/kT)]$$

where I_{B1} is a constant for the depletion region and I_{B2} for the emitter efficiency. For a given well controlled process, it has been widespread experience[8] that I_{B1}, I_{B2} and n listed in the denominator of Equation (20) do not vary appreciably. Thus h_{FE} at a fixed temperature is a function of I_{CO} and V_{BE} only. It follows from Equations (2) and (20) that h_{FE} at a fixed collector current will vary only if I_{CO} varies and:

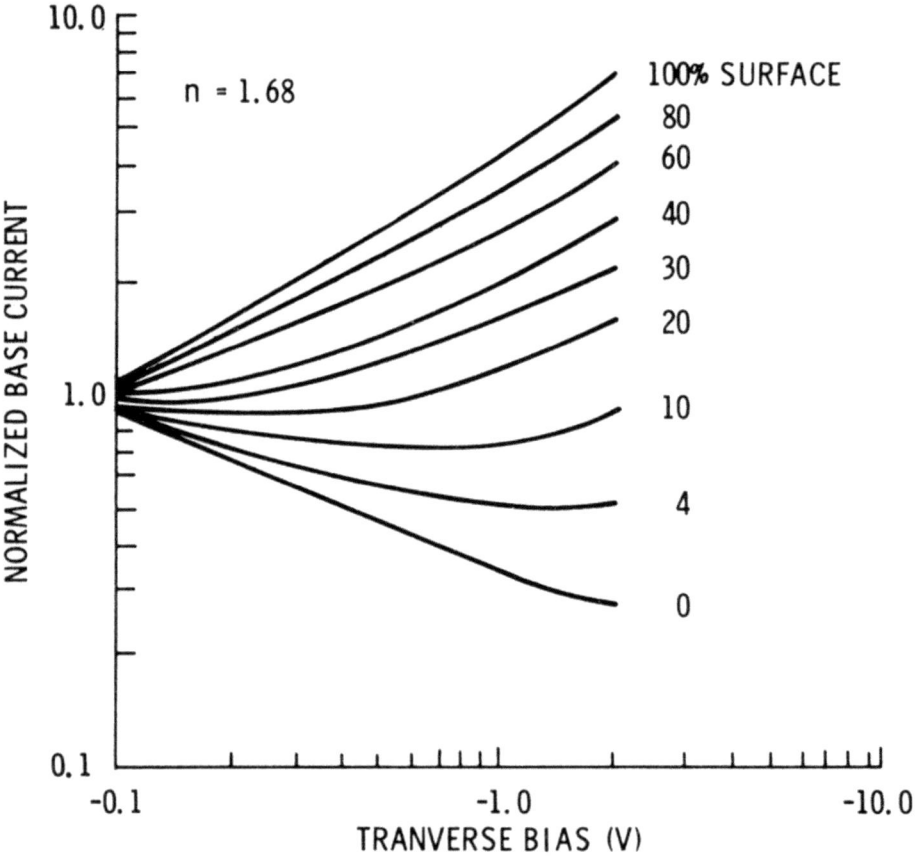

Fig. 6. Variation of normalized base current with transverse voltage for n=1.68.

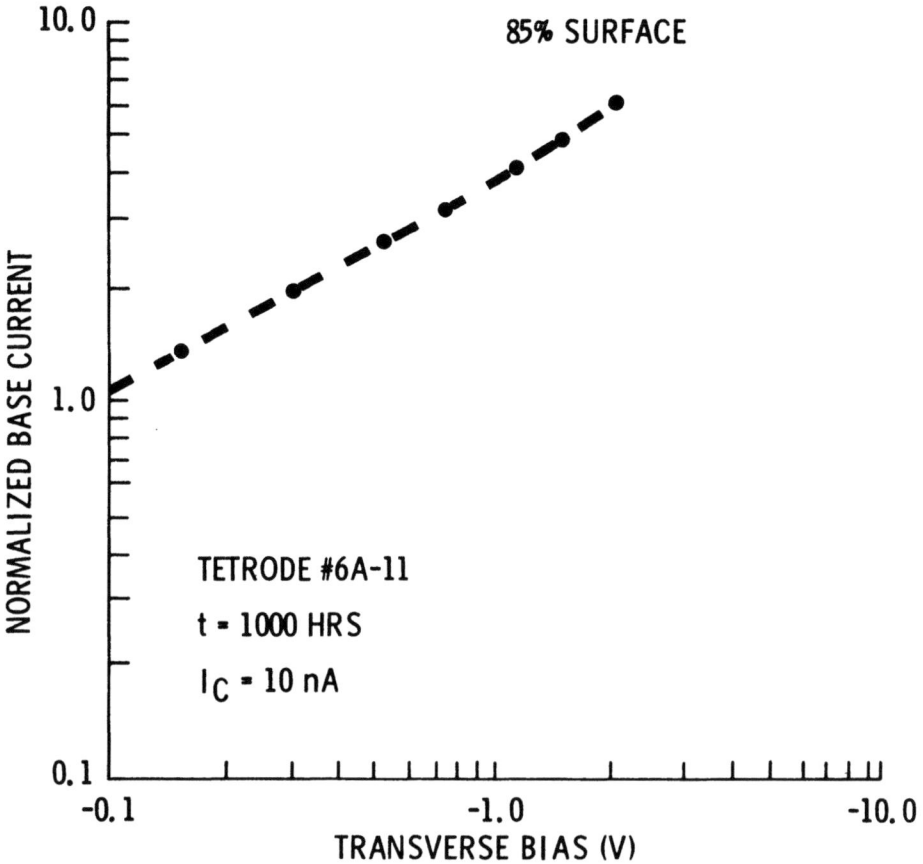

Fig. 7. Experimental tetrode data and model prediction for I_C=10nA indicating 85% surface current for n=1.68.

$$h_{FE} \sim I_{CO} \tag{21}$$

for any and all of the three recombination components considered above. The constant I_{CO} varies inversely with the total density of impurities per unit area in the base under the emitter[9]. Therefore, the current gain is proportional to:

$$h_{FE} \sim 1/\int_{X_{BE}}^{X_{BC}} N(x)\,dx \tag{22}$$

where X_{BE} is the emitter-base junction, X_{BC} the collector-base junction. Equation (22) also assumes the mobility to be a weak function of $N(x)$, which it is indeed over the range of concern.

The transverse base resistance R_{B1}, as measured between B_1 and B_2 in Figure 4, is equal to:

$$R_B = (L_E / W_B P_E \mu q) / \int_{X_{BE}}^{X_{BC}} N(x) dx \qquad (23)$$

where L_E is the emitter width, W_B the base width, P_E the effective emitter length, and μ the base mobility. Combining Equations (22) and (23) leads to:

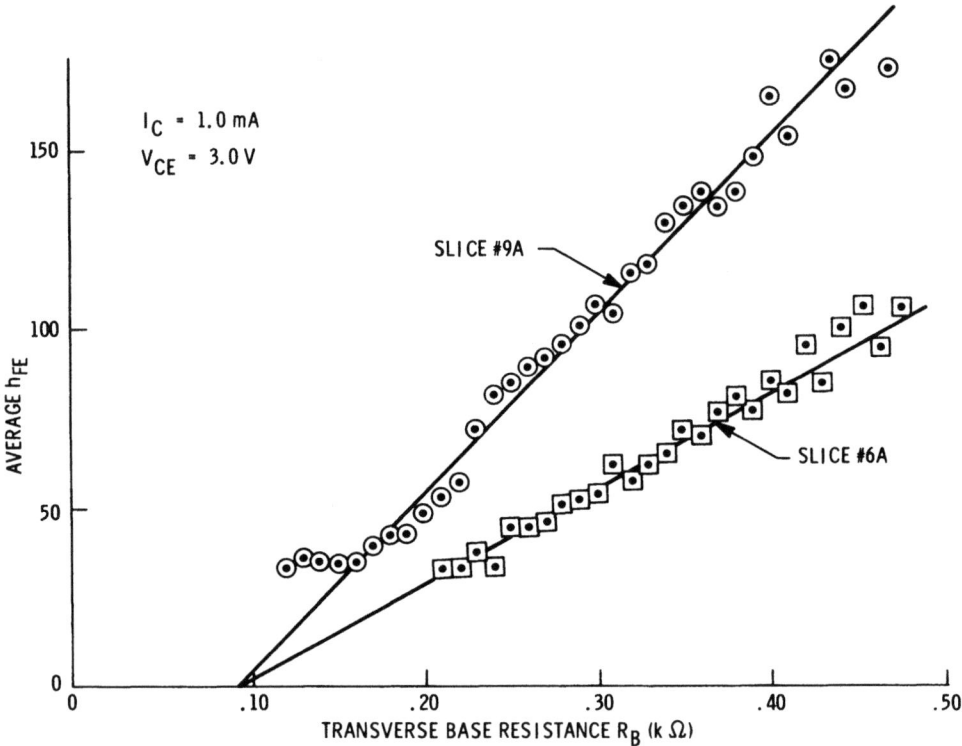

Fig. 8. Current gain h_{FE} as a function of transverse base resistance R_B for stable and degraded devices.

$$h_{FE} = \Gamma R_B \qquad (24)$$

where Γ is a constant at fixed I_C for any well controlled process having a base width less than 10µm.

2.5 Bipolar stability indication through process monitoring.**

It has now been shown that the tetrode transistor can be used to separate surface and bulk components of h_{FE} as well as indicate if the expected relationship between base resistance under the emitter and h_{FE} holds. Both types of measurement can be used as stability indicators.

In fact, the data of Figures 5 and 7 were taken on a device (#6A-11) which failed on high temperature reverse-bias lifetest due to an increase in the surface recombination current. Lifetest data for several samples from a slice (#6A) which degraded plus several samples from a stable slice (#9A) are listed in Table 1. The corresponding plots of h_{FE} versus R_B for the same two slices are shown in Figure 8. The relationship for slice #9A

SAMPLE #	t = 0 (HR) h_{FE}	t = 1000 (HRS)	
		h_{FE}	Δh_{FE} (%)
6A-3	77	68	-12
6A-4	62	55	-11
6A-8	72	66	- 8
*6A-11	56	50	-11
6A-12	74	66	-11
9A-1	87	89	+ 2
9A-3	80	80	0
9A-8	107	106	- 1
9A-9	76	77	+ 1
9A-11	66	67	+ 1

Table 1. Lifetest data high temperature, reverse bias h_{FE} measured at I_C=1mA, V_{CE}=3.0V.

is that typical of the well controlled process; in contrast, the lower h_{FE} at a given R_B for slice #6A indicates an increase in one of the recombination components of base current in Equation (20). Each data point in Figure 8 represents the average h_{FE} for all units having a certain R_B; the total number of units included in one curve of Figure 8 is approximately 500. The intercept of the two lines at $h_{FE}=0$ indicates the sum of the residual external base resistance and the contact resistance; this sum is a function of the process used and a constant for uniform base sheet resistance.

The correlation between Figure 8 and Table 1 verifies that any deviation noticed in the h_{FE}-R_B relationship indicates a potential stability problem for the reasons just discussed. Such potential stability rejects due to h_{FE} instability can be detected with a single measurement of the base resistance under the emitter.

It is important to realize that the constant, Γ, in Equation (24) is a function of the total processing sequence and therefore varies with process changes such as heat treatments, diffusion cycles and surface passivation changes. For instance, it has been found that the choice of surface coverage (sputtered quartz vs. deposited TEOS oxide) can affect Γ.

3. METAL/SILICON CONTACT RESISTANCE

3.1 Metal/semiconductor contact models

A review of the theory and experimental verification of ohmic contacts to semiconductors was presented at the Electrochemical Society Topical Conference[10] in Montreal, Canada in 1968. Theory and experiments of Schottky barriers have been reviewed in detail by Padovani[11] and Duke[12]. The following discussion is restricted to ohmic metal/semiconductor contacts, since they are needed, in one way or another, in all semiconductor devices and circuits. While many aspects of this theory[10] are well understood, the models do not offer a unified description which would allow the prediction and control of important parameters such as the specific contact resistance for various metal/semiconductor combinations.

An ideal ohmic metal/semiconductor contact would behave as though the contact was made from a homogeneous semiconductor material so that the current carriers see flat-band conditions while traversing the "contact" region[13]. In most practical situations, however, the presence of work-function differences between the metal and semiconductor, and surface states results in a surface space-charge region in the semiconductor. This space-charge region and its associated energy barrier can render the contact nonohmic. Under reverse bias conditions, the impedance of the space-charge region will completely dominate the contact impedance. Usually, successful ohmic metal/semiconductor contacts are fabricated on heavily doped semiconductors where the narrow space-charge regions are favorable for tunneling to occur.

Among the newer attempts of modeling ohmic metal/semiconductor contacts, Gossick[14,15] has proposed the concept of nonreflecting potential boundaries for electrons, and has pointed out that the commonly used accumulation contact is an example of a nonreflecting boundary potential. The role of very thin oxide layers in metal-silicon contacts was investigated by Rhoderick and co-workers[16,17,18].

A technique to fabricate nonreflecting boundaries approximating an ideal metal/semiconductor contact has been developed by Schroen[19]. It is based on a thin tunneling barrier provided by a thin dielectric layer between the metal and the semiconductor, which can hold the charge necessary for accommodating the metal/semiconductor work-function difference and for offsetting charged surface states. This eliminates the space-charge in the semiconductor, as shown in Figure 9, and the majority carriers see flat-bands out to the semiconductor-insulator interface where they immediately tunnel through the insulator to the metal, or vice versa. Measurements[20,21] using ohmic aluminum contacts to silicon doped as low as $10^{15} cm^{-3}$ demonstrated specific contact resistances from 10 to 100 times lower than those offered by Pt-Si techniques. This low contact resistance (Al-nSi) is compared in Figure 10 with literature data. The contacts display I-V characteristics which are extremely linear, particularly through the origin; the resistance is that of the sheet resistance of the sample with the contact resistance so small that it is difficult

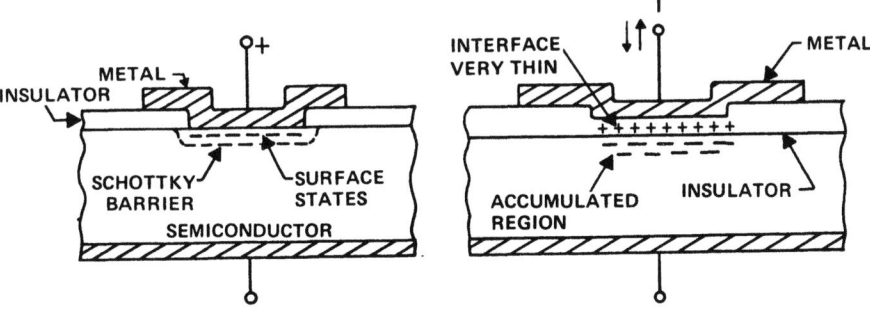

Fig. 9. Ohmic metal-semiconductor contacts.

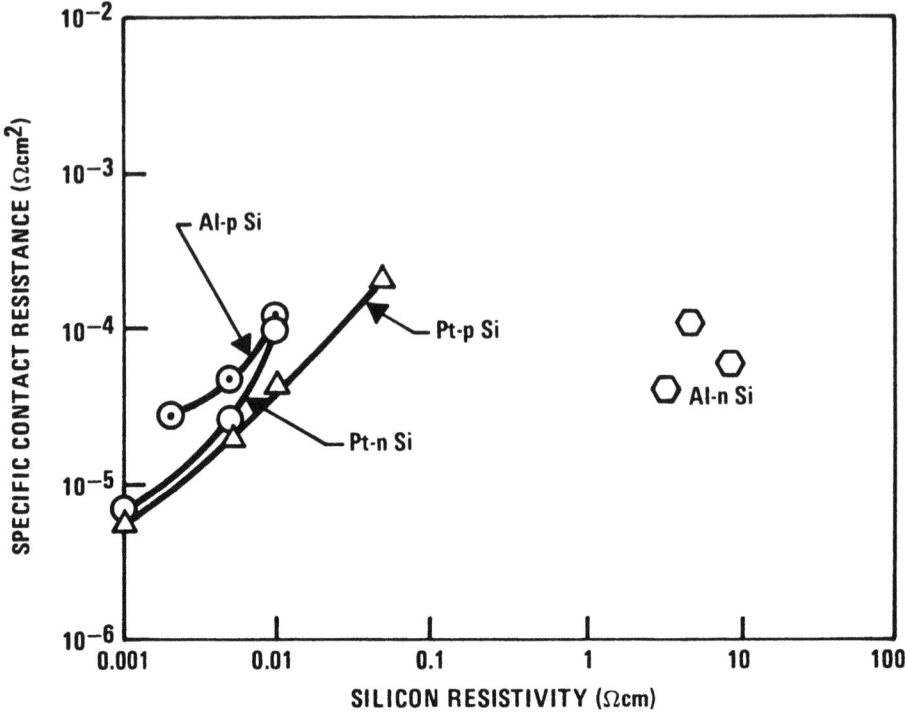

Fig. 10. Metal-silicon contacts.

to measure accuractly. The insulating layer is formed on the surface of a chemically cleaned slice of n-or p-type Si in an oxygen-rich ambient. The thin layer of oxide is grown less than 100 Å thick. Afterwards, it is annealed in hydrogen. Evaporated Al has been used as the contact metal.

3.2 Shockley's test structure for contact resistance

The most notable of several techniques developed for precise measurement of contact and sheet resistances has been proposed by Shockley[22]. Recently, this technique has been used by Yu[23]. Figure 11 shows a schematic diagram and a sketch of the test structure. Basically there is one contact in series with a tapped resistor. The distance from each tap to the contact is known. Thus, the voltage along the diffused resistor can be plotted as a function of distance. The contact resistance can be determined by calculating the intercept, using regression techniques. If L_t is the "transfer length" (in centimeters), R_s (in ohms cm^2) can be

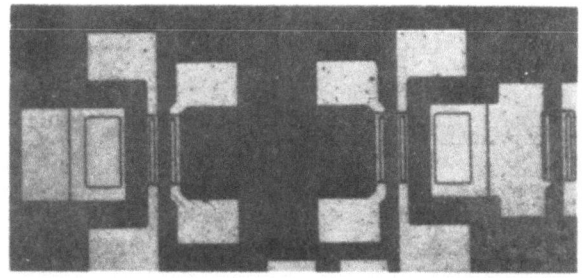

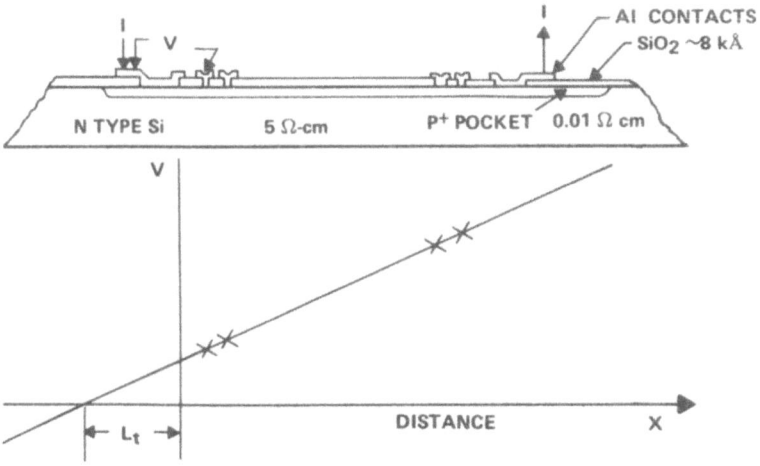

Fig. 11. Shockley's scheme of contact resistance.

obtained by

$$R_c = R_s L_t^2 \tag{25}$$

Since L_t appears squared in Sq. (25), it has to be determined with great precision. The origin X=0 and the locations of the contacts along the X-axis can be found using a microscope of high magnification. The sheet resistance of the P^+ pocket in Figure 11 can be determined from the slope of the line and the geometry of the contacts.

4. MULTILAYER ANALYSIS OF SPREADING RESISTANCE MEASUREMENTS

4.1 Spreading resistance technique

Spreading resistance measurements provide a highly flexible technique for the determination of dopant profiles in semiconductors. However, because each measurement samples a greater depth into the sample than the depth difference between successive measurements, the direct conversion of resistivity readings to dopant concentration values will not yield a correct profile.

The model used to deduce a "true" dopant profile is that of circular contacts to a laterally infinite medium which is partitioned vertically into layers of homogeneous resistivity, each layer corresponding to one spreading resistance measurement point. The analysis is performed by a computer program. Some detail in the development of the program is discussed. The results of this analysis technique compare favorably to profile results of other profiling techniques such as capacitance voltage and incremental sheet resistance on profile types to which they can be applied.

Continuous profiles can be obtained over several orders of magnitude of concentration change and across p-n junctions. The profiles are obtained by stepping the probe points down a bevel lapped at a small angle to the original surface (Fig. 12). Each measurement point is assumed to approximate a measurement on a flat surface formed by removing all material from the sample above the depth of that point on the bevel. By using small angles (17' to 5°) it is possible to obtain profiles in which consecutive points differ in depth by less than 200 Å.

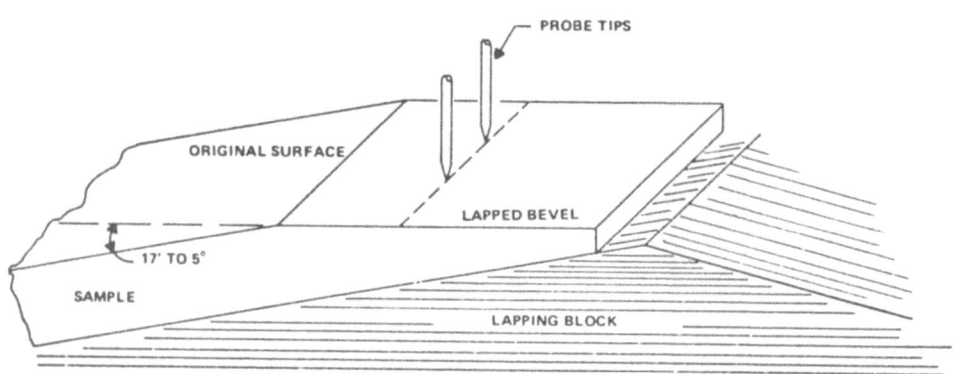

Fig. 12. Spreading resistance probe geometry (scale distorted).

Unfortunately, the precision of the geometry is negated by
the comparatively large volume in the material being profiled
which is sampled by the electrical measurement. This volume may
extend one to several micrometers into the material (which in
shallow devices may involve orders of magnitude of resistivity
change). As a result, the characteristics at a given depth in
the material being profiled affect not one, but several of the
measurement points on the bevel. Thus, the direct result of
spreading resistance measurements is somewhat like a profile of
local averages, rather than of discrete point values.

4.2 Theoretical model

The theoretical model employed has been developed by Lee[24]
and is similar to that utilized by several other authors[25,26,27].
The continuous resistivity profile is approximated by a stack of
layers, each of homogeneous resistivity, with thicknesses equal
to the spacing of the spreading resistance data (Fig. 13). The
potential distribution in this structure can be solved numerically
and employed to derive the actual profile from the originally
measured values. The following discussion is excerpted from Lee[24].

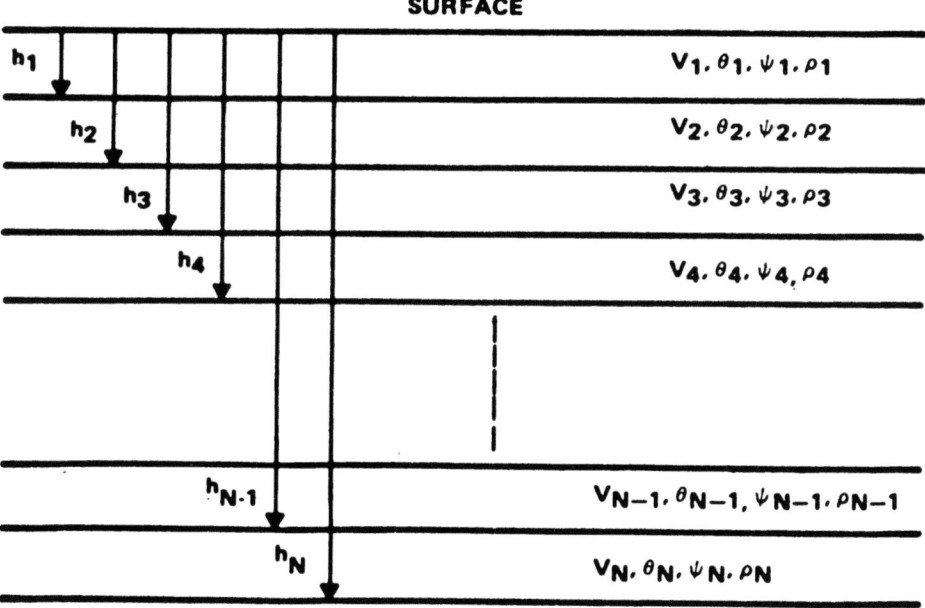

Fig. 13. Multilayer structure.

4.2.1 One probe potential distribution.

The general solution of Laplace's equation in cylindrical coordinates (r, Θ, z) for a circularly symmetric potential such as that resulting from a single probe is[24]:

$$V(r,z) = \int_0^\infty [A(\lambda)J_0(\lambda r)e^{\lambda z} + B(\lambda)J_0(\lambda r)e^{-\lambda z}]\, d\lambda \qquad (26)$$

where V is voltage and A and B are arbitrary functions of the integration variable λ. This can be rewritten, with V_i defined for each layer:

$$V_i(r,z) = K \int_0^\infty e^{-\lambda z} \frac{\sin(\lambda r_0)}{\lambda} J_0(\lambda r)\, d\lambda$$

$$+ K \int_0^\infty \Theta(\lambda) e^{-\lambda z} \frac{\sin(\lambda r_0)}{\lambda} J_0(\lambda r)\, d\lambda \qquad (27)$$

$$+ K \int_0^\infty \psi(\lambda) e^{\lambda z} \frac{\sin(\lambda r_0)}{\lambda} J_0(\lambda r)\, d\lambda$$

where

$$A(\lambda) \equiv K[1+\Theta(\lambda)]\frac{\sin\lambda r_0}{\lambda} \qquad B(\lambda) \equiv K\psi(\lambda)\frac{\sin\lambda r_0}{\lambda}$$

and r_0 is the effective radius of contact of the probe, which is assumed to be circular.

At this point boundary conditions are imposed which assume that current is flowing, which may be inconsistent with Laplace's equation if charge accumulation exists. However, the expected error introduced is not large[28], and the results of the method do agree with other profiling techniques.

The boundary conditions are these: (refer to Fig. 13):

I. No current flows through the surface of the structure beyond the radius of the probe contact:

$$\left.\frac{1}{\rho 1}\frac{\partial V_1}{\partial z}\right|_{\substack{z=0 \\ r=r_o}} = 0$$

II. At each layer interface

  a. V is continuous:

  $$V_i(r,h_i) = V_{i+1}(r,h_i)$$

  b. The current component perpendicular to the interface is continuous:

  $$\left.\frac{1}{\rho_i}\frac{\partial V_i(r,z)}{\partial z}\right|_{z=h_i} = \left.\frac{1}{\rho_{i+1}}\frac{\partial V_{i+1}(r,z)}{\partial z}\right|_{z=h_i}$$

III. Two possible appropriate boundary conditions exist for the bottom layer. One assumes that the last layer is semi-infinite and of homogeneous resistivity. The other assumes that a junction exists beneath the last layer.

In the first case, the boundary condition is that

$$\lim_{z\to\infty} V_N(r,z) = 0$$

In the second case, no current crosses the junction:

$$\left.\frac{1}{\rho_N}\frac{\partial V_N(r,z)}{\partial z}\right|_{z=h_N} = 0$$

When the expression for $V_i(r,z)$ is substituted into these equations, they reduce to the following set of equations. Note that Θ_i and ψ_i are functions of λ only, as previously defined.

$$\Theta_1 - \psi_1 = 0 \tag{28}$$

$$(\theta_i - \theta_{i+1})e^{-2\lambda h_i} + (\psi_i - \psi_{i+1}) = 0 \tag{29}$$

$$(-\theta_i + \beta_i \theta_{i+1})e^{-2\lambda h_i} + (\psi_i - \beta_i \psi_{i+1}) = e^{-2\lambda h_i}(1 - \beta_i) \tag{30}$$

For the first case:

$$\psi_N = 0 \tag{31}$$

For the second case:

$$-\theta_N e^{-2\lambda h_N} + \psi_N = e^{-2\lambda h_N} \tag{32}$$

where

$$\beta_i \equiv \frac{\rho_i}{\rho_{i+1}}$$

This gives 2N equations, which are sufficient for the solution of the N θ's and N ψ's. An additional boundary condition involving the current distribution under the probe contact has been used by some authors to resolve K in Equation (29). However, because the final form of the current distribution will be affected by the second probe of apparatus, in this treatment its resolution is deferred until after the potential distribution of the two probes in combination is determined.

4.2.2 <u>Two probe potential difference measurement</u>. The voltage measured by the spreading resistance apparatus is the potential drop between the two probe points. This derived from a superposition of the one probe solution for each of the two probes. For this solution point contacts are assumed. This is not too unreasonable since the ratio of the probe spacing to the physical contact diameter is 600 to 10, and that of the probe spacing to the effective electrical contact diameter 600 to 3. The measured potential difference between the probes Vm, is

$$V_m = V_a - V_b$$

where V_a and V_b are the potentials at probe a and probe b.

Each potential consists of two components: the potential due to probe a and that due to probe b. Note that the one probe solution coordinates of b relative to a and to b are both (r=D, z=0), where D is the probe spacing. Their coordinates relative to themselves are of course (0,0). That is:

$$Va = V_1(0,0) - V_1(D,0)$$

$$Vb = V_1(D,0) - V_1(0,0)$$

So

$$Vm = Va - Vb = 2[V_1(0,0) - V_1(D,0)] \qquad (33)$$

Noting from Equation (28) that $\theta_1 = \psi_1$, expressing V_1 in the form derived in Equation (27) yields:

$$V_m = K \int_0^\infty (1+2\theta_1)[1-J_o(\lambda D)] \frac{\sin(\lambda r_o) d\lambda}{\lambda}. \qquad (34)$$

4.2.3 <u>Resolution of the proportionately constant K.</u> The boundary condition imposed in order to resolve K is that the measured voltage on a semi-infinite substrate of homogeneous resistivity should equal the voltage measured on that same substrate covered by a finite layer of equal resistivity. That is, a two layer problem on semi-infinite medium:

```
           Surface
 ↓  ┌─────────────────────────┐
 h₁ │   V₁, θ₁, ψ₁, ρ₁        │
 ↓  └─────────────────────────┘   ρ₁=ρ₂
        V₂, θ₂, ψ₂, ρ₂
```

In this case the boundary equation is

$$\theta_1 = \psi_1 = \frac{(\beta_1 - 1) e^{-2\lambda h_1}}{(\beta_1 + 1) - e^{-2\lambda h_1}(\beta_1 - 1)}.$$

Since $\rho_1 = \rho_2, \beta_1 = 1$. So $\theta_1 = \psi_1 = 0$ and

$$V_{m2} = K \int_0^\infty \frac{\sin(\lambda r_o)}{\lambda} [1 - J_o(\lambda D)] \, d\lambda$$

$$= K \int_0^\infty \left[\frac{\pi}{2} - \sin^{-1}\left(\frac{r_o}{D}\right) \right].$$

(35)

On a semi-infinite substrate the measured spreading resistance voltage would be:

$$V_{m1} = I \frac{\rho_1}{2r_o} \tag{36}$$

The defined boundary condition is that $V_{m1} = V_{m2}$, so

$$K \left[\frac{\pi}{2} - \sin^{-1}\left(\frac{r_o}{D}\right) \right] = \frac{I\rho_1}{2r_o} \tag{37}$$

and

$$K = \frac{I\rho_1}{2r_o} \left[\frac{\pi}{2} - \sin^{-1}\left(\frac{r_o}{D}\right) \right]^{-1} \tag{38}$$

This expression could have been developed directly from the model of two discs on a semi-infinite medium. The direct result of this development is:

$$\frac{V}{I} = \frac{\rho}{\pi r_o} \left(\frac{\pi}{2} - \sin^{-1} \frac{r_o}{D} \right) \tag{39}$$

However, it has been traditional to neglect the second term and write

$$\frac{V}{I} = R = \frac{\rho}{2r_o} \tag{40}$$

The approach just shown was used primarily to employ standard formulations as much as possible for familiarity.

It is necessary to keep the $\sin^{-1}(r_0/D)$ term because the iterative technique (Sect. 4.3.1) employed in the solution tends to reinforce this small error on successive iterations resulting in an error several times as large as that in the initial approximation.

4.3 Solution technique

Replacement of K in Equation (35) yields:

$$V_m = \frac{I\rho_1}{2r_0}\left[\frac{\pi}{2} - \sin^{-1}\left(\frac{r_0}{D}\right)\right]^{-1} \int_0^\infty [1+2\theta_1(\lambda)]\frac{\sin(\lambda r_0)}{\lambda}[1-J_0(\lambda D)]\, d\lambda. \quad (41)$$

Previous authors[25,26,27] have chosen to reorder this equation to:

$$2r_0\frac{V_m}{I} = \rho_1\left[\frac{\pi}{2} - \sin^{-1}\left(\frac{r_0}{D}\right)\right]^{-1} \int_0^\infty (1+2\theta_1)\frac{\sin(\lambda r_0)}{\lambda}[1-J_0(\lambda D)]\, d\lambda \quad (42)$$

and to define

$$\rho_0 = 2r_0\frac{V_m}{I}, \quad (43)$$

since ρ_0 would be the correct resistivity if the measurement of Vm and I were taken on a semi-infinite sample of homogeneous resistivity. (This is the usual calibration procedure for a spreading resistance probe.) In addition, the bracketed factor is defined as a "correction factor", so that

$$C = \left\{\frac{\pi}{2} - \sin^{-1}\left(\frac{r_0}{D}\right)\right\}^{-1}\int_0^\infty (1+2\theta)[1-J_0(\lambda D)]\frac{\sin(\lambda r_0)}{\lambda}\, d\lambda, \quad (44)$$

and thus

$$\rho_o = \rho_1 \cdot C \tag{45}$$

As Hu has pointed out[27], this is not an entirely appropriate terminology since "correction factor" implies an arbitrary adjustment on "fudge factor" rather than a logical data reduction scheme. However, in spite of the name, the format is a convenient one and is used here.

4.3.1 <u>Iterative technique</u>. The last equation can be reordered to

$$\rho_1 = \rho_o/C$$

where C is a function of r_o, d, and θ_1. θ_1 is itself a function of h_1 h_2...h_N and p_1 p_2...p_N. In other words, the equation for the "corrected" value of ρ_1 is not explicit in that C is a function of ρ_1. In general the integral factor of C has no analytical solution, and thus the equation cannot be solved explicitly for ρ_1.

The approach to solving for ρ_1 taken here is to define ρ_o as an approximation of ρ_1 and to obtain successfully better approximations of ρ_1 from an iterative solution of 3.3. C may be treated as a function of ρ_1 only.

<u>Simple relaxation</u>: Always converges, but often very slowly in terms of the number of repetitions required. In addition, the only available convergence criterion is comparison of successive values of ρ_1. The iteration is terminated when $\Delta\rho/\rho$ falls below a predefined limit. However, in some cases the convergence is so slow that $\Delta\rho/\rho$ may reach .001 when ρ_1 is still 10% aways from the correct value.

A <u>Newton-Raphson</u> technique, which uses both the last value of ρ_1 and its derivative to compute a new ρ_1, converges 3 to 5 times as fast as relaxation, but suffers the same weakness of convergence criteria and may oscillate in some cases, especially if the input data is not smooth, rather than converge.

Both relaxation and the Newton-Raphson method can be characterized as analytic techniques. Another technique which can be employed is a pragmatic hunting method which converges about one-half as quickly as the Newton-Raphson technique, but converges absolutely and to a known absolute accuracy. The technique is a familiar one, and best described as "<u>bracketing</u>". If a criterion can be found to determine if the correct value of ρ_1 lies between two other values of ρ_1, a search along the ρ_1 axis can be conducted using an arbitrary step size $\Delta\rho$ until the criterion is met. Once this has occurred, the correct value of ρ_1 is bracketed to a known accuracy. By splitting the bracket and retesting either of the two new brackets created for the presence of the correct value the bracket size can be reduced by one-half. Repeating the process results in a one digit accuracy improvement each $3.322 (=\log 2)^{-1}$) iterations.

$$A: \quad \frac{1}{C(\rho_1)} = \left\{ \left[\frac{\pi}{2} - \sin^{-1}\left(\frac{r_o}{D}\right) \right]^{-1} \int_0^\infty (1+2\theta)[1-J_o(\lambda D)] \frac{\sin(\lambda r_o)}{\lambda} d\lambda \right\}^{-1}$$

$$B: \quad \rho_1 = \frac{\rho_o}{C}$$

The relationship of these two equations is depicted qualitatively in Figure 14. Curve B is clearly a straight line. The shape of

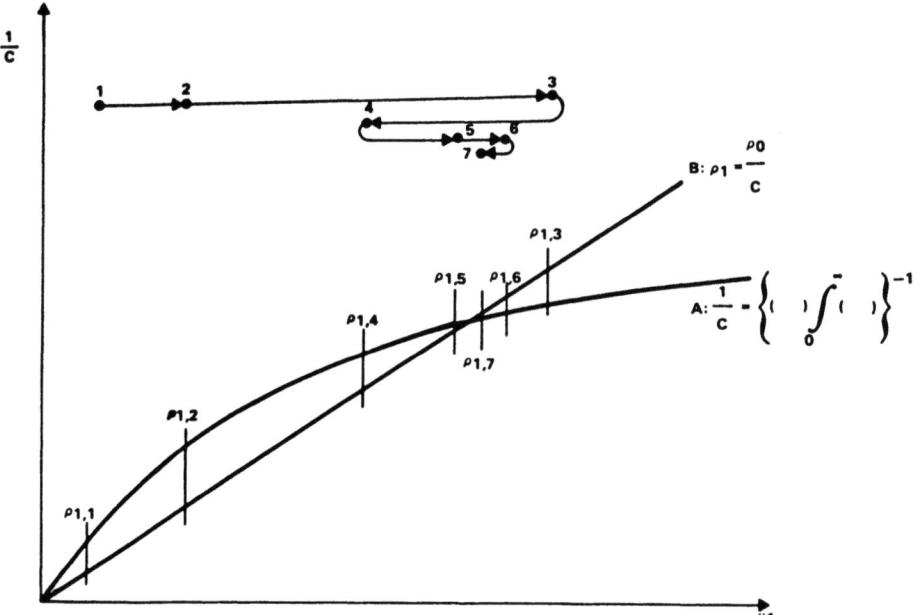

Fig. 14. Qualitative representation of bracketing convergence technique and a typical set of consecutive approximations of ρ.

curve A has been determined from numerical solutions of C (ρ_1) and, except for local curvature reversals associated with noisy data, the shape has been qualitatively the same for every case tested. Since the two curves cross at the correct solution of ρ_1, C_A-C_B will change sign at this point. This is the bracketing criterion discussed previously. The initial search is always conducted upward to avoid the trivial solution at ρ_1=0. Figure 14 also illustrates a hypothetical series of approximations toward the correct value of ρ_1.

The program developed by Lee[24] uses a combination of Newton-Raphson and relaxation techniques to achieve short running time. However, bracketing is selected for use on profiles which are ill-suited to the Newton-Raphson analysis.

4.3.2 **Bootstrap solution.** An iterative solution is applied to only one layer at a time. The method for correcting the entire structure is a bootstrap process proceeding from the deepest to shallowest layer. In other words, the deepest layer is corrected on the basis of the semi-infinite substrate or a junction boundary condition. Then the next is corrected on the basis of the corrected value of the bottom layer and boundary conditions, the next on the basis of these two corrected values, etc. The shallower layers, of course, do not affect the layer being corrected since they were not present at the point on the bevel at which the measurement was made.

4.4 Data pre-processing

When input data is not "smooth," the Multilayer Analysis may magnify input irregularities into the output profile. Figure 15, the original and analyzed profile of an epi-substrate interface, is an example of this effect. The output profile is quite irregular even though the input appears smooth to the eye. The effect occurs when layers beneath the layer being corrected make a major contribution to the measured voltage. When this is true, the program decides that minor variations in the measured voltage must be caused by major ones in the single surface layer since its contribution to the total measured voltage is relatively small. In Figure 15 this is particularly true since the surface layer at the point where irregularity begins is underlain by layers of from one to three orders of magnitude higher conductivity than the surface layer.

The problem is not in the program per se, but in the fact that the theoretical physical model used does not consider the problem of noisy data and also in the fact that although the model is a reasonable approximation of a smooth profile, it is a poor one for the derivative of the profile. The problem, then,

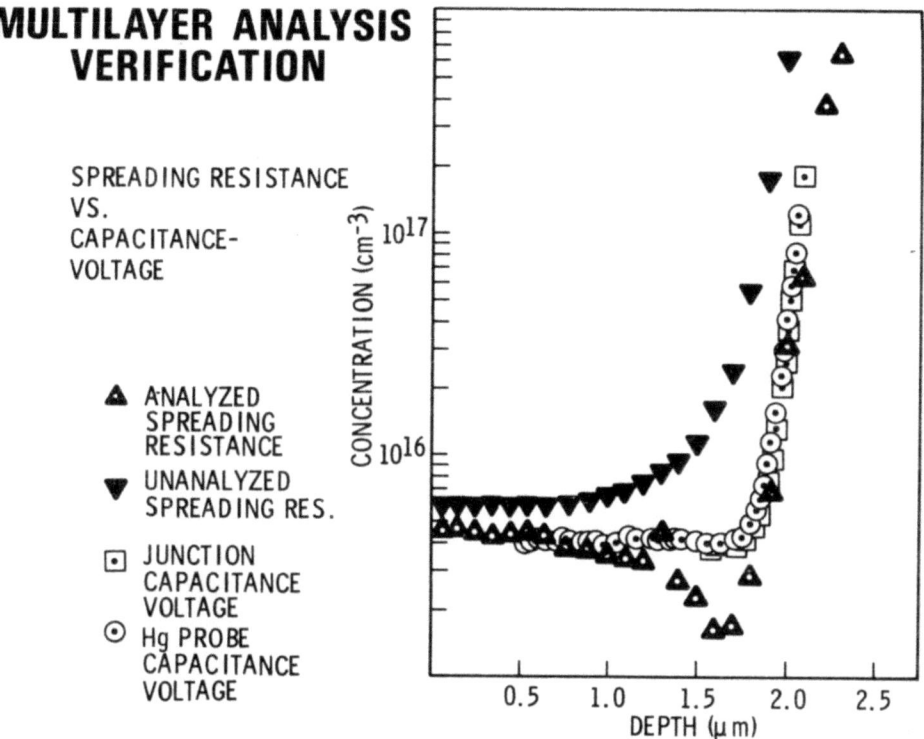

Fig. 15. Multilayer analysis verification.

cannot be solved by any change to the analysis technique short of complete revision. It must be solved at the input data. This is reasonable anyway, since one normally believes that the profile is really smooth and that variations of the type discussed here are indeed noise.

Modifying the input data presents two problems. The first is to define criteria of "smoothness." The second is to smooth the data while avoiding both the loss of information originally contained in the data and the introduction of an arbitrary shape to the data profile from the smoothing technique.

The second problem may not be completely avoidable, and in any case cannot be measured quantitatively, so the best approach is to do as little as possible and accept some irregularity in the output profile if it is not severe.

Since the problem arises from the poor model of the derivative supplied by a step function profile, the basic criterion of smoothness is a locally monotonic set of first differences, which are the finite difference analog of the first derivative. The first difference between two data points is:

$$\Delta_1 V_i \equiv V_{i+1} - V_i \tag{46}$$

(Voltage, V, is normally smoothed since this is the form of the raw data, but R or ρ could be smoothed as well.) No consideration of Δ_h, the depth change, is necessary, since it is constant. The second difference is defined as:

$$\Delta_2 V_i \equiv \Delta_1 V_{i+1} - \Delta_1 V_i \tag{47}$$

Notice that if the set of first differences varies monotonically, then all second differences have the same sign.

Since the first differences must be smooth, it is these values on which the smoothing routine operates. However, data may be too noisy to allow this operation to be performed directly. The routine followed is this:

- The initial data values are smoothed.
- First differences are computed from the smoothed values.
- These first differences are smoothed.
- The final profile is recomputed based on the smoothed differences.

The smoothing procedure referred to is as follows:

- A best fit parabola is generated for each set of five consecutive original points.
- The center point of the five is recomputed from the parabola.

This avoids imposing a functional shape because no two points are adjusted using the same parabola. It also avoids displacing the profile because only initial data is used to generate the smoothed result. That is, the parabolas are fitted only to initial values, not to the smoothed results for previously adjusted points.

Second order smoothing does occur, however, when the first differences computed from a smoothed data set are themselves smoothed. When a smoothed data profile is recomputed from the smoothed differences, displacement of the profile from its original envelope is avoided by allocating the change in the first difference equally to the positions of the data points at each end of the interval. Expressed algebraically for a non-end point:

$$V_{smooth,i} = V_i + \frac{1}{2}(\Delta_1 V_{1\ smooth,i-1} - \Delta_1 V_{1,i-1})$$
$$- \frac{1}{2}(\Delta_1 V_{1\ smooth,i} - \Delta_1 V_{1,i}) \qquad (48)$$

Thus, the resulting profile contains not the smoothed differences, but differences changed in the direction of the smoothed values, but limited by the original data envelope.

Two tests of this technique are particularly important. First, does the profile drift from its original position? And second, does originally smooth data remain smooth? Both can be tested by smoothing profiles generated from various polynomial and transcendental functions whose values and first derivatives are continuous. These profiles are by definition already smooth. It was found that after five iterations of smoothing previously smoothed output, the maximum deviation at any data points was .055% from the original. The results of the first smoothing operation deviated at most .049%. The deviations on subsequent iterations did trend further in the same direction as the first iteration deviations. This and the fact that the greatest parts of the deviations appear on the first iteration suggest that some arbitrary shaping occurs. However, it is small.

At this point, the smoothing program makes no judgment as to the adequacy of the smoothing. In fact, no firm criteria have been developed. The program simply performs the smoothing routine and supplies the initial and smoothed data along with first and second differences. Neither does it attempt to remove grossly erroneous data values from the initial data before smoothing. Occasionally it does change the shape of the profile, but only at very sharp peaks or bends. Obviously it is still necessary to screen the results of smoothing before using them.

4.5 Results

The effect of this analysis on different profile types is illustrated in Figures 15 and 16, but these only verify that the results are qualitative as expected. Some measure of quantitative accuracy can be obtained by comparison of dopant profiles obtained from analyzed spreading resistance data and other profiling techniques. The techniques used were junction and mercury probe capacitance-voltage measurements and incremental sheet resistance. While these techniques cannot be used on all types of profiles,

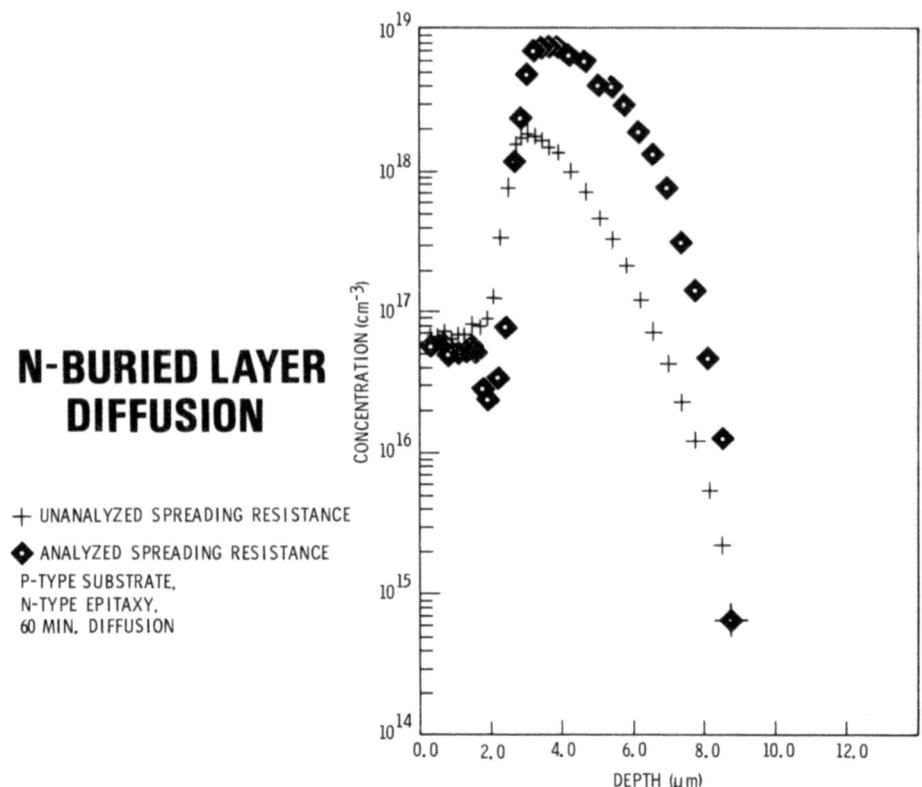

Fig. 16. Original data and result of multilayer analysis for epi-buried layer-junction structure.

they do have a known reliability for specific profile shapes.

The result of such comparisons is general agreement, but certain discrepancies exist. The first is illustrated in

Figure 15, a comparison of analyzed and unanalyzed spreading resistance profiles and capacitance-voltage profiles of an epi-substrate structure. The dip in the analyzed spreading resistance profile at the interface of the flat section of the profile and the steeply rising section is a program artifact. Fortunately, it is one which is easily recognized by an experienced user and thus need not be confusing. The absolute dimensions of the structure involved determined the importance of this defect; the feature is quite prominent in Figure 15, but not nearly so in Figure 16, whose absolute depth is greater.

The effect seems to arise from a discrepancy between the probe model and reality. Recall that the model used for the probe contact was a single circle of uniform potential. However, optical microscopy and SEM of the probe tips and the probe "prints" left on a sample reveal that the contact is actually composed of a large number of smaller contacts over a generally circular area. Furthermore, when a calibration of spreading resistance versus resistivity is attempted using the equation $R=\rho/2r_0$, the value of r_0 obtained is significantly less than the radius of the probe print. Unfortunately, an analytical solution for a model more resembling the probe geometry has not been possible because of the complex boundary conditions involved in a tight array of contacts, either in a density distribution or some arbitrary pattern.

A simple "non-real" model can explain the value of r_0 obtained by the calibration. If the contact is viewed as having a constant voltage, but a limited number of conducting areas in a uniform distribution so that the probability that a contact touches a given point within the limits of the array is δ, then the development of the two probe equation is nearly identical to that of the conducting disc model. The final form is

$$R = \frac{\rho}{2\delta r_D} - \frac{\rho}{\pi \delta r_D} \sin^{-1}\frac{r_D}{D} \frac{\rho}{2\delta r_D}, \qquad (49)$$

where r_D is the radius of the damaged area of the probe point, or the limit of the contact array. Since δ is the probability of contact to an arbitrary point, $\delta = A_C/A_D$, where A_C is the total area of contact and A_D is the area of the region of the array of contacts, or the damaged area. Thus the r_0 measured in calibration is

$$r_0 = r_D \frac{A_C}{A_D} \qquad (50)$$

This allows a solution for A_C, but this is of little help since neither the density nor the size of the individual contacts is revealed in the calibration procedure. Also, observation of probe point hardly suggests uniformity. This model does help to confirm the conjecture that A_C is less than A_D, however.

The value of r_0 given by a R vs ρ calibration is usually called the "effective" radius, and is treated as though it were the radius in a calibration for uniformly doped samples. This is reasonable since the lateral displacement between the points of contact is small compared to the probe to probe distance.

Multilayer analysis, however, is concerned with the potential distribution immediately beneath the probe, so the distribution near individual contacts is important. Figure 17 is a hypothetical representation of the real problem. In a distributed contact array with individual contacts small enough for the potential distribution near each contact to resemble that of an isolated small circular contact, the previously developed contact model indicates that the voltage near the small contact will fall off more quickly than that near a large one.

The single contact probe model then may look too deep in calculating C, and for certain geometries it "overcorrects"; thus, the dip. But the problem is more than one of depth alone; the shape of the potential under the probe is also important. Simply reducing the value of r_0 does indeed remove the dip, but it also moves the rest of the analyzed profile away from the capacitance-voltage data.

A typical comparison of an incremental sheet resistance profile to an analyzed spreading resistance profile of an implanted phosphorus layer is shown in Figure 18. The results are in good agreement on the steeply inclined portion of the profile. There is similarly good agreement shown in Figure 19 for diffused phosphorus. Right under the surface, however, the spreading resistance

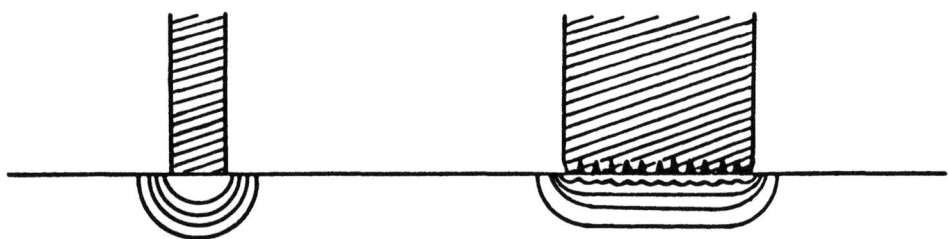

Fig. 17. Hypothetical comparison of single and multiple contact probe voltage equipotentials, showing deeper penetration of single probe field near the surface.

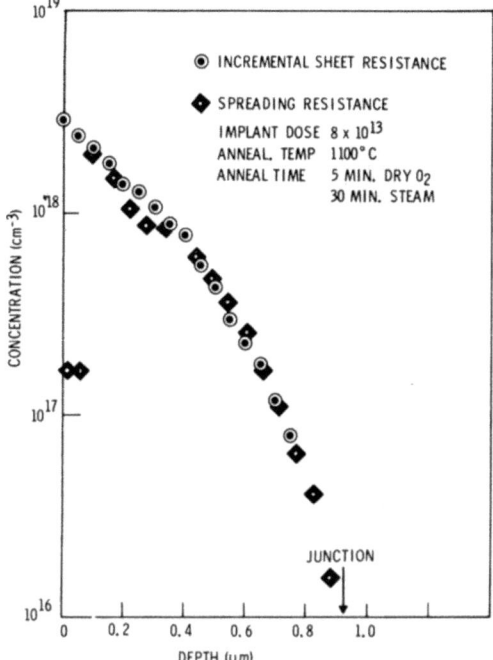

Fig. 18. Phosphorus p^{31} implant profile.

technique seems to suggest a significant decrease in the phosphorus concentration in Figures 18 and 19. The probable origin of this drop is an effective reduction in carrier concentration caused by the formation of a depletion region under the surface generated by the deposition of the protective silicon nitride film[30]. As Figure 19 shows, the depletion region is partially reversed and eliminated, when the silicon nitride film is removed.

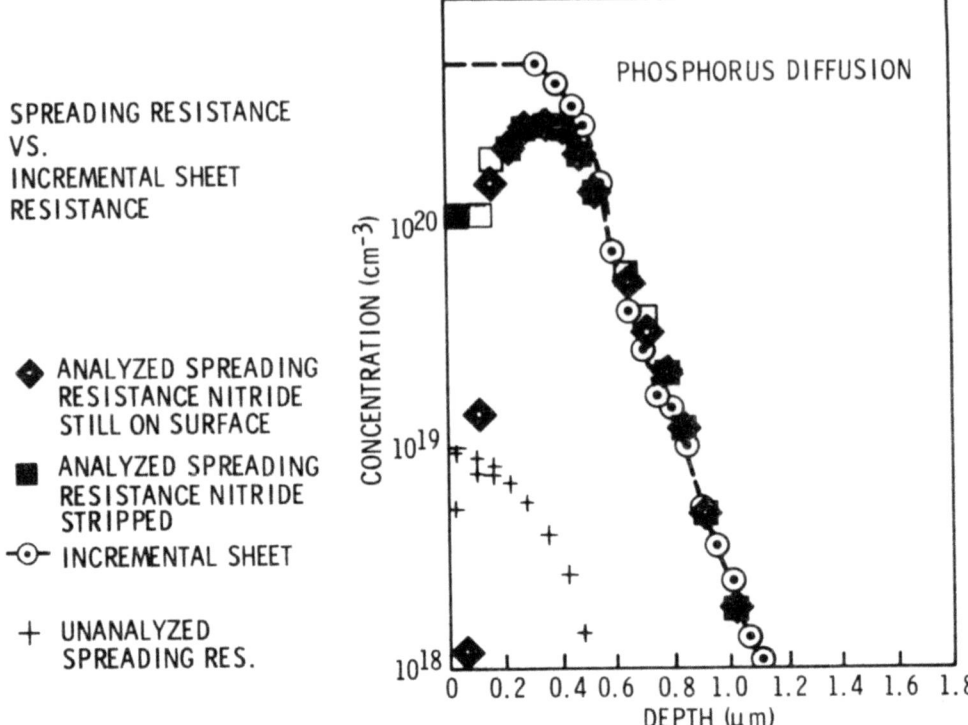

Fig. 19. Multilayer analysis verification.

REFERENCES

1. M.G. Buehler, "Test patterns," National Bureau of Standards SP 400-12, ed. by W.M. Bullis, Washington, D.C., 1975, p. 19.
2. J.G. Aiken. "The bipolar tetrode as a process control tool," Semiconductor Silicon/1973, ed. by H. R. Huff and R. R. Burgess, Electrochem. Soc., Princeton, New Jersey, p. 759.
3. W.H. Schroen, J.G. Aiken, and G.A. Brown, "Reliability improvement by process control," Proc. Tenth Annual IEEE Reliability Physics Symp., Las Vegas, Nevada, 1972, p. 42-48.
4. V.G.K. Reddi, "Influence of surface conditions on silicon planar transistor current gain," Solid-State Electronics 10, 305-335 (1967).
5. P.J. Coppen and W.T. Matzen, "Distribution of recombination current in emitter-base junctions of silicon transistors," IRE Trans. Electron Devices ED-9, 75-81 (Jan. 1962).
6. J.W. Iwerson, A.R. Bray and J.J. Klimac, "Low-current alpha in silicon transistors," IRE Trans. Electron Devices, ED-9, 474 (1962).

7. H.K. Gummell, "A self-consistent iterative scheme for one-dimensional steady state transistor calculations," IEEE Trans. Electron Devices, ED-11, 455 (1964).
8. R. Berry, "Correlation of diffusion process variations with variations in electrical parameters of bipolar transistors," Proc. IEEE 57, 1513-1517 (Sept. 1969).
9. H.K. Gummel, "A self-consistent iterative scheme for one-dimensional steady state transistor calculations," IEEE Trans. Electron Devices, ED-11, 455-465 (Oct. 1964).
10. Ohmic contacts to semiconductors, ed. by B. Schwartz, Electrochem. Soc., New York, 1969. See also C.Y. Chang, Y.K. Fang, and S.M. Sze, "Specific contact resistance of metal semiconductor barriers," Solid-State Electronics, 14, 541-550 (1971).
11. F.A. Padovani, "The voltage-current characteristic of metal-semiconductor contact," Conductors and Semimetals, Vol 7A (Academic Press: New York, 1971), 75-146.
12. C.B. Duke, Tunneling in Solids, (Academic Press: New York, 1969), 30-121.
13. S.M. Sze, Physics of Semiconductor Devices, (Wiley: New York, 1969), 363-424.
14. B.R. Gossick, "Nonreflecting boundary potential," Surface Science, 18, 181-192 (1969).
15. B.R. Gossick, "Electrical characteristics of a metal-semiconductor contact, I and II," Surface Science, 21, 123-135 (1970; 25, 465-490) (1971).
16. M.J. Turner and E.H. Rhoderick, "Metal-silicon Schottky barriers," Solid-State El., 11, 291 (1968).
17. H.C. Card and E.H. Thoderick, "Studies of tunnel MOS diodes I. Interface effects in silicon Schottky diodes," J. Physics D4; 1589 (1971).
18. B.L. Smith and E.H. Rhoderick, "Schottky barriers on p-type silicon," Solid-State El. 14 71 (1971).
19. W.H. Schroen, "Materials quality and process control in integrated circuits manufacture," in Festkoerperprobleme XVII, Vieweg-Verlag, Wiesbaden, 1977.
20. P.B. Ghate, "Failure mechanism studies on multilevel metallization systems for LSI," Final Report RADC-TR-71-186, Sept. 1971, Rome Air Develop. Center, New York.
21. Ch. Ting and Ch. Chen, "A study of the contacts of a diffused resistor," Solid-State El. 14, 433, 1971.
22. W. Shockley, "Research and investigation of inverse epitaxial VHF power transistor," Final Report AL-TDR-64-207 (by R.M. Scarlett), September 1964, Wright-Patterson Air Force Base, Ohio 45433.
23. A.Y.C. Yu, "Electron tunneling and contact resistance of metal-silicon contact barriers," Solid-State Electronics, 13, 239-247 (1970).
24. G.A. Lee, "Multilayer analysis of spreading resistance measurements," Proc. Spreading Resistance Symp., ed. by J.R.

Ehrstein, NBS SP 400-10, Gaithersburg, Md., 1974, P. 75.
25. P.A. Schumann, Jr., and E.E. Gardner, "Spreading resistance correction factors," Solid-State Electronics, $\underline{12}$, pp. 371-375 (1969).
26. T.H. Yeh and K.H. Khokhani, "Multilayer theory of correction factors for spreading resistance measurements," J. Electrochem. Soc. Electrochem. Tech., $\underline{116}$, 10 (1969).
27. S.M. Hu, "Calculation of spreading resistance correction factors," Solid-State Electronics, $\underline{15}$, pp. 809-817 (1972).
28. See also: P.A. Schumann, Jr., and E.E. Gardner, "Multilayer potential distribution," Technical Report 22,404, IBM Components Div., East Fishkill.
29. E.E. Gardner and P.A. Schumann, Jr., "Potential distribution in multi-layered structures," IBM Technical Report 22, 191.
30. W.H. Schroen, G.A. Lee, and F.W. Voltmer, "Comparison of the spreading resistance probe with other silicon characterization techniques," Proc. Spreading Resistance Symp., ed. by J.R. Ehrstein, NSB-SP 400-10, Gaithersburg, Md., 1974, p. 155.

* NOTE: MATERIAL EXCERPTED FROM J.G. AIKEN, "The bipolar tetrode as a process control tool," in SEMICONDUCTOR SILICON /1973, ed. by H.R. MUFF AND R.R. BURGESS, Electrochem. Soc., Princeton, N.J. pp. 759-768.
This information was originally presented at the 143rd Spring Meeting, Chicago, Illinois, 1973, of the Electrochemical Society Inc.
This information reprinted by permission of the publisher, the Electrochemical Society Inc.

** NOTE: MATERIAL EXCERPTED FROM W.H. SCHROEN, J.G. AIKEN, AND G.A. BROWN, "Reliability improvement by process control," Proc. Tenth Annual IEEE Reliability Physics Symp., Las Vegas, Nevada, 1972, pp. 42-48.
© 1972 by THE INSTITUTE OF ELECTRICAL AND ELECTRONICS ENGINEERS, INC.
REPRODUCTION AUTHORIZED BY IEEE.

DEFECT CHARACTERIZATION

H. S. Rupprecht

IBM System Products Division--East Fishkill
Hopewell Junction, New York   12533

ABSTRACT. Defects in materials are a major limitation in
large-scale integration. In this lecture we shall stress
particularly those imperfections that relate to the silicon
substrate and to the passivating dielectric film. A brief
review of the most common defects will be given, and diagnostic
tools and methods that enable one to predict final device yield
at an early processing state will be described.

1. INTRODUCTION

One can advance two major factors that lead to yield losses and
reliability problems in integrated devices. The first concerns
the intrinsic defects in materials, that is, imperfections
caused by the preparation of the substrate by subsequent processing steps.

The second can be termed extrinsic and is related to faults in
the artwork (pattern)-generating process. These faults include
mask layout problems (ground-rule violations and/or design errors)
as well as accidental, statistical defects in the pattern itself.

It is the first category we shall concentrate on in this presentation. These are the defects in the silicon itself as well as
in the dielectric passivation, which lead to (a) soft p-n
junctions, (b) conductive paths between emitter and collectors,
commonly classified as pipes in bipolar devices, and (c) leakage
problems in metal oxide semiconductor (MOS) type transfer and
charge storage devices.

2. SPECIAL TECHNIQUES

Before we start with the actual discussion of the various defects and their characterization, I should like to introduce two techniques which have turned out to be very powerful.

2.1 Technique of Indentation Dislocation Rosettes

The technique of indentation dislocation rosettes (IDR) has been applied by Hu [1] in his defect studies as a means of generating dislocations in a controlled manner and of observing the interaction of those dislocations either with stress regions or with other defects in the crystal.

For this purpose, a diamond stylus is impressed upon a silicon substrate held at about $600°C$. A subsequent heat treatment at $900°C$ for about 30 min permits the defects to propagate sufficiently under the build in stress due to the indentation. The resulting pattern can then be revealed by a chemical preferential etch, e.g., Sirtl etch. Figure 1 depicts schematically the resulting pattern on a (001) surface. A double-column of etch pits extends in a cross-like figure from the indentation center. The etch pits demarcate the intersection of prismatic dislocation loops with the substrate surface. They move along <110> directions on (111) glide planes under shear stress exerted by the indentation center. In a typical experiment the loops penetrate about 10 μm into the bulk (tip of the prismatic trough).

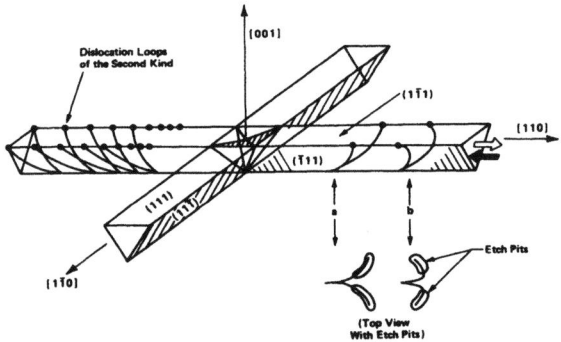

Fig. 1. Indentation dislocation rosette in <100> silicon. Dislocation loops of the second kind are confined to move on the surface of the prisms bounded by {111} and the surface plane. Loop (a) moves in a substrate with a free surface; loop (b) moves in a substrate with a surface film which drags the loop. (Source: Ref. 1)

2.2 Electrolytic Delineation Techniques

For an assessment of a process technology, it is often desirable to map out the geometrical distribution and density of leakage defects. Because of the small size of present-day devices, however, electrical probing of these devices individually is very difficult and cumbersome. The electrolytic mapping technique has proved useful in such applications.

There are essentially two electrolytic techniques described in the literature: the anodic and the cathodic.

2.2.1 The anodic method

In Plantinga's original method [2] the silicon wafer serves as the anode in an electrolytic cell, with Sirtl etch as the electrolyte. The principle is depicted in Fig. 2. In the electrochemical reaction $Si \rightarrow Si^{4+} + 4e$, silicon releases electrons, which either reach the subcollector region via a leakage path formed by a pipe or recombine with holes generated by a defect in the base-junction area. In either case silicon will be preferentially etched away, delineating an electrically active defect.

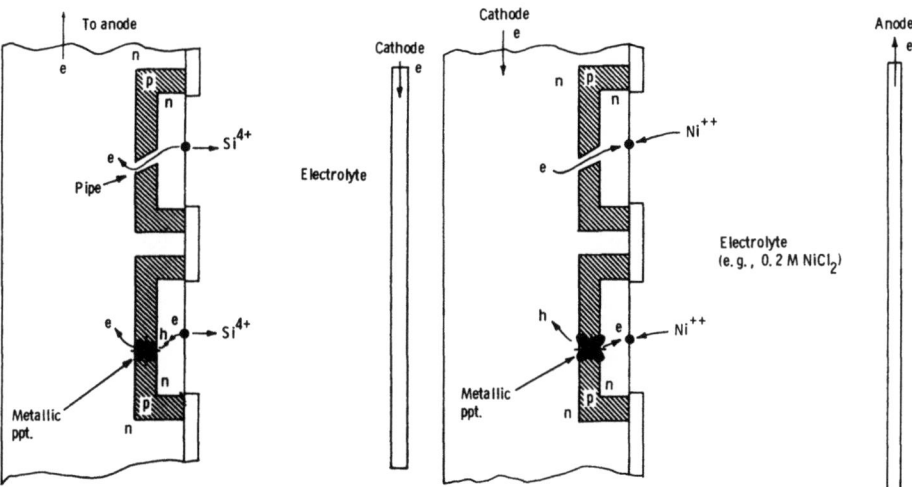

Fig. 2. Anodic etching of leaky transistors. The top transistor is "piped," the bottom one contains a generation-recombination defect. (Source: Ref. 4)

Fig. 3. Cathodic plating of suitable metallic ions, such as nickel, on leaky transistors. The top transistor is "piped," the bottom one contains a generation-recombination defect. (Source: Ref. 4)

In a modification of this method by Seto et al. [3] special Al (deposited under poor vacuum conditions) is plated into the windows of the device structures. Leakage currents cause preferential anodization of the Al, and the defective devices (e.g., emitters) are readily discernible by the color change.

2.2.2 The cathodic method

The cathodic method, described schematically in Fig. 3, was developed by Hu [4] in response to certain shortcomings of the anodic techniques. Here the substrate serves as a cathode in an electrolyte containing $NiCl_2$.

Leakage currents cause a preferential plating of Ni on the wafer surface.

3. THE EFFECT OF GROWN-IN IMPURITIES--OXYGEN AND CARBON

The phenomenon of impurity striations in crystals grown by float zone and Czochralski techniques is well known. It is based upon temperature fluctuations at the melt-solid interface in conjunction with the rotational and vertical motion of the crystal. The question arises whether such "natural" impurities, such as oxygen and carbon, are also incorporated in forms of striae. So far, only carbon has been reported, by Abe et al., to form striations in silicon. This is surprising, as the solubility of oxygen is even higher (2.7×10^{18} atoms/cm^3) than that of carbon (9×10^{17} atoms/cm^3) at the melting point of silicon. It seems to exclude the possibility that oxygen might be responsible for the heterogeneous nucleation of vacancy clusters, leading to striated microdefects, as discussed in Section 4.

We shall describe, in the following, some recent studies of the role of oxygen and its interaction with other crystalline defects.

Hu has carried out IDR experiments on substrates of varying oxygen content and measured the rosette size as a function of oxygen concentration. He has found a marked solution-hardening effect due to oxygen. The critical resolved shear stress increases by a factor of 4 at the maximum oxygen concentration. This is one of the beneficial effects oxygen has in silicon.

On the other hand, oxygen precipitates, as shown by Tan and Tice [5], can give rise to dislocation rosettes (Fig. 4).

Mader and Michel [6,7] have carried out TEM studies on the effect of oxygen imbedded into silicon by means of knock-on collision. (For knock-on collisions, see the lecture on Ion Implantation.)

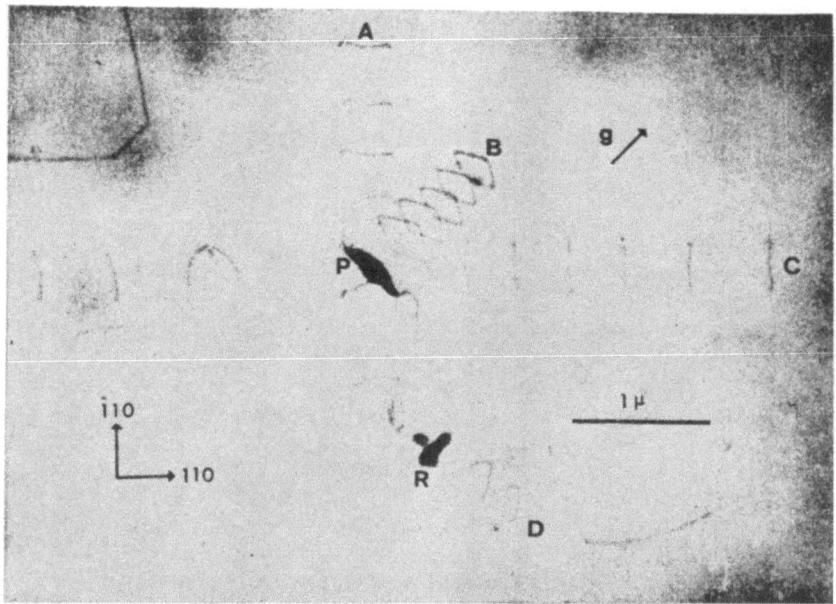

Fig. 4. Prismatic punching of dislocation loops by an SiO_2 precipitate in a silicon matrix. P:ppt with (010) habit plane, &R:ppt with (100) habit plane. Prismatic loops: PA,$(a/2)|\bar{1}10|$ $(\bar{1}10)$; PB,$(a/2)|01\bar{1}|$ $(01\bar{1})$; PC, $(a/2)|110|$ (110); RD,$(a/2)|10\bar{1}|$ $(10\bar{1})$. (Source: Ref. 5)

The annealing behavior of the radiation damage caused by an As implant (80 keV and 10^{16} atoms/cm^2) through a screen oxide of 200 Å SiO_2 deviates markedly from that of an As implant into bare silicon (80 keV and 2×10^{16} atoms/cm^2). In the case of the screen oxide implants (Fig. 5), the resulting dislocation network lies in a plane parallel to the surface, while the dislocations stemming from implants into bare silicon terminate at the surface (Fig 6).

The observations by Mader et al. [6,7] are in agreement with results reported earlier by Chu, Mueller, and Mayer [8]. They observed, in their RBS studies, residual disorder in silicon that had been implanted with As through SiO_2 films and subsequently annealed.

To confirm their suspicion that recoil oxygen is responsible for the retarded annealing behavior, they implanted oxygen directly into bare silicon.

Their results are summarized in Fig. 7.

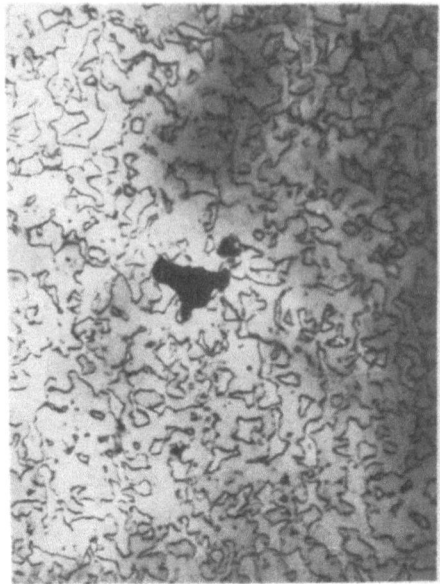 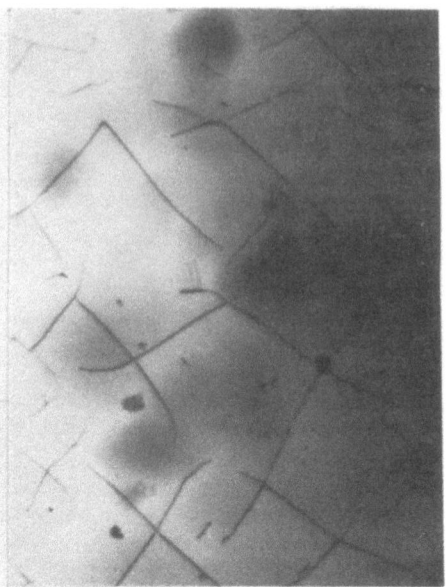

Fig. 5. Meandering dislocation loops with shear character in Si, implanted with 10^{16} As/cm^2 at 80 keV through 200 Å SiO$_2$ and annealed 30 min at 970°C. Magnification 35 000 X, diffraction vector g = [220] vertical. (Source: Ref. 6)

Fig. 6. Stereo TEM pair of dislocation structure in (bare) Si implanted with 2×10^{16} As/cm^2 at 80 keV, and annealed 60 min at 1000°C. Magnification 15 000X, diffraction vector g=[400] vertical. (Source: Ref. 6)

4. SWIRL DEFECTS

The investigation of striations revealed a second type of striae, which could not be attributed to impurities. These striae were discernible in x-ray topographs only after appropriate heat treatments.

The use of copper decoration makes it possible to observe these microdefects by infrared microscopy. They have become known as swirls and are considered to be clusters of microdefects (Fig. 8). At closer scrutiny two kinds of microdefects, as observed by Hu [9], can be distinguished (depicted in Fig. 9): they are usually termed A-clusters (or A-swirl) and B-clusters (or B-swirl) for the large and the small defects, respectively.

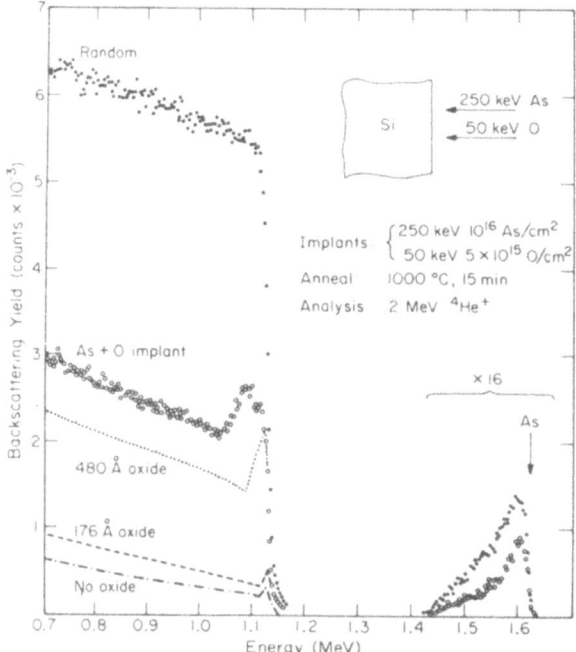

Fig. 7. 2-MeV He* random and aligned backscattering spectra for 250-keV 10^{16} cm^{-2} As implants followed by a 50-keV oxygen implant at a dose of 5 x 10^{15} cm^{-2}. Dotted lines represent the annealing behavior of As through-oxide implants. The dotted line and dashed line represent the aligned spectra for 10^{16} As/cm^2 implanted through 480 and 176 Å of SiO_2, respectively. (Source: Ref. 8)

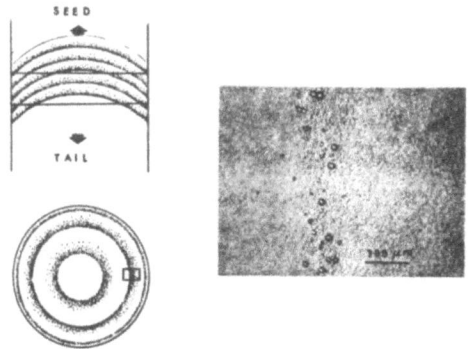

Fig. 8. Schematic illustration (left) showing the distribution of swirl defects in a crystal: left top, longitudinal section: left bottom, transverse section. The micrograph of swirl-defect etch pits on the right (scale—100 μm) corresponds to the framed rectangular area on the left bottom. Etch figures of two distinct sizes can be observed. (Source: Ref. 9)

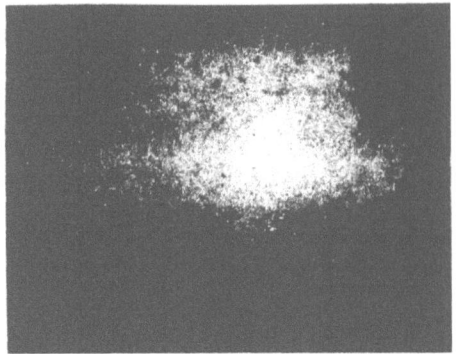

Fig. 9. Infrared micrograph of copper-decorated swirl defects. A-clusters are decorated into rosettes (two in the figure) of copper precipitates, while B-clusters into tiny lumps. Reference scale—50 μm. (Source: Ref. 9)

There seems to be a general consensus that the A-defects are some sort of interstitial dislocation loops. However, the exact nature of the B-swirls is still unresolved. Seeger et al. [10,11] considered silicon self-interstitials to be the dominant point defects at high temperatures which according to Föll and Kolbesen [p.22 of Ref. (9)] suggests that B-clusters are agglomerates of self-interstitials.

5. POINT DEFECTS IN SILICON

For a better understanding of these arguments, it might be advisable to present some of the more recent views on point defects in silicon.

The classical models of interstitial positions in a diamond lattice are depicted in Figs. 10 and 11 [12]. One distinguishes between hexagonal sites (Fig. 10) and tetrahedral sites (Fig. 11).

Such interstitial locations are probably suitable for monovalent ions in silicon such as H^+, Li^+, Na^+, and Cu^+, but are probably unlikely for ions that have a large number of valence electrons, which tend to covalent bonding.

More recently the idea of split self-interstitials has been advanced [13]. Figure 12 shows an electrically neutral configuration, as compared with a single positively charged interstitial, shown in Fig. 13.

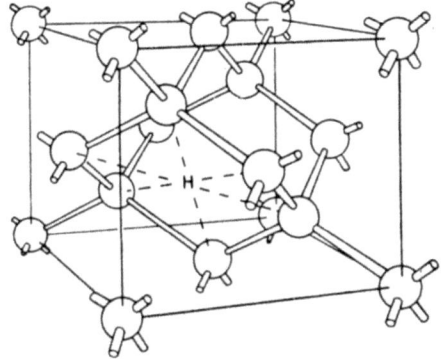

Fig. 10. The hexagonal H interstitial site in the diamond lattice. This site is the center of a six-membered puckered ring in the so-called "chair" configuration. (Source: Ref. 12)

Fig. 11. The terahedral T interstitial site in the diamond lattice. (Source: Ref. 12)

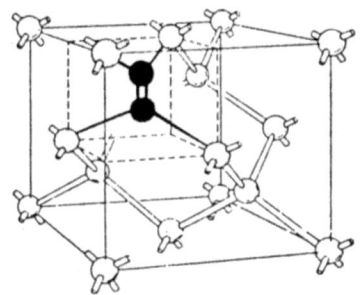
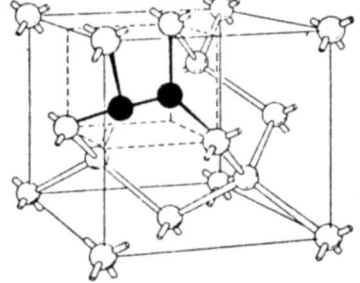

Fig. 12. Electrically neutral <100> split Si self-interstitial. (Source: Ref. 13)

Fig. 13. Singly positively charged <110> split Si self-interstitial. (Source: Ref. 13)

The classical vacancy interpretation [14] is shown in Fig. 14. If one relaxes this configuration by readjusting the position of the atom right on top of the vacancy, one arrives at a semivacancy pair constellation, proposed by Masters [15] (Fig. 15). At high temperatures another combination becomes possible, the divacancy [16] described in Fig. 16.

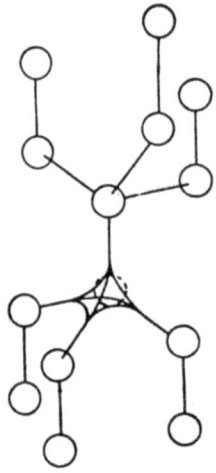

Fig. 14. Immediate environment of vacancy. The empty site is represented by a dashed circle. (Source: Ref. 4)

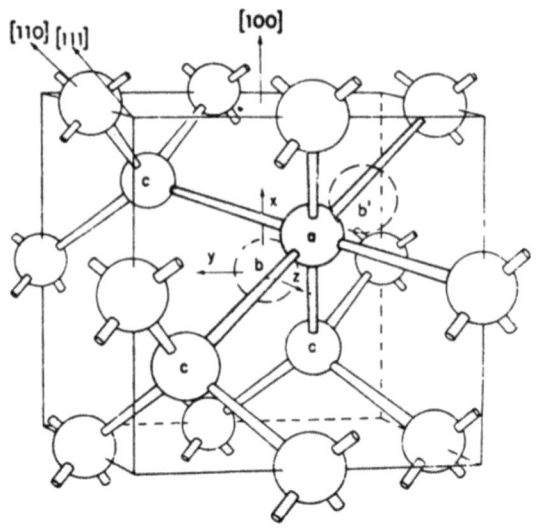

Fig. 15. Model of the semivacancy pair, the proposed configuration of the silicon monovacancy at thermal equilibrium. The indicated crystallographic orientations intersect at the vacant lattice site b. (Source: Ref. 15)

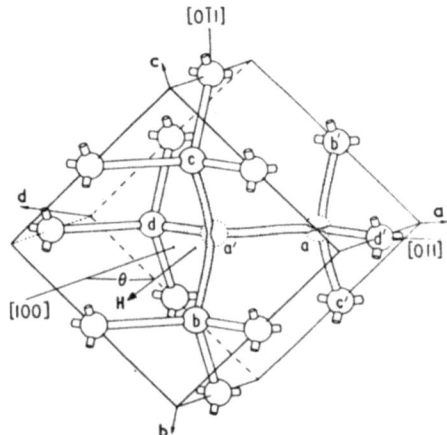

Fig. 16. Model of the divacancy in the orientation ad, showing the crystallographic axes. H denotes the magnetic field direction as varied in the EPR and ENDOR experiments. (Source: Ref. 16)

6. STACKING FAULTS

Stacking faults are due to abrupt changes in the sequence of (111) lattice planes. They occur during (a) crystal growth as a consequence of thermal shock, (b) epitaxial growth in response to interface imperfections, and (c) thermal oxidation.

Stacking faults normally form closed structures [17], as seen in Fig. 17. The partial dislocations, which confine stacking faults, can give rise to low-level leakage.

PAIRS OF STACKING FAULTS

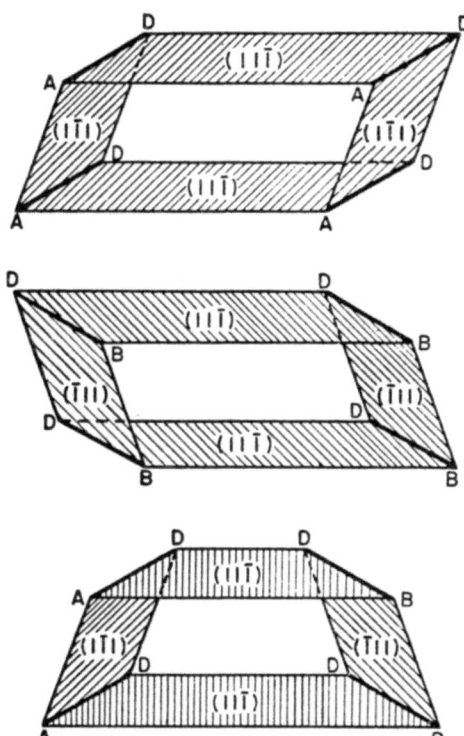

STAIR-ROD DISLOCATIONS AT AD AND BD
WITH BURGERS VECTOR $b=(a/6)[01\bar{1}]$ AND
$b=(a/6)[10\bar{1}]$

Fig. 17. Fault planes in silicon forming closed structures shown schematically. (Source: Ref. 17)

Hu [18] has studied the kinetics of stacking-fault growth during oxidation. Figure 18 depicts the temperature dependence of stacking-fault size for a given oxidation condition. A retrograde behavior has been observed at higher temperatures [9]. The detailed mechanism that leads to the formation is not yet clarified. Hu proposes that during the oxidations self-interstitials are released, which condense into stacking faults. A model advanced by Jaccodine et al. [19] assumes that vacancies are consumed by the oxidation.

To relieve the vacancy undersaturation, Frenkel pairs are formed and the excess interstitials lead to the formation of stacking faults.

7. THERMALLY INDUCED DISLOCATIONS

It is well known that thermal gradients during the heating-up or cooling-down phase can generate slip dislocations. These

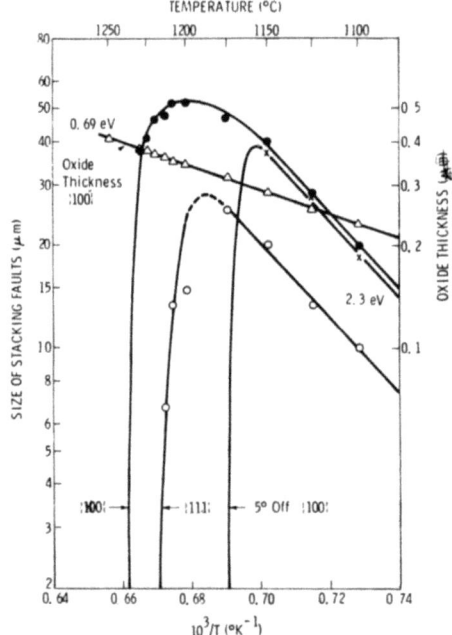

Fig. 18. The kinetics of the growth of oxidation stacking faults in a dry oxygen ambient exhibits an Arrhenius region at lower temperatures, and a retrogrowth region at higher temperatures. (Source: Ref. 9)

slip lines have been observed by many workers in the field
either by preferential chemical etch or by x-ray topography. It
is further known that these slip patterns lead to excess leakage
currents in bipolar devices.

Hu [20] has calculated the slip patterns that one can expect for
high and moderate stress conditions on (001) oriented wafers, as
shown in Fig. 19a and b. At moderate stresses the slip pattern
is confined to the outer zone of the wafer.

This agrees well with the measurements on pipe distributions by
Schwenker and Seto [20, 21]. They applied the aluminum anodization
technique to a special pipe chip test pattern and determined the
pipe density for the center two-thirds of the wafer as well as
for the outer one-third. This test pattern contains devices of
varying perimeter-to-area ratios, enabling one to observe possible
perimeter influence. They found that their results could be
explained by fitting the data on leakage-limited yield to the
relation

$$Y = \exp(-(\ell_A A + \ell_p P)). \tag{1}$$

Y is the yield of nonpiped devices, ℓ_A pipe density per unit
area A, and ℓ_p pipe density per unit perimeter length P. Figure 20
describes their results.

Equation (1) includes a perimeter-related term, $\ell_p P$, whicn takes
into account the fact that the pipes are not distributed over
the emitter area with equal statistical probability. Preferential
etch patterns reveal an increased occurrence of dislocation-derived
defects around the emitter window edge.

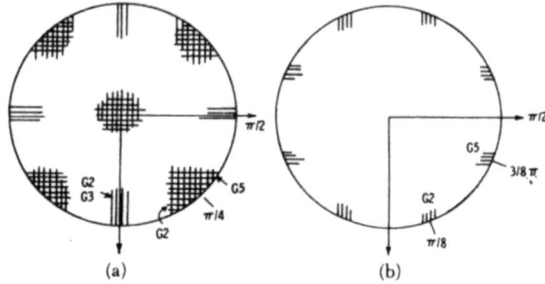

Fig. 19. Schematic illustration of distribution of "slip lines"
in a wafer subjected to a high tangential thermal stress (a) and
to a moderate tangential thermal stress (b). (Source: Ref. 20)

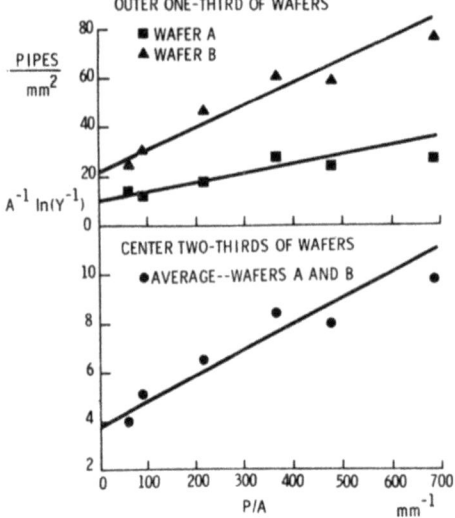

Fig. 20. Logarithm of apparent pipe density versus perimeter-to-area ratio. (Source: Ref. 20)

This phenomenon can be explained in the light of experimental studies by Hu [1] on the propagation of indentation dislocation rosettes at dielectric window edges. Hu observed a retardation of the rosettes when the dislocation movement occurred from an area covered by dielectric into a region of bare silicon surface. As a consequence a dislocation loop formed at the edge.

Figure 21 depicts the situation schematically. On the other hand, the motion of the dislocations is accelerated at the

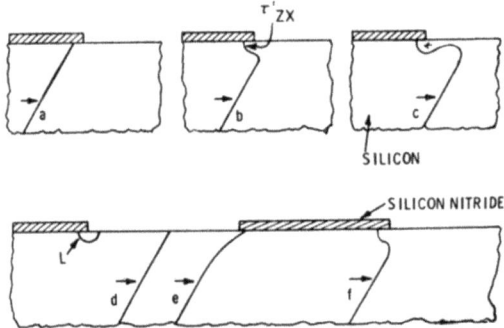

Fig. 21 Mechanism of the formation of emitter edge defects. (Source: Ref. 20)

opposite side of the window, where the local shear stress is
additive to the stress provided by the rosette center. In this
case one obviously does not expect a dislocation loop to be
formed.

On the basis of Hu's model, the pipes originating from a given
slip line should reflect the track of this line and lie always
on the same side of the emitter window. Experimental observations
indeed confirm this behavior.

8. DEFECTS IN THE DIELECTRIC AND AT THE SILICON DIELECTRIC
INTERFACE

Charges in the dielectric give rise to surface depletion of
carriers of the same polarity in the substrate or even surface
inversion layers, which, in turn, greatly increase the junction
leakage in planar device structures.

This fact is well known and has been described by Grove and
Fitzgerald [22] for gated diode structures. The "Grove curves"
(Fig. 22) clearly indicate at least three contributing factors
to the excess leakage:

a. For the surface accumulation regime, the leakage current is
    determined by the generation-recombination mechanism in the

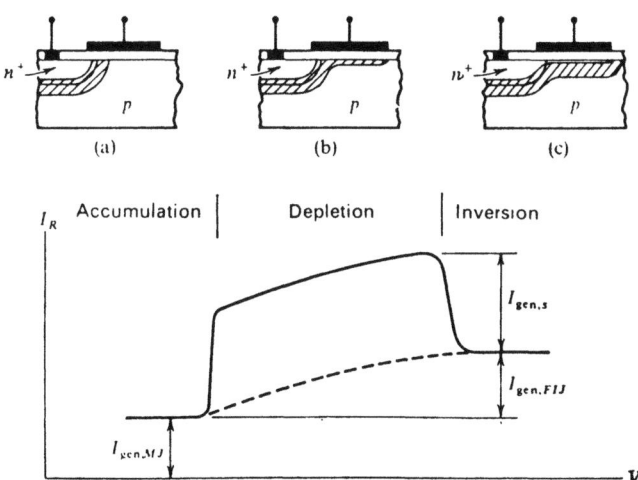

Fig. 22. Illustration of the effect of variation in the nature
of the surface space-charge region on the reverse current of an
n+p diode at a fixed reverse voltage. (Source: Ref. 22)

space-charge region surrounding the total junction area (bulk component).

b. If the surface is depleted in the junction vicinity, an additional component contributes to the generation-recombination, which takes place at the dielectric-silicon interface (surface component).

c. Finally, as the surface vicinity is driven into deep inversion, essentially the junction area is increased and hence the "bulk" component is increased accordingly.

The thermal generation-recombination current is greatly enhanced by the presence of certain defect centers in the silicon (bulk and surface vicinity). These defects can be observed, as described in Section 2, by electrolytic delineation techniques or by generation lifetime tests (MOS recovery measurements).

The charges in the dielectric layer are either <u>extrinsic</u> in nature and caused by impurities such as Na, Al, W, and Ni, which give rise to positive charge centers and can act as recombination centers for electron-hole pairs, or <u>intrinsic</u> defects, such as fast states at the silicon interface (unsaturated bonds) and/or trap centers caused by energetic radiation (x-ray, E-beam, ion implantation, etc.).

Aitken, DiMaria, and Young [23] have recently described, in several papers, a combination of techniques to determine stored charge, charge centroid, and capture cross section of trap centers, using (a) the time dependence of the flatband shift under avalanche injection and (b) photocurrent-induced voltage shifts.

The capture cross section was derived from the relation

$$\Delta V_{FB} = e N_{eff} \frac{L}{\varepsilon \varepsilon_o} (1-\exp(-\sigma\, jt/e)), \qquad (2)$$

where L is oxide thickness, σ capture cross section, j avalanche current, and t time.

$$N_{eff} = \frac{\bar{X}}{L} N_t,$$

where $\bar{X}$ is charge centroid. (See subsection 2.3.2 of lecture on Measurement Techniques.)

They report for implanted tungsten (60 keV, 1×20^{13} atoms/cm^2) in 500 Å of silicon oxide the following results:

σ (cm^2)	N_t (cm^{-2})
1.06x10^{15}	3.21x20^{12}

For 30-keV Al implants into 730 Å of silicon oxide (dose 1x20^{13} atoms/cm^2) the following data are obtained:

σ (cm^2)	N_t (cm^{-2})
1.60·10^{-15}	4.6 ·10^{11}
1.26·10^{-16}	1.14·10^{12}
1.40·10^{-17}	1.4 ·10^{12}
1.26·10^{-18}	5 ·10^{11}

ACKNOWLEDGMENT

I should like to thank S. M. Hu, S. Mader, and R. O. Schwenker for many helpful discussions and for their critique of the manuscript.

REFERENCES

1. S. M. Hu, J. Appl. Phys. 46, 1470 (1975).

2. G. H. Plantinga, IEEE Trans. Electron Devices ED-16, 394 (1969).

3. D. K. Seto, F. Barson, and B. F. Duncan, in Semiconductor Silicon 1973, edited by H. R. Huff and R. R. Burgess, The Electrochemical Society Softbound Symposium Series (Princeton, N. J., 1973), p. 651.

4. S. M. Hu, J. Electrochem. Soc. 124, 578 (1977).

5. W. K. Tice and T. Y. Tan, Appl. Phys. Lett. 28, 564 (1976).

6. S. Mader and A. E. Michel, J. Vac. Sci. Technol. 13, 391 (1976).

7. S. Mader and A. E. Michel, Phys. Status Solidi (a) 33, 793 (1976).

8. W. K. Chu, H. Mueller, and J. W. Mayer, Appl. Phys. Lett. 25, 297 (1974).

9. S. M. Hu, J. Vac. Sci. Technol. 14, 17 (1977).

10. A. Seeger and K. P. Chik, Phys. Status Solidi 29, 455 (1968).

11. A. Seeger and M. L. Swanson, Lattice Defects in Semiconductors, edited by R. R. Hasiguti (University of Tokyo, Tokyo, 1968), p. 93.

12. J. W. Corbett and J. C. Bourgoin, "Defect Creation in Semiconductors," in Point Defects in Solids, edited by J. H. Crawford, Jr., and L. M. Slifkin (Plenum Press, New York and London, 1975), Chap. 1.

13. W. Frank, in Lattice Defects in Semiconductors, 1974, Conference Series Number 23 (The Institute of Physics, London and Bristol, 1975), p. 23.

14. V. B. Glazman and G. S. Myaken'kaya, Sov. Phys. Semicond. 7, 7. 863 (1974).

15. B. J. Masters, Solid State Commun. 9, 283 (1971).

16. J. G. de Wit, C. A. J. Ammerlaan, and E. G. Sieverts, in Lattice Defects in Semiconductors, 1974, Conference Series Number 23 (The Institute of Physics, London and Bristol, 1975), p. 178.

17. Scientific Report No. 1, AFCRL-70-0110, Air Force Contract AF19(628)-68-C-0196, G. H. Schwuttke, Chief Investigator, March 2, 1970.

18. S. M. Hu, Appl. Phys. Lett. 27, 165 (1975)

19. R. J. Jaccodine and C. M. Drum, Appl. Phys. Lett. 8, 29 (1966).

20. S. M. Hu, S. P. Klepner, R. O. Schwenker, and D. K. Seto, J. Appl. Phys. 47, 4098 (1976)

21. R. O. Schwenker and D. K. Seto, Private communication

22. A. S. Grove and D. J. Fitzgerald, Solid-State Electron. 9, (1966).

23. J. M. Aitken, D. J. Maria, and D. R. Young, IEEE Trans. Nucl. Sci. 1526 (1976).

MEASUREMENT TECHNIQUES

H. S. Rupprecht

IBM System Products Division--East Fishkill
Hopewell Junction, New York   12533   USA

ABSTRACT. The development of appropriate diagnostic techniques has played a major role in the advancement of silicon technology. We shall concentrate in this lecture particularly on those techniques that are essential for the development of VLSI, such as methods to determine extremely shallow impurity distributions and techniques to characterize dielectric layers.

1. INTRODUCTION

The development phases through which measurement technology has passed over the years clearly reflect those of solid-state technology as a whole.

In the early days, the main emphasis had been on establishing the bulk properties of materials and their relationship with certain chemical parameters. The measurements had normally been carried out on samples of millimeter or centimeter size.

With the increasing importance of junction effects, however, the interest rapidly shifted away from homogeneous materials. New techniques had to be developed and existing ones modified to deal with inhomogeneous distributions, whose vertical dimensions were now in the order of micrometers and less.

It is the objective of this lecture to discuss particularly some of the more recent developments in the measurement field, which by uniqueness and potential will have an impact on semiconductor technology.

2. ELECTRICAL MEASUREMENTS

We shall subdivide the field of electrical measurements into the following categories:

        Resistivity Type
        Capacitive Techniques
        Photocurrent Methods

We include Hall effect measurements in the first as a special case in which the current-induced transverse potential drop is measured rather than the longitudinal one.

2.1 Resistivity Measurements

2.1.1 Theoretical background

Resistivity measurements had originally been designed to determine bulk parameters, such as resistivity ρ (or its inverse, that is, conductivity), free-carrier concentrations n and p (electrons and holes), and their mobilities μ_n and μ_p, respectively.

These parameters are interrelated by

$$1/\rho = q(n \mu_n + p\mu_p), \quad (1)$$

where q is electronic charge. This relation reduces for n type materials (n>>p) to

$$1/\rho = qn\mu_n.$$

An analogous relation holds for p-type material.

If one deals with nonuniformly doped thin layers, it is convenient to define a new set of parameters (so-called sheet parameters) which correlate with the bulk parameters in the following way. (The relations are given for n-type material):

$$1/\rho_s = \int_0^d \sigma(x) \, dx, \quad (2)$$

where d is the layer thickness,

and

$$n_s = \int_0^d n(x) \, dx. \quad (3)$$

With $1/\rho_s = qn_s \mu_{s\,eff}$ [in analogy to Eq. (1)], one gets

$$\mu_{s\ eff} = \int_0^d n(x)\ \mu(x)\ dx \Big/ \int_0^d n(x)\ dx. \tag{4}$$

For sheet samples the Hall mobility $\mu_{H\ eff}$ and the Hall coefficient $R_{H\ eff}$ can be expressed in terms of the bulk parameters in the following manner:

$$\mu_{H\ eff} = r\int_0^d n(x)\ \mu(x)^2 dx \Big/ \int_0^d n(x)\ \mu(x) dx; \tag{5}$$

$$R_{H\ eff} = \frac{r \cdot d}{q} \int_0^d n(x)\ \mu^2(x) dx \Big/ \Big(\int_0^d n(x)\ \mu(x) dx\Big)^2. \tag{6}$$

The resulting expression for the sheet concentration of free carriers is then

$$n_{s\ Hall} = n_{H\ eff} \cdot d = \frac{\Big(\int_0^d n(x)\ \mu(x) dx\Big)^2}{\int_0^d n(x)\ \mu^2(x) dx}. \tag{7}$$

In deriving Eq. (7) we used the following relations:

(a) $\mu_H = r\ \mu_D$,

where μ_D is the drift mobility as defined in Eq. (1) and r is the Hall factor

(b) $R_{Hs} = \dfrac{r}{n_s}$,

(c) $n_s = n_{H\ eff} \cdot d$.

One notices immediately that the sheet parameters determined in Hall effect measurements can differ considerably from those obtained from sheet-resistivity techniques by way of averaging.

We shall now describe a widely used technique known as incremental sheet conductivity and Hall effect measurements. This method is based upon sectioning the inhomogeneous layer in a coplanar way by means of anodic oxidation and measuring ρ_s and R_{Hs} between the individual sectioning steps. With the subscript i denoting the value of the parameters prior to the ith sectioning step, one obtains the bulk quantities in the removed layer from the following equations:

$$\frac{1}{(\rho_s)_i} - \frac{1}{(\rho_s)_{i+1}} = q\ n_i\ \mu_i\ d_i, \tag{8}$$

where d_i is the thickness of removed layer,

and

$$(R_{Hs})_i/(\rho_s)_i^2 - (R_{Hs})_{i+1}/(\rho_s)_{i+1}^2 = q\, n_i\, \mu_i^2\, d_i. \tag{9}$$

(In practical applications of this technique, it is normally necessary to carry out curve smoothing of the raw data. The scatter, otherwise, could lead to negative values.)

A derivation of Eqs. (8) and (9) is given in Appendices A and B.

The measurements are frequently carried out on van der Pauw type samples.

This technique has been quite successfully employed by many workers in the field to measure the electrical parameters in ion-implanted and diffused layers, but becomes inapplicable for determining the outer profile of a buried layer, diffused or implanted.

Hall effect measurements, however, require additional equipment and are cumbersome to perform. One therefore finds it frequently more convenient to measure the sheet-resistivity only and to determine the impurity concentration by means of Irvin's relations [1], which give $\rho = \rho(N)$ in graphical or tabulated form. They essentially represent Eq. (1).

$$\frac{1}{\rho} = q\, n\, \mu = q\, N\, \mu_{eff}, \tag{10}$$

where N is the total impurity concentration and μ_{eff} is the effective mobility.

If one makes the assumption that all impurity centers are ionized, $n=N$, then, also, the mobilities have to be identical. (The assumption of full ionization is definitely justified at room temperature for impurity concentrations above $10^{15} cm^{-3}$, as can be seen from the temperature dependence of the free-carrier concentration-saturation region.)

The total carrier concentration in the relation $\rho=\rho(N)$ was obtained from chemical data (e.g., radioisotope analysis) or from Hall measurements. In the latter case, a Hall factor $r=1$ was chosen.

Michel [2] has calculated the impurity distribution in boron-implanted structures and subsequently determined the theoretical sheet resistivity by means of Irvin's data. A comparison with the experimental values gives appreciable deviations, particularly in the region from $10^{18} - 10^{19}$ (boron atoms/cm^3). To correct for this discrepancy Michel suggests the use of a Hall factor, which gradually varies from 0.8 at a concentration of 10^{18} (cm^{-3}) to r=1 at a concentration of 10^{19} (cm^{-3}). This correction is shown in Fig. 1, which gives carrier mobility as a function of carrier concentration.

2.1.2 Measurement Tools

The various techniques that have been applied for resistivity measurements are schematically depicted in Fig. 2. They lead from "spider" like samples used for bulk measurements to van der Pauw structures and to four-point probe setups. As the spatial dimensions of impurity distributions steadily decreased, three- and finally two-point spreading resistance probes were employed, which permitted probe spacings of 20 µm. Today, two-point probes are successfully used to sample on bevelled structures

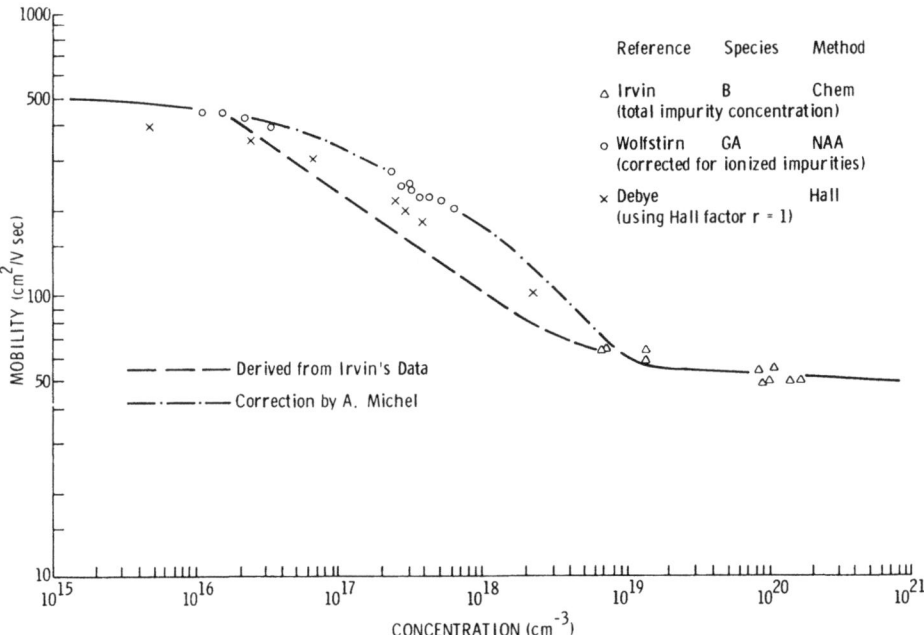

Fig. 1. Relationship between carrier concentration and mobility. (Source: Ref. 2)

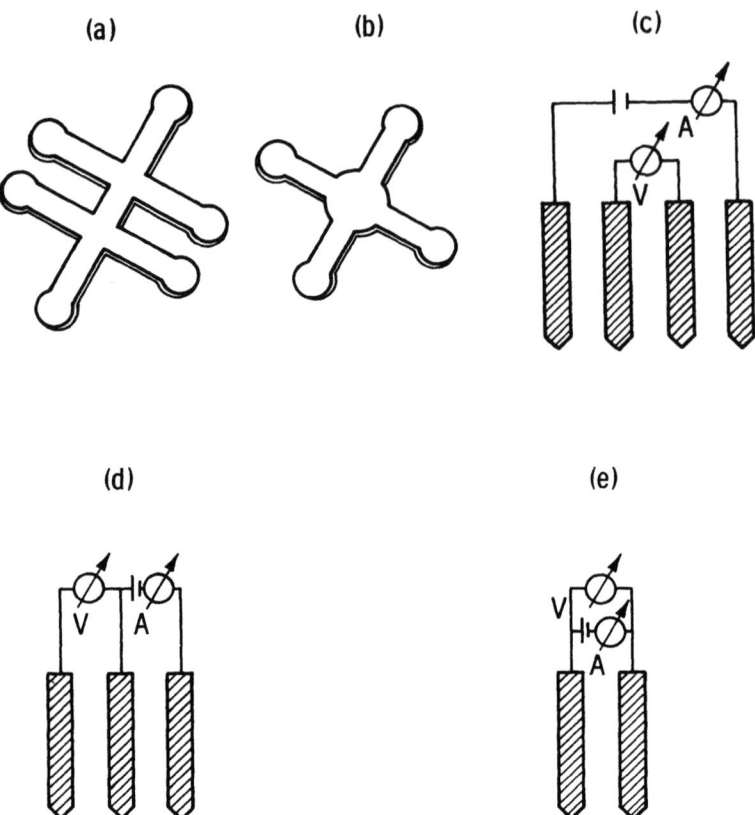

Fig. 2. Typical sample and probe configuration: (a) spider-type samples; (b) van de Pauw samples; (c) 4-point, (d) 3-point, (e) 2-point probe arrangements.

profiles of a few thousand angstroms. The disadvantage of the spreading-resistance technique lies in its extremely cumbersome correction procedure, which is needed to give the true bulk resistivity values in layered structures.

We shall now give a brief description of the main principles of the spreading-resistance method. The spreading resistance, R_{sp}, for a hemispherical probe can be easily derived. For a homogeneously doped sample one finds

$$R_{sp} = V/I = \rho/2\pi a, \qquad (11)$$

where a is the probe radius.

In the case of a disc-like probe one gets

$$R_{sp} = \rho/4a. \tag{12}$$

Schumann et al [3] used an empirical relation for the potential distribution in a layered structure (Figs. 3 and 4) around a disc-shaped probe. The potential in the nth layer is given by

$$V_n(r,z) = \frac{I\rho_1}{2\pi a} \{\int_0^\infty e^{-\ell z} \sin(\ell a) \, J_0(\ell r) \, \ell^{-1} \, d\ell$$

$$+ \int_0^\infty \theta_n(\ell) \, e^{-\ell z} \sin(\ell a) J_0(\ell r) \ell^{-1} \, d\ell \tag{13}$$

$$+ \int_0^\infty \psi_n(\ell) \, e^{\ell z} \sin(\ell a) J_0(\ell r) \ell^{-1} \, d\ell\},$$

where J_ν is the νth Bessel function, and θ_n, ψ_n are integration parameters.

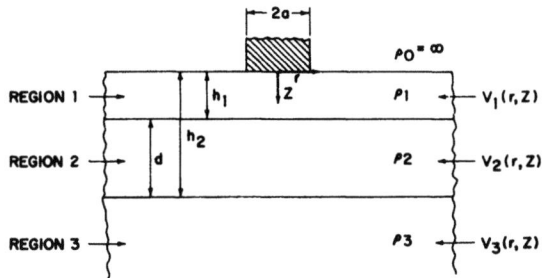

Fig. 3. Geometry of finite area contact on multilayered structure. (Source: Ref. 3)

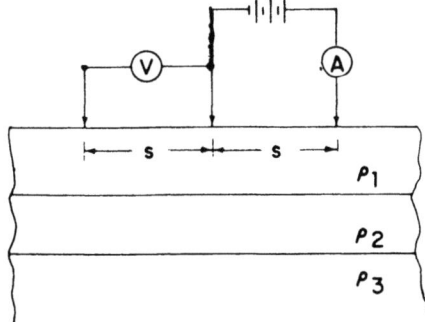

Fig. 4. Three-point spreading resistance probe.

The integration parameters $\theta_n(\ell)$ and $\psi_n(\ell)$ have to be determined from the boundary conditions at the individual layers. These boundary conditions guarantee continuity of current and potential. For a structure of N layers one is left with a 2Nx2N matrix, which has to be solved for a given value of the integration variable ℓ by means of iteration.

Hu [4] suggests that the ρ values obtained from Eq. (12) be used, for assuming a homogeneous medium as a first approximation. This procedure has to be repeated many thousands of times for different ℓ values, to give a sufficiently large set of $\theta_n(\ell)$ values needed for the final evaluation of the measurements.

In the actual experiment, however, only the surface potential distribution is measured, that is,

$$V_1(r,o) = \frac{I\rho_1}{2\pi a} \int_0^\infty (1+2\theta_1(\ell)) \sin(\ell a) \, J_o(\ell r) \, \frac{d\ell}{\ell}, \qquad (14)$$

which has been derived from Eq. (13) for $\theta_1 = \psi_1$; the relation $\theta_1 = \psi_1$ is a consequence of the boundary conditions.

Equation (14) represents the surface potential for r>a; we have also to take into consideration the potential drop due to the disc-like contact (shorting effect).

This "self-potential" is given by

$$V_s = \int_0^a \int_0^{2\pi} V_1(r,o) \, r \, dr \, d\theta \bigg/ \int_0^a \int_0^{2\pi} r \, dr \, d\theta \qquad (15)$$

$$= \frac{I\rho_1}{\pi a^2} \int_0^\infty (1+2\theta_1) \sin(\ell a) \, J_1(\ell a) \, \frac{d\ell}{\ell^2} .$$

With the relations established in Eqs. (14) and (15), we can proceed to determine the measured potential differences between the 3-point probe and 2-point probe arrangements. The situation is schematically depicted in Fig. 5.

Contacts II and III are the current probes, and contact I is the voltage-sensing one. The resulting potentials at each location can be obtained by superposition of the individual contributions. One finds

$$\phi_I = V(s,o) - V(2s,o),$$

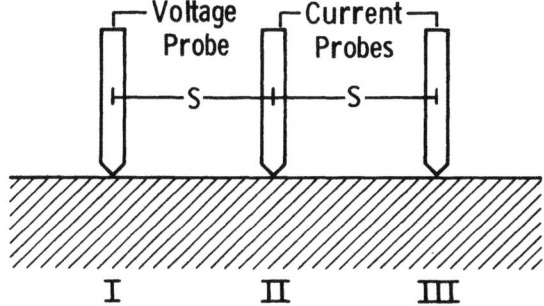

Fig. 5. A schematic of a 3-point probe configuration.

$$\phi_{II} = V_s - V(s,o),$$
$$\phi_{III} = -V_s + V(s,o). \tag{16}$$

Hence the relations for a 3-point probe setup are given by

$$\Delta V = \phi_{II} - \phi_I = V_s + V(2s,o) - 2V(s,o); \tag{17}$$

for a 2-point probe arrangement, one finds

$$\Delta V = \phi_{II} - \phi_{III} = 2\{V_s - V(s,o)\}. \tag{18}$$

We see from Eq. (15) that the probe radius a enters into Eqs. (17) and (18). It is therefore understandable that the measurements are very sensitive to any condition that influences the effective contact area. Figure 6 depicts the results of spreading-resistance experiments on homogeneous samples of known resistivity. One finds, not only deviations from the linear dependence of R_{sp} on ρ, but also dependencies upon crystallographic orientation, type of dopant, and concentration for a given probe loading (weight). It is therefore customary to establish calibration curves that determine effective probe radii for the specific conditions.

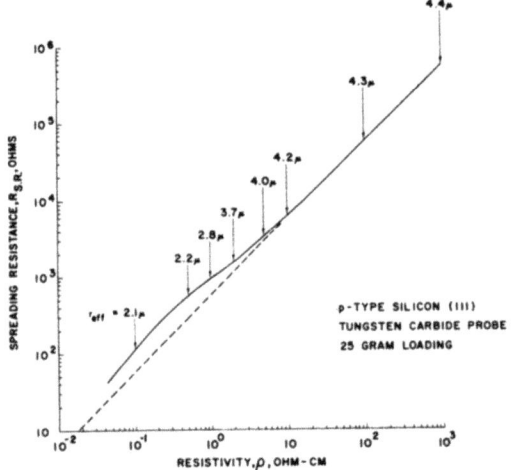

Fig. 6. Typical values of effective radius of contact, a, for p-type silicon. (Source: Ref. 3)

2.2 Capacitive Techniques

2.2.1 C-V profiling methods

In contrast to the techniques discussed in the preceding section, capacitive measurements are capable of providing information about the distribution of ionized impurities rather than free-carrier concentrations. They are commonly used to probe the space-charge regions of p-n junctions, Schottky barriers, and MIS structures.

We shall describe the basic concept and its limitations for the case of abrupt junctions and then extend it to other structures (MIS).

One generally makes the assumption that the space charge in the depletion region consists of ionized impurities only. That means, the free-carrier concentration increases abruptly at the boundary of the space-charge region from zero to the equilibrium value of the bulk material. This is an idealizing assumption and leads to severe complications for space-charge regions smaller than 2 Debye lengths (L_D), as we shall see later.

Let us consider the case, depicted schematically in Fig. 7, of a one-sided abrupt junction.

If we denote the space-charge width by W, a change, dW, will increase the charge in the depletion layer by

$$dQ = qN(W) \, dW, \tag{19}$$

leading to a potential change of

$$dV \sim W \, dE = W \, \frac{dQ}{\varepsilon \varepsilon_o}. \tag{20}$$

From Eqs. (19) and (20) we obtain

$$dV = \frac{q}{2} \frac{N(W)}{\varepsilon \varepsilon_o} dW^2. \tag{21}$$

With the definition for the differential capacitance, $dQ/dV = C_s$, one finds

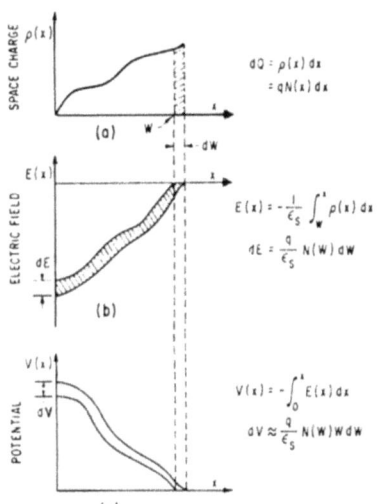

Fig. 7. Schematic illustrations of the (a) charge, (b) electric-field, and (c) potential distribution for a one-sided junction with arbitrary doping, based upon the depletion approximation. (Source: Ref. 5)

$$C_s = \frac{\varepsilon\varepsilon_o}{W} \tag{22}$$

or

$$\frac{1}{C_s^2} = \frac{W^2}{(\varepsilon\varepsilon_o)^2}$$

and

$$\frac{d\, 1/C_s^2}{dV} = \frac{1}{(\varepsilon\varepsilon_o)^2} \frac{dW^2}{dV}. \tag{23}$$

Combining Eqs. (21) and (23) leads to

$$N(W) = \frac{2}{q\varepsilon\varepsilon_o} \left(\frac{d\, 1/C_s^2}{dV}\right)^{-1} = \frac{C_s^3}{q\varepsilon\varepsilon_o} / \frac{dC_s}{dV}. \tag{24}$$

Equation (24) relates the ionized impurity concentration at a depth W to the differential capacitance and its first derivative with respect to applied voltage. It represents a good approximation for $W > 2L_D$. If W becomes smaller than $2L_D$ the contributions of the free-carrier tail give rise to erroneous evaluations. This situation has been discussed in the literature, and various correction procedures have been suggested (6,7). To illustrate the situation, we consider a uniformly doped material and calculate the real differential capacitance by including the free-carrier contribution. The differential capacitance can be derived from an exact solution of the Poisson equation. For details we refer the interested reader to Ref. 7.

The differential capacitance of a uniformly doped p material with doping concentration, N_{Ao}, is given by

$$C_s = \frac{dQ}{d\psi_s} = \frac{\varepsilon\varepsilon_o}{L_D} \frac{1-e^{-x}}{(e^{-x}+x-1)^{1/2}}, \tag{25}$$

where ψ_s represents the surface potential in MIS structures or its equivalent, $V_{build\ in} + V_{appl}$, in abrupt junctions and Schottky devices. x and L_D are defined as

$$x = \beta\psi_s = \frac{q}{kT}\psi_s, \quad (26)$$

and

$$L_D = \left(\frac{2kT\,\varepsilon\varepsilon_o}{q^2 N_{Ao}}\right)^{1/2}. \quad (27)$$

The use of the capacitance, as determined from Eq. (25) in solving Eq. (24) for the impurity concentration, results [Ref. (7)] in

$$N_{eff} = \frac{N_{Ao}}{2} \frac{(1-e^{-x})^3}{(e^{-x}+x-1)e^{-x} - \frac{1}{2}(1-e^{-x})^2}. \quad (28)$$

Figure 8 gives a typical example of the discrepancies one has to expect between the measured value, N_{eff}, and the true value, N_{Ao}.

Several techniques for measuring differential capacitances on MIS structures are described in the literature, such as

    Pulsed high-frequency C-V
    High-frequency C-V
    Second harmonic method.

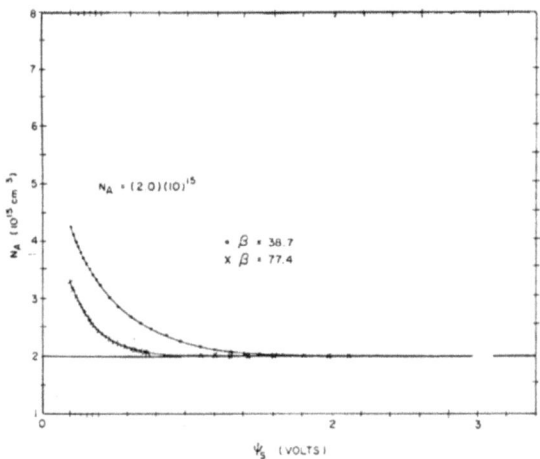

Fig. 8. The calculated impurity concentration as a function of surface potential for $\beta = 38.7$ and 77.4, and an assumed uniform concentration of 2×10^{15} cm^{-3}. (Source: Ref. 7)

Equations (19) through (28) have to be modified, for those applications, to take into account the potential drop across the insulator as well as the oxide capacitance C_{ox}.

$$dV = dV_{ox} + d\psi_s = dQ_s \cdot \frac{D_{ox}}{\varepsilon\varepsilon_o} + d\psi_s \qquad (29)$$

and

$$C = \frac{C_s C_{ox}}{C_{ox} + C_s} . \qquad (30)$$

2.2.2 Generation lifetime measurements

Minority-carrier lifetime measurements have become increasingly important in materials characterization. The frequently used Zerbst [8] technique is cumbersome to perform on a large scale. Fahrner and Schneider [9] have developed a technique which in its basic concept is closely related to pulsed C-V measurements on MIS structures. In the latter case the pulse length has to be adjusted so that no inversion layer can build up. In the generation lifetime measurements the surface layer is already in inversion and is driven by an additional voltage pulse into deeper inversion. The change of the differential capacitance is measured as the inversion layer responds to the additional voltage pulse, dV_A, and finally assumes its new equilibrium configuration. One uses the recovery process to characterize the generation lifetime.

The advantage of this technique over others is that the thermal generation is predominantly determined by bulk phenomena rather than by surface contributions (inversion screening) and the measured generation lifetime is truly a bulk parameter.

Following is the theoretical background of this technique (Ref. 9).

The applied voltage, $V_A + dV_A$, across the MIS structure is given in analogy to Eq. (29) by

$$V_A + dV_A = V_{ox} + \psi_s . \qquad (31)$$

The surface potential, ψ_s, is related to the silicon differential capacitance by Eq. (25). For the depletion approximation (x>1) we find

$$\psi_s = \varepsilon\varepsilon_o \frac{qN_A}{2} \frac{1}{C_s^2} . \qquad (32)$$

With Q_s denoting the space charge due to ionized impurities and Q_i denoting the inversion charge, one gets

$$V_{ox} = (Q_s + Q_i)/C_{ox} \qquad (33)$$

and

$$\frac{dQ_s}{dt} = C_s \cdot \frac{d\psi_s}{dt} \, . \qquad (34)$$

From Eq. (31) we obtain

$$\frac{d(V_A + dV_A)}{dt} = 0 = \frac{d\,V_{ox}}{dt} + \frac{d\psi_s}{dt} \, , \qquad (35)$$

which results, by use of Eqs. (32) - (34), in

$$q\,\varepsilon\varepsilon_o\,\frac{N_A}{2} \cdot d(1/C_s^2)/dt$$
$$+ \frac{C_s}{C_{ox}} q \frac{N_A}{2} \varepsilon\varepsilon_o\, d(1/C_s^2)dt \qquad (36)$$
$$+ (1/C_{ox}) \frac{dQ_i}{dt} = 0 \, .$$

We adopt, for the present, a thermal generation model of the form

$$d\,Q_i/dt = q\,n_i(W - W_f)/2\tau \, . \qquad (37)$$

W_f is the resulting final depletion layer width after the voltage pulse, dV_A, has been applied. The quantities C_f and C_{sf} are defined in an analogous manner. τ is the minority-carrier lifetime.

With Eq. (37), we rewrite Eq. (36):

$$\frac{C_{ox} + C_s}{C_{sf} - C_s} \frac{1}{C_s^2} dC_s = \frac{n_i}{2\tau N_A} \frac{1}{C_{sf}} dt \, . \qquad (38)$$

The high-frequency measurements are normally carried out in such a way that the total MOS capacitance C, as given by Eq. (30), is determined rather than C_{ox} or C_s. It is therefore convenient to express Eq. (38) as a function of C:

$$\frac{1}{C_R^2(1-C_R)} \frac{dC_R}{dt} = \frac{C_f}{C_{ox}} \frac{n_i}{2\tau N_A}, \tag{39}$$

where $C_R = \frac{C}{C_f}$.

The solution of Eq. (39) is given by

$$[\ln \frac{C_R}{1-C_R}] - \frac{1}{C_R} = \frac{C_f}{C_{ox}} \frac{n_i}{2\tau N_A} (t-t_o). \tag{39a}$$

2.3 Photocurrent Methods

2.3.1 Lifetime measurements

H. Reichl and H. Bernt [10] have recently described a technique for measuring minority-carrier lifetime in silicon wafers and thick epitaxial layers. The excess carriers are generated by sinusoidal photo excitation and collected at a Schottky barrier contact. The lifetime of the minority carriers can be determined either from the frequency dependence of the photocurrent or from the phase shift between the photocurrent and the exciting light (Fig. 9).

For the case of infinitely large surface recombination velocity, s, it can be shown that the phase shift, ψ, is related to the lifetime, τ, by the following relation:

$$\psi = 1/2 \arctan(\omega\tau), \tag{40}$$

where ω is the angular frequency of the modulated light.

This relation holds for sample thickness, d, larger than the diffusion length of the minority carriers.

2.3.2 Measurements of insulator bulk trapped charges

Trap centers in dielectric films can have a profound effect upon device performance and reliability. It is therefore of

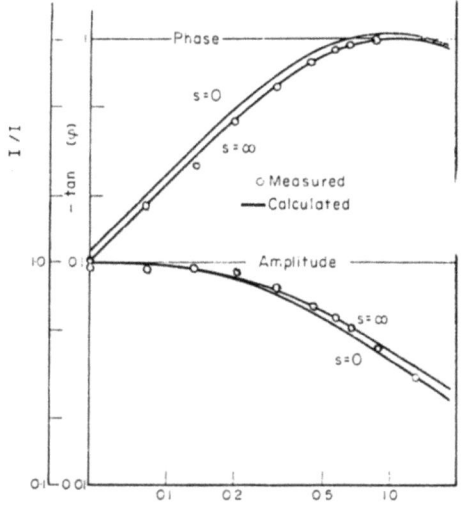

Fig. 9. Phase shift and amplitude of the photocurrent vs modulation frequency. (Source: Ref. 10)

interest to determine the density of trapped charges and their centroids. A method that permits such a characterization has recently been described by DiMaria [11].

It is based upon studies by Powell and Berglund [12,13] of photo injections into MOS structures. In analogy to the barrier-lowering of Schottky devices due to image forces (Ref. 14), Powell and Berglund developed an injection model, which allows for charges in the dielectric layer.

The effect of an image force on the effective barrier height is schematically depicted in Fig. 10.

It can readily be shown that the potential maximum occurs at a position x_o given by

$$x_o = [q/16\pi \,\varepsilon\varepsilon_o\, E(x_o)]^{1/2}, \qquad (41)$$

where $E(x_o)$ is the electric field without the image force contribution.

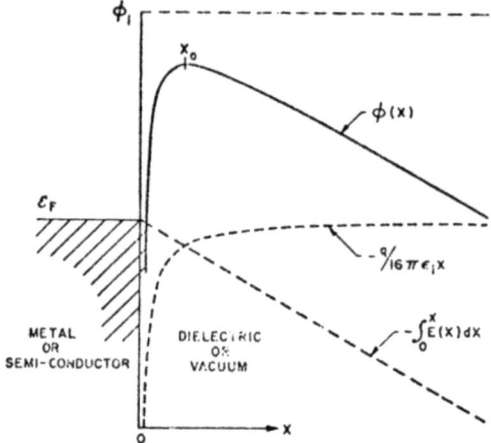

Fig. 10. Potential versus position near a metal-dielectric interface illustrating the contributions to the potential. (Source: Ref. 12)

$$E(x_o) = - \frac{1}{D_{ox}} (V_A + \phi_1 - \phi_2) - \frac{1}{\varepsilon \varepsilon_o} \frac{1}{D_{ox}}$$

$$\cdot \int_0^{D_{ox}} (D_{ox} - x) \rho(x) \, dx + \frac{1}{\varepsilon \varepsilon_o} \int_0^{x_o} \rho(x) \, dx, \quad (42)$$

where D_{ox} is the thickness of the oxide layer.

The photocurrent can be expressed as

$$I = I_o \exp(-x_o/\ell), \quad (43)$$

where ℓ is a characteristic scattering length, I_o is a function of photon energy $\hbar\omega$ and of barrier-lowering $\Delta\phi$, and p is an exponent.

$$I_o = A\hbar\omega (\hbar\omega - \phi_1 + \Delta\phi)^P. \quad (44)$$

DiMaria [11] rewrites Eq. (42) for two cases.

    1. Injection from the aluminum gate (negative voltage) x_o near the aluminum-SiO_2 interface.

2. Injection from the silicon substrate (positive gate voltage) x_o^+ near the silicon-SiO$_2$ interface.

$$E(x_o^-) = \frac{V_A^- - \phi_{MS}^- - \psi_s^-}{D_{ox}} - \frac{1}{\varepsilon\varepsilon_o}(1 - \frac{<\bar{x}>}{D_{ox}})Q + \frac{1}{\varepsilon\varepsilon_o}\int_0^{\bar{x}_o} \rho(x)dx, \quad (45)$$

and

$$E(x_o^+) = \frac{V_A^+ - \phi_{MS}^+ - \psi_s^+}{D_{ox}} + \frac{1}{\varepsilon\varepsilon_o}\frac{<\bar{x}>}{D_{ox}}Q - \frac{1}{\varepsilon\varepsilon_o}\int_0^{D_{ox}} \rho(x)dx, \quad (46)$$

where $\phi_{MS} = \phi_2 - \phi_1$, Ψ_s is the band-bending voltage, and $<\bar{x}>$ is the position of the charge centroid.

For the case in which the trapped charge is located in the bulk of the dielectric film the contribution of the last term in Eqs. (41) and (42) can be neglected. For degenerate substrate doping levels, Ψ_s can also be neglected.

According to Eqs. (41) and (42) the photocurrent depends on $E(x_o)^{1/2}$. Therefore the fields have to be identical before and after bulk charging for a given photocurrent; this means

$$E_a^{\pm}(x_o) = E_b^{\pm}(x_o).$$

For the voltage shifts ΔV_A^- and ΔV_A^+ (Fig. 11), one obtains

$$\Delta V_A^- = V_{A_b}^- - V_{A_a}^- = \frac{1}{\varepsilon\varepsilon_o}(D_{ox} - <\bar{x}>)Q \quad (47)$$

and

$$\Delta V_A^+ = V_{A_b}^+ - V_{A_a}^+ = \frac{-<\bar{x}>}{\varepsilon\varepsilon_o}Q. \quad (48)$$

From Eqs. (47) and (48) one readily gets:

$$Q = \frac{\varepsilon\varepsilon_o}{D_{ox}}(\Delta V_A^- - \Delta V_A^+) \quad (49)$$

and

$$\frac{<\bar{x}>}{D_{ox}} = [1-\Delta V_A^-/\Delta V_A^+]^{-1}. \tag{50}$$

3. MASS ANALYTICAL TECHNIQUES

Amongst the mass-sensitive techniques that have obtained widely accepted use to analyze composition and interface reactions, we shall discuss two relatively new ones:

Rutherford Backscattering (RBS)
Secondary Ion Mass Spectrometry (SIMS)

3.1 Rutherford Backscattering

The RBS technique is capable of providing information on (a) mass species, (b) concentration and (c) depth scale. The major advantages of the method are that those quantities are obtained by absolute measurements and that the profiling does not require layer removal, such as SIMS, Auger, and other surface measurement techniques.

One limitation, however, is the limited sensitivity of the technique. Arsenic in silicon, for instance, can be measured only at concentrations of about 2×10^{-19} cm^{-3} and higher.

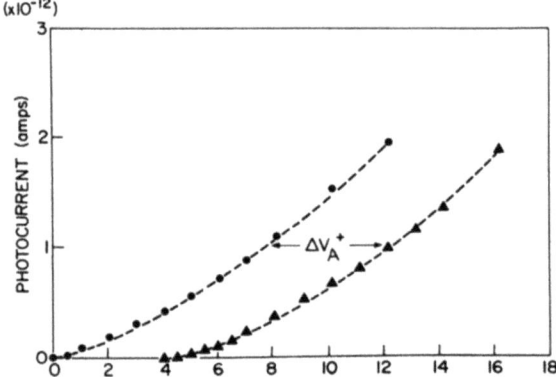

Fig. 11. Photocurrent for a MOWOS structure as a function of applied gate voltage before (solid circles) and after (solid triangles) partial electronic charging of the W layer. Gate voltage polarity was positive and the Si-SiO$_2$ (thermal) is the injecting interface. (Source: Ref. 11)

The experiments are frequently carried out using H^+ or He^+ ions, ranging from an energy of 0.3-3 MeV. Low-energy backscattering is possible by means of electrostatic analyzers.

For a scattering event involving an angle Θ between the original particle direction and the direction after the scattering event (Laboratory coordinates), one finds the following relation:

$$\frac{E'}{E_o} = k_M = \frac{M_1 \cos\Theta + \sqrt{M_2^2 - M_1^2 \sin^2\Theta}}{M_1 + M_2} . \qquad (51)$$

The mass of the scattering target atom (M_2) can therefore be determined from E', E_o, and Θ.

The depth scale for the investigated layer can be obtained from the following considerations (Fig. 12). A projectile entering a target under angle θ_1 with energy E_o has an energy E at depth t according to the relation

$$E = E_o - \int_0^{t/\cos\theta} \left(\frac{dE}{dx}\right) dx . \qquad (52)$$

After a scattering event at depth t resulting in an angle θ_2, the backscattered projectile leaves the target with an energy

$$E_1 = kE - \int_0^{t/\cos\theta_2} \left(\frac{dE}{dx}\right) dx . \qquad (53)$$

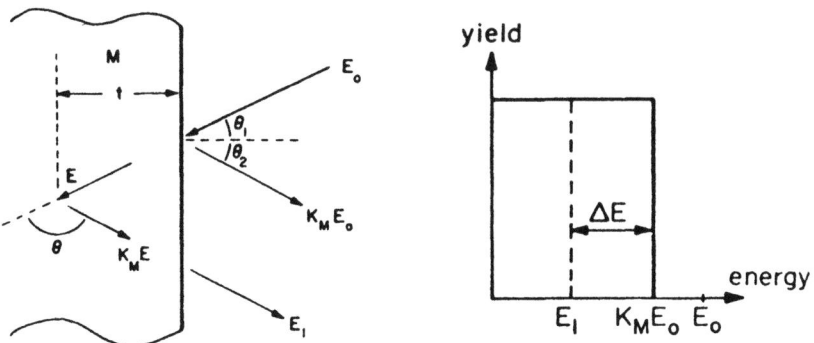

Fig. 12. Backscattering geometry, notation and schematic energy spectrum. (Source: Ref. 15)

We find, therefore, an energy difference $\Delta E = kE_o - E_1$ between a projectile that has been scattered at the surface and a projectile that has been scattered at depth t.

Using Eqs. (52) and (53), we get

$$\Delta E = k \int_0^{t/\cos\theta_1} \left(\frac{dE}{dx}\right) dx + \int_0^{t/\cos\theta_2} \frac{dE}{dx} dx. \qquad (54)$$

In case dE/dx is varying only slowly (thin films), ΔE can be expressed by a linear approximation:

$$\Delta E = \left(\frac{K}{\cos\theta_1} \frac{dE}{dx}\bigg|_{E_o} + \frac{1}{\cos\theta_2} \frac{dE}{dx}\bigg|_{kE_o}\right) t = [s]\, t, \qquad (55)$$

where $[s] = \frac{k}{\cos\theta_1} \frac{dE}{dx}\bigg|_{E_o} + \frac{1}{\cos\theta_2} \frac{dE}{dx}\bigg|_{kE_o}$. $\qquad (56)$

Figure 13 depicts schematically the dependence of dE/dx on particle energy and indicates the errors made by a linear approximation.

The differential cross section $d\sigma/d\Omega$ is an essential parameter for the qualitative analysis and can be obtained from the classical treatment of Rutherford scattering:

$$\frac{d\sigma}{d\Omega} = \frac{Z_1 Z_2 e^2}{2E \sin^2\theta} \cdot \frac{\{\cos\theta + [1 - (\frac{M_1}{M_2}\sin\theta)^2]^{1/2}\}^2}{[1 - (\frac{M_1}{M_2}\sin\theta)^2]^{1/2}}. \qquad (57)$$

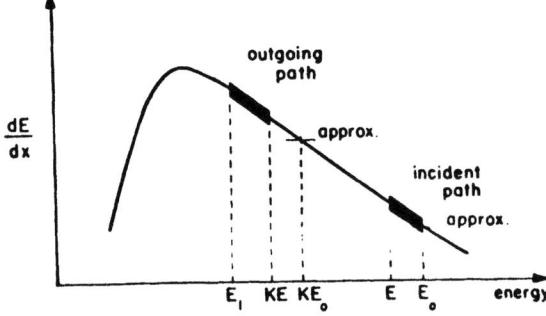

Fig. 13. Schematic diagram of dE/dx versus E and regions used for backscattering analysis. (Source: Ref. 15)

From Eq. (52) one can obtain the average scattering cross section, determined by the acceptance angle, Ω, of the detector arrangement:

$$\sigma = \frac{1}{\Omega} \int \left(\frac{d\sigma}{d\Omega}\right) d\Omega \ . \tag{58}$$

The detector itself has an energy resolution, δE_1, which corresponds to a depth uncertainty of δx, with

$$\delta E_1 = k E_o - E_1 = [s] \, \delta x = N \cdot [\epsilon] \, \delta x, \tag{59}$$

where $[\epsilon] \cdot N = [s]$ and N is the number of target atoms per unit volume.

The height, H, of the spectrum (backscattering yield per channel) depends upon the total number of projectiles, Q, the solid angle of the detector system, Ω, the average differential scattering cross section, Θ, and the total number of target atoms per unit area, $N\delta x$ [with δx corresponding to detector resolution δE_1, Eq. (59)].

One finds

$$H = Q\sigma\Omega N \delta x \tag{60}$$

and, with Eq. (59),

$$H = Q\sigma\Omega N \frac{\delta E_1}{[s]} = Q\sigma\Omega \frac{\delta E_1}{[\epsilon]} \ . \tag{61}$$

We shall now apply these relations to the quantitative analysis of a thin film ($E_o \gg \Delta E$; dE/dx is nearly constant). Therefore,

$$\Delta E = [s]t = N[\epsilon]t \ . \tag{62}$$

If we define the area A of the backscattering yield curve (Fig. 14) as the sum of the counts in each channel, then we find

$$A = H \cdot \frac{\Delta E}{\delta E_1} = Q\sigma\Omega Nt \ . \tag{63}$$

The area under the curve, therefore, is directly proportional to the area density of the atoms in the film.

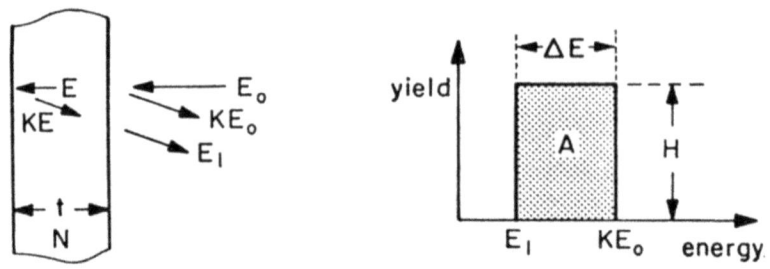

Fig. 14. Thin elemental film analysis. (Source: Ref. 15)

One can avoid tedious calibration procedures to determine Q and Ω for a given experiment by using the known substrate parameters. From Eq. (61) we get, for the substrate yield,

$$Q\Omega = \frac{H_M \, [\varepsilon]_M}{\sigma_M \, \delta E_1} \, , \qquad (64)$$

and, consequently, Eq. (63) can be expressed as

$$A_i = (Nt)_i \, \frac{H_M \, \sigma_i \, [\varepsilon]_M}{\delta E_1 \, \sigma_M} \, . \qquad (65)$$

In a similar way one can measure impurity distributions within the substrate and compositions of the target. For details we refer the interested reader to Ref. 15.

Figure 15 gives a typical example for the compositional analysis of a silicon nitride film prepared at various ratios δ_o of ammonia to silane. Additional examples relating to the application of backscattering in silicon are found in Ref. 17.

3.2 Secondary Ion Mass Spectrometry Measurements

The SIMS technique uses ion sputtering to generate the secondary ions and is therefore a destructive analytical tool in contrast to Rutherford backscattering. It is an excellent method for profiling impurity distributions in silicon, particularly bipolar transistor structures, provided that appropriate precautions are taken.

Being an ion mass-spectrometry technique, the analysis is sensitive only to the ratio of ion mass to charge, M/q; any

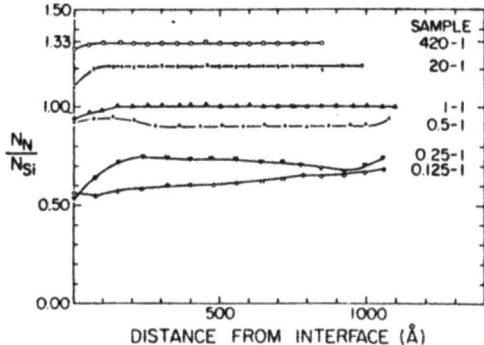

Fig. 15. Distribution of N_N/N_{Si} for samples prepared with different δ_o values. (Source: Ref. 16)

molecule with identical M/q value, therefore, has to interfere with the analysis. This problem can be minimized, however, by selection of a suitable impurity complex (e.g., $X_x Y_y$) rather than the impurity element itself (e.g., X) once the composition or better mass-to-charge ratio of the interfering species is known. Another source of error can arise from the sputtering yield. First, is is well known that the sputtering yield is very sensitive to the matrix itself. Increased sputtering yields have been reported for the initial phases of the sputtering process, in cases, for instance, where the surface was covered by a natural oxide film. Therefore, one frequently uses oxygen ions as the primary species for the sputtering to keep the sputtering yield constant.

Another sputtering-related error can arise from compositional changes due to sputtering itself. Chu and Howard [18] have reported marked differences in the data on compositions of Al-Cu couples when measurements were done either by RBS or by Auger spectroscopy. In the latter case they used sputtering as a means of depth profiling. One would expect that for a homogeneous composition, even under the condition of widely varying sputtering yields of the various constituents, the equilibrium values of the sputtered ions would reflect the exact composition. For extreme gradients in compositions, however, one can see that it will be impossible ever to reach quasi-equilibrium. This fact could lead to major errors in the profile data.

Figure 16 gives a typical SIMS profile for the As and B distribution in an npn transistor. One can clearly discern the dip in the boron profile at the emitter-base junctions and see how

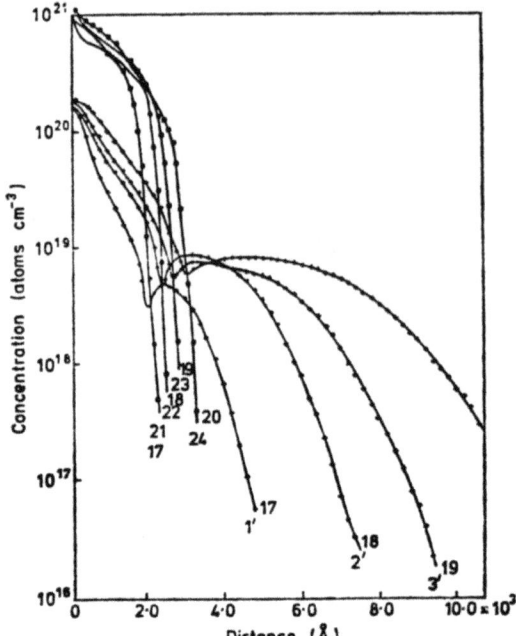

Fig. 16. Arsenic and boron distribution for typical bipolar emitter-base structures. (Source: Ref. 19)

it shifts with increasing diffusion time. The detailed description of this phenomenon is found in Section 7 of the lecture on Diffusion Phenomena in Silicon.

ACKNOWLEDGMENT

I should like to thank my colleagues A. E. Michel, T. H. Yeh, S. M. Hu, and W. K. Chu for many helpful discussions and for their valuable critique of the manuscript.

REFERENCES

1. J. C. Irvin, Bell Syst. Tech. J. 41, 387 (1962).
2. A. E. Michel, Private communication (to be published).
3. P. A. Schumann and E. E. Gardner, J. Electrochem. Soc. 116, 87 (1969). Solid-State Electron. 12, 371 (1969).
4. S. M. Hu, Solid-State Electron 15, 809 (1972).
5. C. P. Wu, E. C. Douglas, and C. E. Mueller, IEEE Trans. Electron Devices, ED-22, 319 (1975).

6. K. Ziegler, E. Klausmann, and S. Kar, Solid-State Electron. 18, 189 (1975).
7. A. R. LeBlanc, D. D. Kleppinger, and J. P. Walsh, J. Electrochem Soc., 119, 1068 (1972).
8. M. Zerbst, Z. Angew. Phys. 22, 1, 30 (1966).
9. W. R. Fahrner and C. P. Schneider, in Technical Report No. 5, ARPA Contract No. DAHC15-72-C-0274, December 1974, p. 33.
10. H. Reichl and H. Bernt, Solid-State Electron 18, 453 (1975).
11. D. J. DiMaria, J. Appl. Phys. 47, 4073 (1976).
12. C. N. Berglund and R. J. Powell, J. Appl. Phys. 42, 573 (1971).
13. R. J. Powell and C. N. Berglund, J. Appl. Phys. 42, 4390 (1971).
14. S. M. Sze, Physics of Semiconductor Devices (Wiley-Interscience, A Division of J. Wiley and Sons, New York-London, 1969), p. 364.
15. W. K. Chu, "Materials Analysis by Nuclear Backscattering" in New Uses of Ion Accelerators, Edited by J. F. Ziegler, (Plenum Press, New York, 1975), Chap. 2C.
16. J. Gyulai, O. Meyer, and J. W. Mayer, J. Appl. Phys. 42, 451 (1971).
17. W. K. Chu, J. W. Mayer, M. A. Nicolet, T. M. Buck, G. Amsel, and EF. Eisen, in Semiconductor Silicon 1973, edited by H. R. Huff and R. R. Burgess, The Electrochemical Society Softbound Symposium Series (Princeton, N. J., (1973), p. 416.
18. W. K. Chu, J. K. Howard, and R. F. Lever, J. Appl. Phys. 47, 4500 (1976).
19. M. Bonis, et al., Abstract 266, p. 658, The Electrochemical Society Extended Abstracts, Fall Meeting, Miami Beach, Florida, 1972.

APPENDIX A: INCREMENTAL SHEET-CONDUCTIVITY MEASUREMENTS

Let us assume that we have a sample as depicted schematically in Fig. A1, which is composed of n layers (Δx_i) of differing carrier concentrations, n_i, and mobilities, μ_i. The current in the ith filament under the influence of an external voltage, U_c, is then given by

$$I_i = \Delta x_i \, W \, q \, n_i \mu_i (U_c/L), \tag{A1}$$

which leads to a conductance for the ith filament of

$$G_i = (\Delta x_i \, W \, q \, n_i \mu_i)/L, \tag{A2}$$

and the sheet conductance ($W/L \equiv 1$) of a structure, consisting of

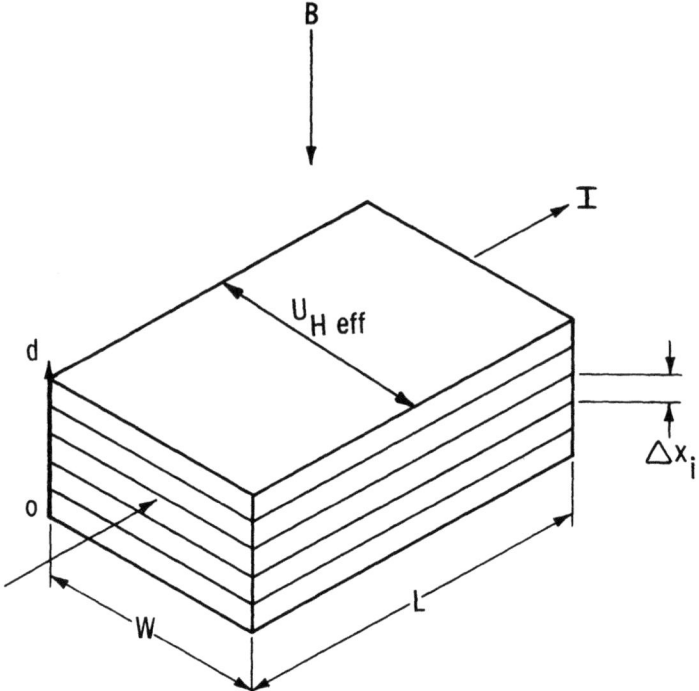

Fig. A1. Schematic of a layered structure.

layers i, i+1,...n, is given by

$$\sigma_{si} = \sum_{j=i}^{n} \Delta x_j \, q \, n_j \, \mu_j \qquad (A3)$$

or, for the limit $\Delta x \to 0$,

$$\sigma_{si} = q \int_0^{d_i} n(x) \, \mu(x) \, dx, \qquad (A4)$$

where

$$d_i = \sum_{j=i}^{n} \Delta x_j \, .$$

After removal of the ith incremental layer (Δx_i), one obtains from the sheet conductivity values, measured before and after the removal,

$$\sigma_{s_i} - \sigma_{s_{i+1}} = \Delta x_i \, q \, n_i \, \mu_i = G_i, \qquad (A5)$$

where σ is the average or effective <u>bulk</u> conductivity in the removed layer,

and for the limit $\Delta x \to 0$

$$\frac{d\,\sigma_s}{dx} = d\,\frac{1/\rho_s}{dx} = \sigma(x). \tag{A6}$$

APPENDIX B: INCREMENTAL HALL EFFECT MEASUREMENTS

The Hall voltage generated in the ith layer for the conditions depicted in Fig. A1 is given by

$$U_{Hi} = (R_{Hi}\,B\,I_i)/\Delta x_i \tag{B1}$$

$$= (\mu_{Hi}/\sigma_i)\,B\,(I_i/\Delta x_i),$$

where $\mu_H = r\,\mu$.

Equation (B1) can be rewritten as

$$U_{Hi} = r\,\mu B \cdot W \cdot (U_c/L). \tag{B2}$$

The Hall voltage, $U_{H\,eff}$, which can be determined experimentally for a layered structure, represents an average value of all Hall voltages, U_{Hi}, of the n layers.

There has to be a transverse current (transverse to the directions of the applied external electric field and of the magnetic field) in each of the n layers because of the incomplete compensation of the Lorentz forces by the Hall field ($U_{H\,eff}/W$).

The compensation current in the ith layer is given by

$$I_i^{comp}/L\Delta x_i = (U_{Hi} - U_{H\,eff}) \cdot \sigma_i/W . \tag{B3}$$

For the total system the compensation currents have to cancel out; that is,

$$\sum_i^n I_i^{comp} = 0 \tag{B4}$$

or

$$\sum_{j=1}^{n} (U_{H_j} - U_{H\,eff}) \sigma_j \Delta x_j = 0 . \tag{B5}$$

For limit $\Delta x_i \to 0$ one therefore obtains

$$\int_0^d U_H(x) \sigma(x) dx = U_{H\,eff} \int_0^d \sigma(x) dx \tag{B6}$$

or

$$U_{H\,eff} = \int_0^d U_H(x) \sigma(x) dx / \int_0^d \sigma(x) dx . \tag{B7}$$

Using Eqs. (A1) and (B2), one finds

$$U_{H\,eff} = \frac{r B (U_c/L) W \int_0^d n(x) \mu^2(x) dx}{\int_0^d n(x) \mu(x) dx} . \tag{B8}$$

We can further define an average mobility, using Eqs. (A1) and (B1):

$$\bar{\mu}_{eff} = U_{H\,eff}/C = \frac{\int_0^d n(x) \mu^2(x) dx}{\int_0^d n(x) \mu(x) dx} , \tag{B9}$$

where $C = rWB(U_c/L)$.

The mobility of the ith layer is obtained from the following electrical parameters, where subscript i denotes the values measured <u>prior</u> to the removal of the ith layer and i+1 denotes the values <u>after</u> the removal:

$$\{U_{H\,eff}^i \cdot \sigma_i - U_{H\,eff}^{i+1} \sigma_{i+1}\}/C$$

$$= q n_i \mu_i^2 \Delta x_i = \mu_i \Delta \sigma_{s_i} \tag{B10}$$

$$= \mu_i \{\sigma_{s_i} - \sigma_{s_{i+1}}\},$$

which can be rewritten as

$$\mu_i = \frac{U_{H\,eff\,i}^i/C - U_{H\,eff}^{i+1}/C \cdot (\sigma_{s,i+1}/\sigma_{s,i})}{1 - \sigma_{s,i+1}/\sigma_{s,i}} . \tag{B11}$$

FUNDAMENTAL LIMITS IN INTEGRATED CIRCUITS

D. Widmann

Research Laboratories, Siemens AG,
D-8000 München 70, F.R. Germany

ABSTRACT. The limits for the size and the operating frequency of silicon-based digital devices and circuits are discussed considering the limiting factors given by material properties, fabrication processes and circuit operation conditions.

1. INTRODUCTION

Since the manufacturing of the first integrated circuits in 1959 the complexity of integrated circuits has about doubled every year. Moore /1/ has pointed out that this growth can be divided in contributions from three fields: Increase of the chip area, reduction of dimensions and device and circuit cleverness (Fig. 1). Considering the evolution of the minimum dimensions of integrated circuits in production (Fig. 2), a reduction of the minimum dimensions by a factor of 2 every 5 years can be stated. Similar improvements have been obtained regarding speed and power consumption of logic circuits.
  The question arises how long the trend toward higher complexity, smaller dimensions, larger chip size and higher speed can continue, and which are the factors limiting further progress.
  The present paper presents simplified considerations from which possible limitations of future integrated circuits can be derived. The analysis will be limited to silicon-based digital circuits with active elements based on pn-junctions. Fundamental limits which are not important because they are far beyond other limits, such as the limitations given by the atomic distance in the silicon lattice, will not be considered.
  In the first part (Sections 2, 3 and 4) the limitations due to material properties and fabrication processes as well as cir-

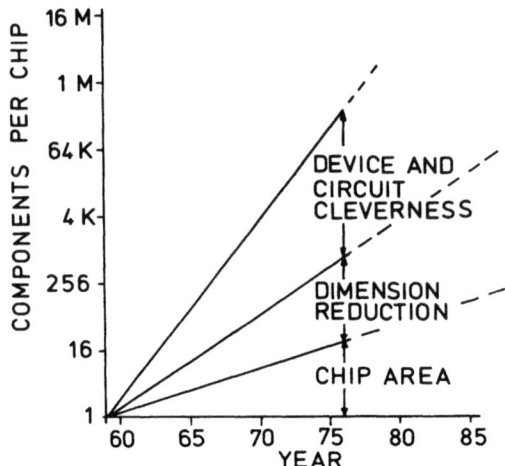

Fig. 1. Evolution of the complexity of integrated circuits in recent years (after Moore /1/). The contributions of increasing chip area, reduction of dimensions and device and circuit cleverness are indicated.

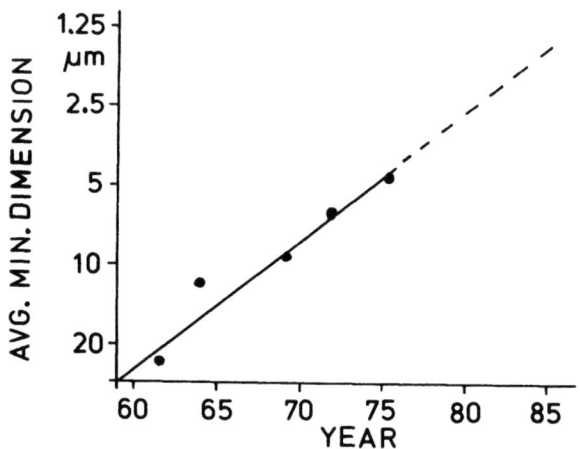

Fig. 2. Evolution of the average minimum dimensions of integrated circuits in production (after Moore /1/).

cuit operation conditions will be discussed. The consequences of these limitations for the size and operating frequency of integrated circuits will be presented in the second part (Sections 5 and 6).

The analysis in the present paper is based on the publications of Hoeneisen and Mead /2, 3/, Wallmark /4/, Keyes /5/, Folberth and Bleher /6/ and Garbrecht and Stein /7/.

2. LIMITATIONS DUE TO MATERIAL PROPERTIES

Some physical properties of the materials involved in silicon technology, i.e. silicon, silicon dioxide and aluminum, give rise to certain limitations. Defects in materials may influence the yield of integrated circuits (see Section 3.4), however it is not reasonable to assume a lower limit for the achievable defect density, since the quality of materials is steadily being improved. Material defects are therefore not considered in the following.

2.1. Band Structure and Fermi Levels in Silicon

The bandgap of silicon, i.e. the energy difference between the lower edge of the conduction band and the upper edge of the valence band, is 1,12 eV at room temperature. The Fermi levels of p-doped and n-doped regions involved in integrated circuits are near the valence band edge and the conduction band edge, respectively. Across a pn-junction a so-called built-in voltage is formed which is the difference between these two Fermi potentials, i.e. the built-in voltage is in the 0.5V - 0.8V range. In order to drive currents in the order of µA or mA across pn-junctions, applied forward voltages near the built-in voltage, which is also called a "diode drop", are required. Also the potential difference which is needed across the source-substrate pn-junction of an MOS transistor to turn on the transistor corresponds to about the built-in voltage.

For applied voltages V below the built-in voltage (this range is called the subthreshold region for MOS transistors) the current density j across the pn-junction is approximately given by

$$j \sim \exp\left(\frac{qV}{kT}\right) \qquad (1).$$

V is positive if it is applied in the forward direction of the pn-junction.

The depletion layer thickness W_{DEPL} of a one-sided n^+p step junction is

$$W_{DEPL} = \sqrt{\frac{2\varepsilon_0 \varepsilon_{si}(V_{BI} - V)}{qN_A}} \qquad (2),$$

where V_{BI} is the built-in voltage and N_A is the impurity concentration in the lower doped p side of the n^+p-junction.

The capacitance of an n^+p-junction is given by

$$C = \sqrt{\frac{q\varepsilon_o \varepsilon_{Si} N_A}{2(V_{BI}-V)}} \qquad (3).$$

For high doping levels the bandgap is supposed to become narrower /8/. At 10^{19} cm^{-3} the bandgap narrowing is about 50 mV and at 10^{20} cm^{-3} it is about 100 mV. This effect reduces the efficiency of highly doped emitters of bipolar transistors.

2.2. Breakdown Voltage of pn-Junctions

If the electric field in a pn-junction reaches a critical limit, breakdown due to avalanche or field emission effects takes place. The critical fields and the breakdown voltages of one-sided step pn-junctions are listed in Table 1 for different doping concentrations in silicon /2/. Increasing the doping concentration at the low-doped side of a pn-junction above about 10^{18} cm^{-3} severely restricts the allowable voltages.

Doping concentration (cm^{-3})	Critical field (Vcm^{-1})	Breakdown voltage (V)
10^{17}	7×10^5	14
10^{18}	1.5×10^6	6
10^{19}	2×10^6	1
10^{20}	6×10^6	0.1

Table 1. Critical electric field and breakdown voltage of plane one-sided step pn-junctions at various doping concentrations in silicon (after /2/).

Junction Curvature (μm)	Doping concentration (cm^{-3})		
	10^{17}	10^{18}	10^{19}
1	1.1	1.0	1.0
0.1	1.9	1.1	1.0
0.01	4.5	1.9	1.1

Table 2. Factors by which the breakdown voltages of curved one-sided step pn-junctions are lowered compared to plane junctions.

If the pn-junctions are curved the critical field is reached at lower applied voltages /9/. In Table 2 the factors are listed by which the breakdown voltages are lowered compared to plane junctions.

2.3. Electron Mobility in Silicon

The mobility μ_e of electrons is defined by the equation:

$$j = qn\mu_e F \qquad (4)$$

where n is the concentration of free electrons and F is the electric field.

At room temperature μ_e is about 1400 $cm^2V^{-1}s^{-1}$ in silicon. This value is reduced
- if the impurity concentration exceeds $10^{16} cm^{-3}$ (at a concentration of $10^{19} cm^{-3}$ the mobility is only 100 $cm^2V^{-1}x^{-1}$),
- if the electric field exceeds 10^4 Vcm^{-1} (the electron drift velocity $v_e = \mu_e F$ then approaches the limit value of about $10^7 cm\ s^{-1}$),
- and if the electrons form an inversion layer (at typical charge densities of 10^{11} to 10^{12} electrons per cm^2 the electron mobility is about 500 $cm^2V^{-1}s^{-1}$).

2.4. Minimum Resistivity of Doped Silicon

The concentration of active dopants in silicon is limited by the solid solubility of the dopants. This doping concentration limit is near $10^{21} cm^{-3}$ for arsenic, phosphorus and boron. The corresponding minimum resistivity of single crystal silicon is about 10^{-4} Ωcm. Polycrystalline silicon films can be doped to a minimum resistivity of about 3×10^{-3} Ωcm.

The contact resistance between aluminum and heavily doped n^+ regions is in the order of $4 \times 10^{-7} \Omega cm^2$.

2.5. Doping Fluctuations in Silicon

If the average number of doping atoms in a given volume is N, then the standard deviation from N can be assumed to be $\sqrt{N}$, according to Shockley /10/.

2.6. Thermal Conductivity of Silicon

The thermal conductivity constant of silicon is k_H = 1,5 $Wcm^{-1}°C^{-1}$. As the volumes in which heat is dissipated during operation of semiconductor devices are generally very small, these parts of the silicon crystal may assume a higher temperature than the average temperature of the silicon chip. For instance, if the area of the emitter of a bipolar transistor is 1 μm x 1 μm, the thermal resistance R_{th} between the transistor and its environment can be calculated to about 3000° per Watt.

The thermal resistance for the heat flow from the front side to the backside of a chip is not critical (< 1° per Watt).

2.7. Minimum Oxide Thickness

The gate insulator thickness of MOS transistors has a lower limit of about 5 nm. Below this thickness electrons can flow through the oxide due to the tunneling effect.

2.8. Maximum Current Density in Conductors

At high current densities in conductors lattice atoms can migrate in the direction of the electron flow. Due to inhomogeneities of the material this electromigration effect may lead to interruptions in the metal interconnections of integrated circuits.

Black /11/ has given the following relation for the median time to failure τ at current density j and temperature T:

$$\tau \sim \frac{wt}{j^2} \exp\left(\frac{E_A}{kT}\right) \qquad (5).$$

W and t are the width and the thickness of the conductor stripe, respectively. E_A is an activation energy.

Scoggan /12/ has shown that the linear dependence of τ on the conductor width w is only valid if w is larger than about the average grain size in the conductor. For smaller widths τ doesn't depend significantly on w.

Using pure aluminum for the interconnections the current density should be kept substantially lower than 10^5Acm^{-2} in order to ensure sufficient reliability. If other metals such as copper are added to aluminum /13/ the allowable current density can be increased by about a factor of 5.

Further improvements may be achieved by using other metallization systems such as Cr-Ag-Cr /11/. The upper limit for the current density in conductors is assumed to be about 10^6Acm^{-2} taking into account elevated temperatures up to about 150°.

3. LIMITATIONS DUE TO FABRICATION PROCESSES

3.1. Deviations from Nominal Values of Parameters

In manufacturing of integrated circuits certain errors of the technological parameters such as film thicknesses, linewidths of structures, edge-to-edge positions and doping concentrations must be taken into account. If a normal distribution of the actual magnitude of a parameter, say the channel length of a MOS transistor, is assumed with a standard deviation σ, the fraction S of the transistors with a channel length L deviating from the mean value L_o by more than a given tolerance ε is obtained from

$$S = 1 - \frac{1}{\sigma}\sqrt{\frac{2}{\pi}} \int_{x=0}^{x=\varepsilon} \exp\left(-\frac{x^2}{2\sigma^2}\right) dx \quad (6).$$

x is the variable indicating the deviation from the mean value L_o of the channel length.

As an example, let us assume a nominal channel length $L_o = 1\mu m$, a standard deviation $\sigma = 0.1~\mu m$ and a required tolerance $\varepsilon \stackrel{\circ}{=} 0.3\mu m$. Using equation (6) the fraction S is 0.27% in this case. This means that one transistor out of 370 transistors can be assumed to have a channel length which is outside the tolerance range.

3.2. Limits of Pattern Generation

Considering optical, x-ray and electron beam methods for pattern generation in resist films the following limits can be assumed for the manufacturable minimum dimensions:
- Exposure with ultraviolet light has a limit at 0.5 ... 1 µm.
- The resolution limit of x-ray exposure is assumed to be given by the range of secondary electrons in the resist. This range is about 40 nm in PMMA resist using a wavelength of 0.8 nm /14/.
- The resolution limit of electron beam exposure is caused by electron backscattering from the substrate. The minimum linewidth achievable with electron beam exposure on silicon wafers is about 0.3 µm at an electron energy of 20 keV and a resist thickness of about 0.5 µm /15/. If the influence of the substrate is reduced by using thin substrates the resolution can be increased. Broers /16/ has recently demonstrated the generation of 8 nm lines and spaces on a 10 nm thick carbon foil.

It is important to note that the limits indicated above refer to plane substrates. If there are steps on the wafer surface the limit values may be somewhat higher.

Although there are pattern definition techniques available which allow the transfer of resist patterns into oxide or metal patterns without any change of the linewidths of the pattern details, it is difficult to delineate grooves whose width is smaller than their depth. Similarly, the width of oxide or polysilicon or metal stripes cannot be made substantially smaller than their thickness.

Overlay precision may be another important factor limiting the packing density of circuit elements. In order to draw full advantage out of the resolution capability of pattern generation systems for densely packed integrated circuits, it is necessary that the edge-to-edge error of two edges belonging to different mask levels is smaller than about half the minimum linewidth. Overlay errors of not more than a few tenths of a micron seem to be very difficult to achieve for full-wafer exposure techniques considering the large number of factors contributing to the overlay error, such as temperature variations, in- plane wafer distortions, linewidth errors, and so on. With step-and-repeat exposure techniques which allow several realignments during exposure of an entire wafer surface an overlay precision of a few tenths of a micron seems to be possible.

3.3. Limitations for Doping Profiles

While doping profiles generated by standard predeposition and drive-in techniques are substantially Gaussian profiles with the maximum doping concentration at the silicon surface, epitaxial growth of doped silicon layers and ion implantation of dopants into silicon substrates offer the possibility of generating a large variety of profiles. The following limitations must however be considered:
- Using standard techniques epitaxial growth of silicon is performed at temperatures around $1000°$ in order to obtain high-quality layers. Solid-state diffusion and out-diffusion of dopants are not negligible at such temperatures. Even with slowly diffusing dopants, such as antimony or arsenic, steep profiles cannot be achieved with transition regions smaller than a few tenths of a micron.
- It has been demonstrated that with molecular beam epitaxy which can be performed at lower temperatures extremely steep profiles are feasible /17/. This technique is however actually in a research stage.
- Due to scattering in the silicon lattice the vertical and lateral spread of ion implanted impurities may be significant. The range of some implanted ions may be substantially different from the so-called projected range R_p /18/. For instance, the pn-junction between a heavily doped implanted region and a moderately doped background region may be distant from the projected range by more than R_p.

Any high-temperature process following the impurity doping process leads to a redistribution of the impurities, which in most cases means a further spread of the impurity atoms.

3.4. Economic Limitations

In practice cost is perhaps the most important limitation on the way toward better performance and/or higher complexity of integrated circuits. For instance, a new pattern generation technique allowing smaller dimensions will not be introduced into production unless this new technique yields cheaper devices or unless the increase in cost is balanced by the gain in performance. An optimum complexity of the integrated circuits to be produced can be attributed to any existing production line depending on the progress in technology know-how.

4. LIMITATIONS DUE TO CIRCUIT OPERATION CONDITIONS

4.1. Temperature of Operation

The temperature range over which an integrated circuit shall be operated is important because most material parameters are dependent on temperature. The "worst-case" conditions which normally refer to the highest operation temperature have to be considered for the layout of the integrated circuit.

For MOS integrated circuits operation at cryogenic temperatures offers some advantages /19/ due to the higher electron mobility and the sharper turn-on characteristics of the MOS transistors.

4.2. Heat Removal

With air cooling generated heat up to about 2 Wcm^{-2} can be removed from an integrated circuit. Using liquid cooling heat removal can be increased to about 20 Wcm^{-2} /5/.

5. MINIMUM SIZE OF SEMICONDUCTOR DEVICES

5.1. Minimum Size of MOS Devices

The electrical properties of MOS transistors with reduced dimensions can be predicted very easily if Dennard's scaling principle /20/ is used. The approach is to scale down all horizontal and vertical dimensions as well as the resistivities and the operating voltages by the scaling factor α. To a first approximation the result is as follows:
- Electric fields and power dissipation per unit area remain unchanged
- Currents and delay times are reduced by the factor α
- The size of the transistors and the power per transistor is reduced by the factor α^2.

$\alpha = 1$ corresponds to the parameters of a conventional n-channel MOS transistor, characterized by the following parameters:

- Minimum dimensions   5 µm
- Gate oxide thickness   0.12 µm
- Source and drain junction depths   1.2 µm
- Substrate doping 3×10^{15} cm^{-3}
- Drain voltage   10 V
- Substrate bias   -5 V

The scaling approach neglects the built-in voltage which varies only slightly during scaling (Section 2.1.). Taking into account the built-in voltage (about 0.6 V), the widths of the depletion layers of the source-substrate and the drain-substrate pn-junctions (equation (2)) establish a lower limit for the distance between both pn-junctions, i.e. for the channel length L of the MOS transistor (punch-through condition):

$$L_{min} \approx W_{DEPL}(\text{Source}) + W_{DEPL}(\text{Drain}) \qquad (7)$$

Using the scaling principle the minimum channel length L_{min} given by equation (7) is reached when

$$L_{min} \approx \text{minimum dimension} \qquad (8)$$

Based on the parameters corresponding to $\alpha = 1$ (see above) this situation is reached for $\alpha \approx 10$. In this case N_A is 3×10^{16} cm^{-3}, which, according to equation (2), means that W_{DEPL} (Source) is 0.2 µm and W_{DEPL} (Drain) is 0.3 µm, so that $L_{min} \cong 0.5$ µm which is also the minimum dimension for $\alpha = 10$. The thickness d_{ox} of the gate oxide (12 nm for $\alpha = 10$) is already near the limit of 5 nm (Section 2.7). Furthermore, the operating voltage (1V) has also approached its limit value (Section 2.1.), because a gate voltage of 1V is just sufficient to cause a potential difference of 0.7V at the corner of the source pn-junction. 0.3V are "lost" as the voltage V_{ox} across the gate oxide, according to the equation

$$V_{ox} = \frac{d_{ox}}{\varepsilon_o \varepsilon_{ox}} \sqrt{2 \varepsilon_o \varepsilon_{Si} q N_A (|2\phi_F| - V_{SUB})} \qquad (9)$$

If the minimum channel length L_{min} shall be reduced below 0.5 µm, N_A must be increased proportional to L_{min}^2 (equation (2)). The gate oxide thickness must be further reduced proportional to L_{min}. Otherwise V_{ox} would increase above 0.3V (equation (9)) which would mean an increase of the gate voltage above 1V.

The following set of parameters for a minimum size MOS transistor seems theoretically possible:

[+] Swanson and Meindl /21/ have predicted a lower limit of 0.1V for the supply voltage of CMOS circuits. Such ICs would be operated in the weak inversion region which means that the non-linearity of the transfer characteristics of inverters is badly degraded /5/.

- Minimum dimensions   0.2 μm
- Gate oxide thickness   5 nm
- Source and drain junction depths   50 nm
- Substrate doping 2×10^{17} cm^{-3}
- Drain voltage   1 V
- Gate voltage   1 V
- Substrate bias   ≈ 0 V

The electric fields in the gate oxide and in the silicon (at the drain corner) are just below the critical fields (Sections 2.2. and 2.7.) under these conditions.

The minimum dimensions of 0.2 μm mean a reduction of the dimensions by a factor of 25 compared to conventional 5 μm dimensions. The packing density of components can thus theoretically be increased by a factor of 625.

If we consider, for instance, the single-transistor cell area (2 components per cell) of a standard dynamic 16K random-access memory in double polysilicon technology with 5 μm dimensions, the average size of a component is about 250 μm^2. Reducing this area by a factor of 625 results in a minimum size per component of 0.4 μm^2. Due to parameter errors (Sections 2.5 and 3.1) the practical minimum size is however assumed to be somewhat larger, possibly 1 μm^2.

Scaled-down MOS transistors show a relatively "softer" turn-on characteristic in the subthreshold region /22/. This may impact device performance. Only if the absolute temperature T is also scaled down /19/, the turn-on characteristic can be made sharper (equation (1)).

5.2. Minimum Size of Bipolar Transistors

Due to the bandgap narrowing effect (Section 2.1.) the doping concentrations in the emitter and base regions of bipolar transistors cannot be made substantially higher than about 10^{20} cm^{-3} and 3×10^{18} cm^{-3}, respectively. Otherwise emitter efficiency would be strongly reduced. Assuming the same doping concentration for the emitter and the collector regions and a minimum operating voltage of 1V as in the case of the MOS transistor the widths of the emitter-base and collector-base junction add to about 0.05 μm (equation (2)). If the same width of 0.05 μm is allowed for the neutral base region, the minimum geometrical base region is about 0.1 μm. The practical limit may be somewhat higher due to parameter errors (Sections 2.5 and 3.1). Also local heating (Section 2.6) must be taken into account. Dissipation of 1 mW in the minimum-size transistor leads to a local temperature rise of about 100°.

The minimum size of a bipolar transistor with base width 0.1 μm has been estimated from geometrical considerations /3/ to be about 200 x 0.1 μm x 0.1 μm = 2 μm^2, assuming square emitter, base and collector contacts and local oxidation techniques. This transistor size is about three orders of magnitude smaller than the size of standard oxide-isolated bipolar transistors with 5-μm dimensions /23/.

5.3. Maximum Complexity of Integrated Circuits

The minimum average size of about 0.4 μm^2 for a component in an MOS circuit has been estimated in Section 5.1. This corresponds to a component density of 2.5×10^8 cm^{-2}. Using this density and assuming a chip size corresponding to the maximum chip size of 7 cm x 7 cm on a 4" wafer a complexity of about 10^{10} components per chip is theoretically possible. The corresponding values for the density and complexity of bipolar circuits are 5×10^7 cm^{-2} and 2.5×10^9, respectively (see Section 5.2.).

The maximum operating frequency of such complex circuits is however drastically limited due to the following factors:
- Signal propagation delay on signal lines
- Limited current density in power supply and signal lines due to electromigration
- Voltage drop on power supply and signal lines
- Limited heat removal.

A compromise must be made between the maximum allowable operation frequency and circuit complexity. This will be discussed in Section 6.

6. LIMITATIONS FOR THE OPERATING FREQUENCY

6.1. Propagation Time of Electrons Through Transistors

The time t_B required for electrons with velocity v to travel through the base of a bipolar transistor with geometrical base width W_B is

$$t_B = \frac{W_B}{v} \quad (10)$$

If the maximum electron drift velocity of 10^7 cms^{-1} (Section 2.3.) is assumed for v*, t_B is 10^{-12} s for the minimum-size bipolar transistor (Section 5.2) with W_B = 0.1 μm.
 Since the minimum-size MOS transistor (Section 5.1.) has a channel length which is approximately twice the minimum base width of the bipolar transistor, the propagation time for electrons through the channel region is about 2×10^{-12} s.

---
*This assumption is reasonable for base widths of about 0.1 μm /6/.

6.2. RC Loading Times and Signal Propagation Delay

The time needed to turn on a transistor in a digital circuit is not only given by the maximum velocity of mobile charges discussed in Section 6.1., but also by the time needed to charge the circuit capacitances and by the propagation velocity of signals on lines.

Let us first consider neighboring transistors in a circuit. The most significant contribution to the load resistance in this case is the base resistance and the channel resistance for bipolar and MOS transistors, respectively.

The base resistance R_B for a bipolar transistor with square emitter and one-sided base contact is

$$R_B \approx \frac{\rho_B}{2 W_{BN}} \qquad (11)$$

ρ_B is the base resistivity and W_{BN} is the width of the neutral base. Taking $\rho_B = 3 \times 10^{-2} \Omega\,cm$ and $W_{BN} = 5 \times 10^{-6}$ cm, according to the minimum-size bipolar transistor, R_B is 3KΩ.

The channel resistance R_C of an n-channel MOS transistor is approximately

$$R_C \approx \frac{V_D}{W N q v} \qquad (12)$$

N is the electron density in the channel, which is between 10^{11} and 10^{12} cm^{-2}. For W = 0.2 μm, V_D = 1V and v = 10^7 cms^{-1} (Section 2.3.) R_C is between 30 and 300KΩ, which is at least an order of magnitude larger than the R_B value estimated above.

The capacitances which must be loaded if neighboring transistors are considered are about 10^{-15} F for the bipolar as well as for the MOS case, because the capacitance of the connecting lines is assumed to be more significant than the intrinsic transistor capacitances. (The collector-base capacitance of the minimum-size bipolar transistor is about 10^{-16} F (equation (3)) and the gate capacitance of the minimum- size MOS transistor is 3 x 10^{-16} F).

Taking the resistance and capacitance values listet above and defining the RC loading time as the time required for a 90% loading of the capacitors, the following minimum loading times t_{LO} can be estimated:

$$t_{LO} = 2.3 R_B C \approx 7 \times 10^{-12} s \text{ (bipolar circuits)} \qquad (13a),$$

$$t_{LO} \approx 2.3 R_C C \approx 7 \times 10^{-11} s \text{ (MOS circuits)} \qquad (13b).$$

Maximum clock frequencies in the order of 10^{11} Hz for bipolar circuits and 10^{10} Hz for MOS circuits are thus possible.

If however complex integrated circuits with relatively long signal lines are taken into account the propagation delay of the signals on the lines may become predominant.

For metal lines the propagation velocity v of signals is in the order of one tenth of the phase velocity of electromagnetic waves in vacuum /24/, i.e. v is about 3×10^9 cm s^{-1}. The signal propagation delay t_{PD} is then

$$\left(\frac{t_{PD}}{s}\right) \approx 3 \times 10^{-10} \left(\frac{l_L}{cm}\right) \qquad (14),$$

where l_L is the length of the line. Let us assume that a time delay of up to one tenth of the cycle time is allowed the maximum allowable clock frequency f_{max} is

$$\left(\frac{f_{max}}{Hz}\right) \approx 3 \times 10^8 \left(\frac{cm}{l_L}\right) \qquad (15).$$

Equation (15) means that a comprise must be made between the maximum complexity and the maximum operating frequency. In Table 3 some values are listed according to equation (15).

Operating frequency (Hz)	Maximum length l_L of line (cm)	Maximum chip area l_L^2 (cm^2)	Maximum number of components* in MOS cir.	Maximum number of components* in bipol.cir.
10^7	3×10^1	10^3	2×10^{11}	5×10^{10}
10^8	3×10^0	10^1	2×10^9	5×10^8
10^9	3×10^{-1}	10^{-1}	2×10^7	5×10^6
10^{10}	3×10^{-2}	10^{-3}	2×10^5	5×10^4
10^{11}	3×10^{-3}	10^{-5}	----	5×10^2

Table 3. Dependence of the maximum complexity of integrated circuits on the maximum operating frequency due to signal propagation delay.

If n* doped lines are used instead of metal lines the maximum operating frequency is substantially lower than with metal lines. To a first approximation the delay time on an n* doped line is $t_D = R_L C_L$ where R_L and C_L are the resistance and the capacitance of the line, respectively. Taking $f_{max} = 1/10 t_D$ as in the previous case of metal lines and assuming the limit parameters discussed in Section 5.1. ($t_L = 0.05$ μm, $\rho_L = 10^{-4}$ Ωcm, $d_{DEPL} = 0.1$ μm) the following relation is found:

---

*For the area of the components the minimum sizes estimated in Sections 5.1. and 5.2. are assumed.

$$\left(\frac{f_{max}}{Hz}\right) = 4 \times 10^4 \left(\frac{cm}{l_L}\right)^2 \qquad (16)$$

Comparison of relations (15) and (16) clearly implies the necessity of metal signal lines for integrated circuits with maximum packing density.

6.3. Operating Fequency Limitations due to Electromigration

As has been pointed out in Section 2.8. the current density in metal lines is restricted to below about 10^6Acm^{-2} due to electromigration.

Assume an aluminum line with thickness $t_L = 0.2$ µm, width $W_L = 0.2$ µm and length l_L running over a SiO_2 film ($d_{ox} = 0.1$ µm) on silicon. If this line is operated with pulses of $U \cong 1V$ at a frequency f and with a rise time of the pulses corresponding to 1/10f, the current I and the current density j in the aluminum conductor during rise time are as follows:

$$\left(\frac{I}{A}\right) = \frac{10 f \varepsilon_o \varepsilon_{ox} l_L W_L U}{d_{ox}} \left(\frac{1}{A}\right) = 8 \times 10^{-12} \left(\frac{f}{Hz}\right)\left(\frac{l_L}{cm}\right) \qquad (17a)$$

$$\left(\frac{j}{Acm^{-2}}\right) = \frac{10 f \varepsilon_o \varepsilon_{ox} l_L U}{d_{ox} t_L} \left(\frac{1}{Acm^{-2}}\right) = 2 \times 10^{-2} \left(\frac{f}{Hz}\right)\left(\frac{l_L}{cm}\right) \qquad (17b)$$

In equations (17) the capacitance of the components connected to the line are not considered. If this is done I and j can assume about 3 times the values given in equations (17).

Inserting the maximum allowable current density of 10^6 Acm^{-2} in equation (17b) yields:

$$\left(\frac{f_{max}}{Hz}\right) = 5 \times 10^7 \left(\frac{cm}{l_L}\right) \qquad (18)$$

This is an even stronger limitation for the operating frequency or the complexity of a circuit than the limitation given by equation (16). In Table 4 some values are listed according to equation (18).

6.4. Operating Frequency Limitations due to the Voltage Drop on Lines

The voltage drop ΔU on aluminum lines 0.2 µm wide and 0.2 µm thick corresponding to minimum-size transistors, is as follows:

Operating frequency (Hz)	Maximum length l_L of line (cm)	Maximum chip area l_L^2 (cm^2)	Maximum number of components in MOS circuits	Maximum number of components in bipol. cir.
10^7	5×10^0	2.5×10^1	6×10^9	10^9
10^8	5×10^{-1}	2.5×10^{-1}	6×10^7	10^7
10^9	5×10^{-2}	2.5×10^{-3}	6×10^5	10^5
10^{10}	5×10^{-3}	2.5×10^{-5}	6×10^3	10^3
10^{11}	5×10^{-4}	2.5×10^{-7}	-----	10^1

Table 4. Dependence of the maximum complexity of integrated circuits on the maximum operating frequency due to current density limitations in metal lines (electromigration).

$$\left(\frac{\Delta U}{V}\right) = 6 \times 10^3 \left(\frac{I}{A}\right)\left(\frac{l_L}{cm}\right) \quad (19)$$

I is the average current in the line. If the assumptions leading to equations (17) are used, I is half the value given in equation (17a), and the following relation is obtained:

$$\left(\frac{\Delta U}{V}\right) = 2.5 \times 10^{-8} \left(\frac{f}{Hz}\right)\left(\frac{l_L}{cm}\right)^2 \quad (20)$$

If the allowable voltage drop is, for instance, 0.1V, the following relation between the maximum allowable operation frequency f_{max} and the length l_L of the line can be calculated:

Operating frequency (Hz)	Maximum length l_L of line (cm)	Maximum chip area l_L^2 (cm^2)	Maximum number of components in MOS cir.	Maximum number of components in bipol. cir.
10^5	6×10^0	4×10^1	10^{10}	2×10^9
10^6	2×10^0	4×10^0	10^9	2×10^8
10^7	6×10^{-1}	4×10^{-1}	10^8	2×10^7
10^8	2×10^{-1}	4×10^{-2}	10^7	2×10^6
10^9	6×10^{-2}	4×10^{-3}	10^6	2×10^5
10^{10}	2×10^{-2}	4×10^{-4}	10^5	2×10^4
10^{11}	6×10^{-3}	4×10^{-5}	10^4	2×10^3

Table 5. Dependence of the maximum complexity of integrated circuits on the maximum operating frequency due to the voltage drop (0.1V allowed) on lines.

$$\left(\frac{f_{max}}{Hz}\right) = 4 \times 10^6 \left(\frac{cm}{l_L}\right)^2 \qquad (21)$$

In Table 5 some values are listed according to equation (21).

Using other metals with lower resisitivity compared to aluminum does not change the result significantly. The only way to arrive at lower resistivities is cooling down of the integrated circuit to cryogenic temperatures.

Doped silicon has a minimum resistivity which is nearly two orders of magnitude higher than the resistivity of aluminum. Accordingly, the constant in equation (21) is about 4×10^4 instead of 4×10^6 which means lower allowable operating frequencies. This points to the desirability for metal interconnection lines for high-complexity circuits.

6.5. Minimum Energy per Binary Operation

The energy which is dissipated during loading or unloading of the circuit capacitances C belonging to a binary operation is

$$E = \frac{1}{2} CU^2 \qquad (22)$$

Assuming the minimum values $C = 10^{-15}$ F (Section 6.2.) and $U = 1V$, E_{min} is 5×10^{-16} J /6/.

6.6. Operating Frequency Limitations due to Limited Heat Removal

We assume that 2 components are needed for performing a binary operation. Using the minimum energy per binary operation E_{min} (Section 6.5.) the average power dissipation P_C of each component is

$$P_C = E_{min} f \qquad (23)$$

f is the operating frequency.

If we assume that the energy E_{min} is dissipated in each logic element during one cycle time the total power dissipation P in the integrated circuit with N_C components is

$$P = N_C E_{min} f \qquad (24)$$

Introducing the chip area A and considering the maximum allowable heat removal per unit area $(P/A)_{max}$ given in Section 4.2., the maximum operating frequency f_{max} can be given as a function of packing density (N_C/A) of components:

$$f_{max} = \frac{1}{E_{min}} \left(\frac{P}{A}\right)_{max} \left(\frac{A}{N_C}\right) \qquad (25).$$

With $E_{min} = 5 \times 10^{-16}$ J and $(P/A)_{max} = 20$ Wcm^{-2} equation (25) can be written as follows:

$$\left(\frac{f_{max}}{Hz}\right) = 4 \times 10^{16} \left(\frac{A}{N_C}\right) \qquad (26).$$

In Table 5 some values are listed according to equation (26).

Operating frequency (Hz)	Maximum density of components (cm^{-2})	Maximum number of components on a 50cm^2 chip
10^7	4×10^9	2×10^{11}
10^8	4×10^8	2×10^{10}
10^9	4×10^7	2×10^9
10^{10}	4×10^6	2×10^8
10^{11}	4×10^5	2×10^7

Table 5. Dependence of the maximum complexity of integrated circuits on the operating frequency due to the limited heat removal of liquid cooling (20Wcm^{-2}). For air cooling the maximum component densities are a factor of 10 smaller (see Section 4.2.).

7. CONCLUSIONS

The packing density of integrated circuits can theoretically be increased by about 3 orders of magnitude. Assuming a maximum chip area of 50 cm^2 the complexity of integrated circuits may be increased by about 5 orders of magnitude compared to present-day integrated circuits. A compromise must however be made between maximum complexity and maximum operating frequency. The most significant limiting factors are the voltage drop on power supply and signal lines and electromigration. Due to the voltage drop the maximum operating frequency of integrated circuits with maximum possible complexity (about 10^{10} components per chip) is only about 100 kHz. At the other end of the scale the maximum complexity of integrated circuits operated with the maximum possible frequency of about 10^{11} Hz is only about 10 components per chip due to the limited current density in metal lines given by electromigration effects.

REFERENCES

1. G.E. Moore, Proc. Int. El. Dev. Meeting, Washington, 11, (1975).
2. B. Hoeneisen and C.A. Mead, Solid State El., 15, 819, (1972)
3. B. Hoeneisen and C.A. Mead, Solid State El., 15, 891, (1972)
4. J.T. Wallmark, Institute of Physics Conf. Ser., London, 25, 133 (1975).
5. R.W. Keyes, Proc. of the IEEE, 63, 740, (1975).
6. O.G. Folberth and J.H. Bleher, Nachrichtentechn. Zeitschrift, 30, 307, (1977).
7. K. Garbrecht and K.-U. Stein, Siemens Forschungs- und Entwicklungsberichte, 5, 312, (1976).
8. R. van Overstraeten et al., IEEE Trans. El. Dev., ED-20, 290, (1973)
9. S.M. Sze and G. Gibbons, Solid State El., 9, 831, (1966).
10. W. Shockley, Solid State El., 2, 35, (1961)
11. J.R. Black, IEEE Trans. El. Dev., ED-16, (1960)
12. G.A. Scoggan et al., 13th Int. Rel. Phys. Symp., Las Vegas, (April 1975)
13. I. Ames et al., IBM J. Res. Develop., 14, 461, (1970)
14. R. Feder et al., J. Vac. Sci. Technol., 12, 1332, (1975)
15. H.I. Smith, Proc. of the IEEE, 62, 1361, (Oct. 1974)
16. A.N. Broers et al., Appl. Phys. Letters, 29, 596, (1976)
17. E. Kasper et al., Appl. Phys., 8, 199, (1975)
18. A. Furukawa et al., Jap. J. Appl. Phys., 11, 134, (1972)
19. F.H. Gaensslen et al., Proc. Int. El. Dev. Meeting, Washington, 43, (1975)
20. R.H. Dennard et al., Proc. Int. El. Dev. Meeting Washington, (1972)
21. R.M. Swanson and J.D. Meindl, Proc. Int. Solid State Circ. Conf., Philadelphia, 110, (1975)
22. R.H. Dennard et al., Proc. Int. El. Dev. Meeting, Washington, 152, (1974)
23. H. Murrmann, Siemens Forschungs- und Entwicklungsberichte, 5, 353, (1976)
24. O.G. Folberth, Int. Elektronische Rundschau, 28, 9 and 29, (1974)

Section III

MODELING OF BIPOLAR DEVICES

REVIEW OF MODELS FOR BIPOLAR TRANSISTORS

H.C. de Graaff

Philips Research Laboratories
Eindhoven, The Netherlands

ABSTRACT. This review gives the basic concepts for bipolar
transistor modelling and discusses the major analytical models
like the Ebers-Moll, charge-control, Gummel-Poon and Linvill
models.

1. INTRODUCTION

A model is a representation of a device or system. There must
exist a great similarity between the model and the device or
system it represents, at least in some meaningful aspects.
This similarity may be, for example, a set of mathematical
equations, describing the behaviour of both the model and the
device being modelled. We can also take the set of mathematical
equations itself as the model, as is done in transistor modelling.
We use models because they are simpler to handle than the real
thing. Mathematical models e.g. allow us to do computer
experiments which are in many cases more easily carried out than
experiments with the real device.
Especially in integrated circuit design these computer experiments
(Computer Aided Circuit Design or CACD for short) can be extremely
time and cost saving. In the aspect of series inductances CACD is
even more realistic than breadboarding. To obtain meaningful
results with CACD one needs simple, yet accurate mathematical
descriptions of the circuit elements as resistors, capacitors,
diodes and transistors.
    Transistor modelling starts with solving, in one way or
another, the basic set of equations for the carrier transport in
semiconductors:

$$J_n = q D_n \frac{\partial n}{\partial x} + q\mu_n nE \qquad (1a)$$

$$J_p = -q D_p \frac{\partial p}{\partial x} + q\mu_p pE \qquad (1b)$$

$$\frac{\partial n}{\partial t} = \frac{1}{q} \frac{\partial J_n}{\partial x} - R \qquad (1c)$$

$$\frac{\partial p}{\partial t} = \frac{1}{q} \frac{\partial J_p}{\partial x} - R \qquad (1d)$$

$$\frac{\partial E}{\partial x} = \frac{q}{\varepsilon}(N_d - N_a + p - n) \qquad (1e)$$

The first two are the current equations for electrons and holes, then the two continuity equations follow and the last one is Poisson's equation. With the help of a computer the set of equations (1) can be solved numerically. As input data we then require the impurity profiles $N_d(x)$ and $N_a(x)$ and appropriate boundary conditions. Although the solutions, obtained in this way, are very accurate in themselves, the overall result largely depends on the accuracy of the input data. Most of the existing numerical solutions are one-dimensional, but two-dimensional solutions are also coming into use [1]. The numerical solutions not only give the terminal behaviour, but also the internal carrier distributions and can therefore be used in explorative and heuristic device research. A second method that can be used to solve the set of equations (1) is to derive analytical expressions for the solutions, preferably in closed form. These solutions are often approximate and for reasons of simplicity one-dimensional. Generally speaking the analytical method is simple, approximate (and less accurate) and requires little computation time. Analytical models are preferred in the design of large complicated integrated circuits because in that case the designer needs a mathematical device model that is simple and consumes little computation time.
In most cases the analytical models are based on the "regional" approach, where the transistor is divided into several adjacent regions of space charge and space charge neutrality. In each region a solution of (1) is obtained and the different regional solutions are connected by means of relevant boundary conditions.
    Some analytical models focus on the terminal behaviour of the device and are well suited to CACD programs (e.g. the Ebers-Moll model) because they can generate the current-voltage relationships needed in circuit design.
    Other, more physical models emphasize the importance of the internal carrier distribution. Typical of these are the Linvill models. The charge control models occupy an intermediate position:

the terminal behaviour is closely related to the internal charge storage.

A mathematical model is not only defined by its equations but also by the coefficients or parameters appearing in these equations.

We must require that the parameters of a model can be determined experimentally and in an unambiguous way. It is also desirable that we have some knowledge about the statistical spread and the influence of temperature rise and ageing.

Moreover, the transistor models should not only confirm the experimental results of the real transistors, they must also be able to predict results under different circumstances. In other words, modelling is not only curve fitting, it also must have forecasting abilities.

As the numerical modelling will be treated elsewhere [2] this review will give only the analytical basic models for bipolar transistors.

2. THE EBERS-MOLL MODEL

2.1. Basic models [3,4]

The physical concept of the basic Ebers-Moll model is that of two diodes connected back to back, as indicated in fig. 1 for an npn transistor.
The additional current sources are driven by the diode currents, which are assumed to have ideal characteristics:

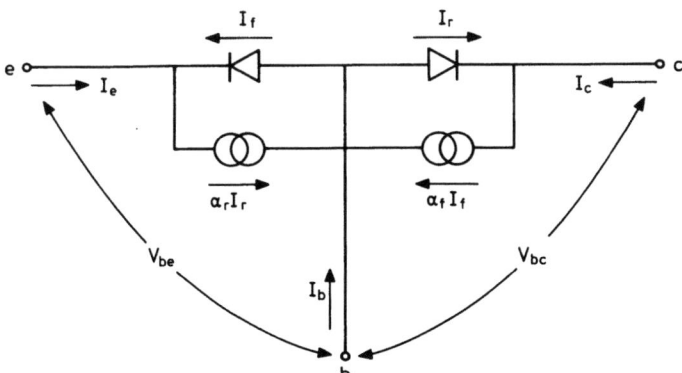

Fig. 1. The basic Ebers-Moll model with two diodes and two current sources, for an npn type of transistors.

$$I_f = I_{fo} \{ \exp(qV_{be}/kT)-1 \} \qquad (1a)$$

$$I_r = I_{ro} \{ \exp(qV_{bc}/kT)-1 \} \qquad (1b)$$

The terminal currents are:

$$I_e = -I_f + \alpha_r I_r \qquad (2a)$$

$$I_b = (1-\alpha_f)I_f + (1-\alpha_r)I_r \qquad (2b)$$

$$I_c = \alpha_f I_f - I_r \qquad (2c)$$

The eqs. (1) and (2) give the relations between the terminal currents and the terminal voltages V_{be} and V_{bc}.
 This model has four parameters, namely the diode current constants I_{fo} and I_{ro} and the forward and reverse current gains α_f and α_r. One can prove that $\alpha_f I_{fo} = \alpha_r I_{ro}$, so only three of them must be determined experimentally.
 This basic model contains many simplifications:
1. the diodes are ideal and not to be used in high-injection cases;
2. the current gain factors α_f and α_r are constants and not functions of current, as they should be;
3. series resistances are omitted;
4. time dependent situations are handled poorly, because all delays and capacitive charging effects are absent, the so-called quasi-static assumption.

The quasi-static approach mentioned in 4., can be improved by making the α's frequency-dependent:

$$\alpha_f = \frac{\alpha_{fo}}{1 + s/\omega_{\alpha f}} \qquad (3a)$$

and

$$\alpha_r = \frac{\alpha_{ro}}{1 + s/\omega_{\alpha r}} \qquad (3b)$$

with $s = j\omega$ or $s = \frac{d}{dt}$.

For normal forward operation ($I_r = 0$) the current gain in grounded emitter configuration is then given by

$$\beta_f = \frac{I_c}{I_b} = \frac{\beta_{fo}}{1 + s/\omega_{\beta f}} \quad (4)$$

where

$$\beta_{fo} = \frac{\alpha_{fo}}{1 - \alpha_{fo}} \quad (5)$$

and

$$\omega_{\beta f} = \frac{\omega_{\alpha f}}{1 + \beta_{fo}} \quad . \quad (6)$$

For the reverse operation ($I_f = 0$) we can derive a similar expression:

$$\beta_r = \frac{\beta_{ro}}{1 + s/\omega_{\beta r}} \quad . \quad (7)$$

Fig. 2 gives β_f as a function of frequency in the complex frequency plane.

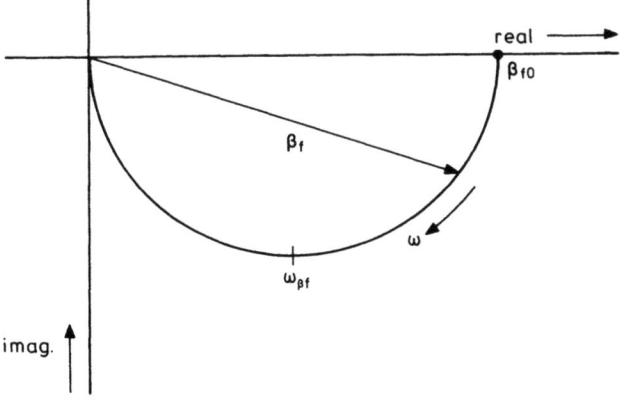

Fig. 2. The grounded emitter current gain β_f as a function of frequency in the complex plane.

The cut-off frequency f_T, at which $|\beta_f| = 1$ is given, according to equation (4), by

$$2\pi f_T = \omega_T \approx \beta_{fo} \cdot \omega_{\beta f} \qquad (8)$$

Eq. (4) also enables us to calculate in a simple way a transient response, e.g. the collector current response to a sudden change in I_b. With s as the differential operator $\frac{d}{dt}$, eq. (4) gives:

$$I_c + \tau_f \frac{dI_c}{dt} = \beta_{fo} I_b \text{ with } \omega_{\beta f} \cdot \tau_f = 1$$

The solution is

$$I_c = \beta_{fo} I_b \{ 1 - \exp(-t/\tau_f) \}$$

Returning to fig. 2 and eq. (4) we can see that the phase angle of β_f at high frequencies is about -90 degrees. However, in reality the phase shift of β_f can exceed this amount. To incorporate this excess phase shift of β_f, α_f is made a higher order function of frequency [5]:

$$\alpha_f = \frac{\alpha_{fo}}{(1+s\tau_1)(1+s\tau_2)} \qquad (9a)$$

and

$$\beta_f = \frac{\beta_{fo}}{1 + \frac{\tau_1+\tau_2}{1-\alpha_{fo}} s + \frac{\tau_1 \tau_2}{1-\alpha_{fo}} s^2} \qquad (9b)$$

These functions have two poles instead of one and the phase angle of β_f at high frequencies is 180 degrees.
It is possible to introduce even more complex functions [6] with poles and zeroes; they give a refined description of the high-frequency behaviour of the E-M model, but the model complexity and the number of parameters to be determined also increase.
Another necessary extension of the basic model is the addition of series resistances and transition layer capacitances (see fig. 3).

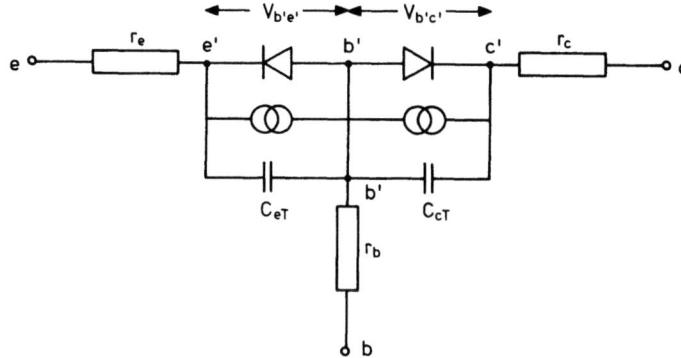

Fig. 3. Basic E-M model with additional series resistances and transition layer capacitances (C_{eT} and C_{cT}).

Note that in fig. 3 the diodes are controlled by the internal junction voltages $V_{b'e'}$ and $V_{b'c'}$ and no longer by the externally applied voltages.

2.2. The Early effect

The Early effect [7] is the effect whereby an increase in collector voltage gives rise to an increase in collector current. This increase in current is brought about by an increase in depletion layer thickness at the reverse biased base-collector junction, which makes the neutral base region thinner.

As an approximation it can be said that the extrapolated (I_c, V_{ce}) characteristics converge to one point, as sketched in Fig. 4a [8,9].
In the model the Early effect can be represented by an extra current source between the internal emitter and collector points and controlled by I_f and V_{ce}. It does not affect the base current. The Early voltage V_A can be written as

$$V_A \approx \beta_{fo} \cdot V_N . \qquad (10)$$

The parameter V_N is determined by the doping profile of the transistor. We know from experiments [10] that V_N is independent

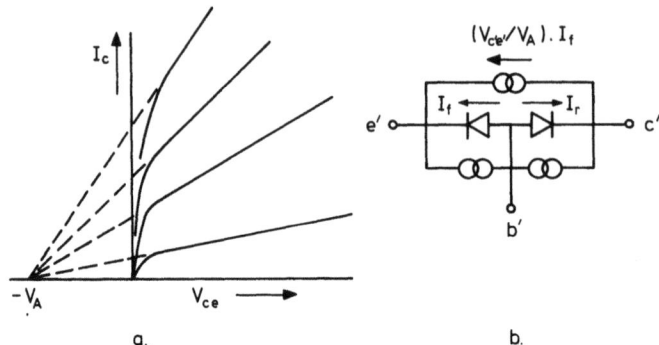

Fig. 4. Influence of the Early effect on the characteristics (a) and its incorporation in the E-M model (b).

of voltage temperature.
The Early effect also makes the current gain β_{fo} voltage-dependent. From the model and eq. (10) we can deduce:

$$\beta_{fo} \approx \frac{\alpha_{fo}}{1-\alpha_{fo}} + \frac{V_{ce}}{V_N} . \qquad (11)$$

2.3. Current crowding and IBIS model

The base current causes a voltage drop in the lateral direction along the base-emitter junction. Therefore the emitter current is not homogeneously distributed over the emitter area, but the current is highest at the emitter edge which is nearest to the base contact. This effect is called current crowding; it increases when the base current increases. The lateral voltage drop in the base region underneath the emitter, which is a non-linear function of base current, can be modelled by a diode [11]:

$$I_b = I_{bo} \{ \exp(qV_{bb'}/kT)-1 \} \qquad (12)$$

with

$$I_{bo} = \frac{4kT}{q} \cdot \frac{2\pi W_b}{\rho_b}$$

where W_b = base thickness

and ρ_b = resistivity of the base material.

Eq. (12) was derived for a cylindrical structure, but it also applies more or less to rectangular geometries [12], but with a different value for I_{bo}.
The circuit representation of the E-M model with current crowding is shown in fig. 5.
This extended version of the E-M model is called the IBIS model [13].

3. THE CHARGE-CONTROL MODEL [14]

The stored minority carrier charge in the neutral base region is of fundamental importance for the charge control model. If we consider again an npn transistor and integrate the continuity equation over the base region, we obtain

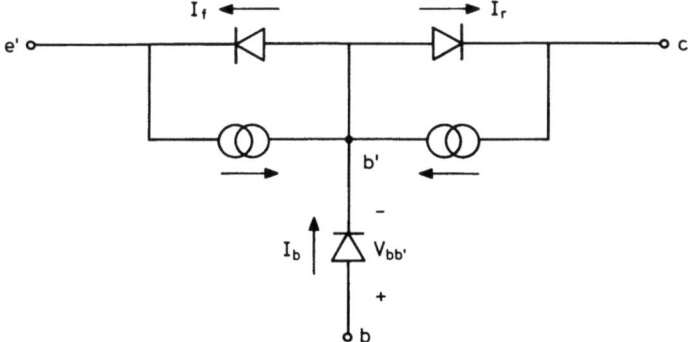

Fig. 5. Basic Ebers-Moll model with an extra diode in the base lead, to represent emitter-base current crowding.

$$\int \nabla \cdot J_n \, dv = \int q \frac{\partial n}{\partial t} dv + \int q R \, dv$$

or

$$I_b = \frac{dQ}{dt} + \frac{Q}{\tau} \tag{13}$$

The term Q/τ represents the recombination, although this recombination does not necessarily take place in the base itself, but also, for example, in the emitter. Eq. (13) states that the total base current supplies the recombination current Q/τ and the charging current dQ/dt. The base charge Q can be split into a forward part Q_f, controlled by the emitter-base junction voltage, and a reverse part Q_r, controlled by the collector-base junction voltage (see fig. 6).
Eq. (13) becomes

$$I_b = \frac{Q_f}{\tau_f} + \frac{dQ_f}{dt} + \frac{Q_r}{\tau_r} + \frac{dQ_r}{dt} \tag{14}$$

with

$$Q_f = Q_{fo} \{ \exp(q V_{b'e'}/kT) - 1 \} \tag{15}$$

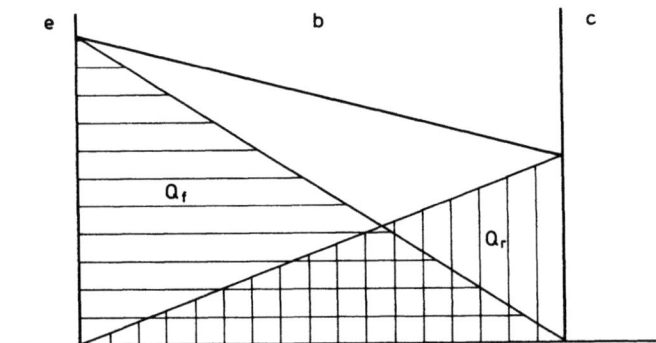

Fig. 6. Sketch of the charge storage in the neutral base region. The fraction Q_f is controlled by $V_{b'e'}$ and the fraction Q_r by $V_{b'c'}$.

and

$$Q_r = Q_{ro} \{ \exp(q V_{b'c'}/kT)-1 \} \qquad (16)$$

The charge storage in the epitaxial collectors of modern transistors is not very well modelled by eq. (16), but this problem is discussed elsewhere in this issue [15].
The charge-control principle states that the collected current is directly proportional to the stored charge of its carriers.

Applying this principle we get for collector and emitter currents:

$$I_c = \beta_{fo} \frac{Q_f}{\tau_f} - (1+\beta_{ro}) \frac{Q_r}{\tau_r} - \frac{dQ_r}{dt} \qquad (17)$$

and

$$I_e = \beta_{ro} \frac{Q_r}{\tau_r} - (1+\beta_{fo}) \frac{Q_f}{\tau_f} - \frac{dQ_f}{dt} . \qquad (18)$$

The circuit representation of eqs. (14) to (18) is given in fig. 7.

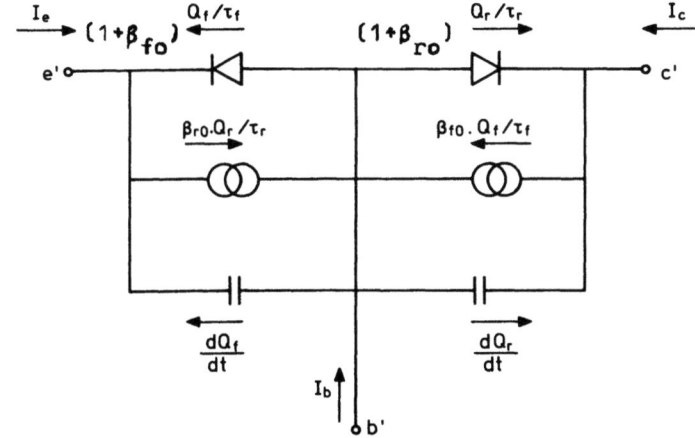

Fig. 7. Circuit diagram of the basic charge-control model.

The capacitances in this circuit are the diffusion capacitances. The charge control model gives approximately the same results as the Ebers-Moll model if

$$\beta_{fo} = \frac{\alpha_{fo}}{1-\alpha_{fo}} = \alpha_{fo} \frac{I_{fo}\tau_f}{Q_{fo}}$$

and

$$\beta_{ro} = \frac{\alpha_{ro}}{1-\alpha_{ro}} = \alpha_{ro} \frac{I_{ro}\tau_r}{Q_{ro}} ,$$

which can be derived from eqs. (2c) and (17) by putting $dQ_r/dt = 0$.
It follows from eqs. (14) and (17) that the grounded emitter current gain in forward operation is

$$\beta_f = \frac{\beta_{fo}}{1+s\tau_f} . \tag{19}$$

Thus the charge-control model again gives a one-pole model for the current gain, predicting a 6 dB/octave fall-off and an f_T given by $\omega_T = \beta_{fo}/\tau_f \approx \omega_{\alpha f}$.
The phase shift of β_f does not exceed -90 degrees, similar to that with the one-pole Ebers-Moll model. We can improve on that by introducing extra delays [16]. For the forward operation ($Q_r = 0$) this takes the following form:

$$I_c(t) = \frac{\beta_{fo}}{\tau_f} Q_f(t-\tau_1) \tag{20}$$

and

$$Q_f(t) = Q_{fo} \left\{ \exp\left[\frac{qV_{b'e'}}{kT}(t-\tau_2)\right] -1 \right\} \tag{21}$$

The expression for the base current, eq. (14), does not change. The extra delay times τ_1 and τ_2 are functions of the built-in electric field in the base [16].
The modification of eq. (20) resembles the higher order functions for the current gain in the Ebers-Moll model (see section 2.1, eq. (9)), whereas eqs. (20) and (21) mean that we have dropped the quasi-static assumption.
With the new equation (20) for I_c we find

$$\beta_f = \frac{\beta_{fo}}{1+j\omega\tau_f} \exp(-j\omega\tau_1). \tag{22}$$

The factor $e^{-j\omega\tau_1}$ provides us with the necessary extra phase shift.

It will be obvious that the equivalent circuit in fig. 7 can be extended, like the Ebers-Moll model, with series resistances, transition layer capacitances, an extra base diode for current crowding and an Early current source.

4. THE SMALL-SIGNAL EQUIVALENT CIRCUIT

From the charge-control equations (14) to (18) we can easily derive the small-signal admittance parameters for the forward mode of operation. In the grounded emitter situation we find:

$$Y_{fe'} = \frac{qI_c}{kT} = g_m \tag{23}$$

$$Y_{ie'} = \frac{g_m}{\beta_{fo}} + j\omega \frac{g_m \tau_f}{\beta_{fo}} \tag{24}$$

and

$$\beta_f = \frac{\beta_{fo}}{1+j\omega\tau_f}.$$

The equivalent circuit is sketched in fig. 8 (part inside the dashed lines). The diffusion capacitance C_{ed} is equal to $g_m \tau_f / \beta_{fo} = g_m/\omega_T$.

The grounded base parameters are obtained from the general relations:

$$Y_{fb'} = -Y_{fe} = -g_m \tag{25}$$

$$Y_{ib'} = (1+\beta_f) Y_{ie} = g_m(1+\frac{1}{\beta_{fo}}) + j \frac{g_m \tau_f}{\beta_{fo}} \tag{26}$$

$$\alpha_f = \frac{\beta_f}{1+\beta_f} = \frac{\alpha_{fo}}{1+j\omega/\omega_{f\alpha}} \tag{27}$$

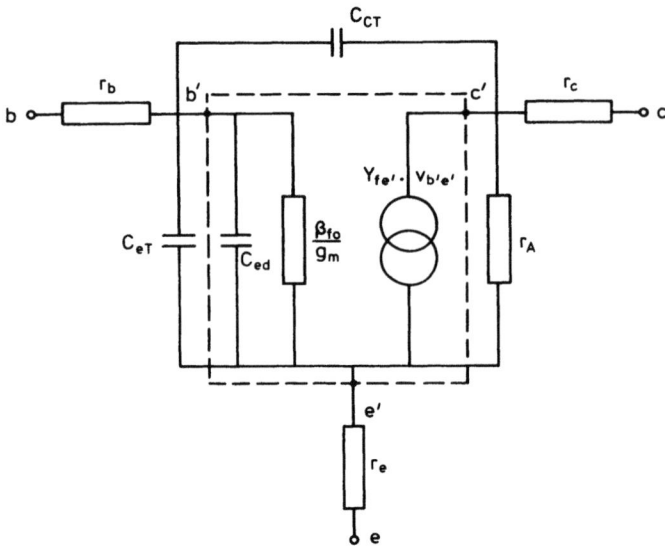

Fig. 8. A.c. small-signal equivalent circuit, derived from the charge-control model. The part inside the dashed lines is the basic or intrinsic part.

where the α cut-off frequency is given by $\omega_{f\alpha} = \dfrac{\beta_{fo}+1}{\tau_f} \approx \omega_T$. Besides the already known insufficient phase shift of β_f, the shortcomings of this simple equivalent circuit are the following:

1. the α cut-off frequency is $\omega_{f\alpha} \approx \omega_T$, whereas in reality it should be higher than ω_T.
2. $Y_{ie'}$ and $Y_{ib'}$ have equal capacitive parts, but in reality the common-base capacitance is lower.
3. the transconductance values $Y_{fb'}$ and $Y_{fe'}$ are purely real, but should have a negative imaginary part too.

All these shortcomings can be remedied by the introduction of the extra time delays τ_1 and τ_2 [16]. By omitting the dependence of the built-in electric field we can write as a first approximation

$$\tau_1 \approx \tau_2 \approx \frac{1}{6\omega_T} \tag{28}$$

Using eqs. (20) and (21) we obtain for the small-signal parameters

$$Y_{fb'} = -Y_{fe} \approx -g_m \exp(-j\frac{\omega}{3\omega_T})$$

$$Y_{ie'} \approx (\frac{g_m}{\beta_{fo}} + j\omega \frac{g_m \tau_f}{\beta_{fo}}) \exp(-j\frac{\omega}{6\omega_T})$$

$$\approx \frac{g_m}{\beta_{fo}} + j\omega \frac{g_m \tau_f}{\beta_{fo}} \tag{29}$$

$$Y_{ib'} \approx g_m (1 + \frac{1}{\beta_{fo}}) + \frac{2}{3} j\omega \frac{g_m \tau_f}{\beta_{fo}}. \tag{30}$$

Comparing eqs. (26) and (30) we can see that the extra delay indeed lowers the grounded base input capacitance. Moreover, from $\alpha = \beta/\beta+1$ we now also find that

$$\omega_{f\alpha} = \frac{6}{5} \omega_T. \tag{31}$$

The equivalent circuit in fig. 8 is extended by incorporating the series resistances (r_e, r_b, r_c) and the transition capacitances (C_{eT}, C_{cT}). The Early effect is presented by the resistance r_A between e' and c'. The complete circuit of fig. 8 is known as a hybrid π circuit.

5. THE GUMMEL-POON MODEL

The Gummel-Poon model is based on the integral charge-control relation [17], which is an extension of the Moll-Ross formula [18] in the sense that the base charge is not constant, but bias dependent. The integral charge-control relation states that the dominant electron current in an npn transistor is given by

$$I_n = (q\, n_i\, A_{em})^2\, D_n\, \frac{\exp(q\, V_{be}/kT) - \exp(q\, V_{be}/kT)}{Q_b} \tag{32}$$

The major assumptions made in the derivation of eq. (32) are: no recombination in the base region (j_n = constant), the use of

low field mobilities and the validity of the Einstein relation.
We can also write for I_n:

$$I_n = I_f - I_r \qquad (33)$$

where

$$I_f = I_s Q_{bo} \frac{\exp(q\,V_{be}/kT)-1}{Q_b} \qquad (34a)$$

and

$$I_r = I_s Q_{bo} \frac{\exp(q\,V_{bc}/kT)-1}{Q_b} \qquad (34b)$$

Note that eqs. (33) and (34) resemble the Ebers-Moll eqs. (1) and (2). In the Gummel-Poon model the base charge Q_b is bias-dependent and modelled as follows [19]:

$$Q_b = Q_{bo} + Q_e + Q_c + B\tau_f I_f + \tau_r I_r \qquad (35)$$

In this equation

Q_{bo} is the fixed base charge, given by the impurity distribution,

Q_e, Q_c are the depletion layer charges of the emitter and collector junctions respectively,

$B\tau_f I_f$ and $\tau_r I_r$ are the minority carrier charge storages, connected with the forward and reverse currents,

B   is a factor which is usually equal to one, but which becomes larger in quasi-saturation (Kirk-effect [20]).

If we substitute eqs. (34a) and (b) into eq. (35) we get a quadratic equation for Q_b, with the solution

$$Q_b = \frac{Q_{bo}+Q_e+Q_c}{2} + [\,(\frac{Q_{bo}+Q_e+Q_c}{2})^2 + I_s Q_{bo} \{\, B\tau_f\,(\exp(qV_{be}/kT)$$
$$-1) + \tau_r\,(\exp(qV_{bc}/kT)-1\,)\,\}\,]^{\frac{1}{2}}. \qquad (36)$$

For the depletion layer charges we can write

$$Q_e = C_{eT} V_{be} \qquad (37a)$$

and

$$Q_c = C_{cT} V_{bc}. \qquad (37b)$$

The transition layer capacitances C_{eT} and C_{cT} are dependent on the junction voltages and often modelled as

$$C_T = \frac{C_o}{(1-\frac{V}{V_o})^n} = \frac{C_o}{\upsilon^n} \qquad (38)$$

with $\upsilon = 1 - \frac{V}{V_o}$.

Eq. (38) has the drawback of a singularity at $V = V_o$ when the forward voltage equals the junction built-in voltage; see fig. 9.
This can be remedied by using for C_T a more complex formula [21] with four instead of three parameters:

$$C_T = \frac{C_o}{(\upsilon^2+b)^{n/2}} \left(1 + \frac{n}{1-n} \cdot \frac{b}{\upsilon^2+b} \right). \qquad (39a)$$

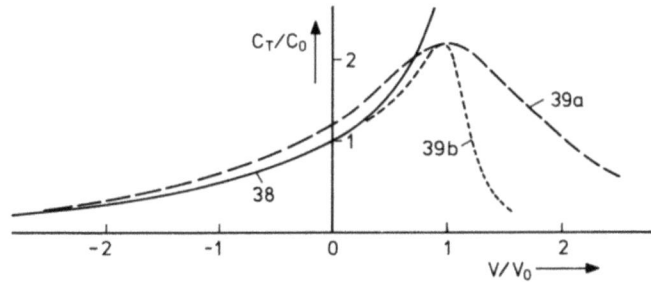

Fig. 9. The transition layer capacitance C_T versus applied voltage V, according to the equations (38), (39a) and (39b).

Eq. (39a) shows symmetry around the point $\upsilon = 0$ (or $V = V_o$). A more rapid decline of C_T for $V \neq V_o$ is obtained with the following expression, which also has four parameters:

$$C_T = \frac{C_o}{(\upsilon^2+K)^{\frac{1}{2}}} \left\{ \frac{\upsilon^2+(\upsilon^2+K)^{\frac{1}{2}}}{2} \right\}^{1-n} \tag{39b}$$

The results of eqs. (38) and (39) are compared in fig. 9. If we reinstate the expressions for Q_e, Q_c and Q_b into eqs. (33) and (34), we are able to express the dominant current I_n as a function of the applied voltages V_{be} and V_{cb}.

For the base current the integral of the continuity equation again holds (cf. eq. 13):

$$I_b = \frac{dQ_b}{dt} + I_{rec}, \tag{40}$$

but the recombination is now modelled as

$$I_{rec} = I_{be} + I_{bc}.$$

In other words, it is split into an emitter and a collector part, with

$$I_{be} = I_1 \{ \exp(qV_{be}/kT)-1 \} + I_2 \{ \exp(qV_{be}/m_e kT)-1 \} \tag{41a}$$

and

$$I_{bc} = I_3 \{ \exp(qV_{bc}/m_c kT)-1 \}. \tag{41b}$$

The total collector and emitter currents are now given by:

$$I_c = I_n - I_{bc} - \tau_r \frac{dI_r}{dt} + C_{cT} \frac{dV_{cb}}{dt} \tag{42}$$

and

$$-I_e = I_n + I_{be} + \tau_f \frac{dI_f}{dt} + C_{eT} \frac{dV_{be}}{dt}. \tag{43}$$

The circuit diagram of this model, completed with series resistances, is given in fig. 10.
The model, as described by eqs. (33) to (43) and represented in fig. 10, accounts for many physical phenomena inside the transistor. Because Q_b is voltage-dependent the effects of high injection in the base ($\tau_f I_f$ becoming larger than Q_{bo}) and quasi-saturation of the collector ($\tau_r I_r > Q_{bo}$) are included. The base push-out or Kirk effect is represented by the factor B, which is a function of I_c and V_{cb} and can become larger than one. The emitter part I_{be} of the base current is modelled by two diodes in parallel, one ideal and one with a non-ideality factor $m_e > 1$. This makes the current gain at low current levels bias-dependent, as it is in reality.

The voltage dependence of the charge Q_c (eq. (37b)) models the Early effect.

The base conductivity modulation at high injection can be accounted for by putting for the base series resistance

$$r_b = r_{ex} + r_b' \frac{Q_{bo}}{Q_b}, \qquad (44)$$

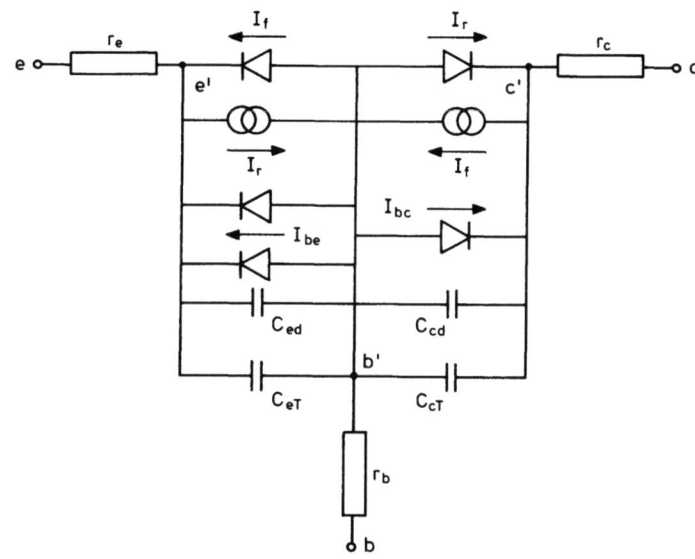

Fig. 10. Circuit representation of the Gummel-Poon model, as given by eqs. (33) to (43), with additional series resistances.

r_{ex} is the resistance of the base region outside the emitter and r'_b is the resistance underneath the emitter, at low injection levels.

The Gummel-Poon model is very accurate, taking many physical effects into account, but it requires a lot of parameters for its characterization, namely four parameters (I_s, Q_{bo}, τ_f, τ_r) for the dominant current, two times four (C_o^s, V_o^b, n and b or K) for the transition layer capacitances, five for the base current (I_1, I_2, I_3, m_e, m_c) and four for the series resistances (r_e, r_{ex}, r'_b, r_c), altogether 21 parameters. Modelling the base push-out factor B requires another four parameters.

As far as the high-frequency behaviour is concerned, it is a one-pole model without an excess phase shift for the current gain.

6. THE LINVILL MODELS

The Linvill models are also based on the regional approach, which divides the transistor into depletion layers with space charge and quasi-neutral regions. It is especially the quasi-neutral region that is modelled, by considering the carrier distribution in that region. The modelling is more physical than electrical.

If, for example, we want to model the neutral base region of an npn transistor we start with the transport and continuity equation for electrons, see section 1. After multiplying by the emitter area we get

$$I_n(x,t) = q A D_n \frac{\partial n}{\partial x} \qquad (45)$$

and

$$\frac{\partial I_n}{\partial x} = q A \frac{\partial n}{\partial t} + q \frac{A}{\tau_n} n(x,t) \qquad (46)$$

The field term is neglected in eq. (45). These two equations are analogous to the equations of a GRC transmission line, but they relate minority current and carrier density instead of current and voltage.

The circuit diagram of the model now looks like a network section with new names for the circuit elements [22]. A base region section of length Δx is characterized by a combinance $H_c \Delta x = qA/\tau \, \Delta x$,

a storance $S \Delta x = qA \Delta x$,

and a diffusance $H_d / \Delta x = qAD_b / \Delta x$.

Figure 11 gives the network section with the new elements. The input and output quantities are the carrier concentrations $n(x,t)$ and the current $I_n(x,t)$.

To model the total base region we can cascade many sections, but the more sections the more complex the model will be.

A one-section model is equivalent to the charge-control model as formulated by eqs. (14) to (18).

Fig. 12 gives a complete transistor model, indicating how the Linvill model for the base is connected to the transistor terminals.

The independent variables for the base lumped network are $n(o,t)$ and $n(w,t)$ and these depend on the junction voltages, e.g. as

$$n(o,t) = n_{po} \{ \exp(qV_{b'e}/kT)-1 \} \tag{47a}$$

and

$$n(w,t) = n_{po} \{ \exp(qV_{b'c}/kT)-1 \} . \tag{47b}$$

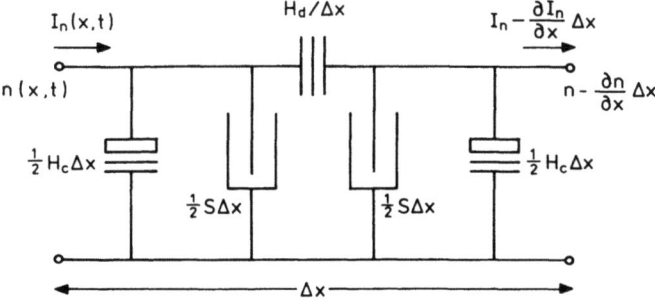

Fig. 11. Linvill model for a section (length Δx) of the quasi-neutral base region, satisfying eqs. (45) and (46).

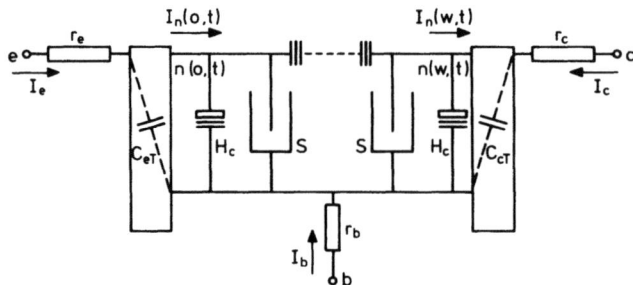

Fig. 12. Complete transistor Linvill model with additional series resistances and transition layer capacitances. The junction voltages $V_{b'e'}$ and $V_{b'c'}$ control the input quantities $n(o,t)$ and $n(w,t)$ of the Linvill network.

The base current is made up of the difference between the input and output currents $I_n(o,t)$ and $I_n(w,t)$ plus the charging currents for the transition layer capacitances C_{eT} and C_{cT}. Therefore it is obvious that

$$-I_e = -I_n(o,t) + C_{eT} \frac{dV_{b'e'}}{dt} \qquad (48a)$$

and

$$-I_c = I_n(w,t) + C_{cT} \frac{dV_{b'c'}}{dt} . \qquad (48b)$$

Eqs. (47) resemble the ideal diode characteristic and are only valid in low injection situations. For high injection situations eqs. (47) have to be modified in the following way:

$$n(o,t) \{ N_a(o) + n(o,t) \} = n_i^2 \{ \exp(qV_{b'e'}/kT)-1 \} \qquad (49a)$$

and

$$n(w,t) \{ N_a(w) + n(o,t) \} = n_i^2 \{ \exp(qV_{b'c'}/kT)-1 \} \qquad (49b)$$

where $N_a(o)$ and $N_a(w)$ are the impurity concentrations at emitter and collector sides.

In some situations it is a serious drawback that the field term is neglected in eq. (45). In the case of a fixed built-in field (E) this can be remedied by connecting an extra current source $qA\mu_n E(n(o)+n(w))/2$ in parallel with the diffusance element, but if the field becomes dependent on the carrier concentrations, this no longer helps.
Thus Linvill modelling is applicable to neutral regions with electric fields that are not bias-dependent.

7. CONCLUDING REMARKS

In the preceding sections we have treated the basic facts of the Ebers-Moll, the Charge-Control, the Gummel-Poon and the Linvill models. Only npn transistors were discussed, but with slight modifications the modelling theories are quite as valid for pnp transistors.
Although many physical effects were included in the transistor models, we felt that avalanche multiplication was beyond the scope of this review because of its modelling complexity.
For each model we can see that making the model more accurate also means making it more complex. In order to refine the high-frequency behaviour of the Ebers-Moll model we had to introduce several extra parameters. The Gummel-Poon model takes care of many specific physical effects, but it requires some 20 to 25 parameters to do so. The Linvill model can be used to simulate transient responses, but if we want to have better results than are obtainable with the charge-control models, we need many lumps or sections.
In each case, therefore, the designer must seek a compromise between accuracy and complexity.

REFERENCES

1. J.W. Slotboom, IEEE Trans. on Electron Devices, ED-20, 669 (1973).
2. A. Wieder, "Numerical models for bipolar devices", this issue.
3. J.J. Ebers and J.L. Moll, Proc. I.R.E., 42, 1761 (1954).
4. J.L. Moll, Proc. I.R.E., 42, 1773 (1954).
5. D.J. Hamilton, F.A. Lindholm and J.A. Narud, Proc. I.E.E.E., 52, 239 (1964).
6. W.L. Engl and J.B. Kioustelidis, Solid-State Electronics, 12, 239 (1969).

7. J.M. Early, Proc. I.R.E., 40, 1401 (1952).
8. F.A. Lindholm and D.J. Hamilton, Proc. I.E.E.E., 59, 1377 (1971).
9. J. Logan, Proc. I.E.E.E., 60, 335 (1972).
10. J. Logan, Proc. I.E.E.E., 60, 78 (1972).
11. G. Rey, Solid-State Electronics 12, 645 (1969).
12. J.R. Hauser, I.E.E.E. Trans. on Electron Devices, ED-11, 238 (1964).
13. G. Rey, K. Lemaire and J.P. Bailbe, l'Onde Electrique, 50, 503 (1970).
14. R. Beaufroy and J.J. Sparkes, Automat. Tel. Eng. Journal 13, 310 (1957).
15. D. Scharfetter, "Bipolar models for IC design", this issue, and
    H.C. de Graaff, "High current density effects in the collector of bipolar transistors", this issue.
16. J. te Winkel, I.E.E.E. Transactions on Electron Devices, ED-20, 389 (1973).
17. H.K. Gummel, Bell Syst. Techn. J. 49, 115 (1970).
18. J.L. Moll and J.M. Ross, Proc. I.R.E., 44, 72 (1956).
19. H.K. Gummel and H.C. Poon, Bell Syst. Techn. J. 49, 827 (1970).
20. C.T. Kirk, I.R.E. Trans. on Electron Devices, ED-19, 164 (1962).
21. H.C. Poon and H.K. Gummel, Proc. I.E.E.E., 57, 2181 (1969).
22. D. Koehler, Bell Syst. Techn. J. 46, 523 (1967) and
    J.G. Linvill, Proc. I.R.E. 46, 1141 (1958).

MEASUREMENTS FOR BIPOLAR DEVICES

P.G.A. Jespers

Université Catholique de Louvain
Microelectronics Lab, Bâtiment Maxwell
1348 Louvain-la-Neuve, Belgium

SUMMARY

DC and low frequency AC parameters

1.1. Basic parameters of the Ebers and Moll model.
1.2. Evaluation of the Ebers and Moll basic parameters.
1.3. The Gummel-Poon model.
    1.3.0. Basic concepts underlying the G-P model.
    1.3.1. Early Effect.
    1.3.2. High injection.
    1.3.3. Review of the G-P model parameters.
    1.3.4. Introduction of reverse feedback in the G-P model.
1.4. Series resistances.
    1.4.0. Collector and emitter series resistances $R_{CC'}$ and $R_{EE'}$.
    1.4.1. Base series resistance $R_{BB'}$ and base crowding effects.

High frequency and dynamic large signal parameters

2.1. Dynamic large signal model of the bipolar transistor.
2.2. Small signal dynamic model (the hybrid π).
2.3. Experimental evaluation of the dynamic parameters.
    2.3.0. Forward transit time τ_F, and depletion capacitances C_E and C_C.
    2.3.1. Reverse transit time τ_R.

2.3.2. Simulation of an invertor stage.

2.4. High frequency measurements.

Conclusion

DC AND LOW FREQUENCY AC PARAMETERS

1.1. Basic parameters of the Ebers and Moll model.

The Ebers and Moll model was introduced in a previous section. The equations may be written as follows:

$$\pm \begin{bmatrix} I_E \\ I_C \end{bmatrix} = \begin{bmatrix} -I_{fo} & \alpha_R I_{ro} \\ \alpha_F I_{fo} & -I_{ro} \end{bmatrix} \cdot \begin{bmatrix} e^{\frac{V_E}{V_T}} - 1 \\ e^{\frac{V_C}{V_T}} - 1 \end{bmatrix} \quad (1)$$

The corresponding equivalent circuit known as the injection model is shown in fig. 1. Currents are positive when entering into the transistor and negative when flowing out of it. Hence in the above expression, the positive sign must be taken for NPN transistors and the negative for PNP's. Voltages V_E and V_C across respectively the emitter and collector junctions are positive when considering forward biased junctions and negative in the opposite case.

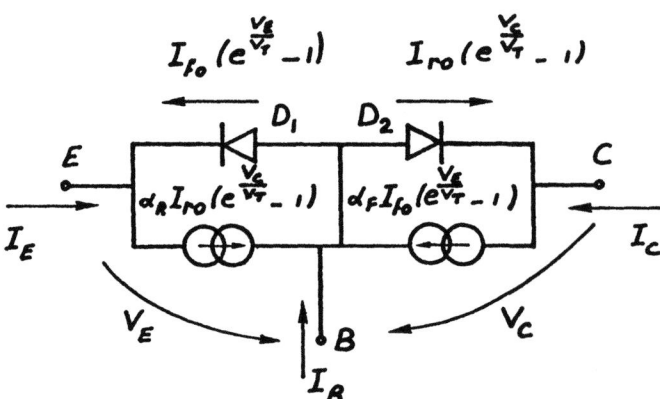

Fig. 1   Ebers and Moll model (injection version).

α_F the common base forward current gain, is given by :

$$\alpha_F = - \frac{I_C}{I_E} \quad \text{considering } V_C = 0 \quad (2)$$

α_R the common base reverse current gain, similarly, is given by :

$$\alpha_R = - \frac{I_E}{I_C} \quad \text{with the condition that } V_E = 0$$

α_F and α_R are both positive quantities.

I_{fo} is the saturation current of the diode D_1.
I_{ro} is the saturation current of the diode D_2.
It can be shown that :

$$\alpha_F I_{fo} = \alpha_R I_{ro} \quad (4)$$

Three quantities only need to be measured experimentally ($\alpha_F I_{fo}$, α_F and α_R) in order to fully characterize the Ebers and Moll model.

Another first order model often considered, is known as the transport version of the bipolar transistor. Instead of two saturation currents, one only is defined :

$$I_S = \alpha_F I_{fo} = \alpha_R I_{ro}$$

and the common emitter current gain β_F is introduced instead of α_F :

$$\beta_F = \frac{I_C}{I_B} \quad \text{with the condition that } V_C = 0$$

Consequently,

$$\beta_F = \frac{\alpha_F}{1-\alpha_F} \quad \text{or} \quad \frac{1}{\alpha_F} = 1 + \frac{1}{\beta_F} \quad (5)$$

Similarly, the reverse current gain, is introduced :

$$\beta_R = \frac{I_E}{I_B}\quad \text{with } V_E = 0$$

thus,

$$\beta_R = \frac{\alpha_R}{1-\alpha_R} \quad \text{or} \quad \frac{1}{\alpha_R} = 1 + \frac{1}{\beta_R} \quad (6)$$

With these notations, expression (1) becomes :

$$\pm \begin{bmatrix} I_E \\ I_C \end{bmatrix} = I_S \begin{bmatrix} -\frac{1}{\beta_F} & \boxed{-1} & \boxed{1} \\ \boxed{1} & \boxed{-1} & -\frac{1}{\beta_R} \end{bmatrix} \cdot \begin{bmatrix} \exp(V_E/V_T)-1 \\ \exp(V_C/V_T)-1 \end{bmatrix} \quad (7)$$

If the ± 1 terms are grouped, the following results is obtained for an NPN transistor :

$$I_E = -(I_F - I_R) - I_F/\beta_F \quad (8a)$$

$$I_C = (I_F - I_R) - I_R/\beta_R \quad (8b)$$

which corresponds to the equivalent circuit of fig. 2, assuming that :

$$I_F = I_S (e^{\frac{V_E}{V_T}} - 1) \quad (9a)$$

and

$$I_R = I_S (e^{\frac{V_C}{V_T}} - 1) \quad (9b)$$

thus : $(I_F - I_R) = I_S (e^{V_E/V_T} - e^{V_C/V_T})$

Again, three parameters $(I_S, \beta_F$ and $\beta_R)$ must be determined experimentally for full characterization of the model.

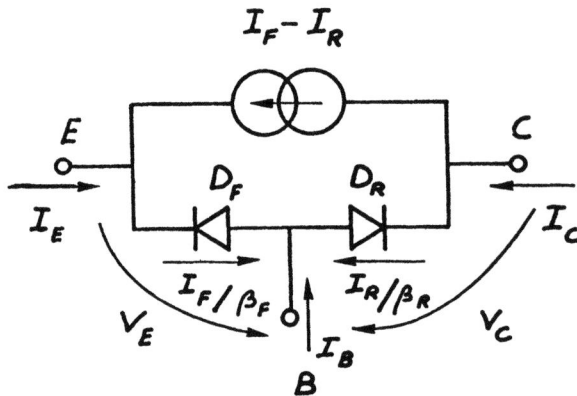

Fig.2 *Ebers and Moll model (transport version).*

The transport version often is preferred because it is simpler than the injection model and better suited for CAD applications. It is a truly common emitter oriented model as results from the example shown in fig. 3 which represents a forward biased bipolar transistor.

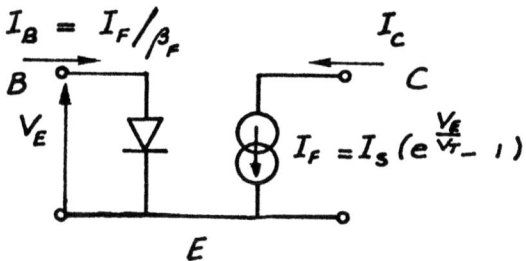

Fig. 3 Ebers and Moll transport version of the common emitter configuration in the active region.

1.2. Evaluation of the Ebers and Moll basic parameters.

The three quantities I_s, β_F and β_R can experimentally be obtained by means of the test set shown in fig. 4. Operational amplifiers are used rather than ammeters for the voltage drop across their input terminals can be kept much smaller; forward biasing of the collecting junctions thus may be avoided.

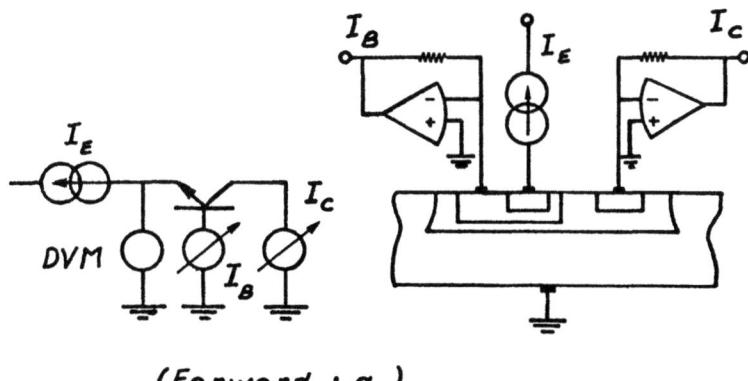

(Forward : a)

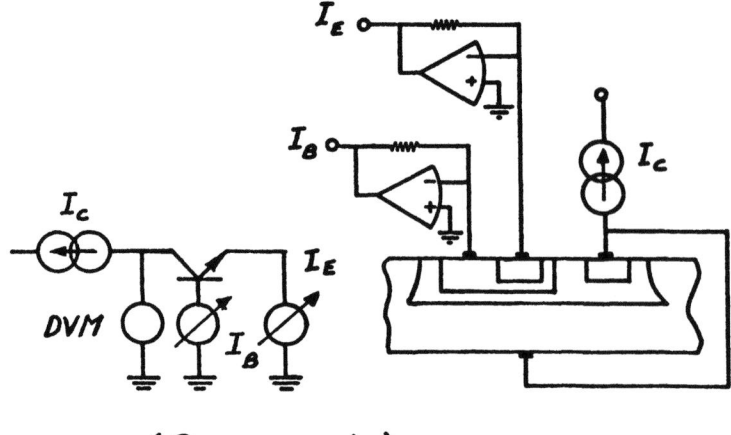

(Reverse : b)

Fig. 4 Experimental set-up for the determination of the collector and base currents versus the emitter voltage in the forward and reverse modes.

A typical plot of ln I_C and ln I_B versus V_E considering transistor 2N5109 of Sescosem is shown in fig. 5. In the same figure, reverse bias conditions are considered (ln I_E and ln I_B versus V_C). The currents I_C and I_E practically coincide over more than 7 decades following a straight line accordingly to expressions (9a) and (9b). The slope of this line should be equal to V_T. A value of 25,2 mV was obtained from the actual plot. The saturation current I_s can be found by extrapolation of this line till V_E is equal to zero Volt. In the present case, I_s is equal to $1.70 \cdot 10^{-15}$ A.

Base current, however, does not obey the first order model as well. Nowhere is it parallel to the collector current. Additional mechanisms must be taken into account in the forward region to explain this, for instance the recombination in the emitter junction which is taking place under very low bias conditions ($V_E \simeq 400$ mV) leading to the following expression of base current [1] :

$$I_B = I_{BO} \cdot \left[\exp(V_E/nV_T) - 1 \right] \qquad (10)$$

with n equal to two.
This is verified below 1 nA in the present example, with a corresponding value of I_{BO} given by $2.9 \cdot 10^{-12}$ A. In the intermediate region, from 10 nA till 10 µA, the slope of the ln $I_B(V_E)$ curve gradually varies, with n changing from 2 to 1. In this region, surface recombination must be considered.

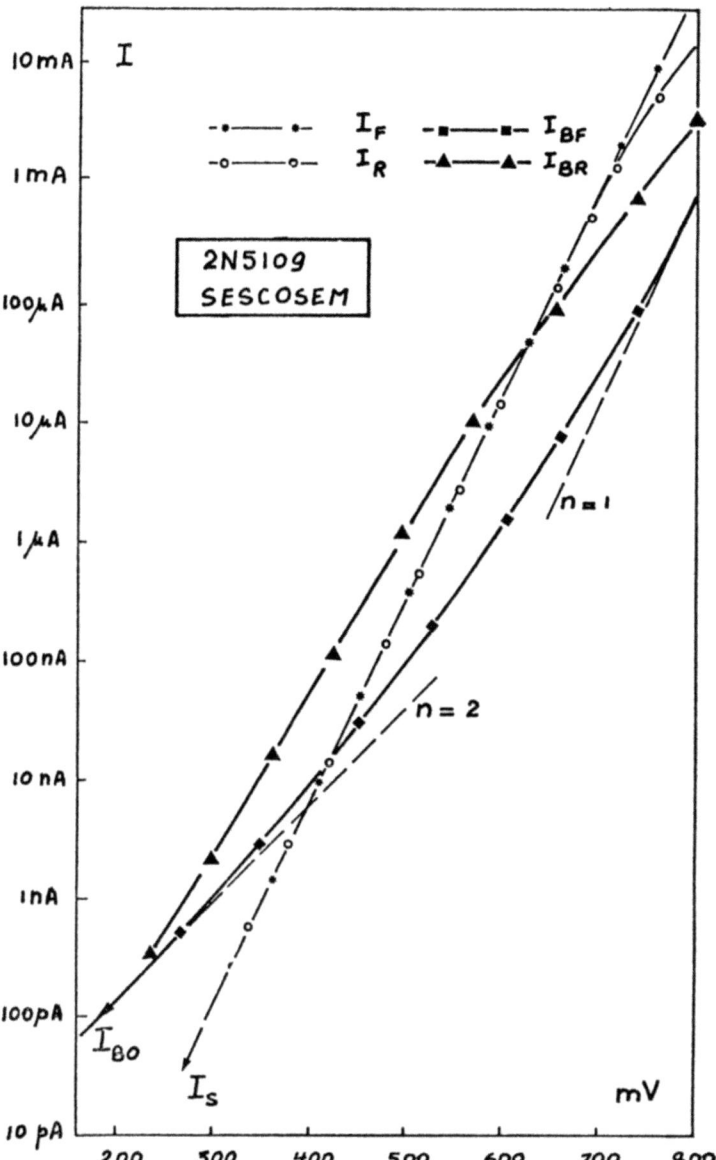

Fig. 5 Forward and reverse characteristics of the 2N5109 transistor (Sescosem).

The reverse base current $I_B(V_C)$ is also influenced by the same recombination mechanisms, but to a much larger extend since the injecting junction not only encompasses the collecting region but also a large oxide region where surface recombination is taking place mainly.

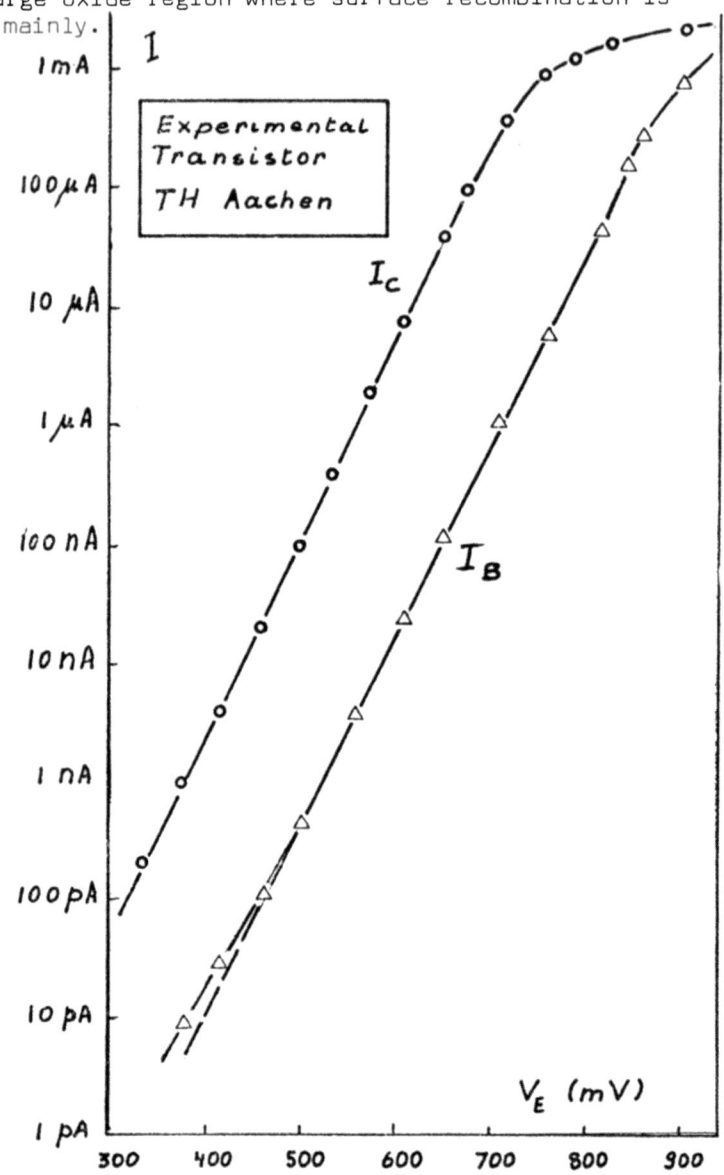

Fig. 6 I_C and I_B versus V_E in the forward mode for an experimental transistor with the lowest possible value of interface states.

Surface recombination is largely dependent upon the technology used. Figure 6 represents a plot similar to the one of figure 5, considering a bipolar transistor for which extreme care has been taken in order to minimize the surface recombination. The Si-SiO$_2$ interface was carefully controlled using a technology which is typical of MOS device fabrication as reported in [2]. It is quite obvious that the simple Ebers and Moll model very well describes the behavior of this last transistor under low injection conditions. This is not true however for the transistor of fig. 5 for which, one may consider current-dependent values for β_F and β_R in order to adequately describe the recombination mechanisms. Fig. 7 represents a plot of both β_F and β_R versus ln I_C. Instead of considering variable β's another method which is often preferred consists to introduce additional diodes - usually one or two max- in parallel with the base diodes with adequate values of saturation current and factor n as in equation (10). This procedure provides sufficient modeling flexibility in order to describe properly the low current behavior of β as well.

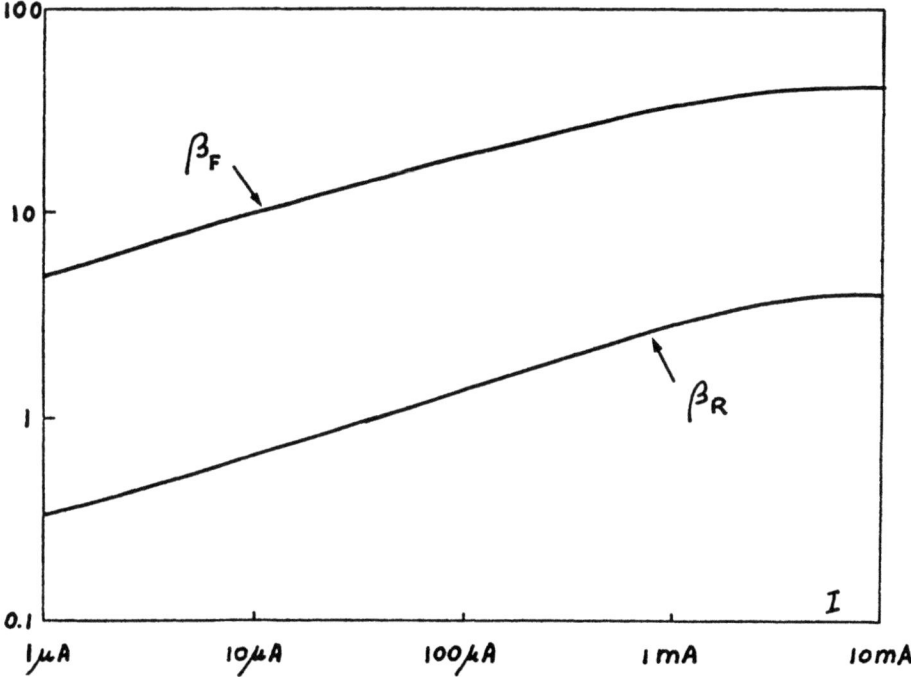

Fig. 7 Forward and reverse current gain versus collector current for the 2N5109 transistor.

For high level injection, other mechanisms must be considered in order to explain the current dependance of β_F and β_R. Some of them will be reviewed later on in this section. Measurements must always be carried out with little power dissipated in the transistor, especially under high reverse bias conditions. The internal temperature of the transistor indeed may raise substantially, reducing the emitter voltage drop approximatively by 2,5 mV per degree centigrade, and causing the $\ln I_C$ and $\ln I_E$ curves progressively to shift from right to left. Since the thermal time constant of most transistors is quite small, only pulsed measurement should be considered. The emitter junction voltage must be measured by means of a calibrated oscilloscope using an accurate offset voltage source in order to achieve both sensitivity and accuracy. A typical oscillogram representing V_E versus time is shown in fig. 8. Duty cycles as low as 10^{-3}, may be required in order to avoid substantial mean temperature raising.

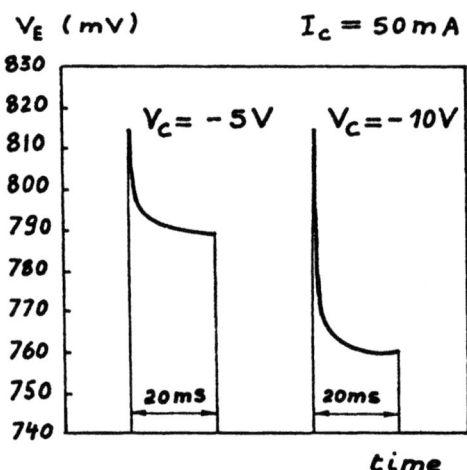

Fig. 8 *Thermal sensitivity of V_{BE} versus time as a function of power dissipated in the collector.*

1.3. The Gummel-Poon model

1.3.0. Basic concepts underlying the G-P model

The Gummel-Poon model, introduced in 1970 [3,4],is based on an extension of the Moll and Ross formula [5]. The basic idea underlying the Moll and Ross formula is the integral charge concept Q which is the charge represented by the total majority carriers per unit area in the neutral emitter, base or collector regions. Under low injection conditions, Q is equal to the integral of the net impurity concentrations in the respective region; thus one may write in the base region, for instance, :

$$Q_B = q \cdot \int_{\text{neutral base}} N_B(x) \, dx \tag{11}$$

Under high level injection however, expression (11) may be violated locally for the minority carrier density may become larger than the net impurity concentration N(x), forcing excess majority carrier to enter into the base in order to maintain quasi-neutrality. When this is the case, the integral base charge Q_B still may be determined by means of expression (11) provided N(x) be replaced by the actual majority carrier density. Similar considerations also hold for the collector region which generally is less doped than the base, thus more sensitive to high level injection conditions.

The Moll-Ross formula is based on the assumption that bulk recombination is negligeable, a more or less valid approximation in the base region, which may be incorrect for the collector region which is much thicker than the base region. This certainly is the case in non epitaxial transistors for which expression (11) may be replaced by :

$$Q_C = q \, N_C \, L_C \tag{12}$$

where N_C represents the uniform collector impurity concentration and L_C the diffusion length in the collector region. For epitaxial, or buried collector transistors, a different situation occurs : the epi-interface layer is supposed to act as a recombination medium characterized with some unknown surface recombination velocity. Experimental values of this recombination velocity can be determined indirectly from inverse current gain measurement.

The equations of the bipolar transistor presented in matrix form, accordingly to the Moll-Ross formula are :

$$\pm \begin{bmatrix} I_E \\ I_C \end{bmatrix} = q^2 n_i^2 A_E \begin{bmatrix} -(D_E/Q_E + D_B/Q_B) & D_B/Q_B \\ D_B/Q_B & -(D_B/Q_B + D_C/Q_C) \end{bmatrix} \cdot \begin{bmatrix} \exp(V_E/V_T) - 1 \\ \exp(V_C/V_T) - 1 \end{bmatrix} \tag{13}$$

with the same notations as in expression (1) and :
$q = 1,6 \, 10^{-19}$ C, the electron charge,

$n_i = 1.45 \; 10^{10}$ cm^{-3}, the Si intrinsic carrier concentration,

A_E : the emitter area.

D_E, D_B and D_C, respectively the emitter, base and collector diffusion constants.

and

Q_E, Q_B, Q_C the respective integral charges defined above.

The fundamental idea of Gummel and Poon is that Q_B is not a constant, but that it is influenced by several factors. For instance Q_B is modulated by the junction voltages V_E and V_C, for any modification of V_E and V_C changes the width of the emitter and junction depletion regions. Consequently, when the reverse bias increases, Q_B decreases as a result of the collector depletion width increase (Early effect). Similarly, when high level injection conditions prevail, Q_B increases for the majority carrier density in the base may exceed the net impurity concentration in order to maintain neutrality within the base (base conductivity modulation).

In the Gummel-Poon model, Q_{BO} is defined as the value of Q_B, when $V_E = V_C = 0$ V. Dividing Q_{BO} by the electron charge q yields the Gummel number "GU" which also is called the "integral base doping".

If, in equation (13), all expressions $1/Q_B$ are replaced by

$$\frac{1}{q_B} \cdot \frac{1}{Q_{BO}}$$

with $q_B = \dfrac{Q_B}{Q_{BO}}$ (relative Gummel number), the following result is obtained :

$$\pm \begin{bmatrix} I_E \\ I_C \end{bmatrix} = \frac{q^2 n_i^2 A_E D_B}{Q_{BO}} \begin{bmatrix} -(\frac{D_E Q_{BO}}{Q_E D_B} + \frac{1}{q_B}) & \frac{1}{q_B} \\ \frac{1}{q_B} & -(\frac{D_C Q_{BO}}{Q_C D_B} + \frac{1}{q_B}) \end{bmatrix} \cdot \begin{bmatrix} \exp(V_E/V_T)-1 \\ \exp(V_C/V_T)-1 \end{bmatrix} \quad (14)$$

Let us define $I_S = \dfrac{q^2 n_i^2 A_E D_B}{Q_{BO}}$ (15a)

and $\beta_{FO} = \dfrac{Q_E D_B}{D_E Q_{BO}}$ (15b)

$\beta_{RO} = \dfrac{Q_C D_B}{D_C Q_{BO}}$ (15c)

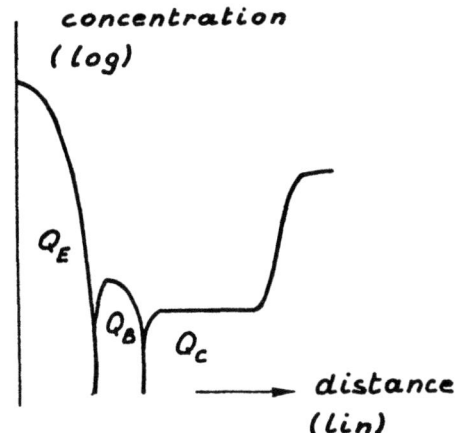

Fig. 9 Integrated charges of the bipolar transistor.

Expression (15b) describes an important result illustrated by fig. 9, which says that the common emitter current gain β_F is proportional to the ratio of the shaded emitter and base areas, assuming D_B and D_E are constants. This statement however needs to be corrected for the high concentration of the emitter introduces band tailing effects which reduce Q_E to some lower value Q_{eff} as will be shown in another section [6].

$$\pm \begin{bmatrix} I_E \\ I_E \end{bmatrix} = I_S \begin{bmatrix} -(\frac{1}{\beta_{FO}} + \frac{1}{q_B}) & \frac{1}{q_B} \\ \frac{1}{q_B} & -(\frac{1}{\beta_{RO}} + \frac{1}{q_B}) \end{bmatrix} \cdot \begin{bmatrix} \exp(V_E/V_T)-1 \\ \exp(V_C/V_T)-1 \end{bmatrix} \quad (15d)$$

Similarly to the expression (9), we define :

$$I_F = \frac{I_S}{q_B} \cdot [\exp(V_E/V_T)-1] \quad (16a)$$

and $\quad I_R = \dfrac{I_S}{q_B} \cdot [\exp(V_C/V_T)-1] \quad (16b)$

or $\quad I_F - I_R = \dfrac{I_S}{q_B} \cdot [\exp(V_E/V_T) - \exp(V_C/V_T)] \quad (16c)$

so that :

$$\pm I_E = -(I_F - I_R) - \frac{I_S}{\beta_{FO}} \cdot [\exp(V_E/V_T)-1] \quad (16d)$$

$$\pm I_C = (I_F - I_R) - \frac{I_S}{\beta_{RO}} \cdot \left[\exp(V_C/V_T) - 1\right] \qquad (16e)$$

and

$$I_B = -(I_E + I_C) = \pm I_S \left[\frac{1}{\beta_{FO}} \cdot (e^{V_E/V_T} - 1) + \frac{1}{\beta_{RO}}(e^{V_C/V_T} - 1)\right] \qquad (16f)$$

The dependance of Q_B from actual bias and current level conditions is modelled accordingly to the Gummel-Poon approach by means of the following expression :

$$Q_B = Q_{BO} + Q_{JE} + Q_{JC} + B \frac{\tau_F I_F}{A_E} + \frac{\tau_R I_R}{A_E} \qquad (17)$$

where :

a) Q_{BO} is the unbiased total base charge per unit area defined above.
b) Q_{JC} and Q_{JE} represent the contributions respectively of the collector and emitter depletion layers per unit area with respect to zero bias.
c) $\frac{\tau_F I_F}{A_E}$ and $\frac{\tau_R I_R}{A_E}$ represent the forward and reverse stored charges in the base per unit area.
d) B is a current depending factor which increases $\tau_F I_F$ in order to take the base charge variation caused by base push-out into account. Obviously, there is no such factor for inverse functioning of double diffused transistors.

When the terms of equation (17) are normalized with respect to Q_{BO}, one has :

$$q_B = 1 + q_{JE} + q_{JC} + B \cdot \frac{I_F}{I_{kF}} + \frac{I_R}{I_{kR}} \qquad (18a)$$

with $I_{kF} = \dfrac{A_E Q_{BO}}{\tau_F}$ and $I_{kR} = \dfrac{A_E Q_{BO}}{\tau_R}$ (18b)

Replacing I_F and I_R by (16a) and (16b) respectively, now gives :

$$q_B = q_1 + \frac{q_2}{q_B}$$

with $q_1 = 1 + q_{JE} + q_{JC}$ (Early effect)

$$q_2 = \frac{I_S}{A_E Q_{BO}} \cdot \left[B\tau_P(e^{V_E/V_T} - 1) + \tau_R(e^{V_C/V_T} - 1)\right]$$

(High injection)

$\qquad\qquad\qquad\qquad\qquad\qquad\qquad\qquad\qquad\qquad\qquad$ (19)

This quadratic expression in q_B can be solved, leading to :

$$q_B = \frac{q_1}{2} + \sqrt{(\frac{q_1}{2})^2 + q_2} \qquad (20)$$

If q_B is reintroduced in I_F and I_R, the Gummel-Poon model of the bipolar transistor is obtained which takes into account a number of effects ignored by the Ebers and Moll first order model. It should be pointed out that the new model is a non-linear one, and consequently that the superposition principle is no more applicable.

In order to fully illustrate the power of the new model, and to derive experimental procedures leading to the knowledge of important parameters, we consider the Early effect and the high injection phenomena separately.

1.3.1. Early Effect

Suppose that low level condition prevails, so that the two last terms of (18) may be deleted. Furthermore, we consider a forward biased transistor and neglect q_{JE} for the emitter depletion layer is very thin, and the contribution of this term indeed is quite small.

Expression (20), thus reduces to :

$$q_B = 1 + q_{JC}$$

The collector depletion capacitance C_C is defined as :

$$C_C = \frac{dQ}{dV_C}$$

where Q represents the charge of the actual collector depletion layer.
Note that :

$$\frac{dQ}{A_C} = - dQ_{JC}$$

where Q/A_C represents the collection depletion charge per unit area. The negative sign is due to the fact that when the collector depletion layer width increases, the base charge Q_B shrinks and vice-versa. One has :

$$Q_{JC} = - \frac{1}{A_C} \int_0^{|V_C|} C_C(V_C) dV_C \qquad (21)$$

Let us call C_o the value of C_C when V_C is equal to zero. We may write :

$$q_{JC} = - \frac{C_o/A_c}{Q_{Bo}} \int_0^{|V_C|} \frac{C_c(V_c)}{C_o} dV_C \qquad (22)$$

Following expressions (16), the collector current I_C (or I_F) now is given by:

$$I_C = \frac{I_S}{1 + q_{JC}} \cdot \exp(V_E/V_T)$$

Taking into account that q_{JC} is always small compared to 1:

$$I_C \simeq I_S \exp(V_E/V_T) \cdot \left[1 + \frac{1}{V_A} \cdot \int_0^{|V_C|} \frac{C_c(V_c)}{C_o} dV_C \right] \qquad (23)$$

with $V_A = \dfrac{Q_{Bo}}{C_o/A_c}$ \qquad (24)

which is called the Early voltage.

The output conductance $\partial I_C/\partial V_C$ along a constant V_E curve is given by:

$$\left. \frac{\partial I_C}{\partial V_C} \right|_{V_E = cte} \simeq \frac{I_S \exp(V_E/V_T)}{V_A} \cdot \frac{C_c(V_c)}{C_o} \qquad (25)$$

This expression leads to the well known conclusion shown in fig. 10, which relates the Early voltage to the output conductance of the common emitter configuration.

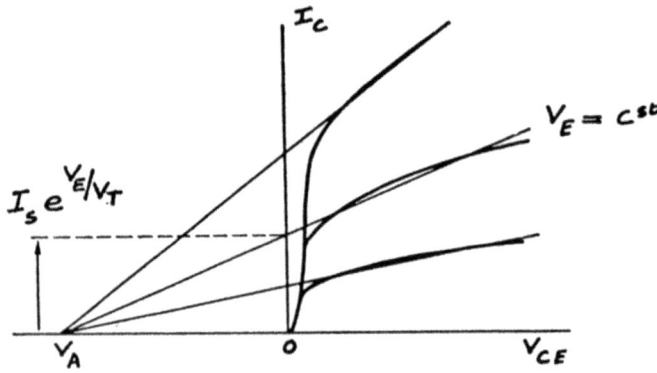

Fig. 10 *Illustration of the Early voltage of a bipolar transistor.*

Expressions (24) and (15a) provide two different manners to measure Q_{BO}, indeed :

a) from (15a)
$$Q_{BO} = \frac{q^2 n_i^2 A_E D_B}{I_S} \quad (26)$$

and

b) from (24)
$$Q_{BO} = V_A \cdot \frac{C_o}{A_C} \quad (27)$$

The current I_S appearing in (26) is found by extrapolating the ln I_C curve versus V_E, as shown previously. A_E is known from the geometry of the transistor, D_B generally is not well known for it is concentration-dependent. Usually a rough estimate of D_B is used, based on the average concentration of the base. This method was proposed by Gummel [7]. Expression (27), seems more convenient because all quantities involved are better known. A_C may be determined from the geometry of the transistor. The collector-base capacitance C_c may be measured by means of a transformer bridge like the Boonton Capacitance Bridge 75D, and V_A can be evaluated by means of a conventional curve tracer, provided that the base of the transistor is controlled by means of a voltage source rather than by a current source. A more refined method consists to compute the integral of C_c/C_o which appears in (23), using for instance a best fit approximation such as :

$$C_c/C_o = (1 + \frac{|V_c|}{\phi})^{-n} \quad (28)$$

which yields :

$$\int_0^{|V_C|} \frac{C_c(V_C)}{C_o} dV_C = \frac{\phi}{1-n} \cdot \left[(1 + \frac{|V_c|}{\phi})^{1-n} - 1 \right] \quad (29)$$

The Early voltage V_A then can be adjusted in order to find the correct $I_c(V_c)$ curve. Careful attention must be given to the temperature sensitivity of the transistor during the measurement since the base is voltage controlled.

Let us illustrate a third procedure. Fig. 11 represents the test set used in order to measure the small signal transistor output impedance at some frequency low enough in order to avoid the "looping" effect visible in most curve tracers. The principle of the apparatus is simple : transistors T_1 and T_2 act as two opposite current sources controlled by the output voltage V_2 of integrator A_5. Op amp. A_4 is a current mirror controlled by the collector current of T_2. Proper adjustment of V_2 makes ΔI equal to zero. This is automatically obtained by means of the loop formed by A_1, A_3 and A_5, provided the switch S has been closed. The opamp A_1 furthermore sets the collector to emitter voltage of the tested device at the desired control voltage.

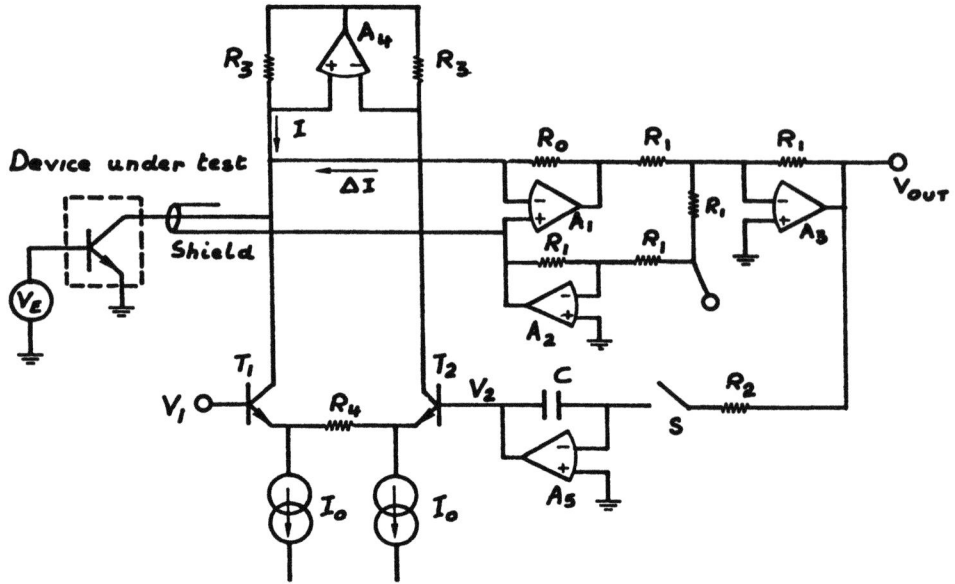

Fig. 11   Experimental set-up for the determination of the high output conductance of a bipolar transistor at low frequencies.

After closure of S long enough in order to reach steady state conditions, the switch is opened again. V_{out} remains normally at zero volt, until a small change ΔV_{CE} produces a small current change ΔI, thus an output voltage $-(R_o \Delta I)$. Knowing V_E and I_S, the Early voltage can be found accordingly to (25) leading to the evaluation of Q_{BO} from (27).

Data obtained from the transistor taken as test sample in fig. 5 are reported in table I considering a value of $I_s \exp(V_E/V_T)$ of 690 µA.
Knowing that C_E/A_E is equal to 21,3 nF/cm^2, Q_{BO} is approximatively equal to 1.8 10^{-6} C/cm^2 and the corresponding Gummel number is 10^{13} cm^{-2}. Introduction of Q_{BO} in (26) furthermore leads to a value of 13 cm^2 s^{-1} for D_B which corresponds to an average hole concentration somewhere between 10^{15} and 10^{16} cm^{-3}.

V_{CE} (V)	V_C (V)	$\dfrac{C_c(V_c)}{C_o}$	g_{oe} (kΩ)	V_A (V)
2	1.3	0.636	166	72.8
4	3.3	0.547	196	74.0
6	5.3	0.502	217	75.2
8	7.3	0.477	227	74.7

Table I

1.3.2. High injection

Starting from equation (18 a et b), let us restrict ourselves to forward bias conditions ($I_R = 0$), and furthermore neglect the contribution of the Early effect ($q_{JE} = q_{JC} = 0$). One has :

$$q_B = 1 + B \frac{I_F}{I_{kF}}$$

Knowing from (16a), that :

$$I_F \simeq \frac{I_s}{q_B} \cdot e^{V_E/V_T}$$

the elimination of q_B between the above equations leads to :

$$V_E = V_T \cdot \ln \left[\frac{I_F}{I_S} \left(1 + B \frac{I_F}{I_{kF}} \right) \right] \quad (30)$$

B is a current dependent factor which is empirically determined in order to simulate base push out. In few cases B may be assimilated to one even under high injection, for instance in such cases where the collector doping concentration exceeds always the base doping (e.g. reverse biased planar transistors, the NPN transistor of I^2L, alloy transistors). High injection effects then reduce to base conductivity modulation, a phenomenon which is called the "Webster effect" [8]. When this happens, expression (30) reduces to :

$$V_E = V_T \cdot \ln \left[\frac{I_F}{I_S} \left(1 + \frac{I_F}{I_{kF}} \right) \right] \quad (31)$$

We see that when $I_F < I_{kF}$, the usual exponential law is obtained :

$$I_F = I_s \cdot \exp(V_E/V_T) \qquad (32)$$

If $I_F > I_{kF}$ one finds :

$$I_F = \sqrt{I_s \, I_{kF}} \cdot \exp(V_E/2V_T) \qquad (33)$$

The current I_{kF}, which is equal to $A_E \, Q_{BO}/\tau_F$ has been given the name of "forward knee current", by Gummel [4] . The corresponding "forward knee voltage" is equal to :

$$V_{kF} = V_T \ln \frac{I_{kF}}{I_S} = V_T \cdot \ln \frac{A_E Q_{BO}}{\tau_F \, I_S} \qquad (34)$$

Since base current computed from (16f) does not depend from the integrated base charge, the $\ln I_C$ and $\ln I_B$ curves versus V_E must be like in fig. 12. The difference between the actual $\ln I_C$ curve and the ideal straight line thus provides a means to measure directly the reduced Gummel factor q_B. This leads to the expression of forward current gain :

$$\beta_F = \frac{\beta_{FO}}{q_B} \qquad (35)$$

resulting from the combination of expressions (16e) and (16f).

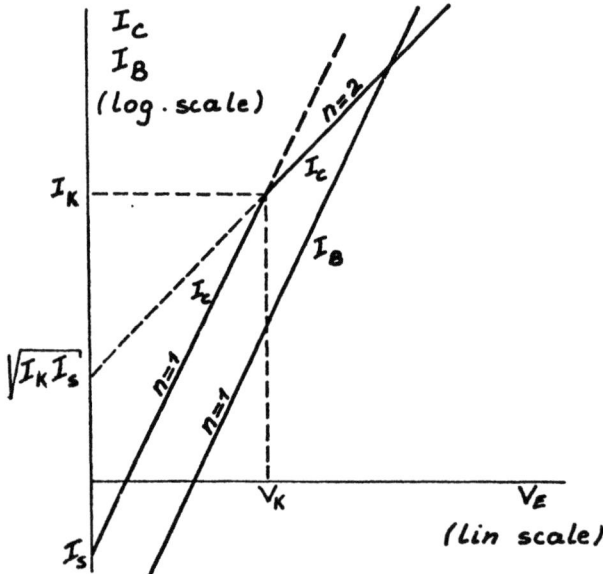

Fig. 12 Illustration of the "Webster effect"

In most transistors, the influence of the factor B usually overrules the Webster effect, for the collector doping of double diffused transistors necessarily is lower than that of the base. Hence conductivity modulation prevails in the collector region before it is experienced in the base. The "Kirk effect" [9] really is the dominating phenomenon. Modellisation of the transistor still can be achieved by means of the Gummel-Poon model, using expression (30) with B ≠ 1. Thus a new knee current is found which is below $A_E Q_{BO}/\tau_F$ and moreover the slope of the ln I_F versus V_E curve is bended by a factor becoming rapidly larger than 2.

Modellisation of B, as proposed by Gummel, is based on the assumption that at the time the collector current exceeds a value I_1 given by :

$$I_1 = \frac{A_E(|V_c| + V_d)}{\rho W_C} \quad (36)$$

the actual base width of the transistor very rapidly extends from its initial value W_B to $(W_B + W_C)$ where W_C represents the width of the thin epitaxial layer between the collector-base junction and the interface between the low doped collector region and the highly doped substrate or buried layer. (ρ is the low doped collector resistivity, V_d the diffusion potential of the collector-base junction which usually is close to .7V, and $|V_c|$ is the absolute value of the reverse voltage applied to the collector-base terminals). Gummel introduced the effective base width W_{eff}, equal to W_B when the collector currents is below I_1, which becomes equal to $(W_B + W_C)$ when I_c exceeds I_1. Knowing that the base transit time varies like the square of the base width, Gummel defines an effective base transit time given by :

$$\tau_{eff} = \left(\frac{W_{eff}}{W_B}\right)^2 \cdot \tau_F = B \cdot \tau_F \quad (37)$$

Various expressions of B can be considered. The one proposed by Gummel and Poon is a function of I_c and I_1 plus an additional parameter I_2 which must be determined empirically in order to fit experimental data.

Once more let us consider the transistor of fig. 5. The "knee current" I_{kF} accordingly to (18b) is given by :

$$I_{kF} = \frac{A_E Q_{BO}}{\tau_F} = 0.60 \text{ A.}$$

knowing that the emitter area A_E is equal to $3,76 \cdot 10^{-5}$ cm^2 and that τ_F is of the order of 10^{-10} s (this will be reported later on in this section). I_{kF} is much larger than the actual "knee point"; computing I_1, indeed, one finds :

$$I_1 = 0.034 \; (|V_C| + 0.7) \; A$$

(with ρ = 2.7 Ωcm, W_C = 4.1μ , V_d = 0.7 V). Considering a series of values of V_{CE}, the following table is obtained :

V_{CE} (V)	V_C (V)	I_1 (A)
0,7	0	.029
1,7	1	.058
2,7	2	.092
3,7	3	.126
5,7	5	.194
17,7	17	.600

Table II

Base push-out overrules thus completely base conductivity modulation. Examination of fig. 5 shows indeed that beyond 30 mA, the actual I_F curve no more coïncides with the ideal $I_S \exp(V_E/V_T)$ curve, thus the "knee current" in the present case is approximatively equal to 30 mA. Of course increasing V_C, as done in normal circumstances, substantially improves I_{kF}.

1.3.3. Review of Gummel-Poon model parameters

The Gummel-Poon model requires a rather large number of parameters which may be categorized in five groups totalizing an ensemble of 21 data.

Group n° 1 contains 4 parameters : the "knee" parameters I_k and V_k plus the forward base transit time τ_F and the reverse base transit time τ_R , or preferably the ratio r_t equal to τ_R/τ_F. Remember that I_k, τ_F and Q_{BO} are linked (18) and that Q_{BO} may be determined from I_s.

Group n° 2 contains 5 parameters describing the base current: the ideal base saturation current I_s/β_{FO}, and the forward non ideal low base current parameters, e.g. I_{BO} and n, plus the reverse base current parameters defined by another couple of values of I_{BO} and n.

Group n° 3 is relative to the emitter capacitance and contains four parameters describing the current dependance of C_E. Instead of expression (28), the Gummel-Poon model uses a more elaborate equation which is better suited for CAD applications.

Group n° 4 contains 4 parameters describing the collector depletion capacitance in the same manner as group n° 3.

Group n° 5 describes base push-out by means of 4 parameters, among which are I_1 (36), an additional parameter I_2, the ratio r_W or (W_C/W_B) and a so called "push-out exponent" which is used to adjust the B function in order to fit the experimental data.

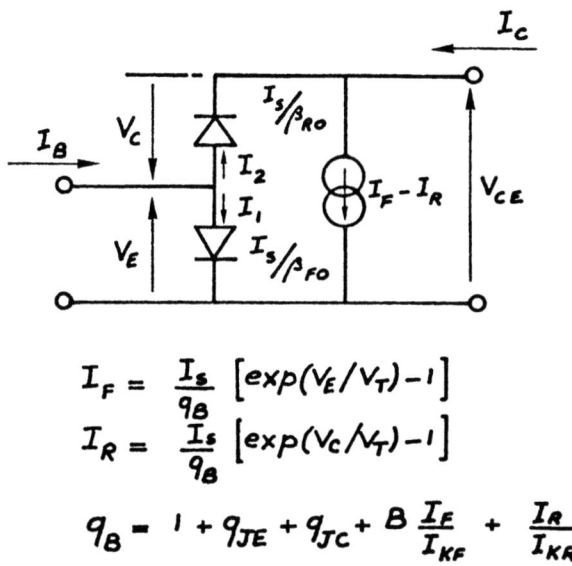

$$I_F = \frac{I_s}{q_B}\left[\exp(V_E/V_T)-1\right]$$

$$I_R = \frac{I_s}{q_B}\left[\exp(V_C/V_T)-1\right]$$

$$q_B = 1 + q_{JE} + q_{JC} + B\frac{I_F}{I_{KF}} + \frac{I_R}{I_{KR}}$$

Fig. 13 Gummel-Poon model.

1.3.4. Introduction of reverse feedback in the Gummel-Poon model

Fig. 14 shows a circuit representation of the Gummel-Poon model. Although representative of many important transistor characteristics, it does not describe reverse feedback. Hence the parameter h_{re} is always equal to zero. Fig. 14 shows an attempt to introduce reverse feedback considering the case of a forward biased transistor only. Instead of a single input diode with a saturation current I_s/β_{FO}, we consider the parallel combination of two diodes. One diode has a constant saturation current I_s/β_{F1} like in the original Gummel-Poon model. The second diode has a q_B dependant saturation current : $I_s/q_B\beta_{F2}$. In this manner, the input node is influenced by the output voltage V_{CE} and reverse feedback is introduced.

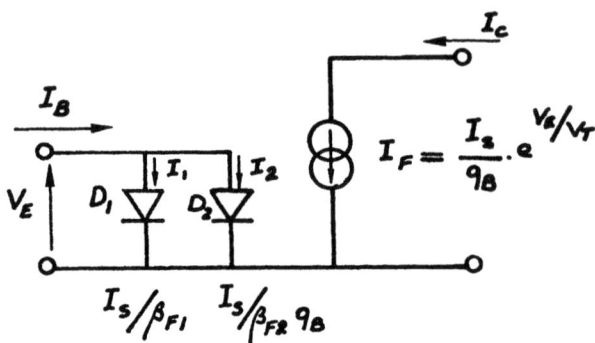

Fig. 14. Introduction of reverse feedback in the Gummel-Poon model (only the active region is considered).

One may also interpret fig. 14 as follows : emitter efficiency is modelled by D_1 like in the original transport model and recombination is modelled by D_2 for the recombination current is strictly proportional to the collector current. In other words, the two input diodes may be considered as separate symbolic representations of both mechanisms controlling the current gain of the transistor. If we can find experimentally the values of β_{F1} and β_{F2}, some insight of the relative contributions of both mechanisms may be obtained.

Let us consider a transistor in the common emitter configuration with its base terminal connected to a voltage source, thus with shorted input terminals. Consider a small collector voltage change v_{CE} and suppose we measure in the same time the small signal currents i_C and i_B. From (16a) :

$$i_C = \frac{dI_C}{dq_B}\bigg|_{V_E=cte} = -\frac{I_C}{q_B}$$

Similarly, one has :

$$i_B = \frac{dI_B}{dq_B}\bigg|_{V_E=cte} = -\frac{I_2}{q_B} \qquad (39)$$

where I_2 represents the DC current flowing into diode D_2. Consequently :

$$\frac{i_C}{i_B} = \frac{dI_C}{dI_B}\bigg|_{V_E=cte} = \frac{I_C}{I_2} = \beta_{F2} \qquad (40)$$

The ratio of i_C/i_B thus provides a direct measurement of β_{F2}. However since β_{F2} in most cases is much larger than β_F, the current i_B may be quite small, of the order of a few nA, and the measurement quite difficult to perform. A better approach is to measure the output impedance of the transistor first with open and second with shorted input terminals and to determine h_{re} from expression :

$$h_{re} = (h_{oe} - g_{oe}^{-1}) \cdot \frac{V_T}{I_E} \tag{41}$$

where h_{oe} is the small signal output conductance with open circuit terminals and g_{oe}^{-1} is the small signal output conductance with shorted input terminals.
The difference $(h_{oe} - g_{oe}^{-1})$ can accurately be measured at very low frequency (e.g. : 10^0 Hz) by means of the test set shown in fig. 11.
Furthermore, since :

$$I_B = I_1 + I_2 = I_s(\frac{1}{\beta_{F1}} + \frac{1}{q_B \beta_{F2}}) \cdot \exp(V_E/V_T)$$

the total derivative of the above expression, considering I_B constant, leads to :

$$\frac{I_B}{V_T} dV_E - \frac{I_2}{q_B} dq_B = 0$$

or :

$$\frac{I_B}{V_T} \frac{dV_E}{dV_{CE}} = \frac{I_2}{q_B} \cdot \frac{dq_B}{dV_{CE}}$$

From (22), we find that :

$$h_{re} = \frac{dV_E}{dV_{CE}}\bigg|_{I_B=cte} = -\frac{1}{q_B} \frac{I_2}{I_B} \frac{V_T}{V_A} \cdot \frac{C_c(V_c)}{C_0}$$

Since h_{re} is known from (41), the ratio $\frac{I_2}{I_B}$ can be determined by means of the above expression.
Furthermore, expression :

$$\frac{I_2}{I_B} = \frac{1/q_B \beta_{F2}}{1/\beta_{F1} + 1/q_B \beta_{F2}}$$

leads to the final result :

$$\frac{\beta_{F2}}{\beta_F} = 1 + \frac{1}{q_B}\left[\frac{1}{q_B} \frac{V_T}{V_A} \frac{1}{|h_{re}|} \frac{C_c}{C_0} - 1\right] \tag{42}$$

Considering again the 2N5109 transistor, the following experimental data were obtained with the same nominal current $I_s \exp(V_E/V_T)$ as in table 5 (the frequency was 10 Hz).

V_{CE} (V)	$h_{oe} - g_{oe}^{-1}$ (Ω^{-1})	h_{re}	$1/q_B$	C_c/C_o	β_{F2}/β_F
2	$-1,0 \cdot 10^{-6}$	$-3,77 \cdot 10^{-5}$	1,014	0,696	6,6
4	$-1,0 \cdot 10^{-6}$	$-3,77 \cdot 10^{-5}$	1,032	0,570	5,6
6	$-1,0 \cdot 10^{-6}$	$-3,77 \cdot 10^{-5}$	1,048	0,520	5,3
8	$-1,0 \cdot 10^{-6}$	$-3,77 \cdot 10^{-5}$	1,064	0,478	5,0

Table III

Knowing that β_F is equal to 31, we find that β_{F2} is comprised between 174 and 155 for V_{CE} varying between 4 and 8 V.

The assumption that β_{F2} is representative of the recombination mechanisms is confirmed by the fact that a transistor like the one considered in fig. 6 exhibits a very large value of β_{F2}. Practically no output conductance difference is measurable indeed whether the base terminal is connected to a current or a voltage source. Considering a value of 100 µA for $I_s \exp(V_E/V_T)$, the Early voltage was found approximatively, equal to 50 V and the difference in output conductances was found below $5 \cdot 10^{-8} \Omega^{-1}$, with an absolute output conductance between 2 and $1,5 \cdot 10^{-6} \Omega^{-1}$ for V_{CE} comprised between 2 and 8 V respectively. The resulting ratio β_{F2}/β_F therefore must be larger than 50, leading to a value of β_{F2} larger than 10.000.

Additional confirmation of the fact that β_{F2} may describe the recombination mechanisms can be found in the frequency dependance of β_{F2}. Measurements carried out on another high gain transistor (BC 109) led to the conclusion that β_{F2} tends to increase with frequency, in the range from 0 to 100 Hz by approximatively 3 dB per octave. This is the typical slope associated with the recombination process in bipolar transistors [10]. Hence attention always must be given to specify the frequency at which β_{F2} is measured.

1.4. Series resistances

1.4.0. Collector and emitter series resistances $R_{CC'}$ and $R_{EE'}$

The low doped collector region introduces a series resistance $R_{CC'}$, which is responsible for the base push-out effect mentioned above. The current I_1, defined by expression (36), is an important parameter in this respect for it sets the upper limit wherefrom the collector junction becomes forward biased. Below I_1,

R_{CC}, is not a very important quantity since it represents only a relatively small resistance in series with the high output resistance of the transistor. At high frequencies however, R_{CC}, combined with the collector to substrate capacitance introduces an additional time constant which may substantially lower the transistor performances.

One may determine R_{CC}, under low injection conditions by comparing the actual transistor characteristics in the saturation mode with those computed from the Ebers and Moll model. Consider for instance the circuit shown in fig. 13 and let us eliminate V_E, V_C, I_1 and I_2 in order to obtain the following expression :

$$I_C = I_B \cdot \frac{\beta_{RO} \cdot (e^{V_{CE}/V_T} - 1) - 1}{1 + \frac{\beta_{RO}}{\beta_{FO}} \cdot e^{V_{CE}/V_T}} \tag{43}$$

The first order derivative of I_C with respect to V_{CE} is given by :

$$\frac{dI_C}{dV_{CE}} = \frac{I_B}{V_T} \frac{\frac{\beta_{RO}}{\beta_{FO}} \cdot e^{V_{CE}/V_T} \cdot (1 + \beta_{RO} + \beta_{FO})}{1 + \frac{\beta_{RO}}{\beta_{FO}} e^{V_{CE}/V_T}} \tag{44}$$

When V_{CE} is taken equal to $V_T \ln \beta_{FO}/\beta_{RO}$, the second derivative of expression (43) becomes equal to zero, and the above expression reduces to :

$$\left. \frac{dI_C}{dV_{CE}} \right|_{V_{CE}=V_T \ln \frac{\beta_{FO}}{\beta_{RO}}} = \frac{I_B}{4V_T} (1 + \beta_{RO} + \beta_{FO}) \tag{45}$$

Considering that :

$$\beta_{FO} >> \beta_{RO} \quad \text{and also} >> 1$$

expression (45) is approximatively equal to the reciproqual of $4 h_{ib}$. Moreover introduction of $V_T \cdot \ln \beta_{FO}/\beta_{RO}$ into (43), leads to the conclusion that :

$$\left. \frac{I_C}{I_B} \right|_{V_{CE} = V_T \ln \beta_{FO}/\beta_{RO}} = \frac{\beta_{FO} - (1+\beta_{RO})}{2} \simeq \frac{\beta_{FO}}{2} \tag{46}$$

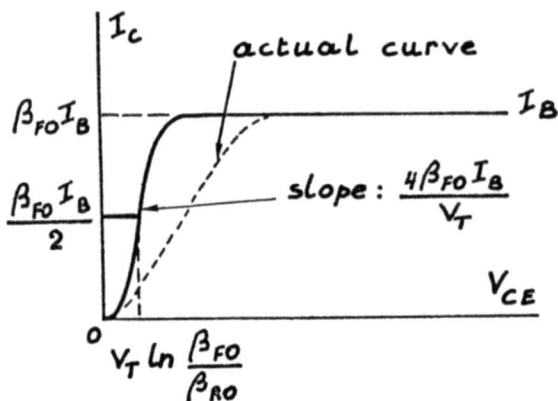

Fig. 15. The series collector resistance may be determined from the difference between the actual saturation characteristic and the theoretical.

A simple means to evaluate $R_{CC'}$, based on the above considerations is illustrated in fig. 15. It simply consists to determine $R_{CC'}$, from the difference between the theoretical $I_C(V_{CE})$ curve and the actual physical curve. Since this method is based upon the assumption that β_{FO} and β_{RO} are constant quantities, one may prefer to follow a slightly different procedure for which either graphical or analytical procedures may be considered. In both cases one starts from a given value of the current flowing through the forward base diode in order to determine V_E first and therefrom the current flowing through the reverse base diode, knowing I_B. Once this current is obtained V_C is known, thus V_{CE} and I_C also, for it results directly from the knowledge V_E and V_C.

This method is illustrated in fig. 16 which represents the theoretical and experimental saturation characteristics of the 2N5109 transistor. The good agreement between curves obviously indicates that $R_{CC'}$ is quite small and that in this case it may be neglected for low injection conditions.

The same method may be proposed for the determination of the emitter series resistance $R_{EE'}$, but due the very small value of this resistance, another procedure may be better suited as summarized in fig. 17 [11]. It consists to plot the open collector DC voltage of the device under test versus base current and to determine $R_{EE'}$ from the slope of the straight line which is obtained.

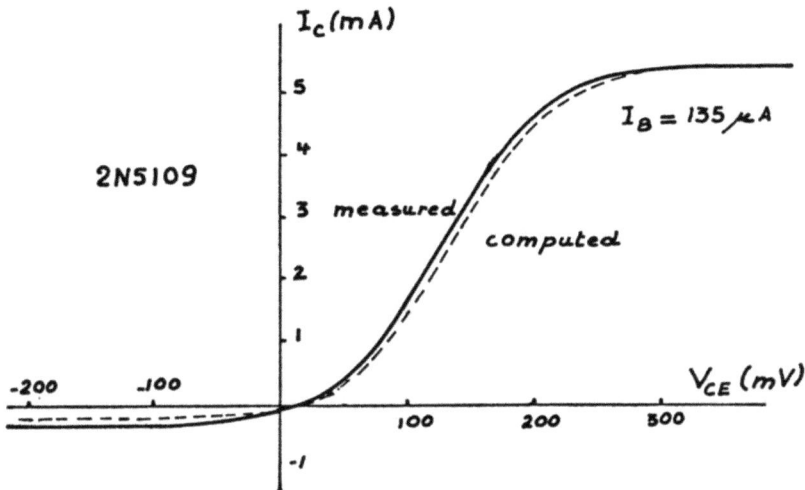

Fig. 16. Comparison between the theoretical and experimental saturation curves of the 2N5109 transistor.

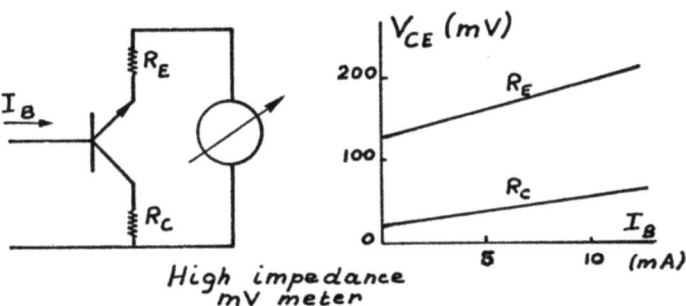

Fig. 17. Circuit for the measurement of emitter and collector series resistances.

Indeed, the elimination of I_F/I_R between the equations :

$$\frac{I_R}{\beta_R} = I_F - I_R$$

and $V_{CE} = V_T \ln \dfrac{I_F}{I_R}$

leads to the well known result :

$$V_{CE} = V_T \ln\left(1 + \frac{1}{\beta_R}\right) \quad (47)$$

so that the introduction of R_E in series with the emitter changes expression (47) into :

$$V_{CE} = V_T \ln\left(1 + \frac{1}{\beta_R}\right) + R_E I_B \quad (48)$$

A very high impedance DC voltmeter of course is needed in order to carry out this measurement properly as well as the assumption that β_R does not vary too much with I_B. In regions where β_R is a strong function of current, there can be considerable departure from the ideal straight line.

Naturally this method is also applicable to the evaluation of $R_{CC'}$.

1.4.1. Base_resistance_and_base_crowding_effects

The extrinsic base resistance $R_{BB'}$ is substantially more difficult to model than the collector and emitter series resistances for it is current and frequency dependant. It is composed of two parts : one is the passive base which corresponds to the bulk resistance between the base contact and the actual active base region, the second is the equivalent resistance of the active base region where base crowding is taking place. The second which dominates has been thoroughly studied in [1] and a first order model proposed in which the current dependance of $R_{BB'}$ is modelled by means of a diode with a very large saturation current :

$$I_B = I_{SB} (\exp V_{BB'}/V_T - 1) \quad (49)$$

with :

$$I_{SB} = \frac{\lambda V_T}{R}$$

λ is equal to 3 for transistors having a rectangular base provided with one base contact, and 4 for the circular geometry, R is equal to $\rho l/Wh$ (ρ : mean resistivity of the active layer base; l, W and h respectively : length, width and height of the active base).

Such representation of the extrinsic base resistance leads to the following small signal resistance :

$$r_{BB'} = \frac{R/3}{1 + I_B/I_{SB}} \quad (50)$$

which is obtained by derivation of expression (49).

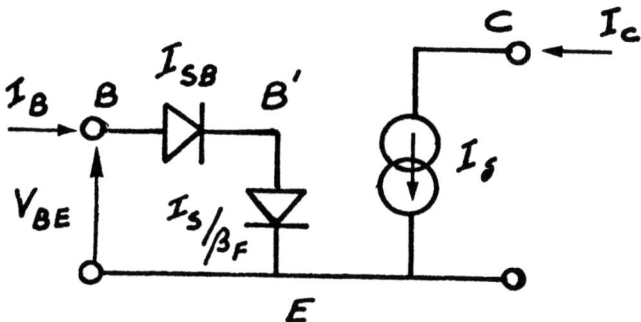

Fig. 18. *The extrinsic base resistance can be simulated by means of an extra diode having a very large saturation current (ref. 1).*

Before we describe several measuring techniques suitable for the evaluation of $r_{BB'}$, let us examine the influence of the diode-like behavior of $r_{BB'}$ on the DC characteristics of the transistor. Calling V_{BE} the actual base to emitter voltage which differs from V_E only by the voltage drop $V_{BB'}$ across the extrinsic base diode D_B^E (fig. 18), one has :

$$I_B \approx I_{SB} \cdot e^{\frac{V_{BB'}}{V_T}} \approx \frac{I_S}{\beta_F} \cdot e^{\frac{V_E}{V_T}}$$

It is easy to show that :

$$\left. \begin{array}{l} V_{BE} = 2 V_T \cdot \ln \dfrac{I_C}{\sqrt{I_S I_{SB} \beta_F}} \\ \text{or} \quad I_C = \sqrt{I_S I_{SB} \beta_F} \cdot \exp(V_{BE}/2V_T) \end{array} \right\} \qquad (51)$$

These expressions are valid only if $V_{BB'} > V_T$. Below this, the usual $I_S \cdot \exp(V_E/V_T)$ expression prevails. A plot of expressions (51) is represented in fig. 19, clearly showing the bending of the $\ln I_C$ and $\ln I_B$ curves when V_{BE} becomes larger than $2V_T \ln \sqrt{\beta_F I_{SB}/I_S}$. Notice that this effect is very similar to the one predicted by Gummel-Poon at high injection level. However in the G-P model, I_B remains unaffected while in the present situation I_B and I_C undergo the same bending.

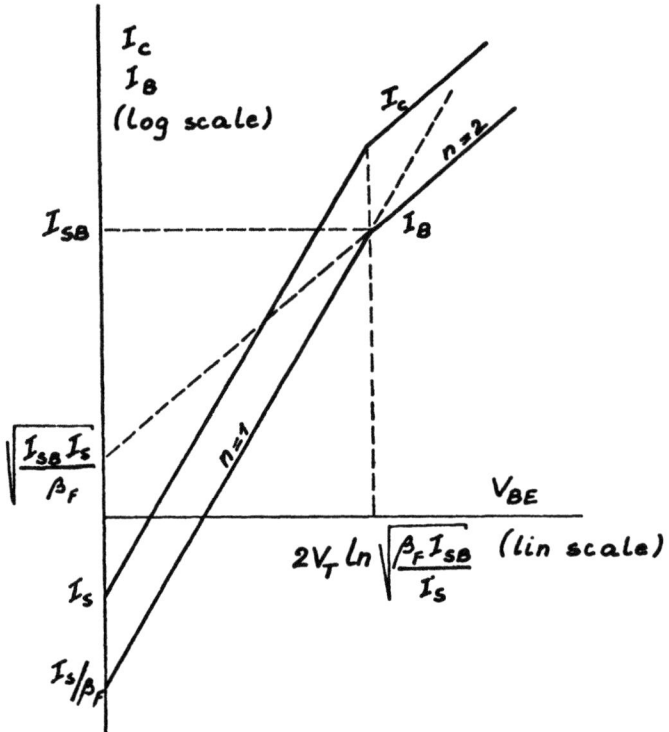

Fig. 19. *The influence of the series diode representing the extrinsic base resistance upon the I_C and I_B curves versus V_E.*

Since the extrinsic base resistance only modifies the base to emitter voltage, the forward current gain β_F of course remains constant. However, when the base current is increasing lateral injection grows faster than vertical injection for the extrinsic base resistance preferantially forward biases the contour of the emitter region. Hence, we consider the equivalent circuit of fig. 20, where I_{SL} represents the saturation current of the diode which is supposed to simulate lateral injection. The forward current gain again decreases as is clearly visible in the right part of fig. 20, and the overal behavior becomes very similar to the one considered by Gummel and Poon.

In practice, few modern transistors exhibit the above described effect for $r_{BB'}$ is always very small since the emitter periphery is made as large as possible. For instance, in the 2N5109 transistor which was considered before, 16 emitters are connected in parallel, each emitter stripe being 3μ wide and 56μ long. The base is obtained by two different diffusion steps, one lowly doped P type layer .9 μ thick with a sheet resistivity of 140 Ω per sq (the actual base region) and one highly doped P type grid surrounding the washed emitters, which is 1.2 μ thick and has a sheet resistivity of 10 Ω per sq. In such case the corresponding "knee" current (or voltage) is well above the G-P "knee-point point", and no visible indication of extrinsic base resistance can be found in the plot of gig. 5.

This does not mean that the small signal extrinsic resistance may always be neglected. In high frequency applications, it plays an important role and techniques enabling to measure $r_{BB'}$ are absolutely necessary. Among the various techniques, the one proposed by Rey & Leturcq [1] will be presented first.

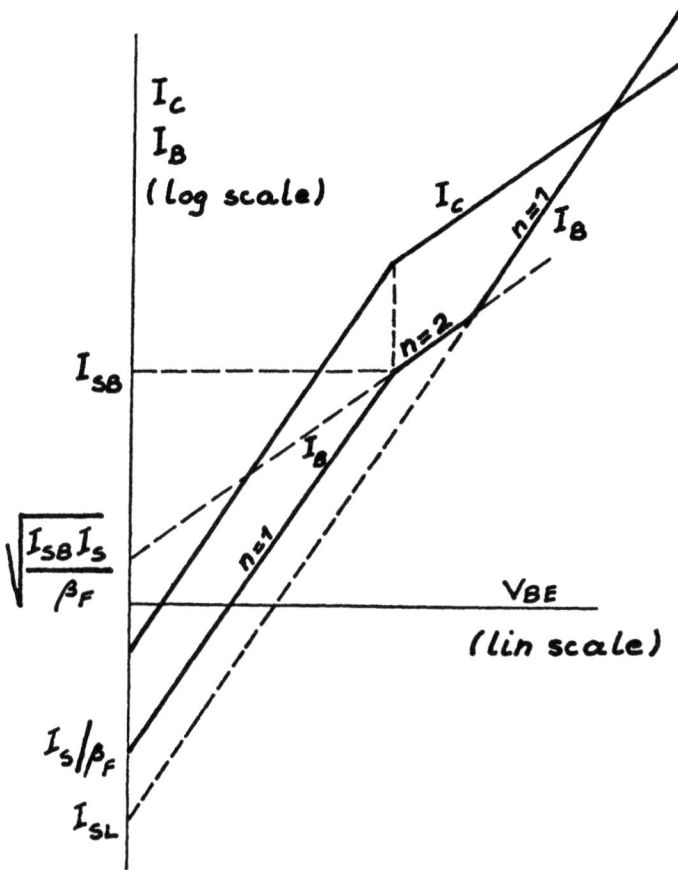

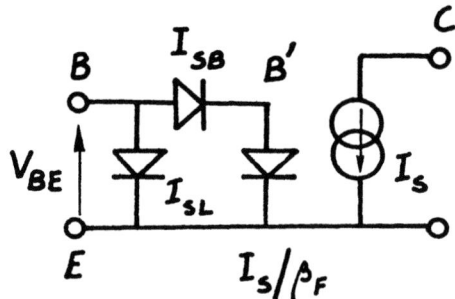

Fig. 20. When lateral injection is included in the Rey and Leturcq model, the current gain fall-off can be simulated at high collector currents.

a) Determination of $r_{BB'}$ by means of a series base resistor

The experimental set up is shown in fig. 21. The principle of the method consists to measure the voltage drop across $r_{BB'}$ caused by injection of the capacitive current flowing through C_C from the collector to the base terminal. Instead of measuring the actual collector current, a two step procedure is preferred which is based on the introduction of an additional small resistor R (e.g. 100Ω) in series with $r_{BB'}$. This resistor may be shorted by means of the switch S.
First S is closed and the voltage drop across $r_{BB'}$, namely $v_{e1} = j\omega C_C \cdot r_{BB'} \cdot v_C$ is measured through the forward biased emitter terminal by means of a high impedance selective voltmeter. Then S is opened and a new voltage signal is obtained equal to :

$$v_{e2} = j\omega C_C \, r_{BB'} \cdot v_C + j\omega (C_C + C_L) \, R v_C$$

After elimination of v_C between the expressions of v_{e1} and v_{e2}, one finds :

$$r_{BB'} = R(1 + \frac{C_L}{C_C}) \cdot \frac{v_{e1}}{v_{e2} - v_{e1}} \tag{52}$$

In this expression C_C represents the part of the collector to base junction capacitance which is facing the active base region, while C_L represents the remaining part. Thus, one may replace $(1 + C_L/C_C)$ by A_E/A_C. The value of $r_{BB'}$ now may be found knowing v_{e1} and v_{e2}, R and the geometry of the transistor. In order to perform the measurement correctly, attention must be given to the following points :

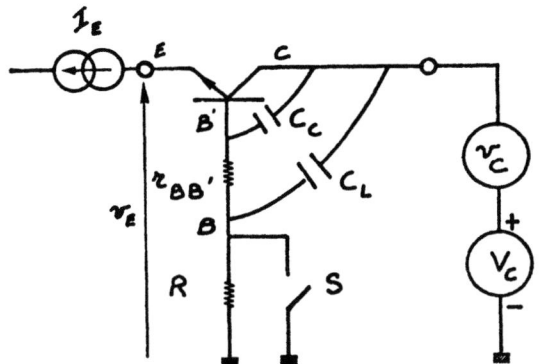

Fig. 21. *Determination of $r_{BB'}$ by means of the voltage drop across this extrinsic base resistance caused by the capacitive collector current flowing through C_C.*

- the frequency of the measurement must be kept low enough in order to make $(R + r_{BB'})$ much smaller than the impedance of $(C_C + C_L)$
- it must be large enough in order to overrule the Early effect.

An easy way to check experimentally the validity of the above conditions consists to change the frequency of the sine wave signal source v_c and to verify that v_e does vary proportionnally.

A typical $r_{BB'}$ plot versus I_C is shown at fig. 22 for the 2N5109 transistor which was considered before.

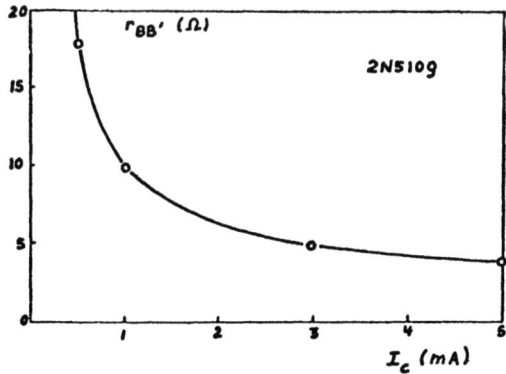

Fig. 22. *$r_{BB'}$ versus I_C for the 2N5109 transistor as it results from the method shown in fig. 21.*

b) Circle diagram

This high frequency method will be described later in this section in the microwave measurements paragraph.

c) Phase cancellation method

This method, proposed by Sansen and Meyer [12] in 1972, is based on the frequency behavior of the common base input impedance versus collector current. It can be shown that the input impedance has one pole and one zero, and that phase cancellation may be obtained by proper adjustment of the collector current. When this is the case, the input impedance of the emitter terminal is simply given by $(r_{BB'} + r_E)$ wherefrom $r_{BB'}$ can be derived easily. The procedure is quite simple and less time-consuming than the previous ones. It is required to utilize a measuring set which enables to measure both phase and amplitude of the input impedance.

d) Noise measurements

From the noise theory of bipolar transistors, it is well known that the extrinsic base resistance contributes substantially to the overal noise figure. An estimate of this resistance may be obtained as reported by several authors [13,14].

One may for instance determine $(r_{BB'} + r_{EE'})$ from the theoretical expression of the noise figure:

$$F = 1 + \frac{1}{\beta_F}\left[1 + \frac{qI_E}{kT}(r_{BB'} + r_{EE'})\right]$$

$$+ \frac{1}{R_S}\{(1+\frac{1}{\beta_F})(r_{BB'}+r_{EE'}+\frac{1}{2}\frac{kT}{2qI_E}) + \frac{qI_E}{kT}\frac{(r_{BB'}+r_{EE'})^2}{2\beta_F}\}$$

$$+ R_S \frac{qI_E}{kT}\frac{1}{\beta_F} \qquad (53)$$

The measurement should be carried out at such a frequency to avoid the 1/f noise, e.g. 100 kHz. Accordingly to H.C. de Graaff, the values of $R_{BB'}$ determined by this method compare quite favourably with those obtained by other methods.

HIGH FREQUENCY AND DYNAMIC LARGE SIGNAL PARAMETERS

2.1. <u>Dynamic large signal model of the bipolar transistor</u>

Bipolar transistors dynamic models are based generally on the assumption that the currents I_F and I_R, are strictly proportional to the respective forward and reverse charges stored in the base. It was already stated in expression (17), that :

$$\left. \begin{array}{l} Q_F = \tau_F \, I_F \\ \text{and} \quad Q_R = \tau_R \, I_R \end{array} \right\} \quad (54)$$

Any modification of collector or emitter currents requires thus a change of the stored charge in the base controlled by means of the base current I_B :

$$\left. \begin{array}{l} I_B = \dot{Q}_F \text{ in the forward direction} \\ \text{or} \quad I_B = \dot{Q}_R \text{ in the reverse direction} \end{array} \right\} \quad (55)$$

This first order model was introduced by several authors [15,16].

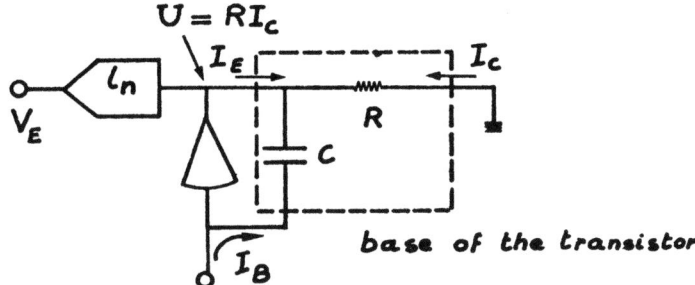

Fig. 23. *A simple analogue simulation of the charge control model of the bipolar transistor operating in the active region.*

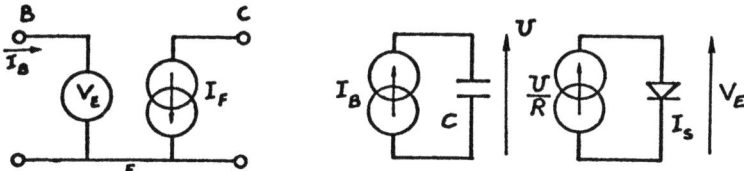

Fig. 24. *Another simulation circuit of the charge control model of the bipolar transistor operating in the active region.*

Fig. 23 represents an analogue circuit which simulates the dynamic behavior of the base of a froward biased transistor controlled by means of its base current (the RC network is the simplest representation of the current and continuity equations within the base). The circuit represented in fig. 24 is strictly equivalent to the previous one, but the opamp has been replaced by current controlled sources. I_B, the base current, controls the charging of capacitor C, and the voltage across this capacitor in turn determines I_C and V_E as in the previous network.

One may write in both cases :

$$U = \frac{1}{C} \int_0^t I_B \, dt \tag{56}$$

and $\quad V_E = V_T \cdot \ln \frac{U/R}{I_S} \tag{57}$

Therefrom

$$V_E = V_T \cdot \ln \left(\frac{\omega_F}{I_S} \int_0^t I_B \, dt\right) \tag{58}$$

calling

$$\omega_F = (RC)^{-1} = \tau_F^{-1} \tag{59}$$

Furthermore :

$$I_C = \frac{U}{R} = \omega_F \int_0^t I_B \, dt = \frac{Q_F}{\tau_F} \tag{60}$$

Above equations describe the large signal dynamic behavior of a forward biased transistor having infinite current gain. Hence, in order to complete our analysis, we must introduce base recombination, or emitter efficiency, or both. Base recombination can be simulated by means of a resistor R_B in parallel with the capacitor C. Emitter efficiency is simulated by means of an additional diode with saturation current I_s/β_F placed in parallel with the controlled voltage source V_E (fig. 25). Hence, the circuit resumes to the network of fig. 3 under steady state conditions (I_1 equal to zero).

Under reverse bias conditions, similar conclusions may be drawn introducing Q_R, ω_R or τ_R. Since the charge-current relations are linear, the superposition principle applies. This leads to the network of fig. 26 which represents the dynamic equivalent network of the transport version of the Ebers and Moll equations. Notice that the junction transition capacitances C_E and C_C are also included in the last model.

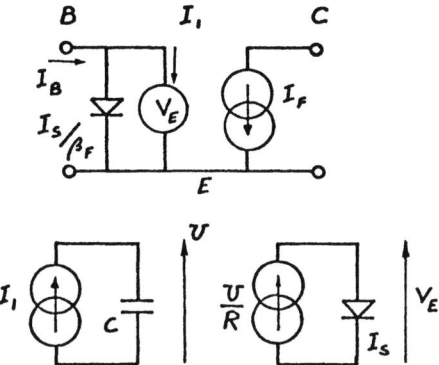

Fig. 25. The introduction of a diode with saturation current I_s/β_F models the emitter efficiency of the bipolar transistor (active region only).

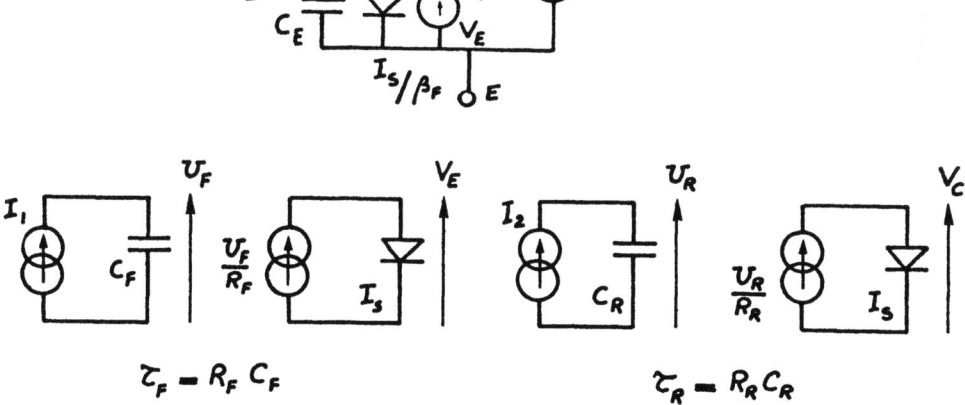

Fig. 26. Full dynamic large signal model of the bipolar transistor (this model is an extension of the transport version of the Ebers and Moll model).

2.2. Small signal dynamic model (the hybrid Π)

We restrict ourselves to the common emitter configuration under forward bias and small signal conditions. The network of fig. 26 resumes to the one of fig. 27. The voltage source v_E being controlled by i_1 (via u_F), one may write :

$$i_1 = j\omega C_F \, u_F = j \frac{\omega}{\omega_F} i_F$$

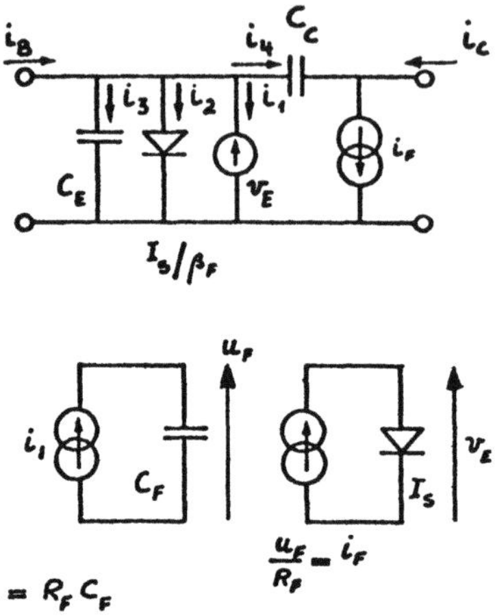

Fig. 27. Small signal equivalent of the model represented in fig. 26 for the active region only.

After differentiation of expression (57) we obtain :

$$v_E = V_T \frac{u_F}{U} = V_T \frac{i_F}{I_F}$$

thus :

$$i_1 = j \frac{\omega}{\omega_F} \frac{I_F}{V_T} v_E \qquad (61)$$

Hence, the branch containing the controlled voltage source, may be replaced by a capacitance C_D, called the diffusion capacitance, such as

$$C_D = \frac{1}{\omega_F} \frac{I_F}{V_T} \qquad (62)$$

C_D is a current depending capacitor, although C_F is not.

The branch containing the diode with saturation current I_S/β_F leads to the expression :

$$i_2 \simeq \frac{1}{\beta_F} \cdot \frac{I_F}{V_T} \cdot v_E \qquad (63)$$

or $\quad i_F = \frac{I_F}{V_T} \cdot v_E \qquad (64)$

If the two capacitances C_E and C_D are grouped in order to form C_π, the hybrid Π equivalent circuit is nearly obtained. Only $r_{BB'}$ should be added in order to obtain the well known circuit represented in fig. 28.

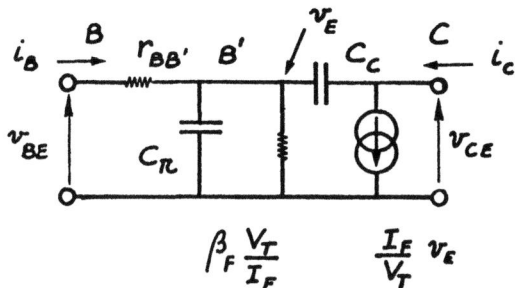

Fig. 28. The hybrid Π network can be derived directly from the model shown in fig. 27.

2.3. Experimental evaluation of the dynamic parameters

2.3.0. Forward transit time τ_F, and depletion capacitance C_E and C_C

The forward transit time τ_F, as well as the sum of the transistor capacitances C_E and C_C may be derived from the frequency behavior of the current gain β_F with shorted output terminals.

Since : $i_B = i_1 + i_2 + i_3 + i_4$

we express each of these currents as a function of i_F rather than v_E, using expression (64).

Thus : $i_1 = j \frac{\omega}{\omega_F} \cdot i_F \quad$ from (61)

$\qquad i_2 = \frac{1}{\beta_F} \cdot i_F \quad$ from (63)

$\qquad i_3 = j\omega C_E v_E = j\omega C_E \frac{V_T}{I_F} i_F$

and $\quad i_4 = j\omega C_C \frac{V_T}{I_F} i_F$

Hence

$$i_B = \left\{ \frac{1}{\beta_F} + j\omega \left[\frac{1}{\omega_F} + (C_E + C_C) \frac{V_T}{I_F} \right] \right\} i_c \qquad (65)$$

The transition pulsation then is defined as :

$$\frac{1}{\omega_T} = \frac{1}{\omega_F} + (C_E + C_C) \frac{V_T}{I_F} \qquad (66)$$

Assuming β_F is large, ω_T is defined as the pulsation for which the common emitter short circuit current gain is equal to one. The frequency f_T can be measured easily versus I_C and plotted as the reciprocal of ω_T versus the reciprocal of I_C in order to obtain the straight line described by expression (66). This line intersects the vertical axis at the point $1/\omega_F$ or τ_F (fig. 29). At the same time the sum $(C_E + C_C)$ may be found from the slope of the straight line accordingly to the expression :

$$tg\,\alpha = (C_C + C_E) \frac{V_T}{I_F} \qquad (67)$$

Knowing C_C from the previous measurements (section 1.3.1) it is easy to find C_E, the depletion capacitance of the forward biased emitter junction.

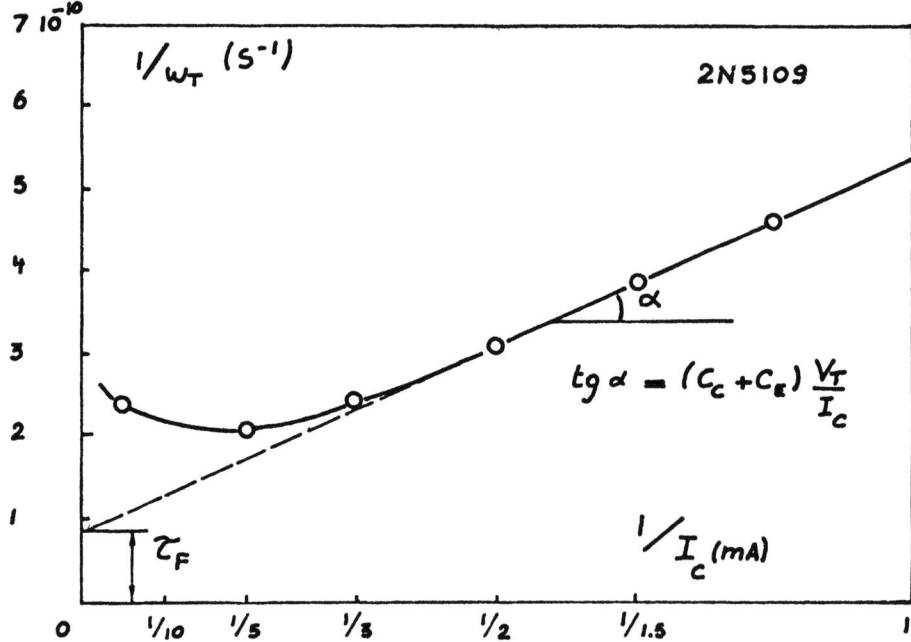

Fig. 29. Determination of the forward transit time τ_F and the depletion capacitance $(C_C + C_E)$ from the current dependance of ω_T

The plot of fig. 29, which was obtained for the 2N5109 transistor, shows that when I_F (or I_C) becomes very large, the experimental curve departs substantially from the ideal straight line. The phenomenon once more provides an illustration of the base widening effect considered in section (1.3.2) under expression (37). When I_F becomes larger than I_1 (see expression (36)), the base moves into the thin lowly doped epi layer of the collector and storage of carriers rapidly increases within the extended base, thus Q_F (or C_F) increases, as well as τ_F. Hence the determination of τ_F must be based upon the extrapolation of the straight line as shown in fig. 29, in order to always restrict I_F below I_1.

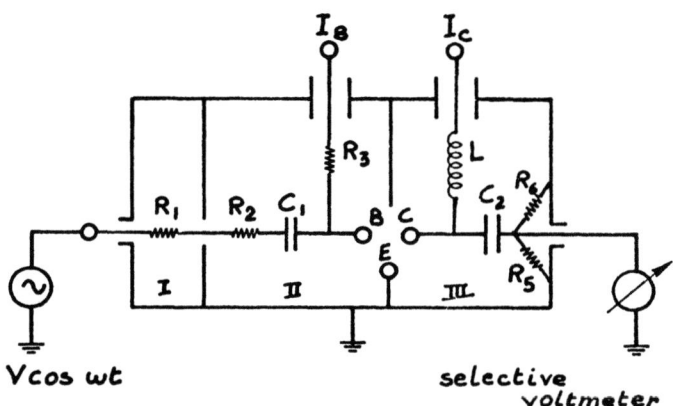

Fig. 30. Simple experimental set-up for the determination of the transition frequency f_T.

Let us discuss now the experimental test set up used for the measurement of f_T. The vast majority of modern transistors exhibit values of f_T which are of the order of 300 to 600 MHz. Some, especially those having base and emitter thicknesses of a few tenths of a micron, have f_T's as high as a few GhZ. Obviously the determination of f_T cannot be performed directly. Current gain measurements must be carried out at some lower frequencies for it is almost impossible to realize good current sources at such high frequencies. Usually one measures the common emitter current gain at some lower fixed frequency, e.g. 100 MHz, using the set up which is shown in fig. 30. Here a relatively good current source is achieved by means of the input voltage generator and two or more series resistances (R_1, R_2) of a few kΩ. Adequate screening is an absolute requirement in order to avoid direct radiation form the input terminal to the base of the transistor. In fig. 30 double screening is achieved by placing R_1 and R_2 into

two different compartiments.
The transistor is inserted in the middle between box II and box III, in order to avoid by-passing by means of direct coupling. The transistor operating point is fixed by means of the DC bias circuitry formed by R_3 and L which is used in order to block the HF component. C_1 and C_2 provide DC isolation between HF and DC circuitry. At the output terminal the collector is almost grounded through the low impedance resistance formed by means of R_5 and R_6 in parallel (e.g. 10Ω). It is very important to keep this impedance as low as possible, in order to minimize the voltage gain, otherwise substantial decrease of f_T may be experienced due to the Miller effect (the reverse feedback path associated with C_c). The measurement should be carried out as follows :

a) short the terminals B and C, leaving E open, in order to determine the unity gain overal transfer function of the system.
b) remove the short circuit, and insert the bipolar transistor in order to measure the actual current gain by means of the ratio of the output signals obtained under (b) and (a).

If we assume that the frequency at which the measurement occurs is higher than the f_T/β_F but still lower than f_T, the transition frequency may be determined from the simple 6 dB per octave law described in expression (65). In other words :

$$f_T = f_{measure} \times \text{current gain}$$

Repeating the measurement for different currents I_c with constant V_{CE}, leads to the curve shown in fig. 29, wherefrom τ_F as well as $(C_E + C_c)$ are computed.

2.3.1. Reverse transit time τ_R

The above considered procedure can be extended to the case of the reverse biased transistor to find τ_R and $(C_C + C_E)$. This method however often leads to substantial errors. Indeed, in the forward direction, the high emitter doping prevents significative charge storage to occur in the emitter. Most of the charge therefore finds itself restricted within the base and the proposed model adequately describes the dynamic behavior of the transistor. In the reverse mode, the actual "emitter" comprises the lowly doped epi-layer of the collector where substantial charge storage occurs [17] (the same problem also exists in I^2L structures). Since the measurement described above is based on the transfer of charges through the base, chances exist that the evaluation of the reverse f_T does not determine τ_R correctly.

An alternative method is described hereunder in which the determination of τ_R is obtained indirectly from data derived from the saturated mode of operation of the invertor stage. It is well known that the collector current of an overdriven transistor does not decay immediately after the direction of the base current has been reversed. A delay t_s, called the "desaturation time" is needed indeed before the collector current effectively begins to decrease. During this time, removal of excess charge is taking place by means of the imposed base current till, the transistor again enters into the active region. The time constant governing this mechanism depends on τ_F as well as on τ_R, since both forward and reverse conditions coexist in the saturated mode. Straightforward analysis of the charge control model leads to the well known expression of this time constant [18] :

$$\tau_S = \frac{\tau_R + \tau_F(1 + 1/\beta_F)}{1/\beta_R + 1/\beta_F(1 + 1\beta_F)} \tag{68}$$

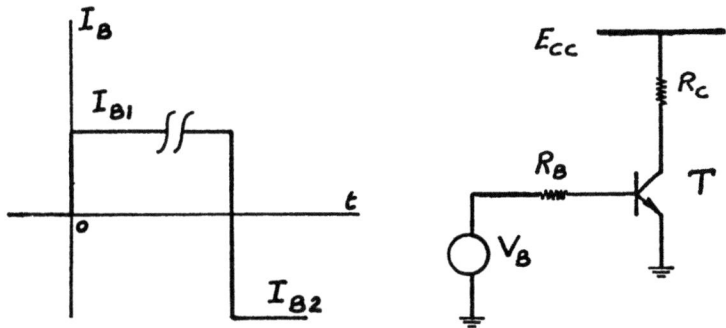

Fig. 31. *Determination of the switching times of a simple bipolar invertor stage.*

We define (fig. 31) :

I_{CMAX} as E_{cc}/R_L
and I_{BO} as I_{CMAX}/β_F

The switch-on overdrive factor now is introduced,

$$n = \frac{I_{B1}}{I_{BO}} \qquad n > 0$$

and the switch-off overdrive factor m :

$$m = \frac{I_{B2}}{I_{BO}} \qquad m < 0$$

With these notations, the desaturation time t_S can be computed from the following expression [18] :

$$t_s = \tau_s \cdot \ln \frac{n-m}{1-m} \tag{69}$$

provided the initial base current step is at least five times longer than τ_S.

Plotting t_s versus $\ln[(n-m)/(1-m)]$ for a number of different test conditions with variable n and m factors, thus yields a set of points on a straight line wherefrom τ_S can easily be evaluated. Such a plot, reproduced in fig. 32 for the 2N5109 transistor, shows that close agreement indeed is obtained in practice. The corresponding value of τ_S is equal to 185 ns and τ_R determined from (68) and the previously obtained values of τ_F, β_F and τ_R, is equal to 49 ns. This is 79% higher than the value which would be obtained by means of the method described in § 2.30. It should be pointed out that in contrast with other methods, the evaluation of τ_S based on t_s does not depend on the actual values of transition capacitances for the junction voltages remain practically constant during the whole desaturation time.

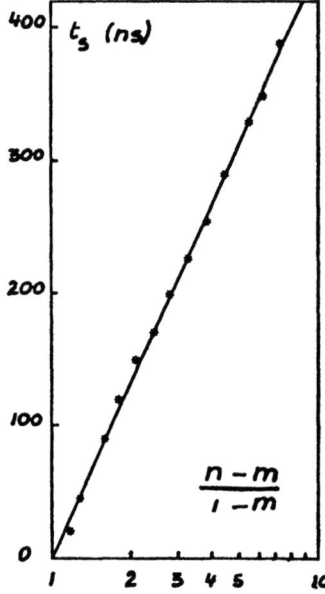

Fig. 32. *Determination of τ_s from the desaturation time measurement.*

2.3.2. Simulation of an invertor stage

This section has been mainly devoted up till now to the description of acquisition methods of the basic bipolar transistor parameters. In order to illustrate their implementation in a modeling process, a very simple example will be presented : the simulation of an invertor stage similar to the one shown in fig. 31. The transistor is the 2N5109 and the model the one represented in fig. 26. A survey of 2N5109 parameter is given hereunder.

a) β_F is supposed to be constant and equal to 40. Since the range of interest of forward collector current is restricted from 1 to 15 mA.
b) $\beta_R = 4$
c) $I_S = 1.70 \ 10^{-15}$ A
d) $\tau_F = 0.85 \ 10^{-10}$ s

R_F is arbitrarily choosen equal to 100 Ω, whereas C_F is equal to .85 pF in order to obtain the correct τ_F.

e) $\tau_R = 4.9 \ 10^{-8}$ s.
With R_R equal to R_F, C_R is given by 490 pF.

f) $C_C = C_{CO} \cdot (1 - \dfrac{V_C}{\phi_C})^{-n}$

with $C_{CO} = 6$ pF
$\phi_C = 0.8$ V
and $n = 0.5$

g) $C_E = C_{EO} (1 - \dfrac{V_E}{\phi_E})^{-n}$

$C_{EO} = 2.3$ pF
$\phi_E = 0.8$ V
and $n = 0.5$

h) $R_{BB'} = 5 \ \Omega$

Furthermore : $E_{cc} = 5$ V, $R_L = 500 \Omega$, $R_B = 2.2$ kΩ, $V_{Bon} = 5$ V and $V_{Boff} = -2$ V with rise and fall times equal to 6 ns. The calculated plots are shown in fig. 33 and fig. 34. The saturation mechanism of the transistor is clearly visible in the graph representing Q_R. Fig. 35 shows the experimental verification closely following the theoretical curves.

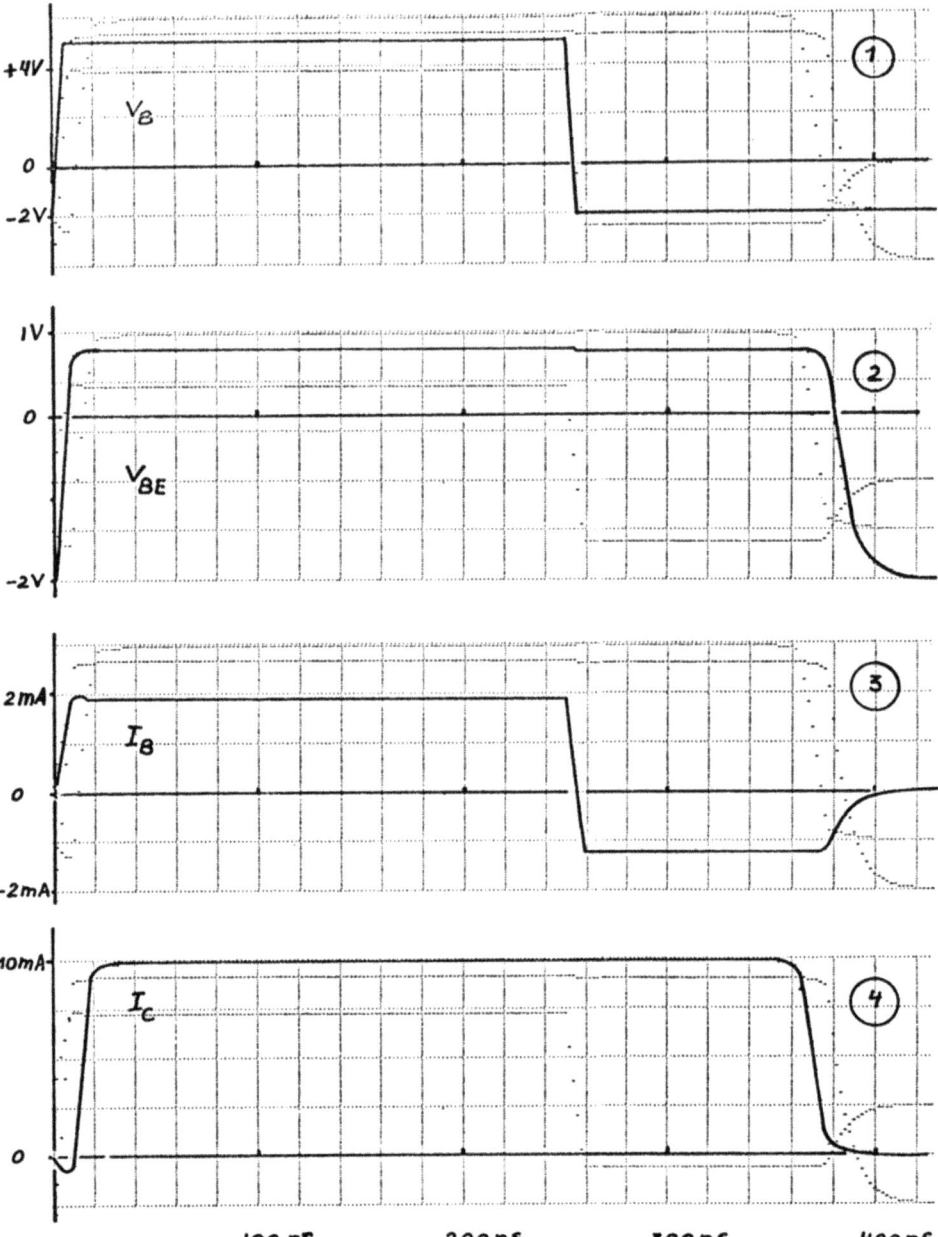

Fig. 33. C.A.D. simulation of the inverter stage shown in fig.31.
1) Base control voltage V_B
2) Base voltage V_{BE}
3) Base current I_B
4) Collector current I_c

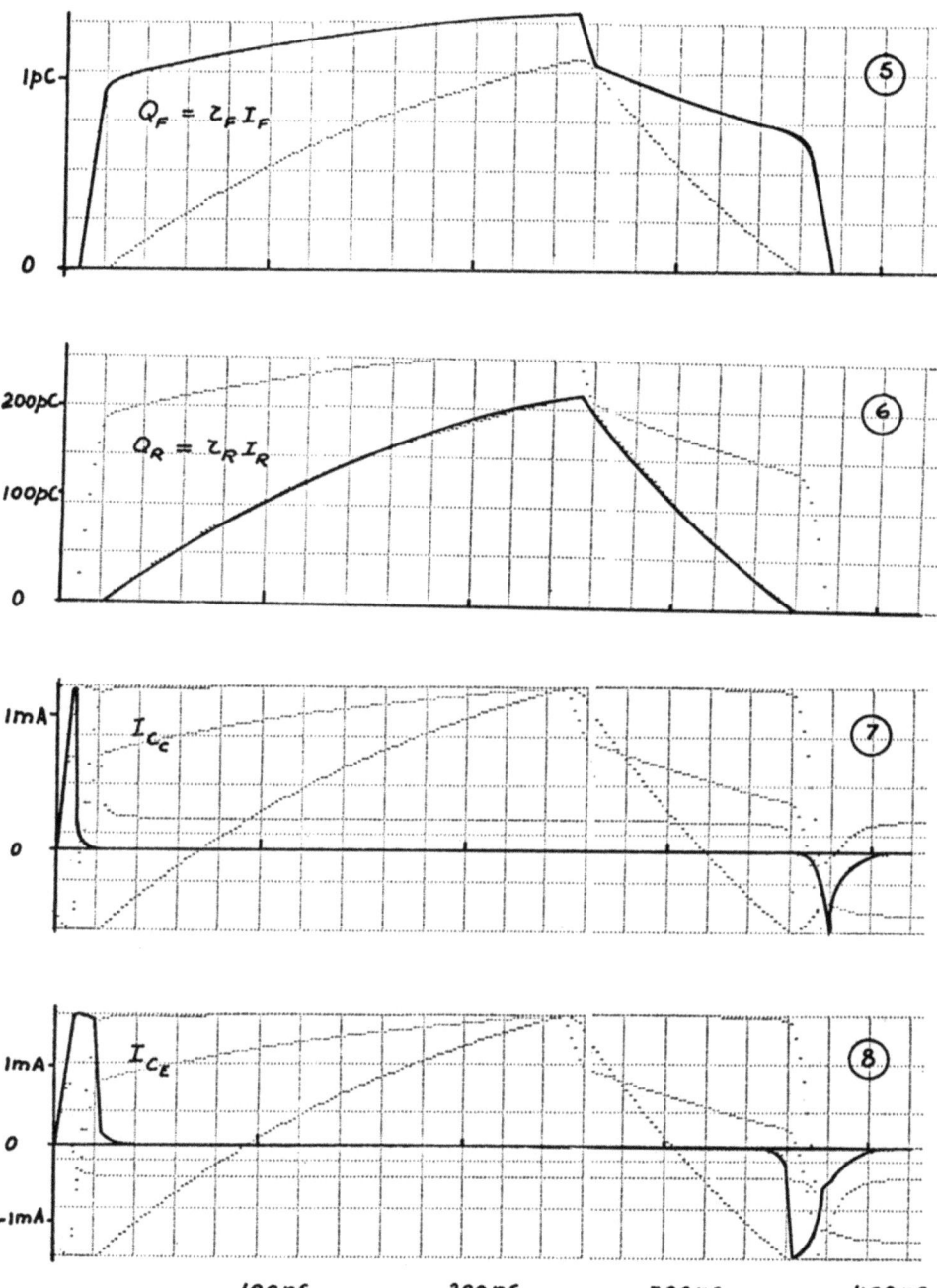

Fig. 34. C.A.D. simulation of the inverter stage shown in fig.31.
 5) Forward base charge Q_F
 6) Reverse base charge Q_R
 7) Current flowing in the emitter transition capacitance
 8) Current flowing in the collector transition capacitan-

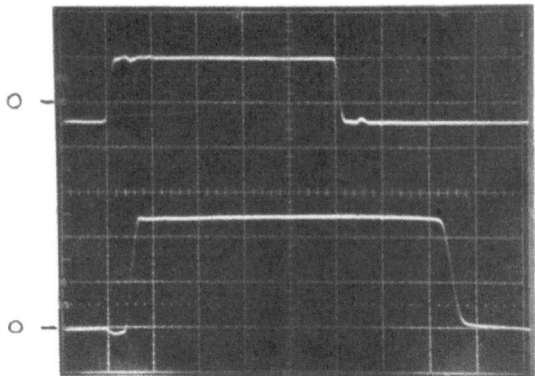

Fig. 35. Experimental verification of the curves represented in fig. 33.
Upper trace : 5 V per div. Control voltage V_B
Lower trace : 4 mA per div. Collector current I_c
Horizontal : 50 ns per div.
Probe capacitance (2pF) was not included in the CAD simulation.

2.4. High frequency measurements

With the exception of transit time determinations, few high frequency measurements have been proposed up till now in this section. A trend however exists to increasingly use scattering parameter measurements. The most obvious advantage is the fact that the transistor terminals always are loaded by means of broadband low impedances. In this manner spurious oscillations are almost eliminated and transistors with f_T's as high as a few GHz can be tested without problems even under DC conditions. Parameter acquisition at several GHz however may be more questionnable for number of additionnal parameters are introduced which seriously complicate the interpretation of the results, unless the goal to achieve precisely is to evaluate these parameters.

Very high frequency measurements provide a unique means to determine stray inductances of discrete transistors. The packaging plays an important role in this respect. Minimum perturbation of course occurs when the transistor chip is mounted directly in a microstrip structure with short bondings. However, even so, the stray inductance of the common lead of the transistor may cause unwanted feedback affecting an important parameter such as the input impedance. This effect can be minimized using three port measurements instead of two port for the stray inductance of a bonding path not used for grounding is somewhat smaller than that of a ground connection.

Furthermore, the reference impedance, in general 50 Ω, overrules the inductance so that the impedances in series with each terminal are better defined. Moreover an additional advantage of three port techniques is the fact that among the nine parameters which it is possible to measure four only are independent. Hence cross-checking is possible accordingly to the equations :

$$\sum_{i=1}^{3} s_{ij} = 1$$

and

$$\sum_{j=1}^{3} s_{ij} = 1$$

Comparative measurements of the two port parameter s_{11} (the input reflection coefficient) with the same parameter derived from three port measurements (s_{ij}), accordingly to the expression

$$s_{11} = s_{11}^* - \frac{s_{31}^* s_{13}^*}{1 + s_{33}^*}$$

cast some light on the unavoidable common lead inductance which is often in the order of a few tenths of a nH.

When a transistor is measured within its package the influence of the stray inductances becomes in general more important. For illustration, we consider the hybrid Π network of the 2N5109 transistor with a collector current of 1 mA and a corresponding value of $r_{BB'}$ of 10 Ω. All other parameters are the same as before with C_c equal to 3 pF due to the reverse biasing of the collector and f_T equal to 1.87 GHz.

Fig. 36 shows the computed s_{ie} parameter first with no inductance in the emitter lead and second with 0.1 nH in series with the ground terminal. In order to investigate also the influence of the current gain excess phase in the vicinity of the transition frequency, two more curves are added in fig. 36 accordingly to the model proposed by Pritchard [19] . This model is the same as the well known hybrid Π except for the output current source which is multiplied by the excess phase factor $\exp(-j \frac{\omega}{\omega_T} m)$. Appropriate values of m usually rank between 0.1 and 0.5. The following comments may be derived from fig. 36 :

a) even an inductance as small as 0.1 nH drastically changes the behavior of the input impedance.
b) excess phase only modifies the inductive part of the input impedance to an extend which is strongly influenced by the actual value of m. With m equal to 0.5 instability may occur.

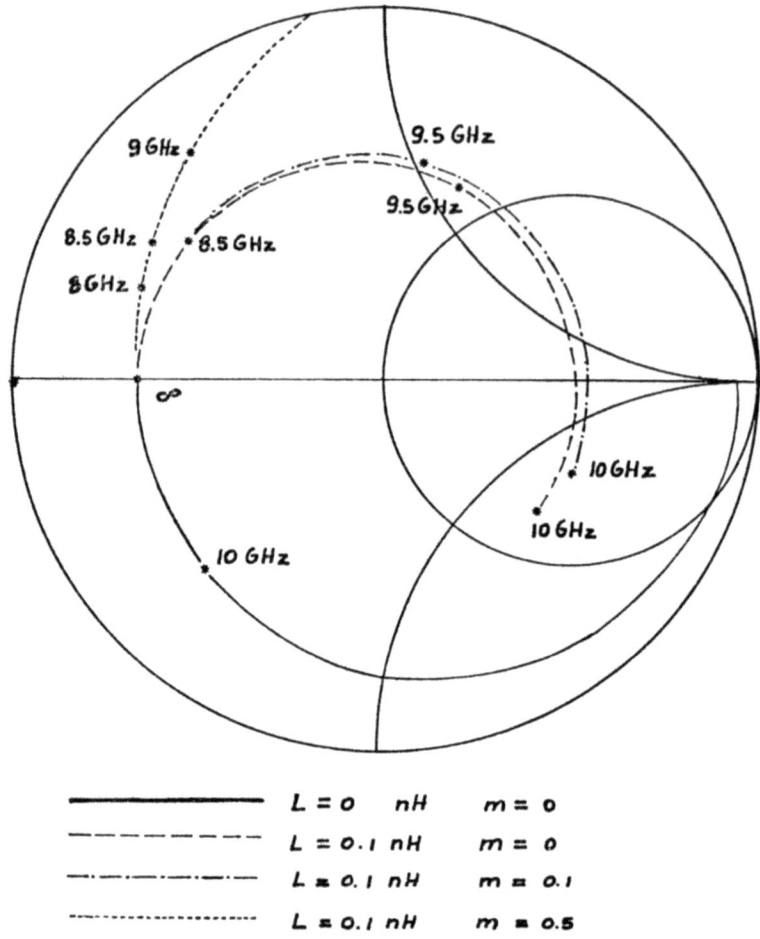

Fig. 36. *Theoretical curves of the input impedance of the 2N5109 transistor showing the influence of the emitter stray inductance and collector current extra phase shift.*

With an emitter inductor if 1 nH, which is typical of an encapsulated transistor, the effect on the input impedance is somewhat different as shown in fig. 37.

We now see that :
a) the shift toward the inductive behavior occurs sooner, somewhere between 1 and 1.5 GHz.
b) the effect of excess phase is much less sensitive than before.

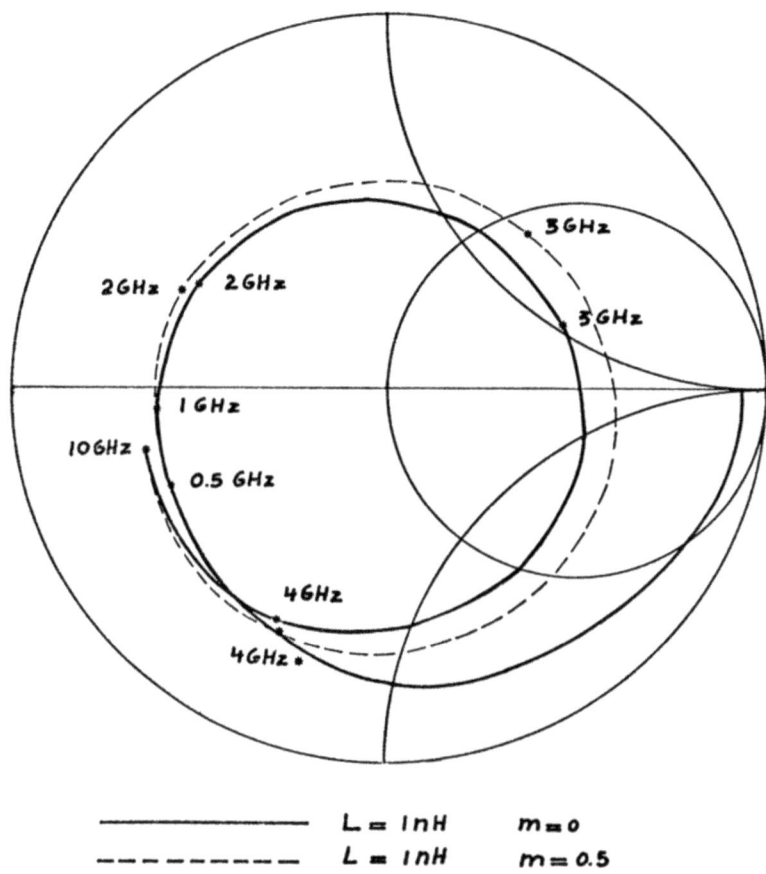

Fig. 37. Theoretical curves of the input impedance of the 2N5109 transistor with a 1 nH emitter series inductor for an encapsulated transistor.

At the point where s_{ie} crosses the real axis the input impedance becomes real. With no emitter inductor, s_{ie} becomes real at infinite frequency for the hybrid Π circuit then resumes to $r_{BB'}$. An alternative method of evaluation of $r_{BB'}$ is based on this property and known as the circle diagram method [20]. Fig. 36 and 37 prove that this method is more or less valid even when the input inductance becomes inductive but that an increasing error should be expected when the emitter stray inductance is larger than a few tenths of a nH. Since the circle method necessarily implies measurements at very high frequencies, this effect should not be overlooked nor should the base stray inductance be omitted in the modelisation process.

CONCLUSION

We have described in this section methods for the determination of bipolar transistor parameters. The transport version of the Ebers and Moll model should be preferred to the older injection model, since it is simpler, common emitter oriented, and its basic parameters are easy to measure. Moreover the transport version is a simplified version of the Gummel-Poon model.

The Gummel-Poon model requires a large number of parameters and describes adequately many phenomena which are not included in the Ebers and Moll model, such as : Early effect, high injection effects, non linear transition capacitances and low current imperfections of the base current. The Gummel number GU provides an objective measure of the departure from the ideal transistor characteristics. The ln I_C and ln I_B versus V_E and V_C curves provide many important informations with respect to the Gummel-Poon model.
The extrinsic base resistance evaluation probably is one of the most controversive measurements. $R_{BB'}$ is a non-linear frequency-dependent component. Several methods of evaluation were reviewed. It is unlikely that any measurement method of $R_{BB'}$ provides a satisfactorily model for a transistor controlled by a voltage source. The extrinsic collector and emitter resistances can be measured fairly well, but the extrinsic collector resistance becomes quite difficult to model when high injection is taking place and base push-out is becoming the dominant factor.

The most effective small signal dynamic model of the bipolar transistor is the hybrid Π for it is a good compromise between accuracy and complexity. The classical transit time measurement, based on the $1/\omega_T$ versus $1/I_C$ curve, is sufficient for the forward direction but other methods are more suitable for the inverted mode.

A large signal dynamic model has been described where the current dependant diffusion capacitance is simulated by means of a constant capacitor and a non linear algebraic expression. This model is not used in most of the existing computer programs but perhaps it may provide a better physical insight of what is happening in the transistor.

Very high frequency measurements, such as S parameter measurements, ments, should be considered only for preventing oscillations in the case of microwave transistors. They also provide a good means to measure stray inductances, but their use for parameter extraction should be carefully considered.

Acknowledgments

The author wants to thank Mr. GRUNBERG of Sescosem (France) for permission to publish physical and geometrical data concerning the 2N5109 transistor. He also wants to express his gratitude to Professor W. Engl of the Technical University Aachen (Germany) who provided the experimental transistor used in fig. 6. The contributions of MM. A. Sevrin, Duquesne, Boussingault, Adant and Cambier are gratefully acknowledged for the measurements described in this section.

References

1 Théorie Approfondie du Transistor Bipolaire.
by G. Rey and P. Leturcq,
Masson (1972), France.

2 The Influence of Fixed Interface Charges on the Current-Gain Falloff of Planar N-P-N Transistors,
by W.M. Werner.
Journal of the Electrochemical Soc. (Solid-State Science and Technology), Vol 123, n° 4, p. 540-543, April 1976.

3 A Charge Control Relation for Bipolar Transistors,
by H.K. Gummel,
The Bell Systel Technical Journal, p. 115-120, January 1970.

4 An Integral Charge Control Model of Bipolar Transistors,
by H.K. Gummel and H.C. Poon,
The Bell System Technical Journal, p. 827-851, May-June 1970.

5 The Dependence of Transistor Parameters on the Distribution of Base Layer Resistivity,
by J.L. Moll and I.M. Ross, Proc. I.R.E., Vol. 44, p. 72-78, January 1956.

6 The Influence of Heavy Doping on the Emitter Efficiency of a Bipolar Transistor,
by H. De Man,
IEEE Transactions on Electron Devices, Vol ED-18, n° 10, p. 833-835, Oct. 1971.

7 Measurement of the Number of Impurities in the Base Layer of a Transistor,
by H.K. Gummel, Proc. of the IRE, p. 834, April 1961.

8   On the Variation of Junction-Transistor Current Amplification Factor with Emitter Current,
    by W.M. Webster,
    Proc. of the IRE, Vol. 42, p. 914-920, June 1954.

9   A Theory of Transistor Cutoff Frequency (f_T) Falloff at High Current Densities,
    C.T. Kirk,
    IEEE Transactions on Electron Devices, Vol ED-9, p. 164-174, March 1962.

10 Physics and Technology of Semiconductor Devices,
    by A.S. Grove,
    J. Wiley (1967), New-York.

11 Accurate Measurement of Emitter and Collector Series Resistances in Transistors,
    by B. Kulke and S.L. Miller,
    Proc. of the IRE, Vol. 45, p. 90, 1957.

12 Characterization and Measurement of the Base and Emitter Resistances of Bipolar Transistors,
    by W.M.C. Sansen and R.G. Meyer,
    IEEE Journal of Solid State Circuits, Vol SC-7, n° 6, p. 492-498, Dec. 1972.

13 Comparison of Noise-Measured and Theoretical Value of Base Spreading Resistance for an Interdigitated Transistor,
    by K.F. Knoht and R.T. Urwin,
    Electronics Letters, Vol. 11, N° 7, p. 147-148, April 1975.

14 Personal communication,
    H.C. de Graaff.

15 Large Signal Behavior of Junction Transistors,
    by Ebers J.J. and J.L. Moll,
    Proc. IRE, Vol. 42, p. 1761-1762, Dec. 1954.

16 Large Signal Transient Response of Junction Transistors,
    by J.L. Moll,
    Proc. IRE, Vol. 42, p. 1773-1784, Dec. 1954.

17 The Minority Carrier Storage Effect in the Collector Region and the Storage Time of Transistors,
    by. H. Ianai, T. Sugana, K. Tada,
    Journal of the Institute Electr. Commun. Engineers of Japan, Vol. 46, p. 577-582, April 1963.

18 Electronic Principles, Physics, Models and Circuits,
    by Paul E. Gray and Campbell L. Searle,
    J. Wiley, 1967.

19 Electrical Characteristics of Transistors,
    by R.L. Pritchard,
    Mc. Graw Hill, 1967.

20 Power Gain of Transistors at High Frequencies,
    by J. Lindmayer,
    Solid-State Electron., Vol. 5, p. 171-175, Jan. 1962.

Bipolar Transistor MODEL for IC Design

D. Scharfetter

Bell Laboratories
Murray Hill, New Jersey 07974

Introduction

This paper presents a review of the Gummel-Poon model[1,2] for bipolar transistors as implemented in the SPICE[3] general purpose circuit analysis program. In section II certain deficiencies in the model are discussed and an extended Gummel-Poon model is presented which removes the deficiencies.

I. The Spice Gummel-Poon Model

The Gummel-Poon model[1] for bipolar transistors is an extension of the transport formulation (β's instead of α's) of the basic Ebers-Moll description of transistor action. The transported current I_{CC} depends explicitly on internal junction voltages V_{BE} and V_{BC}. The recombination currents I_{BE} and I_{BC} also depend explicitly on the junction voltage and constitute the base current, I_B. The collector and emitter currents I_C and I_E are given by

$$I_C = I_{CC} - I_{BC} \qquad \qquad 1$$

$$I_E = -I_{CC} - I_{BE.} \qquad \qquad 2$$

where the sign convention is that positive current flows into the device.

The Gummel-Poon model can be described at this point by allowing:

a) Two recombination current components for both I_{BE} and I_{BC}; b) A normalized conductivity modulated and junction depletion controlled base change (Gummel Charge),[2] Q_{BN}.

The above effects result in a) current gain β fall-off for low current, b) current gain fall-off for high current, c) finite output conductance (Early effect), d) f_T fall-off at high currents and e) the flexibility to model a conductivity modulated base resistance.

The SPICE Gummel-Poon model equations are reviewed in the following section.

SPICE Gummel-Poon Model NGP and PGP (npn and pnp)

$$I_{BE} = I_1 E^{V_{BE}/V_T} + I_2 e^{V_{BE}/N_e V_T} \qquad 3$$

$$I_{BC} = I_3 e^{V_{BC}/V_T} + I_4 e^{V_{BC}/N_c V_T} \qquad 4$$

$$I_{CC} = I_S (e^{V_{BE}/V_T} - e^{V_{BC}/V_T})/Q_{BN} \qquad 5$$

$$Q_{BN} = 1 + Q_{BE} + Q_{BC}$$

The terms Q_{BE} and Q_{BC} represent junction and diffusion capacitance charge-storage elements. The junction capacitance charge-storage element is the integral of the junction capacitance normalized to zero bias value, i.e.

$$Q_{BE_j} = \frac{1}{V_B} \int_0^{V_{BE}} C_{JE}(V) \frac{dV}{C_{JE_0}}, \qquad 6$$

Where V_B is a model parameter, chosen to fit the (inverse) output conductance; similarly

$$Q_{BC_j} = \frac{1}{V_A} \int_0^{V_{BC}} C_{JC}(V) \frac{dV}{C_{JC_0}}. \qquad 7$$

The diffusion capacitance charge storage element is given simply by

$$Q_{BE_D} = I_S(e^{V_B E/V_T} - 1)/Q_B/I_K \qquad 8$$

or

$$Q_{BE_D} = I_C/I_K, \qquad 9$$

where I_K (knee current) is a model parameter, Eq. (9) was the original formulation,[1] while Eq. (8) allows Q_B to be obtained explicitly from junction voltage values, and is for all practical purposes identical.[3] The concequences of Eq. (9) is that high level gain (I_C/I_B) will always fall-off exactly as $1/I_C$, (ratio of Eq. (5) to (3) as $V_{BE} \to \infty$). Also the delay time $\frac{dQ_B}{dI_C}$ is given by

$$\frac{dQ_B}{dI_C} = C_{JE} \frac{dQ_{BEj}}{dV_{BE}} \frac{dV_{BE}}{dI_C} + T_F \qquad 10$$

where T_F physically is related to I_K, but kept general to allow modeling flexibility. Eq. (10) yields the familiar delay-time $(1/f_T)$ versus $1/I_C$ characteristic, which intercepts the delay time axis at T_F for infinite current and has a slope of

$$\frac{d}{d(1/I_C)} \text{ of } \left(\frac{1}{2\pi f_T}\right) = C_{JE} * V_T \qquad 11$$

for small currents.

The original SPICE program recognized this "non-modeling" of f_T fall-off and "patched-up" the ac model by "modulating" the diffusion capacitance by Q_{BN} i.e.

$$\frac{dQ_B}{dI_c} = C_{JE} \frac{dQ_{BE_J}}{dV_{BE}} \frac{dV_{BE}}{dT_C} + T_F \cdot Q_{BN} \qquad 12$$

this gives the proper qualitative effect with extreme parameter economy (no additional parameters!). This is the SPICE NGP and PGP model. Also base resistance can be easily modulated by Q_B i.e.

$$r_b = RB/QB. \qquad 13$$

II. Extended Gummel-Poon Model

The deficiencies of the NGP(PGP) model are:
1) Beta can only roll off as $1/I_C$
2) Both peak gain and peak f_T must be modeled by one parameter, IK.
3) No provision for modeling forward-biased junction capacitance (see Eq. 11) i.e. $C_J = C_{Jo}/(1-V/P)^m$ in reverse, $C_J = C_{Jo} mV/P$ in forward bias.

In particular the high level Kirk effect[4] and forward biased junction capacitance are ignored. The forward biased junction capacitance deficiency is easily removed, and will be discussed later. The inadequate modeling of the Kirk effort, which results in deficiencies 1) and 2) above, was resolved by use of detailed device simulations.[5,6,7]

Shown in Figures 1-4 are solutions for carriers at two levels of bias, low and high. Note that at high bias base widening is occuring and base charge (diffusion capacitance) is increasing more rapidly than accounted for by Eq. (9). In order to develop a more accurate model for the diffusion capacitance, the base charge, as evaluated from the simulations, was characterized as emitter and collector voltage controlled capacitance, as defined in Fig. 5. When these results are plotted, as shown in Fig. 6, it is apparent that the diffusion capacitance at high bias varies simply as an exponential with applied junction voltage, with a model parameter n_D, i.e.

$$Q_{BE_D} = K(e^{V_{BE}/n_D V_{VT}} - 1) \qquad 14$$

This simple expression allows great flexibility in removing the discrepancies in the NGP and PGP models. For example, for n_D equal to 2, the asymptotic behavior is identical to that predicted by the old model. For n_D equal to 1, extreme gain fall-off ($\frac{d\beta}{dI_C} \to -\infty$) is obtained. Also, peak Beta and peak f_T can occur at independent values of current, and finally as the result derives from the physical behavior of transistors, the dc and ac characteristics are properly linked.

The constant K in Eq. 14) should be chosen to set $Q_{BE_D} = I_C/I_K$, at $I_C = I_K$ to maintain upward compatibility with the NGP and PGP models. Also note in particular that the delay time $\frac{dQ_B}{dI_C}$ is directly derived from the normalized base charge, as with the NGP and PGP models, the charge-storage elements are unnormalized by the parameter C_{JE} and T_F. (C_{JC} and T_R for inverse operation).

Removal of deficiency 3) on the value of forward biased junction capacitance, is easily accomplished by the addition on an additional model parameter.[8] This additional parameter should be chosen to yield a value of forward biased capacitance which forces a match to the observed forward delay time characteristic.

III. Summary and Results

The basic Gummel-Poon model as implemented in SPICE has been reviewed and an extension proposed which readily models the base charge in high level injection. By use of device simulation programs it was found that the component of base charge associated with diffusion capacitance at high bias can be modeled by a simple exponential in junction voltage. Shown in Fig. 7 are current gain and delay time as predicted by the model. Model parameter values were extracted from the same simulations as illustrated in Figures 1-6.

References

1. H. K. Gummel, H. C. Poon, "An Integral Charge Control Model of Bipolar Transistors," BSTJ, vol. 49, pp. 827-852, May-June 1970.
2. H. K. Gummel, "A Charge Control Relation for Bipolar Transistors," BSTJ, vol. 49, pp. 115-120, Jan. 1970.
3. L. W. Nagle, D. O. Pederson, "SPICE Simulation Program with Integrated Circuit Emphasis," Midwest Symposium on Circuit Theory, Waterloo, Ontario, April 12, 1973.
4. C. T. Kirk, Jr., "A Theory of Transistor Cutoff Frequency (f_t) Falloff at High Current Densities," IRE Trans. on Electron Devices, ED-9, No. 2, pp. 164-174, March 1962.
5. J. G. Ruch, D. L. Scharfetter, "Characterization of Bipolar Devices," 1973 IEDM, Washington, D.C., December 1973.
6. D. L. Scharfetter, H. K. Gummel, "Large - Signal Analysis of a Silicon Read Diode Oscillator, TEEE Trans. El. Devices, vol. ED-16, pp. 64-77, Jan. 1969.
7. J. G. Ruch, H. C. Poon, W. J. McCalla, D. L. Scharfetter, "Automated Transistor Characterization and Parameter Generation for Linear Integrated Circuit Design," 1974 ISSCC, Philadelphia, Pa., Feb. 13-15, 1974.
8. H. C. Poon, H. K. Gummel, "Modeling of Emitter Capacitance," Proc. of IEEE, vol. 57, No. 12, pp. 2181-2182, Dec. 1969.

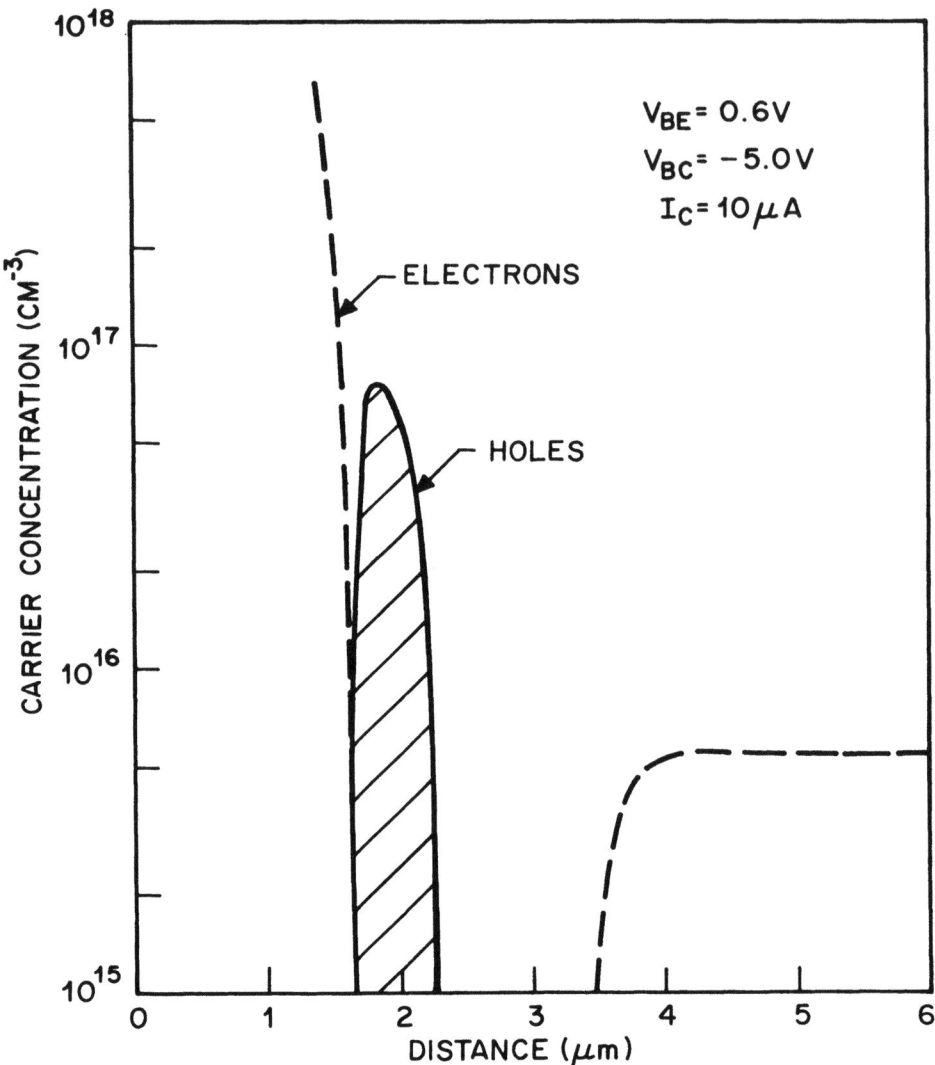

FIG. 1 HOLE AND ELECTRON CONCENTRATIONS, AT LOW BIAS

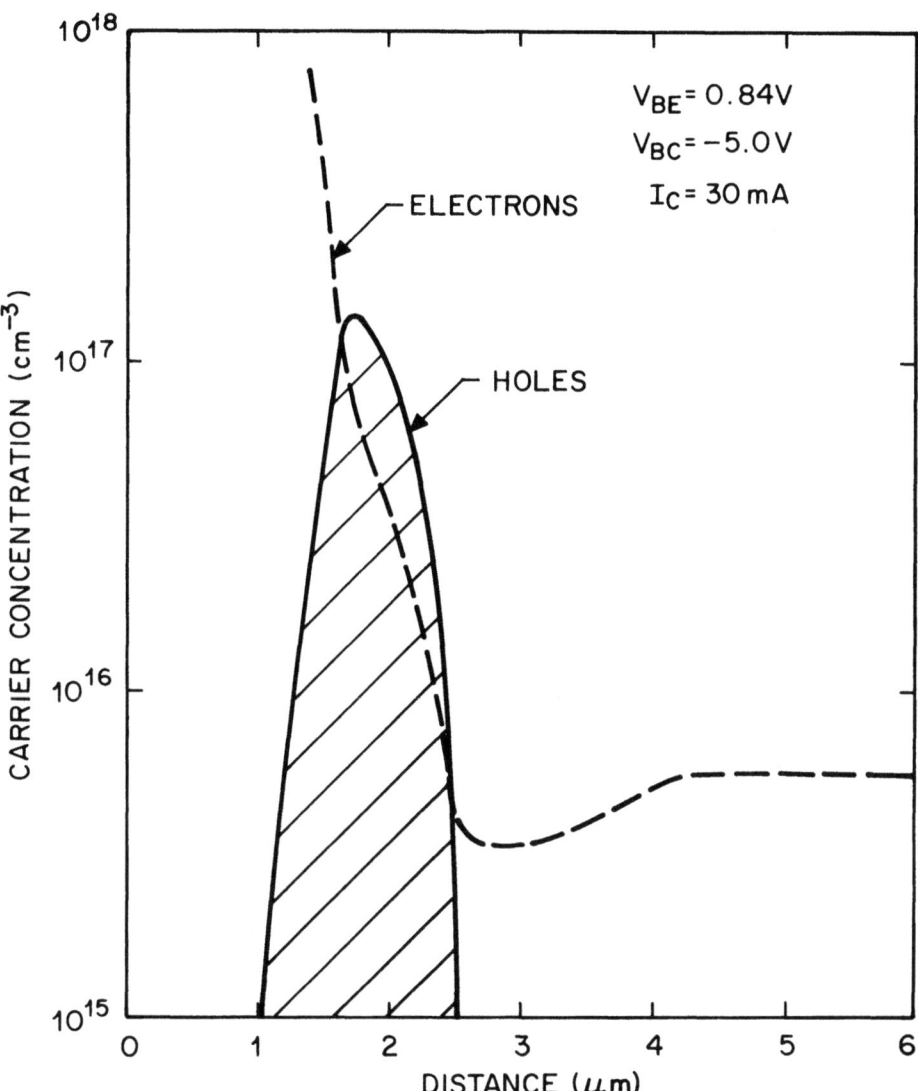

FIG. 2 HOLE AND ELECTRON CONCENTRATIONS, AT HIGH BIAS

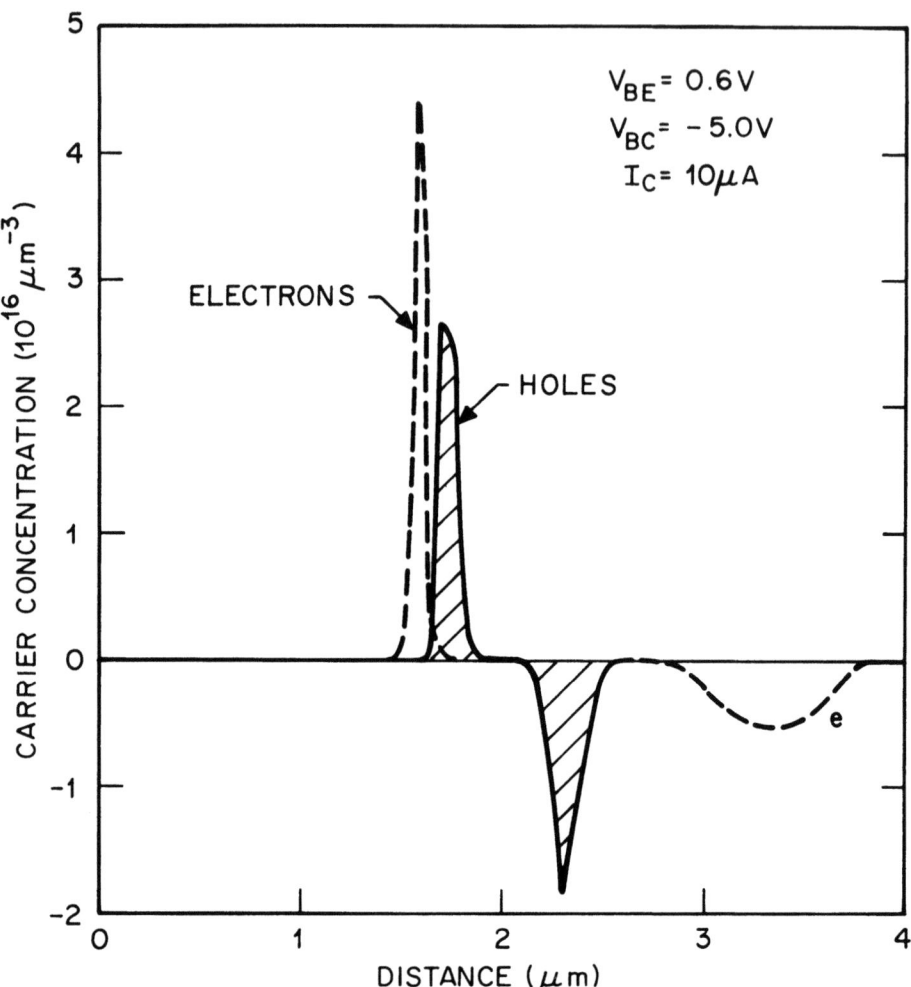

FIG. 3 HOLE AND ELECTRON CONCENTRATIONS IN EXCESS OF ZERO BIAS VALUES, AT LOW BIAS

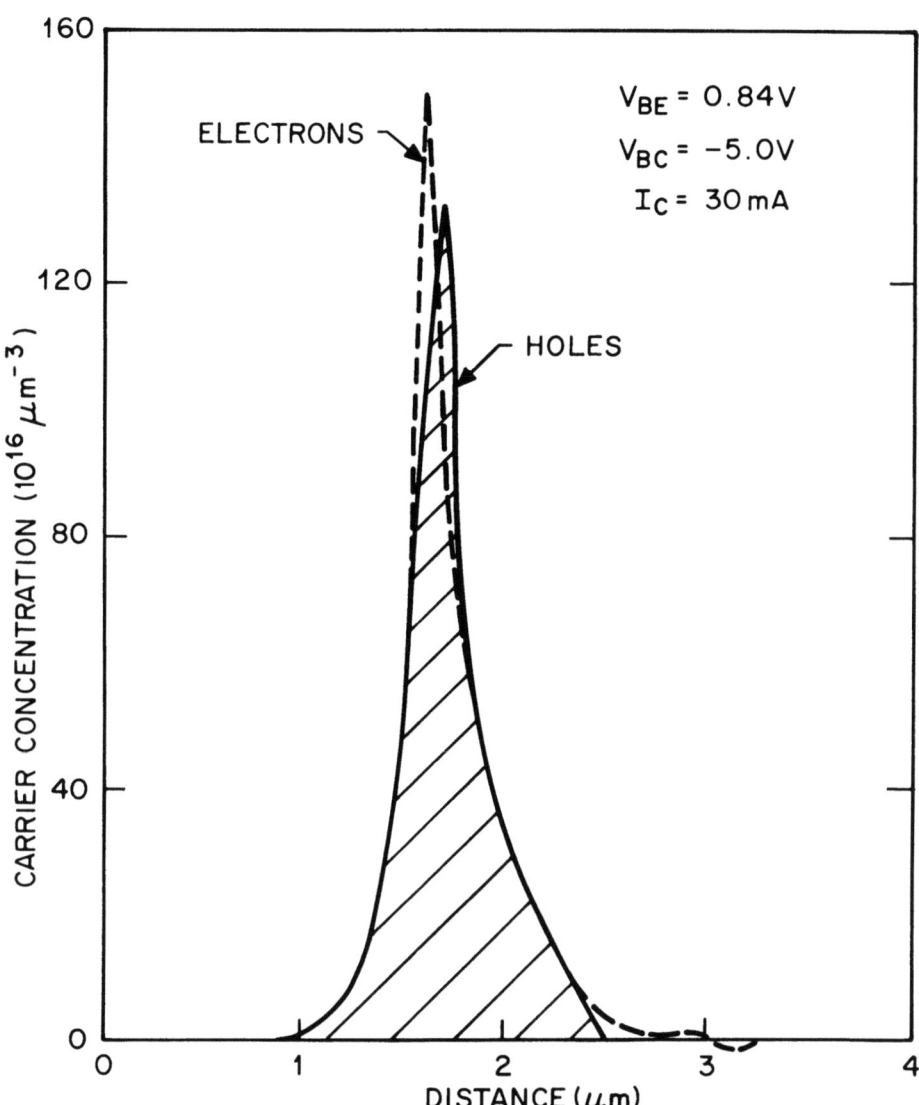

FIG. 4 HOLE AND ELECTRON CONCENTRATIONS IN EXCESS OF ZERO BIAS VALUES, AT HIGH BIAS

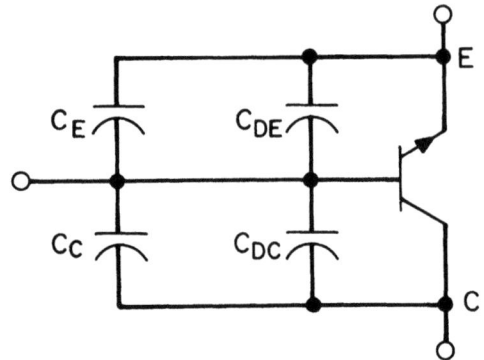

$$I_\alpha = \frac{I_S Q_0 \left(e^{\frac{V_{BE}}{V_T}} - e^{\frac{V_{BC}}{V_T}}\right)}{(Q_0 + Q_E + Q_C + Q_{DE} + Q_{DC})}$$

$$C_E = \left.\frac{\partial Q_E}{\partial V_{BE}}\right|_{V_{BC}} \qquad C_C = \left.\frac{\partial Q_C}{\partial V_{BC}}\right|_{V_{BE}}$$

$$C_{DE} = \left.\frac{\partial Q_{DE}}{\partial V_{BE}}\right|_{V_{BC}} \qquad C_{DC} = \left.\frac{\partial Q_{DC}}{\partial V_{BC}}\right|_{V_{BE}}$$

FIG. 5 DEFINITION OF FUNTION AND DIFFUSION CAPACITANCES

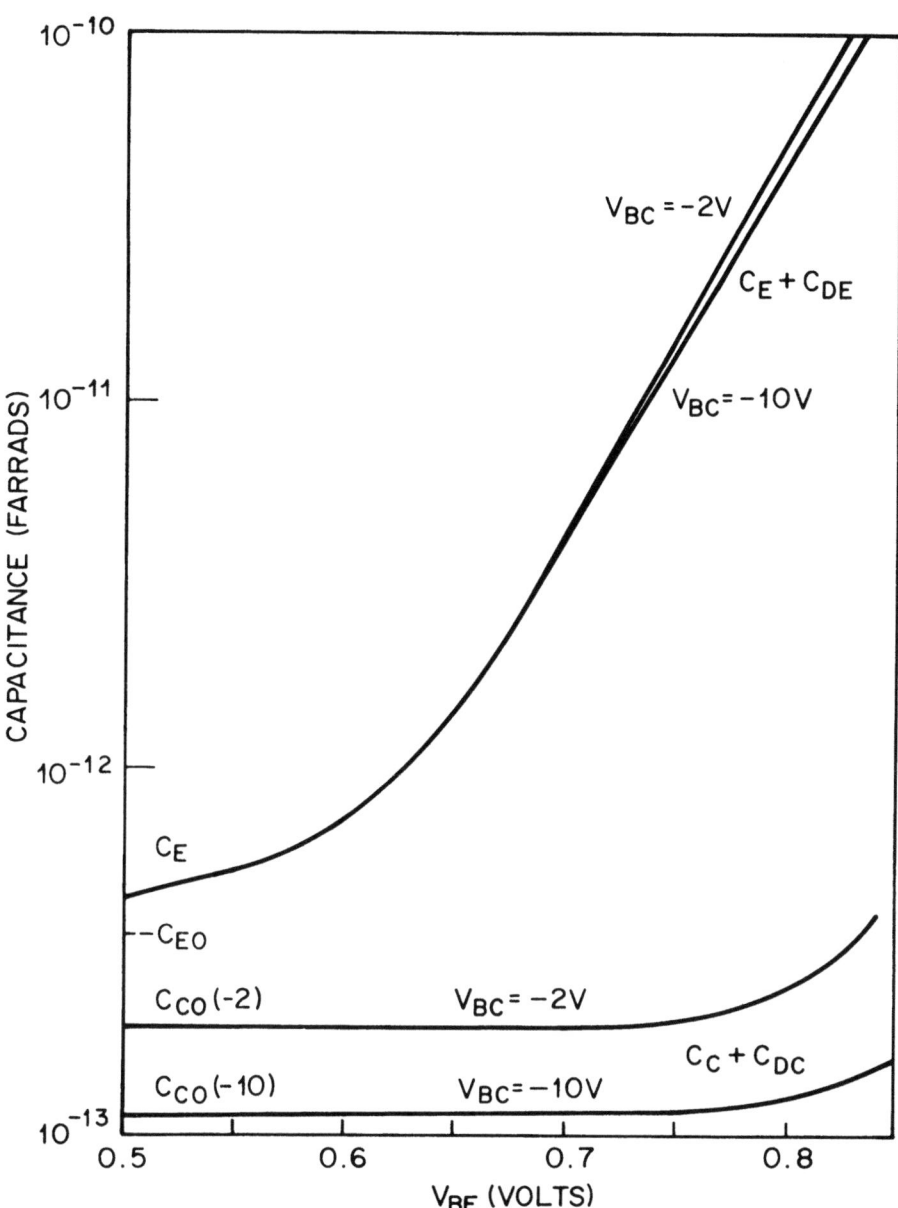

FIG. 6 JUNCTION AND DIFFUSION CAPACITANCES VS V_{BE}

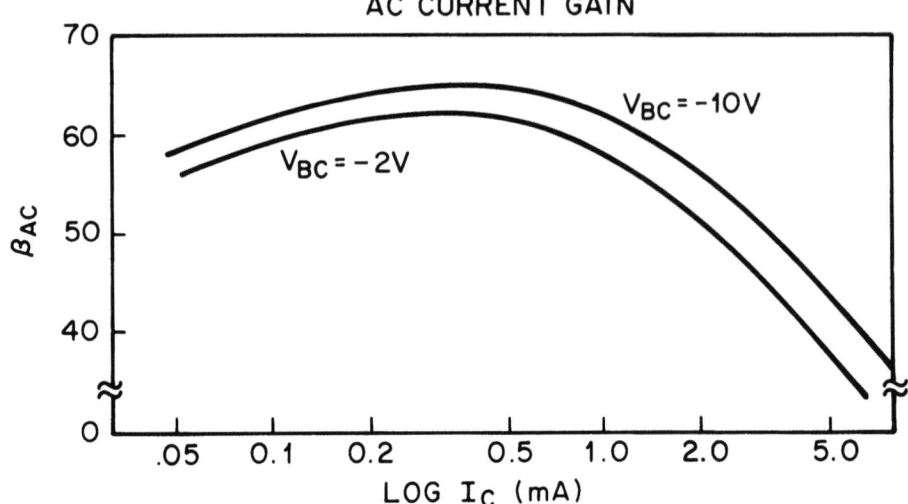

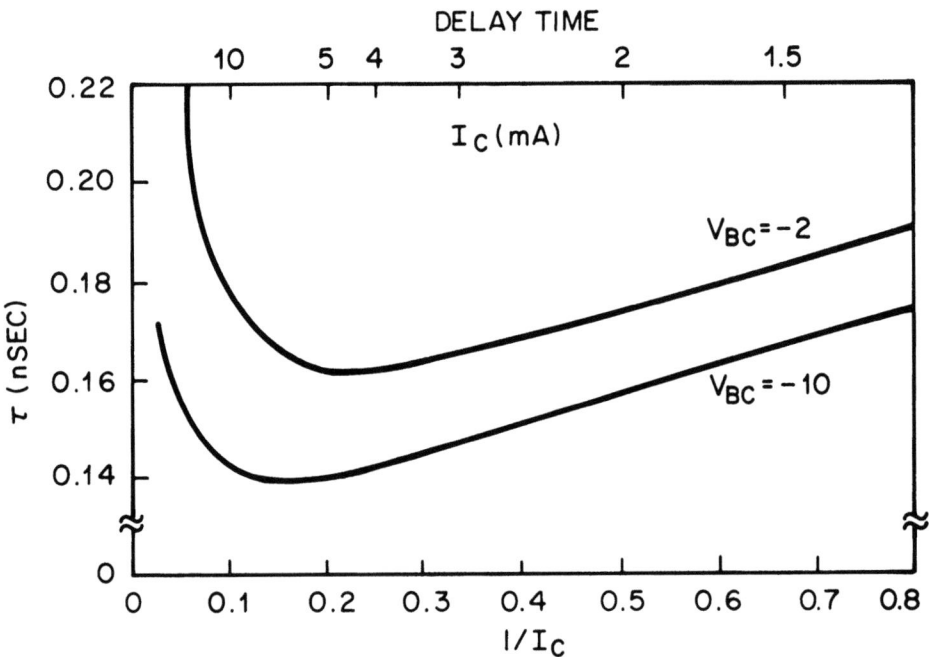

FIG. 7 AC TRANSISTOR CHARACTERISTICS

MODELING OF BIPOLAR DEVICES

W.L. Engl, O. Manck, A.W. Wieder

Institut für Theoretische Elektrotechnik
Technische Hochschule Aachen, Germany

I. NUMERICAL MODELS

1. Basic Equations and Physical Models

In semiconductors the transport of electrons in the conductivity band and likewise the transport of holes in the valence band is described by a continuity equation for each type of carriers:

$$\frac{\partial n}{\partial t} - \frac{1}{q} \nabla \cdot \vec{J}_n = -R \tag{1.1}$$

$$\frac{\partial p}{\partial t} + \frac{1}{q} \nabla \cdot \vec{J}_p = -R \tag{1.2}$$

where R is the recombination rate. The current densities of electrons and holes are proportional to the gradients of the corresponding quasi Fermi potentials or imrefs

$$\vec{J}_n = -q\mu_n n \nabla \phi_n, \tag{1.3}$$

$$\vec{J}_p = -q\mu_p p \nabla \phi_p, \tag{1.4}$$

where µ denotes mobility. Carrier density, imref and energy level within the crystal are related by an integral relation

$$n = \int_{E \approx E_c}^{\infty} \frac{D_c(E)dE}{1 + \exp((E + q\phi_n)/kT)}, \qquad (1.5)$$

$$p = \int_{-\infty}^{E \approx E_v} \frac{D_v(E)dE}{1 + \exp(-(E + q\phi_p)/kT)}. \qquad (1.6)$$

$D_{c,v}$ is the effective density of states in the conductivity and valence band. In order to get symmetrical expressions for n and p, the electrostatic potential ψ in the crystal is referenced to the intrinsic Fermi level E_i ($\psi = -E_i/q$). The potential distribution can be calculated by solving Poisson's equation:

$$\nabla^2 \psi = \frac{q}{\varepsilon}(n-p-N), \qquad (1.7)$$

where $N = N_D^+ - N_A^-$ is the electrically active net impurity concentration. If the imrefs $\phi_{n,p}$ are at most a few thermal voltages $V_T = kT/q$ apart from the band edges the integral relations (1.5,6) can be simplified and yield the Boltzmann approximation:

$$n = n_i \exp((\psi - \phi_n)/V_T), \qquad (1.8a)$$

$$p = n_i \exp((\phi_p - \psi)/V_T), \qquad (1.8b)$$

where $n_i^2 = N_c N_v \exp(-(E_c - E_v)/kT)$ defines the intrinsic concentration. The effective densities of states are given by the constants:

$$N_c = \int_0^{\infty} D_c(E+E_c) \exp(-E/kT)dE,$$

$$N_v = \int_{-\infty}^{0} D_v(E+E_v) \exp(E/kT)dE. \qquad (1.9)$$

The set of equations (1.1-1.8) in combination with boundary conditions determines the electrical performance of semiconductor devices. The dimensions of these devices are such, that up to frequencies of the order of 10^{10} Hz the wavelength of an electromagnetic wave in matter is still large compared to the device dimensions. Hence the quasistatic formulation of the equations is justified.

Band gap narrowing caused by heavy doping effects can be taken into account by a doping dependent intrinsic density in (1.8), which yields:

$$n = n_{in}(N) \exp((\psi-\phi_n)/V_T),$$
$$p = n_{ip}(N) \exp((\phi_p-\psi)/V_T). \qquad (1.10)$$

The product of carrier densities in thermal equilibrium defines an effective intrinsic concentration, which can be measured as a function of impurity concentration [2,3]:

$$n_{ie}^2 = n_{in} n_{ip} = n_i^2 \exp(-\Delta E_g/kT). \qquad (1.11)$$

Relation (1.10) also implies the quasi-neutrality condition $n-p-N \approx 0$ to hold, which is in general true for heavily doped regions. One more condition is needed to determine the factors of the product in (1.11) separately. To yield this condition we consider the position of the Fermi level in heavily doped regions. Numerical calculations of the thermal equilibrium in such a region assuming hopping conductivity in the impurity band [1] indicate that the Fermi energy hardly exceeds the energy level of the dopands $E_{A,D}$. The reason for this is, that the density of states in the impurity band increases strongly as a function of impurity concentration. The electrostatic potential as a function of e.g. donator concentration can be approximated by fitting numerical results:

$$\psi = \begin{cases} V_T \ln(N/n_i) & \text{for } N \leq N_1 \\ V_T \left[\ln(\frac{N}{n_i}) + a\left(1-\left(1-\frac{\ln(N/N_1)}{\ln(N_2/N_1)}\right)^\alpha\right)\right] & \text{for } N_1 \leq N \leq N_2 \end{cases} \qquad (1.12)$$

$N_1 = 10^{17}$ cm^{-3}, $N_2 = 10^{21}$ cm^{-3}, $a = (E_D-E_i)/kT - \ln(N_2/n_i)$,
$\alpha = \ln(N_2/N_1)$ and $\ln(\frac{N}{n_i}) \approx \sinh^{-1}(\frac{N}{n_i})$.

This approximation also holds in the non equilibrium state, if the minority carrier concentration is small compared to the concentration of majorities, which is normally true in heavily doped regions. The effective intrinsic concentration turns out to be the most important quantity for describing heavy doping effects. In regions where these effects become pronounced, the majority carrier density is fixed by the doping level, thus the minority carrier density in thermal equilibrium increases with n_{ie}^2 corresponding to (1.11). Therefore minority carrier transport as well as recombination is enhanced likewise. The increase in current can be described by a fictitious electric field [4]:

$$\vec{E}^* = 2V_T \, \nabla \ln(n_{ie}/n_i), \qquad (1.13)$$

acting only on minority carriers in quasi neutral regions.

Mobility is determined by different scattering mechanisms [5,6]. Impurity scattering decreases the mobility as a function of the absolute impurity concentration $N_t = N_A + N_D$. Experimental results are fitted with an expression of the form:

$$\mu(N_t) = \mu_{min} + \frac{\mu_{max} - \mu_{min}}{1 + (N_t/N_{Ref})^\gamma} \qquad 0.5<\gamma<1 \qquad (1.14)$$

Since the velocity in a high-field region is essentially independent of impurity concentration the combined dependence can be approximated by:

$$\mu(N_t, E) = \frac{\mu(N_t)}{\left(1 + \left(\frac{E \cdot \mu(N_t)}{v_S}\right)^\beta\right)^{1/\beta}}. \qquad (1.15)$$

v_S is the saturation velocity and $\beta = 1$ gives the hole and $\beta = 2$ the electron mobility, respectively. However, writing the gradient of quasi-Fermi potentials in (1.15) instead of the electric field strength seems to be more reasonable since scattering should be a function of carrier velocity. Furthermore, the imref gradient automatically gives the correct mobility with respect to current flow. In case of extremely high injection levels, occuring in PIN-diodes and thyristors electron-hole scattering reducing the carrier mobility may prevail ion scattering especially in lowly doped regions [7,8]. The effect of ion and electron-hole scattering on mobility can be simulated by an effective doping N_t^* which is plugged into equation (1.14):

$$N_t^* = (1-b)(N_D + N_A) + b(n+p) \qquad (1.16)$$

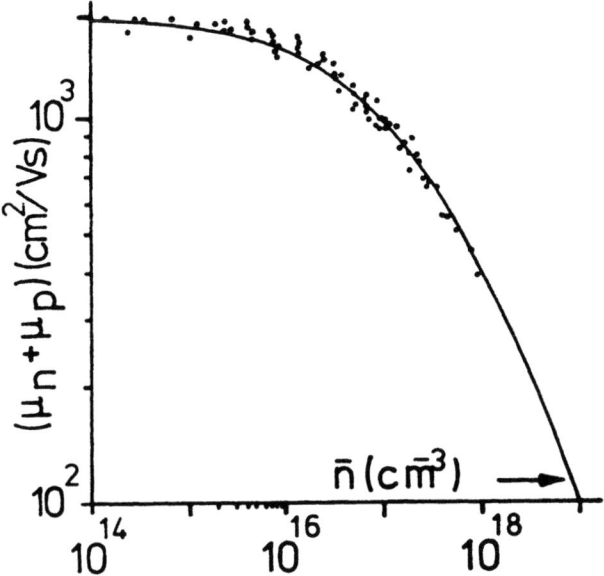

Fig. 1. Mobility as function of carrier concentration for high injection conditions. Measurement [7,8] and approximation corresponding to (1.16).

The parameter not yet determined is used to match the experimental results in Fig. 1 (b = 0.33).

The recombination rate is described by a Shockley-Read-Hall term:

$$R_{SRH} = \frac{np - n_{ie}^2}{\tau_p(n+n_t) + \tau_n(p+p_t)} \qquad (1.17)$$

$$n_t = n_{ie} \exp((E_t-E_i)/kT)$$

$$p_t = n_{ie} \exp((E_i-E_t)/kT)$$

Interface recombination is treated in an analogous way by replacing lifetimes by the reciprocal values of the recombination velocities. Experimental data for the SRH-model have been established to a lesser extent, compared to data required for the simulation of transport mechanisms. Particular knowledge is required for electron and hole lifetimes, the corresponding recombination velocities and the energy distribution of traps. These parameters are not all known a priori with sufficient accuracy and thus in part have the character of fitting parameters. In (1.17) we use a lifetime relation first proposed by Scharfetter [9]:

$$\tau = \tau_{min} + \frac{(\tau_{max} - \tau_{min})}{1 + N_t/N_{Ref}} \qquad (1.18)$$

The lifetime varies between its maximum value in intrinsic material and a minimum value depending on fabrication processes. The energy level of the recombination trap can be fitted from the slope of the h_{FE}/I_C-curves. Another important mechanism in silicon is Auger recombination which describes the energy transfer to a third carrier:

$$R_{AU} = (C_n n + C_p p) \cdot (np - n_{ie}^2) \; . \qquad (1.19)$$

It becomes significant in regions of heavy doping or high injection. The coefficients measured by different authors do not differ very much [10,11,12]. Hence, they can be introduced into the calculation as a priori parameters.

Avalanche generation

$$G_{AV} = \frac{1}{q} (\alpha_n J_n + \alpha_p J_p) \qquad (1.20)$$

occurs when high reverse voltages are applied. The published ionization coefficients depend somewhat on technology [13,14,15]. Finally carrier generation caused from outside must be considered if optoelectronic devices or measuring set ups for determining lifetime are to be simulated.

2. Numerical Algorithms

Exact analytical solutions of the set of partial differential equations (1.1 - 1.8) are not feasible, hence numerical methods are mandatory. In the course of this procedure the differential equations have to be transformed into a set of finite difference equations. These nonlinear algebraic equations have then to be linearized before finally the linear system can be solved. An iterative process of the last two steps converges to the solution of the nonlinear difference equations. Each of the three subsequent steps is covered by standard mathematics in the one-dimensional case. Multidimensional problems however cannot be treated by standard approximations, thus special numerical algorithms become necessary, the development of which was guided by the physical understanding of bipolar devices. This is especially important for finite difference approximations.

Since constant space charge density leads to a parabolic function for the potential, the standard Taylor series method is adequate to discretisize Poisson's equation. For the particular (i, j) grid point (Fig. 2) one obtains

$$\nabla^2 \psi|_{ij} \to \frac{2}{h_i h_{i+1}(h_i+h_{i+1})} \qquad (2.1)$$

$$\cdot \{h_{i+1}\psi_{i-1,j} - (h_i+h_{i+1})\psi_{i,j} + h_i \psi_{i+1,j}\}$$

$$+ \frac{2}{k_j k_{j+1}(k_j+k_{j+1})}$$

$$\cdot \{k_{j+1}\psi_{i,j-1} - (k_j+k_{j+1})\psi_{i,j} + k_j \psi_{i,j+1}\} .$$

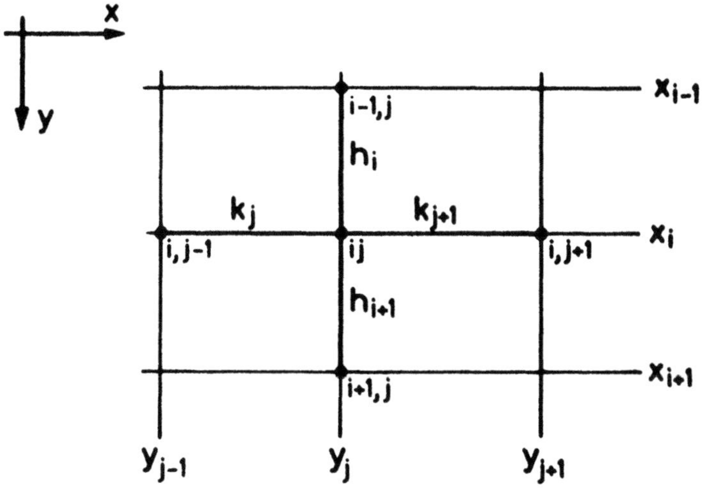

Fig. 2. Typical two-dimensional grid cell.

For the continuity equations, the analogous approach for the quasi Fermi potentials would imply exponential current density distributions in the neighborhood of each grid point. This would certainly be a poor approximation, since the current densities $\vec{J}_p$ and $\vec{J}_n$ can vary only weakly with position. Therefore, a different discretization scheme is used which was originally proposed by Gummel [16]. The scheme has previously proven to be the very key for the two-dimensional semiconductor device modeling enabling one to drastically reduce the number of necessary grid points, while maintaining numerical accuracy. The basic idea is to assume constant hole and electron current densities and a constant electric field between pairs of grid points. Equations (1.3) and (1.4) can then be integrated, yielding for the normalized current

$$J_{px}^{i+1/2,j} = -\mu_p^{i+1/2,j} \frac{\psi^{i+1,j} - \psi^{i,j}}{\exp(\psi^{i+1,j}) - \exp(\psi^{i,j})} \quad (2.2)$$

$$(\exp(\phi_p^{i+1,j}) - \exp(\phi_p^{i,j})).$$

The subscript $i + 1/2, j$ denotes average values in the interval $x_i < x < x_{i+1}$, $y = y_j$. The divergence is approximated by making use of the customary Taylor series approach, yielding for (1.2)

$$\nabla \cdot \vec{J}_p\big|_{ij} \to \frac{J_{px}^{i+1/2,j} - J_{px}^{i-1/2,j}}{\frac{1}{2}(h_i + h_{i+1})} + \frac{J_{py}^{i,j+1/2} - J_{py}^{i,j-1/2}}{\frac{1}{2}(k_j + k_{j+1})} \quad (2.3)$$

Numerical calculations have shown that SRH recombination is concentrated at or in the neighborhood of PN junctions, hence the assumption of constant recombination between neighboring grid points is a poor approach to discretize the continuity equation and hence this is not even allowed in the one-dimensional case. The local integration of the recombination rate assuming constant imrefs and a linear potential variation between neighboring grid points overcomes these difficulties in the bulk as well as at the interfaces.

A solution vector $\vec{\zeta}$ is defined, representing the unknown values of ϕ_p, ϕ_n and ψ at the grid points

$$\vec{\zeta} = (\phi_p^{11}, \phi_n^{11}, \psi^{11}, \phi_p^{21}, \phi_n^{21}, \psi^{21}, \ldots, \phi_p^{IJ}, \phi_n^{IJ}, \psi^{IJ}) \quad (2.4)$$

where the subscripts I and J denote the number of grid lines parallel to the x- and y-axis, respectively. The system of difference equations can be arranged in the vector equation

$$\vec{F}(\vec{\zeta}) = (f_{11}(\vec{\zeta}), g_{11}(\vec{\zeta}), h_{11}(\vec{\zeta}), f_{21}(\vec{\zeta}), \ldots, f_{IJ}(\vec{\zeta}), g_{IJ}(\vec{\zeta}), h_{IJ}(\vec{\zeta}))$$
$$= 0 \quad (2.5)$$

Here, f_{ij}, g_{ij} and h_{ij} stand for the difference equations in the (ij) grid point derived from the continuity equations for holes and electrons and Poisson's equations, respectively. Thus for each grid point, three difference equations in the variables ϕ_n, ϕ_p and ψ are obtained.

The system of nonlinear difference equations can be solved by
different methods. Most simply one treats each differential equation separately and succesively, and repeats this procedure until
a desired accuracy is reached. This successive procedure is advantageous for multidimensional calculations as it saves storage
and converges quite good. However, it is restricted to low and
medium injection levels where the variables are not strongly
coupled.

In case of high level injection such a procedure fails, at
least in this simple formulation and the set of equations has to
be solved simultaneously. For one-dimensional calculations this
concept is the most efficient for all injection levels, because
in this case the solution of the linearized system is easily obtained and enables one to fully take advantage of the quadaratic
convergence of the Newton-Raphson linearization. For two-dimensional calculations the simultaneous solution of the linearized
system requires a large effort hence combinations of the successive
and simultaneous approach become meaningful.

The solution of (2.5) is found by Newtonian iteration. Assuming that an approximation $\vec{\zeta}^0$ of $\vec{\zeta}$ is available, one obtains a better approximation by the process

$$F'(\vec{\zeta}^k)\delta\vec{\zeta}^{k+1} = -\vec{F}(\vec{\zeta}^k), \quad k = 0,1,2,\ldots, \qquad (2.6)$$

$$\vec{\zeta}^{k+1} = \vec{\zeta}^k + \delta\vec{\zeta}^{k+1}$$

where $F'(\vec{\zeta}^k)$ is the Jacobian matrix, and $\delta\vec{\zeta}^{k+1}$ is a correction
vector in the iteration step (k+1).

The convergence of this iteration process is quadratic, provided the linearized systems are solved exactly. The linearized
system is solved best by Gaussian elimination methods for one-dimensional calculations. For the two-dimensional problem this
method proves to be successful too, provided that the successive
procedure can be used and small corrections (2.6) allow to take
advantage of the quadratic convergence of the Newtonian cycle.
In case of large corrections and for large systems to be solved
simultaneously a second iterative process is necessary. The best
results have been obtained with the successive line overrelaxation (SLOR) method applied to vertical or lateral grid lines.

The large amount of storage for one grid point, nearly 60
elements in the simultaneous SLOR method, requires careful programming. Storing only the actually needed information in the
core, problems with 8000 grid points can be solved in a 128 k
machine with 48 bits words. Computing time last not least depends
on the quality of the starting solution, which approximates the
solution to be found by analytical means. On a Telefunken TR 440
(time for addition, multiplication and division are 1.8 µs,
3.8 µs and 12 µs, respectively) computing time for each grid point
and bias condition is in the order of .05 s for one - and in the

order of .5 s for two-dimensional calculations.

Numerical accuracy can be checked by summing up all terminal currents. This sum must vanish with the order of permitted inaccuracy of a certain specified terminal current.

3. One- and two-dimensional examples.

No analytical solution is available for a one-dimensional thyristor structure. Various characteristics calculated from a numerical simulation show how avalanche generation effects the switching voltage of the device (fig. 3) and how Auger recombination and electron-hole scattering influence the current in the forward conducting region (fig. 4) [17].

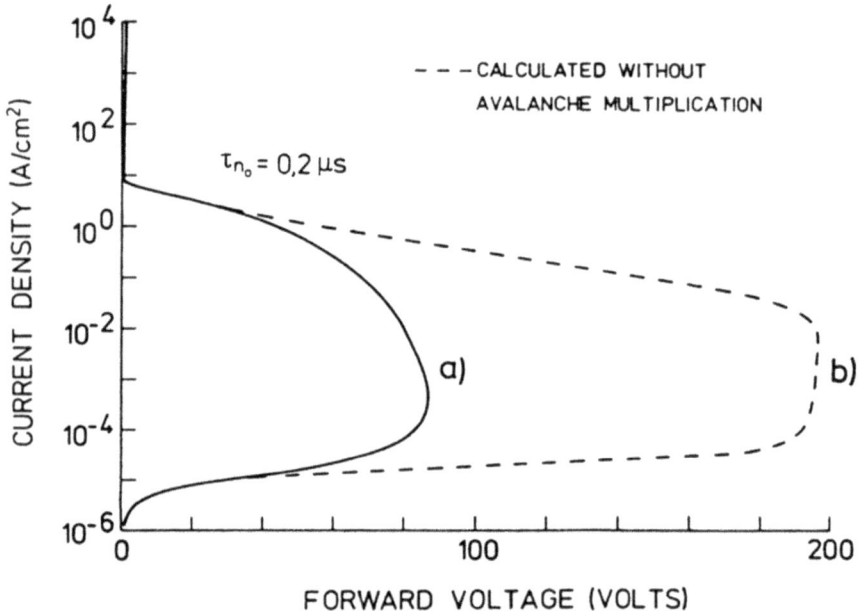

Fig. 3. One-dimensional simulation of a planar thyristor. Calculated forward I-V characteristic
a) triggered by avalanche multiplication,
b) triggered by punch through.

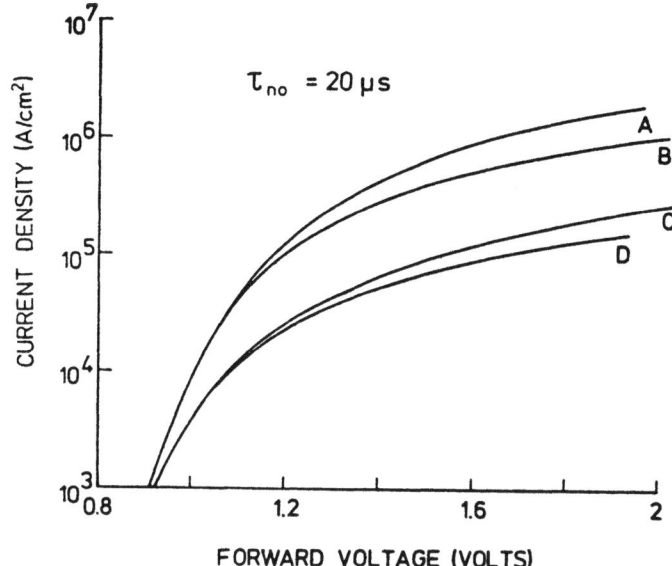

Fig. 4. High current region: A) with SRH recombination and field and doping dependent mobilities, B) as A) but with Auger recombination, C) as A) but with carrier-carrier scattering, D) combining B) and C).

Well known effects of the bipolar transistor are conductivity modulation in the base, sidewall injection, emitter crowding, base widening and spreading of emitter current towards the collector. For high level injection, these effects are strongly coupled by reciprocal action and analytical predictions fail. Figures 5 and 6 show respectively the potential distribution over the two-dimensional transistor cross section at 700 mV and at 900 mV emitter base voltage [18, 19]. At 700 mV the metallurgical junctions still coincide with the electrical and the base resistance is given by the doping. In contrary, the appreciable widened base at 900 mV is formed completely by moveable carriers. The collector is flooded and the base collector junction shifted towards the vicinity of the collector contact. The spreading in the collector region is rather small.

The part of the collector beneath the base contact is almost free of injected carriers, if the ratio of base resistance of the active base is small compared to that of the inactive base.

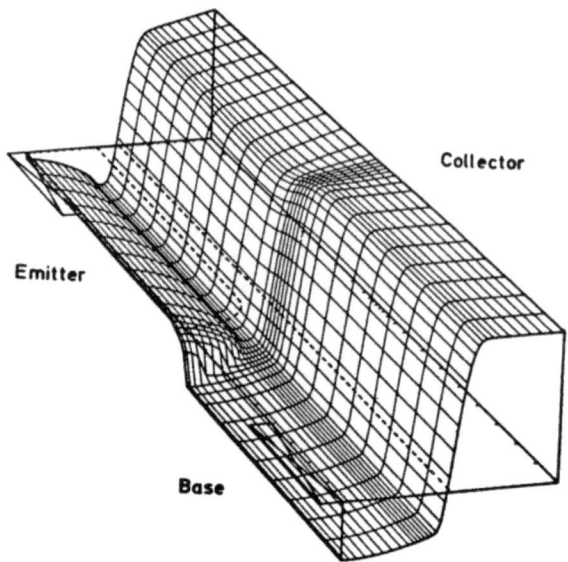

Fig. 5. Potential distribution of a NPN-transistor. Dashed lines represent metallurgical junctions. Bias conditions: V_{BE} = 700 mV, V_{CB} = 1 V.

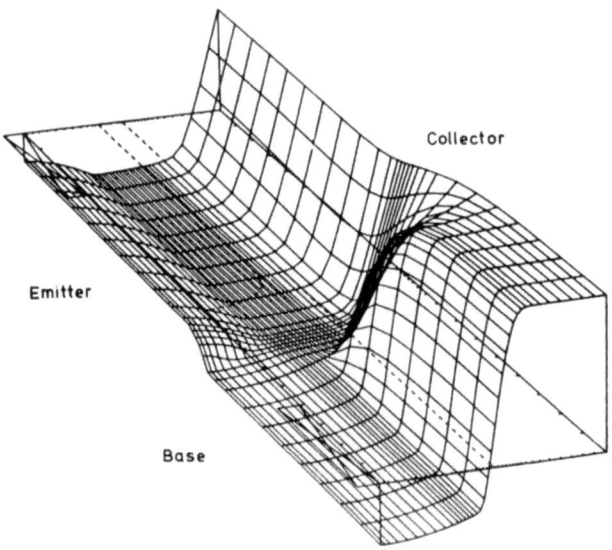

Fig. 6. For reference see figure 5. Bias conditions: V_{BE} = 900 mV, V_{CB} = 1 V.

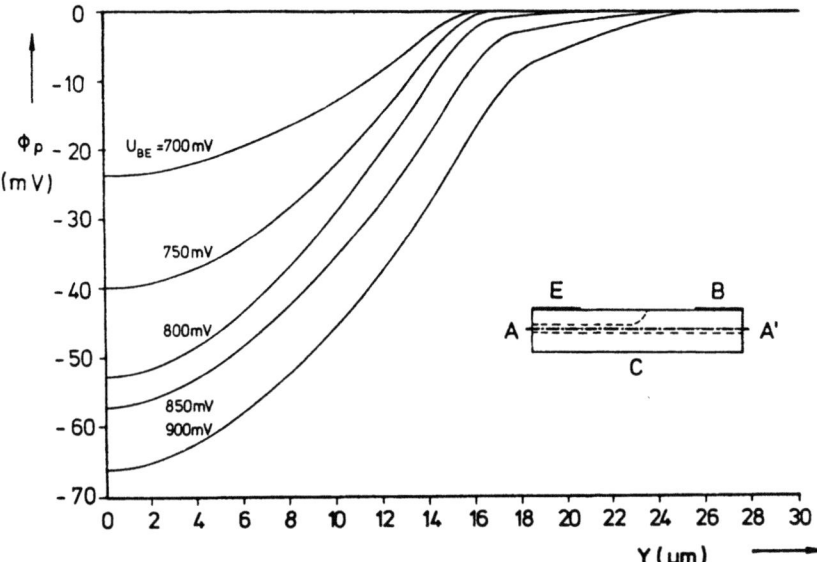

Fig. 7. Drop of the quasi Fermi potential of holes in the active and inactive base, $V_{CB} = 1$ V.

Fig. 7 demonstrates the influence of emitter crowding. The voltage drop in the base is negligible at low injection level and reaches almost 1 V_T at V_{BE} = 700 mV giving rise to emitter crowding. However, base push out prevents crowding from further monotonically increase. Even stabilization of the crowding level may occur ($\Delta\phi_p(x = 0) \approx \Delta\phi_p(x = 15 \mu m)$ in the range between V_{BE} = 800 and 900 mV).

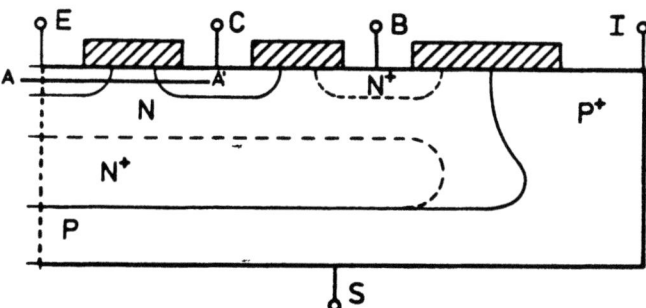

Fig. 8. Cross-section of a lateral PNP-transistor. Inner solid lines represent PN- and dashed line NN$^+$-junctions

For a lateral PNP-transistor (figure 8) with buried layer, isolation diffusion and substrate, figure 9 shows the two-dimensional current distribution [20]. Over the xy-plane the log. of the absolute value of hole current density is displayed for high level injection. The emitter current spreads essentially over the whole epi-layer thickness while flowing towards the collector contact. One also notices the limited screening effect of the buried layer, since an appreciable part of the current flows vertically through the buried layer into the substrate. It becomes likewise apparent that the collector cannot absorb the lateral current totally. A small fraction flows on laterally, partially to the contact of the isolation region and partially down to the substrate. Figure 10 gives impurity, electron- and hole concentration over a lateral cross-section through emitter, base and collector in a linear scale. For high injection levels ($V_{EB} > 600$ mV) both carrier densities increase uniformly, but their difference still equals the impurity concentration thus maintaining quasi-neutrality in the base for all injection levels.

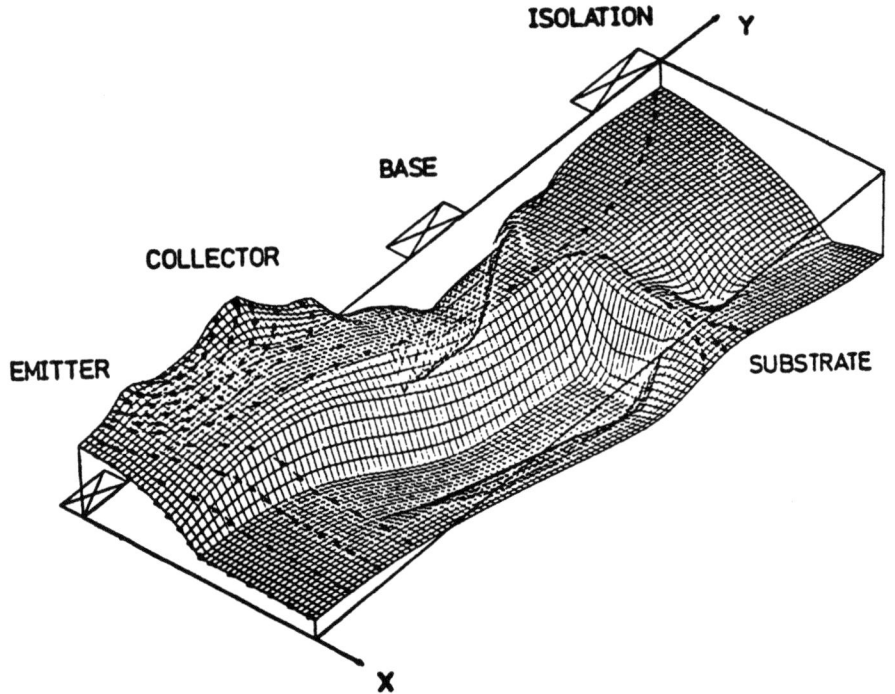

Fig. 9. Absolute value of hole current density of a lateral PNP-transistor (logarithmic plot). Crossed rectangles in the reference plane refer to contact widths. Bias conditions: V_{EB} = 850 mV, V_{CB} = -1 V.

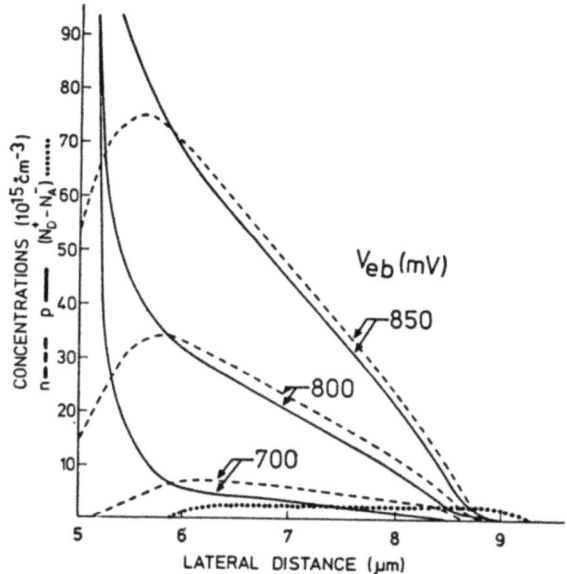

Fig. 10. Doping profile (dotted line), electron (dashed lines) and hole concentrations (solid lines) over the lateral cross-section $\overline{A\,A'}$ (see figure 8). Bias conditions: $V_{CB} = -1$ V, $V_{SB} = -1$ V.

The transient solution for the NPN-transistor [21] differs largely from the one-dimensional vertical behaviour of the steady state solution mentioned before. In fig. 11 and 12 the base emitter voltage is switched from 900 mV to zero Volt.

Figure 6 shows the initial potential distribution for a forward bias of 900 mV. Immediately after switching the potential in the emitter and the active base region has been raised by the switching voltage (figure 11). The potential difference between the active and inactive base builds up a gradient causing discharge. Following the discharge of the sidewall transistor a wave starts to travel laterally through the active base region (figure 12).

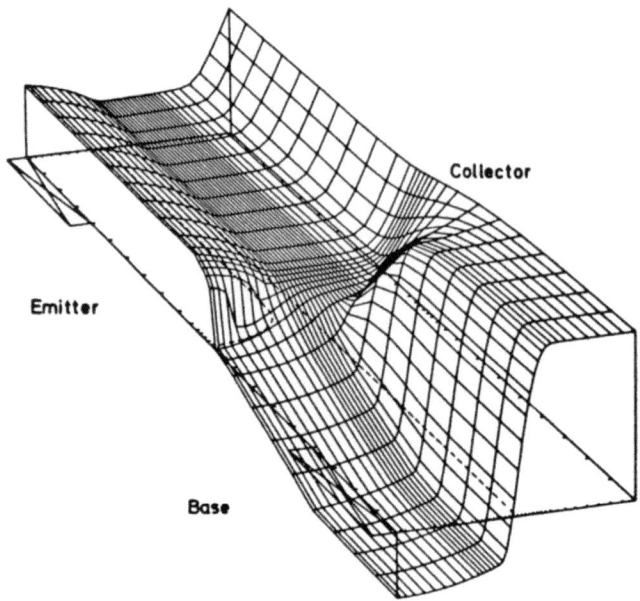

Fig. 11. Potential immediately after switching the NPN-transistor out of saturation, $t = 10^{-10}$ s.

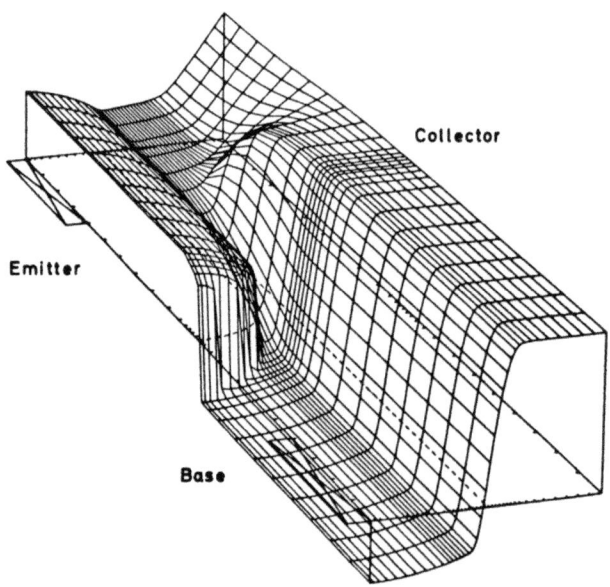

Fig. 12. Potential at $t = 4 \times 10^{-8}$ s.

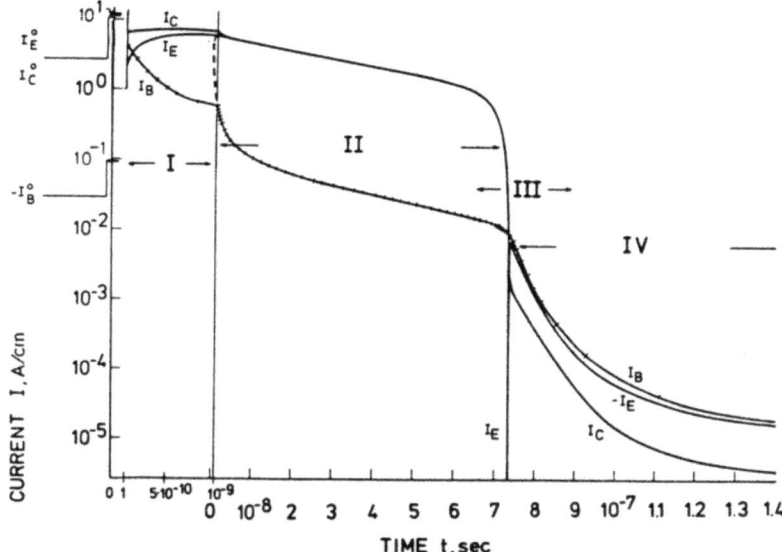

Fig. 13. Terminal currents versus time. Time region I-IV can be associated with different mechanisms. In regions I, the time scale is enlarged by 40 x. I_E^0, I_B^0, I_C^0 are the initial values.

After the lateral wave has reached the emitter center most of the stored charge is discharged. Lateral flow of holes to the base contact rules the switching mechanism. The corresponding terminal characteristics are shown in figure 13. The dominating lateral wave is associated with time region two.

4. Effective modeling.

We deal only with models which deduce terminal characteristics from physical and geometrical data. Especially we do not treat curve fitting techniques to describe measured characteristics. The computational effort of exact numerical analysis is rather high, particularly if more than one dimension is considered, and the accuracy of the basic physical data is in many cases relatively low. Hence obviously one should search for ways to solve the device equations with reduced effort, tolerating some sacrifice in accuracy. It is further desirable to have a hierarchy of models, which allows the user to select the most appropriate for his application. Such a set of numerical models can also be used profitably to save computer time, since the result at one level can be used as a starting solution for the next higher level of the model hierarchy.

The lowest level is given by analytical solutions. A solution of this kind for a one-dimensional npn transistor in steady state will be treated in chap. II. There, the significance of the underlying assumptions will be explained in detail. For a bipolar device one starts by calculating the Gummel numbers from the doping profile along one-dimensional cross sections. From these numbers and the corresponding injecting areas one gets the transfer currents. Their sum approximates the terminal current if the following conditions hold: The forward biasing voltage causes low level injection. Recombination and generation currents are small compared to transfer currents. The reverse bias does not lead to punchthrough and Ohmic resistances can be neglected. With respect to the distribution of internal fields these conditions correspond to: Majority carrier imrefs are constant, quasineutrality determines the potential and transfer current fixes the minority carrier density. These restrictions are not unrealistic. The operation of many devices which perform digital functions can be estimated this way. Computation time is negligible, but to pin down the cross sections and corresponding injection areas correctly is sometimes difficult.

Which improvements are being made next, respectively which assumptions are being dropped first, depends on the operating conditions of the device. Dropping quasineutrality in the junction regions requires to solve Poisson's equation there. To achieve this ϕ_n and ϕ_p are assumed to have the constant potential values which result from neglecting Ohmic resistances in the quasineutral regions. At the edges of the junction regions the potential is continued by its quasineutral values. The resulting potential differs very little from the exact solution unless high level injection exists. The Gummel numbers calculated from the potential have the same degree of accuracy. In these numbers the Early effect is now also regarded for. SRH generation and recombination which occurs at low level injection almost completely in the junction region can now be easily integrated numerically, there. Also depletion layer capacitances can be calculated correctly. However two-dimensional current distributions and recombination in quasineutral regions are not taken care of. The computational effort is very little as long as the potential problem is sufficiently characterized by one-dimensional cross section. A two-dimensional solution is also not too expensive, since the extension of the space charge region is only a small fraction of the total device volume.

The next step is to solve for the minority carrier imref exactly, still assuming a constant majority carrier imref and using the aforementioned solution for the electrostatic potential. Recombination in quasineutral regions and at the surface as well as the distribution of the transfer current is described correctly with this solution, provided Ohmic resistances and high injection effects do not yet come into play. Computer time increases a good deal because of the relatively large region in which the equations must be solved. To summarize the numerical procedure at this stage: Poisson's equation has to be solved only once in the small space charge regions and the two transport equations have to be solved likewise in subregions of the device, the union of which is only as large as the whole device structure. The rank of this model in the hierarchy can be described such, that according to our experience it is well suited to describe device behaviour up to a forward bias of 700 mV and a reverse bias below avalanche breakdown or punchthrough and it still can be implemented economically in a circuit analysis program.

If high injection effects and Ohmic resistances have to be incorporated in the model the numerical effort increases considerably, since the three differential equations governing the device can no longer be decoupled. From the assumptions made before, only quasineutrality still holds and it now holds within forward biased junction regions, too. The latter extension does not buy too much, however. The two transport equations have to be solved simultaneously, mostly in the whole device region. Instead of a simultaneous solution, successive solutions of the two equations may suffice.

A large reverse bias at a junction requires in the respective region a simultaneous solution of Poisson's equation and the transport equation of that carrier which carries the dominating part of the current there.

The last two members of the hierarchy are models which solve two of the three basic equations. In comparison with a solution of the complete system the savings are a factor 2 to 4 depending on convergence. In general, a succeeding exact solution changes only the internal distributions a little with no influence on terminal characteristics. Besides this general model hierarchy, certain devices offer the possibility of additional simplifications. A MOS model almost equivalent with the exact solution treats only the potential equation two-dimensionally [22].

So far the exact numerical models were aimed at the device designer and the experimentally fitted analytical models at the circuit designer. However, if optimization with respect to performance and tolerances is the goal, device and circuit design have to be merged and models based on technological parameters have to be implemented in circuit simulation programs. Particularly basic circuit cells which are being repeated many times on the same chip justify an elaborate treatment and likewise requires a combination of analog and digital functions, e.g. amplifiers and I^2L logic, a detailed consideration of impurity profiles. At the lowest level of the model hierarchy this is always possible. Especially the one-dimensional exact solution can be implemented without leading to excessive computation time [23]. A circuit with 9 transistors all of which have been simulated by the three basic equations, required 17 sec CPU-time on a TR 440 computer for one DC operating point. In general, not all transistors of a circuit need to be modelled with such completeness.

Two and three dimensional effects in a device can also be modelled with this type of a program in a very efficient way by coupling one-dimensional devices with elements being controlled by internal field variables, e.g. carrier concentrations. The choice of the number of coupling elements allows any desired trade off between computational effort and accuracy.

REFERENCES

1. Morgan, T.N., Broadening of Impurity Bands in Heavily Doped Silicon, Phys. Rev. 139 (1965) 343.
2. Vol'fson, A.A., Subashiev, V.K., Fundamental Absorption Edge of Silicon Heavily Doped with Donors or Acceptor Impurities, Sov. Phys.-Semicond. 1 (1967) 327.
3. Slotboom, J.W., The pn-Product in Silicon, Solid State Electronics, 20 (1977) 279.
4. De Man, H.J., The Influence of Heavy Doping on the Emitter Efficiency of a Bipolar Transistor, IEEE Trans. Electron Devices ED-18 (1971) 833.
5. Caughey, D.M., Thomas, R.E., Carrier Mobilities in Silicon Empirically Related to Doping and Field, Proc. IEEE 55 (1967) 2192.
6. Jacobini, C., et al., A Review of some Charge Transport Properties of Silicon, Solid State Electron. 20 (1977) 77.
7. Dannhäuser, F., Die Abhängigkeit der Trägerbeweglichkeit in Silizium von der Konzentration der freien Ladungsträger, I, Solid State Electron. 15 (1972) 1371.
8. Krausse, J., Die Abhängigkeit der Trägerbeweglichkeit in Silizium von der Konzentration der freien Ladungsträger, II, Solid State Electron. 15 (1972) 1377.
9. Scharfetter, D.L., Measured Dependence of Lifetime upon Defect Density and Temperature in Depletion Layers of Epitaxial Silicon Diodes, Solid State Dev. Research Conf., Santa Barbara, 1967.
10. Beck. J.D., Conradt, R., Auger Rekombination in Si, Solid State Comm. 13 (1973) 93.
11. Burtscher, J., Dannhäuser, F., Krauss, J., Die Rekombination in Thyristoren und Gleichrichtern aus Silizium: ihr Einfluss auf die Durchlasskennlinie und das Freiwerdezeitverhalten, Solid State Electron. 18 (1975) 35.
12. Nilsson, N.G., The Influence of Auger Recombination on the Forward Characteristic of Semiconductor Power Rectifiers at High Current Densities, Solid State Electron. 16 (1973) 681.
13. Van Overstraeten, R., De Man, H., Measurement of the Ionization Rates in Diffused Silicon p-n Junctions, Solid State Electron. 13 (1970) 583.
14. Lee, C.A., Logan, R.A., et al., Ionization Rates of Holes and Electrons in Silicon, Phys. Rev., 134 (1964) A 761.
15. Grant, W.N., Electron and Hole Ionization Rates in Epitaxial Silicon at High Electric Fields, Solid State Electron. 16 (1973) 1189.
16. Gummel, H.K., A Self-Consistent Iterative Scheme for One-Dimensional Steady State Transistor Calculations, IEEE Trans. Electron Devices ED-11 (1964) 455.
17. Anheier, W., Engl, W., Manck, O., Wieder, A.W., Rigorous Numerical Analysis of a Planar Thyristor, IEEE Int. Electron Devices Meeting, Washington D.C. (1975) Dig. Techn. Pap. 363.

18. Heimeier, H.H., A Two-Dimensional Numerical Analysis of a Silicon NPN Transistor, IEEE Trans. Electron Devices ED-20 (1973) 708.
19. Manck, O., Heimeier, H.H., Engl, W., High Injection in a Two-Dimensional Transistor, IEEE Trans. Electron Devices ED-21 (1974) 403.
20. Wieder, W.W., Manck, O., Engl, W., Two-Dimensional Analysis of a Monolithic PNP Transistor, IEEE Int. Electron Devices Meeting, Washington D.C. (1974) Dig. Tech. Pap. 414.
21. Manck, O., Engl., W.L., Two-Dimensional Computer Simulation for Switching a Bipolar Transistor out of Saturation, IEEE Trans. Electron Devices ED-22 (1975) 339.
22. De la Moneda, Threshold Voltage from Numerical Solution of the Two-Dimensional MOS Transistor, IEEE Trans. Circuit Theor. CT-20 (1973) 666.
23. Laur, R., Strohband, H.P., Numerical Modeling Technique for Computer Aided Circuit Design, IEEE Int. Symp. Circuits and Systems, Munich (1976) Proceedings 247.

II. NUMERICALLY BASED ANALYTICAL STEADY STATE TRANSISTOR THEORY

1. The Transfer Relation

The numerical solutions of the semiconductor device equations result in distributions of potential, carrier concentrations, and related values like current densities and others. These calculations not only give an understanding of device performance and internal mechanisms but also give valuable hints for technological improvements and limits. Furthermore approximations enabling analytical solutions are suggested and likewise simplifying approximations can be justified. This will be demonstrated by a one dimensional double diffused NPN-transistor. The boron base diffused into the epitaxial layer ($2.5 \times 10^{15} cm^{-3}$) has a depth of 3.0 µm and a surface concentration of $4 \times 10^{18} cm^{-3}$. The phosphorous diffusion for the emitter has a surface concentration of $8 \times 10^{20} cm^{-3}$ and reaches down to 2 µm. It reduces the Gummel number (1.2) to $1.2 \times 10^{11} cm^{-4} s$. The doping profile has been approximated by Gaussian and error functions.

The electrical performance of bipolar transistors can be described by recombination-generation mechanisms in the bulk as well as at the different interfaces and by what one may call transfer mechanisms. The transfer current is carried by minorities and passes its respective region without significant recombination loss.

By means of numerically exact solutions Gummel [1] found the model for the transfer current flowing from the emitter into the collector. It is defined for electrons through-passing a p-region

$$J_T = q n_i^2 \frac{\exp(V_{BE}/V_T) - \exp(V_{BC}/V_T)}{\int_{\text{p-region}} \frac{p}{D_n} \frac{n_i^2}{n_{ie}^2} dx} \tag{1.1}$$

The current is proportional to the exponent of the biasing voltage and inversely proportional to the charge of the majorities stored in this region. In case of low level injection the majority carrier density is fixed by the doping, thus the transfer current is characterized by only one quantity depending on device structure, the Gummel number:

$$G = \int_{\text{p-region}} \frac{N_A}{D_n} \frac{n_i^2}{n_{ie}^2} dx \tag{1.2}$$

1.1 Derivation of the Transfer Relation

The electron current density J_n can be subdivided into two parts:

$$J_n = J_n^{(0)} + J_n^{(1)} \qquad (1.3)$$

$J_n^{(0)}$ is the divergencefree part, hence is constant in the one dimensional case, whereas for $J_n^{(1)}$ equation (I.1.1) yields:

$$-\frac{1}{q}\frac{d}{dx} J_n^{(1)} = -R \qquad (1.4)$$

Assuming a sufficiently large current gain $h_{FE} = I_C/I_B$ for a transistor ($h_{FE} > 10$) the condition

$$J_n^{(1)} \ll J_n^{(0)} \qquad (1.5)$$

is fulfilled everywhere in the transistor. With (1.5) in mind, we can interprete (1.3) as being the first two terms of a perturbation analysis. The zero order approximation is the recombinationless transfer current density, which is being calculated first. After having solved the basic equations for $n^{(0)}$, $p^{(0)}$ the recombination rate $R^{(0)}$ is calculated from (I.1.17) and plugged into (1.4) which finally yields the first order correction term $J_n^{(1)}$. In deriving the transfer relation the first basic assumption is, that for electron current density it suffices to determine the transfer term. Hence we drop the superscripts and get

$$-J_n = J_T = q\mu_n n \frac{d\phi_n}{dx} = q\mu_n n_{in} \exp((\psi-\phi_n)/V_T) \frac{d\phi_n}{dx} =$$

$$= -qn_{in} D_n \exp(\psi/V_T) \frac{d}{dx} \exp(-\phi_n/V_T) = \text{const.} \quad , \qquad (1.6)$$

which can also be given in integral form:

$$-J_n = J_T = -q \frac{\exp(-\phi_n(x)/V_T) - \exp(-\phi_n(x_o)/V_T)}{\displaystyle\int_{x_o}^{x} \frac{\exp(-\psi/V_T)}{n_{in} D_n} dx'} \qquad (1.7)$$

The arbitrary integration limits in (1.7) may coincide with the emitter contact ($x_o = 0$) and the collector contact ($x = x_{cc}$), respectively. Thus in common emitter configuration $\phi_n(0) \stackrel{cc}{=} 0$, $\phi_n(x_{cc}) = V_{CE}$ and J_T could be calculated, if the electrostatic potential ψ were known.

$$J_T = q \frac{1 - \exp(-V_{CE}/V_T)}{\displaystyle\int_0^{x_{cc}} \frac{\exp(-\psi/V_T)}{n_{in} D_n} dx} \qquad (1.8)$$

Also (1.7) could be solved for $\phi_n(x)$ and hence $n(x)$ could be determined. The problem at hand is to find the potential ψ, which is governed by Poisson's equation (I.1.7) containing besides $\phi_n(x)$ the likewise unknown $\phi_p(x)$.

The transport equations (I.1.3), (I.1.4) can be understood as the general form of Ohm's law for electrons and holes, respectively:

$$\vec{J}_{n,p} = \sigma_{n,p}(\vec{E}+\vec{E}^{(i)}_{n,p}); \quad \sigma_{n,p} = q\mu_{n,p}\begin{Bmatrix} n; \\ p; \end{Bmatrix} \quad \vec{E}+\vec{E}^{(i)}_{n,p} = -\nabla\phi_{n,p} \quad . \qquad (1.9)$$

The current is caused by the sum of the electric field $\vec{E}$ and the impressed electric field $\vec{E}^{(i)}_{n,p}$. This sum is given by the gradient of the respective quasi Fermi potential. In thermal equilibrium the majority carrier conductivity in extrinsic material is determined by the impurity concentration and is large in comparison with the minority carrier conductivity. It is assumed that this is still true for small deviation from equilibrium i.e. for low level injection. Hence the semiconductor may be considered as an ideal conductor for majority carriers, which then flow through their respective region without any drop in imref.

Under forward bias the region of constant majority carrier imrefs $\phi_n(x)$ and $\phi_p(x)$, respectively ranges beyond the metallurgical emitter junction $x_{j,E}$. There the electron density drops some orders of magnitude causing a large diffusion current, which must be nearly compensated by a drift current in order to maintain a constant transfer current. This implies that thermal equilibrium is almost preserved within the junction and likewise is the Boltzmann relation $\psi(x) \propto \ln n(x)$, yielding $\phi_n(x) = $ const.

Let us consider for a moment the emitter and base region of a transistor as a short base diode with an Ohmic contact in place of the collector junction. According to the foregoing argument we may apply elementary diode theory in order to make an estimation for

$$J_T \cong \frac{q D_n}{x_{j,C} - x_{j,E}} n(x_{sp,E}) \quad ,$$

where $x_{sp,E}$ denotes the p edge of the emitter space charge region. Comparing this with (1.9) we get:

$$\frac{d\phi_n}{dx} \cong \frac{V_T}{x_{j,C} - x_{j,E}} \frac{n(x_{sp,E})}{n(x)} \quad . \tag{1.10}$$

Since $\phi_n(x)$ raises to V_{BE} at the "contact" over a distance approximately equal to $x_{j,C} - x_{j,E}$, we conclude that $n(x)$ may drop well below $n(x_{j,E})$ before $\phi_n(x)$ starts to raise at $x_{n,E}$. Under low level conditions the concentration ratio in (1.10) is independent of the applied voltage, hence the curves for $\phi_n(x)$ must coincide within the base region. A similar argument holds for $\phi_p(x)$, whose level is V_{BE} at the "diode contact" and stays constant until it starts to drop to zero across the junction at $x = x_{p,E}$.

Thus we have an overlapping region $x_{p,E} < x < x_{n,E}$ of constant imrefs, which is called transition region and which encloses $x_{j,E}$ for low level injection (fig.1). Within the transition region the imrefs are split by the applied voltage V_{BE}:

$$\phi_p(x) - \phi_n(x) = V_{BE} \; , \quad x_{p,E} < x < x_{n,E} \quad . \tag{1.11}$$

Approaching high injection levels the assumption of constant imrefs breaks down, however it is to be expected, that the more extrinsic the region is, the higher is the permitted injection level (compare $\phi_n(x)$ in the emitter for normal mode of operation fig.1 with $\phi_n(x)$ in the collector for the inverse mode fig.2).

At the reverse biased junction no such transition region exists, because $\phi_n(x)$ reaches V_{CE} at $x_{n,C}$ beyond $x_{j,C}$. Carriers crossing the collector space charge region travel at saturation velocity over some distance therein and hence $n(x)$ cannot fall below a certain constant value, which is determined by J_T. This in turn leads to a limited and constant slope of $\phi_n(x)$ along that distance.

In conclusion we can firstly shrink the integration interval in (1.7) to $x_{n,E} < x < x_{n,C}$. Secondly we have a constant majority carrier imref throughout the base region, which implies a Boltzmann relation there:

$$\phi_p(x) = \text{const} = V_{BE} = \psi(x) + V_T \ln \frac{p(x)}{n_{ip}}, \; x_{p,E} < x < x_{p,C} \tag{1.12}$$

Thirdly we can replace the lower integration limit $x_{n,E}$ by $x_{p,E}$ and express with (I.1.10) the potential ψ with (1.12) by the majority carrier concentration p if $x_{n,C}$ is likewise replaced by $x_{p,C}$. The latter replacement is not exact, since no transition region exists at the collector but for reverse voltages

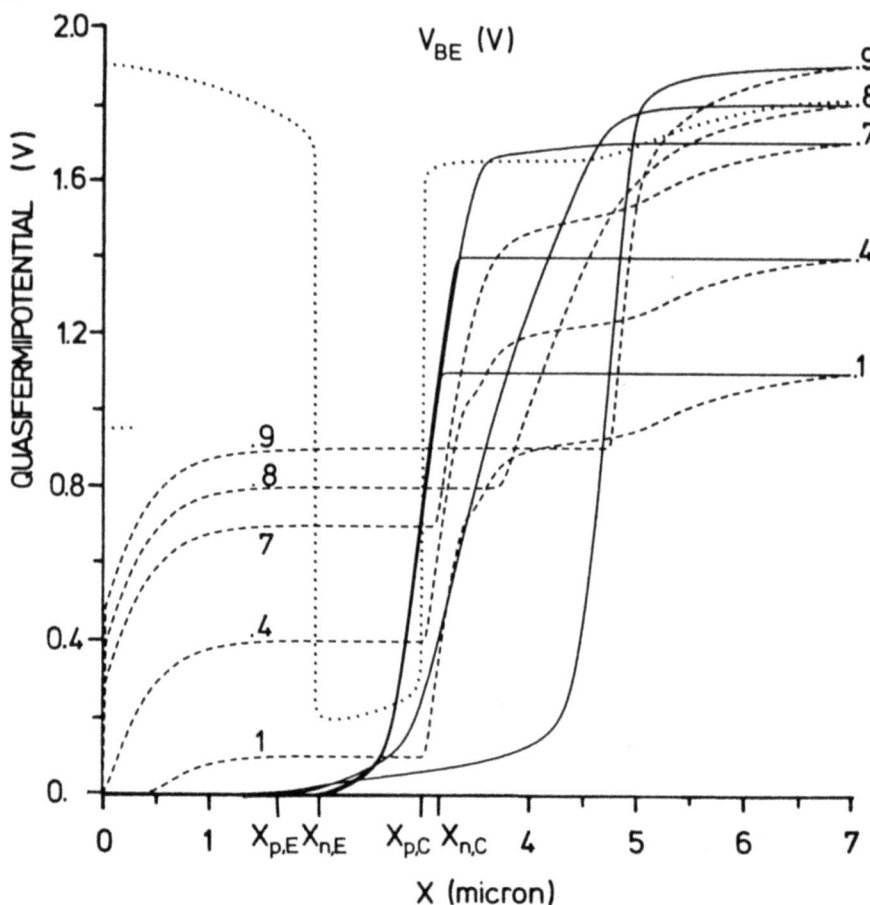

Fig. 1: NPN-transistor in normal mode. Quasi Fermi potential of electrons (——) and holes (---) vs base emitter voltage. V_{CB} = const = 1V. The dotted line represents the net doping profile ($\sinh^{-1}(N)$). The limits of the regions $\phi_{n,p}$ = const are indicated for low injection conditions.

$$\exp(-\phi_n(x_{n,C})/V_T) = \exp(-V_{CE}/V_T), \quad \exp(-\phi_n(x_{p,C})/V_T) \ll 1 \qquad (1.13)$$

holds and hence the second term in the numerator of (1.7) can be neglected. Preserving this term a symmetrical form of the transfer relation results:

$$J_T = qn_i^2 \frac{\exp(V_{BE}/V_T) - \exp(V_{BC}/V_T)}{\int_{x_{p,E}}^{x_{p,C}} \frac{n_i^2}{n_{ie}^2} \frac{p}{D_n} dx} \qquad (1.14)$$

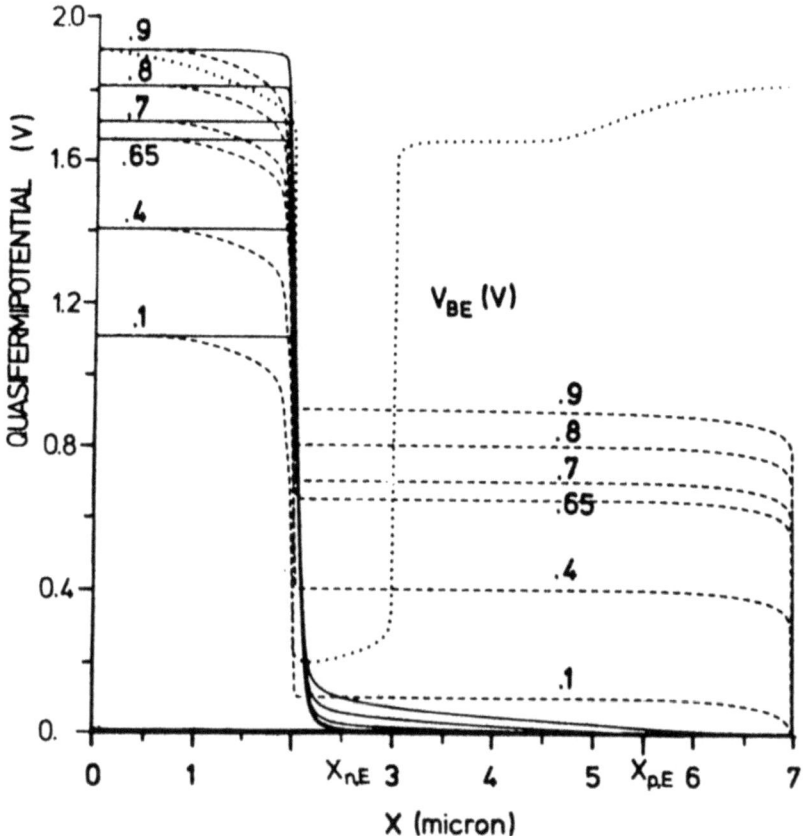

Fig. 2: Same as Fig. 1, but in inverse mode.

which gives the transfer current for the normal and inverse mode of operation. The current characteristic is for low level injection identical in both cases. The influence of reverse V_{BC} is only contained in the denominator integral.

Despite the intuitive insight which is gained from (1.14) it must be recalled from (1.8) that in general the potential distribution is required in order to calculate the transfer current. Hence Poisson's equation must be solved. The only known exact solution of this nonlinear equation in closed form is the solution which expresses charge neutrality $n - p - N = 0$ and which holds for $N(x)$, $\phi_n(x)$, $\phi_p(x)$ = const. In graded regions this solution is still applicable if the doping is almost balanced by mobile carriers, such that the logarithmic dependance (I.1.10) of the electrostatic potential on carrier density distribution outweighs the dependance generated by the second derivative of ψ. Hence, if the unbalance condition

$$\rho_{unb} = \frac{\varepsilon \left| \frac{d^2\psi}{dx^2} \right|}{q \cdot \max(n,p)} \ll 1 \qquad (1.15)$$

is met within a certain region, then quasineutrality holds in this region:

$$n - p - N \cong 0, \qquad (1.16)$$

and the electrostatic potential ψ can be expressed by the imrefs:

$$\psi = \phi_n + V_T \ln(n/n_{in}) = \phi_p - V_T \ln(p/n_{ip}) ; \qquad (1.17)$$

$$\left.\begin{matrix}n\\p\end{matrix}\right\} = \pm \frac{N}{2} + \sqrt{\frac{N^2}{4} + n_{ie}^2 \exp((\phi_p - \phi_n)/V_T)} \quad.$$

Fig.3 shows ρ_{unb} calculated from the numerically exact solution. The transistor is quasineutral except in space charge regions around the junctions. With increasing injection quasineutrality holds increasingly better and at high level injection after base push out has occurred only small space charge regions remain within the collector region. A numerical value $\rho_{unb} \leq 10\%$ seems to be reasonable in order to define quasineutrality. An analytical estimation for ρ_{unb} is also possible, using a Gaussian doping profile with characteristic length L as a testprofile. Calculating $\frac{d^2\psi}{dx^2}$ from (1.17) yields for low injection levels $\max(n,p) = N$:

$$\rho_{unb} = \left(\frac{L_D^{(e)}}{L}\right)^2 ; \quad L_D^{(e)} = \frac{L_D}{\sqrt{N}} , \qquad (1.18)$$

where L_D and $L_D^{(e)}$ are the intrinsic and extrinsic Debye lengths, respectively. For example, the 10% limit of ρ_{unb} results from L = 0.5 μm and N = 10^{15}cm^{-3}.

The quasineutral expression (1.17) suggests to define the high injection onset by the two terms under the square root becoming equal:

$$V^{HI} = (\phi_p - \phi_n)^{HI} = 2 V_T \ln (|N|/2n_{ie}) \qquad (1.19)$$

Since V^{HI} can be identified with V_{BE} in the region defined in (1.11) it can be estimated from (1.19) where high injection effects are to be expected locally.

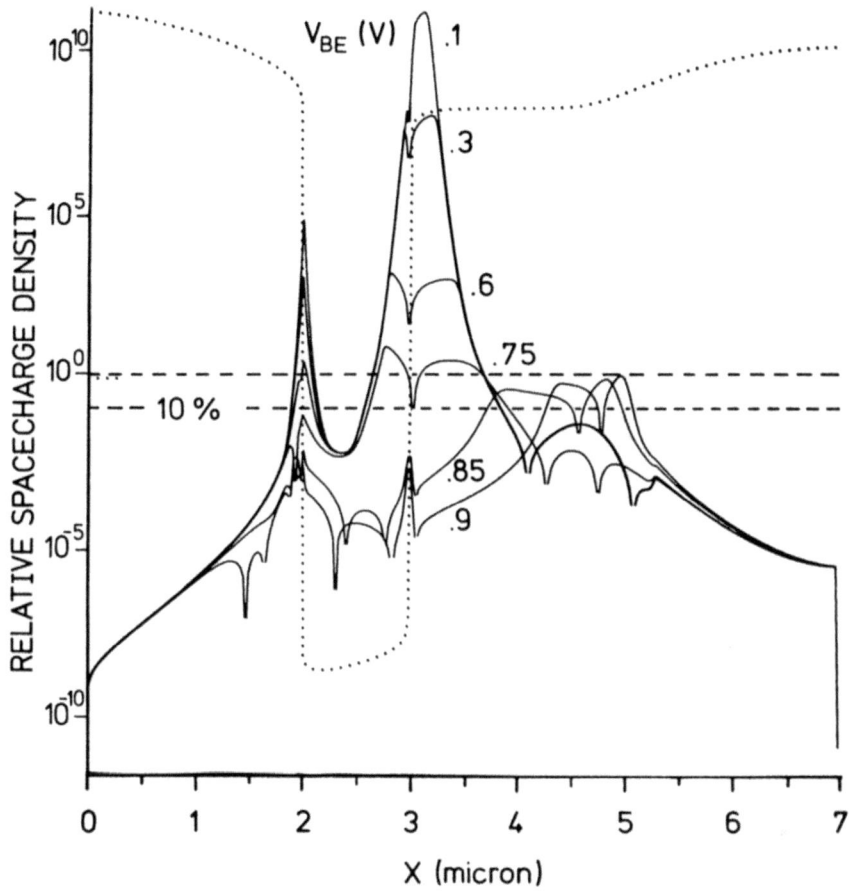

Fig. 3: Relative space charge ρ_{unb} vs base emitter voltage (1.15).
V_{CB} = const = 1V.

Quasineutrality yields from (1.17) for low level injection the potential $\psi = \phi_p - V_T \ln(N_A/n_{ip})$. Plugging it into (1.8) and, recalling (1.12), gives the denominator integral, which we also get more simply from (1.14) by $p(x) \cong N_A(x)$:

$$G = \int_{x_{p,E}}^{x_{p,C}} \frac{n_i^2 N_A}{n_{ie}^2 D_N} dx \quad . \tag{1.20}$$

Furthermore $x_{p,E}$ can be replaced by $x_{j,E}$ for low level injection and $x_{p,C}$ by $x_{j,C}$ at least for small V_{BC} and hence G becomes the Gummel number, a technological quantity which can be directly evaluated from the doping profile in the base region. As long as the hole density can be approximated by the acceptor doping the injection factor m equals one.

In the high injection case the hole density increases quasi-neutrally with the injected electron density, as soon as $\phi_p - \phi_n$ exceeds locally the onset voltage (1.19). From (1.17) we then have:

$$n \cong p \cong n_{ie} \exp((\phi_p - \phi_n)/2V_T). \tag{1.21}$$

With (1.21) and (1.12) equation (1.6) yields again an integral form:

$$J_T = 2q \frac{\exp(V_{BE}/2V_T) - \exp(V_{BC}/2V_T)}{\displaystyle\int_{x_{p,E}}^{x_{p,C}} \frac{dx}{D_n\, n_{ie}}} \tag{1.22}$$

To justify this expression we must assume the injection level to be high enough that $D_n(n)$ can be replaced by its constant limes, otherwise no closed expression is obtainable. For a transistor in normal mode the upper integration limit no longer coincides approximately with the junction, but has been shifted deeply into the collector region (fig.1). Notice also the difference with respect to the inverse mode (fig.2.). Comparing (1.14) with (1.22) we see that the injection factor m approaches two in agreement with the numerical solution (fig.4). However, m = 2 cannot be verified in a two dimensional structure since the internal base emitter voltage decreases continously with increasing distance from the base contact due to base resistance and hence the collector current characteristic is smoothed.

Corresponding to the calculation of the transfer current we also get the minority carrier imref in the base region from (1.7) and (1.12). To express $\phi_n(x)$ conveniently we define the quantity

$$Q(x) = \int_{x_{p,E}}^{x} \frac{n_i^2\, p}{n_{ie}^2\, D_n} dx' \tag{1.23}$$

which is proportional to the base charge accumulated between $x_{p,E}$ and any cross section x within the base, if $n_{ie}^2 D_n$ were a constant and which we call for this reason the modified base charge.

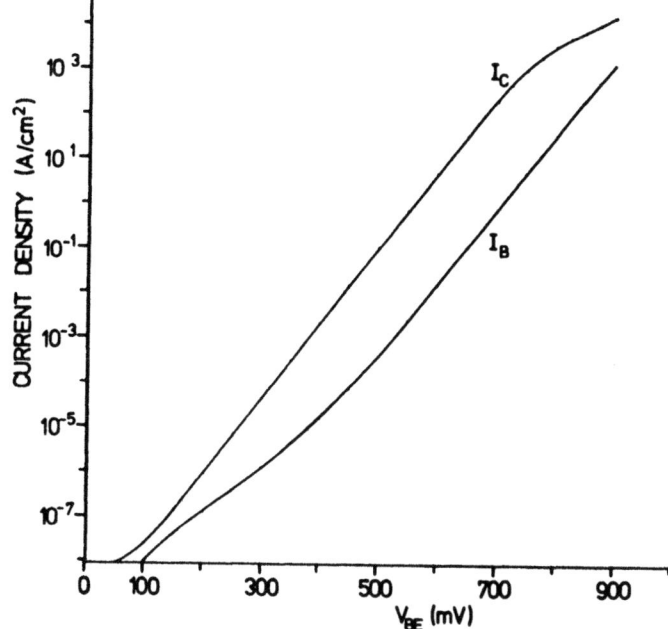

Fig. 4: Terminal characteristics in the normal mode.

We then have

$$\phi_n(x) = - V_T \ln(1 - \frac{Q(x)}{Q(x_{p,C})} (1 - \exp(- V_{CE}/V_T))). \quad (1.24)$$

Since $Q(x)$ is not dependent on V_{BE} for low injection levels the same holds for $\phi_n(x)$ and the respective curves of $\phi_n(x)$ coincide in this range (fig.1,2). Finally (1.24) and (1.12) yield the electron density in the base:

$$n(x) = \exp(V_{BE}/V_T) \frac{n_{ie}^2}{p(x)} (1 - \frac{Q(x)}{Q(x_{p,C})}(1 - \exp(- V_{CE}/V_T))) \quad (1.25)$$

As long as $p \cong N_A$ holds $n(x)$ consists of a factor depending exponentially on the injection voltage, and of a spatial distribution composed of the solution $n_{ie}^2/N_A(x)$ for thermal equilibrium and a term which declines towards the base collector junction. For $V_{BC} < 0$ this term is characterized by the ratio of the remaining modified base charge between x and $x_{p,C}$ and the total modified base charge (fig.5).

Applying (1.25) for the inverse mode of operation we have to replace V_{BE}/V_{BC}, V_{CE}/V_{EC} and we have to count x in the direction from the collector to the emitter. A doping profile which is unsymmetrical with respect to the base center yields highly different

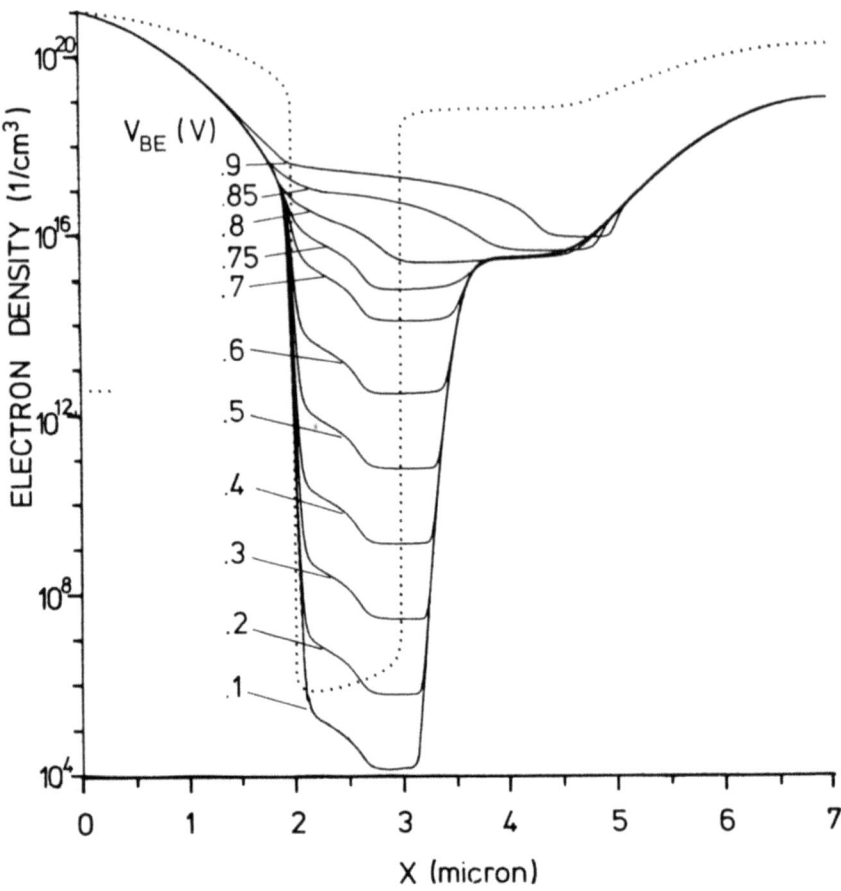

Fig. 5: Electron density. Bias condition as in Fig. 1.

electron density distributions for the normal and inverse mode. Looking into the base from a forward biased junction we notice electron stagnation in front of the maximum doping and a fast decay behind it. Hence an unsymmetrical doping profile yields different values of the stored minority charge according to the mode of operation. In steady state however the doping profile influences only the position of the integration limits, especially with respect to their dependence on the reverse bias voltage.

In analogy to (1.23) we define the quantity

$$P(x) = \int_{x_{p,E}}^{x} \frac{n_i^2 \, dx}{n_{ie} D_n} \qquad (1.26)$$

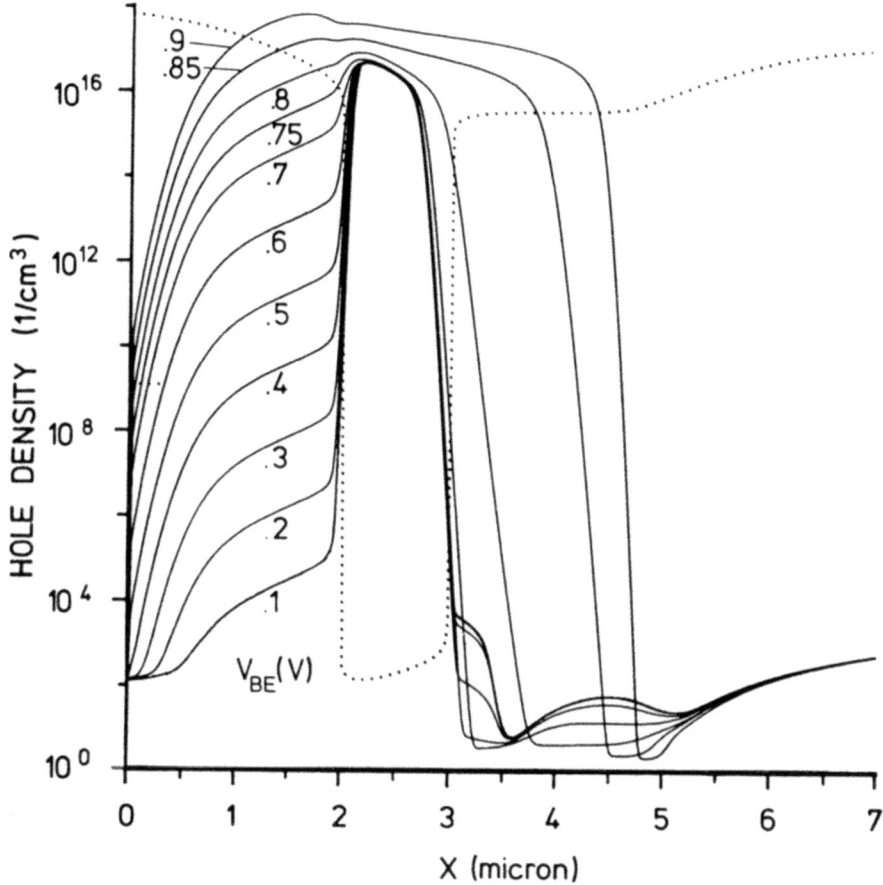

Fig. 6: Hole density. Bias condition as in Fig. 1.

for the high injection case and get from (1.6), (1.12) and (1.22):

$$\phi_n(x) = -2V_T \ln\left(1 - \frac{P(x)}{P(x_{p,c})}\left(1 - \exp(-V_{CE}/2V_T)\right)\right) \quad (1.27)$$

$$n(x) \cong p(x) = \exp(V_{BE}/2V_T)\, n_{ie}\left(1 - \frac{P(x)}{P(x_{p,c})}\left(1 - \exp(-V_{CE}/2V_T)\right)\right) \quad (1.28)$$

The wellknown linear carrier distribution is given by (1.28), if the product $n_{ie}D_n$ is constant over the range of integration. This is the case for uniform doping and if the change of $\mu_n(n(x)) \propto D_n(n(x))$

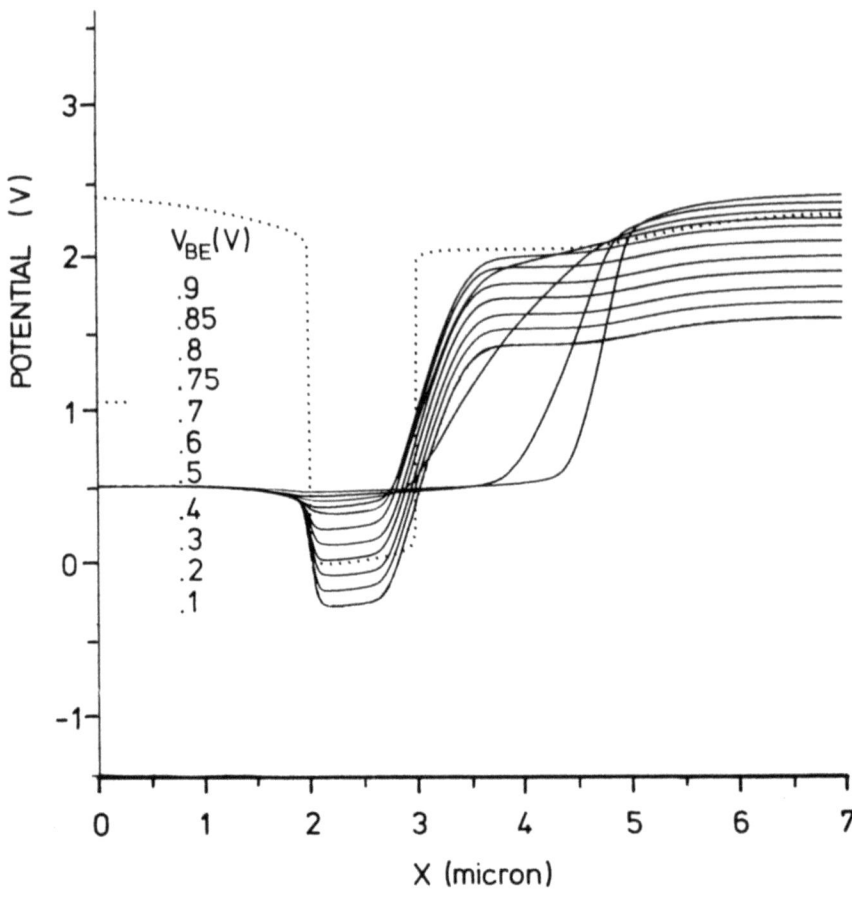

Fig. 7: Potential distribution. Bias condition as in Fig. 1.

is nearly compensated by an inverse change of $n_{ie}(N(x))$. Fortunately enough this happens quite often (fig.8).

1.2 The Influence of the Base Collector Voltage.

Next we shall determine the dependance of the transfer current J_T on the reverse collector bias V_{BC}. We restrict ourselves to low injection levels where the injected electrons carrying the impressed transfer current through the depletion region do not alter the space charge there. The contribution of $p(x)$ to the integral $Q(x_p,c)$ in the denominator of J_T vanishes beyond the edge $x_{sp,C}$ of the collector depletion region (fig.6):

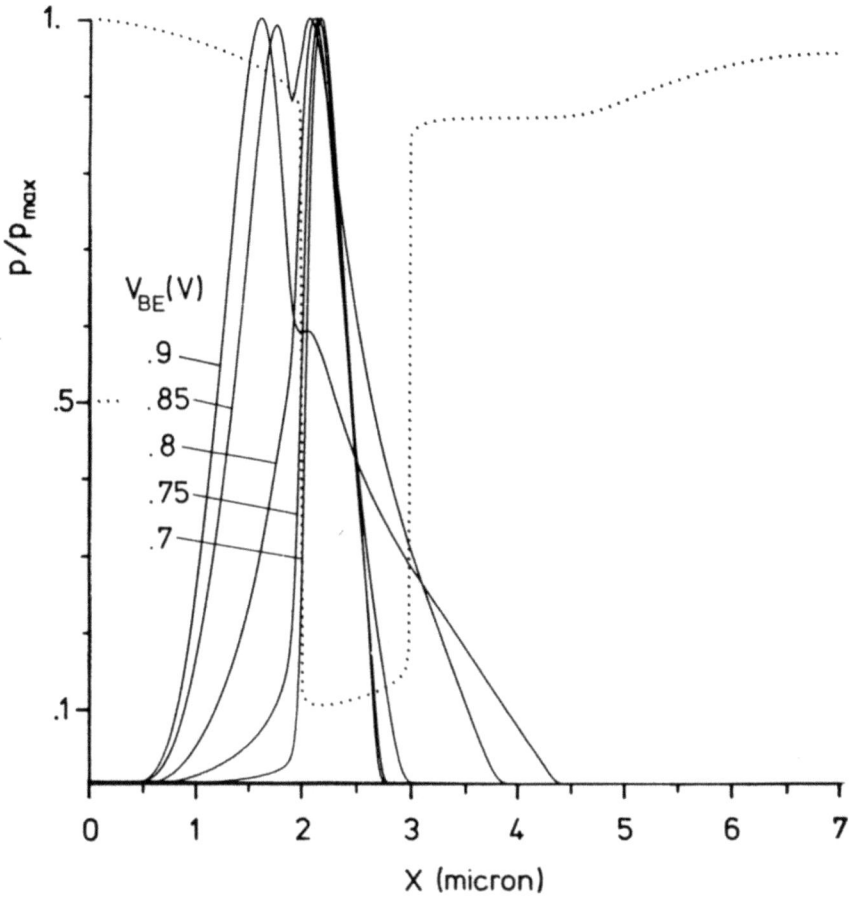

Fig. 8: Linear plot of the relative hole density. Bias condition as in Fig. 1.

$$Q(x_{p,C}) = \int_{x_{p,E}}^{x_{p,C}} \frac{n_i^2 \, p}{n_{ie}^2 D_n} \, dx = \int_{x_{p,E}}^{x_{sp,C}} \frac{n_i^2 N_A}{n_{ie}^2 D_n} \, dx \qquad (1.29)$$

This edge is shifted as a function of V_{BC} resulting in:

$$\frac{\partial Q(x_{p,C})}{\partial V_{BC}} = \frac{n_i^2 N_A(x_{sp,C})}{n_{ie}^2(x_{sp,C}) D_n(x_{sp,C})} \frac{dx_{sp,C}}{dV_{BC}} = \frac{n_i^2}{n_{ie}^2 D_n q} C_{BC} : = \frac{C_{BC}}{a(x_{sp,C})}$$

(1.30)

using the relation $C_{BC} dV_{BC} = q N_A(x_{sp,C}) dx_{sp,C}$. The differential conductance $\left|\frac{\partial J_T}{\partial V_{BC}}\right|$ can now be related to the collector space charge capacitance by (1.30):

$$\frac{\partial J_T}{\partial V_{BC}} = -\frac{C_{BC}}{aqn_i^2 \exp(V_{BE}/V_T)} J_T^2 \quad , \tag{1.31}$$

and hence the dependence of J_T on V_{BC} is given by:

$$J_T(V_{BC}) = \frac{J_T(0)}{1 + \frac{1}{Q(x_j,C)} \int_0^{V_{BC}} \frac{C_{BC}}{a(x_{sp,C})} dV'_{BC}} \tag{1.32}$$

$Q(x_j,C)$ can be identified with the Gummel number and $a(x_{sp,C}(V_{BC}))$ can be approximately treated as a constant, since $a(x)$ is only weakly dependant on x [2]. The singularity in (1.32) expresses punch through. An explicit form of (1.32) can be given for a transistor with an epitaxial collector yielding a highly unsymmetrical base collector junction, which will be approximated as an abrupt junction:

$$C_{BC} = \sqrt{\frac{\varepsilon q N_{epi}}{2(V_D - V_{BC})}} = C^o_{BC} \frac{1}{\sqrt{1 - V_{BC}/V_D}} \quad . \tag{1.33}$$

From (1.32) and (1.33) follows:

$$J_T(V_{BC}) = \frac{q n_i^2 \exp(V_{BE}/V_T)}{Q(x_j,C) - \frac{2 C^o_{BC} V_D}{a}(\sqrt{1 - V_{BC}/V_D} - 1)} \tag{1.34}$$

Finally, the transfer current can be identified with the collector current if for small V_{BC}, SRH-generation and for large V_{BC}, avalanche generation in the space charge region is sufficiently small.

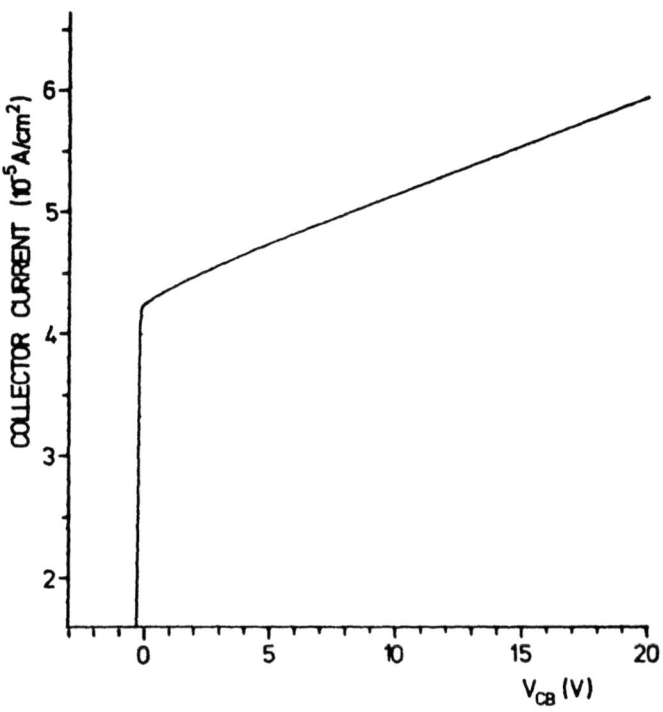

Fig. 9: Collector current vs collector base voltage. V_{BE} = 300 mV.

Plugging numerical values into (1.34) leads to a characteristic which is usually fitted by the Early line (fig.9). However, the analytical expression suggests as definition of the Early voltage:

$$J_T(V_{BC} = -V_{EA}) = 2 J_T(V_{BC} = 0). \qquad (1.35)$$

which gives:

$$V_{EA} = \frac{aQ(x_{j,C})}{4C^o_{BC}} \left(\frac{aQ(x_{j,C})}{4C^o_{BC}V_D} + 2 \right) \simeq \frac{1}{4}\left(\frac{aQ(x_{j,C})}{2C^o_{BC}V_D}\right)^2 V_D \qquad (1.36)$$

The Early voltage is proportional to the square of the ratio of the stored modified base charge in equilibrium to the charge pushed aside from the collector junction by the diffusion voltage.

2. The Recombination Current

One of the main causes of base current J_B is recombination in the bulk [3]. It even completely determines J_B in the one dimensional structure under investigation. For low injection levels SRH-recombination with one trapping level in the band gap prevails and leads to an injection factor m > 1 in the J_B vs V_{BE} characteristic (fig.4). Recombination is then essentially restricted to the emitter transition region $x_{p,E} \le x \le x_{n,E}$ as can be seen from the numerical result (fig.10), which displays the recombination rate R(x), and even better from the linear plot in fig.11. In the transition region $\phi_n(x) = 0$ and $\phi_p(x) = V_{BE}$, only the potential ψ is unknown and we thus have:

$$R(\psi) = \frac{n_{ie}^2 \exp(V_{BE}/V_T)}{\tau_p(n_{in}\exp(\psi/V_T)+n_t) + \tau_n(n_{ip}\exp((V_{BE}-\psi)/V_T)+p_t)} \qquad (2.1)$$

Since the numerator is a constant and the denominator is exponentially dependant on $\pm \psi$, the recombination rate $R(\psi)$ has a pronounced extreme at

$$\psi = \frac{V_{BE}}{2} - \frac{V_T}{2} \ln(\tau_p/\tau_n) \quad , \qquad (2.2)$$

with a maximum value of:

$$R_{max} = \frac{n_i^2 \exp(V_{BE}/V_T)}{\sqrt{\tau_n\tau_p}(2n_i \exp(V_{BE}/2V_T) + n_t\sqrt{\tau_p/\tau_n} + p_t\sqrt{\tau_n/\tau_p})} \qquad (2.3)$$

Here, $n_{ie}(x) = \text{const} = n_i$, $\tau_{n,p}(x) = \text{const}$ was assumed. At the point x_R where the recombination rate reaches its maximum (2.2) yields the carrier densities:

$$n(x_R) = \sqrt{\tau_n/\tau_p}\; n_i \exp(V_{BE}/2V_T) \qquad (2.4)$$

$$p(x_R) = \sqrt{\tau_p/\tau_n}\; n_i \exp(V_{BE}/2V_T) \qquad (2.5)$$

For $\tau_n = \tau_p$ we have $n(x_R) = p(x_R)$ and hence x_R coincides with the junction x_j and is not too far apart from it for $\tau_n/\tau_p = 10$. R_{max} increases with m = 2, if n_t and p_t, respectively are small compared to $n_i\exp(V_{BE}/2V_T)$. The results (2.3) to (2.5) can be inter-

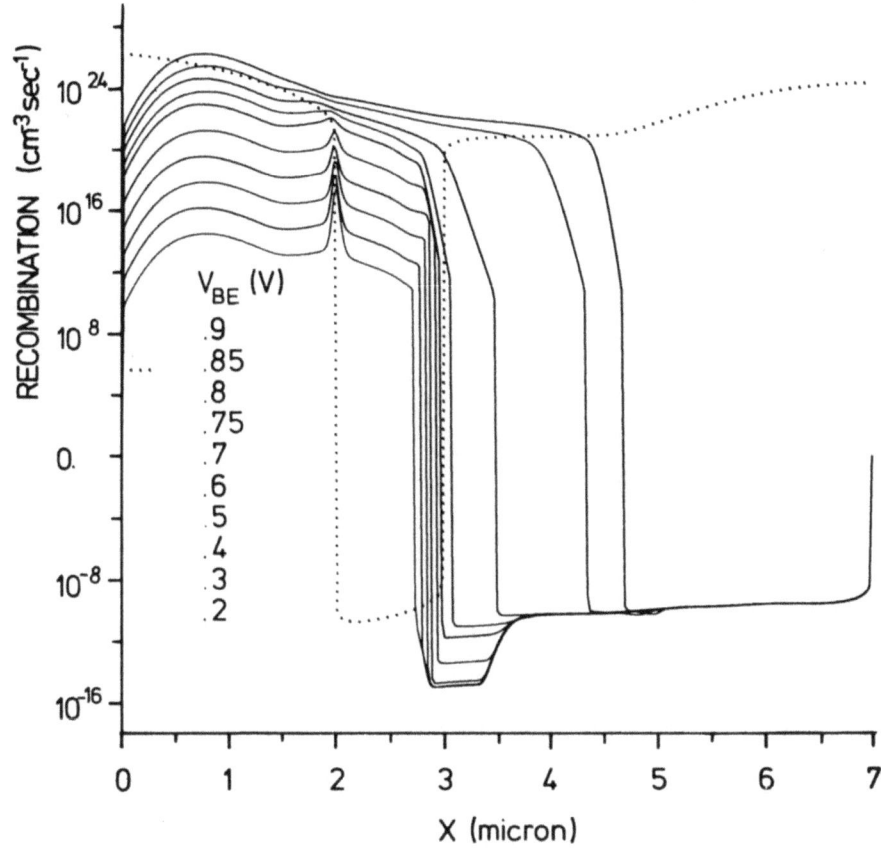

Fig. 10: Total recombination density R, plotted as $\sinh^{-1}(R)/\ln 10$. Bias condition as in fig.1.

preted quite easily: The numerator n·p of the SRH-expression is only weakly dependent on x. Thus the denominator $\tau_p n + \tau_n p$ determines the recombination rate. It has its minimum at $\tau_p n = \tau_n p$ because of n·p = const.

Corresponding to the increase of the majority carrier concentration towards the edge of the quasineutral region the SRH-recombination decreases and thus may finally be neglected in the quasineutral region. There m has approached one, since the denominator equals the impurity concentration. In conclusion, the base current caused in the transition region is composed of a part with a corresponding recombination rate which is very large, but restricted to a small volume around x_R and which has an injection factor m = 2,

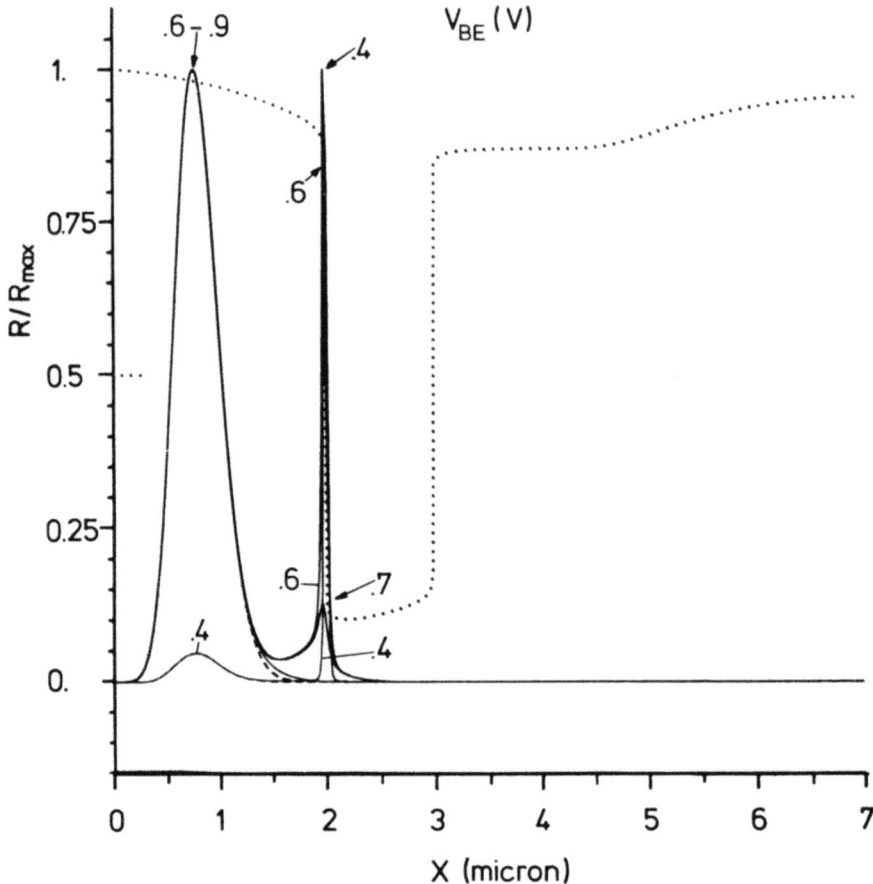

Fig. 11: Linear plot of the relative recombination density. Dashed line represents the Auger component.

and other parts with decreasing rate and decreasing m spread over the rest of the transition region. This composition leads to m = 1.4 for the base current in the low injection range (fig.4).

The recombination rate in the transition region can be estimated with a linear approximation of ψ in this region centered at $x = x_R$:

$$\psi = \psi(x_R) - E_R(x-x_R) \quad . \qquad (2.6)$$

Here, we had to assume $\tau_{n,p}$ = const, which certainly holds in this small range. We then have

$$J_{B,trans} = q \int_{x_{p,E}}^{x_{n,E}} R(x)dx = \frac{q}{E_R} \int_{\psi(x_{n,E})}^{\psi(x_{p,E})} R(\psi)d\psi =$$

$$= \frac{2n_i^2 \, q \, V_T \, \exp(V_{BE}/V_T)}{E_R \sqrt{\Delta}} \tan^{-1}\left(\frac{\sqrt{\Delta}}{\tau_p n_t + \tau_n p_t}\right) ; \quad (2.7)$$

$$\Delta = 4\tau_n\tau_p n_i^2 \exp(V_{BE}/V_T) - (\tau_p n_t + \tau_n p_t)^2 .$$

The condition for (2.7) to hold is.

$$\exp(\psi(x_{n,E})/V_T) \ll \exp(\psi(x_{p,E})/V_T) \text{ and } \exp((V_{BE}-\psi(x_{p,E}))/V_T) \ll 1 \quad (2.8)$$

restricting (2.7) to $V_{BE} < 500$ mV. We can simplify (2.8), if the trap level is located in the middle of the bandgap:

$$J_{B,trans} = \frac{\pi \, q \, V_T \, n_i}{2\sqrt{\tau_n\tau_p} \, E_R} \exp(V_{BE}/2V_T) \quad (2.9)$$

Comparing this result with (2.3) and defining a recombination length L_{SRH}:

$$L_{SRH} := \frac{\pi \, V_T}{E_R} \quad (2.10)$$

we can write:

$$J_B = J_{B,trans} := q \, R_{max} \cdot L_{SRH} \quad (2.11)$$

The increase of J_B with $m < 2$ can then be explained as an increase of the recombination length with V_{BE}.

Fitting numerical results for an almost linear base emitter junction gives:

$$E_R^{(lin)}(V_{BE}) = E_R^o \left(1 - \frac{V_{BE}}{V^o}\right)^{0.7}, \quad (2.12)$$

where E^o denotes the fieldstrength in thermal equilibrium and $V^o = 780$ mV. The exponent .7 compares well with 2/3 for an ideal linear junction. The result according to (2.9) and (2.12) is in good agreement with exact numerical values up to V_{BE} = 500 mV. Also of practical interest is the result for an unsymmetrical abrupt layer. In this case we have:

$$E_R^{(abr)}(V_{BE}) = \sqrt{\frac{q\,N_{epi}}{\varepsilon}(V_T \ln(\frac{N_{epi}^2}{n_i^2}\frac{\tau_n}{\tau_p}) - V_{BE})} \qquad (2.13)$$

Relation (2.9) and (2.13) offer a possibility to estimate carrier lifetimes in a junction.

For higher injection levels approximately from $V_{BE} \geq 500$ mV the Auger band to band recombination with energy transfer by a third carrier is dominating and recombination takes mainly place in the emitter region (figs.10,11). A numerical solution of the transport equation for holes is required, taking the doping profile and Auger recombination into account and hence no analytical expression for the base current can be given. The following interpretation of the numerical results seems possible. The decay of the local lifetime with $1/N_D^2$ causes a drain for holes at $N_D \cong 10^{20} cm^{-3}$ and the remaining still more highly doped part of the emitter region is without any effect on the hole behavior.

REFERENCES

1. Gummel, H.K., A Self-Consistent Iterative Scheme for One Dimensional Steady State Transistor Calculations, IEEE Trans. Electron Devices ED-11, (1964) 455.
2. Heimeier, H.H., Berger, H.H., Evaluation of Electron Injection, IEEE J. Solid State Circ., SC-12, (1977) 205.
3. Werner, W., The Influence of Fixed Interface Charges on the Current Gain Fall Off of Planar n-p-n Transistor, J. Electrochem. Soc. 123 (1976) 540.

HIGH CURRENT DENSITY EFFECTS IN THE COLLECTOR OF BIPOLAR
TRANSISTORS

H.C. de Graaff

Philips Research Laboratories
Eindhoven, The Netherlands

ABSTRACT. In this review the behaviour of the base-collector
junction of transistors with lightly doped, epitaxial collector
regions is discussed. We have distinguished the following modes
of operation: depletion, scattering-limited drift velocity (SLDV)
and quasi-saturation (or injection). The influence of these
various modes on electrical quantities like h_{FE} and f_T is
discussed.

1. INTRODUCTION

In this paper high current density effects in the lightly
doped collector region of bipolar transistors will be discussed.
The base-collector junction in the Ebers-Moll and related
models is represented by an ideal diode. For alloy transistors
this was a reasonable approximation but for planar transistors
with epitaxial collector regions this model fails in the
quasi-saturation. In that situation the terminal voltage V_{cb}
is a reverse bias, but internally the base-collector junction
is forward biased. This is caused by the internal voltage drop
across the epitaxial collector region. That voltage drop can be
ohmic [1] or non-ohmic [2], but in both cases a minority
carrier charge is stored in the epitaxial collector region,
because this region is more lightly doped than the base. The
charge storage in the lightly doped collector in quasi-
-saturation has an important influence on the current gain, the
cut-off frequency f_T and the collector capacitance.

2. HIGH CURRENT DENSITY COLLECTOR EFFECTS

The treatment in this section uses a one-dimensional d.c. analysis. Two-dimensional current spreading in the collector region is approximated by the introduction of a collector spreading resistance r_c [3] . We further assume that the base is much more heavily doped than the homogeneously doped collector region, so the relevant phenomena take place on the collector side of the junction. In the calculations of the charge distributions we will neglect recombination. Furthermore we will use Boltzmann statistics and assume complete ionization of the impurities.

Fig. 1 gives a cross-section of an npn transistor with an epitaxial collector region of width W and constant impurity concentration N_d. The analysis pertains to the active part underneath the emitter.

We can distinguish two situations: saturation and non-saturation, or injection and depletion. In the depletion situation the base-collector junction is reverse-biased internally. The typical field and majority carrier distributions are sketched in fig. 2a. Fig. 2b gives these typical distributions in the injection situation, where the base-collector junction is forward-biased internally.

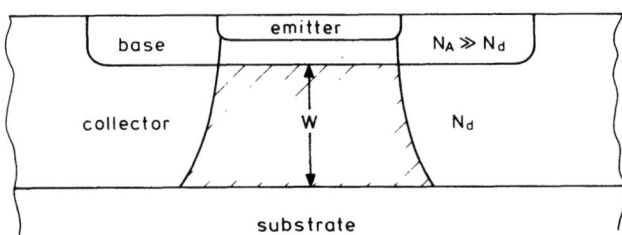

Fig. 1. Cross-sectional view of an npn transistor. The lightly doped epitaxial collector layer has a width W and a constant doping concentration N_d. The shaded area is the active transistor part.

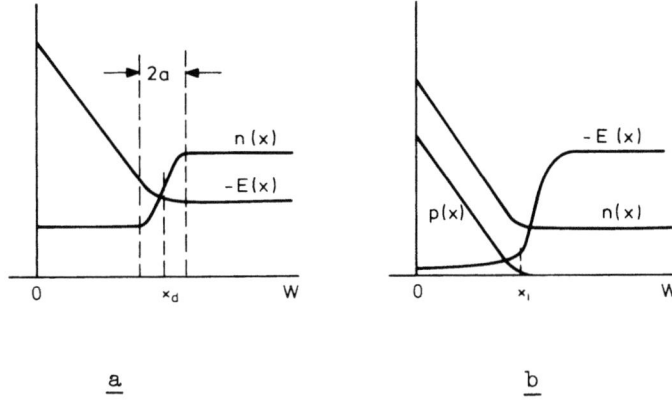

Fig. 2. Typical field and carrier distributions in the collector epilayer:
a) depletion mode, depletion region from $0-x_d$;
b) injection or quasi-saturation mode, injection region from $0-x_i$.

2.1. Depletion

The drift velocity is a function of the electric field. For electrons in silicon we have $v_{dr} = \mu_n E$, where the mobility μ_n is constant up to a field strength E_2 of about 2500 V/cm. Above a field strength E_1 of about 15000 V/cm the drift velocity saturates at a value of approximately 10^7 cm/s (see fig. 3).
Thus we can distinguish three cases:
  ohmic behaviour for $E < E_2$,
  tepid carrier behaviour for $E_2 < E < E_1$,
  hot carrier behaviour for $E > E_1$.
Ryder [4] has modelled this behaviour as follows:

$$v_{dr} = \mu_n \cdot E \qquad \text{for } E < E_2,$$
$$v_{dr} = \mu_n \sqrt{E_2 E} \qquad \text{for } E_2 < E < E_1, \qquad (1)$$
$$v_{dr} = v_{lim} = \mu_n \sqrt{E_2 E_1} \qquad \text{for } E > E_1.$$

The velocity v_{lim} is the scattering-limited drift velocity (SLDV). Assuming a constant velocity v_{lim} in the depletion layer and ohmic behaviour in the remaining part of the lightly doped collector, Kirk has found for the depletion layer width x_d [1]:

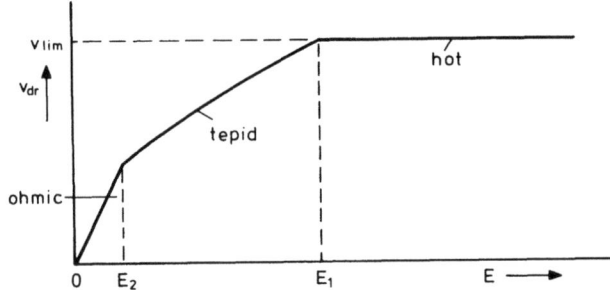

Fig. 3. Drift velocity v_{dr} versus electric field E. In silicon $E_1 \approx 15000$ V/cm and $E_2 \approx 2500$ V/cm, while $v_{lim} \approx 10^7$ cm/s.

$$x_d = \left(\frac{2\varepsilon}{qN_d} \cdot \frac{V_{cb} + V_o - \rho W J_c}{1 - J_c/J_1} \right)^{\frac{1}{2}} \tag{2}$$

Here V_{cb} is the externally applied voltage,
W the width of the lightly doped epitaxial collector layer,
$\rho = \frac{1}{q\mu_n N_d}$, the resistivity of the epitaxial collector,
$J_1 = qN_d v_{lim}$, a critical current density,
ε the permittivity ($\approx 10^{-12}$ F/cm^2 for silicon) and
V_o the built-in voltage of the base-collector junction.

The term $(1-J_c/J_1)$ in the denominator in eq. (2) represents the influence of the mobile majority carriers (electrons in this case) on the total space charge. The mobile carriers compensate the fixed ionized impurities charge. This effect tends to increase x_d with increasing current.
The ohmic voltage drop, expressed by the term in the numerator of eq. (2) tends to decrease x_d with increasing current. What actually happens depends on V_{cb}:
for $V_{cb} + V_o > \rho W J_1$, x_d increases and for
$V_{cb} + V_o < \rho W J_1$, x_d decreases with increasing current.

Eq. (2) can be used for $J_c < J_1$ and $0 < x_d < W$. It is based on a theory which neglects the tepid carrier behaviour and the presence of an intermediate zone around the depletion layer edge. If we do recognize the tepid carrier behaviour we have to modify the voltage drop term $\rho W J_c$ in eq. (2) into $\rho W J_c \cdot K(J_c)$. The factor K is a function of current and given by [5]:

$$K = 1 \quad \text{for } J_c \leqslant J_2 = \frac{E_2}{\rho}$$
$$\text{and} \quad K = \frac{J_c}{J_2} \quad \text{for } J_2 < J_c < J_1. \tag{3}$$

The field strength E_2 corresponds in Ryder's model with the onset of the tepid carrier behaviour.

With this correction eq. (2) becomes

$$x_d = \left(\frac{2\varepsilon}{qN_d} \frac{V_{cb} + V_o - K\rho W J_c}{1 - J_c/J_1} \right)^{\frac{1}{2}}. \tag{4}$$

Furthermore the depletion layer edge is not abrupt, but there exists an intermediate layer with a gradual change from space charge to neutrality (see fig. 2a). The width (2a) of this intermediate region depends on the current only and it has been shown that this dependence can be written as [5]:

$$a = \frac{\varepsilon E_1}{qN_d} \frac{1 - K\rho J_c E_1}{1 - J_c/J_1}. \tag{5}$$

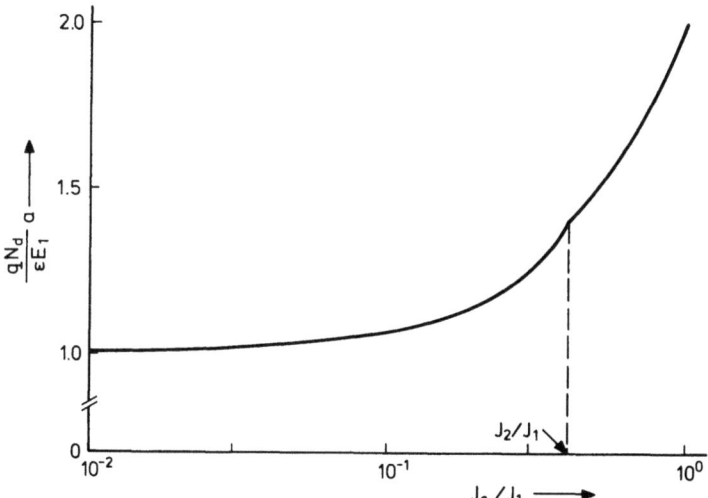

Fig. 4. The half-width (a) of the intermediate region in the depletion mode, normalized to $\varepsilon E_1/qN_d$, as a function of J_c/J_1.

Fig. 4 shows $\frac{qN_d}{\varepsilon E_1} \cdot a$ as a function of J_c/J_1.

Fig. 5 gives an illustration of a typical depletion situation. The collector doping concentration $N_d = 10^{15}$ cm^{-3}, $\rho = 4,6$ Ωcm, $J_1 = 1372$ A/cm^2, $J_2 = 555$ A/cm^2 and $W = 6$ µm. For $V_{cb} + V_o = 6,3$ volts and $J_c = 293$ A/cm^2 we find, with the help of eqs. (3) and (4), $x_d = 3,05$ µm and from eq. (5) it follows that the width of the intermediate region $2a = 2,32$ µm. In the region $0 < x < x_d - a$ the carriers move with the scattering-limited drift velocity because $E > E_1$; their concentration is J_c/qv_{lim}. In the greater part of the intermediate region the carriers show tepid behaviour and for $x > x_d + a$ the conduction is ohmic because $J_c > J_2$ (cf. eq. (3)).

The results of the depletion model are compared with numerical calculations [6] (fully drawn lines in fig. 5).

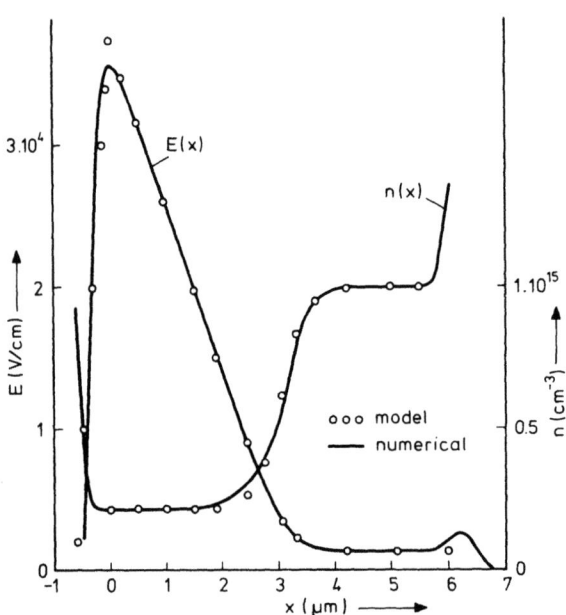

Fig. 5. Field and electron distribution in a typical depletion situation. $N_d = 10^{15}$ cm^{-3}, $W = 6$ µm, $J_c = 293$ A/cm^2 and $V_{cb} + V_o = 6,3$ volts.

2.2. The SLDV mode of operation

If the electric field strength is greater than E_1 in the entire (epitaxial) collector region the carriers move everywhere in this region with the scattering-limited drift velocity (SLDV) and the slope of the electric field is independent of position for $0 < x < W$:

$$\frac{dE}{dx} = \frac{qN_d}{\varepsilon}(1-J_c/J_1) . \qquad (6)$$

The applied V_{cb} finally determines the actual field distribution:

$$-E(x) = \frac{V_{cb}+V_o}{W} - \frac{qN_d}{\varepsilon}(1-J_c/J_1)(x-\tfrac{1}{2}W) . \qquad (7)$$

The electron density is constant throughout the lightly doped collector region and given by:

$$n = \frac{J_c}{qv_{lim}} = \frac{J_c}{J_1}N_d . \qquad (8)$$

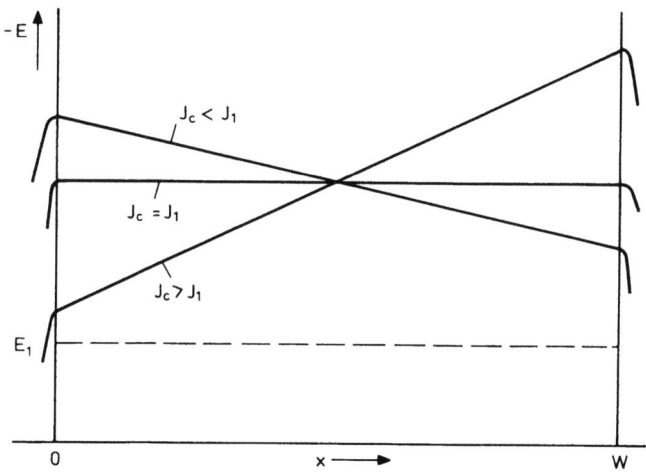

Fig. 6. Field distribution in the SLDV mode of operation for a given V_{cb}. The field gradient changes with current.

Fig. 6 gives the field distributions for a constant V_{cb} but various current densities. For $J_c = J_1$ the mobile carriers just compensate the fixed charges and the total space charge is zero.

2.3. The (J_c, V_{cb}) plane

In sections 2.1 and 2.2 we have discussed the depletion and SLDV modes of operation and we will now establish the boundaries of these models in the (J_c, V_{cb}) plane.
The depletion model is valid for $J_c < J_1$ and $0 < x_d < W$. By putting $x_d = 0$ we get from eq. (4) the boundary

$$V_{cb} + V_o = K(J_c) \cdot \rho W J_c \tag{9}$$

and by putting $x_d = W$ we get

$$V_{cb} + V_o = K(J_c) \cdot \rho W J_c + \frac{qN_d}{2\varepsilon} W^2 (1-J_c/J_1) \tag{10}$$

The SLDV-model ceases to be valid when the minimum of $|E(x)|$ drops below the E_1-value, which can occur at $x = 0$ or at $x = W$. The boundaries found in this way, are:

$$V_{cb} + V_o = E_1 W \pm \frac{qN_d}{2\varepsilon} W^2 (1-J_c/J_1) \tag{11}$$

The plus sign refers to $J_c \leq J_1$, the minus sign to $J_c > J_1$.
The boundaries given by eqs. (9), (10) and (11) are sketched in fig. 7. Left of the boundaries $x_d = 0$ and $-E(0) = E_1$ the base-collector junction is forward-biased internally and the injection model (see section 2.4) must be used.
Between the SLDV- and depletion areas of the (J_c, V_{cb}) plane in fig. 7 there exists some piece of no man's land which is mainly governed by tepid carrier behaviour and not covered by our simplified models. It is possible to give a complete and consistent set of models [5], covering this area too, but that is beyond the scope of this paper.

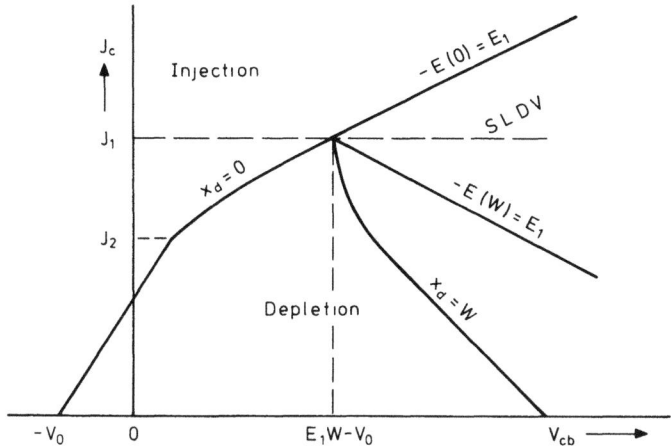

Fig. 7. Division of the (J_c, V_{cb}) plane into three main areas: depletion, SLDV and injection.

2.4. The injection mode of operation

The injection mode of operation is characterised by the base-collector junction being forward-biased internally. In the literature this mode of operation is often called quasi-saturation. Because the collector region is relatively lightly doped, holes are injected from the base into the collector. This injection is counteracted by the electric field, brought about by the dominant emitter-collector current. Under these circumstances we can divide the epitaxial collector region into an injection region $(0 < x < x_i)$, where the injected hole charge is stored [7] and the remaining part, where the carrier behaviour can be ohmic, tepid or hot; see fig. 2b.

For the calculation of the charge distribution in the injection region the following is assumed [8]:

1. quasi-neutrality exists in the injection region;

$$p + N_d = n \qquad (12)$$

2. recombination is neglected, so we can put

$$J_p \approx 0 . \qquad (13)$$

Because of eq. (13) we can solve the hole current equation for the electric field and substitute the result into the electron current equation. Together with the Einstein relation $qD_n = \mu_n kT$ and eq. (12) we then get the following first order differential equation for holes:

$$qD_n(2 + \frac{N_d}{p}) \frac{dp}{dx} = J_n \approx J_c. \tag{14}$$

If we choose as boundary condition [2]

$$p(x_i) = n_i ,$$

the solution of eq. (14) is

$$x_i = \frac{2qD_n}{J_c} \{ p(o) - n_i + \tfrac{1}{2}N_d \ln \frac{p(o)}{n_i} \}. \tag{15}$$

p(o) is the hole concentration at the metallurgical junction. In fig. 8 we have sketched, as a function of position, the quasi-fermi levels for holes and electrons $\varphi_p(x)$ and $\varphi_n(x)$ plus the electrostatic potential $\psi(x)$.
The quasi-fermi levels are defined by the following equations:

$$n = n_i \exp \{ \frac{q}{kT} (\psi - \varphi_n) \} \tag{16a}$$

and

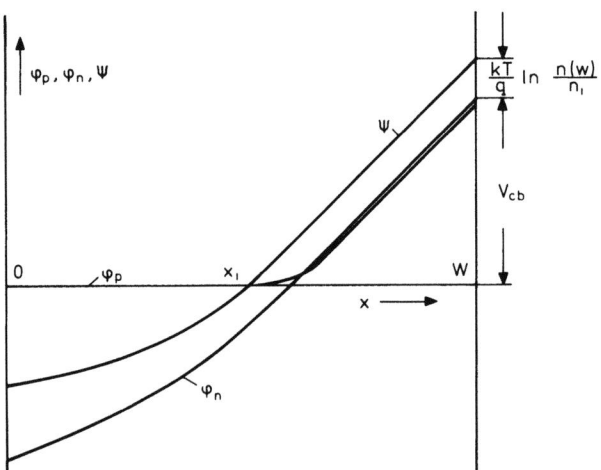

Fig. 8. The quasi-fermi levels (φ_p, φ_n) and the electrostatic potential (ψ) as a function of position in the collector in case of quasi-saturation.

$$p = n_i \exp\left\{ \frac{q}{kT}(\varphi_p - \psi) \right\}. \tag{16b}$$

For $0 < x < x_i$ (injection region) the $\varphi_p = 0 =$ constant. This follows from the current equation for holes, which can be written as

$$J_p = -q\mu_p p \frac{d\varphi_p}{dx}$$

We have assumed $J_p = 0$, but in the injection region $p \neq 0$, so we must have $d\varphi_p/dx = 0$ and $\varphi_p =$ constant.

The boundary condition at the end of the epitaxial region $(x = W)$ is taken as

$$n(W) = N_d \tag{17a}$$

with $J_c < J_1$ for ohmic and tepid carrier behaviour, and as

$$n(W) = \frac{J_c}{J_1} \cdot N_d \tag{17b}$$

with $J_c > J_1$ for hot carrier behaviour.

In the first case the electric field in the end region $(x_i < x < W)$ is constant and equal to $K(J_c) \cdot \rho J_c$, where $K(J_c)$ is given by eq. (3). Because the applied V_{cb} is equal to the difference in quasi-fermi levels, it follows from eqs. (16a) and (17a) that

$$\psi(W) = V_{cb} + \frac{kT}{q} \ln \frac{N_d}{n_i} \tag{18}$$

Further we have, from $\frac{d\psi}{dx} = K\rho J_c$,

$$\psi(W) - \psi(x_i) = \psi(W) = K\rho J_c(W - x_i) \tag{19}$$

Combining eqs. (18) and (19) we have established a relation between x_i, J_c and V_{cb}:

$$K\rho J_c (W - x_i) = V_{cb} + \frac{kT}{q} \ln \frac{n(W)}{n_i} \tag{20}$$

Eliminating x_i from eqs. (15) and (20) leads to:

$$2 \frac{p(o)}{N_d} + \ln \frac{p(o)}{N_d} = \frac{q}{kT} \left\{ \rho W J_c - \frac{V_{cb} + \tfrac{1}{2}V_o}{K(J_c)} - \tfrac{1}{2}V_o \right\} \tag{21}$$

with $V_o \approx 2\frac{kT}{q} \ln N_d/n_i$, the built-in voltage of the base-collector junction. For purely ohmic carrier behaviour $K(J_c) = 1$ and the right-hand side of eq. (21) becomes $(\rho W J_c - V_{cb} - V_o)$.

In the second case the electrons move with the scattering--limited drift velocity through the end region and we have

$$-\frac{dE}{dx} = \frac{qN_d}{\varepsilon}(\frac{J_c}{J_1} - 1)$$

$$-E(x) = E(x_i) + \frac{qN_d}{\varepsilon}(\frac{J_c}{J_1} - 1)(x-x_i)$$

and $\psi(W) = (W-x_i) E(x_i) + \frac{qN_d}{2\varepsilon}(\frac{J_c}{J_1} - 1)(W-x_i)^2$. (22)

From eqs. (17b), (18) and (22) we obtain:

$$(W-x_i) E(x_i) + \frac{qN_d}{2\varepsilon}(\frac{J_c}{J_1} - 1)(W-x_i)^2 = V_{cb} + \frac{kT}{q} \ln \frac{J_c}{J_1} \cdot \frac{N_d}{n_i} \quad (23)$$

Eq. (23) relates x_i with V_{cb} and J_c in the hot carrier case. Elimination of x_i with the help of eq. (15) then gives p(o) as a function of V_{cb} and J_c, similar to eq. (21), but more complicated, reason why it is omitted here.

Theoretically one should take for $E(x_i)$ the value E_1 (see fig. 3), but computer simulations [6] show that a better fit is obtained if $E(x_i) \approx 8 \times 10^3$ V/cm is taken. This is because the model does not take care of an intermediate zone between injection and hot carrier region.

Fig. 9 gives an example of a base-collector junction in quasi-saturation. It is the same structure as in fig. 5, but now with $V_{cb} + V_o = 7$ volts and $J_c = 1920$ A/cm. Here we have $J_c > J_1$, with hot carrier behaviour in the end region.

2.5. The influence of the stored collector charge on the device performance

In section 2.4 we have calculated the minority carrier distribution by means of a one-dimensional analysis for the injection or quasi-saturation mode of operation. By means of eq. (21) we can calculate the total charge storage in the collector:

$$Q_r = qA_e \int_0^{x_i} p\,dx \approx \frac{q^2 A_e^2}{I_c} D_n\, p(o) \{ p(o) + N_d \} \quad (24)$$

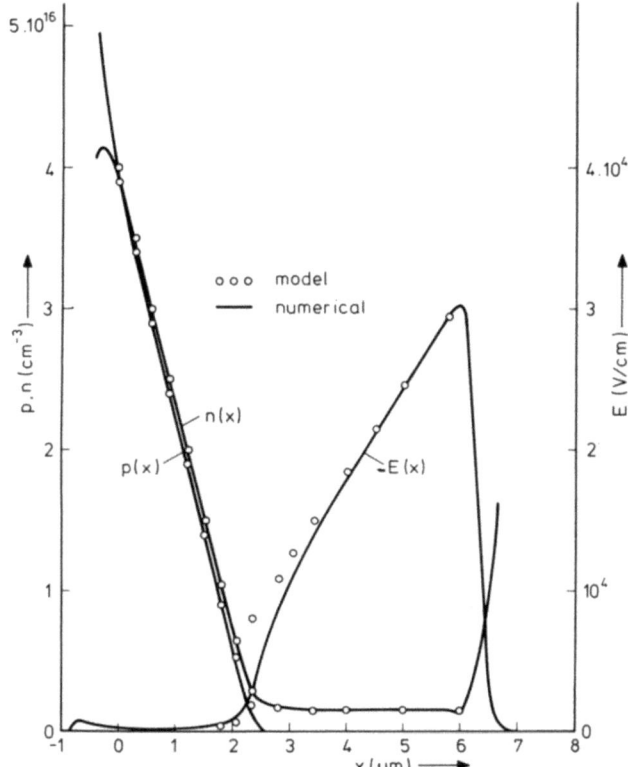

Fig. 9. Field and carrier distributions in quasi-saturation. Same structure as in fig. 5, but with $V_{cb} + V_o = 7$ volts and $J_c = 1920$ A/cm^2.

where $I_c = J_c A_e$ and A_e is the effective emitter area. If we limit ourselves for reasons of simplicity to the purely ohmic carrier behaviour we find by means of eqs. (21) and (24) for low injection levels ($p(o) < N_d$):

$$Q_r \cdot I_c \approx D_n (qN_d A_e)^2 \exp \left\{ \frac{q(r_c I_c - V_{cb} - V_o)}{kT} \right\} \quad (25a)$$

and for high injection levels ($p(o) > N_d$):

$$Q_r \cdot I_c \approx D_n (qN_d A_e)^2 \left(\frac{q}{2kT}\right)^2 (r_c I_c - V_{cb} - V_o)^2 \quad (25b)$$

Note that the one-dimensional resistance ρW has been replaced by the collector spreading resistance $r_c = \rho W/A_e$.
The main influence of the stored charge Q_r is on the current gain h_{FE}, the cut-off frequency f_T and the collector diffusion capacitance C_{cd} [9-12].

2.5.1. The current gain h_{FE}

The h_{FE} decreases in quasi-saturation because the base current I_b increases and the collector current I_c decreases. The I_b increase is due to the extra recombination in the collector, given by Q_r/τ_r where τ_r is the effective lifetime in the epitaxial collector region.
The charge Q_r must be added to the base charge and thus diminishes the collector current, as can be seen from the integral charge-control relation [13].
Which of the two effects dominates the h_{FE}-fall off in quasi-saturation depends on the value of τ_r and the fixed base charge Q_{bo}.
For $\tau_r > 10^{-6}$ s in most practical cases the h_{FE} decreases because I_c decreases, but for $\tau_r < 10^{-7}$ s the increase of I_b becomes very important.
We can sometimes distinguish two regions in the (I_c, V_{ce}) characteristics (see fig. 10).

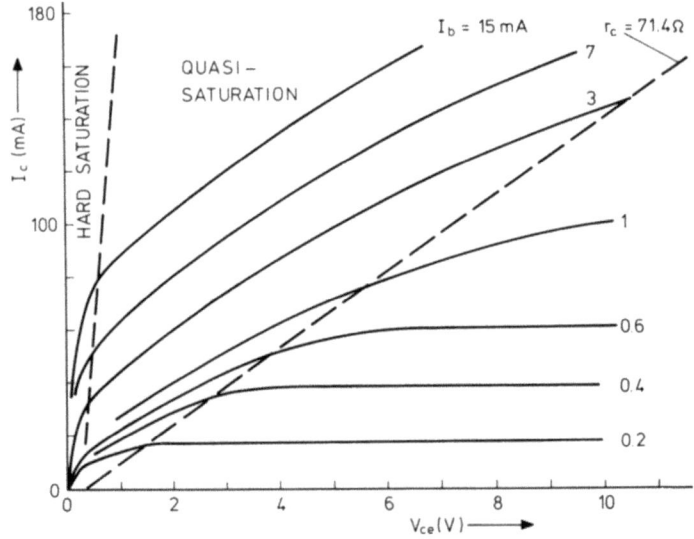

Fig. 10. (I_c, V_{ce}) characteristics with a two-region saturation: hard saturation and quasi-saturation.

Apart from the quasi-saturation we also have the hard saturation, when the entire base-collector junction is forward-biased ($V_{cb} < -400$ mV) and not only the part underneath the emitter.

2.5.2. f_T and diffusion capacitance

The charge Q_r will decrease the cut-off frequency f_T and it will give rise to a collector diffusion capacitance C_{cd}. The contribution of the collector charge Q_r to the total delay between emitter and collector is

$$\left(\frac{dQ_r}{dI_c}\right)_{V_{ce}} = \left(\frac{\partial Q_r}{\partial I_c}\right)_{V_{cb}} + \left(\frac{\partial Q_r}{\partial V_{cb}}\right)_{I_c} \cdot \frac{dV_{cb}}{dI_c} \quad (26)$$

$$= \Delta \tau_b + C_{cd} \cdot r_o,$$

where

$$\Delta \tau_b = \left(\frac{\partial Q_r}{\partial I_c}\right)_{V_{cb}},$$

$$C_{cd} = -\left(\frac{\partial Q_r}{\partial V_{cb}}\right)_{I_c}$$

and

$$r_o = \frac{kT}{qI_c}.$$

In eq. (26) Q_r is considered as a function of I_c and V_{cb}, which is given by eq. (25). Using eq. (25b) for very high currents ($I_c \to \infty$) we get

$$\Delta \tau_{b\,\infty} \approx -\frac{Q_r}{I_c} + \tfrac{1}{2} D_n (qN_dA_e)^2 \left(\frac{q}{kT}\right)^2 r_c^2 = \frac{W^2}{4D_n}$$

and

$$C_{cd\,\infty} \approx \tfrac{1}{2} \frac{q}{kT} (qN_dA_e) W.$$

So at very high injection levels C_{cd} and $f_T (\approx 1/2\,\pi\Delta\tau_b)$ become constant.
Figs. 11 and 12 give experimental results.

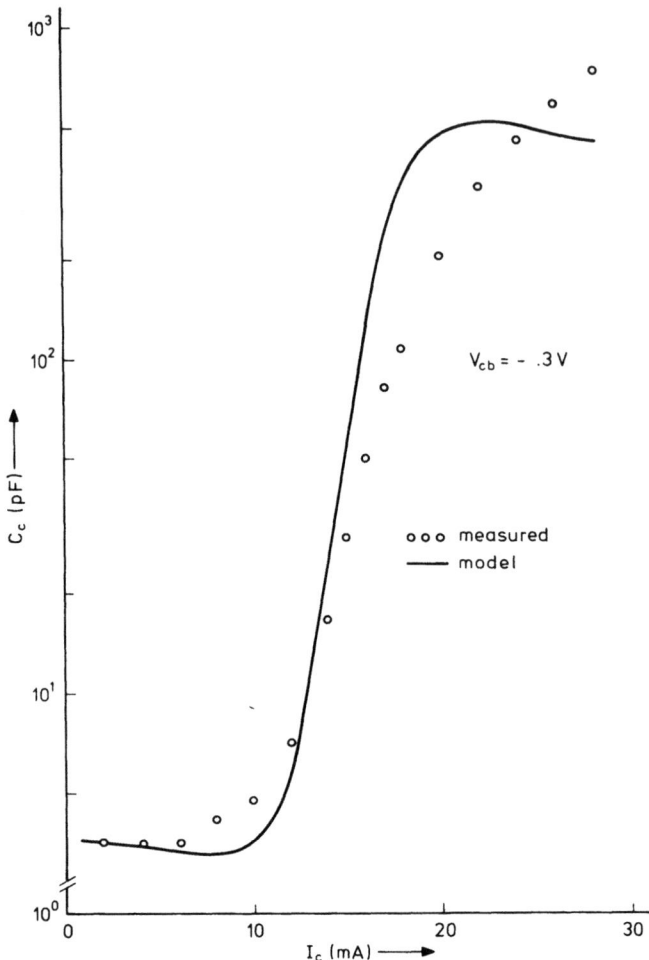

Fig. 11. Total collector capacitance as a function of I_c for a given value of V_{cb}. At high injection the C-value levels off.

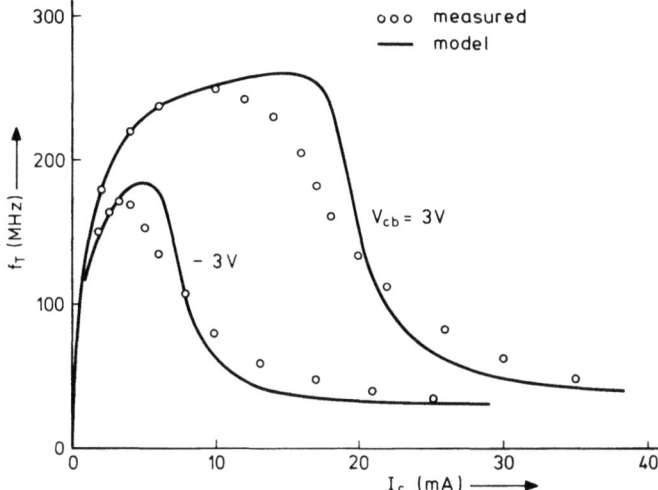

Fig. 12. f_T versus I_c, for two different values of V_{cb}. At high collector injection levels f_T becomes constant.

2.6. Measuring the onset of quasi-saturation

Eqs. (9) and (11) of section 2.3. give the boundary in the (J_c, V_{cb}) plane between the injection and the depletion or SLDV mode of operation (see also fig. 7). This boundary consists of three parts, corresponding to the ohmic, tepid and hot carrier modes. Usually the complete boundary cannot be measured d.c.-wise in one and the same transistor because the dissipation will be too high.

Low collector doping and thin epilayers however, enhance the hot carrier flow and for $N_d \cdot W < 10^{11}$ cm^{-2} this is the only possibility [5,14].

Measuring the onset of collector injection or quasi-saturation can be based on the low-frequency third-order harmonic distortion. This distortion strongly depends on the third derivative d^3Q_r/dI_c^3 of the stored charge in the collector, which is a complicated function of I_c and V_{cb} (see fig. 13). It can be shown [15] that the third harmonic in the collector current has a maximum at the boundaries given by eqs. (9) and (11).

Fig. 14 gives the measurement set-up. The twin-T filter suppresses the third harmonics of the signal generator. The third harmonic output is measured as a function of the d.c. current I_c with the d.c. voltage V_{cb} as a parameter: see fig. 15. The first maximum is caused by the emitter-base junction and is independent

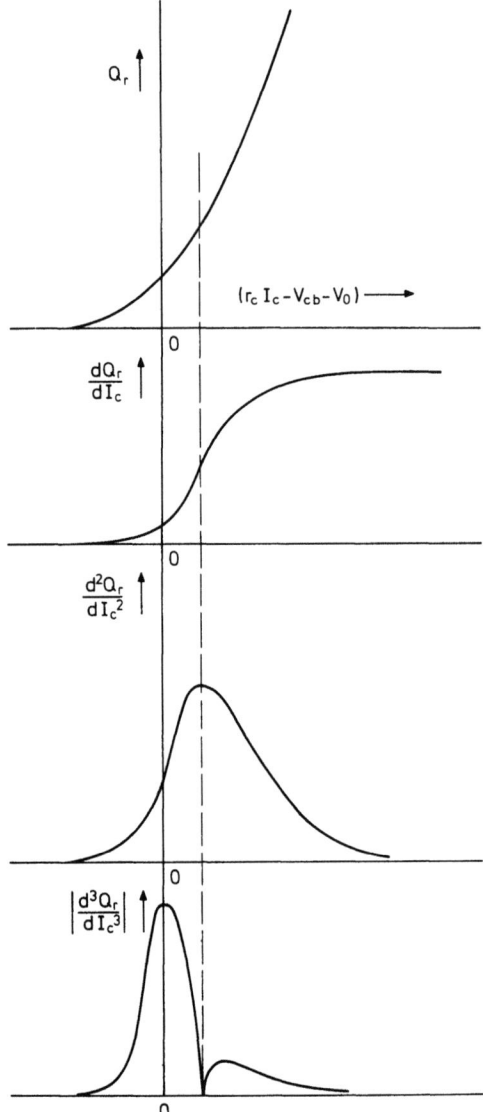

Fig. 13. The stored collector charge Q_r and its derivatives as a function of $(I_c r_c - V_{cb} - V_0)$, showing that $d^3 Q_r / dI_c^3$ has a maximum for $I_c r_c - V_{cb} - V_0 = 0$

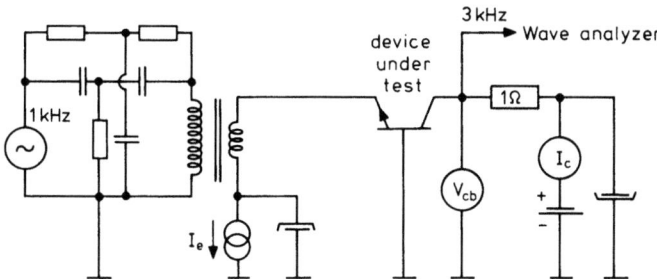

Fig. 14. The measurement set-up.

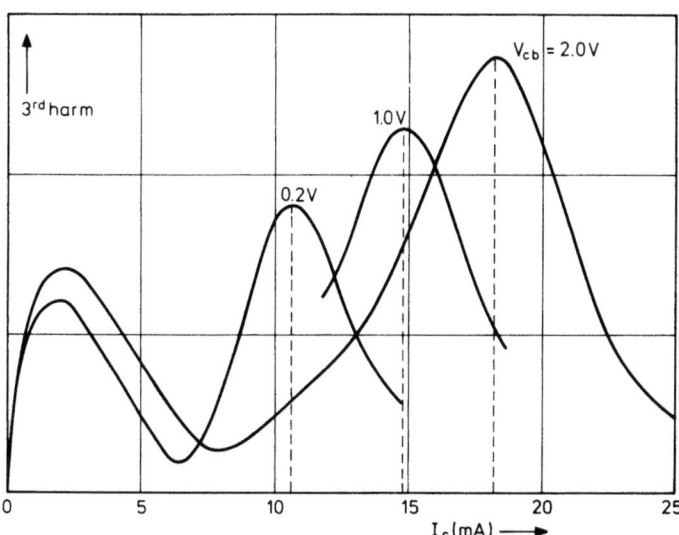

Fig. 15. The third harmonic output (arbitrary scale) as a function of bias current (I_c) with V_{cb} as a parameter.

of V_{cb}.
The second maximum, however, is due to the collector charge Q_r, and the current at which it appears strongly depends on V_{cb}.
In this way the onset of quasi-saturation can be measured and some experimental results are given in fig. 16.

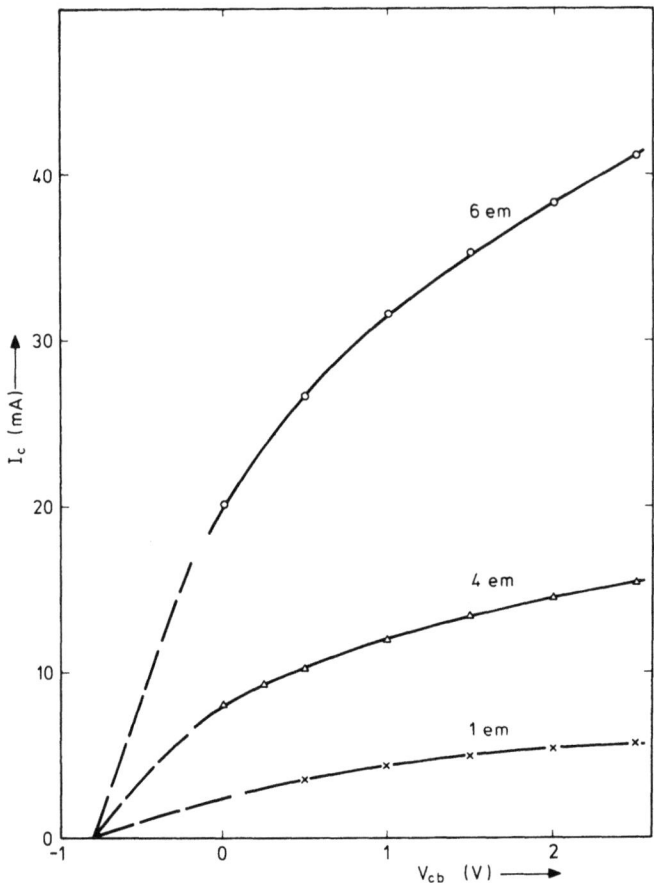

Fig. 16a. Experimentally determined quasi-saturation boundaries of microwave transistors with 1, 4 and 6 emitter stripes. $N_d \cdot W = 5 \times 10^{11}$ cm^{-2}, tepid carrier flow.

In fig. 16a the boundaries of three microwaves transistors are given. All had a $N_d \cdot W$ product equal to 5×10^{11} cm^{-2} but differed in the number of emitter stripes. We can see here the tepid carrier behaviour (cf. eq. (9)).

In fig. 16b the $N_d \cdot W$ product is 6×10^{10} cm^{-2} and the measurements reveal a hot carrier behaviour (cf. eq. (11)).

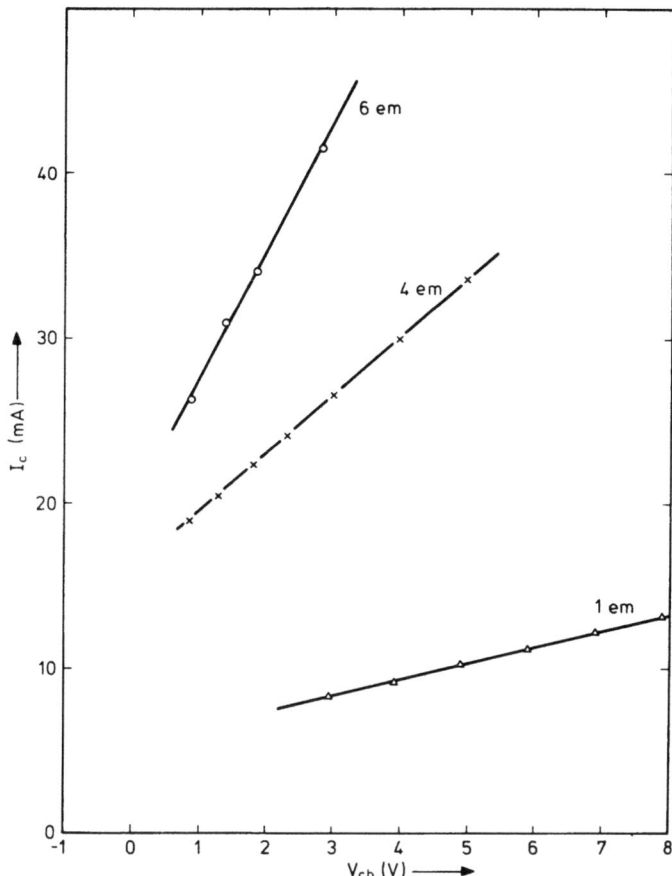

Fig. 16b. As fig. 16a, but for $N_d \cdot W = 6 \times 10^{10}$ cm^{-2}, hot carrier flow.

2.7. Circuit models for quasi-saturation

It is well-known that the classical Ebers-Moll model [13] is not very exact in forecasting the quasi-saturation behaviour of lightly doped epitaxial collector structures. If we want to improve the Ebers-Moll model in this respect we have to add a few elements [16], namely a fixed collector series resistance and an extra controlled voltage source. Then we get a diagram as in fig. 17. The resistor r_c is the unmodulated collector series

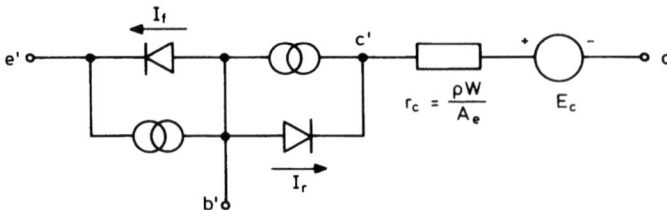

Fig. 17. The Ebers-Moll model with additional elements r_c and E_c to simulate the quasi-saturation behaviour of lightly doped collectors.

resistance: $r_c = \rho W/A_e$. For the Ebers-Moll part we have the usual equations

$$I_f = I_{fo} \{ \exp (qV_{b'e'}/kT) -1 \}$$

and

$$I_r = I_{ro} \{ \exp (qV_{b'c'}/kT) -1 \}.$$

For the controlled voltage source E_c Gwyn et al.[16] have proposed

$$E_c = \frac{I_c r_c}{C + I_c r_c} (I_c r_c - V_{cb}) \tag{27}$$

where C is a constant of about ·2 to ·8 volts. The quasi-saturation behaviour is here very much determined by the constant C.

Another possibility is to express E_c as a function of I_r. A formula of that kind, based on the theory of section 2.5, is the following [17]:

$$E_c \approx \frac{kT}{q} [2 \frac{n_i}{N_d} \sqrt{\frac{I_{ro}}{I_{fo}}} + \ln \{ \frac{N_d}{n_i} \cdot \frac{I_{ro}}{I_{fo}} \sqrt{\frac{I_{fo}}{I_r}} \}]. \tag{28}$$

Eq. (28) assumes high injection into the collector ($p(o) > N_d$). In fig. 18 this model is compared with measured characteristics for a BC107.
In addition to the Ebers-Moll parameters two other parameters are needed: r_c and N_d/n_i.
Still another possibility is to omit E_c and to take r_c as a modulated series resistance:

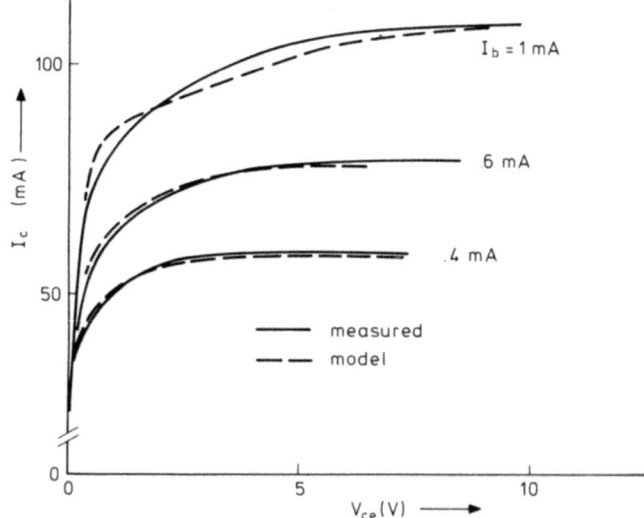

Fig. 18. Measured and calculated (I_c, V_{ce}) characteristics of the BC107.

$$r_c = \frac{\rho(W-x_i)}{A_e}.$$

The width x_i of the injection region is calculated by means of a Linvill-type network, coupled to the Ebers-Moll model [18,19].

3. CONCLUSIONS

Taking the complete dependence of the drift velocity on electric field (ohmic, tepid and hot) into account, the d.c. field and carrier distributions in the lightly doped collector are calculated analytically. In this way a division of the (I_c, V_{cb}) plane can be given and the adequate transistor model for each point of operation chosen. The total charge stored in the collector, influences the current gain, collector capacitance and cut-off frequency in quasi-saturation.

The theory can be incorporated in an Ebers-Moll like-model, which has a collector series resistance and a controlled voltage source as extra elements.

REFERENCES

1. C.T. Kirk, I.R.E. Trans. ED-9, 164 (1962).
2. J.A. Pals and H.C. de Graaff, Philips Res. Rep. 24, 53 (1969).
3. H.C. de Graaff, Philips Res. Rep. 24, 34 (1969).
4. E.J. Ryder, Phys. Rev. 90, 766 (1953).
5. H.C. de Graaff, Solid-State Electr. 16, 587 (1973).
6. J.W. Slotboom, I.E.E.E. Trans. Ed-20, 669 (1973).
7. L.A. Hahn, I.E.E.E. Trans. ED-16, 654 (1969).
8. J.R.A. Beale and J.A.G. Slatter, Solid-State Electr. 11, 241 (1968).
9. R. Kumar and L.P. Hunter, I.E.E.E. Trans. ED-22, 52 (1975).
10. R. Kumar and L.P. Hunter, I.E.E.E. Trans. ED-22, 1031 (1975).
11. L.E. Clark, I.E.E.E. ED-16, 113 (1969).
12. R.J. Whittier and D.A. Tremere, I.E.E.E. Trans. ED-16, 39 (1969).
13. H.C. de Graaff, Review of Models for Bipolar Transistors, this issue.
14. D.L. Bowler and F.A. Lindholm, I.E.E.E. Trans. ED-20, 257 (1973).
15. H.C. de Graaff and R.J. van der Wal, Solid-State Electr. 17, 1187 (1974).
16. C.W. Gwyn, G.G. Summers and W.T. Corbett, I.E.E.E. Trans. on Nucl. Sci., NS-17, 63 (1970).
17. H. Lindeman, to be published.
18. P.P. Wang and F.H. Branin Jr., I.E.E.E. Power Electr. Spec. Conf., Pasadena, P.E.S.C. 73 Record, 80 (1973).
19. F.A. Perner, I.E.D.M, Washington, Technical Digest, 418 (1974).

EMITTER EFFECTS IN BIPOLAR TRANSISTORS

H.C. de Graaff

Philips Research Laboratories,
Eindhoven - The Netherlands

ABSTRACT. In heavily doped emitter regions bandgap narrowing and lifetime reduction result in the fact that the base current is one or two orders of magnitude larger than expected from the ratio of the total number of impurities in emitter and base. Experiments show that the above mentioned effects become important for impurity concentrations larger than 10^{17} cm^{-3}. Both effects are incorporated in a general theory for emitter efficiency and two recently developed emitter structures (LEC and polysil) are discussed on this basis.

1. INTRODUCTION

The d.c. current gain h_{FE} of a bipolar transistor is a function of the collector current (I_c). At low current levels h_{FE} increases with current, then reaches a maximum and decreases at still higher current levels. This decrease is caused by high injection in the base and/or collector saturation. The low level current gain is dominated by recombination at the surface and in the emitter-base space charge region. In this current range the base current (I_b) usually increases with exp ($q\ V_{be}/mkT$) with the non-ideality factor m between 1 and 2 [1].

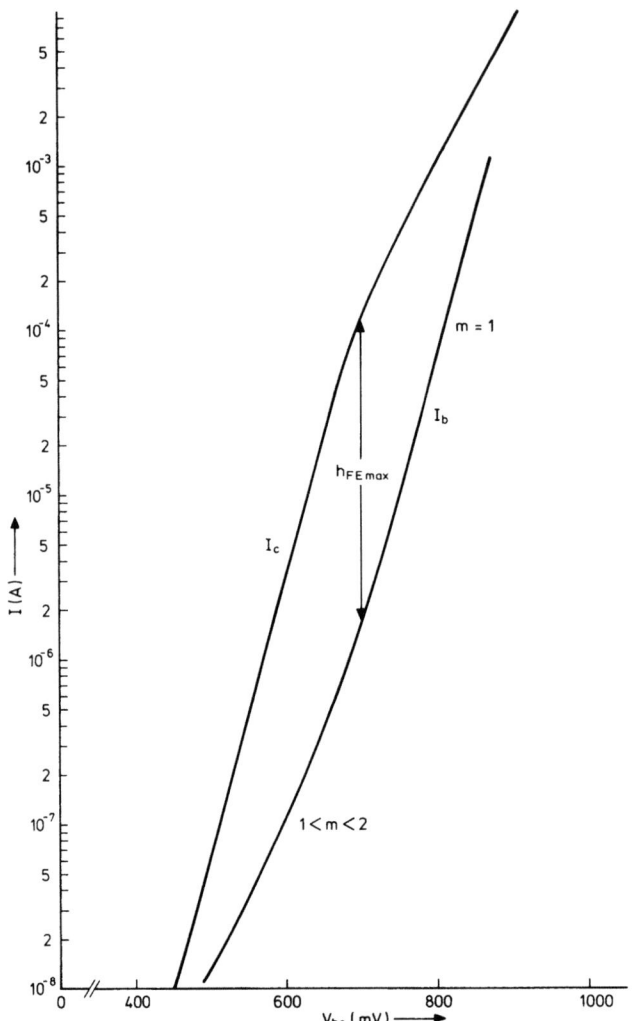

Fig. 1. Sketch of I_c and I_b versus V_{be}. m is the non-ideality factor for the base current.

The collector current increases with $\exp(qV_{be}/kT)$ at low levels and gradually goes to an increase with $\exp(qV_{be}/2kT)$ at high levels. Fig. 1 shows I_c and I_b schematically as functions of the applied base-emitter voltage (V_{be}). This figure clearly reveals the above mentioned h_{FE}-behaviour. The maximum value $h_{FE\,max}$ is determined mainly by the emitter efficiency η, defined as

$$\eta = \frac{J_n}{J_p + J_n} \qquad (1)$$

for an npn transistor.
If the recombination in the bulk regions of emitter and base is neglected, the classical theory predicts that $h_{FE\,max}$ is determined by the total number of impurities in base and emitter, devided by the respective minority carrier diffusion constants [2]:

$$h_{FE\,max} = \int_{em.} \frac{N_d}{D_p}dx \, / \int_{base} \frac{N_a}{D_n}dx. \qquad (2)$$

The current gains found in practice are usually much lower than eq.(2) predicts. This has two main reasons:
a. The high impurity concentration present in the emitter reduces the bandgap and increases the intrinsic concentration n_i. This increases, in turn the minority carrier concentration in the emitter and thus the base current [3].
b. The minority carrier lifetime decreases at high impurity concentrations and may become very low in the emitter. This too increases the base current [4].

In general, the impurity concentration is not constant in the emitter and we have the following complex situation. The intrinsic concentration (n_i) and lifetime (τ) are functions of impurity concentration (N) and, therefore, of position. Moreover, we have a built-in electric field $E(x)$ and the diffusion constant D_p also depends on $N(x)$.
We will treat these various aspects in the following sections and we will start with the intrinsic concentration.

2. BANDGAP NARROWING AND PN-PRODUCT

In this section we will first give some theoretical background and then describe a few experiments with bipolar transistors.

2.1. Theoretical background

The electron and hole concentrations are given by [5]:

$$n = \int_{-\infty}^{+\infty} \rho_c(E) \frac{dE}{1 + \exp\{(E - E_f)/kT\}} \qquad (3)$$

$$p = \int_{-\infty}^{+\infty} \rho_v(E) \frac{dE}{1 + \exp\{(E_f - E)/kT\}} \qquad (4)$$

Here ρ_c and ρ_v are the density-of-states functions for the conduction and valence bands and E_f is the Fermi level. E_f can be found by imposing the neutrality condition:

$$p - n + N = 0 \qquad (5)$$

where N is the (donor) impurity concentration. In non-degenerate material we have [5]:

$$\rho_c(E) = \frac{4\pi}{h^3}(2m_e)^{3/2} (E)^{\frac{1}{2}} \qquad (6)$$

and

$$\rho_v(E) = \frac{4\pi}{h^3}(2m_h)^{3/2} (-E-E_g)^{\frac{1}{2}}. \qquad (7)$$

Spin degeneracy is taken into account, m_e and m_h are the the effective masses and E_g is the bandgap (see fig. 2).

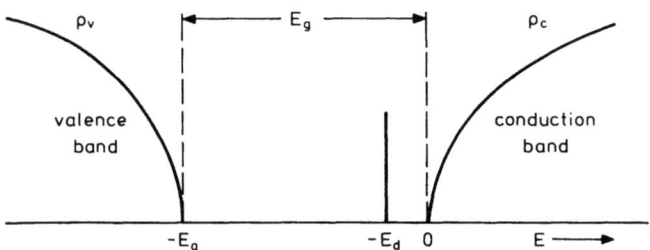

Fig. 2. The density-of-state functions ρ_c and ρ_v as a function of energy (E). The bottom of the conduction band is chosen as E=0. The donors have a single level (E_d).

In intrinsic material N=0 and $p=n=n_{io}$. From eqs.(3)-(7) it follows that the pn-product in intrinsic material is given by

$$pn = n_{io}^2 = 4(\frac{2\pi kT}{h^2})^3 (m_e m_h)^{3/2} \exp(-E_g/kT). \qquad (8)$$

The bandgap E_g is a function of temperature [6] and above 250 K it decreases linearly. This makes it possible to rewrite eq. (8) as [7]:

$$n_{io}^2(T) = C \cdot T^3 \exp(-qV_{go}/kT) \qquad (9)$$

with $C = 9.61 \times 10^{32}$ cm^{-6}K^{-3}
and $V_{go} = 1.206$ volt.

The energy qV_{go} is found by extrapolating the linear part of E_g versus T towards T=0.

At high impurity densities two things happen, namely the density-of-states functions acquire tails and the impurity band broadens, see fig.3.

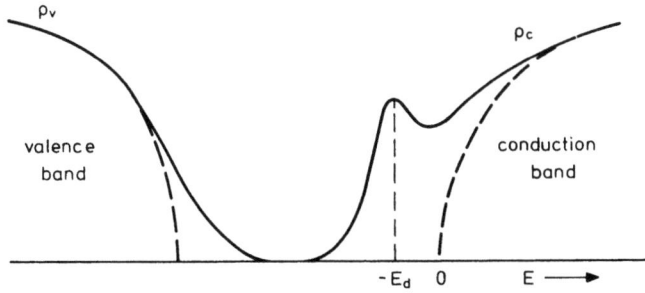

Fig. 3. Density-of-state functions in heavily doped material, showing band tailing and impurity band broadening.

It will be clear from fig. 3 that in this situation the effective bandgap will be narrower than in the intrinsic case. If this bandgap narrowing is denoted by ΔV_{go}, we can write for n_i^2 as a function of temperature and impurity concentration [8]:

$$n_i^2(T,N) = n_{io}^2(T)\exp\left\{\frac{q\Delta V_{go}(N)}{kT}\right\} \quad (10)$$

The bandgap narrowing ΔV_{go} increases with N because both the impurity band broadening and the band edge tailing increase with increasing N.

2.2 Experiments

Bandgap narrowing in silicon can be measured by means of optical absorption[9], and also in bipolar transistors with heavily doped base regions [10].
In the latter case one starts with the expression for the low level collector current:

$$I_c = I_o \exp(qV_{be}/kT) \quad (11a)$$

$$\text{with } I_o = A_e \frac{\mu_n kT n_i^2}{Q_{bo}} \quad (11b)$$

Q_{bo} is the total number of holes in the base/cm^2; at low level injections it is a constant.
From eqs. (9) and (10) it follows that I_c can be written as:

$$I_c = \text{const.} \times T^4 \cdot \mu_n(T)\exp\{\frac{q}{kT}(V_{be} - V_g)\}. \quad (12)$$

differentiating eq. (12) to T and keeping I_c constant gives:

$$V_g - T\frac{dV_g}{dT} = V_{be} - T\frac{dV_{be}}{dT} - \frac{kT}{q}\left\{4 + T\frac{d(\ln\mu_n)}{dT}\right\}. \quad (13)$$

V_g is the narrowed bandgap: $V_g = V_{gi} - \Delta V_{go}$.
The experimental procedure consists in measuring V_{be} as a function of T for I_c = constant in the range 150-420 K, with small

temperatures steps, thus determining the quantity $V_{be} - \frac{\Delta V_{be}}{\Delta T}$ in eq.(13). The sheet resistance of the base underneath the emitter is also measured in the same temperature range and $d(\ln \mu_p)/dT$ is determined from this. By assuming that

$$\frac{d(\ln \mu_p)}{dT} = \frac{d(\ln \mu_n)}{dT} \qquad (14)$$

the right-hand side of eq.(13) is known completely and therefore

$$V_g - T\frac{\Delta V_g}{\Delta T}.$$

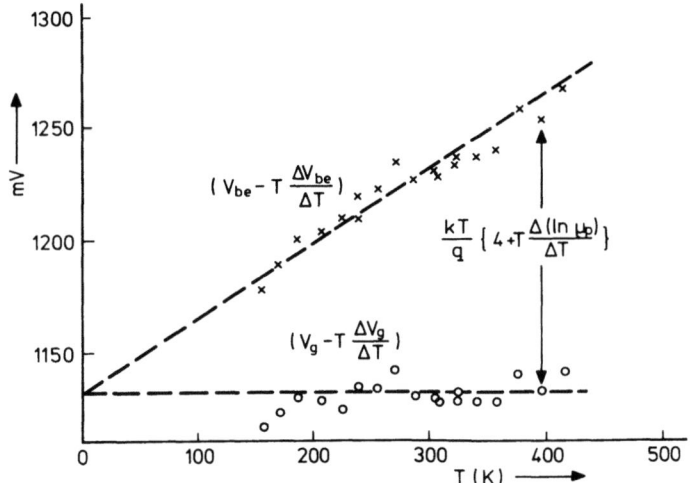

Fig. 4. The quantities $(V_{be} - T\frac{\Delta V_{be}}{\Delta T})$ and $(V_g - T\frac{\Delta V_g}{\Delta T})$ versus temperature. Base doping $N_A = 1 \times 10^{19}$ cm^{-3}.

Above 250 K, where V_g decreases linearly with T, $(V_g - T \frac{\Delta V_g}{\Delta T})$ equals the new extrapolated value of the bandgap and therefore ΔV_{go} is known. Fig. 4 gives the experimental results, for a base doping of 1×10^{19} cm^{-3}; $\Delta V_{go} = 76$ mV in this case. Another procedure is to use eq.(11b) at room temperature. From I_C versus V_{be} and the base sheet resistance underneath the emitter we determine I_o and Q_{bo} and then calculate n_i^2. Using eq.(10) finally gives us ΔV_{go} again. Here we have assumed that the electron mobility (μ_n) has the majority carrier value in n-type material with the same doping as in the base. This assumption is similar to the one in eq.(14).

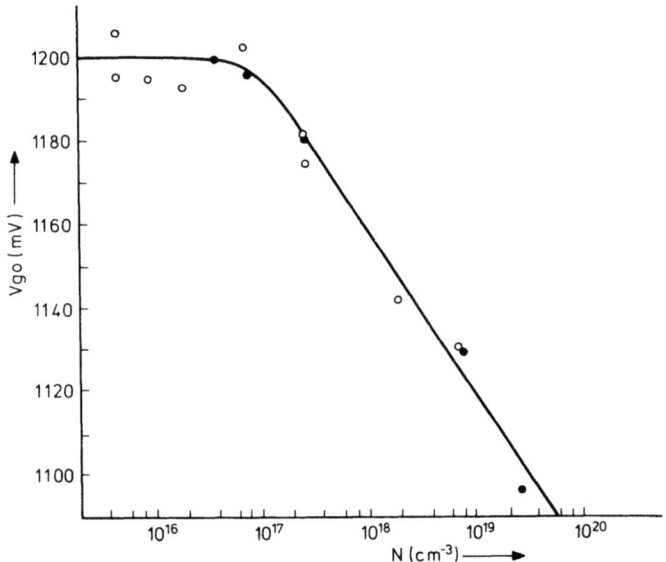

Fig.5. The measured extrapolated bandgap value V_{go} versus impurity concentration (N).

The results of fig. 5 were obtained by using various base dopings, ranging from 10^{15} to 2.5×10^{19} cm^{-3}. These results can be fitted by

$$\Delta V_{go}(N) = V_1 \left\{ \ln \frac{N}{N_o} + \sqrt{(\ln \frac{N}{N_o})^2 + a} \right\} \quad (15)$$

with $V_1 = 9$ mV,
$N_o^1 = 10^{17}$ cm^{-3},
$a = .5$.

As we can see, bandgap narrowing becomes important above impurity concentrations of 1×10^{17} cm^{-3}. From the optical absorption measurements [9] we learn that bandgap narrowing starts for $N > 1 \times 10^{19}$ cm^{-3}. However, for transistors we will use the results shown in fig. 5.

3. DOPING-DEPENDENT LIFETIME

In this section we will consider two recombination mechanisms, namely the well-known Shockly-Read-Hall (SHR) and the Auger recombination.
In the SHR-process electrons and holes recombine via traps with energy levels somewhere in the middle of the bandgap [11].
The recombination rate is given by

$$R_{SHR} = \frac{pn - n_i^2}{\tau_{po}(n + n_1) + \tau_{no}(p + p_1)} . \quad (16)$$

where τ_{po} and τ_{no} are constants, determined by the trap density and the capture cross-section (e.g. $\tau_{po} = 1/\sigma_p v_{th} N_t$).
The energy level of the traps determines n_1 and p_1; for midband levels $n_1 = p_1 = n_i$.
In the Auger process electrons and holes recombine directly, with an energy transfer to a third particle. It is the inverse of avalanche multiplication. The recombination rate can be writter as [12]:

$$R_A = (C_n n + C_p p)(pn - n_i^2). \quad (17)$$

Values of C_n and C_p for silicon, found experimentally [13], are $C_n \approx 2.9 \times 10^{-31}$ cm^6/s and $C_p \approx 1.2 \times 10^{-31}$ cm^6/s.

In n-type emitters we have $n=N_d \gg p$ and the total recombination rate R is given by

$$R = p\left(\frac{1}{\tau_{po}} + \frac{1}{\tau_A}\right) \qquad (18a)$$

with $\tau_A = \dfrac{1}{C_n N_d^2}$ \qquad (18b)

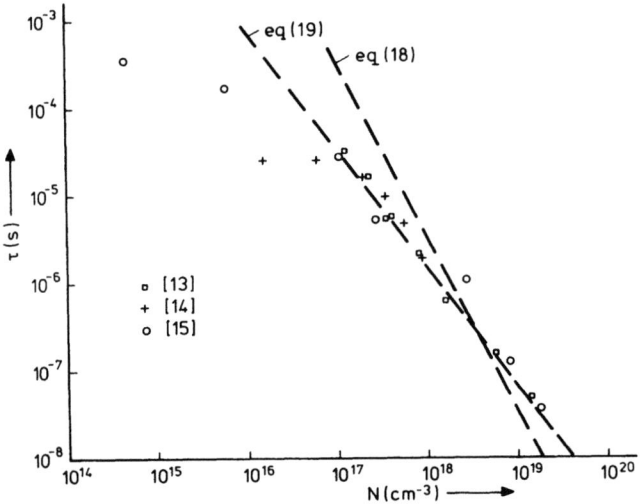

Fig. 6. Lifetime of holes in n-type Si as a function of doping.

The overall lifetime $\tau(= \dfrac{\tau_{po}}{1 + \tau_{po} C_n N_d^2})$ is a function of the impurity concentration, as shown in fig. 6. However, a better fit of the experimental values [13-15] in fig. 6 is obtained with

$$\tau_A^{-1} = 2.25 \times 10^{-19} \cdot N_d^{1.36} \qquad (19)$$

4. GENERAL THEORY OF EMITTER EFFICIENCY

The density of the dominant current component in an npn transistor is given by

$$J_n = \dfrac{q\, n_{io}^2}{G_b} \exp(q\, V_{be}/kT), \qquad (20)$$

provided there is no high injection in the base. G_b is called the (base) Gummel number [16] and determined by the base doping profile:

$$G_b = \int_0^{W_b} \left(\dfrac{n_{io}}{n_i}\right)^2 \cdot \dfrac{N_a(x)}{D_n} dx \qquad (21)$$

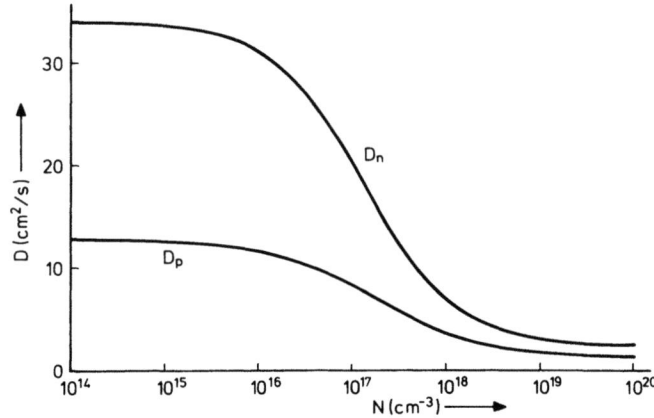

Fig. 7. Minority carrier diffusion constants as a function doping, obtained from experimental mobility data for majority carriers.

For base dopings higher than 10^{17} cm^{-3} the factor $(\frac{n_{io}}{n_i})^2$ is smaller than 1 (cf. eqs. (10) and (15)).
The diffusion constant D_n also depends on the impurity concentration, as is shown in fig. 7. These curves were obtained from experimental values of the mobility for holes [17] and electrons [18] by using the Einstein relation $D = \frac{kT}{q}\mu$.
The density of the ideal component (with m=1, see fig.1) of the base current is now written as

$$J_p = \frac{q\, n_{io}^2}{G_e} \exp(q\, V_{be}/kT). \qquad (22)$$

Here we have introduced the figure of merit (G_e) for the emitter or the emitter Gummel number, in analogy with eq.(20). If we have e.g. a $G_b = 5 \times 10^{11}$ s cm^{-4} and a $G_e = 5 \times 10^{13}$ s cm^{-4} then we may expect $h_{FE\,max}$ to be $G_e/G_b = 100$.
Although we will not give the derivation here, it can be shown [19] that

$$G_e = \frac{1}{s} \left| (\frac{n_{io}}{n_i})^2 \cdot N_d(x) \cdot g(x) \right|_{x=0} + \int_0^{W_e} (\frac{n_{io}}{n_i})^2 \cdot \frac{N_d}{D_p} \cdot g(x)\,dx \qquad (23)$$

in which s is the emitter contact recombination velocity and $g(x)$ is a weighting function for the emitter doping profile.
The first term in eq.(23) arises from the recombination at the emitter contact; for an infinite recombination velocity s it vanishes and eq.(23) resembles eq.(21) but for the weighting function $g(x)$. For Al emitter contacts we have found [19] $s \approx 3 \times 10^5$ cm/s.
The weighting function $g(x)$ is defined as the ratio of the minority carrier current densities at a certain point in the emitter and at the emitter-base junction:

$$g(x) = \frac{J_p(x)}{J_p(W_e)}. \qquad (24)$$

$g(x)$ is sketched in fig. 8, together with a doping profile $N_d(x)$ and its reduced doping profile $\{\frac{n_{io}}{n_i(x)}\}^2 \cdot N_d(x)$, divided by $D_p(x)$.

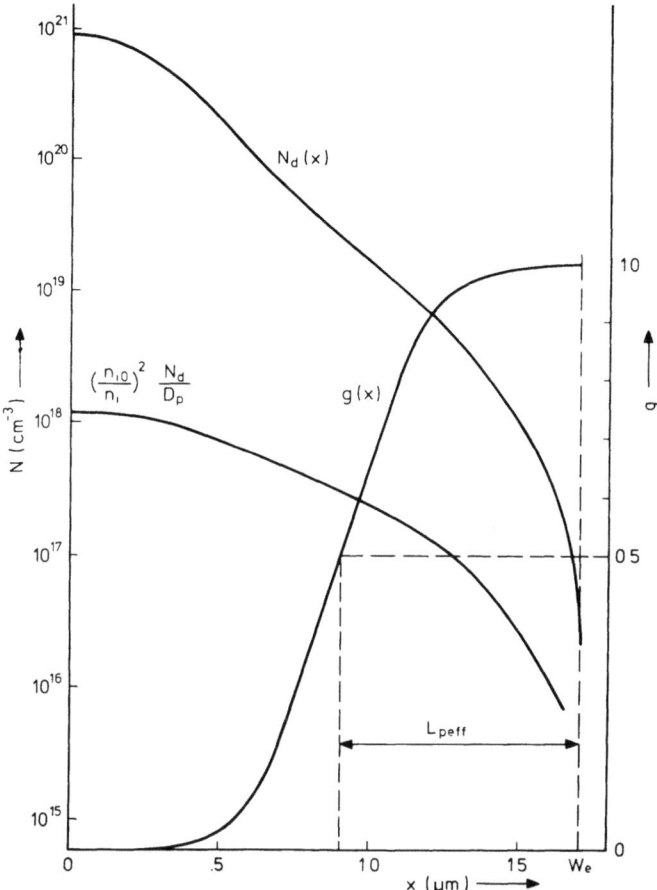

Fig. 8. Emitter doping profile $N_d(x)$, reduced profile $(\frac{n_{io}^2}{n_i^2} N_d)$ divided by D_p and weighting function $g(x)$.

The function $g(x)$ represents the influence of the emitter bulk recombination on G_e: if this recombination can be neglected $g(x) = 1$ throughout the emitter. From computer simulations it is found that $g(x)$ can be modelled as [19]:

$$g(x) \approx \exp\left\{-\left(\frac{W_e-x}{1.096\,L_{p\,eff}}\right)^4\right\}. \qquad (25)$$

At a distance $L_{p\,eff}$ from the junction $g(x) = 0.5$. $L_{p\,eff}$ is a kind of effective diffusion-recombination length and it strongly depends on bandgap narrowing $(n_i(x))$, Auger recombination $(\tau(x))$ and the built-in electric field $E(x)$, or in short on the doping profile $N_d(x)$.
For gaussian profiles, characterised by

$$N_d(x) = N(0)\exp\left\{-\left(\frac{x}{L}\right)^2\right\} \qquad (26)$$

computer simulations have revealed that $L_{p\,eff}/W_e$ is a function of $N(0)$ only, as depicted in fig. 9.

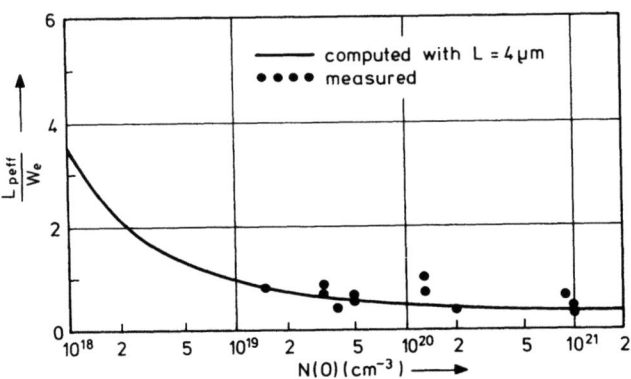

Fig. 9. The quantity $L_{p\,eff}/W_e$ versus the surface concentration $N(0)$ for gaussian profiles. The dots are experimental values.

These computer simulation incorporate the results of section 2, eqs.(10) and (15) and of section 3, eq.(19).
If the impurity concentration in the emitter is increased, the reduced doping profile is also increased, but $L_{p\ eff}$ decreases, which is why the integral in eq.(23) changes relatively little. This means, in practice, that for all kinds of emitters G_e has the same order of magnitude ($G_e \sim 1\text{-}5 \times 10^{13}$ s cm^{-4}), a fact that has already been found experimentally [20].
From fig. 8 it can be seen that if $W_e \gg L_{p\ eff}$, $g(o) = 0$ and the contact recombination velocity is of no importance. This is because the bulk recombination is so strong that the minority carriers do not reach the contact. However, if $W_e \ll L_{p\ eff}$, $g(0) = 1$, the bulk recombination is negligible and the value of s becomes very important.
For gaussian profiles, as given by eq.(26), we can easily prove, by using eq.(23), that G_e increases with the charcateristic length L. In more general terms this means that G_e increases with the emitter width W_e, provided that the doping profile is adapted to ensure also an increase of $L_{p\ eff}$. In this way, emitters can be made with $W_e \approx 15$ µm and $G_e^p \approx 1 \times 10^{14}$ s cm^{-4}.

5. SPECIAL EMITTER STRUCTURES

5.1. The LEC transistor [21,22]

The bipolar transistor with Low Emitter Concentration has an emitter with a normal n$^+$ part and a lightly doped epitaxial region, see fig. 10. This epitaxial emitter region forms a sort of buffer zone between the heavily doped n$^+$ region and the base.

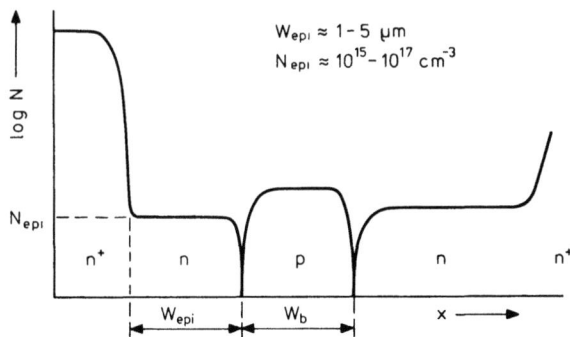

Fig. 10. Typical doping profile of an LEC transistor.

This has the advantage that the fixed base charge can be made with strict tolerances, even for lightly doped bases ($G_b \approx 10^{11}$ s cm^{-4}).
With a lightly doped base we can attain a high collector current at a given V_{be} (see eq.(20)) and the $h_{FE\ max} \approx G_e/G_b$ will also be high.
The drawback, however, is that a lightly doped base will suffer from high injection effects at relatively low current levels. The emitter figure G_e has the normal value ($\approx 10^{13}$ s cm^{-4}) in LEC transistors. This can be explained with the help of fig. 11, which shows the recombination rate R as a function of position for various values of V_{be}. The base current $I_b = \int R\ dx$.
We can distinguish four components:
A. recombination in the neutral base,
B. recombination in the space charge region around the junction,
C. recombination of the charge, stored in the lightly doped emitter region and
D. recombination in the n^+ emitter.

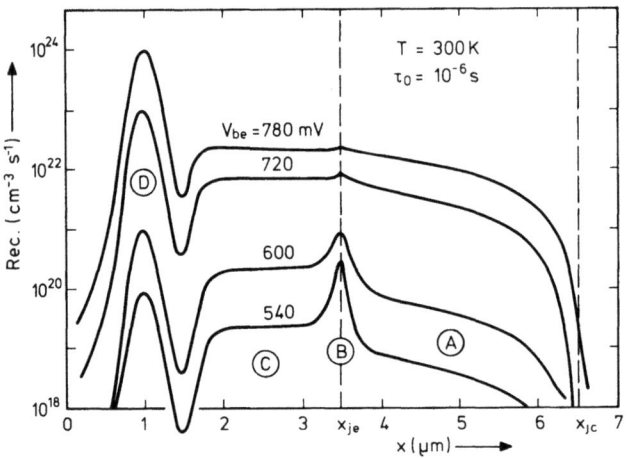

Fig. 11. Computer simulation of the recombination rate in an LEC transistor. The parts A,B,C and D are related to the four components of the base current.

For $V_{be} > 700$ mV, component D will dominate and it is this component that usually determines the base current in the vicinity of $h_{FE\ max}$. The actual value of component D depends on the n^+ doping profile and this is more or less the same as in a double diffused transistor. So it is not surprising to find the same G_e- values for LEC and double diffused transistors [22].

5.2. The Polysil emitter

The polysil emitter [23] is made by diffusinf the dopant (phosphorus or arsenic) through a layer (width ≈ 1 μm) of polycrystalline silicon.
The complete emitter then consists of a very thin (.1 μm) monocrystalline region, with a polycrystalline layer on top. The estimated surface concentration of the impurities in the monocrystalline part is about 1×10^{20} cm^{-3} or lower.
Applying eq.(23) to this monocrystalline part, and noting that $W_e \ll L_{p\ eff}$ and therefore $g(x) = 1$, we get

$$G_e = \frac{1}{s} \left| \left(\frac{n_{io}}{n_i}\right)^2 N_d \right|_{x=0} + \int_0^{W_e} \left(\frac{n_{io}}{n_i}\right)^2 \frac{N_d}{D_p} dx.$$

The value of the integral is about 3×10^{12} s cm^{-4}, but the G_e of polysil emitters is about 2×10^{14} s cm^{-4}. That is one order of magnitude larger than with conventionally made emitters. From the foregoing it will be clear that the contribution of the monocrystalline region to G_e is negligible and that recombination must take place in the polysilicon and/or at the interface.

6. CONCLUDING REMARKS

In this review some recently acquired new insights on emitter efficiency and current gain are discussed. In heavily doped regions such as emitters usually are, two phenomena are of great importance: bandgap narrowing and Auger recombination. Experimental data enable us to formulate these phenomena quantitatively and to incorporate them in a more general theory of emitter efficiency.
Finally some new, special emitter structures such as LEC and polysil are discussed and assessed against theory.

REFERENCES

1. S.M. Sze, Physics of Semiconductor Devices, p. 269. Wiley, New York (1969).
2. J. Lindmayer and Ch.Y. Wrigley, Fundamentals of Semiconductor Devices p. 128. Van Nostrand, New York (1965).
3. H.J.J. de Man, IEEE Trans ED-$\underline{18}$, 833 (1971).
4. W.W. Sheng, IEEE Trans. ED-$\underline{22}$, 25 (1975).
5. R.A. Smith, Semiconductors, p. 74. Cambridge University Press, Cambridge (1961).
6. G.G. Mc Farlane, J.P. Mc Lean, J.E. Quarrington and V. Roberts, Phys. Rev. $\underline{111}$, 1245 (1958).
7. E.H. Putley and W.H. Mitchell, Proc. Phys. Soc. London, $\underline{A72}$, 193 (1958).
8. J.W. Slotboom, Solid-State Electr. $\underline{20}$, (1977).
9. A.A. Vol'fson and V.K. Subashiev, Sov.Phys. Semiconductors $\underline{1}$, 327 (1967).
10. J.W. Slotboom and H.C. de Graaf, Solid-State Electr. $\underline{19}$, 857 (1976).
11. W. Shockley and W.T. Read Jr., Phys. Rev. $\underline{87}$, 835 (1952).
12. R. Conradt, Festkörperprobleme, Bd. XII, $\underline{449}$, (1972).
13. J.D. Beck and R. Conradt, Solid-State Comm., $\underline{13}$, 93 (1973).
14. J.Krausse, Solid-State Electr. $\underline{17}$, 427 (1974).
15. D. Kendall, Conf. Phys. and Appl. of Li Diffused Si, NASA (1969).
16. H.K. Gummel, Proc. I.R.E. $\underline{49}$, 834 (1961).
17. Wagner, J. Electrochem. Soc. $\underline{119}$, 1570 (1972).
18. G. Baccarani and P. Ostoja, SOlid-State Electr. $\underline{18}$, 579 (1975).
19. H.C. de Graaff, J.W. Slotboom and A. Schmitz, Solid-State Electr., to be published.
20. J. Burtscher, F. Dannhäuser and J. Krausse, Solid-State Electr. $\underline{18}$, 35 (1975).
21. H. Yagi and T. Tsuyuki, Suppl. to J. Jap. Soc. Appl. Phys. $\underline{44}$, 279 (1975).
22. H.C. de Graaff and J.W. SLotboom, Solid-State Electr. $\underline{19}$, 809 (1976).
23. J. Graul, A. Glasl and H. Murrmann, IEEE J. Solid-State Circ. SC-$\underline{11}$, 491 (1976).

BIPOLAR MODELS FOR STATISTICAL IC DESIGN*

R.W. Dutton, D.A. Divekar

Stanford University Integrated Circuits Laboratory
Stanford, California 94305

ABSTRACT. The Gummel-Poon model is reviewed with emphasis directed toward automated measurement. An algorithm for model parameter extraction is given and measured and simulated results are compared. A statistical sample of area-scaled transistors across a wafer has been characterized. Factor Analysis is used to help reduce the data and define a statistical model. Computer programs for Process and Device Simulation are used to facilitate understanding of the observed statistical variations and parameter correlations.

1. INTRODUCTION

The intent of this discussion is to define and develop the modeling tools required for process-oriented statistical IC design. Statistical design covers the spectrum from the fabrication process design through characterization and data reduction capabilities for test devices. The objective is to build device and circuit models directly from process design specifications. As a result, computer simulation can reflect process variations on circuit performance in a way which aids the engineer in "design centering" for improved performance and yield. The following sections define terminology, give the results of an experimental study and illustrate a method for developing model statistics directly from process modeling.

---

* This work was sponsored initially through Hewlett-Packard Corporate Engineering and continuing support of DARPA contract DAA-B07-75-C-1344 is gratefully acknowledged.

2. AUTOMATED DEVICE PARAMETER EXTRACTION

2.1 Model formulation

A specific bipolar-junction transistor model to be considered in this paper is that described by Gummel and Poon [1]. The basis for the model formulation will be developed. A model used in SPICE which is a modification of the Gummel-Poon model will be described, and then the parameter determination will be discussed. Figure 1a shows the conventional Ebers-Moll [2] bipolar-junction transistor model. The terminal currents are described in terms of the injected components of the diode currents at the base-emitter and base-collector junctions. Unfortunately, the model cannot, in this form, account for components of current other than $e^{qV/kT}$ dependence. The model shown in Figure 1b is a modified form of the Ebers-Moll model [1,3,4]. The essential difference in the formulation from that of the Ebers-Moll model is that the transport components of current (i.e., currents traversing the base region) are taken as reference currents, and the base-emitter and base-collector diodes are separated. In the simplest case, the diode relationships can be ideal ($e^{qV/kT}$ dependent), and the results reduce identically to those of the Ebers-Moll equations. In the more general case, however, the effect of space-charge recombination, basewidth modulation, and high current-level effects can be incorporated by appropriate modification of the diode and the transport model to represent the actual device performance in a more physical way.

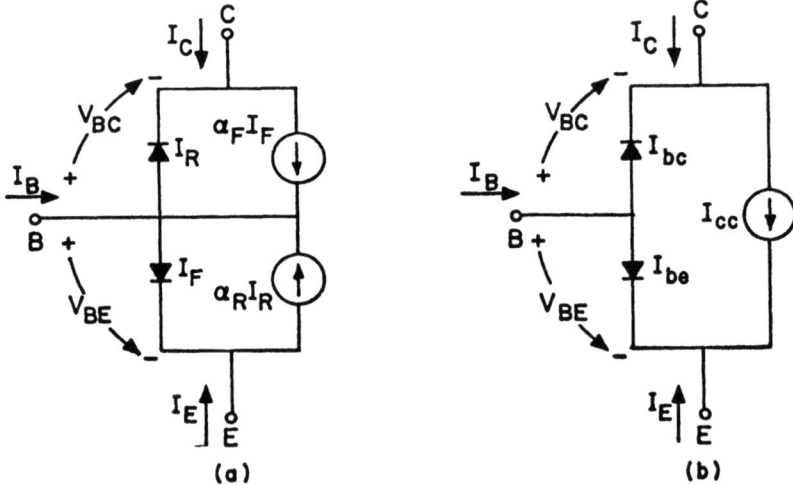

Fig. 1a. The classical Ebers-Moll bipolar model [2].
Fig. 1b. The modified transport formulation [1,3,4].

The Ebers-Moll equations are of the form:

$$I_E = -I_F + \alpha_R I_R \tag{1}$$

$$I_C = \alpha_F I_F - I_R \tag{2}$$

where

$$I_F = I_{ES}\left(e^{qV_{BE}/kT} - 1\right) \tag{3}$$

and

$$I_R = I_{CS}\left(e^{qV_{BC}/kT} - 1\right) \tag{4}$$

with the constraint $\alpha_F I_{ES} = \alpha_R I_{CS}$. Let us define $I_S = \alpha_F I_{ES} = \alpha_R I_{CS}$. We also know that $1/\alpha_F = 1 + 1/\beta_F$ and $1/\alpha_R = 1 + 1/\beta_R$. Using the above equations, the Ebers-Moll equations can be written as

$$I_E = -I_S(1 + 1/\beta_F)\left(e^{qV_{BE}/kT} - 1\right) + I_S\left(e^{qV_{BC}/kT} - 1\right)$$

$$I_C = I_S\left(e^{qV_{BE}/kT} - 1\right) - I_S(1 + 1/\beta_R)\left(e^{qV_{BC}/kT} - 1\right)$$

or

$$I_E = -I_S\left(e^{qV_{BE}/kT} - e^{qV_{BC}/kT}\right) - I_{S/\beta_F}\left(e^{qV_{BE}/kT} - 1\right)$$

and

$$I_C = I_S\left(e^{qV_{BE}/kT} - e^{qV_{BC}/kT}\right) - I_{S/\beta_R}\left(e^{qV_{BC}/kT} - 1\right)$$

From the above equations, the terminal currents can be written as

$$I_E = -I_{cc} - I_{be} \tag{5}$$

$$I_C = I_{cc} - I_{bc} \tag{6}$$

$$I_B = I_{be} + I_{bc} \tag{7}$$

where

$$I_{cc} = I_S\left(e^{qV_{BE}/kT} - e^{qV_{BC}/kT}\right) \tag{8}$$

$$I_{be} = I_{S/\beta_F}\left(e^{qV_{BE}/kT} - 1\right) \tag{9}$$

$$I_{bc} = I_{S/\beta_R}\left(e^{qV_{BC}/kT} - 1\right) \tag{10}$$

The emitter and collector currents are separated into the dominant component I_{CC} and the base-current components. This allows us to use different voltage dependences for the individual components, thus giving an improved representation of the physical processes in the transistor. In the Gummel-Poon model, Eqs. (6) and (7) are retained and Eqs. (8) through (10) are replaced by relations describing physical processes in the transistor. The component I_{CC} (which includes a high-level injection approximation) can be obtained from the charge-control relation [5] which is derived from basic physical considerations as

$$I_{cc} = I_S Q_{bo} \frac{\left(e^{qV_{BE}/kT} - e^{qV_{BC}/kT}\right)}{Q_b} \tag{11}$$

where Q_b is base charge or charge of carriers which communicate with the base terminal—electrons for pnp transistor and holes for npn transistor. The "o" subscript indicates the value of Q_b under zero bias conditions. The Q_b term is voltage and current dependent and represents changes in transport phenomena due to basewidth modulation and conductivity modulation in the base. The transport model as shown in Figure 1b can be represented by the two basic equations:

$$I_C = I_S Q_{bo} \frac{\left(e^{qV_{BE}/kT} - e^{qV_{BC}/kT}\right)}{Q_b} - I_{bc} \tag{12}$$

and

$$I_B = I_{be} + I_{bc} \tag{13}$$

The dominant component I_{cc} may be separated into a forward and reverse component as

$$I_{cc} = I_S Q_{bo} \frac{\left(e^{qV_{BE}/kT} - 1\right)}{Q_b} - I_S Q_{bo} \frac{\left(e^{qV_{BC}/kT} - 1\right)}{Q_b}$$

$$= \hat{I}_F - \hat{I}_R \tag{14}$$

where $\hat{I}_F$ and $\hat{I}_R$ are the reference transport components in the modified formulation.

At this point, it is worthwhile to note an important difference between the injection and the transport models. If the emitter junction is forward biased with $V_{BC} = 0$, from the injection model it follows from (1) and (4) that

$$I_F = -I_E = \underset{\underset{\text{component}}{\uparrow}}{I_C} + \underset{\underset{\text{component}}{\uparrow}}{I_B} \tag{15}$$

for the transport model we use (10), (12), and (14) to get

$$\hat{I}_F = \underset{\underset{\text{ideal component}}{\uparrow}}{I_C} \tag{16}$$

Equations (15) and (16) show the fundamental difference [3] between the two models. The reference current $\hat{I}_F$ for the transport model represents an ideal component of current in the sense that collector current and emitter-base voltage are related by simply $e^{qV_{BE}/kT}$. The reference current I_F for the injection model, on the other hand, is made up of two components of currents of which I_B, the base current, is nonideal. It is always easier to work with components approximating the ideal theoretical behavior. For the bipolar transistor, this ideal transport behavior is observed experimentally. For this reason, the transport model is preferred over the injection model.

Equations (12) and (13) form the basis of the Gummel-Poon model. The base-current components I_{be} and I_{bc} depend strongly on the recombination properties of the structure and cannot be readily calculated from first principles. On the other hand, it is possible to calculate Q_b as a function of V_{BE} and V_{BC} knowing the doping profile since it is nearly independent of recombination properties. However, this process may be difficult, hence normally it is preferable to approximate Q_b to simple algebraic or algorithmic representations which give acceptable accuracy depending on the desired model complexity. In the model originally proposed by Gummel-Poon [1], the term Q_b is represented by the combination of two terms q_1 and q_2. The term q_2 represents the excess base charge or the current-dependent charge associated with the diffusion capacitances. This charge can also depend on the base push-out effects represented by B. The term q_1 consists of the zero-bias charge in the base and the charge associated with the junction capacitances. The contribution from emitter-base and collector-base junction capacitances is represented as a four-dimensional vector P. Recombination in the transistor is represented by base current which is a sum of terms which are exponential in the junction voltages. The forward base current I_{be} is described by two components, one ideal and the other nonideal. For the reverse base current I_{bc}, a single nonideal component is considered to be adequate. Appropriate parameters are introduced to represent Q, I_{be}, and I_{bc} as described above in the basic Eqs. (12) and (13). The complete description of the original Gummel-Poon model is given by the following set of equations:

$$i_c = I_c/I_k \qquad V_k = \frac{q(V_{BE})}{kT} I_k$$

$$i_b = I_b/I_k$$

$$V_e = (qV_{BE}/kT) - V_k \qquad e_k = e^{-V_k}$$

$$V_c = (qV_{BC}/kT) - V_k \qquad e_{ke} = e^{-V_k/n_e}$$

$$q_b = Q_b/(I_k \tau_f) \qquad e_{kc} = e^{-V_k/n_c}$$

$$f_{(V,P)} = P_3 \left\{ \frac{1}{\left(1 + P_4\right)^{P_2}} + \frac{(V/P_1 - 1)}{\left[(V/P_1 - 1)^2 + P_4\right]^{P_2}} \right\}$$

$$i_{be} = i_1 \left(e^{V_e} - e_k \right) + i_2 \left(e^{V_e/n_e} - e_{ke} \right)$$

$$i_{bc} = i_3 \left(e^{V_c/n_c} - e_{kc} \right)$$

$$i_b = -i_{be} - i_{bc}$$

$$i_4 = i_c + (V_{oc} - V_c - V_k)/V_{rp}$$

$$B = \left\{ 1 + r_w \frac{\left[\left(i_4^2 + r_p\right)^{1/2} - i_4\right]^{n_p}}{4\left(i_c^2 + r_p\right)} \right\}$$

$$q_1 = 1 + f(V_e + V_k, P_e) + f(V_c + V_k, P_e)$$

$$q_2 = B\left(e^{V_e} - e_k\right) + r_t\left(e^{V_c} - e_k\right)$$

$$q_b = q_1/2 + \left[\left(q_1/2\right)^2 + q_2\right]^{1/2}$$

$$i_c = -\frac{\left(e^{V_e} - e^{V_c}\right)}{q_b} + i_{bc}$$

The above set of equations contains 21 parameters.

The model used in SPICE is a modification of this model [8]. The starting point is the same for the model implementation [Eqs. (12) and (13)]. The normalized base charge Q_b/Q_{bo} is represented by Q_B. The SPICE Q_B term also consists of two components, Q_1 and Q_2,

$$Q_B = 1/2\left[Q_1 + \sqrt{Q_1^2 + 4Q_2}\right] \quad (17)$$

Instead of representing Q_1 as a complicated function of voltage and a four-dimensional vector, the junction capacitances are assumed to be constant (only for approximating changes in transport current). The charge associated with junction capacitance, therefore, will be directly proportional to the junction voltages. The parameters V_A and V_B are the constants of proportionality and give rise to the finite output conductance,

$$Q_1 = 1 + \frac{V_{BC}}{V_A} + \frac{V_{BE}}{V_B} \quad (18)$$

Q_2 is the current-dependent charge contributed by the diffusion capacitances of the two junctions. It is assumed that there is no space charge in the collector body, which is equivalent to assuming that base push-out effects are neglected and that $B = 1$ in the Gummel-Poon equations. The equation for Q_2, therefore, can be written as

$$Q_2 = \frac{I_S}{I_k}\left[e^{qV_{BE}/kT} - 1\right] + \frac{I_S}{I_{kr}}\left[e^{qV_{BC}/kT} - 1\right] \tag{19}$$

The terms I_k and I_{kr} determine the conditions for the onset of high-level effects; they are referred to as the "knee" currents. Each of the diode-current components I_{be} and I_{bc} is represented by an ideal and a nonideal component. The equations for I_C and I_B therefore become

$$I_C = \frac{I_S}{Q_B}\left(e^{qV_{BE}/kT} - e^{qV_{BC}/kT}\right) - \frac{I_S}{\beta_{RM}}\left(e^{qV_{BC}/kT} - 1\right)$$
$$- C_4 I_S \left(e^{qV_{BC}/n_c kT} - 1\right) \tag{20}$$

$$I_B = \frac{I_S}{\beta_{FM}}\left(e^{qV_{BE}/kT} - 1\right) + C_2 I_S \left(e^{qV_{BE}/n_e kT} - 1\right)$$
$$+ \frac{I_S}{\beta_{RM}}\left(e^{qV_{BC}/kT} - 1\right) 1 + C_4 I_S \left(e^{qV_{BC}/n_c kT} - 1\right) \tag{21}$$

The coefficients n_e and n_c allow adjustment of the exponential slopes for the nonideal base-current dependences and C_2 and C_4 set the intercept coefficients. Equations (17) through (21) represent the bipolar junction transistor model used in SPICE [8]. This model contains 11 parameters -- I_S, β_{FM}, β_{RM}, C_2, C_4, n_e, n_c, I_k, I_{kr}, V_A, and V_B.

To illustrate the essential features of this implementation of the Gummel-Poon model, consider log I_C, log I_B vs V_{BE} characteristics for constant V_{BC} for a transistor, as shown in Figure 2. At very low forward bias, the component Q_2 can be neglected. In this case, $Q_B \simeq Q_1$ and, for moderate forward bias,

$$I_C = \frac{I_S}{1 + \frac{V_{BC}}{V_A} + \frac{V_{BE}}{V_B}} e^{qV_{BE}/kT} \simeq I_S e^{qV_{BE}/kT}\left(1 - \frac{V_{BC}}{V_A}\right) \tag{22}$$

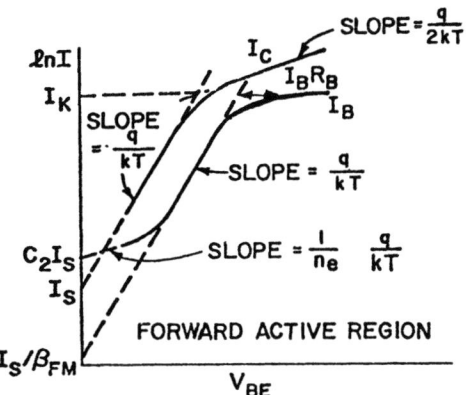

Fig. 2. Logarithm of I_C and I_B vs V_{BE} for a constant V_{BC}. The model parameters are those used in SPICE [8].

Thus, at low forward bias, collector current has an ideal component with a slope of q/kT and the finite output conductance is determined by the parameter V_A, the early voltage. Measurements of I_C at several values of V_{BC} are needed to determine V_A. At low forward bias, and hence at low currents, recombination in the space-charge layers is significant and the curve for I_B has a nonideal slope of $q/n_e kT$. At high-current levels and for low values of V_{BC}, the component Q_2 predominates and Q_1 can be assumed equal to 1. Thus,

$$Q_B = \frac{Q_1}{2}\left(1 + \sqrt{1 + 4\frac{Q_2}{Q_1^2}}\right)$$

$$\approx 1/2\left(1 + \sqrt{1 + 4Q_2}\right) \text{ for } Q_1 \simeq 1 \quad (23)$$

In case of low-level injection, $\frac{4I_S}{I_k} e^{qV_{BE}/kT} \ll 1$, and the collector current follows the ideal law:

$$I_C \simeq I_S e^{qV_{BE}/kT} \quad (24)$$

For high level injection, $\frac{4I_S}{I_k} e^{qV_{BE}/kT} \gg 1$ and the collector current becomes "nonideal."

$$I_C = \frac{I_S}{Q_B} e^{qV_{BE}/kT} \approx \frac{I_S e^{qV_{BE}/kT}}{\sqrt{\frac{I_S}{I_k} e^{qV_{BE}/kT}}} = \sqrt{I_S I_k}\, e^{qV_{BE}/2kT} \qquad (25)$$

At high current levels, therefore, the curve for I_C has a slope of $q/2kT$. For the I_B curve, after a certain value of current, the ideal component dominates and the curve has a slope of q/kT. Again at high-current levels, the voltage drop in the extrinsic base resistance R_B starts to dominate and the curves for I_B and I_C deviate from their straight-line behavior on a semilogarithmic plot. I_B does not rise as steeply as before and, hence, the slope of the curve becomes less than q/kT.

2.2 Measurement techniques

As explained above, log I_C, log I_B vs V_{BE} curves depict all the physical phenomena represented by the model. The parameters for the model can be determined from the curves, as is described below. First consider the curve for I_C. By extrapolating the region with the slope of q/kT, we obtain the value of I_S. As seen above, in regions where high-level injection dominates, the curve has a slope of $q/2kT$. If it was possible to assume that low-level and high-level effects could be distinguished by regions, a value of I_k could be obtained by the intersection of the two asymptotes with slopes q/kT and $q/2kT$. In most of the transistors, however, the two effects overlap and some trial and error is required to produce the correct parameter. For the I_B curve, by extrapolating the region with a slope of $q/n_e kT$ at low-current levels, we can determine the value of C_2 and, by determining this slope, we can find n_e. The parameter β_{FM} is determined by the regions of the two curves when they are parallel to each other with a slope of q/kT. In this region, both I_C and I_B have dominant ideal components and the ratio of I_C and I_B in this region gives β_{FM} which can be determined from the I_B curve by extrapolating the region with a slope of q/kT, as shown in Figure 2. We can obtain an estimate of the value of R_B from the deviation of the I_B curve from its ideal slope of q/kT at high-current levels.

In addition to determining the parameters for the SPICE model, values of I_B with I_C as a parameter can be obtained. If we plot the ratio of respective values of I_C and I_B against I_C, we obtain the β vs I_C curve for that particular device. Typically, the curve is in the form displayed in Figure 3.

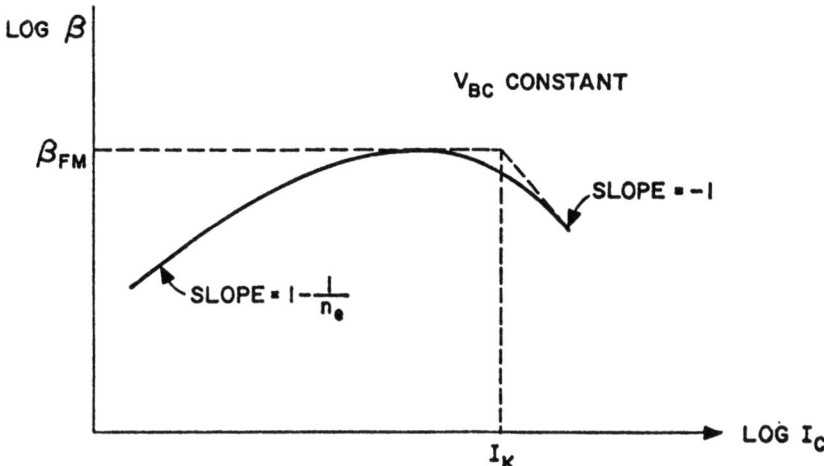

Fig. 3. A log-log plot of β vs I_C.

Figure 2 has illustrated the typical I_C and I_B dependence for a bilopar transistor for forward active bias conditions. We have seen that we can determine the Gummel-Poon parameters from these curves. However, to obtain a good estimation of these parameters, care and measurement accuracy are required. The parameters I_S and C_2 are determined by extrapolating the curves at lower currents. In order to have an accurate estimate of the intercepts, it is necessary to take measurements at as low currents as possible. On the other hand, a good estimate of I_k and R_B requires that measurements be taken at high values of currents. Thus, it is necessary to take measurements over a wide range of current values. In order to measure the Gummel-Poon parameters, efficient measurement and data-processing capabilities are required.

Consider the possible techniques for measuring the log I_C log I_B vs V_{BE} characteristics at constant V_{BC}. Since a transistor is a three-terminal device, we have three terminal currents I_B, I_C, I_E and three voltages V_{BE}, V_{CE}, V_{BC}. The characteristics, as shown in Figure 3, are to be measured at constant V_{BC}. This is appropriate for the determination of the parameters for the Gummel-Poon model. Thus, we can set V_{CE} or V_{BC} to the desired value. Having chosen one parameter as an independent variable (i.e., V_{BE}, I_C, I_B or I_E) and by changing this variable in steps, the rest of the variables can be measured. Thus, for any measurement point, we are setting two variables independently,

(a) V_{BC} or V_{CE}

(b) V_{BE}, I_C, I_B or I_E

If we assume that we can set any one of (a) and any one of (b) independently, then there are eight possible ways of measuring the desired characteristics.

Consider group (a) first. The two possibilities are to (i) set V_{BC} and (ii) set V_{CE}. Since the measurement is to be done at constant V_{BC}, V_{BC} does not give rise to any additional problems. Now consider setting V_{CE}. We know that, for a transistor, $V_{BC} = V_{CE} - V_{BE}$. V_{BE} varies over a range of 0 V to 0.8 V during the measurement. Hence V_{BC} also varies by the same amount. Therefore, if we wish to keep V_{BC} constant, V_{CE} must also be varied by using some feedback arrangement. The arrangement may not become complicated in this case since V_{CE} has to be varied just according to the variation in V_{BE}.

Now consider group (b). There are four possibilities in this case: (i) set V_{BE}, (ii) set I_C, (iii) set I_B, and (iv) set I_E. Consider the case when V_{BE} is controlled. Since, at low-current levels, I_C is less than I_B (see Figure 3), a current measurement limitation will be imposed by I_C. When V_{BE} is zero, I_C should equal I_S. We can vary V_{BE} in steps of 1 mV. When V_{BE} is equal to 1 mV, I_C will ideally be equal to $I_S e^{1/26}$ which is approximately 1.041 of I_S. Thus the minimum I_C we should have to measure is on the order of I_s. If we can measure currents less than this, then the V_{BE} source is a limitation. But if it is not possible to measure currents of this order, then the minimum voltage given by the voltage source is adequate and does not restrict our measurement capability.

Now consider the case when one of the three terminal currents is controlled. If we assume that measurement of current is not a limitation, then the minimum current we need to measure depends on which current we set. This is illustrated in Figure 4. If we set I_C, then all measured values of current will be greater than the source current because at low-current levels, I_C is the smallest of all the three currents. If we set I_B, the measurements must extend to lower current values that the source value. The most stringent measurement requirements occur if we set I_E since all current values are less than I_E. If we can easily measure currents lower than the minimum source value, setting I_E is the best solution. If our minimum measurable current is equal to or greater than the minimum source current, it does not matter which current is set. The choice can be made from other considerations. It should be noted that, if any of the three currents is being set, there is not a direct control over the variation in V_{BE}. Therefore, if it is required to take readings at equal V_{BE} steps, some kind of a feedback must be used to adjust the currents accordingly.

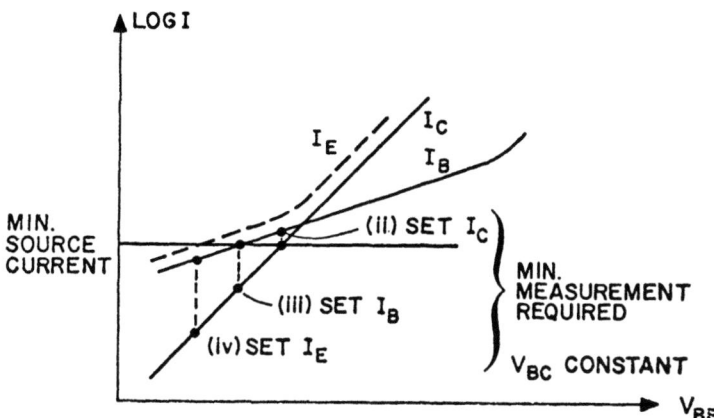

Fig. 4. The comparison of measurement requirements as a function of minimum source current for conditions (ii), (iii) and (iv).

Up to this point, we have not considered the safety of the device under test. It should not be damaged during the testing process. One way to protect it is to control the power dissipated during the testing procedure. By setting V_{BE} as in (1), there is no direct control over the power dissipation because I_C and I_E are exponentially dependent on V_{BE} and small changes in V_{BE} can cause large changes in I_C or I_E. If I_B is set as in (iii) and by knowing the h_{FE} of the device approximately, one can have a loose control over the power dissipation. If I_C or I_E is set as in (ii) or (iv), there is a direct control over the power dissipated in the device under test.

Using the criteria described in the above discussion, we can choose a method that can be used to measure the log I_C, log I_B vs V_{BE} characteristics. The discussion is summarized in Table 1.

Table 1

Constraint (b) ↓ \ Constraint (a) →	(i) Set V_{CE}	(ii) Set V_{CB}
(iii) Set I_B	Feedback required to keep V_{CB} constant. Loose control over power dissipation. Measure I_C, V_{BE} in between (ii) and (iii). Lowest current I_E in between (ii) and (iii). Feedback required to get equal V_{BE} steps	No feedback to set V_{CB}. Loose control over power dissipation. Measure I_C, V_{BE}. Lowest current I_E in between (ii) and (iii). Feedback required to get equal V_{BE} steps
(ii) Set I_C	Feedback required to keep V_{CB} constant. Direct control over power dissipation. Measure I_B or I_E and V_{BE}. Lowest current limited by min. source current. Feedback required to get equal V_{BE} steps	No feedback to set V_{CB}. Direct control over power dissipation. Measure I_B or I_E and V_{BE}. Lowest current limited by min. source current. Feedback required to get equal V_{BE} steps
(iv) Set I_E	Feedback required to keep V_{CB} constant. Direct control over power dissipation. Measure I_B or I_C and V_{BE}. Allows measurement at lowest possible current between (i), (ii) and (iii). Feedback to get equal V_{BE} steps	No feedback to set V_{CB}. Direct control over power dissipation. Measure I_B or I_C and V_{BE}. Allows measurement at lowest possible current between (i), (ii) and (iii). Feedback to get equal V_{BE} steps
(i) Set V_{BE}	Feedback required to keep V_{CB} constant. No direct control over power dissipation. Measure any two of I_B, I_C, I_E. Lowest current determined by measurement capability or V_{BE} source. No feedback to get equal V_{BE} steps	No feedback to set V_{CB}. No direct control over power dissipation. Measure two of I_B, I_C, I_E. Lowest current determined by measurement capability or V_{BE} source. No feedback to get equal V_{BE} steps

Based on typical equipment constraints, the minimum source current and minimum measurable current are equal. Therefore, the lowest current which can be measured is the same for all the methods. If V_{CE} is set along with any of the three currents, two feedback loops are required -- one to keep V_{CB} constant and one to vary V_{BE} in equal steps. On the other hand, if V_{CB} is set instead of V_{CE} and if any of the three currents is also fixed, we need only one feedback loop. If V_{BE} is set, then a feedback loop is not needed.

Therefore, we have to consider only the second column in Table 1 since a single feedback loop is the most desirable. Since we would like to have direct control over power dissipation in the device, then control using I_B and V_{BE} are ruled out. Both of the remaining choices, to control I_C or I_E, require feedback to get equal V_{BE} steps. It is not possible to set I_C and V_{CB} simultaneously using only voltage and current sources. Therefore, the choice to set V_{CB} and I_E is taken as the desired one for the present measurement system. Figure 5 shows a

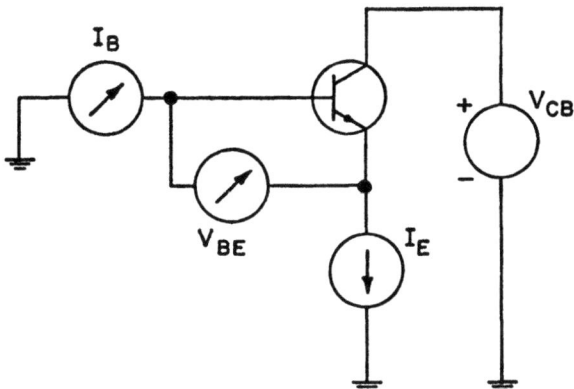

Fig. 5. A measurement configuration used to control V_{CB} and I_E. I_B and V_{BE} are measured parameters.

practical implementation of this method. As discussed earlier, this method places the most stringent requirements on current measurement and gives direct control over the power dissipation.

2.3 Model parameter extraction

This section describes the procedure developed for determining the dc Gummel-Poon parameters for a bipolar junction transistor. The procedure consists of measuring the I_C, I_B vs. V_{BE} characteristics of a device and extracting the parameters using curve-fitting techniques.

Figure 6 shows the actual configuration used for the measurement. V_{CB} is set to the desired value and I_E is varied in steps to cover the entire range of interest. V_{BE} and I_B are measured and I_C is calculated knowing I_E.

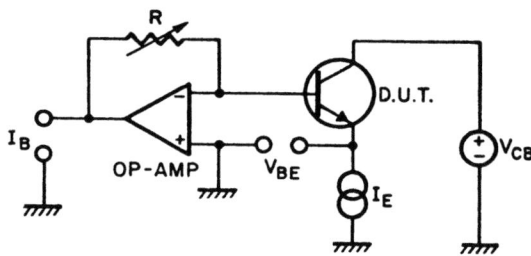

Fig. 6. Actual configuration based on Figure 5.

The above configuration gives direct control over the power dissipated in the device and also allows measurements at very low currents. In this configuration, one current and one voltage are to be measured. The current measurement consists of measuring the voltage drop across a known resistor. This calls for two voltage measurements which can be done using a single DVM and a multiplexer.

In order to determine the parameters with reasonable accuracy, measurements have to be taken over a wide current range. With the available current sources, output currents ranging from 100 nA

full-scale to 100 mA full-scale can be obtained in seven ranges.*
For acceptable accuracy, it is advisable not to go more than a
decade below the full-scale current on each range. Thus, I_E can be
varied over the range of 10 nA to 100 mA, covering seven decades
of current variation. Typical measured data are shown in Figures
7 and 8, for a device from the kitchip [7].

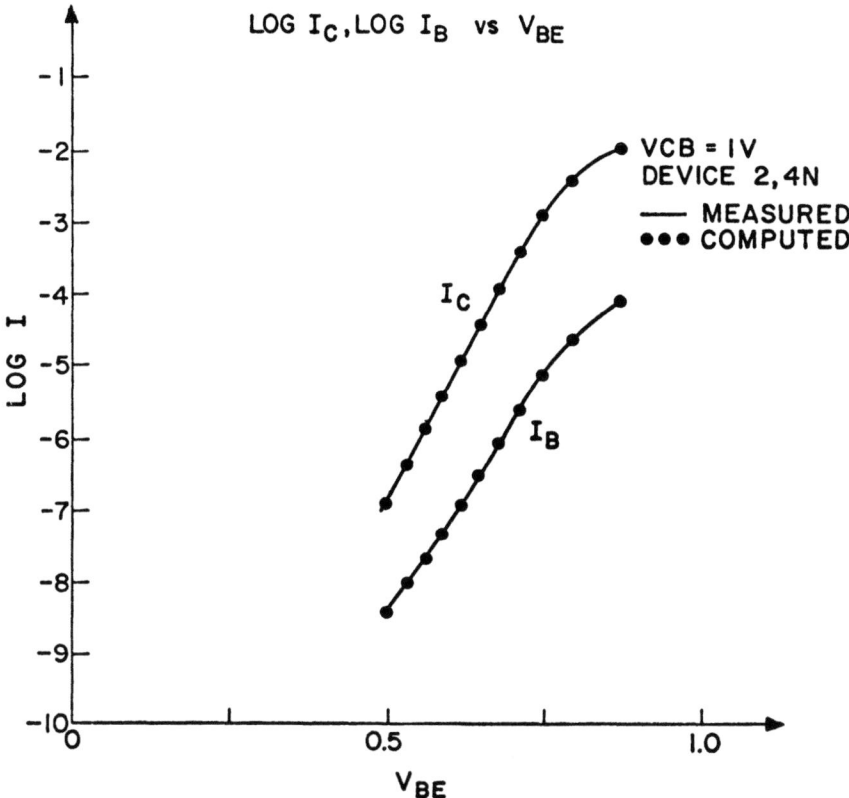

Fig. 7. Measured log I vs V_{BE} data and computed values.

---

* The low-current constraints can be improved with recently announced commercial instruments.

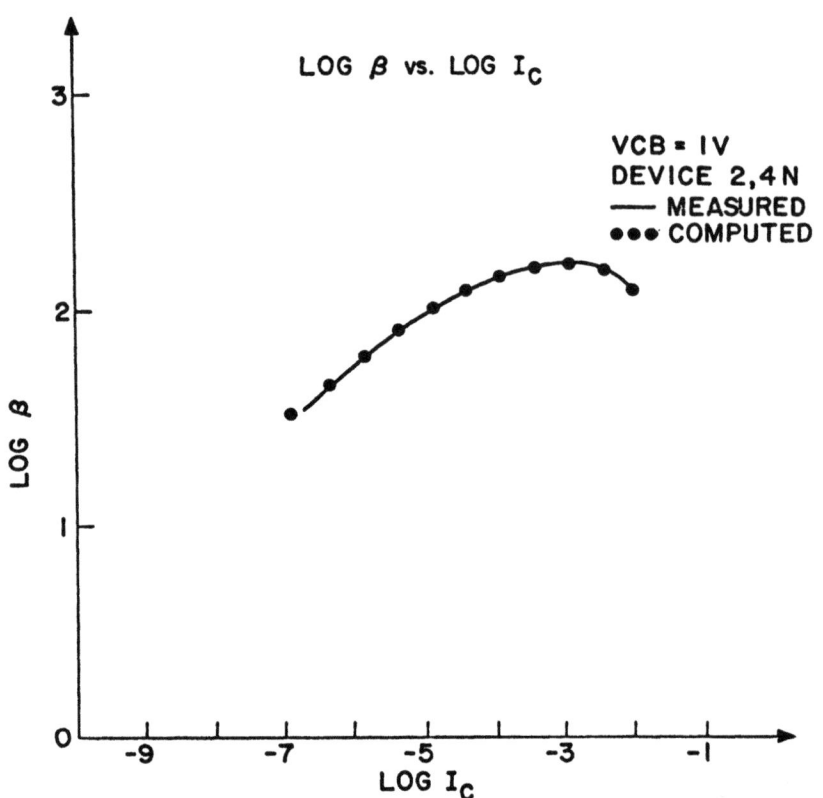

Fig. 8. Measured log β vs log I_C data and computed values.

The program for finding the Gummel-Poon parameters uses the above measured data. Since the measurements are taken starting from low currents and going towards high currents, the data are assumed to be ordered and no reordering is done.

The dc equations for I_C and I_B are given in Eqs. (20) and (21). The forward parameters can be determined from the equations if the measurements are taken at $V_{CB} = 0$. However, the last two terms in both the equations are negligible even at small reverse V_{CB}. Measurements on many devices show that these terms are of the order of tens of pA which is very small compared to the measured values of I_C and I_B. Thus, for the forward parameters, Eqs. (20) and (21) can be written as

$$I_C = \frac{I_S}{Q_B} e^{V_{BE}/V_T} \tag{26}$$

$$I_B = \frac{I_S}{\beta_{FM}} e^{V_{BE}/V_T} + C_2 I_S e^{V_{BE}/n_e V_T} \tag{27}$$

where $V_T = kT/q$. It is convenient to rewrite the equation for I_B as

$$I_B = I_3 e^{V_{BE}/V_T} + I_4 e^{V_{BE}/n_e V_T} \tag{28}$$

where

I_3, I_4, and n_e are the parameters to be determined.

In the low-current region, the second term in Eq. (28) which represents a nonideal component is dominant and, in the high-current range, the ideal component (the first term) dominates. Therefore, five points are taken in the low-current region and initial estimates of I_4 and n_e are obtained using a linear least-squares fit for these points. The equation used is

$$\log I_B = \frac{1}{n_e V_T} \cdot V_{BE} + \log I_4 \tag{29}$$

At this stage, the fit assumes all of I_B is due to the non-ideal term. Knowing I_C and I_B, the β vs I_C curve can be computed. The maximum value of β serves as a good initial estimate of β_{FM}. As explained later, a few points at low I_C can be used to get an initial guess for I_S. The starting value of I_3 can be computed from I_S and β_{FM}.

Typically, five points at low I_B and several points around the maximum β are used for the following iterations. Having estimated I_3, the ideal component of current can be calculated. Subtracting this from the total I_B, the nonideal fraction of I_B can now be separated for the points at low I_B. The estimates of I_4 and n_e are used to calculate a new nonideal component and in turn the ideal component can be easily computed using the points around the maximum β. The estimate I_3 is now corrected using nonlinear least-squares curve fitting. These iterations are continued until consistent values are obtained for the parameters I_3, I_4, and n_e.

At high values of I_B, the effect of R_B becomes important. Figure 9 shows I_B as a function of V_{BE} in this region. For a particular value of V_{BE}, the base current I_B' is calculated using the parameters extracted up to this point, assuming that all V_{BE} appears across the base-emitter junction. From Figure 9 it can be observed that

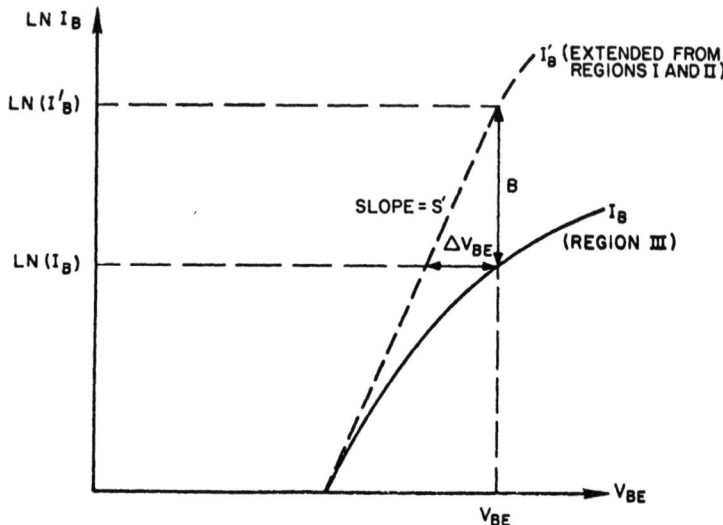

Fig. 9. Plots of log I_B vs V_{BE}, measured and simulated as used for extraction of R_B.

$$\Delta V_{BE} = \frac{B}{S'} \qquad (30)$$

Thus, knowing the value of I_B from the measurements, R_B can be determined from the relationship:

$$\Delta V_{BE} = I_B R_B \qquad (31)$$

Using this value of R_B, the measured values of V_{BE} can be corrected for the $I_B R_B$ drop. These corrected values can now be used to calculate I_3, I_4, and n_e. These new values in turn are used to find a new R_B. By continuing this iteration, consistent values for I_3, I_4, and n_e can be obtained. Taking these values as final, the final value of R_B is found.

All the four parameters from the I_B curve (viz. I_3, I_4, n_e, and R_B) are now known. Using this R_B, $I_B R_B$ correction is applied to all the measured values of V_{BE}.

Now the corrected I_C vs V_{BE} curve is known and there are two parameters to be determined -- I_S and I_k. At low-current levels, the effect of I_k is negligible. Five points are used in this current range to get an initial guess for I_S using linear least squares. The following equation is used:

$$\log I_C = \frac{1}{V_T} \cdot V_{BE} + \log I_S \tag{32}$$

Knowing I_S, the high-level injection approximation is used at the highest I_C to get a starting value of I_k,

$$I_C = \sqrt{I_S I_k} \; e^{V_{BE}/2V_T} \tag{33}$$

Using these starting values, a nonlinear least-squares fit is done for the entire I_C curve to obtain the final values of I_S and I_k.

From the values of I_S, I_3, and I_4, the parameters β_{FM} and C_2 can be easily computed. This gives the six forward dc parameters I_S, I_k, β_{FM}, C_2, n_e, and RB.

The comparison program uses the parameter values obtained from the above procedure to calculate the values of I_C and I_B for the measured values of V_{BE} using the theoretical Gummel-Poon equations. The measured and calculated curves for I_C, I_B, and β are plotted. This gives an indication of how good a fit the calculated parameters achieve. A similar procedure is used for reverse parameters. In this case we get another four dc parameters -- I_{kr}, β_{RM}, C_4, and n_c.

A typical device characterization process can be described as follows: the measurement program is used to get the desired I_C, I_B vs V_{BE} characteristics; from this data, the β vs I_C curve can be computed. Figures 7 and 8 are typical curves for a device in the forward region. These data are used in conjunction with the parameter determination program to get the Gummel-Poon parameters. Parameter values for the above device are listed in Table 2. The measured and theoretical curves are compared using the calculated Gummel-Poon parameters. Figures 7 and 8 show the comparison for the typical device in the forward region.

Table 2

Gummel-Poon Parameters Extracted from the Measured Data of Figure 7.

I_S = 5.1416 x 10^{-16} A

I_k = 16.95 mA

β_{FM} = 216.859

n_e = 1.5242

C_2 = 19.2442

R_B = 60.85 Ohms (average)

R_E = 5.476 Ohms

Comparison of measured and computed characteristics:

RMS error in β = 1.0722%
RMS error in I_B = 0.5792%
RMS error in I_C = 0.9953%

2.4 Parameter extraction summary

This section has defined the basic Gummel-Poon model equations. Measurement configurations and an algorithm for parameter extraction have been illustrated which give suitable model parameter resolution for IC statistical design. The next section illustrates the use of such data in developing a statistical design model.

3. DEVICE MODEL PARAMETER CORRELATIONS

3.1 Statistical models

In statistical modeling, it is necessary to simulate a large number of active devices like bipolar junction transistors manufactured from a particular fabrication process [6]. It is possible to obtain physically unrealistic combinations of model parameters if all the parameters are chosen randomly from their distributions. Therefore, such combinations will not represent physical effects of the practical fabrication process. As a simple example, consider the two parameters β_F and I_S where β_F is the forward beta and I_S is the intercept current for I_C. The scatter diagram for the measured values of these parameters is shown in Figure 10 (the devices and the parameter measurement are described in Section 3.2) along with the least-squares straight-line fit to this data.

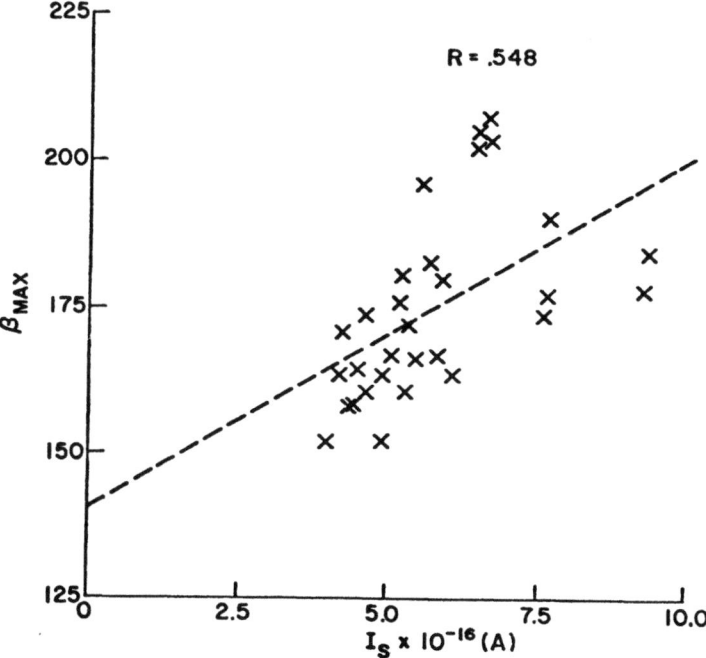

Fig. 10. Scatter diagram of β_F amd I_S.

The straight line has a positive slope due to the positive value of the correlation coefficient (correlation coefficient is explained in Section 3.3). If both parameters are selected at random, it is possible to obtain a maximum value of β_F for a minimum value of I_S which is physically unrealizable. In practice, a large number of parameters are involved and choosing them randomly assumes that no correlations exist. Therefore, to simulate the parameters with realistic correlations, it becomes necessary to find a minimum set of independent parameters from which the rest of the parameters are calculated taking into account their correlations with the independent parameters.

A preliminary study is now reported which uses a sample of devices from a single wafer to establish the basis for the choice of the independent parameters. The model parameters for the Gummel-Poon model are used. Some statistical data-reduction techniques are used to identify the underlying factors which aid in the search of independent parameters. They can also be used to relate the fabrication process parameters to the electrical parameters of the device.

A general description of the devices and the model used for this analysis is given. The "correlation coefficient" is then described. Next, its use in the first step of the statistical analysis is explained. Finally, the results from a statistical data-reduction procedure are described following a brief introduction of the method. The significance of these results is discussed.

3.2 Parameter measurement

In this analysis, the bipolar transistors available in a general-purpose IC chip called "kitchip" are used. This kitchip comprises a large number of transistors and other components and is fabricated in the Stanford Integrated Circuits Laboratory using a planar epitaxial process, the details of which are described elsewhere [7]. The kitchip has bipolar transistors with three different geometric configurations as shown in Figure 11. The "standard npn" transistors are used for the statistical study. However, the other area scaled devices were used to separate area-sensitive factors from profile-dominated effects. In addition, the large area device data most clearly illustrate correlations for BC junction capacitance since measurement errors are minimized.

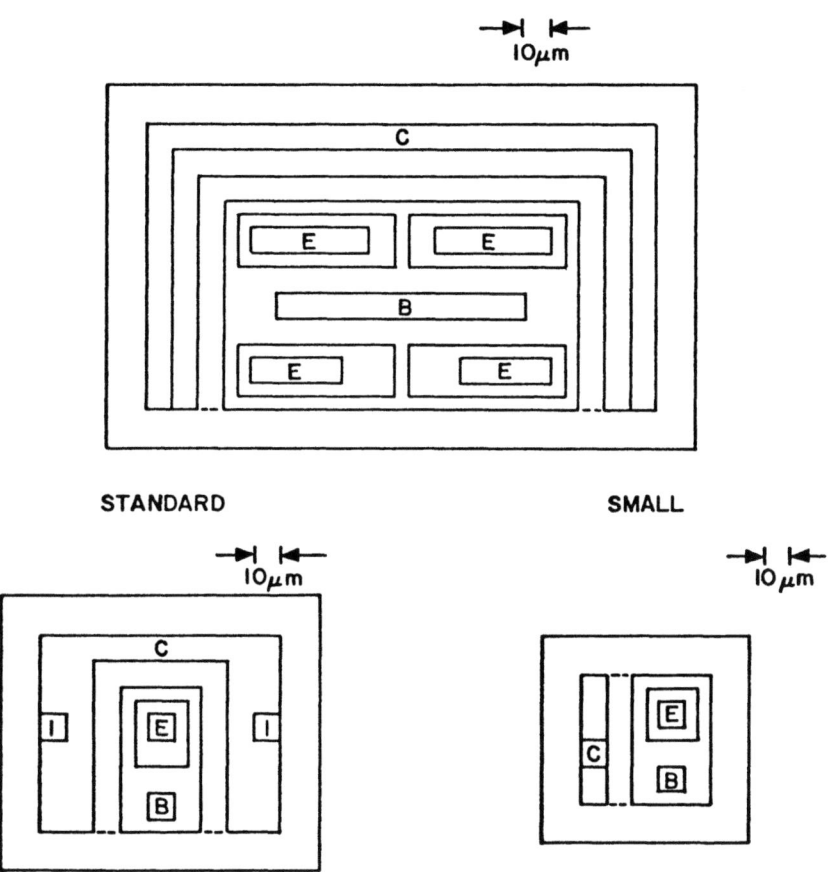

Fig. 11. Transistor geometries.

The model parameters for the Gummel-Poon model as implemented in SPICE [8] are used in this study. For quick referenc, the simplified dc Gummel-Poon model equations for the forward active mode of operation are shown in Figure 12 along with their graphical interpretation. The equation used to model the junction capacitance is also shown in Figure 12.

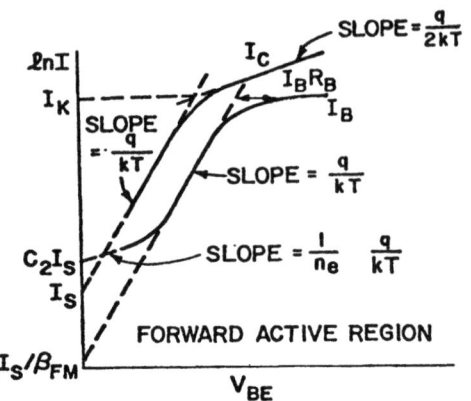

$$I_C = \frac{I_S}{Q_B} e^{V_{BE}/V_T}$$

$$I_B = \frac{I_S}{\beta_F}\left(e^{V_{BE}/V_T} - 1\right) + C_2 I_S \left(e^{V_{BE}/n_e V_T} - 1\right)$$

$$Q_B = \frac{1}{2}\left[Q_1 + \sqrt{Q_1^2 + 4Q_2}\right]$$

$$Q_1 = 1 + \frac{V_{BC}}{V_A} \qquad \text{Capacitance:} \quad C = \frac{C_0}{\left(1 - \frac{V}{\phi}\right)^m}$$

$$Q_2 = \frac{I_S}{I_K}\left(e^{V_{BE}/V_T} - 1\right) \qquad \text{where } V_T = \frac{kT}{q}$$

Fig. 12. Simplified Gummel-Poon model.

The DC parameters I_S, I_K, β_F, C_2, n_e, and R_B are derived from the measured I_C, I_B vs V_{BE} data. These measurements were made at a constant V_{CB} of 1 V and the emitter current was varied from 1 μA to 100 mA. Figure 13a shows the comparison of the measured values of I_C and I_B with their calculated values using the parameters extracted from the measurements. The fit is within ± 1%. β_R was similarly measured at V_{BE} = 1 V and in the collector current range of 1 μA to 100 mA. The Early voltage V_A was measured at constant V_{BE} and the remaining dc parameters R_C and R_E were measured using suitable measurement configurations [9]. The pinched base resistance R_{PB} was obtained from the linear portion of its voltage-current characteristics near the origin for a resistor 10 μm wide and 140 μm long. The junction capacitances were measured using a digital LCR meter over a range of reverse-bias voltages and the parameters C_0, ϕ, and m were extracted. Again, the calculated values agree with the measured values within ± 1%, as shown in

DC: Forward - I_S, I_K, β_F, C_2, n_e, R_B, V_A

Reverse - β_R

R_{PB}

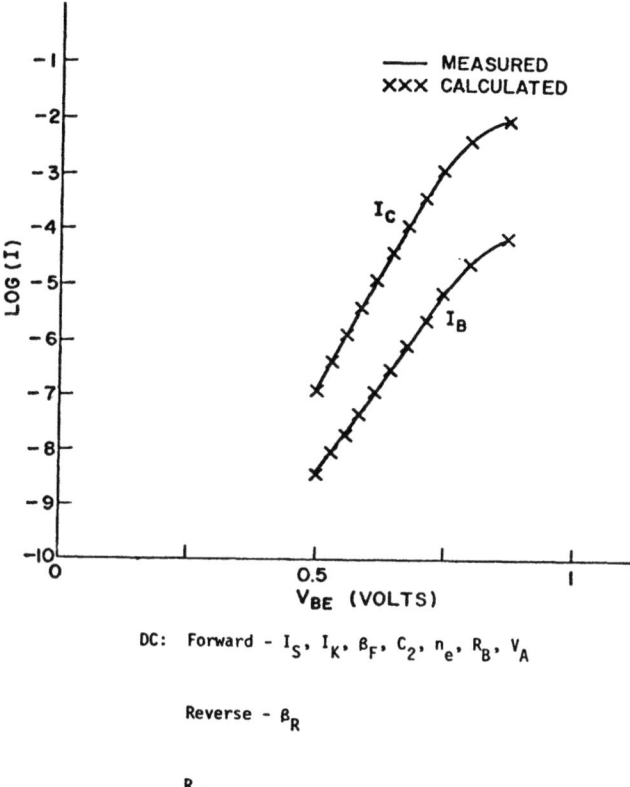

Fig. 13. Comparison of measured characteristics with the characteristics calculated from extracted parameters.
(a) I_C and I_B

Figure 13b. The f_T was calculated using the measured s-parameters. A constant V_{CE} of 3 V was used during these measurements, and the maximum value of f_T was determined after varying the collector current from 50 μA to 50 mA.

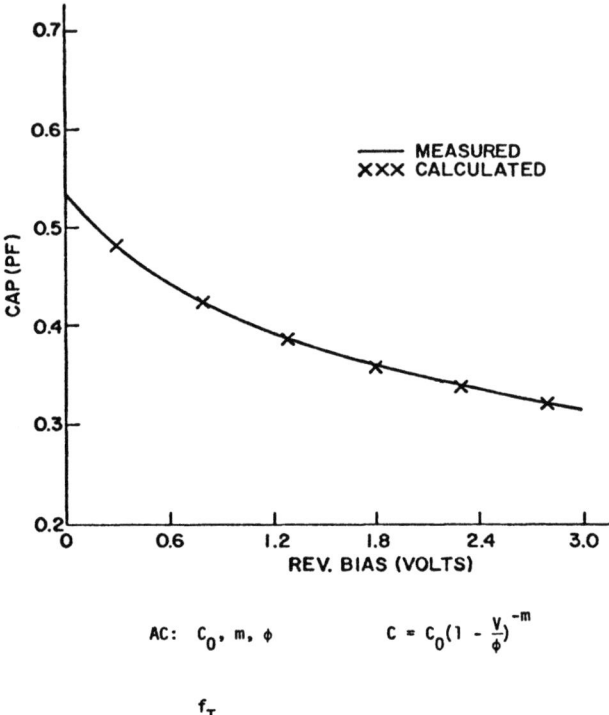

Fig. 13. Comparison of measured characteristics with the characteristics calculated from extrated parameters.
(b) Emitter-base capacitance

After extracting the model parameters for all the devices, the matrix of correlation coefficients described in the next section is computed.

3.3 Correlation coefficient

Correlation between any two parameters can be examined by observing their scatter plot, such as the one shown in Figure 10. With the exception of cases where correlation is very strong, it is often difficult to judge the degree of correlation by observing the scatter plot. This difficulty is overcome by computing a correlation coefficient [10] which gives a quantitative measure of the correlation. This is a dimensionless quantity ranging between -1 and +1. In the simplest form it assumes that the relationship between the two parameters is linear and the measure of this correlation is given by

$$R = \frac{\Sigma(X - \overline{X})(Y - \overline{Y})}{\sqrt{\Sigma(X - \overline{X})^2 \cdot \Sigma(Y - \overline{Y})^2}} \qquad (34)$$

In this expression, X and Y are the paired values of the two parameters and $\overline{X}$ and $\overline{Y}$ are their mean values, respectively. When the points on the scatter plot lie on a perfect straight line, the value of R is -1 or +1, depending on the slope of the line, indicating a perfect correlation. The value of R is zero if the points on the scatter plot lie on a circle. A zero value of R corresponds to absolutely no correlation. Intermediate values of R result when the points lie within an ellipse and their interpretation becomes complex. The correlation between the two parameters is said to be "moderate" if the value of R is around 0.5.

Though a value of zero for R indicates no corelation, it is possible to get nonzero values for R when two completely independent, and hence uncorrelated, sets of parameters are considered. This is due to the finite size of the sample as well as some errors in the measurement. Therefore, the concept of a "significance level" [10] is useful in order to decide whether the two parameters under consideration are really related to each other in the statistical sense. For a particular value of R, the significance level indicates whether this value of R really arises due to the relationship between the two parameters or it is just a random value resulting from some other causes. Significance level is a function of the number of observations, the desired probability, and the correlation coefficient itself. As a general guideline, the significance level for 35 observations can be calculated to be R = 0.28 for a probability of 0.1. Thus, if a correlation coefficient of 0.29 is obtained based on the measurements on 35 devices, then the probability that these two parameters are completely unrelated is less than 10% or, in other words, the probability that there indeed is some relationship between these two parameters is greater than 90%.

In practice, measured correlation coefficients differ from their ideal values of 0 or 1. The significance level becomes useful in such cases to interpret and judge the importance of the measured correlation coefficient.

3.4 Correlation results

In order to study the interdependencies of various parameters, the correlation matrix consisting of the linear correlation coefficients was computed. This matrix consists of correlation coefficients of each parameter with the remaining parameters. Some amount of caution must be exercised in interpreting this matrix. Physical relationships between various parameters have to be taken into account as well as the significance levels of the coefficients. For the data used in this analyis, the parameters had a sufficient spread to give a significance level of about 0.28.

The model parameters under consideration can be divided into two major subdivisions. The first group consists of parameters which depend strongly on collector doping. R_C, I_k, C_{OCB}, m_{CB}, and ϕ_{CB} comprise this group. The correlation matrix of these parameters for the "large"* transistor is shown in Figure 14. Only the upper

	R_C	I_K	C_{OCB}	m_{CB}	ϕ_{CB}
R_C	1.0	-.4	-.7	.45	.75
I_K		1.0	.47	-.21	-.6
C_{OCB}			1.0	-.67	-.56
m_{CB}				1.0	.59
ϕ_{CB}					1.0

Fig. 14. Correlation matrix of parameters depending on collector doping.

(*See next page)

triangle of the matrix is shown since a correlation matrix is symmetric. These parameters show good correlations among themselves but do not correlate well with the other parameters. Within this group, since each parameter shows good correlations with the remaining parameters, it is sufficient to choose only one independent parameter. R_C appears to be a suitable choice because it is easier to measure and depends directly on the collector doping. Thus for this group of parameters, R_C can be selected randomly and the remaining parameters can be computed from their correlations with R_C.

The second group of parameters depends on the doping levels in the emitter and base, which in turn are related to each other, as will be shown later. Figure 15 gives the correlation matrix for the parameters in this group. Out of these 13 parameters, β_F, β_R, R_{PB}, f_T, I_S, R_B and V_A depend directly on the integrated base doping. Parameters n_e, C_2, C_{OEB}, ϕ_{EB}, and m_{EB} depend on the properties of the emitter-base junction and its depletion layer which are a function of the integrated dopings in the base as well as the emitter. R_E depends on the integrated emitter doping but correlates with other parameters due to the fact that the integrated dopings in the base and the emitter depend strongly on each other. Only linear correlation coefficients are considered in this matrix. It is also possible to consider other nonlinear functions. For example, I_S depends on the integrated base doping whereas V_A depends on its reciprocal. Therefore, instead of considering the correlation coefficient between I_S and V_A, the coefficient between I_S and $1/V_A$ can be computed. But it was observed that this does not affect the magnitudes of the correlation coefficients as well as the results of the analysis described later to a considerable extent.

Due to the large number of parameters involved and the number of correlation coefficients which exceed the significance level, it is not possible to identify a few parameters as independent parameter by inspection. Therefore, a statistical data-reduction method called "Factor analysis" [11,12] is employed to understand the pattern in the observed data. A brief introductory description of factor analysis is given in the next section.

---

* The correlation coefficients for the "large" transistor are given since the measurement accuracy is poor for BC capacitance of the "standard" transistor due to its small value. The mean value of C_{OCB} is 0.37 pF for "standard" and 1.38 pF for "large" transistors, respectively.

	β_F	β_R	R_{PB}	f_T	I_S	R_E	n_e	R_B	V_A	C_2	C_{OEB}	ϕ_{EB}	m_{EB}
β_F	1.0	.73	.26	.34	.55	-.15	-.06	.81	-.6	.01	.22	-.13	.6
β_R		1.0	.76	.57	.9	.38	.32	.57	-.88	.51	-.24	.25	.66
R_{PB}			1.0	.56	.83	.8	.67	.2	-.7	.71	-.7	.62	.31
f_T				1.0	.6	.4	.44	.28	-.54	.43	-.55	.41	.28
I_S					1.0	.54	.58	.56	-.81	.77	-.44	.35	.59
R_E						1.0	.66	-.2	-.38	.64	-.78	.68	.04
n_e							1.0	.11	-.25	.82	-.72	.66	.04
R_B								1.0	-.42	.23	.15	-.22	.55
V_A									1.0	-.46	.23	-.2	-.56
C_2										1.0	-.6	.45	.26
C_{OEB}											1.0	-.7	-.07
ϕ_{EB}												1.0	-.21
m_{EB}													1.0

Fig. 15. Correlation matrix of parameters depending on emitter and base doping.

3.5 Factor analysis

Factor analysis attempts to explain observed relations among numerous variables in terms of simpler relations. Its single most distinctive characteristic is its data-reduction capability. Given an array of correlation coefficients for a set of variables, factor-analytic techniques enables the examination of some underlying pattern of relationships so that the data can be rearranged or reduced to a smaller set of factors or components that may be taken as source variable accounting for the observed interrelations in the data [11,12].

The starting point is the square symmetric correlation matrix between n variables. The process of factor analysis is designed to resolve this n x n correlation matrix into an n x k factor matrix, where the number of factors k is usually much smaller than n, the number of variables. This relation is shown in Figure 16 in a simplified form which shows that the factor matrix has to be the one which, when postmultiplied by its transpose, restores the correlation matrix. It is assumed that the observed variable is influenced by various determinants, only some of which are shared by other variables in the set. The part of the variable that is not shared is called "unique," while the part influenced by the shared determinants is called "common". Under this assumption,

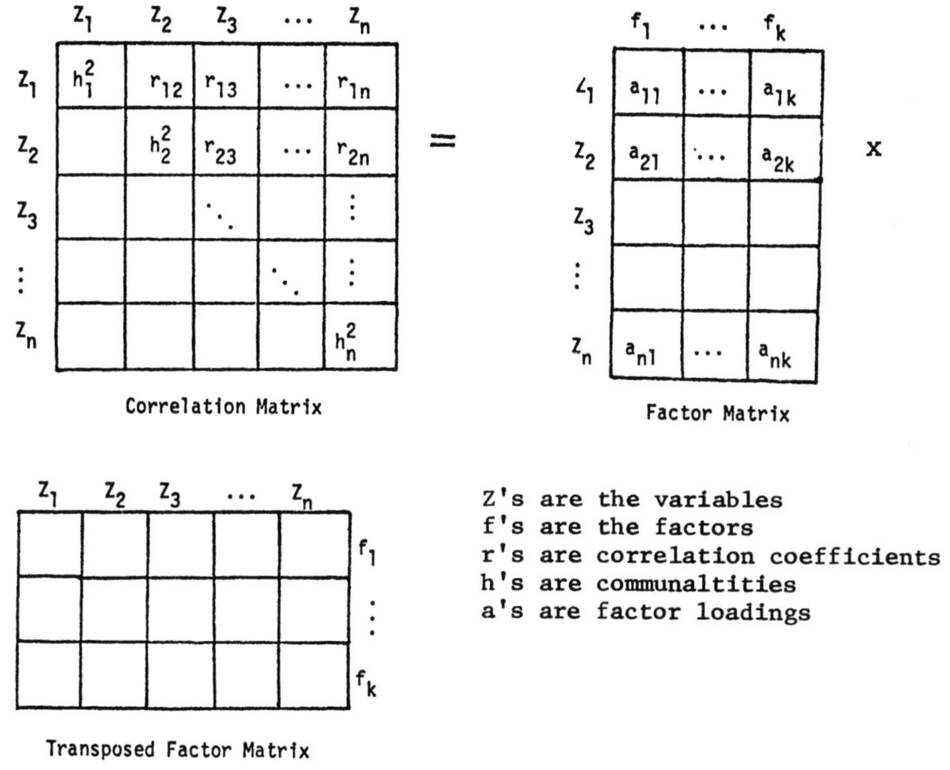

Z's are the variables
f's are the factors
r's are correlation coefficients
h's are communaltities
a's are factor loadings

Fig. 16. Factor analysis.

the unique part of a variable does not contribute to relationships among variables. It also follows from the preceding assumption that the observed correlations must be the result of the correlated variables sharing some of the common determinants.

The basic model, discussed so far, can be expressed compactly by the following equation:

$$z_j = a_{j1}F_1 + a_{j2}F_2 + \ldots a_{jk}F_k + d_j U_j \qquad j = 1, 2, \ldots, n \tag{35}$$

where z_j = variable j in standardized form

F_i = hypothetical factors obtained by taking linear combinations of the n variables

U_j = unique factor for variable j

a_{ji} = standardized multiple-regression coefficient of variable j on factor i called "factor loading"

d_j = standardized regression coefficient in variable j on unique factor j

The factors F_i are assumed orthogonal to each other.

The method of factor analysis proceeds as follows. Initially, the estimates of the "communalities" or the amount of the variance of the variable accounted for by the common factors together are substituted for the diagonal elements of the correlation matrix. After factoring this matrix, a new resultant set of communalities is obtained. These iterations are continued until consistent values are obtained for the communalities. This gives the "factor pattern" matrix consisting of the factor loadings a_{ji}. If a coefficient a_{ji} is large, then the variable j is said to load heavily on factor i. These factors are called "unrotated factors" and the variables tend to load heavily on more than one factor. This is indicated schematically in Figure 17. These factors are then rotated so that each variable is accounted for by as few factors as possible. The rotated factor loadings are conceptually simpler to interpret than the unrotated ones. After achieving the desired rotation, the factors are expressed in terms of the original variables by obtaining the "factor score coefficients":

$$F_i = fsc_{1i}z_1 + fsc_{2i}z_2 + \ldots + fsc_{ni}z_n$$

$$i = 1, 2, \ldots, k \tag{36}$$

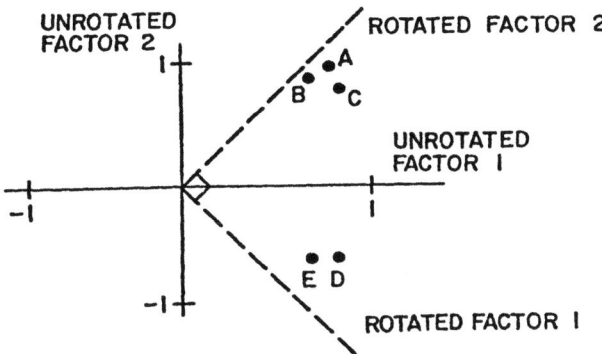

Fig. 17. Factor analysis.

where fsc_i are the factor score coefficients. The various factors can be given some interpretation by observing the factor score coefficients. Usually, for each factor, only a few of these coefficients will be high. A value less than about 0.25 is considered negligible for these coefficients. Thus, by starting with n variables, a large amount of variation in the data is explained by only k factors which in turn can be interpreted in terms of only a few original variables.

This method of factor analysis is applied to the measured device parameters and the results are discussed in the next section.

3.6 Factor-analysis results

The results of a factor analysis performed on the 13 measured parameters are summarized in Figure 18. The communalities of all the parameters are quite high (the maximum possible value is 1), indicating that a large amount of the variance is accounted for by the common factors. Two factors explain about 75% of the variation in the measured data. From the rotated factor matrix, it can be seen that most of the parameters load heavily on either one factor or the other. For example, β_F loads heavily on the second factor whereas n_e loads heavily on the first. The factor score coefficients can be used to interpret these factors. β_F and I_S contribute significantly to the first factor with a slightly smaller contribution from R_{PB}. The main contributors to the second factor are I_S and β_F. Thus I_S, β_F, and R_{PB} emerge as the important parameters which account for a large variation in the data. Since I_S and R_{PB} show extremely good correlation (correlation coefficient of 0.83), only I_S and β_F need be considered as the controlling parameters.

I_S and β_F are given by the following first-order expressions:

$$I_S = \frac{q\, A_e\, n_i^2}{GU_B} \quad \text{and} \quad \beta_F = \frac{GU_E}{GU_B} \tag{37}$$

where q = magnitude of electronic charge

n_i = intrinsic carrier concentration

A_e = area of emitter-base junction

and

$$GU_B = \int_{\text{base}} \frac{N_A(x)}{D_n}\, dx$$

		Varimax Rotated Factor Matrix		Factor Score Coefficients	
Variable	Communality	Factor 1	Factor 2	Factor 1	Factor 2
B_F	0.84919	-0.19441	0.90078	-0.41657	0.73176
I_S	0.97396	0.58512	0.79473	0.46071	1.02441
B_R	0.92962	0.37133	0.88979	-0.06316	0.31098
R_{PB}	0.88897	0.82592	0.45477	0.34406	-0.38530
R_B	0.58364	-0.10090	0.75727	0.03518	-0.38729
C_2	0.63912	0.73307	0.31896	-0.20220	-0.05936
n_e	0.67575	0.81208	0.12758	0.18974	0.01756
R_E	0.74614	0.86345	0.02448	0.13847	-0.27596
v_A	0.77172	-0.33588	-0.81173	0.04999	0.03341
f_T	0.40636	0.46403	0.43707	0.03523	-0.12817
C_{OEB}	0.75137	-0.86646	0.02495	-0.15841	-0.34843
m_{EB}	0.52186	0.06999	0.71900	-0.01874	-0.24472
ϕ_{EB}	0.53441	0.72542	-0.09042	0.12258	-0.29329

Factor	Eigenvalue	Pct. of Var.	Cum. Pct.
1	6.59371	50.7	50.7
2	3.20481	24.7	75.4

Fig. 18. Results of factor analysis using measured parameters.

$$GU_E = \int_{\text{emitter}} \frac{N_D(x)}{D_P}$$

N_A, N_D = acceptor and donor concentrations

D_n, D_P = electron and hole diffusion constants

x = distance

GU_B and GU_E are the Gummel numbers in the base and emitter, respectively, which are the measures of the total charge in these regions. Thus, factor analysis suggests that the variation in measured parameters is due to the variations in GU_B, GU_E, and A_e. To isolate the variation caused by the emitter junction area A_e, correlation of I_S between devices located on the same chip but having different emitter areas is considered, as shown in Figure 19. A correlation coefficient of R indicates that the proportion of variance explained because of a common factor is R^2. Therefore, it can be concluded that about 87% of the variance in I_S is explained by the base Gummel number since that is the common factor in this case. Thus, most of the observed variation in I_S is due to GU_B and not due to the variation in A_E. Hence, it can be concluded from the results of the factor analysis that most of the variations in the measured parameters are controlled by GU_E and GU_B. This agrees with the physical interpretation since all the parameters are related to the doping levels in the emitter and base.

The two Gummel numbers GU_E and GU_B depend on each other very strongly because of the effects of emitter push. The depth of the emitter-base junction also affects the depth of the base-collector junction, thus affecting the total doping in the base. GU_B and GU_E are estimated from I_S and β_F and their correlation coefficient is computed. A value of 0.94 is obtained for this coefficient, showing the strong dependence of the two Gummel numbers. Hence, only one of them needs to be considered as an independent variable. Based on these considerations, the parameter I_S suggests itself as a choice for the independent parameter. I_S depends only on GU_B unlike β_F which depends on both GU_E as well as GU_B. Also, most of the measured parameters depend directly on the GU_B to varying degrees. It is easier to compute I_S from the measurement of I_C for a couple of values of V_{BE}, whereas measurements need to be taken over a wide current range to compute β_F.

Thus, for the 13 measured parameters which depend on base and emitter doping, about 75% of the variation is due to common underlying factors and can be accounted for by choosing I_S as the independent parameter.

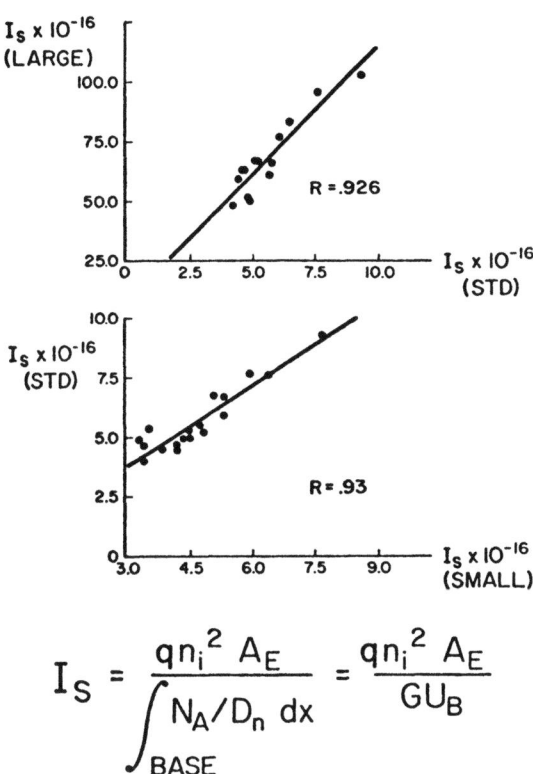

Fig. 19. Scatter diagrams of I_S for area scaled devices.

3.7 Discussion

Using factor analysis, I_S was chosen as the independent parameter for the set of 13 measured parameters. In order to verify this choice, linear regression equations were formed from the measured data for each of these parameters with I_S as the independent variable. These equations are summarized below:

$$\beta_F = 5.98\ I_S + 139.99$$

$$\beta_R = 0.19\ I_S + 1.2$$

$$R_{PB} = 26.39\ I_S + 23.19 \quad (K\Omega)$$

$$R_B = 45.27\ I_S + 759.85 \quad (\Omega)$$

$$C_2 = 30.63\ I_S - 136.85$$

$$n_e = 0.06\ I_S + 1.2$$

$$R_E = 0.28\ I_S + 3.86 \quad (\Omega)$$

$$V_A = -10.95\ I_S + 106.53 \quad (V)$$

$$f_T = 15.87\ I_S + 582.67 \quad (MHz)$$

$$C_{OEB} = -0.01\ I_S + 0.5933 \quad (pF)$$

$$m_{EB} = 0.01\ I_S + 0.31$$

$$\phi_{EB} = 0.01\ I_S + 0.71 \quad (V)$$

where I_S is a multiple of 10^{-16}.

A device from the test wafer was selected and completely characterized. With its measured value of I_S as the starting point, the remaining parameters were calculated using the regression equations. These calculated values were then compared with the measured values as shown in Figure 20. Most of the calculated parameters were within about 10 to 15% of their corresponding measured parameters. Therefore, these parameters should model the device characteristics with reasonable accuracy.

3.8 Parameter correlation summary

The measured device parameters can be divided into two groups. The parameters in the first group depend on the collector doping, whereas the parameters in the second group depend on base or emitter doping. In order to produce a proper statistical model,

Parameter	Calculated	Measured	% Error
I_S		5.14×10^{-16}	
β_F	170.73	167.0	2.23
β_R	2.13	2.2	-1.1
R_{PB} (k-ohms)	158.84	176.64	-10.1
R_B (ohms)	78.1	61.79	26.3
C_2	20.59	19.24	7.0
n_e	1.51	1.52	-0.76
R_E (ohms)	5.3	5.48	-3.3
V_A (volts)	50.25	53.84	-6.67
f_T (MHz)	664.24	660.47	0.57
C_{OEB} (pf)	0.54	0.53	1.89
m_{EB}	0.36	0.359	0.67
ϕ_{EB} (volts)	0.76	0.8	-5.0

Fig. 20. Comparison of measured parameters with parameters computed from I_S.

it is necessary to choose one parameter from each group at random and then the remaining parameters should be computed making use of the correlations. A large number of parameters can be effectively controlled by choosing I_S as the independent variable.

It may also be possible to use this approach to speed up the characterization of the devices. After making all the measurements on some of the devices, the correlations between the parameters can be established. Then for the remaining devices from the same batch, it is only necessary to measure I_S, which is very easy to compute from one or two measurements of collector current with a low forward bias on the emitter-base junction. Once I_S is known, the remaining parameters can be calculated. These calculated parameters will model the device characteristics with reasonable accuracy.

4. CORRELATION OF ELECTRICAL AND FABRICATION PROCESS PARAMETERS

This final section draws on the data presented in Section 3 as well as computer simulation programs discussed elsewhere [13,14,15]. The results of Section 3 rest substantially on empirical data. In this section, the data will be examined with the intent of evolving methods to generate the same data directly from process specifications. To achieve this objective, two simulation tools are required and corroborative measurements of both impurity profiles and electrical device statistics. The simulation tools will be briefly described. These programs are SUPREM and mini-SITCAP. The first program simulates impurity profiles directly from process specifications for a multistep process [13]. The mini-SITCAP program [14,15] uses impurity profiles and device geometry information to calculate device electrical parameters. The final section compares measured profile and device statistics with those generated using SUPREM and mini-SITCAP.

The spectrum of applications for process-oriented device simulation is listed in Figure 21. The circuit designer has greatest interest in the right-most functions and the device and technology specialist might focus on the left-most functions. In the following sections, the basic simulation tools are described as well as their application for studying device parameters correlations.

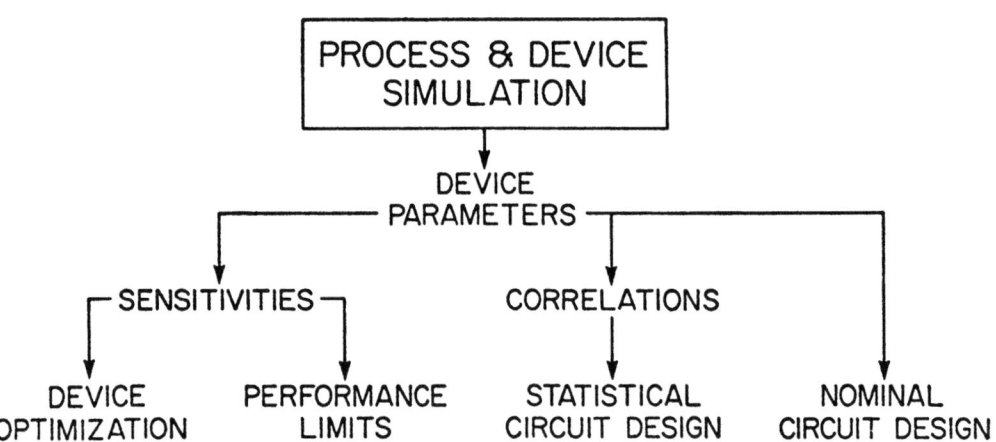

Fig. 21. Applications of process and device simulations.

4.1 PRocess Engineering Models (SUPREM)

The first step in device parameter calculation is to accurately predict device impurity distributions resulting from a given fabrication process. For process simulation, we have implemented process engineering models into a program we call SUPREM -- Stanford University's program for IC PRocess Engineering Models. The program accepts fabrication information identical to run-sheet specifications and outputs impurity distributions. Figure 22 illustrates an input and output for a boron deposition followed by a drive-in sequence with oxide redistribution. The figure also shows measured data generated using spreading resistance [16,17]. The simulated results represent a numerical solution of a moving boundary oxidation which transfers 70% of the boron into the oxide. Details of this algorithm and program documentation are given elsehwere [13, 18].

Each "step" specification in Figure 22 has an associated set of "model" specifications. These models define the details of the process kinetics used for simulation. For example, DIF contains diffusion data for boron, and STM contains oxidation and impurity segregation data for a steam environment. Figure 22 references the default models denoted by the zero suffixes. The user can input other parameters by including model cards with typical parameters, as outlined in the Appendix. A more complete description is given in the user's guide [18] and in a following discussion.

Figure 22 represents only a portion of a complete SUPREM input file. In addition to moving boundary oxidation and diffusion, the program can simulate epitaxial growth, ion implantation, and several other conditions for impurity deposition or etching. The program uses moving boundary solutions for oxidation and epitaxy. In addition, the nonuniform grid features allow for user-specified resolution so that detailed information can be simulated throughout the structure, even with multiple layers. At present, the program is limited to one dimension although two-dimensional simulation is planned.

INPUT--

    COMM BASE PREDEP
    STEP   TYPE = DIFF, TEMP = 950, TIME = 30, ELEM = B, MODE = DIFO

    COMM BASE DRIVE-IN
    STEP   TYPE = OXID, TEMP = 1000, TIME = 45, MODE = DRY1

    STEP   TYPE = OXID, TEMP = 1000, TIME = 60, MODE = WET1

    STEP   TYPE = OXID, TEMP = 1000, TIME = 5, MODE = DRY2

    STEP   TYPE = DRIVE, TEMP = 1000, TIME = 10, MODE = DIF1

Fig. 22. SUPREM input and output for boron predeposition and drive-in.

4.2 Process-oriented device simulation

The use of SUPREM for complete process modeling has been tested for a standard double-diffused epitaxial bipolar process. Simulated and measured impurity profiles are shown in Figure 23 for a structure which includes a buried subcollector. To study the effects of process perturbations on device parameters, a simulation program was used to calculate directly electrical parameters using the data shown in Figure 23. Mini-SITCAP [15] accepts vertical profile and surface geometry information as inputs for parameter calculation of bipolar devices. The program outputs electrical device parameters. Figure 23 shows schematically the data flow and output set of model parameters. The output format shown in Figure 23 corresponds to the Gummel-Poon model as implemented in SPICE [8] and described in Section 2. Table 2 shows a comparison of several SUPREM/SITCAP calculated parameters compared with measurements for the nominal case process and device design.

The mini-SITCAP program implements standard engineering approximations for device design and scaling, along with a detailed one-dimensional numerical solution in the neutral and space-change regions [14]. The significant features which make the program well suited for application in this study are input compatibility with SUPREM and outputs which can be compared directly with measured device parameters and distributions. The next section describes results of a study which applies SUPREM and SITCAP.

4.3 Device parameter distributions and correlations

The application of mini-SITCAP in calculating a nominal set of device parameters has previously been reported [14,15]. Inaccuracies and ambiguities become apparent in these simulations as a result of over-simplification in specifying impurity distributions and in not accounting for the spread in measured device parameters. To overcome these limitations, results of additional efforts are discussed in this section. First, SUPREM profile simulation is now used as input for the device simulation. This capability allows current simulation of oxide redistribution of boron and buried layer out-diffusion. Second, profile and device statistics have been studied to identify the controlling effects which can enable effective modeling of parameter distributions.

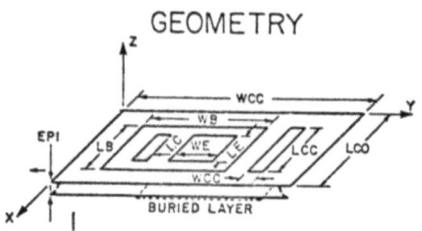
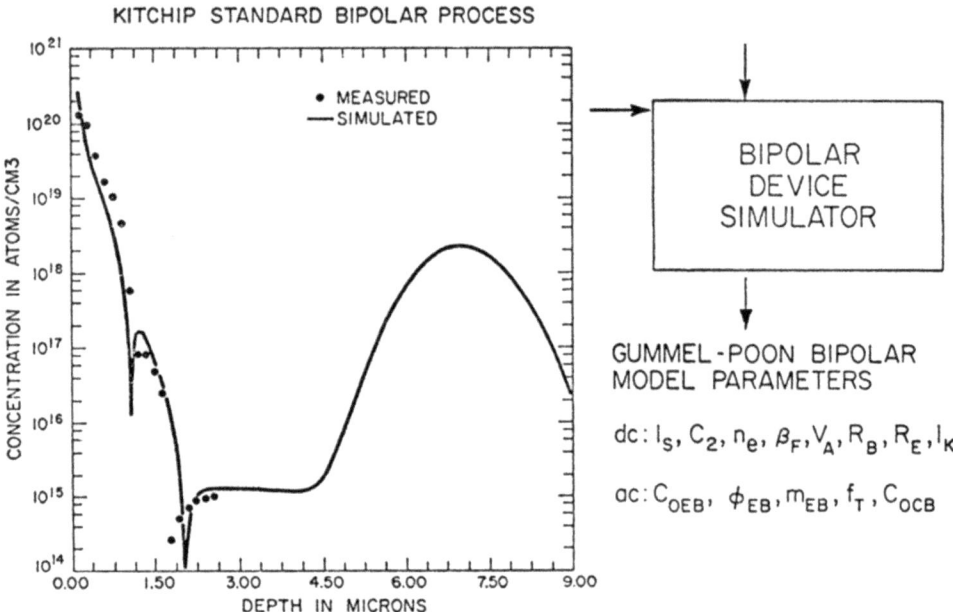

Fig. 23. Input and output parameters for SITCAP.

The interface of SUPREM and mini-SITCAP has made it possible to improve the programs independently. This separates the improvements in technology modeling from specific device simulation capabilities. Although SUPREM presently has rudimentary models for anomalous diffusion and coupled species -- for example, As and As-B -- further developments will soon upgrade these models [18]. The device simulator can similarly be adapted to a specific set of technological and topological constraints.

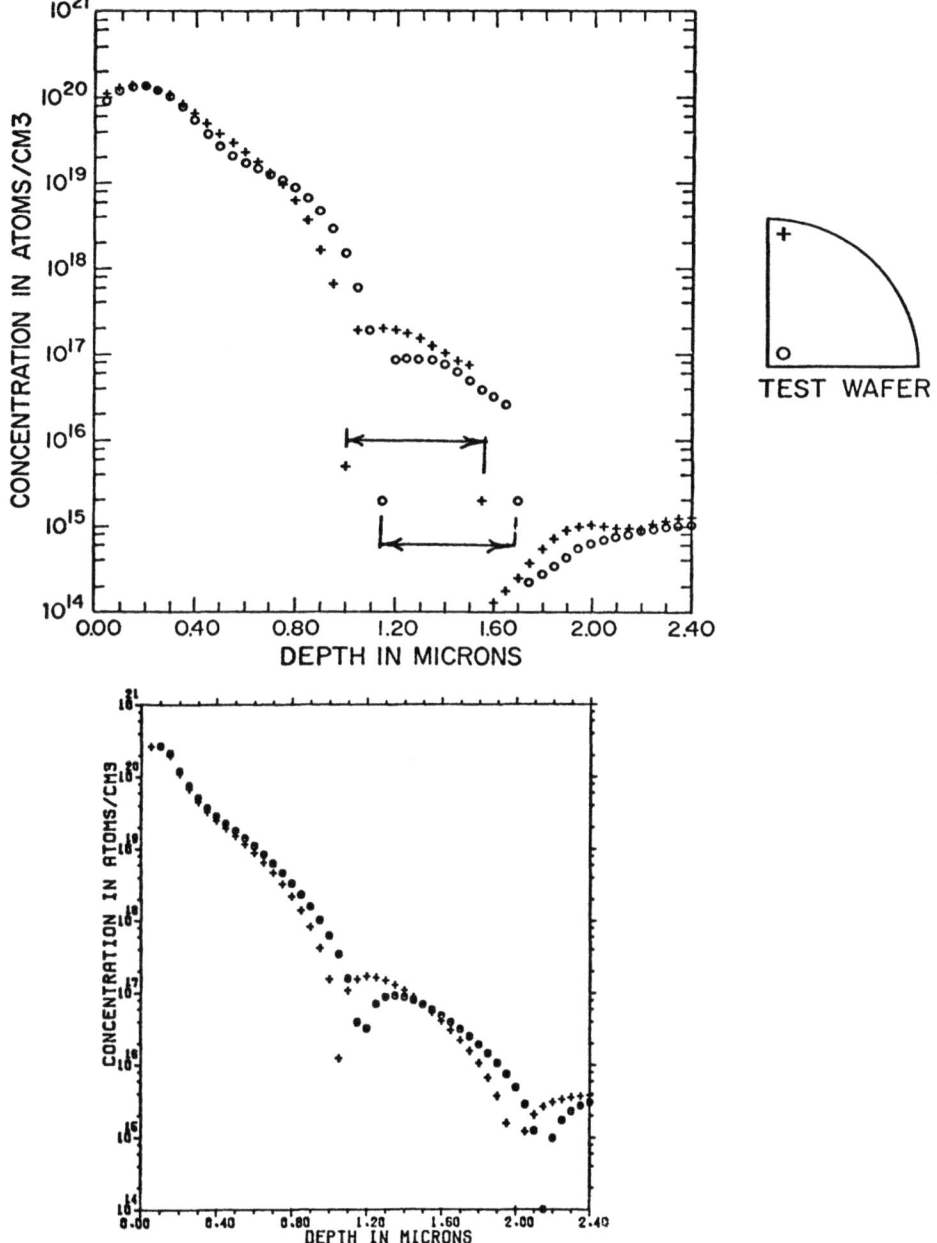

Fig. 24. (a) Measured impurity profiles across the test wafer.
(b) Corresponding impurity profiles used as SITCAP input to compute electrical parameters.

The SUPREM and SITCAP programs have been used to study the effects which control parameter distributions and correlation between parameters for a bipolar process. Parameter characterization was discussed in Section 3 for a wafer with test devices and a process-control wafer for the same fabrication sequence. Bipolar devices of three different emitter areas were characterized for die on a two-inch wafer. The complete set of SPICE Gummel-Poon model parameters [8] was obtained. Figure 24a shows the profile spread for emitter and base diffusions measured using spreading resistance. Figure 24b shows the corresponding simulated profiles used as SITCAP input. The process parameter variations required to achieve

Table 2.

Parameter	SUPREM/SITCAP	Measured
I_S(A)	7.15×10^{-16}	7.64×10^{-16}
β_F	184.00	174.00
$R_C(\Omega)$	34.50	47.83
V_A(V)	20.20	17.67
C_{OCB}(PF)	0.391	0.36
ϕ_{CB}(V)	0.57	0.51
C_{OEB}(PF)	0.50	0.52
ϕ_{EB}(V)	0.721	0.77
τ_F(PSEC)	131.00	179.00

Comparison of simulated and measured electrical parameters.

the simulated results and the calculated variation in junction depths and integrated dopings are shown in Table 3. The second column gives the corresponding variations in the measured parameters. The physical parameters associated with temperature, concentration, and diffusion are consistent with observed junction-depth variations. For this process a phosphorus emitter diffusion was used. The deposition was operated near the knee of solid solubility so that gradients across the wafer were nonnegligible. In addition, the emitter push was found to be a dominant factor in controlling both base and emitter parameters. As a result, the emitter and base Gummel numbers, GU_E and GU_B, are strongly correlated. The profiles were simulated by user-specified alterations in diffusion coefficients, accounting for cooperative diffusion of phosphorus and boron in a manner similar to that given by Nakamura [19]. An improved species-coupling is presently being incorporated directly into SUPREM.

The electrical device parameters have been studied to identify dominant trends controlling parameter distributions and correlations. Figure 25a shows the measured distribution of I_S across the wafer for the standard emitter area device. Using the simulated profile spread from Figure 25b and SITCAP simulation, the dashed upper and lower bounds on Figure 25a were generated. The result indicates that, using simulated profiles which bound the measured profiles, the electrical parameter I_S can be correctly bracketed. The cause of the distribution in I_S is clarified by considering the scatter plots for the I_S values for the area scaled devices given in Section 3. The data are replotted in Figure 25b and 25c. The coefficient R indicated the degree of correlation as described in Section 3. The plots show that I_S values are highly correlated for area scaled devices. Using the relationship for I_S given by

$$I_S = \frac{qA_e n_i^2}{\int_{base} N_A/D_n \, dx} = \frac{qA_e n_i^2}{GU_B}$$

and the correlation coefficient for Figure 25b and 25c, it follows that 87% of the observed variations can be accounted for from variations in GU_B. The dependence of other parameters on I_S and the base Gummel number will now be demonstrated.

As stated earlier, the interdependence of base and emitter profiles is substantial. This is demonstrated by considering the parametric dependence for beta. First-order theory predicts that maximum beta is controlled by the following proportionality:

Table 3.

Parameter	Variation for Simulation	Measured Variation	
Temperatures	± 2°C		
Surface Concentrations	± 16%		
Diffusion Coefficients	± 16%		
XJEB	± 5.41%	± 7%	Spreading resistance measurements on the test wafer
XJBC	± 3.86%	± 4.62%	
Emitter Sheet Resistance	± 2.95%	± 3.4%	
Pinched Base Sheet Resistance	±19.14%	± 29%	
I_S	±36.8%	± 40%	Device measurements on test chip
C_{OEB}	± 9.2%	± 8.3%	

Comparison of simulated and measured parameter variations.

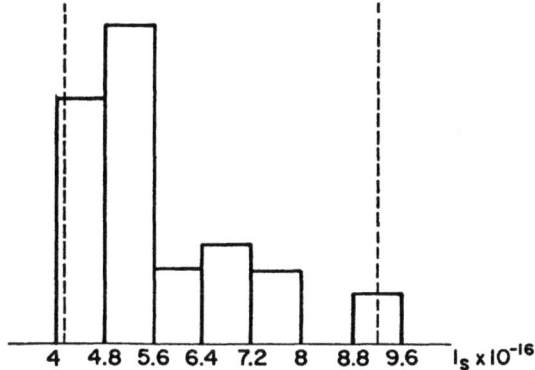
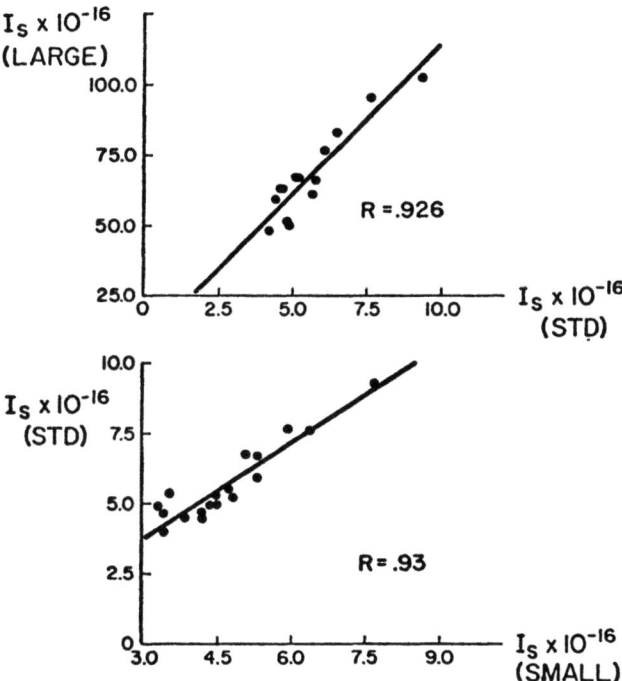

Fig. 25. (a) Histogram of measured I_S and limits generated by profiles of Fig. 4 (b).
(b), (c) Scatter plots of I_S for these three area scaled transistors.

$$\beta_F \propto \frac{GU_E}{GU_B} \qquad \text{where } GU_E = \int_{\text{emitter}} N_D/D_P \, dx$$

Figure 26 shows the measured scatter plot for β_F vs I_S. The correlation coefficient of 0.55 is well above the significance level of 0.28. Using $1/I_S$ to estimate GU_B and β_F/I_S to estimate GU_E, a correlation coefficient of 0.94 is obtained for GU_B and GU_E. This strong coupling of GU_E and GU_B suggests that many emitter-base transistor parameters should be correlated with I_S because of its dependence on GU_B. Results of a factor analysis of all correlations for the forward-mode Gummel-Poon parameters show that 75% of the observed variations can be represented by a linear combination of two factors [20]. Each factor is predominately controlled by I_S and β_F. As a result, Table 4 gives the set of parameter specifications obtained by applying linear regression to the measured parameters in terms of a single independent parameter I_S. I_S is an appropriate independent parameter based on the results of factor analysis and is a readily measurable quantity.

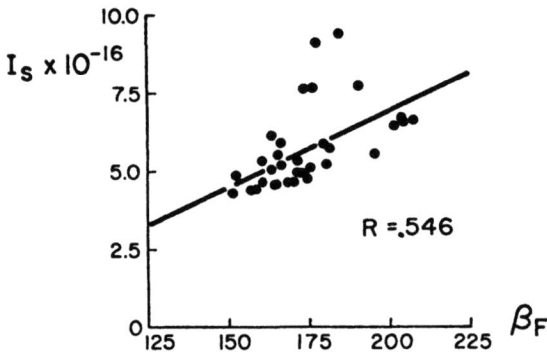

Fig. 26. Measured scatter plot on β_F and I_S.

Table 4.

β_F	$= 5.98\ I_S + 139.99$
β_R	$= 0.19\ I_S + 1.2$
R_{PB}	$= 26.39\ I_S + 23.19$
R_B	$= 45.27\ I_S + 759.85$
C_2	$= 30.63\ I_S + 136.85$
n_e	$= 0.06\ I_S + 1.2$
R_E	$= 0.28\ I_S + 3.86$
V_A	$= -10.95\ I_S + 106.53$
f_T	$= 15.87\ I_S + 582.67$
C_{OEB}	$= -0.01\ I_S + 0.5933$
M_{EB}	$= 0.01\ I_S + 0.31$
ϕ_{EB}	$= 0.01\ I_S + 0.71$

(I_S is a multiple of 10^{-16})

Linear regression equations for device parameters with I_S as the independent parameter.

4.4 Electrical and process parameter correlation summary

The results presented in this section indicate that process modeling and device simulation can help to identify process steps which control device parameters. In addition, the statistical analysis has identified key factors controlling parameter correlations. The strong dependence of model parameters for this process on GU_B and GU_E was shown using factor analysis. The correlation of β_F and I_S confirms process-simulation results which indicate the strong coupling of emitter and base profiles. Further developments in coupled-species modeling in SUPREM will make possible direct prediction of parameter correlations from the process description data.

Capabilities of computer tools for process modeling and device simulation have been described in this section. Using SUPREM, critical processing steps have been identified for a bipolar process, and SITCAP simulation correctly brackets the observed distribution of I_S. A simplified model for forward-mode device parameters has been constructed using only I_S as the independent parameter.

5. REFERENCES

1. H.K. Gummel and H.C. Poon, The Bell System Tech. J., May-Jun 1970, pp. 827-852.

2. J.J. Ebers and J.L. Moll, Proc. IRE (42), 12, Dec 1954, pp. 1761-1772.

3. J. Logan, The Bell System Tech J. (50), 4, Apr 1971, pp. 1106-1147.

4. W.J. McCalla and W.G. Howard, IEEE J. of Solid-State Circuits, Feb 1971, pp. 15-19.

5. H.K. Gummel, The Bell System Tech. J., Jan 1970, pp. 115-120.

6. P. Balaban and J.J. Golembeski, IEEE Trans. on Circuits and Systems (CAS-22), 2, Feb 1975, pp. 100-108.

7. S.R. Combs and J.D. Meindl, paper presented at 3rd International Symposium on Biotelemetry, Asilomar, California, May 1976.

8. L.W. Nagel and D.O. Pederson, paper presented at 19th Midwest Symposium on Circuit Theory, Waterloo, Ontario, 12 Apr 1973.

9. I. Getreu, Electronics (47), 23, Nov 1974, pp. 137-143.

10. G.W. Snedecor and W.G. Cochran, Statistical Methods (sixth edition), Iowa State University Press 1967.

11. E.H. Bryant and W.R. Atchley (eds), Multivariate Statistical Methods: Within Group Covariation, Dowden, Hutchinson and Ross, Inc., 1975.

12. N.H. Nie, C.H. Hull, J.G. Jenkins, K. Steinbrenner, and D.H. Bent, Statistical Package for the Social Sciences (second edition), McGraw-Hill Book Co., 1975.

13. R.W. Dutton, A.G. Gonzalez, R.D. Rung, and D.A. Antoniadis, paper presented at Electrochem. Soc. Meeting, Philadelphia, May 1977.

14. H.J. De Man and R. Mertens, Tham 9.2, ISSCC, Feb 1973, pp. 104, 105, 205.

15. A.G. Gonzalez, S.R. Combs, R.W. Gill, and R.W. Dutton, Paper D-5268, International Electronics Convention, 25-29 Aug 1975, Sydney, Australia.

16. D.C. D'Avanzo and R.W. Dutton, <u>Proc. of Electrochem. Soc.</u> (76-2), Las Vegas, Oct 1976, pp. 769-771.

17. D.C. D'Avanzo, D.C. Rung, and R.W. Dutton, "Spreading Resistance for Impurity Profiles," TR No. 5013-2, Stanford Electronics Laboratories, Stanford University, Stanford, California, Feb 1977.

18. S.E. Hansen, D.A. Antoniadis, A.G. Gonzalez, M. Rodoni, and R.W. Dutton, "SUPREM I - A Program for IC Process Engineering Models," SU SEL 77-006, Stanford Electronics Laboratories, Stanford University, Stanford, California, 1977.

19. H. Nakamura, S. Ohyama, and C. Tadachi, <u>J. Electrochem. Soc.</u> (121), <u>10</u>, Oct 1974, pp. 1377-1381.

20. D.A. Divekar, R.W. Dutton, and W.J. McCalla, paper presented at IEEE Conf. on Circuits and Systems, Apr 1977.

21. B.L. Crowder, J.F. Ziegler, and F.F. Morehead, "The Application of Ion Implantation to the Study of Diffusion of Boron in Silicon," in <u>Ion Implantation in Semiconductors and Other Materials</u> (edited by B.L. Crowder), Plenum Press, New York, 1973, pp. 267-274.

APPENDIX

Although for most uses the internal SUPREM data (default values) should be adequate, it is possible to alter those values to externally furnished ones by means of "model" data. To illustrate this point and to demonstrate the basic physical models used in the present version of SUPREM, a complete description for a steam oxidation of a boron-doped sample of silicon is

  STEP TYPE=OXID, TEMP=1100, TIME=15, MODE=SIM2, MODE=DIF3

  MODEL NAME=SIM2, LRTE=0.1, PRTE=0.01

  MODEL NAME=DIF3, ELEM=B, OXDC=3.1E8, OXEA=4.06
                          SIDC=1.9E8, SIEA=3.02, NI=1.4E19,
                                                 BETA=3
                          SEGR=4.3E3, SGEA=1.135

In terms of the above symbolic names, the physical model for oxide growth is

$$X_o^2 + AX_o = B(t + \tau)$$

where X_o is the oxide thickness, $A = $ PRTE/LRTE, $B = $ PRTE, t is the time, and τ is internally determined based on the oxide thickness at the start of the present process step.

The diffusion coefficients are

$$D_{SiO_2} = \text{OXDC} \exp[-\text{OXEA}/kT]$$

$$D_{Si} = D_i \left[1 + \frac{1}{\sqrt{1 + 4\left(\frac{n_i}{C}\right)^2}} \right] \left[\frac{1 + \beta \frac{C}{n_i}}{1 + \beta} \right]$$

where $D_i = $ SIDC $\exp[-$ SIEA$/kT]$ is the low-concentration diffusivity, C is the impurity concentration, $n_i = $ NI is the intrinsic carrier concentration at the process temperature, and $\beta = $ beta is a factor that determines the effect of charged vacancies on diffusivity [21]. Finally, the segregation coefficient, m, is defined in SUPREM as

$$m = \frac{C(\phi)_{Si}}{C(\phi)_{SiO_2}} = \text{SEGR} \exp[-\text{SGEA}/kT]$$

SURVEY OF I²L MODELLING

F.M. Klaassen

Philips Research Laboratories
Eindhoven, The Netherlands

INTRODUCTION. A standard I²L inverter cell [1,2] has been made up by merging a multicollector npn switching transistor, which operates in the inverse mode, and a lateral pnp transistor, which provides base current to the switching transistor or sinks collector current from an adjacent gate (fig. 1).
In order to avoid unwanted lateral hole diffusion in the epitaxial layer a cell is often surrounded by an N^+ collar. As is well known, I²L offers a high packing density (250 gates/mm² for a 7-μm process) and a low dissipation, which can be easily varied over orders of magnitude.

In order to extract a circuit model or to analyze the electronic properties of an I²L gate, a physical understanding of the switching transistor current gain and cut-off frequency is essential. Therefore we shall first discuss some results of device physics [3-11] that relate to or may be applied to I²L. Particular attention will be paid to the various recombination mechanisms in a cell. From these results a d.c. model of a

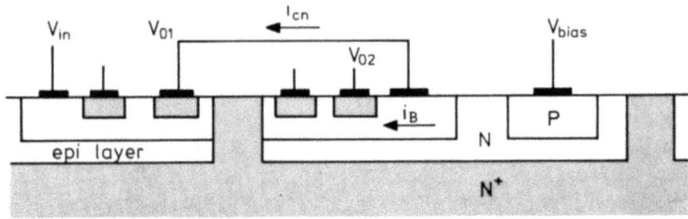

Fig. 1. Cross-section of integrated injection logic.

standard gate will be constructed by breaking up the switching transistor into several parts. This is followed by a discussion of parameters which represent charge storage in the cell, and whose value is needed for a dynamic model. The results will be compared with measurements.

After introducing the model, we next discuss some results of transient analysis, which have been obtained either from a circuit simulation program or by approximate methods [7]. Some measured results will also be given.

Finally we shall discuss several modified gates, which have been developed for improving the packing density (e.g. substrate-fed logic [19,22] and also the switching speed (oxide isolated I^2L [14-17] or Schottky I^2L [21,23].

1. DEVICE PHYSICS FOR I^2L

Unlike the normal mode of transistor operation, the collector current of the npn switching transistor is not only determined by the base dope, but also by a hole charge injected into the epitaxial layer between base and emitter buried layer. It has been shown that for the inverse mode of operation the distribution of these injected holes is practically uniform in the epitaxial layer [7,26]. Using this result the collector current can be expressed in terms similar to well known classical expressions (see next section).

More complications arise in describing the base current, as several factors contribute to the value of this quantity. By varying the diffusion and contact areas of the base and the collector, and using special gated structures, several workers have investigated the base current of a gate [3-6]. Although varying in order of magnitude due to process differences, three major components have been established: hole recombination in the buried layer, electron recombination at the base contact and at the oxide covered surface.

Although these components can be expressed in well known terms (compare next section), it is worth paying attention to some special points. Firstly, for the base area as well as the buried layer the pn-product n_i^2 is much larger than the classical value owing to high dope effects [8,9]. This is shown in fig. 2 [9].
Using the results of this figure the classical equations for injection may be corrected.

In the buried N^+ layer the recombination is not only affected by the above high doping effect, but also by an Auger process. As a consequence of both effects together, the recombination losses do not decrease with increasing doping; after passing a minimum they even increase at high doping values [11]. This is shown in fig. 3.

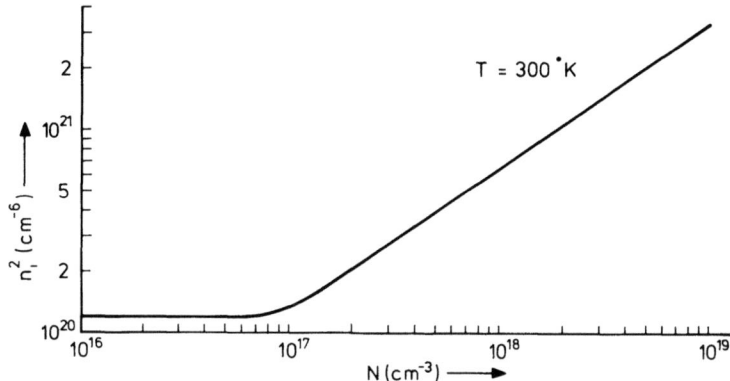

Fig. 2. The pn-product as a function of doping [9].

Fortunately the usual value for the dope of the buried layer is close to the minimum recombination value. This has been assumed to be the case in the next section.

Another parameter often used to express the recombination of minority carriers at the Si-SiO$_2$ interface is the surface recombination velocity (v_{ss}). As is well known, this quantity is strongly increased by surface band bending. After correction of the measured results on an I^{2}L structure for the above phenomena, a value for v_{ss} has been found [6], which is still unreasonably high.

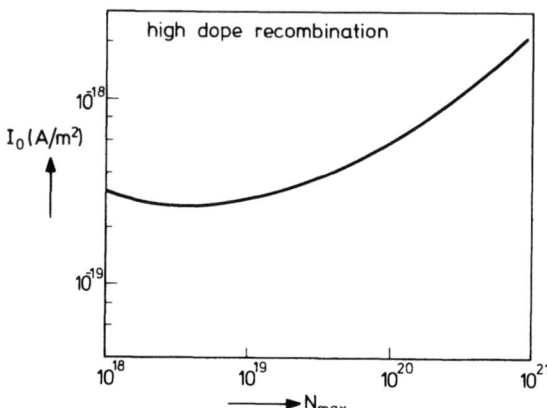

Fig. 3. Recombination current as a function of doping, taking into account high doping effect and Auger processes.

The last parameter we will discuss in this section is essential to an understanding the switching behaviour of an I²L gate. This parameter is the cut-off frequency f_T of the npn switching transistor, which is a measure of the stored charges. At low current levels the difference in f_T between the normal and inverse mode is small because charging and discharging of depletion areas now prevail. However, at medium and high collector current densities the f_T value of the inverse mode is far less than that of the normal mode [7]. This is again caused by storage of holes in the epitaxial layer, which (for the standard gate) exceeds the electron storage in the base by an order of magnitude. This inequality has some advantage for modelling. The value of the cut-off frequency for the downward mode of the npn transistor and for the pnp injector do not have to be known exactly, since these parameters make hardly any contribution to the switching speed. In the section on a dynamic model we will elaborate on the above picture.

2. D.C. MODEL OF A STANDARD GATE

Fig. 4c. shows a dc model based on the layout of a four-collector standard gate (fig. 4b) where the switching transistor is divided into five equal parts.

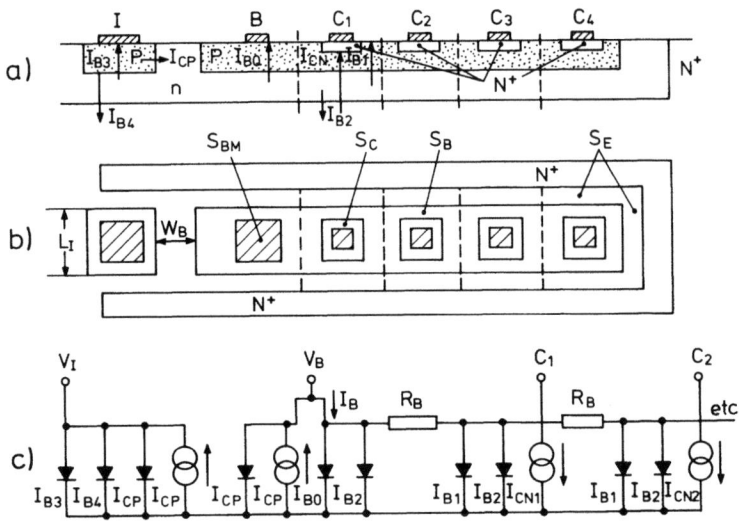

Fig. 4. Cross-section (a), layout (b) and dc model of a four-collector gate.

This model only contains major current components.
In similarity to the Moll-Ross relation the collector current is given by

$$I_{CN} \approx \frac{qD_n n_i^2 S_c \exp qV/kT}{\int N_B' dx + pd_{epi}'} , \qquad (1)$$

where $\int N_B' dx$ is the integral active base dope below the collector area S_c and the hole density in the lower epilayer (thickness d_{epi}', dope N_{epi}) is given by Boltzmann's relation

$$p = [\tfrac{1}{4} N_{epi}^2 + n_i^2 \exp qV/kT]^{\tfrac{1}{2}} - \tfrac{1}{2} N_{epi} . \qquad (2)$$

All other symbols have their usual meaning. Eq. (1) applies both for low and high currents, since high injection which is included in (2) usually prevails in the epilayer only.
The electron current of the parasitic base contact diode is given by

$$I_{BO} \approx [S_{BM} + \gamma(S_B - S_{BM})] \frac{qD_n n_i^2(N_B)}{\int N_B dx} \exp qV/kT, \qquad (3)$$

where the first part represents the current flowing to the contact (area S_{BM}) and the second part represents the current recombining at the oxide interface. $\int N_B dx$ applies to the passive base. For a good quality oxide $\gamma = 0.10$ [4]. It is worth noting that for the usual base doping density the value of the pn-product $n_i^2(N_B)$ is already a factor of 8 larger than the classical value owing to high dope effects (compare fig. 2).
The electron current recombining at the interface surrounding a collector area (I_{B1}) is fairly represented by an equation similar to the second part of eq. (3), if an area ($S_B - S_C$) is taken instead of ($S_B - S_{BM}$).
The hole current that flows to the buried layer and the N^+ collar surrounding a gate is given by

$$I_{B2} = S_E \times 1.1 \times 10^{-19} \exp qV/mkT . \qquad (4)$$

Since the holes are also flowing to the N^+ collar the collecting area $S_E > S_B$. Obviously the numerical constant depends on the dope of the buried layer. However, for dopes exceeding a value of $10^{19}/cm^3$, the variation is small as enforced by high dope effects and Auger recombination [10,11]. Values of m = 1.15 are encountered in practice, causing the upward current gain to depend on the current level.

Although all above current equations represent a quasi--twodimensional model, this approximation is justified. From a numerical analysis [12] it can be concluded that spreading effects are unimportant in I^2L.

Owing to the high value of the downward current gain, the loss current to the collector area can be neglected.

For the pnp collector current we have the well-known relation

$$I_{CP} \approx S' q D_p p/W_B , \qquad (5)$$

in which the effective collector area $S' \approx (D + \frac{1}{2} d'_{epi})L_I$; here two-dimensional current flow is more or less represented by adding a term $\frac{1}{2} d'_{epi}$ to the base diffusion depth D. L_I is the length and W_B the base width of the injector.

In similarity to the npn transistor, the injector loss currents I_{B3} and I_{B4} are given by eq. (3) and eq. (4), provided that the appropriate areas are substituted. Finally a base resistance R_B has been inserted between the multicollector sections, whose value is mainly determined by current paths around the collectors. With all components introduced above the following measurable quantities have been defined [1,7]:

a) the upward current gain of the j-th npn collector

$$\beta_{nj} = I_{CNj}/I_B , \qquad (6)$$

where I_B is that part of the base current that flows to the npn section (compare fig. 1c).

b) the current gain of a gate according to the definition of ref. 1.

$$\beta_j = \frac{I_{CNj}}{I_{in}} = \frac{I_{CNj}}{I_B + I_{CP}} \qquad (7)$$

where the denominator is the total base current flowing to a gate, when the injector is short-circuited. Since the recombination current in the base of the injector is small compared to I_B and I_{CP}, the above current gains can be measured easily from the scheme of fig. 5.

c) the current gain of the injector

$$\alpha_{inj} \approx I_{CP}/(I_{CP}+I_{B3}+I_{B4}) , \qquad (8)$$

where I_{B3} and I_{B4} are the total recombination current under the injector diffusion area.

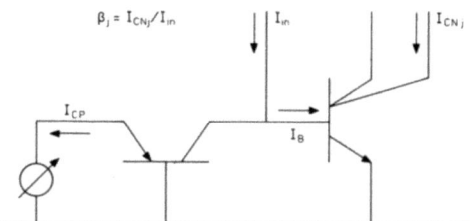

Fig. 5. Usual definition of current gain for I^2L.

In fig. 6 the collector current of the closest and most distant npn transistor and that of the pnp injector have been plotted as a function of the bias condition.
In another figure (7) the current gains according to the above equations have been plotted for a standard gate (15 x 15 μm² - collector areas and 5 μm - clearances).

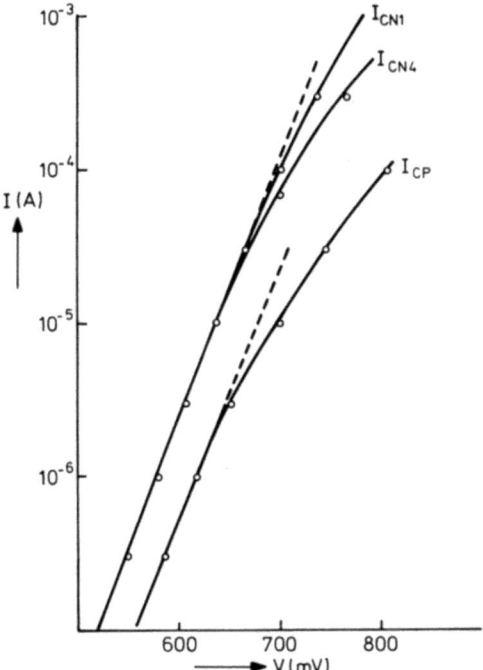

Fig. 6. I-V characteristics for a gate according to fig. 4.

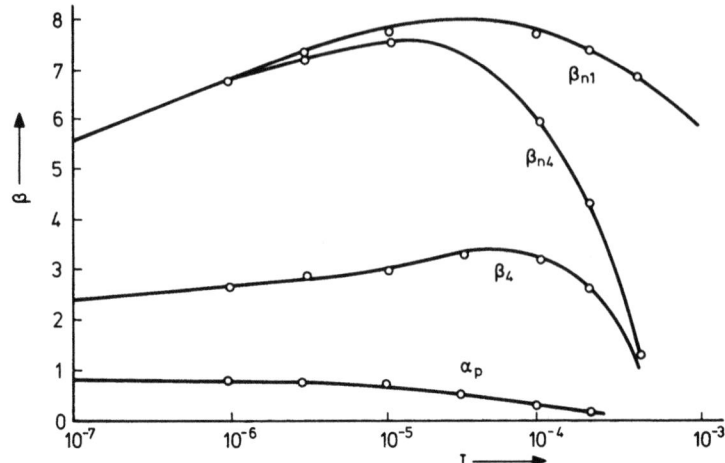

Fig. 7. Current gains corresponding to the results of fig. 6.

For the computation the following data were used:
$D = 2$ μm, $d_{epi} = 2$ μm, $W_B = 7$ μm, $N_{epi} = 5 \times 10^{15}/cm^3$,
$\int N_B dx = 4 \times 10^{14}/cm^2$, $\int N_B^p dx = 5 \times 10^{12}/cm^2$, $R_B = 350$ Ω.
From this figure we observe a fair agreement between computed and measured values. The combined influence of base resistance and high injection conditions on the transistor properties is clearly illustrated. Owing to the low dope of the epitaxial layer the pnp transistor comes already into high injection condition at low current densities. This causes a rapid fall-off of α_p in fig. 7. For the npn transistors the same effect causes the β values to decrease at higher currents. The collector most distant to the base shows the largest decrease owing to the large base resistance.

The value of the current gain β_j is extremely important for logic design owing to its relation to the noise margin ΔV (compare fig. 8).

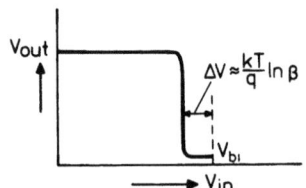

Fig. 8. Definition of noise margin.

It has been shown previously [7] that

$$\Delta V \approx (kT/q) \ln \beta_j . \qquad (9)$$

Requiring that ΔV should be approximately $\frac{kT}{q}$ Volt we have as a necessary condition for the current gain of a gate

$$\beta_j \geq 2 . \qquad (10)$$

This condition should also be met for the most distant collector and even when the current gain is affected by high injection effects and a base resistance.

The model as depicted in fig. 4c may be used to optimize specific gate configurations and to verify whether conditions as mentioned above are met. A procedure similar to the one discussed above for obtaining the process parameters of the model has been given elsehwere [24].

3. A DYNAMIC MODEL

Since, according to fig. 7, the current gains are nearly constant in a large current range, it is allowed to simplify the model for transient analysis. On the other hand depletion and storage capacitances have to be added. In particular at medium and high current levels storage of holes in the epilayer is important and (for standard I^2L) exceeds electron storage by an order of magnitude. It is therefore advantageous to express the collector current in terms of the hole storage, provided that the phase difference between this hole charge and the electron base charge is not large.

$$I_{CN} = S_B \, p \, d'_{epi} / \tau_t \qquad (11)$$

From eq. (11), using eq. (1) upon eliminating of I_{CN}, we obtain for the time constant

$$\tau_t = \frac{(F+1) p \, d'_{epi} \, [\int N_B dx + p \, d'_{epi}]}{D_n n_i^2 (S_C/S_B) \exp q V/kT} , \qquad (12)$$

where F is the number of collectors or fanout. In the case of low injection $pd' < \int N_B dx$, eq. (12) may be approximated by

$$\tau_t = \frac{S_B}{S_C} \frac{(F+1) d'_{epi} \int N_B dx}{D_n N_{epi}} \qquad (13)$$

Equation (11) will be recognized as charge control relation. Experimentally τ_t can be found from the unity gain cut-off frequency f_T as a function of the collector current I_c.

$$2\pi f_T = [\tau_t + kT(C_{eb}+C_{cb})/qI_c]^{-1}, \quad (14)$$

where C_{eb} and C_{cb} are the emitter and collector depletion capacitances. In fig. 9 f_T is plotted for a transistor of the same gate as used in fig. 4.
For $I_c < 10^{-5}$ the influence of the depletion capacitances dominates, for $I_c > 10^{-4}$ f_T is mainly determined by τ_t. The value of f_T according to approximation (12) is indicated by a dotted line. Due to the large value of the ratio between the total base area $((F+1)S_B)$ and a collector area (S_c) and also to the relatively large values of epitaxial layer thickness and base width in the case considered, the f_T value is very low. By tightening up several critical quantities, f_T values up to 30 MHz have been reached, even for a standard gate [7].

Unfortunately the f_T variation at high currents complicates a computation of the delay time, since the majority of transient analysis programs require a constant value of τ_t. However, a fair solution to this problem can be found either by using a computed value of τ_t according to eq. (13) or by using a measured f_T value in the appropriate high current range. Although the cut-off frequencies of the npn transistor in the normal downward mode and those of the pnp transistor are required for numerical calculation as well, approximative values are sufficient, since the charges associated with both quantities are much smaller

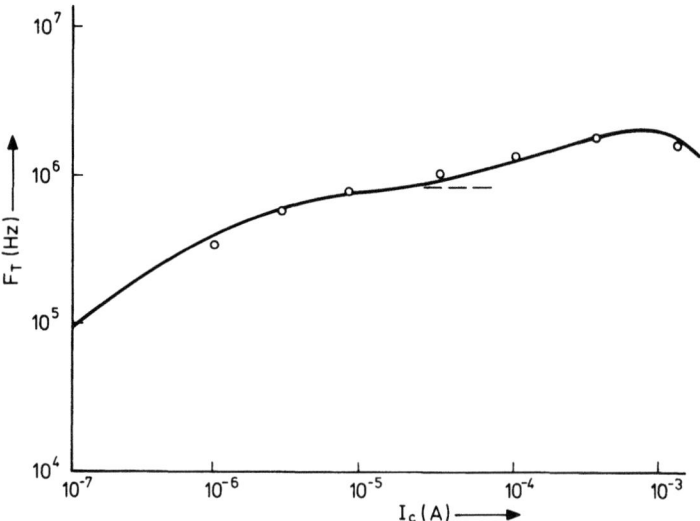

Fig. 9. The cut-off frequency of an npn transistor from fig. 4 as a function of the collector current. The dots represent measured values.

than the hole charge in the epilayer.

In fig. 10 a simple dynamic model is shown. The storage capacitances used are usually defined by

$$C_T = (q/kT)I_c \tau_t \,. \tag{15}$$

Storage under the injector area is neglected as this charge remains constant during switching of an inverter. Since a dynamic model based on the parameters discussed is strictly one-dimensional, an improved model has been suggested by using more distributed transistor sections [13,27]. Generally the improvement is not large.

4. RESULTS OF TRANSIENT ANALYSIS

Fig. 11 shows a comparison between a computer calculation and the measured delay time per gate as obtained from a ring oscillator consisting of a cascade of gates.
It may be concluded that the agreement is excellent. The slight increase of the delay time at high values of injector current is due to base resistance effects. Another figure (nr. 12) indicates more specifically the influence of the base resistance on the delay time at high currents.
In such a case a computer simulation is etremely useful. However, for a configuration where the influence of base resistance has been minimized (e.g. by arranging all collectors parallel to the injector rail or using an extra base diffusion), a first order estimate of the delay time can be made, which gives clear insight into the design aspects.

Applying a charge control analysis to the base nodes of the switching transistors it has been shown [7] that the delay time at low injector currents is given by

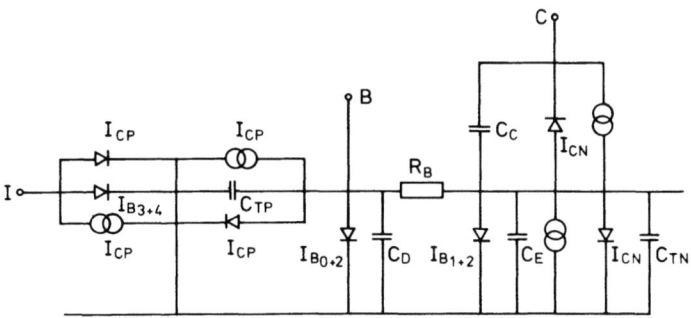

Fig. 10. A dynamic model for I^2L.

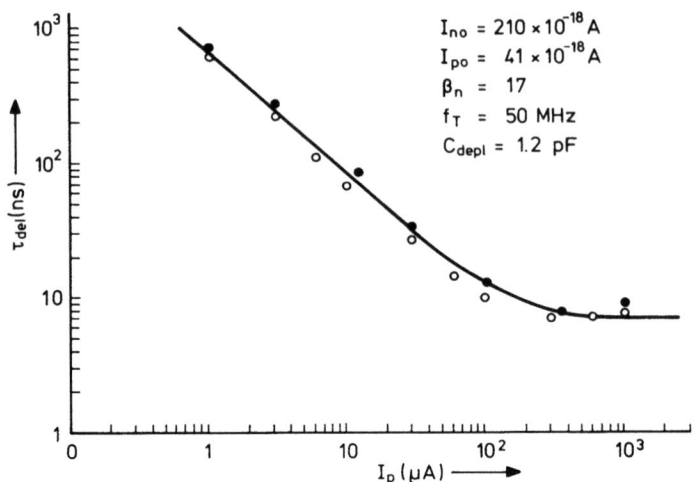

Fig. 11. Propagation delay time as a function of injection current.

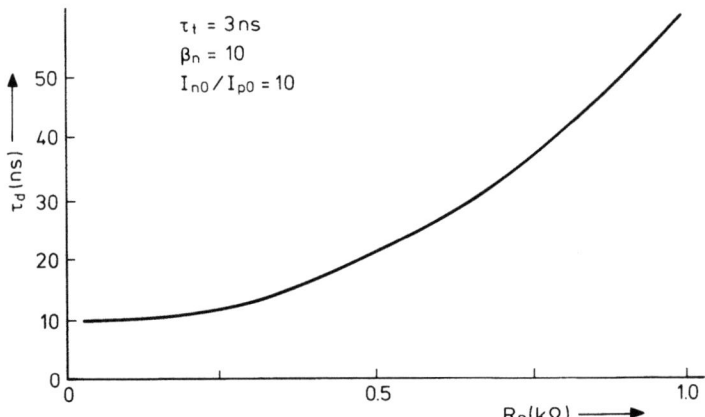

Fig. 12. Influence of base resistance on minimum delay time.

$$\tau_d \approx \frac{[C_{eb}+(F+2)C_{cb}] V_{bi} \tau}{2I_p} \qquad (16)$$

where V_{bi} is the bias voltage. As is well known, the delay time varies with the inverse of the injector current.

On the other hand, when the injector current is sufficiently high, the delay time is determined by the storage of the active charges. This causes τ_d to become independent of the current level applied. Then

$$\tau_d \approx \frac{\beta_j \tau}{2\beta_{nj}} \ln \frac{2\tau_d + (\beta_j/\beta_{nj})\tau}{(\beta_j/\beta_{nj}) - \tau_t} \qquad (17)$$

where τ is the effective lifetime of holes in the epitaxial layer. Generally τ_d is only a weak function of τ. In fig. 13 we have plotted the delay time vs the ratio β_j/β_{nj} with the npn cut-off frequency as a parameter.

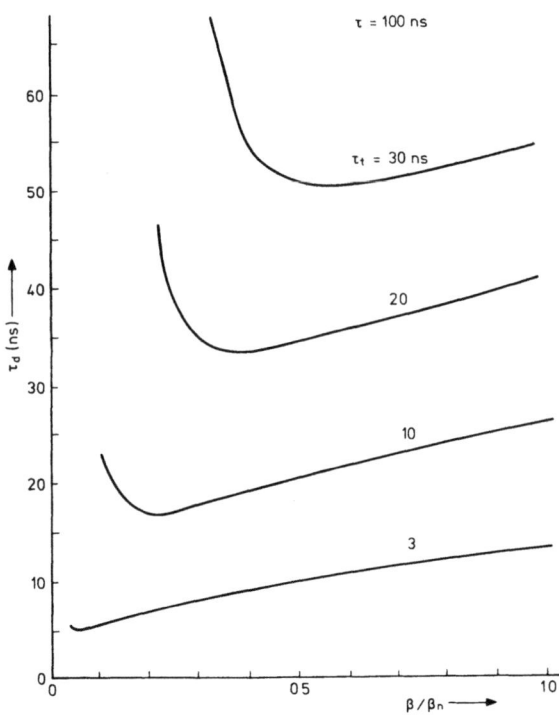

Fig. 13. Delay time as a function of the current gain ratio with "transit time" τ_t as a parameter.

Obviously the minimum delay time may be reduced to a certain point by decreasing the ratio β_j/β_{nj}, (which can be achieved by increasing the injector rail) or by reducing the remaining epilayer and realizing more shallow transistor structures. Since the effective β_j is only a weak function of the fan-out, whereas (compare eq. 13) f_T decreases linearly with the number of collectors, the delay time will increase with the fan-out.

5. MODIFIED GATE STRUCTURES

In this section we discuss several modifications of the standard gate configuration, which on the one hand are technologically more complicated, but on the other hand reveal improved switching speed or higher packing density. For each gate we shall also briefly discuss relevant improvements in properties or some associated draw-backs.

5.1. Oxide-isolated I^2L

In order to increase the packing density and also the switching speed, I^2L has been made in oxide-isolated, shallow epitaxial layers. This can be achieved on n-type [14,15,16] as well on p-type layers [17].

A cross-section is shown in fig. 14a.

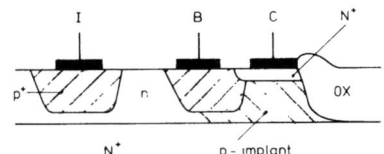

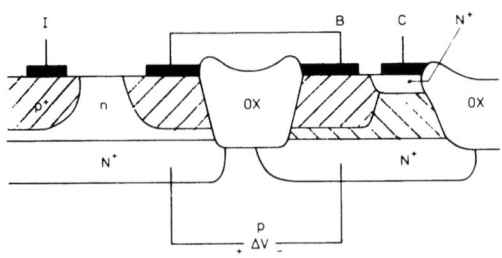

Fig. 14. Oxide-isolated I^2L gate with saturated injector (a) and nonsaturated injector (b).

The epitaxial layer is n-type with a thickness of typically 1.2 μm. Compared to a standard process the diffusion of the injector and the extrinsic base of the switching transistor is an additional step. This has the advantage of providing a better injection efficiency and a lower base resistance.

Qualitatively this gate behaves like the standard version, except for the switching speed. At low currents this speed will increase due to the reduction of the depletion capacitances by the oxide isolation. However one should realize that the low base resistance process step opposes a small depletion capacitance. Moreover the maximum switching speed of a shallow epitaxial version also behaves differently. As the charge stored in the remaining epitaxial layer has been considerably reduced, the charge stored in the base of the pnp transistor becomes a limiting factor. A first-order estimate of the switching time based on the charge control analysis yields [18]

$$\tau_d \approx (I_{po}/I_{no})^{\frac{1}{2}} \cdot (2\pi f_{TP})^{-1} \; , \tag{18}$$

where I_{no} and I_{po} are the current constants of the switching transistor and the injector respectively. It should be emphasized that f_{TP} is the inverse cut-off frequency of the injector with the total base area of the switching transistor acting as emitter. With present designs (I_{po}/I_{no} = 0.10 and $f_{TP} \approx$ 10 MHz) a delay time of 4 ns per gate is achieved [15]. However, since the fan-out is only weakly coupled to f_{TP}, its influence on the delay time is fairly small.

In order to eliminate the influence of the saturation charge of the injector on the delay time, it is interesting to consider an I^2L version, in which a small reverse bias on the bc junction of the injector can be applied. This is possible in a configuration where the injector (with a minor loss in packing density) is no longer merged with the switching transistor [14] (fig. 14b). Now the main charge storage is again in the npn transistor. However, compared to the standard gate, the charge is no longer reduced by a factor (β_i/β_{nj}), but the full injector current charges the switching transistor. As a result the delay time is approximately given by [18]

$$\tau_d \approx \beta_{nj}^{\frac{1}{2}} (4\pi f_{TN})^{-1} \; , \tag{19}$$

where β_{nj} is the intrinsic current gain of collector j of a multicollector switching transistor and f_{TN} is the npn cut-off frequency. With $\beta_{nj} \approx$ 20 and $f_{TN} \approx$ 200 MHz, a value of approximately 3 ns has been realized for τ_d [14].

An interesting slightly different I^2L configuration has been realized on a p-epitaxial layer, though an extra diffusion step

is necessary. This structure is shown in fig. 15.
The injector has been realized by means of two successive implantations through one window. In order to prevent the p-layer from inversion and to decrease the npn base resistance the p-type injector implantation has also been used to surround the shallow n^+ collectors. Compared to previous structures this one offers the advantage of more freedom to design the injector, as there is no longer any need to compromise between injector current gain and charge storage in the epilayer. Moreover the use of a higher injector base doping will reduce the injector current gain fall-off. Since the npn transistor is the main source of charge storage this version will behave like the standard gate. Here too, the advantage of local oxidation for reducing the depletion capacitances is partly annulled by the low resistivity base layer.

Instead of using a conventional downward diffused base, fast oxide-isolated I^2L versions have also been made by making use of an upward diffused base from an extra p-type buried layer [25]. In these structures hole charge in the npn-sections has been almost completely eliminated.

5.2. Substrate-fed logic (SFL)

By using the combination of N^+ buried layer and p-type substrate as injector for an inversely operating npn switching transistor, the packing density of standard I^2L can be improved [19]. This is illustrated in fig. 16.

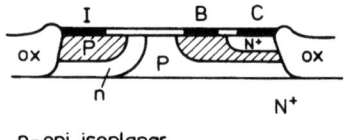

p-epi isoplanar

Fig. 15. Oxide-isolated I^2L made on p-epitaxial layer.

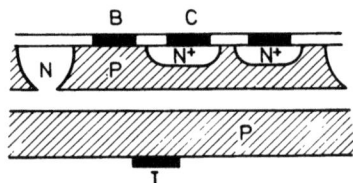

substrate-fed I^2L

Fig. 16. Substrate-fed I^2L.

Injector rails are no longer needed. Apart from the improved
density, additional advantages are the higher current gain of
the injector and somewhat smaller depletion capacitances due
to the low doping of the p-epitaxial base layer. However, this
configuration is only suitable for operation at low currents.
The high base resistance, the relatively thick epilayer and in
particular the high emitter resistance will cause problems at
increased current rates. For instance, with an epilayer
resistivity of 10 kΩ, and keeping in mind that $I_{no}/I_{po} \approx 10$, the
buried layer sheet resistivity should be about 300 Ω. This
value may cause problems at currents exceeding 50 μA.

Instead of using the substrate, an injector may also be
realized by using a second buried layer of the p-type slightly
displaced below the conventional n-type buried layer [22].

5.3. Schottky I^2L

In order to improve the switching speed, two Schottky I^2L
versions have been proposed. In the first one a Schottky diode
has been inserted between a collector of a switching transistor
and the base of another gate [19,20]. (Compare fig. 17).
As a result of the presence of this diode the logic swing of a
gate is reduced by a factor of 2 or 3 (depending on the Schottky
diode current constant), although the preceding npn transistor
may still operate in heavy saturation. Due to the reduction of
the voltage excursion, the delay time at low currents has been
reported to be reduced by a factor of 2. A drawback of the
structure is the rather low value of the downward current gain,
which is a fundamental limitation of a pure Schottky emitter.
This might cause current hogging problems [1]. Moreover at
higher currents excess charge is stored in the epitaxial collector area. Since this charge can only be removed from the
collector at a limited current rate, the minimum delay time will
be rather high.

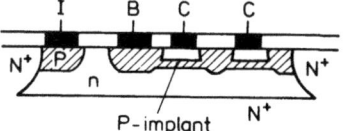

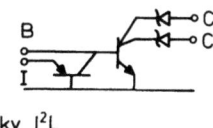

schottky I^2L

Fig. 17. Implementation of Schottky diodes in I^2L.

In a second proposed Schottky I^2L structure (fig. 18) an injector feeds a true Schottky-collector switching transistor, which should preferably be of the pnp type [21].
For a gate configuration as shown together with the basic gate of fig. 18 it can be readily shown that the effective current gain for a four-collector gate with three collectors practically floating is given by

$$\beta_{eff} \approx \frac{(\beta_i + 7/4)\beta_n}{3\beta_n + 4\beta_i + 4} \tag{20}$$

where β_n and β_i are respectively the upward and downward current gain of the Schottky transistors. For safe operation β_i should have a value larger than 6. However such a value cannot be achieved by a pure Schottky collector, but might be reached by introducing a weak, shallow p-layer beneath the metal collector [23].
Depending on the electron barrier at the collector, not only is the logic signal swing reduced but excess charge storage due to saturation is avoided. A favourable speed-power product is therefore to be expected.
From a straightforward calculation we have found that the delay time per gate will be given by

$$\tau_d \approx \tau_t . \tag{21}$$

Comparing this result to the standard results (17), a considerable improvement in speed has to be expected.

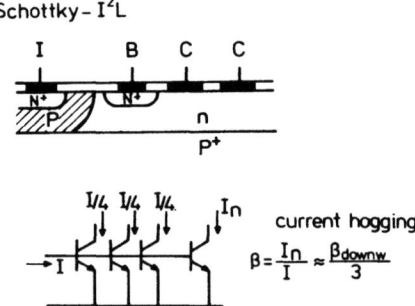

Fig. 18. An I^2L-gate with a Schottky-collector switching transistor. The four-collector configuration illustrates a current hogging problem.

REFERENCES

1. H.H. Berger and S.K. Wiedman, IEEE J.S.S.C. SC-7, p. 340 (1972).
2. K. Hart and A. Slob, IEEE J.S.S.C. SC-7, p. 346 (1972).
3. H.H. Berger, IEEE, J.S.S.C. SC-9, p. 218 (1974).
4. H. Wulms, Digest ISSCC, p. 92 (1976).
5. M. Hillen a.o., ESSDERC, München p. 26 (1976).
6. W. Mattheus a.o., ESSDERC, München, p. 28 (1976).
7. F.M. Klaassen, IEEE, Trans.El.Dev. ED-22, p. 145 (1975).
8. R. Mertens, R. van Overstraeten and H. de Man, IEEE Trans.El.Dev. ED-20. p. 290 (1973).
9. J.W. Slotboom and H.C. de Graaff, Solid State Electr. Vol. 19, p. 857, (1976).
10. J. Burtscher a.o., Sol.St.El., 18, p. 35 (1975).
11. J.W. Slotboom, Solid St.El., 20, p. 167 (1977).
12. R.C. Jaeger, Digest ISSCC, p. 96 (1976).
13. J. Lohstroh, Digest ISSCC, p. 94 (1976).
14. C. Mulder, H. Wulms, IEEE J.S.S.C. SC-11, p. 379 (1976).
15. W.B. Sander a.o., Digest ISSCC, p. 182 (1976).
16. R. Muller a.o., ESSCIRC-Canterbury, p. 10 (1975).
17. R.E. Crippen a.o., Digest ISSCC, p. 30 (1976).
18. F.M. Klaassen, IEEE J.S.S.C. SC-12, April (1977).
19. V. Blatt a.o., IEEE J.S.S.C. SC-10, p. 336 (1975).
20. F.W. Hewlett, IEEE J.S.S.C. SC-10, p. 343 (1975).
21. S. Blackstone and R. Mertens, ESSCIRC-Toulouse p. 66 (1976).
22. T. Nakano a.o., Digest IEDM, p. 555 (1975).
23. F.W. Hewlett a.o., Digest IEDM, p. 304 (1976).
24. J.L. Dunkley a.o., Digest IEDM, p. 312 (1976).
25. J.D. McGreivy, Digest IEDM, p. 308 (1976).
26. H.C. de Graaff and J.W. Slotboom, Sol.St.Electr. Vol. 19, p. 809 (1976).
27. W. Mattheus a.o., Digest ISSCC, p. 44 (1977).

Section IV

MODELING OF MOS DEVICES

REVIEW OF PHYSICAL MODELS FOR MOS TRANSISTORS

F.M. Klaassen

Philips Research Laboratories
Eindhoven, The Netherlands

INTRODUCTION. In common with the description of other devices a physical model of the MOS transistor is a trade-off between simplicity and accuracy. Initially the gradual channel approximation as a simple analytic, one-dimensional theory was also accurate, since earlier MOS transistors had a long channel. However with the progress of technology the device became gradually smaller, and problems arose due to velocity saturation effects and a two-dimensionally shaped potential distribution. In fact only a numerical solution method achieves sufficient accuracy here, but simplicity has been lost. As a compromise many approximative analytic descriptions have been published, mostly with a view to application in a circuit analysis program.

In this review we first present the complete gradual channel approximation including the substrate space charge effect and compare the result with simplified models sometimes used in CAD. After a discussion of the failure of the above models, several modifications to the standard model are introduced, which take into account a field-dependent mobility and the influence of the drain depletion layer. The influence of mobile carrier diffusion, the effect of temperature, drain corner breakdown and a linear high-frequency model are also discussed. For the sake of completeness we next present the modelling of sub-threshold behaviour.

After the above review of practical analytic models we present results of two-dimensional numerical analysis which are used as a check on previously discussed CAD-type device models. Finally an outline is given of two rather separate subjects, a complicated, yet analytic model based on uniform space charge density and the noise characteristics of the MOST.

Through out this paper only the n-channel MOST is considered. By changing the appropriate signs all results also apply to p-channel devices.

1. STATIC DRAIN CHARACTERISTICS

1.1. Threshold voltage

Generally four principal charge components control the behaviour of the MOS transistor. As illustrated in fig. 1 there are: the gate charge q_G, the inversion or channel charge q_N, the substrate charge q_B, and the equivalent interface charge q_S. These are related by the charge neutrality condition

$$q_G + q_N + q_B + q_S = 0 . \tag{1}$$

When only a small voltage difference is present between source and drain, the field in the insulator is normal to the surface and consequently the following relation applies to q_G

$$q_G = C_{ox}(V_G - \varphi_{ms} - \Delta V_s), \tag{2}$$

where $C_{ox} = \varepsilon_{ox}/t_{ox}$ is the insulator capacitance per unit area, V_G is the potential of the gate, ΔV_s the potential of the semiconductor surface or the band bending and φ_{ms} is the contact potential between gate material and the semiconductor. All potentials V are taken with respect to the neutral substrate (compare fig. 1).

Following Shockley's depletion approximation the depletion charge of the substrate is given by

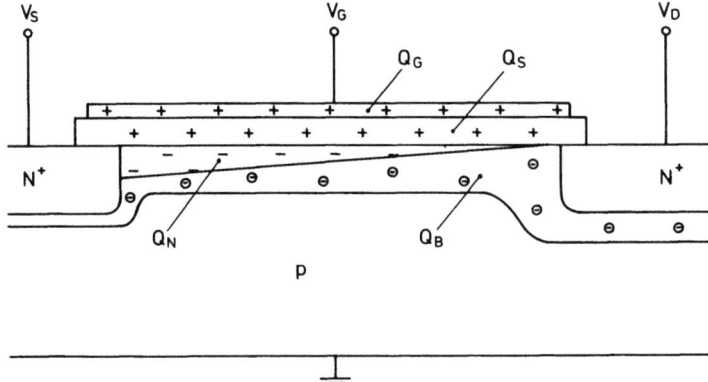

Fig. 1. Charge components in n-channel MOS transistor.

$$q_B = -(2\varepsilon_s qN\Delta V_s)^{\frac{1}{2}},\qquad(3)$$

where N is the substrate doping.

Usually a threshold voltage V_T is defined [1,2] as the gate potential, which causes a band bending $2\varphi_F$, where $\varphi_F(N)$ is the substrate Fermi potential. In this condition the inversion charge may be neglected in eq. (1) and we obtain from eqs. (1), (2) and (3)

$$V_T = \varphi_{ms} + 2\varphi_F - q_s/C_{ox} + k(2\varphi_F)^{\frac{1}{2}},\qquad(4)$$

where the so-called body factor k is given by

$$k = \frac{t_{ox}}{\varepsilon_{ox}}(2q\varepsilon_s N)^{\frac{1}{2}}.\qquad(5)$$

In relation (4) the sum of the first three right hand terms is usually denoted by $V_{TX} = 2\varphi_F + V_{FB}$, where $V_{FB} = (\varphi_{ms} - q_s/C_{ox})$ is often referred as flat band voltage.
In present-day technology devices q_s has a value between $2 \times 10^{-8}/cm^2$ (< 100 >-material) and $8 \times 10^{-8}/cm^2$ (< 111 >-material), φ_{ms} varies between -1V (Al-gate) and 0 (Si-gate) and k varies between .5 and 2.5 (in C-MOS technology).

When a voltage difference V_S is applied between source and substrate, as is often the case with an integrated circuit, a generalized threshold voltage V_{TS} may be derived from eq. (4).

$$V_{TS} = V_S + V_{TX} + k(V_S + 2\varphi_F)^{\frac{1}{2}}.\qquad(6)$$

Although the current for an applied gate voltage $V_G < V_{TS}$ is not strictly zero (compare section 4), for many applications V_{TS} may be considered as a separation between the passive and active mode.
In fig. 2 a range of practical values of V_T has been plotted for the case $V_{FB} = 0$.

1.2. Drain current

In the active mode the mobile inversion charge is often computed under the following assumptions:
the field in the channel is perpendicular to the semiconductor surface (gradual channel approximation); an increase of V_G no longer has influence on band bending; current transport takes place by drift alone with a field-independent mobility. In this case the d.c. channel current at a position x in the channel is given by [1,3]

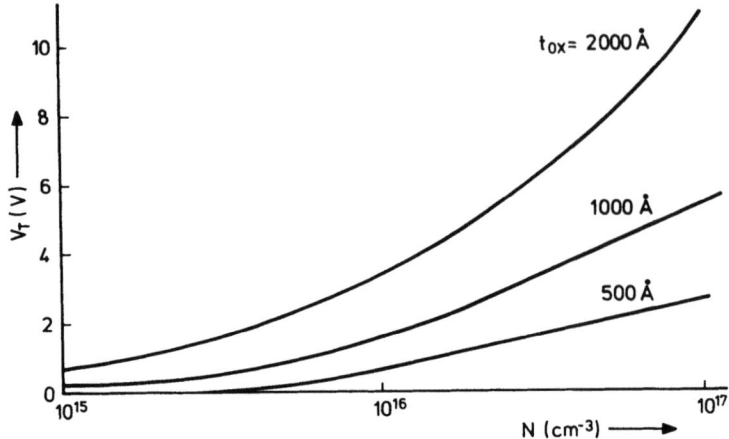

Fig. 2. Threshold voltage as a function of < 100 > substrate doping.

$$I = W\mu_n C_{ox} [\ V_G - V_{TX} - V(x) - k\ \{\ V(x) + 2\varphi_F\ \}^{\frac{1}{2}}\ ]\ \frac{dV}{dx}, \tag{7}$$

where W is the gate width (compare fig. 3) and μ_n the mobility. Solving eq. (7) it is possible to express the drain current as a function, which is symmetrical with respect to source and drain [4]:

$$I_D = \frac{\beta}{2} [\ f(V_G, V_S) - f(V_G, V_D)\ ], \tag{8}$$

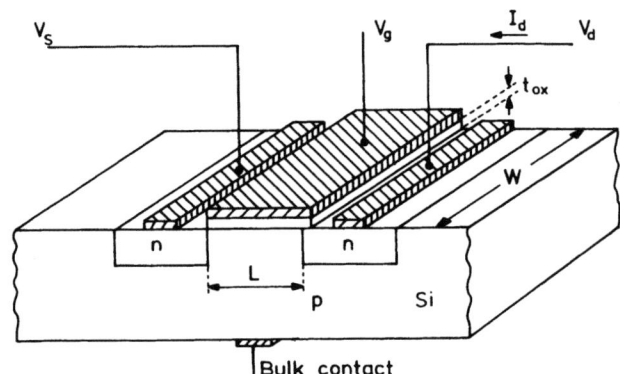

Fig. 3. A MOS structure.

where
$$f(V_G, V) = (V_G - V_{TX} - V)^2 - 4/3 k (V + 2\varphi_F)^{3/2} \tag{9}$$
and
$$\beta = \mu_n \varepsilon_{ox} W / t_{ox} L. \tag{10}$$

L is the channel length.
For drain voltages such that
$$V_G \leqslant V_{TD} = V_D + V_{TX} + k(V_D + 2\varphi_F)^{\frac{1}{2}} \tag{11}$$
or, alternatively,
$$V_D \geqslant V_{DP} = -2\varphi_F + [\, -\frac{k}{2} + (\frac{k^2}{4} + V_G - V_{TX} + 2\varphi_F)^{\frac{1}{2}}\,]^2 \tag{12}$$

the above equations lose their validity, since in condition (11) the mobile inversion charge disappears at the drain end of the channel (the so-called pinch-off region) and the field lines are no longer normal to the surface. In this case the channel current is assumed to be no longer controlled by an increase of the drain potential, as the carriers move through the depleted region either by diffusion or with a saturated drift velocity. The current then has a saturated value, which is given by eq. (8) provided that V_D is replaced by V_{DP}

$$I_{DSS} = \frac{\beta}{2} [\, f(V_G, V_S) - f(V_G, V_{DP})\,]. \tag{13}$$

Formally this result is correct since according to eq. (8) through (11) $dI_D/dV_D \rightarrow 0$ for $V_D \rightarrow V_{DP}$. Consequently the drain conductance is zero in the saturation region.

For practical purposes the above equations are often considered to be rather unwieldy and it is therefore worth considering whether a simpler approximation can be obtained. This can be achieved by expanding the most right-hand term of eq. (7) around $V = V_S$

$$(V + 2\varphi_F)^{\frac{1}{2}} \approx (V_S + 2\varphi_F)^{\frac{1}{2}} + \frac{V - V_S}{2(V_S + 2\varphi_F)^{\frac{1}{2}}}.$$

When the above expansion is substituted in eq. (7) we obtain an approximation for eq. (8)

$$I_D = \beta [\, (V_G - V_{TS})(V_D - V_S) - \tfrac{1}{2}(1 + \gamma)(V_D - V_S)^2\,], \tag{14}$$

where

$$\gamma = k/2(V_S+2\varphi_F)^{\frac{1}{2}}.$$

To this approximation corresponds a saturation voltage

$$V_{DP} = V_S + (V_G-V_{TS})/1+\gamma \ . \tag{15}$$

When eq. (15) is substituted in eq. (14), the saturated drain current I_{DSS} varies with the square of (V_G-V_{TS}), a behaviour often found in practice.

In some models even eq. (14) is considered too complicated, and γ is taken to be zero. In particular for processes with k-values > 1 this approach is too crude. Actually the expansion taken above is only justified for a device operating with a large reverse bias voltage V_S. For low values of V_S however, the expansion overestimates the influence of the substrate charge upon the drain current.
Therefore as a fair compromise γ can be shown to have a value [6]

$$\gamma \approx 0.4 \ k \ (V_S+2\varphi_F)^{-\frac{1}{2}} \ . \tag{16}$$

In fig. 4 a comparison is made between the drain characteristics

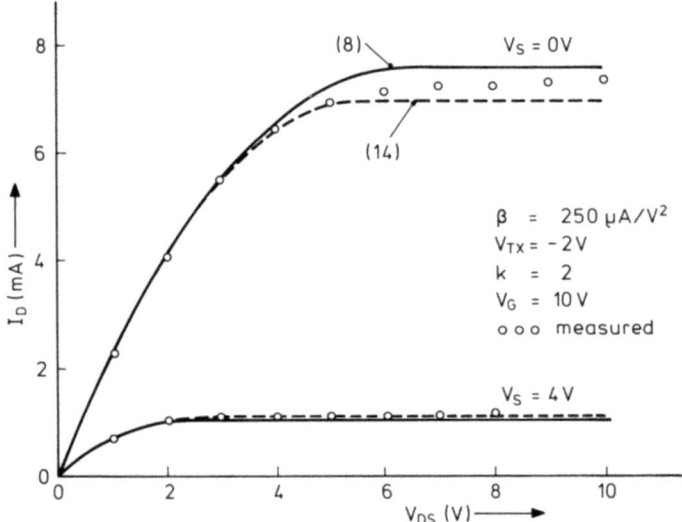

Fig. 4. Simulated drain current vs drain-source voltage.

as given by eqs (8) through (13) and the approximation (14) through (16) for a process with k=2. For $V_S = 0$ the differences are within 5%, for $V_S = 4V$ even negligible.

1.3. Transconductance

The device shown in fig. 3 is a four-terminal device, and in general exhibits a transconductance both from the gate and the substrate. For practical reasons we discuss here only the gate transconductance. Usually the bulk transconductance is much smaller.

The gate transconductance, which is defined by

$$g_m = (\partial I_D / \partial V_G) \mid V_D, V_S,$$

can be evaluated from eq. (8). For the linear region $(V_D < V_{DP})$

$$g_m = \beta(V_D - V_S)$$

and for the saturation region $(V_D \geq V_{DP})$

$$g_{ms} = \beta(V_{DP} - V_S) . \tag{17}$$

From (12) and (17) it will be noted that the transconductance in the linear region is independent of the gate-source voltage. However, under saturation conditions both V_G and V_S affect the transconductance. According to (15) the saturation transconductance may be approximated by

$$g_{ms} \approx \frac{\beta}{1+\gamma} (V_G - V_{TS}) . \tag{18}$$

When the channel length L is not very short, g_{ms} is found experimentally to show a linear behaviour.

1.4. Limitations of the simple model

In figs 5 and 6 a comparison has been made between measured drain current and gate transconductance and computed values according to the above model. The deviations observed clearly illustrate the shortcomings of the model. Whereas both devices exhibit the same calculated characteristics owing to the same W/L ratio, the measured values differ considerably. Only at moderate channel lengths (L > 10 µm) do the absolute values of I_D not differ considerably but the saturated output conductance has a non-zero value. This is mainly caused by the drain

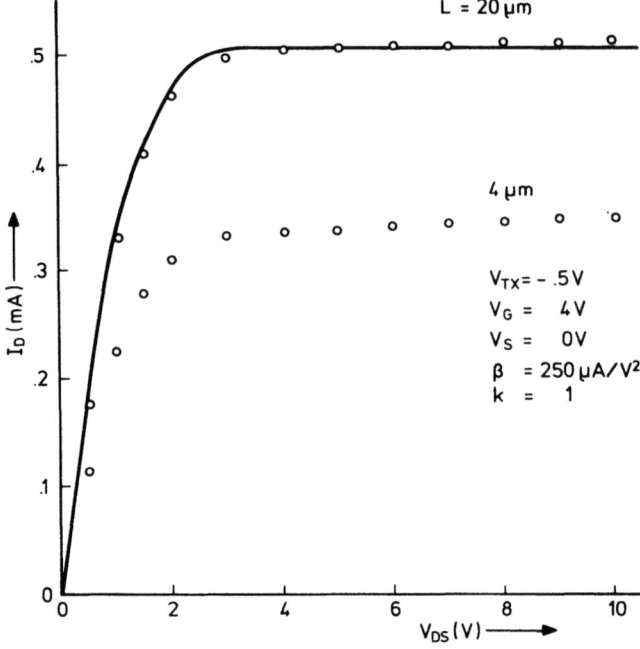

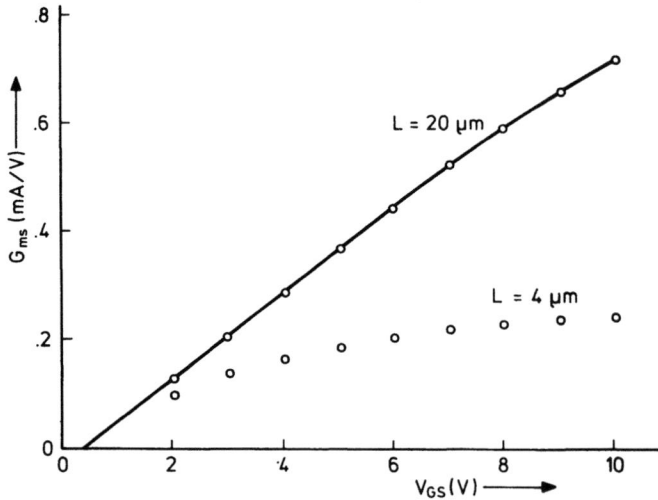

Fig. 5+6. Comparison of simulated (drawnlines) and measured results (dots).

depletion layer, which increases upon an increase of the drain
voltage and causes the pinch-off point, at which the field in
the insulator changes direction, to move towards the source.
As the effective channel length becomes shorter, the saturated
drain current increases slightly. In general, the current
increase can only be calculated precisely by solving the two-
dimensional Poisson equation for the drain-channel area [8].
Fortunately some useful analytical approximations are
available [5,7].

When inspecting the transconductance, it can be concluded that for moderate channel lengths the measured values
are only slightly lower at high gate voltages, however for
$L < 5$ µm large deviations occur. Two factors are responsible
for this result. The mobility has been shown to be field
dependent [9,10] and a series resistance associated with the
source will also affect the transconductance. From measured
results of the characteristics both effects can be modelled
successfully.

Since the free carrier density decreases drastically in
the pinched-off region, it may be questioned whether a diffusion
component can be neglected. Although, in general, the answer is
positive, we next discuss a calculation including the diffusion
current [11].

Finally, for short channel lengths ($L < 3$ µm) the threshold
voltage no longer satisfies the simple relation (6). This is due
to the fact that the one-dimensional formula for the gate-
-induced substrate charge no longer holds, since part of q_B is
controlled by the source and drain potential. This problem is
treated in another chapter (Advanced physical MOS models by
G. Merckel).

2. MODIFICATIONS OF DRAIN CHARACTERISTICS

In order to correct the failure of the simple model we first
discuss in this section several attempts to modify the classical
expression for the drain current by taking into account the
field dependence of the mobility and the channel shrinkage
effect. Next we present a calculation for I_D including the
diffusion current. Finally two inconveniencies often met in
practice, will be discussed: the temperature sensitivity of
I_D and drain corner breakdown.

2.1. Field-dependent mobility

From both the field effect and the Hall effect it has
been found that the free carrier mobility depends on the normal
field. Fig. 7 gives a plot of a room temperature result [9].

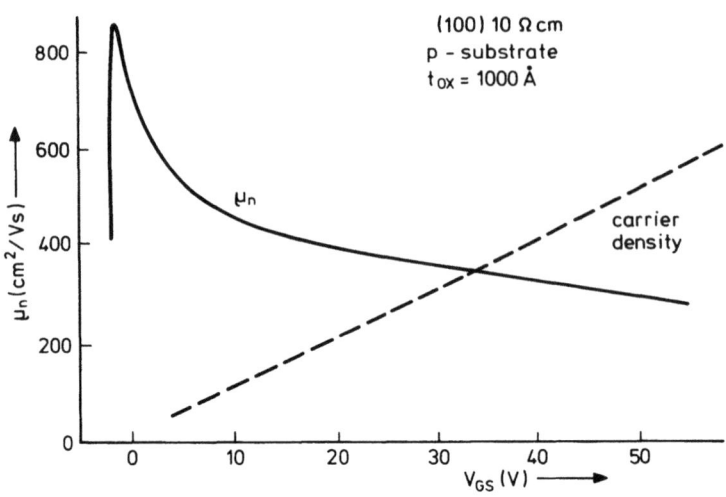

Fig. 7. Hall mobility at 22°.

After a steep rise in the sub threshold region (probably due to inhomogeneities in the conductivity [30] the mobility reaches a sharp maximum (usually lower than the bulk value) and decreases gradually in the strong inversion mode.
At low temperatures often a second peak occurs. From several experiments it was concluded that the mobility decrease does not depend on the substrate doping or the substrate bias and very little on the surface charge and the crystal orientation. The normal field at the surface is the only important parameter. Although a classical theory for surface scattering in a linear potential well supports this result qualitatively ($\mu/\mu_B \sim E_N^{-1}$), serious discrepancies appear from more recent calculations, which take into account that the conduction band is quantized in the surface well and assume a far more likely scattering mechanism by surface phonons or by surface roughness.

As is well known, at high lateral fields the mobility also decreases owing to velocity saturation. This is shown experimentally in fig. 8 [12]. For < 100 >-material the electron velocity saturates at fields of $E_x \approx 2 \times 10^4$ V/cm with an ultimate room temperature velocity $v_{ss} = 6.5 \times 10^6$ cm/s. This result is independent of the inversion density and the substrate doping. For other crystal orientations v_{ss} becomes considerably lower.

Fortunately the field-dependent mobility can be modelled by a relation that satisfies both simplicity and accuracy [6]

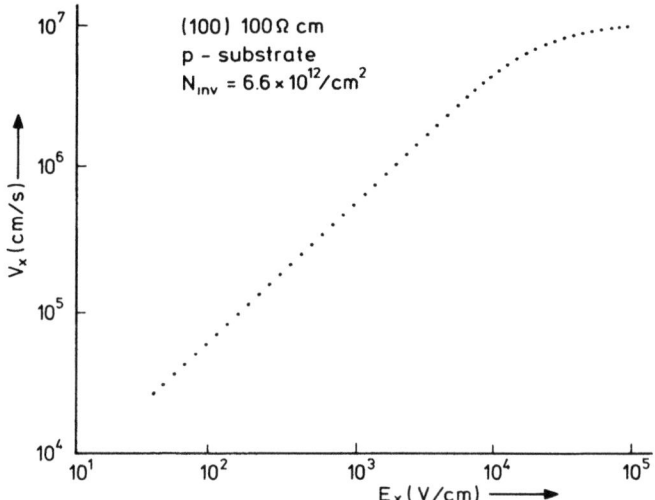

Fig. 8. Drift velocity vs field.

$$\mu = \frac{\mu_o}{1+\Theta(V_G-V_{TS})+E/E_c} \quad . \tag{19}$$

In this equation μ_o is the low-field mobility and Θ and E_c are empirically fitted parameters.
When the above relation is substituted in eq. (8), we obtain for the drain current

$$I_D = \frac{\beta\,[\,f(V_G,V_S) - f(V_G,V_D)]}{2\,[\,1+\Theta(V_G-V_{TS})+(V_D-V_S)/LE_c]} \tag{20}$$

A simpler approach is often adopted by using the following approximation for the carrier velocity [13]

$$v = \frac{\mu_o E}{1+\Theta(V_G-V_{TS})} \qquad \text{for } E < E_c$$

$$v \approx \frac{\mu_o E_c}{1+\Theta(V_G-V_{TS})} = v_{ss} \qquad \text{for } E \geq E_c \quad . \tag{21}$$

Obviously this approach leads to a "hard" saturation of the characteristics. The saturated drain current is given by

$$I_{DSS} = \frac{\beta LE_c}{1+\Theta(V_G-V_{TS})}\,[\,V_G-V_{TX}-V'_{DP}-k(V'_{DP}+2\varphi_F)^{\frac{1}{2}}\,] \quad . \tag{22}$$

Comparison of this value with the normal value (8) for the ohmic region has yielded a quartic equation for the saturation voltage, which can be evaluated only numerically [5]. As this result is not very useful for circuit analysis, several attempts have been made to obtain a direct approximative solution.

When we use the "soft saturation" approach as given by (20) it is reasonable to assume that this relation holds in that part of the channel area where the carriers have not yet attained a velocity v_{ss}. In fact fewer problems arise from fre carrier pinch-off than in the constant mobility approach. In that case saturation occurs at an internal saturation voltage, which satisfies the condition

$$\frac{dI_D}{dV_D} \Big|_{V_{DSS}} = 0$$

Assuming moreover that $V'_{DP} = V_{DP} - \Delta V$, where V_{DP} is the constant-mobility saturation voltage and ΔV is only a correction factor, it can be shown [6] that

$$V'_{DP} \approx V_{DP} - \frac{(V_{DP} - V_S)^2}{2LE_c [\ 1+\Theta(V_G-V_{TS}) + (V_{DP}-V_S)/LE_c\ ]} \quad (23)$$

We have found that this saturation voltage deviates less than 3% from the value obtained numerically from eq. (20), up to values $(V'_{DP}-V_S)/LE_c = 1$.

When the influence of the substrate depletion charge on the drain current is modelled according to the result (14), it is much easier to derive an expression for the saturation voltage. Here we discuss only the result for the "hard saturation" approach (21). From a comparison of the drain current, as given by (14) for the ohmic region, with the saturation value

$$I_{DSS} = \frac{\beta LE_c}{1+\Theta(V_G-V_{TS})} [\ V_G-V_{TS}-(1+\gamma)(V'_{DP}-V_S)\ ]$$

it has been shown [5] that

$$V'_{DP} \approx V_S + (\frac{V_G-V_{TS}}{1+\gamma}) + (LE_c) - [\ (\frac{V_G-V_{TS}}{1+\gamma})^2 + (LE_c)^2\ ]^{\frac{1}{2}}. \quad (24)$$

The above approximation is remarkably close to the result (23), which was obtained in a quite different way. This is shown in fig. 9, which compares both results with the constant mobility saturation voltage as a function of the channel length.

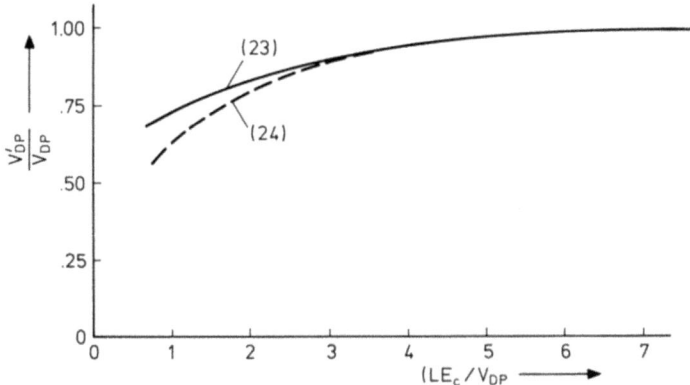

Fig. 9. Decrease of saturation voltage owing to velocity saturation.

Of course, the current expressions as discussed above can only be used for drain voltages lower than the boundary values (23) or (24). For higher values the influence of the drain depletion area has to be taken into account.

2.2. Finite output resistance

As we will discuss later on, two-dimensional numerical analysis of the MOST has shown that an analytical treatment of the drain depletion layer near the pinched-off channel area is very complicated and hard to implement in a MOST model. All models proposed and in use are rather crude approximations. The simplest approach, for instance, postulates the validity of the equations (13) or (20) for the saturation region ($V_D \geqslant V_{DP}$), provided that V_D in the f-function is replaced by V_{DP} and that the geometrical channel length L in the factor β is replaced by an effective channel length (L-l) [14].
The argument behind this approach is that the internal saturation voltage, which causes pinch-off or current saturation, does not vary much on a further increase of drain voltage, but the pinched-off region moves slightly towards the source (thickness l). Moreover, when the normal field is smaller than the lateral field, Poisson's equation may be solved one-dimensionally with the result

$$l \approx \alpha(V_D - V_{DP})^{\frac{1}{2}}, \tag{25}$$

where $\alpha \approx (2\varepsilon_s/qN)^{\frac{1}{2}}$.

For a channel length $L > 5$ μm, the above approximation, when substituted in the current equations, describes the imperfect saturating drain current rather well at high drain voltages. However, for a close fit, the constant α has to be determined empirically rather than with the above theoretical value [6,7]. Furthermore the fact that eq. (25) gives zero output resistance for $V_D = V_{DP}$ will cause numerical difficulties in circuit analysis. Several smoothing correction functions for this problem have been proposed.

The pinched off area can be treated more accurately by using the following assumptions [5]:

1. The pinch off point is characterized by a potential V_{DP} and a lateral field E_c.
2. Mobile carriers between this point and the drain are pushed away from the surface, and are spread uniformly in this region over a depth ranging from δ at the pinch-off point to δ_j for the drain. The following set of equations then applies to the drain-channel space charge layer (normal field neglected)

$$\frac{d^2V}{dx^2} = -\varepsilon_s^{-1}(qN + J/v_{SS})$$

$$J = I_{DSS}/Wy$$

$$y = (\delta_j - \delta)x/1 + \delta \qquad (26)$$

where y is the normal direction and x the lateral one. When the above equations are used for calculating the space-charge layer thickness l and the result is substituted in the relation $I_D = I_{DSS}(1 - l/L)^{-1}$, an analytic V(I) expression is obtained for the saturation region

$$V_D - V_{DP} = \frac{L^2}{\alpha}(1 - \frac{I_{DSS}}{I_D})^2 [1 + \frac{2I_{DSS}}{qNWv_s\delta_j}(\ln \frac{\delta_j}{\delta} - 1)] + LE_c(1 - \frac{I_{DSS}}{I_D}) \qquad (27)$$

This formula gives an accurate value for the drain current beyond saturation, even at channel lengths of 4 μm. This is shown in fig. 10 [5]. The mean width for δ was taken to be 100 Å.

Unfortunately the above result is rather complicated. However, for not too large current densities, it is possible to

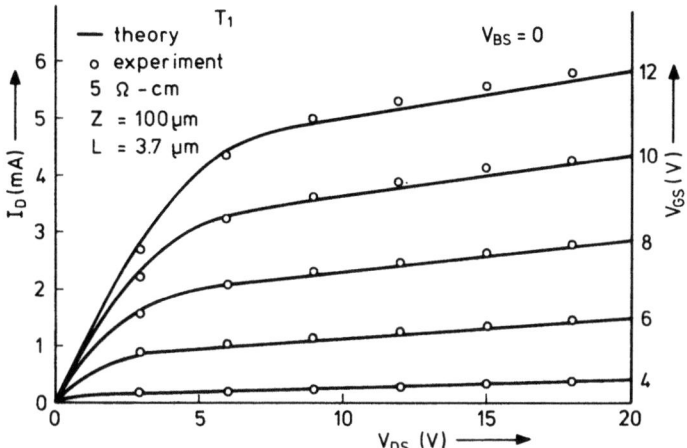

Fig. 10. Calculated and measured values of drain current.

approximate more closely to the expression for the space charge width l, which precedes the calculation of eq. (27). The result [5] is given by

$$l \approx \alpha^2 \left\{ \left[\left(\frac{E_c}{2} \right)^2 + \frac{(V_D - V_{DP})}{2\alpha^2} \right]^{\frac{1}{2}} - \frac{E_c}{2} \right\} . \quad (28)$$

The above expression, which may be considered as a generalization of the previous result (25), does not cause numerical difficulties at $V_D \rightarrow V_{DP}$. Although the output conductance that results from substituting (28) in the current equation has a discontinuity at this voltage, the calculated values compare rather well with measured values, in particular when α is used as an empirical parameter.

2.3. Diffusion current

In the previous discussion only the mobile carrier drift current was assumed to contribute to the drain current. However, near the pinched off region, where the inversion charge density decreases rapidly, the diffusion current might also become an important factor. It can be shown, however, that the contribution to the drain current is small [11].

When the current transport is by drift and diffusion, the drain current is given by

$$I_D = qWD_n \int_0^{y_i} n(x,y) dy \cdot \frac{d\xi}{dx} , \quad (29)$$

where D_n is the diffusion constant, ξ is the electron quasi-Fermi level normalized to kT/q and the electron density has to be integrated in the normal direction up to a point y_i, where its density equals n_i.
Furthermore, assuming that the quasi-Fermi level of majority carriers is constant, while ξ is constant in the y-direction only, the solution of Poisson's equation can be used to transform y in terms of the normalized band bending Δu

$$\frac{d\Delta u}{dy} = F(u_F, \xi, \Delta u)/L_D ,$$

where L_D is the intrinsic Debye length, u_F is the normalized substrate Fermi level and

$$F = [\ \exp(\Delta u - \xi - u_F) + \exp(u_F - \Delta u) + (\Delta u - 1)\exp u_F -$$
$$(\Delta u + \exp - \xi)\exp - u_F\ ]^{\frac{1}{2}} . \tag{30}$$

When the above result is substituted in (29), the drain current reads

$$I_D = \frac{qWD_n n_i L_D}{L} \int_{U_S}^{U_D} d\xi \int_{u_F + U_S}^{\Delta u_s} [\ \frac{\exp(\Delta u - \xi - u_F)}{F(\Delta u, \xi, u_F)}\ ]\ d\Delta u, \tag{31}$$

where Δu_s is the normalized surface potential.
In addition, Poisson's equation provides the relationship between Δu_s and the applied gate to substrate potential, namely

$$U_G' = q/kT(V_G - V_{FB}) = \Delta u_s + (\varepsilon_s t_{ox}/\varepsilon_{ox} L_D) F(u_F, \xi, \Delta u_s) . \tag{32}$$

For strong inversion the F-function can be further approximated by

$$F \approx [\ \exp(\Delta u_s - \xi - u_F) + \Delta u_s \exp u_F\ ]^{\frac{1}{2}} . \tag{33}$$

Once the terminal voltages are given, the amount of band bending is found from (32) and (33) and the drain current can be calculated from the integration indicated in (31).
From a numerical evaluation two conclusions can be drawn. The diffusion current usually contributes only a few percent to the total current. In contrast to the previous results, eq. (31) remains valid in the pinched off region, as the inclusion of the diffusion current automatically leads to complete saturation. Of course, due to the channel shrinkage effect a finite output resistance occurs beyond saturation.

2.4. Temperature dependence

When we consider the drain current as given by (14), two factors cause the current to change with temperature [15]. These are the mobility μ (being part of β) and the Fermi potential $2\varphi_F$, which determines V_T (compare (4)). Often the resulting temperature sensitivity is expressed in an effective input voltage drift dV_{GS}/dT, when I_D is kept constant. Formally this voltage drift is defined by

$$-\frac{\partial I_D}{\partial \mu}\frac{d\mu}{dT} + \frac{\partial I_D}{\partial V_{GS}}\frac{dV_{GS}}{dT} + \frac{\partial I_D}{\partial \varphi_F}\frac{d\varphi_F}{dT} = 0 .$$

Neglecting the influence of γ on the temperature dependence, dV_{GS}/dT can easily be found for the saturation condition from (14). Making use of the empirical relation $\mu \sim T^{-1.5}$ and the well known relation $\varphi_F = kT/q \ln N/n_i$, the temperature drift becomes

$$\frac{dV_{GS}}{dT} \approx \frac{3(V_{GS}-V_T)}{4T} - (1 + \frac{k}{2(2\varphi_F)^{\frac{1}{2}}})(\frac{E_g/2 - \varphi_F}{T}) . \tag{34}$$

Since the first and second terms are both positive, it is evident that a zero temperature drift point exists. This is shown in fig. 11, which compares (34) with an experimental result. Unfortunately this condition is not suitable for practical use in linear circuits.

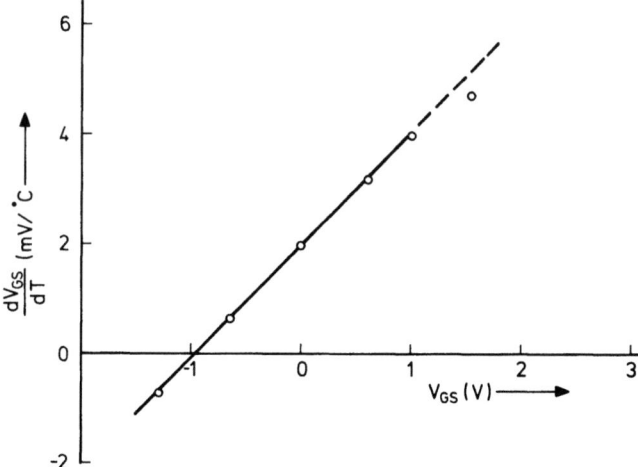

Fig. 11. The temperature drift as a function of bias condition.

2.5. Drain voltage breakdown

Usually the drain voltage breakdown of a MOST is lower than the value according to junction breakdown. This is due to the fact that the gate forces the equipotential lines round the drain to bend back to the drain corner near the interface. Consequently multiplication effects first start here. From a two-dimensional numerical solution of Poisson's equation in the space charge layer it has been concluded [16] that the derivative of the normal field at the surface in the normal direction y is fairly described by

$$\frac{\partial E_{sy}}{\partial y} \approx \frac{V_s - V_G}{(\epsilon_s/\epsilon_{ox}) t_{ox} h} ,$$

where V_s is the surface potential and h is an empirical constant of 1.5 μm. Using this result to calculate the lateral field at the drain corner E_{xc} with the aid of Poisson's equation, and also assuming that breakdown occurs for $E_{xc} = E_{cr} \approx 6 \times 10^5$ V/cm, the drain breakdown voltage becomes

$$V_{BR} = V_G + E_{cr} (\frac{\epsilon_s}{\epsilon_{ox}} t_{ox} h)^{\frac{1}{2}} . \qquad (35)$$

A linear dependence on V_G and a square root dependence on the oxide thickness is indeed found experimentally. This is shown in fig. 12.

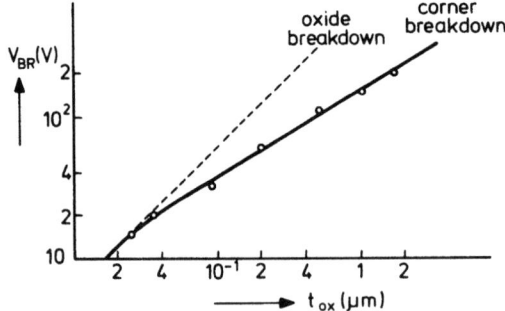

Fig. 12. Drain breakdown vs oxide thickness ($V_G = 0$).

3. CHARGE CHARACTERISTICS

3.1. Gate capacitances

When the gate voltage of a MOST is increased, it takes some time before a new equilibrium is established. An extra charge has to be built up. A representation of these time-dependent effects is given in fig. 13 by the inclusion of capacitances or their corresponding charges.
In this dynamic model only the capacitance C_{GB}, C_{GS} and C_{GD} are related to the intrinsic device structure; all other capacitances (C_{GSO} and C_{GDO}, which represent the gate overlap; C_{BS} and C_{BD}, which represent the source and drain depletion charge) are parisitics which are not discussed here. In nearly all physically based MOS models the value of the gate capacitances is derived by differentiating the gate charge Q_G to the voltages V_G, V_{GS} and V_{GD}.
When the MOST is in the passive mode ($V_G < V_{TS}$), only the capacitance C_{GB} is important. Since the mobile charge may be neglected the capacitance may be derived from differentiation of the substrate charge

$$Q_B = \varepsilon_s E_s = \frac{\varepsilon_s kT}{qL_D} F(u_F, \xi, \Delta u_s) \quad , \qquad (36)$$

where F is given in eq. (30).
Although the result agrees remarkably well with experimental data, a considerable amount of computer time is required for evaluation. For the study of surface states in MOS capacitors this is not a drawback, but for transistors a simpler modelling is required. This can be achieved by realizing that C_{GB} is the series capacitance of the oxide layer and the substrate depletion layer (width = $(2\varepsilon_s \Delta V_s/qN)^{\frac{1}{2}}$)

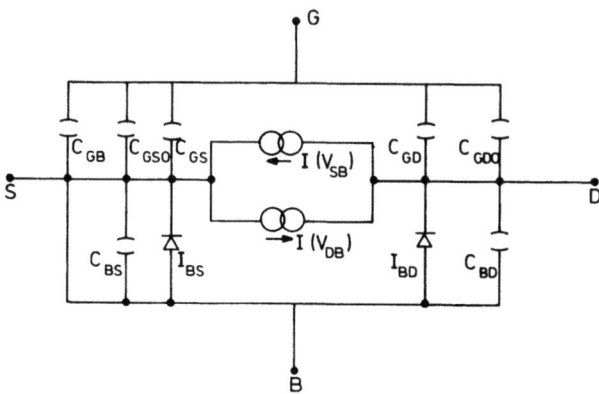

Fig. 13. A dynamic MOS model.

$$C_{GB} = \frac{C_{ox}C_B}{C_{ox}+C_B} = [\ 1 + \frac{2\Delta V_s^{\frac{1}{2}}}{k}\ ]^{-1} \cdot C_{ox}\ .$$

Solving ΔV_s from the equations (1) through (3) we have

$$C_{GB} = C_{ox}[\ 1 + \frac{4}{k^2}(V_G - V_{FB})\ ]^{-\frac{1}{2}}\ . \tag{37}$$

Although the above result is inaccurate at $V_G = V_{FB}$ owing to the failure of the depletion approximation at this bias voltage, this has hardly any practical consequence for computing circuit delay times. For $V_G > V_{TS}$, at which the active mode prevails, the gate-substrate capacitance is usually taken constant with V_{TS} substituted for V_G in eq. (37).

For this mode the active capacitances can be derived from the expression for the total free carrier charge Q_N. Generally this charge is given by

$$Q_N = W\,C_{ox} \int_0^L [\ V_G - V_{TX} - V - k(V+2\varphi_F)^{\frac{1}{2}}\ ]\ dx\ . \tag{38}$$

This integral can be solved by changing the variable dx for dV by making use of (7). Unfortunately the result is complicated and differentiation leads to expression [17] for the capacitances that require a considerable amount of computer time. In fig. 14 it is shown that a simple but accurate modelling for Q_N can be achieved with the relations [6]

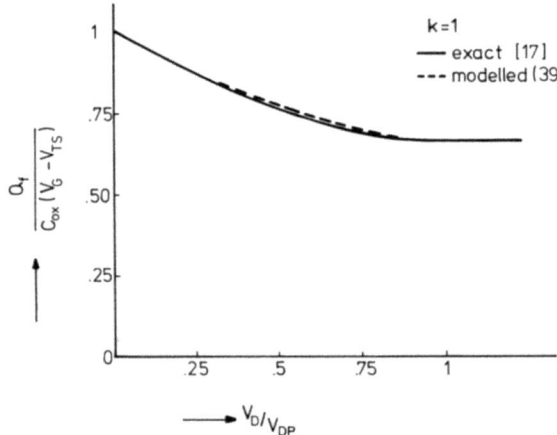

Fig. 14. The total free-carrier charge as a function of V_{GS} and V_{DS}.

$$Q_N = \frac{2}{3} C_{ox} (V_1+V_2 - \frac{V_1 V_2}{V_1+V_2}) \; , \tag{39}$$

where

$$V_1 = V_G - V_{TS}$$
$$V_2 = V_G - V_{TD}$$

Actually this result may be considered as a generalization of an exact expression for Q_N, which is obtained by taking k=0 in (38) [4]. In this case (39) is also found, but V_2 reads

$$V_2 = V_{GD} - V_{TX} - k(V_S + 2\varphi_F)^{\frac{1}{2}} \; .$$

Differentiating the above expression with respect to V_{GS} and V_{GD} the following capacitance values are obtained

$$C_{GS} = \frac{2}{3} C_{ox} [1 - \frac{V_2^2}{(V_1+V_2)^2}] \tag{40}$$

$$C_{GD} = \frac{2}{3} C_{ox} [1 - \frac{V_1^2}{(V_1+V_2)^2}] \; . \tag{41}$$

Since V_2 becomes zero when V_D approaches V_{DP}, the values of C_{GS} and C_{GD}, in agreement with experimental results, draw near to values of $2/3\, C_{ox}$ and 0, respectively. This is not the case with many MOS models, which use (40) and (41), but on the other hand make use of $V_2 = (V_{GD}-V_{TS})$ [4,25]. For low values of V_{DS}, $V_1 \approx V_2$ and the well-known result $C_{GS}=C_{GD}=\frac{1}{2}C_{ox}$ is obtained.

3.2. Admittances

In linear circuits not only the gate capacitances but also the gate conductances of the MOST may be important. In general, these properties are determined both by (7) and the continuity equation

$$\frac{\partial I}{\partial x} = W C_{ox} \frac{d}{dt} [V_G - V_{TX} - V - k(V+2\varphi_F)^{\frac{1}{2}}] \; . \tag{42}$$

Taking the approximation as discussed with eq. (14) and using the small signal approximation

$$I(x,t) = I_o + \tilde{i}(x,\omega)\exp j\omega t$$
$$V_G-(1+\gamma)V = u_o(x) + \tilde{u}(x,\omega)\exp j\omega t \quad,$$

eq. (42) leads to a second-order differential equation, for which numerical values of the y-parameters have been obtained [18].

For the practical case where the transistor operates in saturation and γ is small (low substrate dope), a simple closed form expression for the y-parameters can be obtained by making use of a series expansion of $\tilde{i}(x,\omega)$ in terms of the angular frequency ω [19]. By iterative integration of equations (14) and (42) it has been shown [20] that

$$y_{11} = \frac{\Delta \tilde{i}_G}{\Delta \tilde{v}_{GS}} \bigg|_{v_{DS}=0}$$

and

$$y_{21} = \frac{\Delta \tilde{i}_D}{\Delta \tilde{v}_{GS}} \bigg|_{v_{DS}=0}$$

are given by

$$y_{11} = y_{21} \times (2\omega/3\omega_o)j$$
$$y_{21} = g_{ms}[1 + (4\omega/15\omega_o)j]^{-1} \quad (43)$$

where the gain-bandwidth product follows from

$$\omega_o \approx L^2/\mu_n(V_G-V_{TS}) \quad . \quad (44)$$

These expressions are accurate within 3 percent up to a value ω_o. For higher frequencies (outside the practical range) more accurate values can be achieved by taking higher order terms into account. Since the other y-parameters are mainly determined by parasitics, we do not discuss them here.

4. SUB-THRESHOLD CURRENT

For gate voltages between flatband and threshold the drain current is essentially determined by the amount of band bending at the source. In general, its value may be calculated with the theory outlined in section 2.3 and for which several functional relations can be simplified in weak inversion. To a good approximation the F-function for this region ($2 < \Delta u < 2u_F$)

reads [21,22]

$$F \approx [(\Delta u-1)\exp u_F]^{\frac{1}{2}}$$

Comparing this result with eq. (36), we observe that in the sub-threshold region the normal field E_s in the semiconductor is proportional to the square root of the surface band bending. When the above expression is substituted in (31) it can be shown [22] that the current for not too short channels is given by

$$I_{ST} = \frac{WL_B q D_n n_s}{L} [\ 1-\exp(-qV_{DS}/kT]\ , \qquad (45)$$

where the minority carrier density at the source is given by

$$n_s \approx \frac{n_i[\exp(b-1)u_F-1]}{[2(U_S+bu_F-1)]^{\frac{1}{2}}} \qquad (46)$$

and the normalized band bending parameter b is defined by the source condition

$$\Delta u_S = U_S + b|u_F| \ .$$

In the threshold region $1 < b < 2$.
Equation (45) may be interpreted as a diffusion current, which results from an injection of minority carriers from the source caused by surface band bending, and which flows to a depth from the surface equal to the extrinsic Debye length $L_B = [\varepsilon_s kT/q^2 N]^{\frac{1}{2}}$. The equilibrium surface concentration n_s increases exponentially with the band bending.

The relation between gate voltage and band bending given by (32) also simplifies for the region discussed. Since the factor 1 in the F-function approximation can be omitted with negligible error, eq. (32) yields

$$V_G - V_{FB} = V_S + (\frac{qu_F}{kT})\ b + k\ (V_S + \frac{qu_F b}{kT})^{\frac{1}{2}}. \qquad (47)$$

In fig. 15 some experimental results are compared with the calculated current according to eq. (45) and (47). For a channel length > 5 µm the agreement is fair. From the same equations we obtain for the slope $\varepsilon = d(\ln I_{ST})dV_{GS}$

$$\varepsilon \approx [\ 1 + \frac{k}{2}\ (V_S + \frac{qu_F b}{kT})^{-\frac{1}{2}}]^{-1}\ .$$

In general, then, the exponential slope depends on V_{GS} through the band bending term, but this dependence becomes weaker as a substrate bias is applied. Also, it is noted that ε decreases

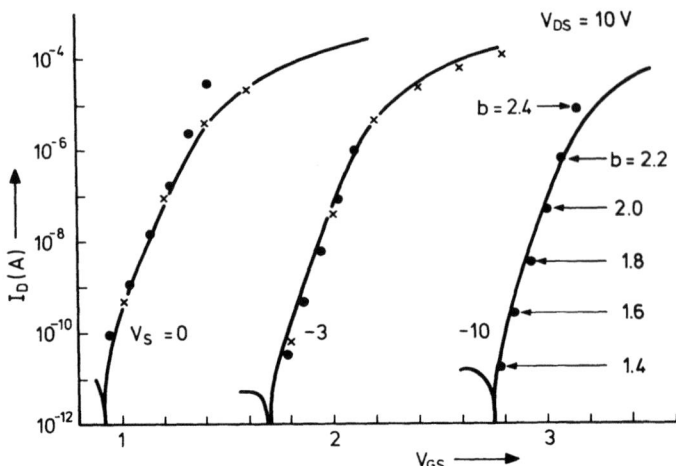

Fig. 15. The sub-threshold current vs gate voltage.[22].

with k, which means that the range of gate voltage between weak and strong inversion increases as the parameter $t_{ox} \cdot \sqrt{N}$ increases.

For short channels deviations from the above theory occur at higher values of the drain potential, which affects in this case the surface bending at the source. A more quantitative discussion is given elsewhere.

5. RESULTS OF TWO-DIMENSIONAL ANALYSIS

Due to the two-dimensional nature of the device, in particular at shorter channel length, all one-dimensional analytic modelling, as discussed previously, depends on simplifications whose validity is difficult to prove. Of course comparison of the results with experiments is a fair proof, but not always conclusive. An alternative can be found in comparison with results from a numerical method, which solves Poisson's equation and the continuity equations for mobile carriers in a two-dimensional device. In general, these methods also allow the use of a field-dependent mobility and doping profiles. Several approaches with their results have been published [8,23]. In this section we discuss some relevant results and compare them with those given before.

Figures 16 through 19 give the distribution of potential, free carrier density and current density of a 4 µm-channel MOST under the following conditions: $V_{GS} = 1.5$ V, $V_{DS} = 4V$, $V_{SB} = 0$, $N = 2 \times 10^{15}/cm^3$, $t_{ox} = 1000$ Å, $q_s = 2 \times 10^{-8}$ C/cm². Moreover a field-dependent mobility according to (19) and realistic doping profiles for source and drain have been used. The metallurgical junctions are represented by drawn lines in fig. 16.

From the plots we first draw some preliminary conclusions:
a. A real pinch-off point is not manifest. Only at $x = 3.25$ μm has the current density at the surface decreased by a factor two, E_x has reached a value $E_c = 2 \times 10^4$ V/cm and the carrier density, after having reached a saddle point value, starts to decrease strongly towards the drain (fig. 18).
b. At $x = 3.45$ μm E_y changes sign (fig. 16).
c. In the "pinch-off" area (between 3.25 and 3.65 μm) the total number of free carriers $\int n(x,y)dy$ remains constant (fig. 17) owing to velocity saturation, but the carriers are pushed away from the surface (fig. 19).
d. Due to the high mobile carrier density ($> 10^{16}/cm^3$) at a depth of 0.15 μm from the surface at the drain boundary, the potential distibution shows a bump (fig. 16).

The gradual channel approximation as applied to the "ohmic" region $x < 3.25$ μm, appears to be reasonably valid from the following observations.
a. From the distribution of potential (fig. 16) and field (not shown here) $E_y > E_x$.
b. The free carrier density corresponding to eqs. (14) through (24) (dotted line) compares well with the numerical values (fig. 17).
c. The saturation voltage as given by (23) fits the potential value at $x = 3.25$ μm (fig. 17).
d. The value of the drain current according to (20) agrees with the numerical value provided that channel shortening is taken into account.

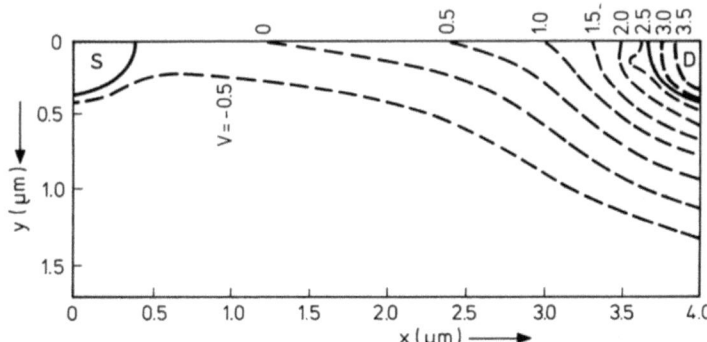

Fig. 16. Potential distribution in the semiconductor.

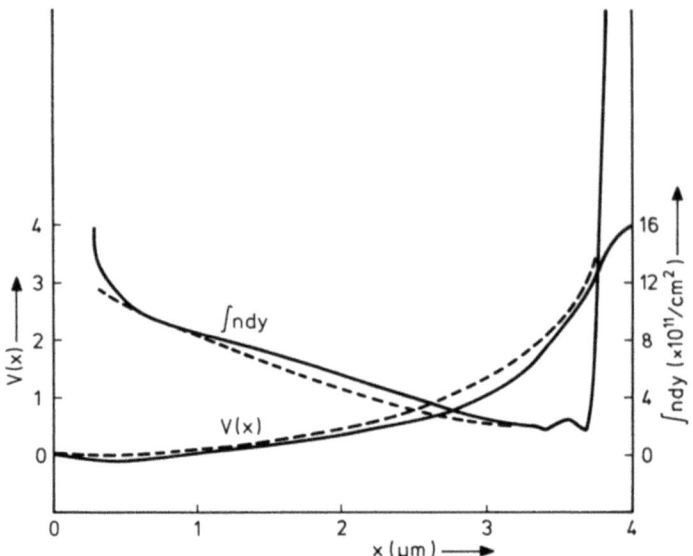

Fig. 17. Distribution of potential and free carriers along the channel.

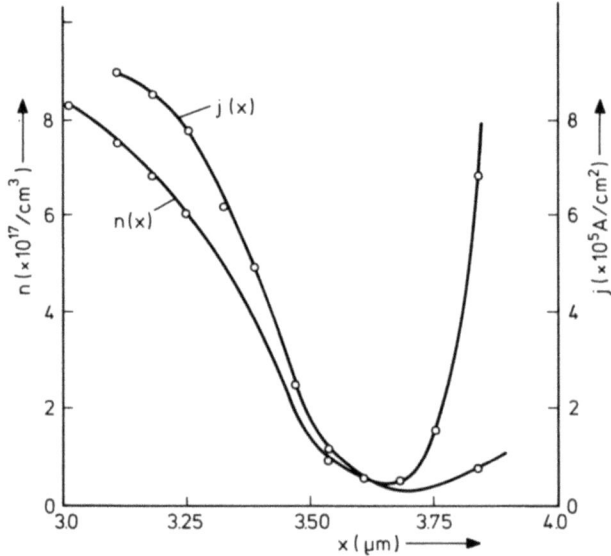

Fig. 18. Free carrier density and current density at the surface.

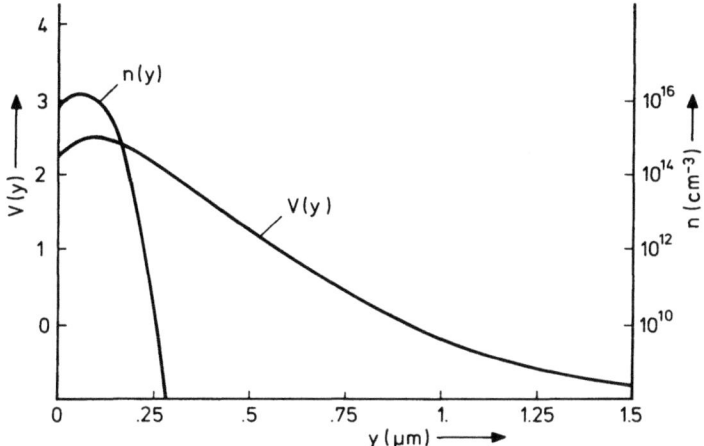

Fig. 19. Potential and free carrier density in the normal direction at x = 3.5 µm.

However the assumptions usually made to arrive at a practical analytical description of the saturation behaviour (compare section 2.2) are not generally justified.
For instance we observe that:
a. the carriers are not distributed uniformly through the space charge region (fig. 19).
b. although the normal field close to the drain becomes small compared with the lateral field, the assumption used in (26) $dEx/dx \gg dEy/dy$ is violated at x = 3.45 µm, where the normal field changes sign.
Therefore it may be concluded that the agreement between experiment and theory of section 2.2 is probably caused by the fact that the integration routine applied is not affected much by point b and also that the drain conductance is not a very critical quantity.

6. A MODEL BASED ON UNIFORM SPACE CHARGE DENSITY

Recently an interesting model has been proposed, which claims to give better agreement with second order effects observed experimentally [26] . It is mainly based on the following assumptions.
a. The space charge layer of mobile carriers and substrate depletion charge, which according to quantum statistics is shown in fig. 20, can be modelled by a single space charge layer, which is uniform normal to the surface.

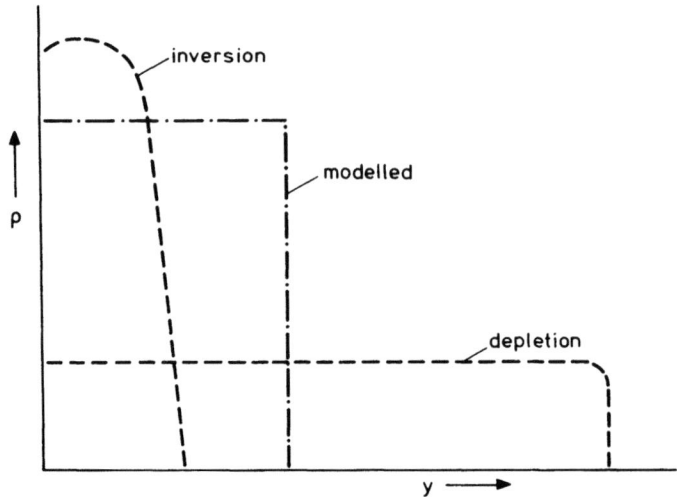

Fig. 20. Space charge normal to the surface.

b. The mobile part of this layer is transported by drift and diffusion with a constant mobility.
c. The channel is split into two regions; to the first part the gradual channel approximation applies, the current in the drain region is obtained from an elegant analytical solution of the two-dimensional Poisson equation.

The first assumptions automatically lead to a parabolic potential distribution normal to the surface. The width of the space charge layer follows from the continuity condition of the displacement at the interface. When the immobile charge in this layer is subtracted from the totally induced charge, the mobile charge is shown to be given by

$$Q_m = -\frac{kT}{q} \frac{\varepsilon_{ox}}{t_{ox} a} \frac{\eta^2 + \eta/a - U_G}{\eta} \tag{48}$$

where

$$\eta = (U_G - \Delta u_s) \tag{49}$$

and

$$a = \left(\frac{kT\varepsilon_{ox}^2}{2q^2 t_{ox}^2 \varepsilon_s N} \right)^{\frac{1}{2}}$$

By elaborating the second assumption, the drain current follows from

$$I_D = -WD_n Q_m \frac{d\xi}{dx} = -(\frac{kT}{q})(\frac{D_n \varepsilon_{ox} W}{La^2})[\frac{\eta^2}{2} + (2a+a^{-1})\eta + (1-U_G)\ln\eta]_{\eta_S}^{\eta_D}, \qquad (50)$$

where the potential ξ, which may be considered as an analogon of the quasi-Fermi potential, is given by

$$\eta^2 + \eta/a - U_G = \exp[-\eta/a + U_G - U_S - \xi - 2u_F]. \qquad (51)$$

Since η_S and η_D can be obtained numerically from eq. (51) by putting ξ equal to U_S and U_D respectively, the current is known from eq. (50).

It is claimed that this model without using the mobility degradation at high values of V_G (compare (19)), provides better agreement between V_G anomalies in current, transconductance and gate capacitances than previous models. Although this is an interesting result, some questions remain with respect to this model's general validity and applicability.

7. NOISE SOURCES

In linear circuits MOS transistors are sometimes used as low noise input stages. In general, this is only useful at high frequencies, where the device is limited by thermal noise; at frequencies $< 10^6$ Hz excess noise of the $1/f$ type dominates. Usually the fluctuating drain current $< \Delta i_D^2 >$ is expressed as a frequency spectrum $< \Delta i_D^2 > = \int_0^\infty S_i(f)df$, the spectral density of which can be measured. In the case of thermal noise it has been shown [27] that

$$S_i(f) = 4kT\beta \{ \int_{V_S}^{V_D} g^2(V)dV / \int_{V_S}^{V_D} g(V)dV,$$

where $g(V)$ is the term between brackets in the current expression (7). Using the approximation leading to the result (14), it is easily shown that

$$S_i(f) = \frac{8}{3}(1+\gamma)kTg_{ms}. \qquad (52)$$

This result may be recognized as a generalized Nyquist formula.

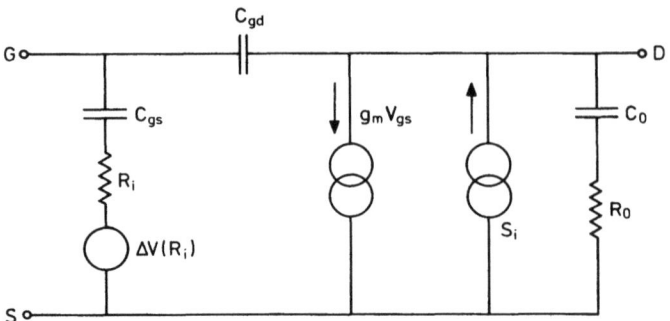

Fig. 21. Noise equivalent circuit of the MOS transistor.

The low frequency noise is mainly generated by surface states. For a clean-technology device the spectral density can be modelled [28] by

$$S_i(f) \approx \frac{\alpha(N_{ss}) g_{ms} I_{DSS}}{C_g f} \quad , \tag{53}$$

where $C_g = WL\, \varepsilon_{ox}/t_{ox}$ and α is a factor proportional to the surface state density N_{SS}. For $N_{SS} \approx 10^{10}/cm^2$ (<100>-material), $\alpha \approx 10^{-4}$.
In devices processed by the ion implantation technique an excess noise term with a $f^{-\frac{1}{2}}$ dependence sometimes appears, which can extend up to very high frequencies. By means of special annealing procedures this component can be reduced [29].

The above noise sources can be incorporated in a noise circuit model as shown in fig. 21. Besides the drain current fluctuations the thermal noise voltage $<\Delta V^2>$ of the input conductance has also been added to this figure. At high frequencies the resistance $R_{in} = 2/5\, g_{ms}^{-1}$, which can be derived from eq. (43), also determines the noise figure.

The author is obliged to E.F. Stikvoort (Philips Research Labs) for providing the numerical results of section 5.

REFERENCES

1. H.K.J. Ihantola and J.L. Moll. Sol. St. Electr. 7 (1964) 423.
2. C.T. Sah and H.C. Pao, I.E.E.E. Trans ED-13 (1966) 393.
3. J.A. van Nielen and O.W. Memelink, Phil. Res. Rep. 22 (1967) 35.
4. J.E. Meyer, RCA Rev. 32 (1971) 42.
5. G. Merckel a.o., IEEE Trans. ED-19 (1972) 681.
6. F.M. Klaassen, Phil. Res. Rep. 31 (1976) 71.
7. D. Frohman-Bentchkowski, IEEE Trans ED-16 (1969) 108.
8. D. van Dorpe a.o., Sol. St. Electr. 15 (1972) 547.
9. N. Murphy a.o., Sol. St. Electr. 12 (1969) 775.
10. F.F. Fang and A.B. Fowler, Phys. Rev. 169 (1968) 619.
11. H.C. Pao and C.T. Sah, Sol. St. Electr. 9 (1966) 927.
12. F.F. Fang and A.B. Fowler, J. of Appl. Phys. 41, (1970) 1825.
13. G. Baum and H. Beneking, IEEE Trans ED-17 (1970) 481.
14. V. Reddi and C.T. Sah, IEEE Trans ED-12 (1965) 139.
15. R.S.C. Cobbold, Electr. Lett. 2 (1966) 190.
16. H.C. de Graaff, Phil. Res. Rep. 25 (1970) 21.
17. R.S.C. Cobbold, Theory and Applications of FET's, Wiley-Interscience (1970).
18. J.A. Geurst, Solid State Electr. 9 (1966) 129.
19. F.M. Klaassen, IEEE Trans ED-14 (1967) 368.
20. J.A. van Nielen, Solid St. Electr. 12 (1969) 826.
21. M.B. Barron, Sol. St. El. 15 (1972) 293.
22. R.R. Troutman, IEEE J. Sol. St. Circ. 9 (1974) 55.
23. D.P. Kennedy and P.C. Murly, IBM J. Res. Dev. 17 (1973) 2.
24. D.A. Hodges and H. Schickman, IEEE J. Sol. St. Circ. 3 (1968) 285.
25. T.K. Young and R.W. Dutton, Mini-Msinc, Techn.Rep. Stanford Univ. (1976).
26. J. El-Mansy and A.R. Boothroyd, Techn. Dig. IEDM (1974) 35.
27. F.M. Klaassen and J. Prins, Phils. Res. Rep. 22 (1967) 505.
28. F.M. Klaassen, IEEE Trans ED-18 (1971) 887.
29. K.L. Wang, Dig. IEDM (1976) paper 14.5.
30. J. Koomen, Sol. St. Electr. 16 (1973) 801.

CHARACTERIZATION AND MEASUREMENTS OF MOST DEVICES

F.M. Klaassen

Philips Research Laboratories
Eindhoven - The Netherlands

INTRODUCTION. In a previous chapter [1] several general purpose, physically based, large-signal and special MOST models have been introduced. Generally these models are characterized by a set of mathematical relations between bias conditions, material constants, dimensions and process parameters.
The latter quantities notably include:
the threshold voltage V_{TS}, which separates the active from the passive mode;
the body factor k, by which V_{TS} increases with substrate bias;
the gain factor β, expressed in A/V^2;
the mobility degradation factor θ_1, which is important at high gate voltages;
the current reduction factor θ_2, related to velocity saturation, and
the channel length modulation constant α, which is a measure of the saturated drain resistance.
To a large extent the accuracy of the model depends on the determination of the above parameters. The present chapter describes the methods used to measure these quantities and some other device figures that characterize the MOST.
  First the procedures widely used for determining the d.c. drain current parameters are discussed and considered in rational succession. Nearly all methods considered are graphical and physically oriented. It is shown that more automated methods are needed.
Next the applicability of the approximations for the intrinsic gate capacitances is checked. This is followed by the determination of small signal admittance parameters. Finally two special topics are considered: noise and cross-modulation. The

results of both properties demonstrate the shortcomings of all available models.

1. THRESHOLD VOLTAGE AND BODY FACTOR

One of the key parameters characterizing MOS transistors in the active mode is the threshold voltage (V_{TS}). In general, this parameter depends on the substrate bias (V_{SB}), the oxide thickness and the substrate doping. Except for some implanted substrate layers [2,3] the threshold voltage satisfies the relation (compare eq. (6) of ref. 1)

$$V_{TS} = V_{TX} + k (V_{SB} + 2\varphi_F)^{\frac{1}{2}}, \qquad (1)$$

where V_{TX} and k (known as the body factor) are process parameters and $2\varphi_F$(k) is the diffusion potential, which depends very weakly on k.

Usually V_{TS} is determined by extrapolation from a plot of the square root of the saturated drain current vs the applied gate-source voltage (V_{GS}) [4]. This procedure finds some justification in the approximation for the saturated drain current, which reads (compare eqs (14) and (15) of ref. 1)

$$I_{DSS} \approx \frac{\beta(V_{GS}-V_{TS})^2}{[1+k\gamma^{-1}(V_{SB}+2\varphi_F)^{\frac{1}{2}}][1+\Theta_1(V_{GS}-V_{TS})]}, \qquad (2)$$

where β is the gain factor, Θ a mobility correction factor and γ a modelling factor, which varies between 2.0 and 2.5. Fig. 1a gives a result for an enhancement transistor, fig. 1b for a depletion transistor. Except for the tails at low values of V_{GS}, $I_{DSS}^{\frac{1}{2}}$ vs V_{GS} exhibits to some extent a linear dependence. However, this result is rather accidental owing to the two non-linear terms in the denominator, which compensate each other. Consequently the linear range is often quite small, particularly in devices built on low-resistivity substrates.

A better method, though requiring a more sensitive current reading is to measure the drain current at low drain-source voltage (V_{DS}). In this case the substrate depletion charge does not complicate the current expression and we simply have

$$g_D = \frac{I_D}{V_{DS}} = \frac{\beta(V_{GS}-V_{TS})}{1+\Theta_1(V_{GS}-V_{TS})}. \qquad (3)$$

As Θ_1 is usually small, a first-order value of V_{TS} is obtained from an extrapolation from the plot g_D vs V_{GS}.

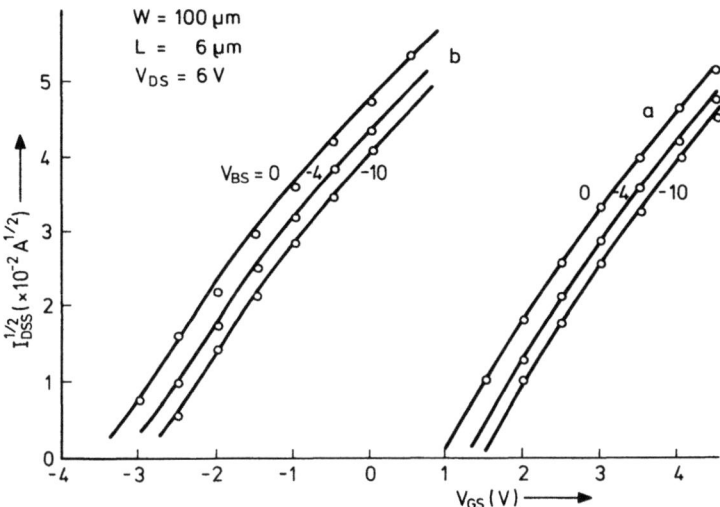

Fig. 1. Square root of drain current vs gate voltage.

This is shown in fig. 2. A second-order correction for V_{TS} can be made once β and θ_1 have been determined. This will be discussed in the next section.

As expected, the V_T values found from fig. 2 often differ from those of fig. 1. However, this deviation is seldom more than the statistical variation on a wafer.

From the values obtained for V_{TS} at different values of the substrate bias, V_{TX} and the body factor k are determined by

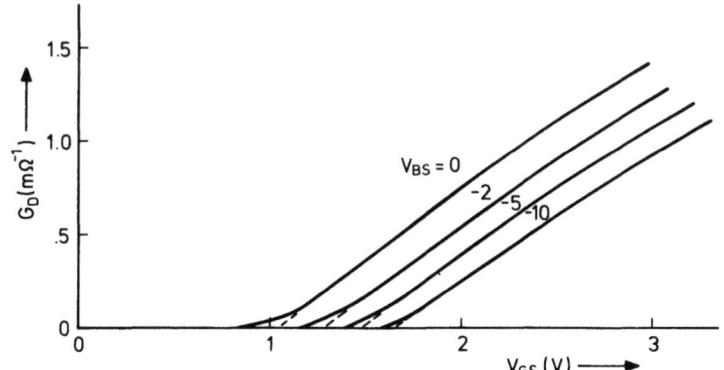

Fig. 2. Drain conductance (V_{DS} = 50 mV) vs gate voltage.

plotting V_{TS} vs $(V_{SB} + 2\varphi_F)^{\frac{1}{2}}$ and approximating $2\varphi_F \approx 0.7$ V.
This is shown in fig. 3. Here, too, a second-order correction for
k is easily obtained by using the first value of k to correct
the value of $2\varphi_F(k) = $ const. $\times \ln(kC_{ox})$.

2. GAIN FACTOR AND MOBILITY REDUCTION

When the threshold voltage has been determined from the
channel conductivity, the gain factor β and the mobility
degradation factor Θ can easily be deduced from a plot suggested
from a rearrangement of (3)

$$\frac{1}{g_D} = \frac{\Theta_1}{\beta} + \frac{1}{\beta(V_{GS}-V_{TS})} . \tag{4}$$

Of course, a more appropriate way to determine the mobility
decrease at high gate fields can be found in a measurement of
the Hall effect [5,6]. However, as this technique requires
special configurations, only possible for long channels, it
is not suitable for MOST process characterization.
In fig. 4 the measured values from fig. 2 have been used for
plotting g_D^{-1} vs $(V_{GS}-V_{TS})^{-1}$. The values of Θ_1 and β are obtained
from a least squares fit. For n-channel devices higher values
of Θ_1 are usually found than for p-channels, particularly in
implanted depletion devices (compare fig. 1b). As far as the
latter devices are concerned this result is caused by the fact
that at low values of V_{GS} the current path extends more into the

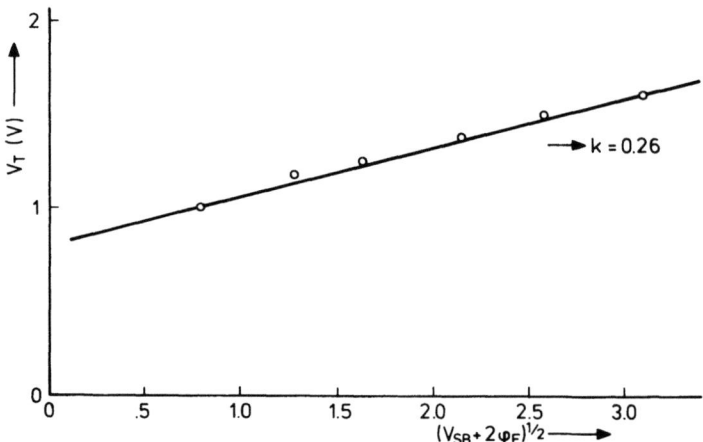

Fig. 3. Threshold voltage vs bulk voltage.

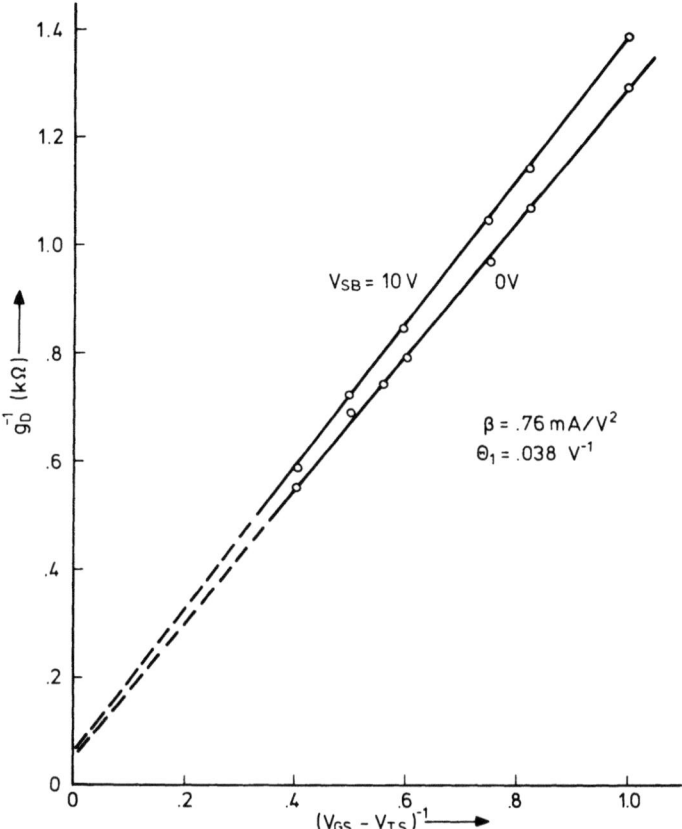

Fig. 4. Drain conductance vs gate voltage.

bulk than it does at higher voltages. Therefore the mobility decreases from bulk value to a value typical of inversion layers [6]. In many cases, however, the value of Θ_1 strongly suggests that a source series resistance is responsible for the non-linear behaviour of g_D or $I_{DSS}^{\frac{1}{2}}$. Although a standard MOST model does not exactly cover the two complications discussed, it has been found that the Θ correction is satisfactory for CAD use.

As mentioned before, the values of β and Θ_1 can be used again to make a second-order correction for V_{TS}. From another rearrangement of (3) we have

$$V_{GS} = V_{TS} + [\ \beta/g_D - \Theta_1]^{-1}. \tag{5}$$

In practice one correction loop has been shown to be sufficient.

As $\beta = \beta_o$ W/L, where $\beta_o \approx 25$ µA/V², or 7 µA/V² for [100] n-channel and p-channel devices, respectively, a measurement of the gain factor can be used to determine the uncertainty ΔL in the effective channel length L. In general ΔL is more or less a constant for a specific process, due to underetching of the gate oxide and lateral outdiffusion of source and drain. ΔL can be found by measuring β as a function of the channel length, as specified by the mask. This is shown in fig. 5 for a Mo gate process, where β^{-1} has been plotted vs L.

The procedures outlined above may be critized on the grounds that they yield parameters only from a small part of the characteristics, close to the plane V_{DS} being zero. Therefore it would be more satisfactory if these parameters, and also a parameter representing velocity saturation, could be determined from the total characteristics. Since such a procedure can only be handled properly by a more automated method, we do not treat such an approach here, but leave this to another chapter "A MOST model for CAD with automated parameter determination".

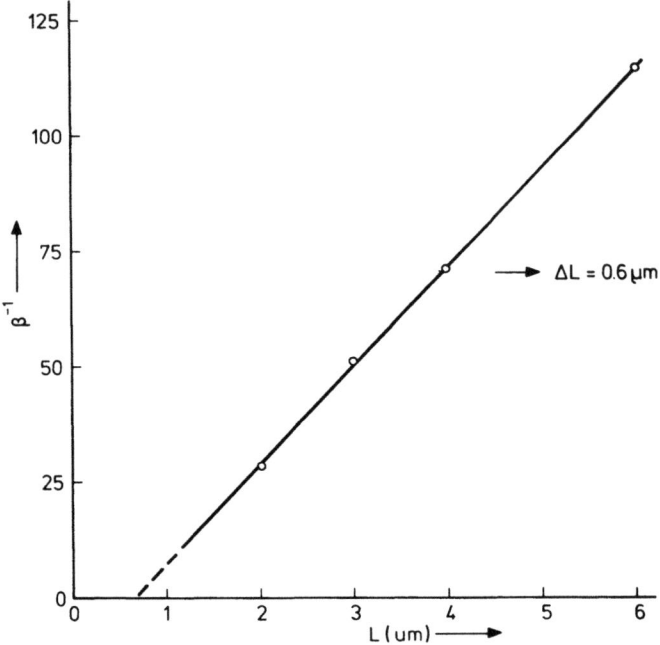

Fig. 5. Gain factor vs channel length (on mask).

Of course, by using a special technique, the effect of velocity saturation can be measured; for example the pulsed bridge technique [7] has been used for measuring the drain current. In order to avoid complications of non-linear field distribution in the channel due to saturation, special devices with very thick oxides were used. A less accurate method consists in measuring the saturated transconductance. When the substrate doping is not too high, this quantity is given to a good approximation (comp. 17) and (19) of ref. 1) by

$$g_{ms} \approx \frac{\beta V_{DP}}{1+\Theta_1(V_G-V_{TS})+\Theta_2 V_{DP}}, \qquad (6)$$

where V_{DP} is the drain voltage at which the transconductance saturates, $\Theta_2 = (LE_C)^{-1}$ and E_C is the field at which velocity saturation starts. The value of Θ_2 can then be obtained from a plot of g_{ms}^{-1} vs V_{DP}^{-1} according to

$$\frac{\beta}{g_{ms}} \approx \Theta_2 + \frac{1+\Theta_1(V_{GS}-V_{TS})}{V_{DP}} \qquad (7)$$

In fig. 6 some results are shown for a MOST with a channel length of 3.5 μm. From the minimum a value $\Theta_2 = .18$ V^{-1} and a corresponding value $E_c = 1.6 \times 10^4$ V/cm have been obtained. These values differ not much from those given in the literature [7].

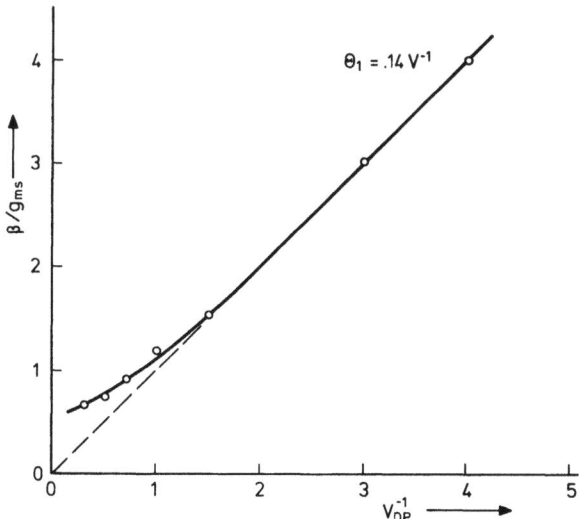

Fig. 6. Transconductance vs saturation voltage.

3. DRAIN RESISTANCE

Obviously a study of the drain output resistance is the best way to investigate experimentally the effect of the drain depletion layer n the saturation region of the transistor. The slight increase in drain current with drain voltage is fairly insensitive to model variations. Following the mechanism for current saturation, as discussed in section 22 of ref. 1, the output resistance can be derived from eq. (27);

$$V_D = \frac{L^2}{\alpha^2}(1 - \frac{I_{DSS}}{I_D})[2\frac{I_{DSS}}{I_D^2} + (1 + \frac{I_{DSS}}{I_D})\frac{2\ln y_j/\delta}{qNWy_j V_s}]$$
$$+ \frac{1}{\Theta_2}\frac{I_{DSS}}{I_D^2} \tag{8}$$

where I_{DSS} is the drain curent value at a drain voltage equal to the internal saturation voltage V_{DP}, I_D is the actual drain current, y_j is the drain diffusion depth, δ is the channel thickness at pinch-off, V_s is the saturated drift velocity and $\alpha = (2\varepsilon_s/qN)^{\frac{1}{2}}$.

The validity of the above expression has been fully checked [8] for devices with channel lengths ranging from 3 to 25 μm, an oxide thickness ranging from 900 to 1500 Å and made on substrates between 1 and 5 Ωcm. The value of y_j was 3.5 μm. As an example we present in fig. 7 results for the curent and drain resistance related to 3.7 μm-channel MOST. Although the fit between theory and measurements is closer than 5 percent, it might be advisable to consider the factors α and Θ_2 as process parameters rather than as known constant. The reason for this

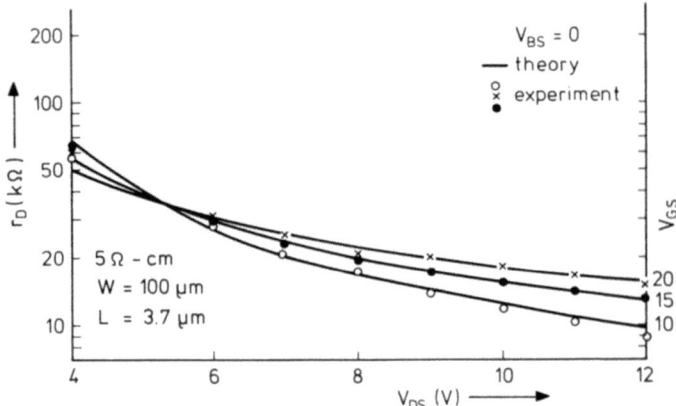

Fig. 7. Dynamic drain resistance.

approach lies in the invalidity of the assumptions leading to eq. (8). It has been shown [9] that in devices with a thin gate oxide, built on a high-resistivity substrate, channel shortening also may be caused by the fringing field between gate electrode and drain area. From qualitative measurements it was concluded that this effect becomes important when 20 Ωcm material is used.

As discussed in ref. 1, a practical expression for the effective channel length has been derived for a device with L > 4 μm (compare (28) of ref. 1). Using this expression we obtain the following approximation for the drain resistance

$$r_D \approx \frac{I_{DSS}}{I_D^2} [\; \Theta_2^2 + 2 \frac{V_D - V_{DP}}{\alpha^2 L^2} \;]^{\frac{1}{2}} \tag{9}$$

Observing this result we conclude that $(r_d I_D)^2$ should vary linearly with V_{DS} [10]. In general, this behaviour is found experimentally, as shown in fig. 8 for a 10 Ωcm, 700 Å gate-oxide, 3 μm-channel device. The value of α that can be derived

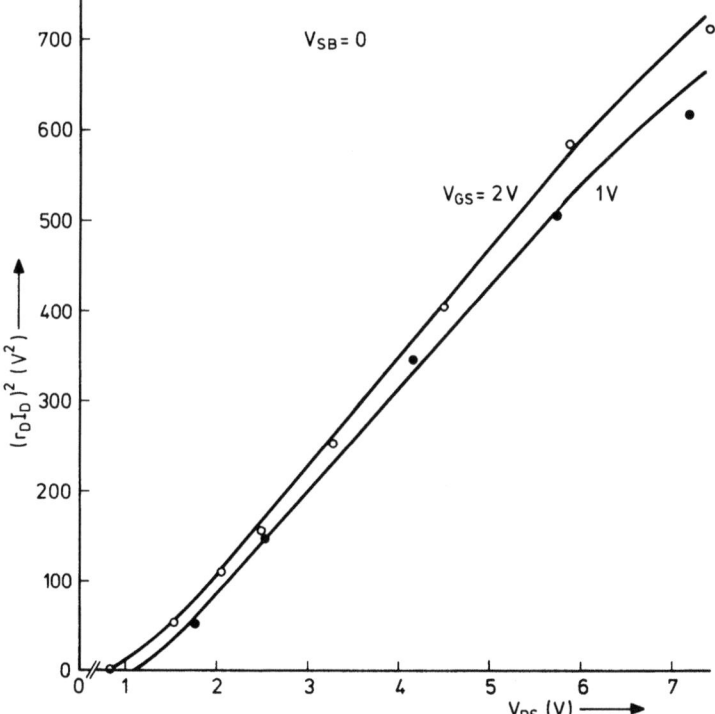

Fig. 8. Variation of $(r_D I_D)^2$ product.

from this plot is somewhat larger than according to the drain depletion effect.

4. CAPACITANCES

In section 3.1 of ref. 1 it was shown that all three intrinsic MOST capacitances C_{GS}, C_{GD} and C_{GB} can only be derived theoretically by approximate modelling.
Fortunately these capacitances, although non-linear, do not vary largely with bias conditions [4]. Moreover the dynamic behaviour of the device is also largely influenced by parasitics, such as junction capacitance, gate-drain overlap and wiring capacitance. On the one hand this has the implication that a model error does not have a drastic effect upon time delay or gain-bandwidth product; on the other hand, determination of the intrinsic capacitances becomes inaccurate.

By making use of a three-terminal capacitance bridge and carefully neutralizing parasitics due to overlap, bonding and header, all three intrinsic MOST capacitances have been determined experimentally for a self-aligned device with L = 3.2 µm, W = 1000 µm, V_T = 0.40 V, V_{FB} = -.35 V and t_{ox} = 700 Å. The results (fully drawn lines) are shown in fig. 9a and fig. 9b at various bias conditions. Together with these results the values

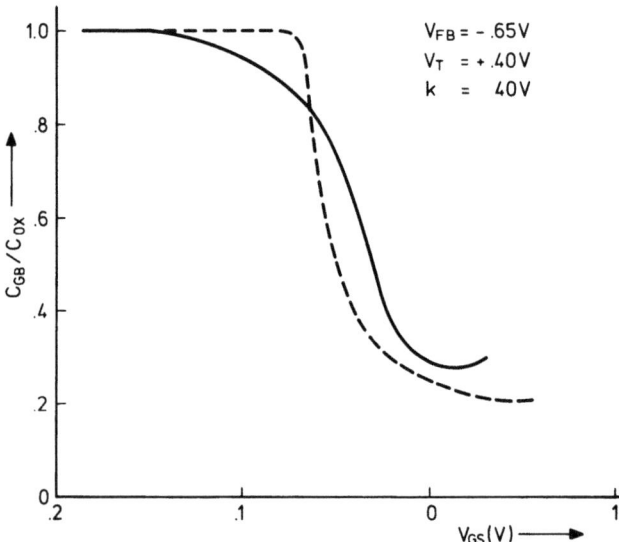

Fig. 9a. Gate-bulk capacitance.

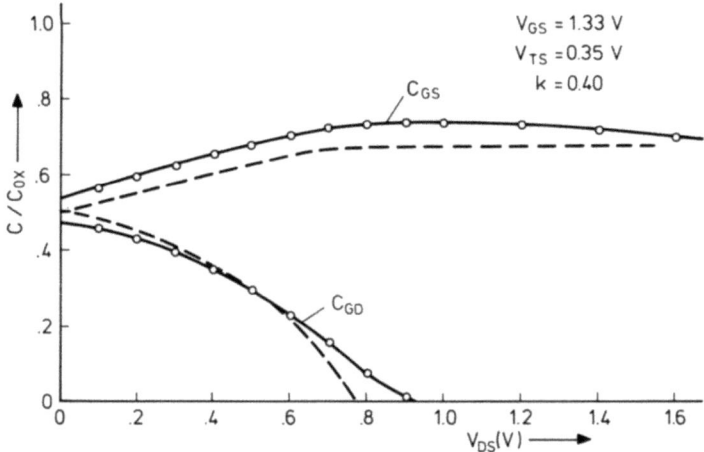

Fig. 9b. Gate-source and gate-drain capacitance.

according to the modelling (compare eqs (37), (40) and (41) of ref. 1) have also been drawn (dotted lines). As can be observed, the agreement is not very good. At $V_{GS} < V_{FB}$ the capacitance C_{GB} exhibits a considerable deviation. As the device in circuits is never in this region or only for a short time, this deviation has no practical consequences. The same applies to the small deviations of the other capacitances in the linear region. In saturation the gate-source capacitance becomes gradually smaller owing to channel shrinkage, an effect which can be modelled properly by using the effective channel length (compare eq. (28) of ref. 1).

5. ADMITTANCES

In high-frequency linear circuits the admittance parameters in the saturation region are also important, since these control the power gain. As was discussed in section 3.2, only the input admittance y_{11} and the transfer admittance y_{21} are determined by the intrinsic device. To a good approximation the above parameters are given by [11]

$$y_{21} \approx g_{ms} (1 + 4/15 xj\, \omega/\omega_o)^{-1}$$
$$y_{11} \approx 2/3\, j\, \frac{\omega}{\omega_o} \times y_{21}$$
(10)

where g_{ms} is the low-frequency transconductance and ω_o is the cut-off frequency as determined by $\omega_o = L^2/\mu_n(V_G-V_{TS})$.

Generally the parameters considered can be measured directly with an admittance bridge or indirectly by measuring the S-parameters with a network analyser. In fig. 10a the results for the real and imaginary parts of y_{11} and y_{21} have been plotted as a function of the frequency. For completeness the parasitic value for the feedback and the output admittance have also been given. The above results were measured for a MOST with L = 5 μm, W = 700 μm and t_{ox} = 1100 Å. As discussed in ref. 1, the agreement between (10) and the measurements is very good. A value $f_o = 1.8 \times 10^9$ Hz can be derived from the measurements. This value compares well with the bias conditions.

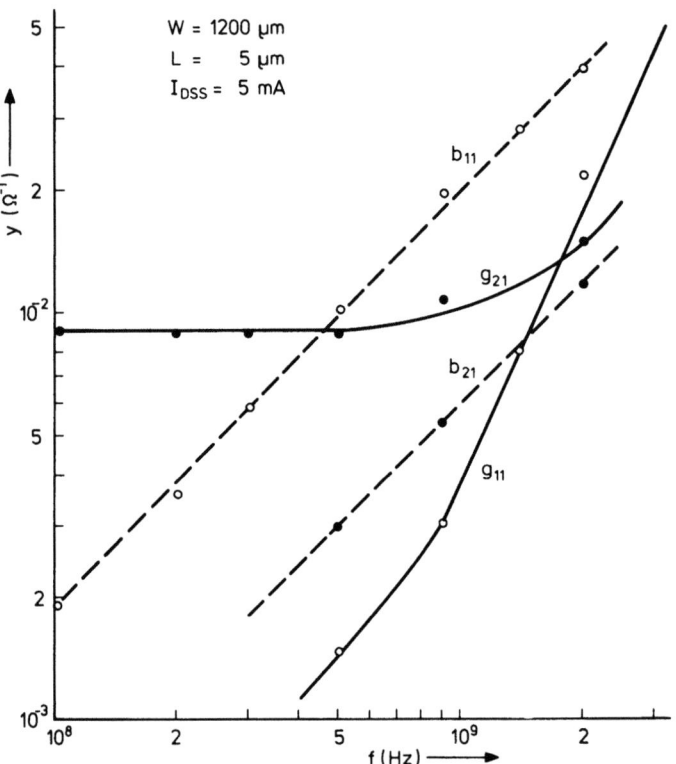

Fig. 10. Real and imaginary parts of admittances vs frequency.

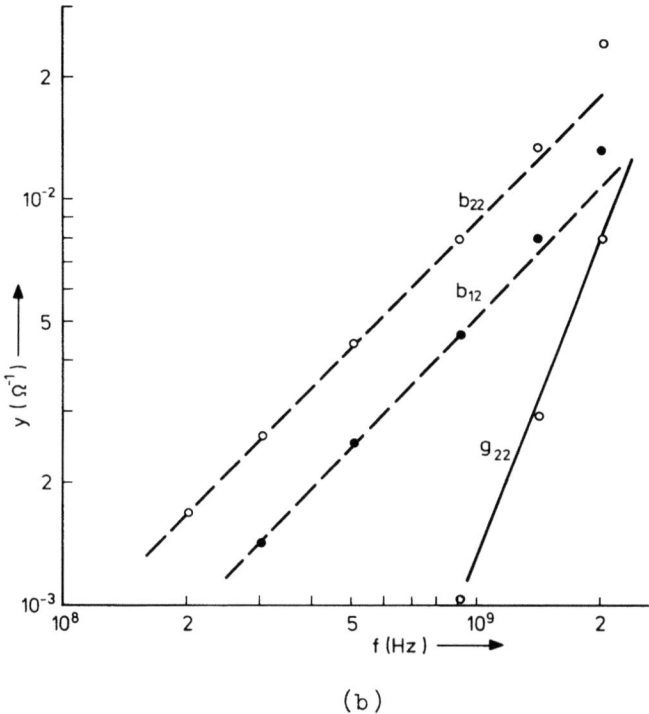

(b)

As the above expressions have been derived for a situation where velocity saturation does not occur, it has to be expected that deviations will occur between (10) and measurements in the case of very short channels [12].

6. NOISE

Fig. 11 gives the noise spectrum of the drain current as a function of the bias conditions [13]. As discussed in ref. 1, two different mechanisms are responsible for this result. At low frequencies the 1/f noise dominates, at higher frequencies thermal noise is the fundamental limit. These results refer to a long-channel device built on a high resistivity substrate. In this case the 1/f noise is relatively low (compare eq. (53) of ref. 1 and the white noise level exactly satisfies

$$S_i \approx \frac{8}{3} kT\, g_{ms}. \tag{11}$$

However in short-channel devices, which are usually employed for input stages of tuners, the 1/f noise is much larger owing to the

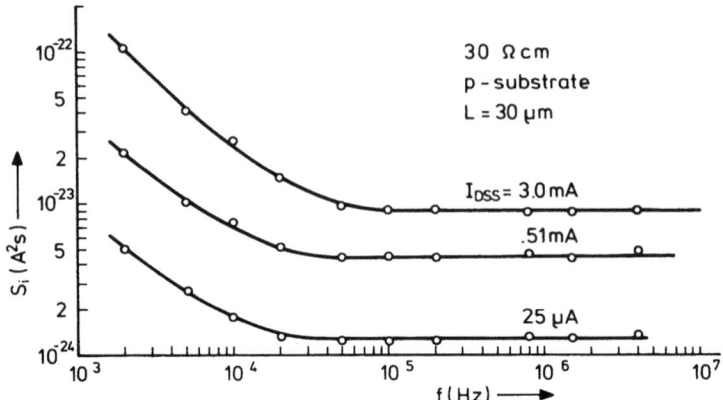

Fig. 11. Noise spectrum of drain current.

smaller gate capacitance, moreover the white noie level may be larger than (11), although seldomly more than a factor of 3. This might be due to an increased lattice temperature accompanied by velocity saturation [14] or to excess surface states as will occur in an implantation process [15].

At high frequencies the noise is often expressed in terms of a noise figure, which is the ratio between the noise measured at the output of the device and the amplified noise of the signal source. By proper matching of this source to the input admittance a minimum noise figure (F_m) is obtained. In fig. 12 the measured value of this quantity is plotted as a function of the frequency. The device is a dual-gate MOST, which is used in commercial tuners. At very high frequencies F_m increases with frequency owing to capacitive coupling between channel noise and the input (compare fig. 21 of ref. 1).

7. CROSS-MODULATION

In all experiments discussed above, the MOST models as given in ref. 1 have largely demonstrated their usefulness. This is also true when these models are used in digital circuit simulation. As an example of cases where the models fail or are less satisfactory, some results of cross-modulation characteristics are exhibited in fig. 13 [16]. As can be observed, the agreement between measurement and computation is only qualitative. This difference is all the more remarkable as the computer result was obtained from a 5-parameter model for the saturated drain current, in which the subthreshold region, the square-law region

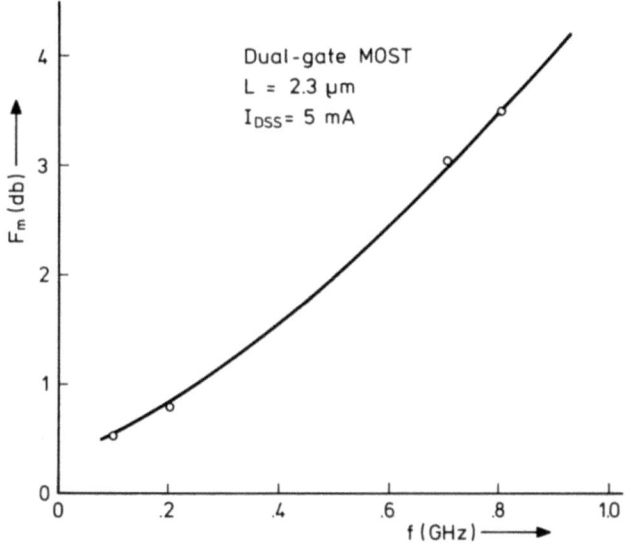

Fig. 12. High-frequency noise factor of MOS tetrode.

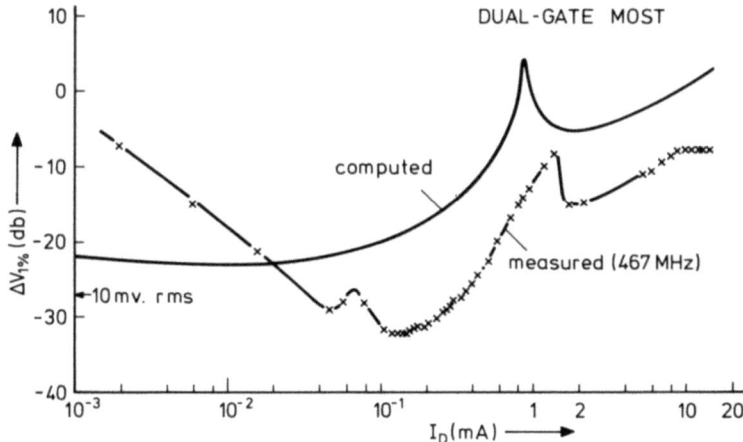

Fig. 13. Input signal for 1% cross modulation as a function of drain current.

and the mobility degradation region were matched together with an empirical optimization program. Obviously a model able to describe higher order effects is needed.

ACKNOWLEDGEMENT

The author wishes to express his gratitude to Mr. J.A. van Nielen for providing the admittance measurements.

REFERENCES

1. F.M. Klaassen, Review of physical MOST models, chapter
2. G. Doucet a.o., Sol.St.Electr. 19 (1976) 191.
3. J. Huang and G. Taylor, Trans. IEEE ED-22 (1975) 995.
4. R.S.C. Cobbold, Theory and applications of field effect transistors, Wiley Interscience.
5. N. Murphy a.o., Sol.St.Electr. 12 (1969) 775.
6. E.P. Jacobs a.o., Techn.Dig. ESSDERC, München 1976.
7. F.F. Fang and A.B. Fowler, J. of Appl.Phys. 41 (1970) 1825.
8. G. Merckel a.o., IEEE Trans. ED-19 (1972) 681.
9. D. Frohman-Bentchkowski, IEEE Trans. ED-16 (1969) 108.
10. P. Rossel a.o., Sol.St.Electr. 19 (1976) 51.
11. J.A. van Nielen, Sol.St.Electr. 12 (1969) 826.
12. T. Okabe a.o., Proc.Conf.Sol.St.Dev. (Tokyo, 1975).
13. F.M. Klaassen and P.A. Hart, Phil.Techn.Rev. 31 (1971) 224.
14. F.M. Klaassen, IEEE Trans. ED-17 (1970) 858.
15. K.L. Wang, Techn.Dig. IEDM (1976) 335.
16. T.G. Mihran, IEEE Trans. ED-22 (1975) 982.

SURFACE CHARACTERIZATION - C-V TECHNIQUE

G. Merckel, J. De Pontcharra

Centre d'Etudes Nucléaires de Grenoble, LETI/MEA
85 X 38041 Grenoble Cedex, France.

ABSTRACT. C(V) curves are dependent on the insulating layer, semiconducting layer, and their interfaces. After a brief review of the physics of MIS structures, some methods of measurement are described.

1. IDEAL MIS DEVICE

1.1 Basic mechanisms

The device has a METAL-INSULATOR-SEMICONDUCTOR structure.
The insulator is a perfect dielectric (no internal charge, no current) and the work function difference ϕ_{MS} between the metal and semiconductor is assumed to be zero.

We briefly recall the mechanism of formation of a space charge region in the semiconductor as a function of its surface potential, ϕ_S. This potential appears if a voltage is applied between the metal and the semiconductor [1] [2]. The relation between the potential, ϕ, and the charge density, $\rho(x)$, in the semiconductor is given by Poisson's equation :

$$\frac{d^2\phi}{dx^2} + \frac{\rho(x)}{\varepsilon_S} = 0 \qquad (1)$$

with $\rho(x) = q \left[N_D - N_A + p(x) - n(x) \right]$

ε_S is the absolute permittivity of the semiconductor, N_D and N_A are the concentrations of donors and acceptors in the semiconductor, $p(x)$ and $n(x)$ are the concentrations of holes and electrons.

$$p(x) = p_0 \exp\left(-\frac{q\phi}{kT}\right) \quad (2) \quad \text{and} \quad n(x) = n_0 \exp\left(\frac{q\phi}{kT}\right) \quad (3)$$

p_0 is the equilibrium concentration of holes in the neutral bulk
n_0 is the equilibrium concentration of electrons in the neutral bulk.

Integrating (1), gives the electric field, $E(x)$, and the potential $\phi(x)$, in the semiconductor [1] [2].
At the surface of the semiconductor, the field, E_S, and the total charge density, Q_S, are given by Gauss law :

$$Q_S = \varepsilon_S E_S$$

Fig. 1 and Fig. 2 illustrate the dependence of the charge distribution on gate voltage, V_G, and surface potential, ϕ_S (P type Si)

- a) $\phi_S = 0$, ($V_G = 0$). Flat band condition.
  The energy bands are flat and the charge Q_S is zero.

- b) $\phi_S < 0$, ($V_G < 0$). Accumulation.
  The energy bands are bent upwards. The holes are attracted toward the surface and there is an accumulation of holes. The charge Q_S is positive.

- c) $\phi_S > 0$, ($V_G > 0$). Depletion.
  The holes are pushed into the bulk. The charge Q_S is mainly composed of the negative fixed charges Q_B (ionized acceptors). The space charge region extends to a distance x_D which depends on the doping, N_A, and the surface potential, ϕ_S. Assuming that there are no free carriers in the space charge region (depletion approx.), one has :

$$\phi_S = \frac{q N_A x_D^2}{2 \varepsilon_S}$$

- d) $\phi_S > 0$, ($V_G \gg 0$). Inversion.
  The electric field is sufficient to attract electrons to the surface (charge Q_n). The concentration of electrons exceed the concentration of the fixed charges, and strong inversion occurs for $\phi_S \simeq 2\phi_F$ (see Fig. 2),

  with $\phi_F = \frac{kT}{q} \text{Log} (N_A/n_i)$.

If V_G is increased, all additional charges brought to the metallic gate will cause a corresponding increase in $n(x)$ and there will be no more increase in x_D which attains a maximum value, x_{Dmax} :

$$x_{Dmax} \simeq \left(\frac{2 \varepsilon_S}{q N_A} 2 \phi_F\right)^{1/2}$$

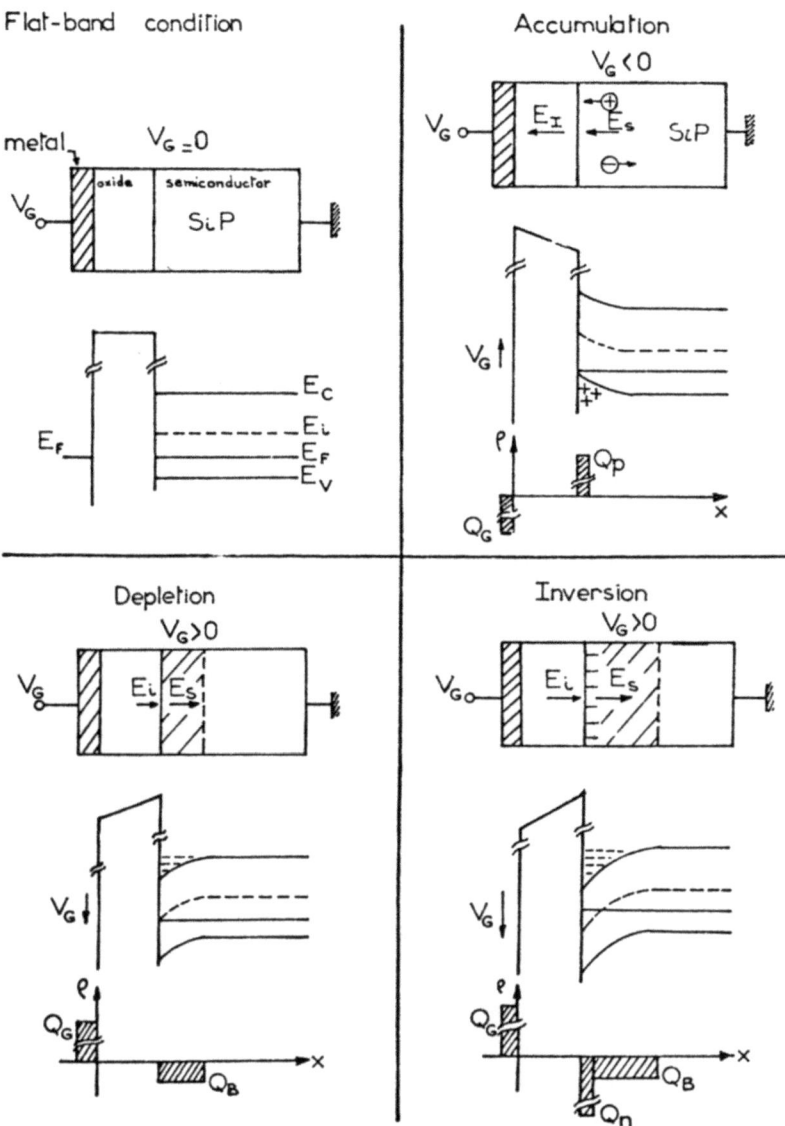

Fig. 1. Ideal MIS device – semiconductor P-type
Band and charge distribution diagrams
- flat-band condition $V_G = V_{FB} = 0$
- accumulation $V_G < 0$
- depletion $V_G > 0$
- inversion $V_G > 0$

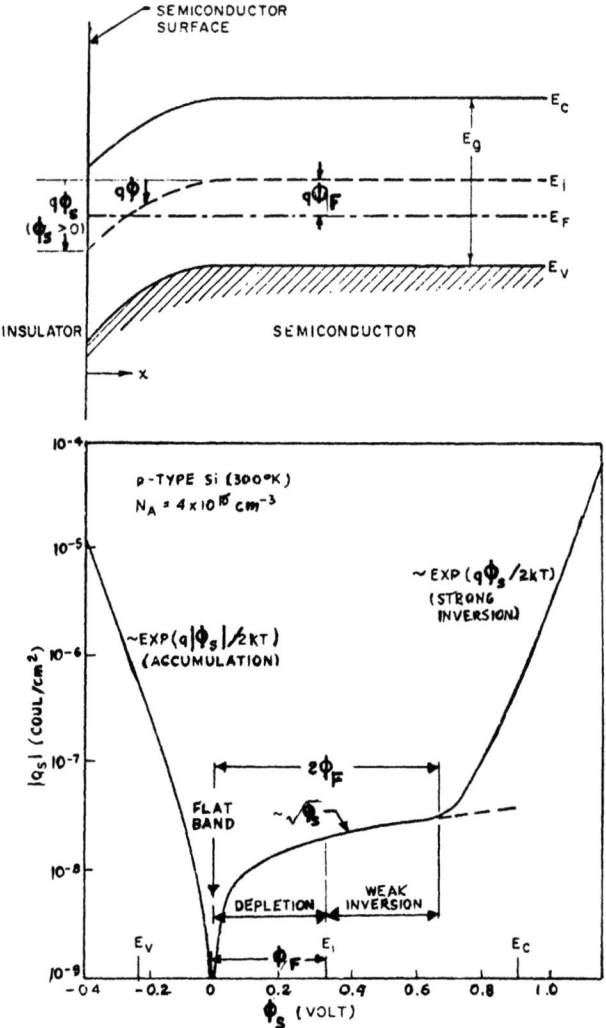

Fig. 2. Surface charge density, Q_S, versus surface potential ϕ_S. Band structure. P-type semiconductor. (After SZE [1])

The charge in the semiconductor is the sum of the electron charge, Q_n, in the inversion layer, and the charge, $Q_B = -qN_A x_{Dmax}$, of the ionized acceptors.

1.2 C(V) curves [1]

In the absence of any work-function differences, the applied voltage, V_G, appears partly across the insulator (V_I) and partly across the silicon ($\emptyset_S$), thus :

$$V_G = V_I + \emptyset_S$$

As shown in Fig. 3, the total capacitance per unit area, C, is a series combination of the insulator (oxide) capacitance, C_I ($= \varepsilon_I/x_I$), and the silicon space-charge capacitance C_S ($= \varepsilon_S/x_D$) : $1/C = 1/C_I + 1/C_S$.

For a negative voltage, we have an accumulation of holes and therefore a very high differential capacitance C_S. As a result, the total capacitance is close to the insulator capacitance. As the negative voltage is reduced sufficiently, the depletion region appears, C_S decreases, and the total capacitance decreases.
The capacitance goes through a minimum an then increases again as the inversion layer forms at the surface. This minimum corresponds to the maximum space charge layer width, x_{Dmax}.

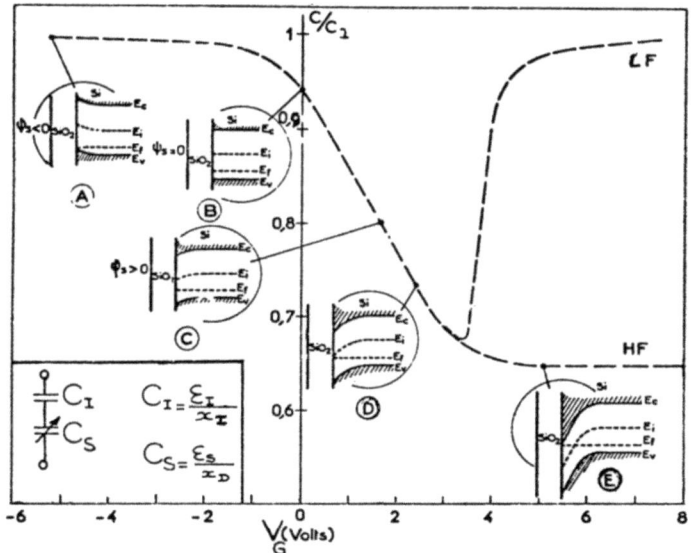

Fig. 3. Ideal MIS capacitance. C(V) plot.

It depends on the doping level and oxide thickness as shown in Fig. 4. The increase of the capacitance is dependent on the ability of the electron concentration, in the inversion layer, to follow the applied ac signal. At low frequencies (LF), the generation-recombination rates of electrons leads to a charge exchange with the inversion layer in step with the applied ac signal. The surface potential is constant and the measured capacitance is C_I. At flat-band condition, the semiconductor capacitance, C_{SFB}, is given by [1] :

$$C_{SFB} = \left(\frac{q^2 N_A \varepsilon_S}{kT}\right)^{1/2}$$

The transition frequency from L.F. to H.F. response depends on the rate of generation-recombination in the semiconductor and can be modified by increasing this rate by illumination or heating (Fig. 5).

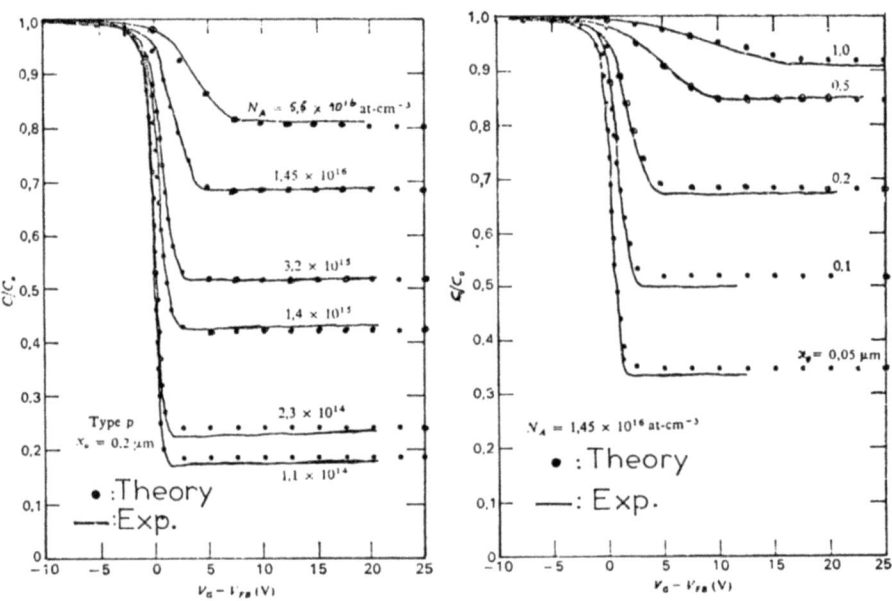

Fig. 4. Variation on C(V) plots with variation on doping and oxide thickness. (After GROVE [2])

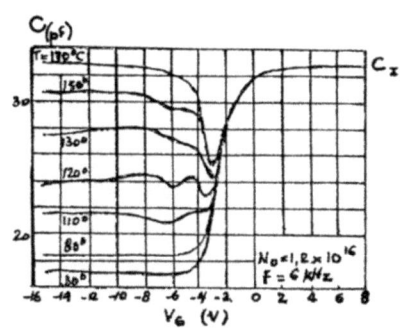 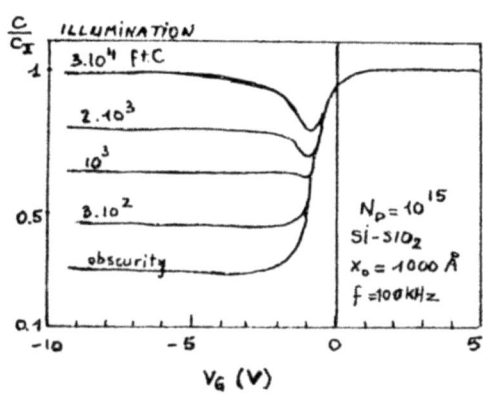

Fig. 5. Temperature and illumination effects. (After GROVE [2])

2. REAL MIS CAPACITANCE

2.1 Differences with ideal MIS capacitance

This structure can be differentiated from the ideal one by the existence of : [1] [2] [5]

a) a work function difference, ϕ_{MS}, between metal and semiconductor. The flat-band condition is $V_G = V_{FB} = \phi_{MS}$.

b) an oxide charge Q_O or $\rho(x)$.

$$V_{FB} = -\frac{Q_O}{C_I} \quad \text{or} \quad -\frac{1}{C_I} \int_0^{x_I} \frac{x}{x_I} \rho(x) \, dx$$

c) surface states, N_{SS}. The following three cases may be possible : - the occupancy of surface states is independent of applied voltage. The C(V) plot gives a measure of Q_{SS} (= q N_{SS}).

$$V_{FB} = -\frac{Q_{SS}}{C_I}$$

- the occupancy depends on the applied voltage
- the surface states follow the measurement ac signal.

d) some imperfections in oxide like leakage current.

Table 1 and Figure 6 show the deviations on C(V) characteristics.

DIFFERENCE WITH IDEAL MIS		INFLUENCE ON HF C(V)
Work-function difference	$V_{FB} = \phi_{MS}$	Parallel shift along the voltage axis by an amount ϕ_{MS}.
Mobile or fixed density of charge, $\rho(x)$, in oxide	$V_{FB} = -\frac{1}{C_I}\int_0^{x_I}\frac{x}{x_I}\rho(x)dx$	Parallel shift along the voltage axis dependent on time, voltage or temperature if the charge is due to mobile ions.
INTERFACE STATES :		
Fixed charge Q_{SS} independent of applied voltage	$V_{FB} = -\frac{Q_{SS}}{C_I}$	Parallel shift along the voltage axis.
Dependent on dc bias		Distorsion and displacement. C_{min} remain unchanged.
Dependent on ac and dc bias		Distorsion, displacement and change in C_{min}.
Leakage in oxide		Error in C measurement.
Surface conduction		Error in C measurement

TABLE 1

REAL MIS CAPACITANCE (see Fig. 6)

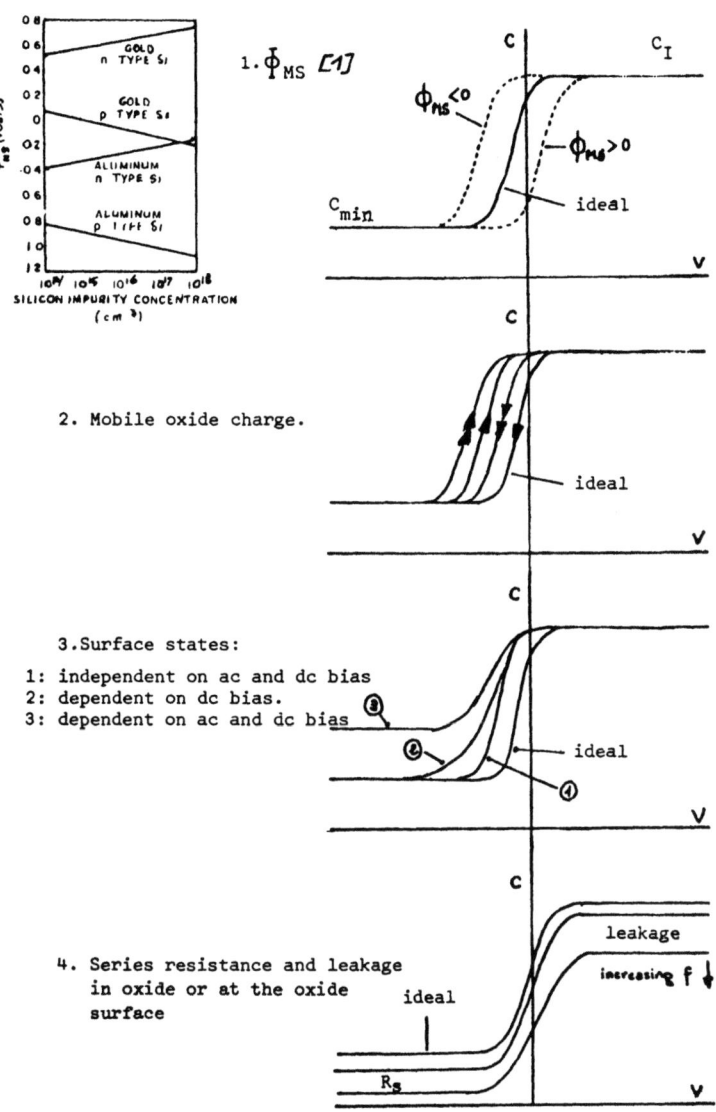

Fig. 6. Real MIS capacitance. Deviations on C(V) plots.

When interface states follow ac signal, no information can be
obtained in a simple manner. The frequency must be increased to
a few MHz to eliminate the fast interface state contribution.
In this case the flat-band voltage is :

$$V_{FB} = \phi_{MS} - \frac{1}{C_I} \int_0^{x_I} \frac{x}{x_I} \rho(x)dx - \frac{Q_{SS}}{C_I}$$

2.2 Equivalent circuit

The contribution of the surface states in the capacitance can be
expressed by a capacitance C_{SS} and a resistance R_{SS} in parallel
with the depletion capacitance C_S, Fig. 7. This circuit can be
represented by a frequency dependent capacitance, C_P, and
conductance, G_P, in parallel.

$$C_P = C_S + \frac{C_{SS}}{1+\omega^2\tau^2} \qquad (6)$$

$$\frac{G_P}{\omega} = \frac{C_{SS}\,\omega\tau}{1+\omega^2\tau^2} \qquad (7)$$

with $\tau = R_{SS} C_{SS}$, lifetime of surface states.

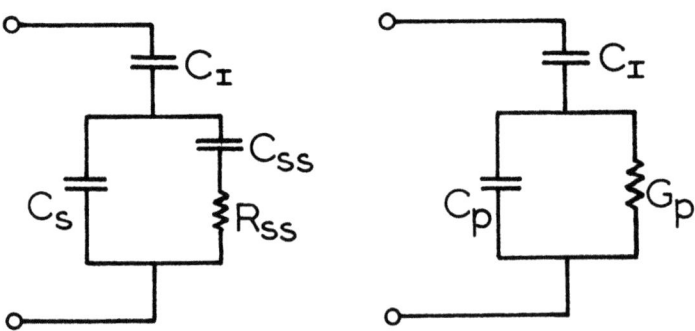

Fig. 7. Equivalent circuit.

3. CHARACTERIZATION OF INTERFACE STATES

Several methods are known for determining the interface state density

$$N_{SS} = \frac{1}{q}\left(\frac{\partial Q_{SS}}{\partial \phi_S}\right) \quad \text{states/cm}^2/\text{eV}.$$

All of those assume that the depletion approximation is valid and that the semiconductor is uniformly doped.

3.1 Differentiation method [6]

Conditions : measurement at high frequencies for eliminating the capacitance C_{SS}. The displacement ΔV along the voltage axis is

$$\phi_{MS} - \left(\frac{Q_{SS}}{C_I}\right)$$

Pratical procedure : an ideal C(V) curve is plotted with the same parameters as the experimental curve (C_I and C_{min}). The shift gives Q_{SS}/C_I for each value of V_G, therefore of ϕ_S. A graphical differentiation gives

$$N_{SS} = \frac{1}{q}\left(\frac{\partial Q_{SS}}{\partial \phi_S}\right)_{V_G}$$

Limitations : - only the integral of N_{SS} is measured
- the frequency must be so high that no surface states can follow ac signal
- non uniform doping profile appears as an apparent N_{SS}.

This method is useful for MIS capacitances with N_{SS} larger than 10^{11} states/cm²/eV. If the measurements are carefully done, the sensitivity is of the order of 5.10^9 states/cm²/eV. [7].

3.2 Integration method [8]

The measurement is made at low frequency. The surface states are in equilibrium with applied voltage, thus

$$V_G - V_{FB} = V_I + \phi_S \tag{8}$$

$$C_I dV_I = C dV_G \tag{9}$$

C is the total measured capacitance.

(8) and (9) give $\dfrac{d\phi_S}{dV_G} = 1 - \dfrac{C}{C_I}$ and $\dfrac{d\phi_S}{dV_I} = \dfrac{C_I}{C} - 1$ (10)

Integrating (10) gives ϕ_S :

$$\phi_S(V_{G1}) - \phi_S(V_{G2}) = -\int_{V_{G1}}^{V_{G2}} \left[1 - \dfrac{C}{C_I}\right] dV_G \qquad (11)$$

A theoretical curve $\dfrac{C_S}{C_I} = f(\phi_S)$ is calculated by computer [9]. Comparison with experimental curve in accumulation region gives $\phi_S(V_I)$.

The charge equation is :

$$Q_M(V_G) = Q_S(\phi_S) + Q_{SS}(\phi_S) \qquad (12)$$

Differentiation of (12) gives [8] [9] [10]

$$\dfrac{d\phi_S}{dV_I} = \dfrac{\varepsilon_i}{x_I} \left[\dfrac{dQ_S}{d\phi_S} + q\, N_{SS}(\phi_S)\right]^{-1} \qquad (13)$$

Equations (10) and (11) gives an experimental curve $\dfrac{d\phi_S}{dV_I} = f(\phi_S)$ and the comparison with the ideal curve gives N_{SS}.

Note : The capacitance C can be measured by the current when a slow ramp voltage is applied (quasi-static method, Q_{SM})

$I(t) = C \dfrac{dV}{dt}$. The slope dV/dt is constant and the measurement of I gives C [9].

This method is useful for N_{SS} lower than 10^{13} states/cm²/eV, if $1 - C/C_I$ is measured with a small error, i.e. for ϕ_S far from conduction or valence bands.

3.3 Conductance method [11]

The conductance is directly related to surface states. That was not the case for capacitance methods.
Using equation (7), the admittance of MIS capacitor is measured. At a given applied voltage, G_P/ω is measured as a function of frequency. $G_P/\omega = f(\omega\tau)$ rises to a maximum value ($= C_{SS}/2$) if $\omega\tau = 1$. Thus N_{SS} is given by $N_{SS} = C_{SS}/qA$.

A is the gate area.

So, one obtains $N_{SS}(\phi_S)$, through $V_G(\phi_S)$.

This method is accurate for N_{SS} below 10^{11} states/cm²/eV.

3.4 Temperature method [12]

The principle is based on the variation of Fermi level with temperature. This leads to a charge exchange with interface states. The variation of V_{FB} with T gives a measure of the charge of interface states in the upper part of the band-gap for n-type, and in lower part for p-type semiconductors. Varying T, the HF C(V) plot gives the change in V_{FB}, i.e. in Q_{SS} ($\emptyset_S$). Then,

$$N_{SS} = \frac{1}{q} \frac{\partial Q_{SS}}{\partial \emptyset_S}$$

This method gives N_{SS} close to the band edge and is independent of oxide charge. The frequency must be high so that no surface states contribute to H.F. capacitance measurements.

Fig. 8 shows an example for three crystal orientations in Si-SiO₂ structure. For p-type, N_{SS} can only be measured near the valence band, for n-type near the conduction band.

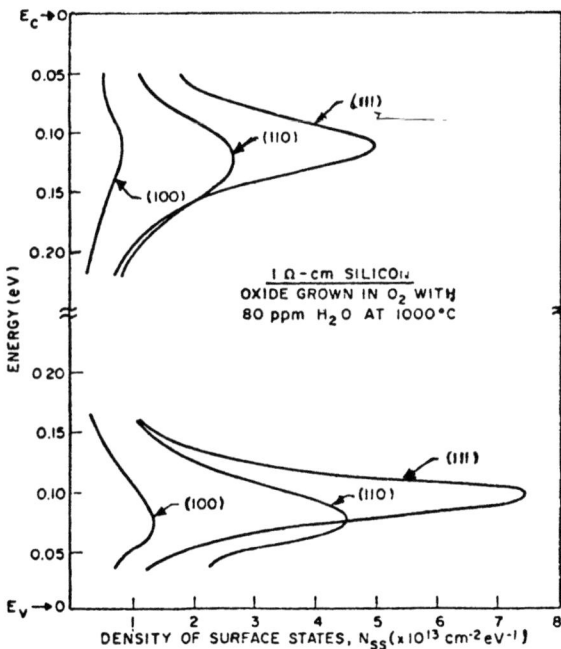

Fig. 8. Temperature method. Variation of N_{SS} with crystal orientation. (After SZE [1])

4. EXPERIMENTAL RESULTS

Figure 9 shows the effect of annealing and crystal orientation on Q_{SS} measured by HF C(V) (SiO_2 - Si).

In Figure 10, one presents a comparison between charge states in MIS capacitors with In_2O_3 and Al gates (QSM method).

Figure 11 shows the effect of annealing on interface states, at 550°C and with increasing annealing times (conductance method and LF capacitance method).

Fig. 12, 13, 14 refer to the conductance method applied to samples before and after irradiation. The effect of annealing is also shown.

The density N_{SS} decays after annealing, the conditions are indicated below the figures.

The samples are <111> oriented, and are irradiated under zero applied voltage.

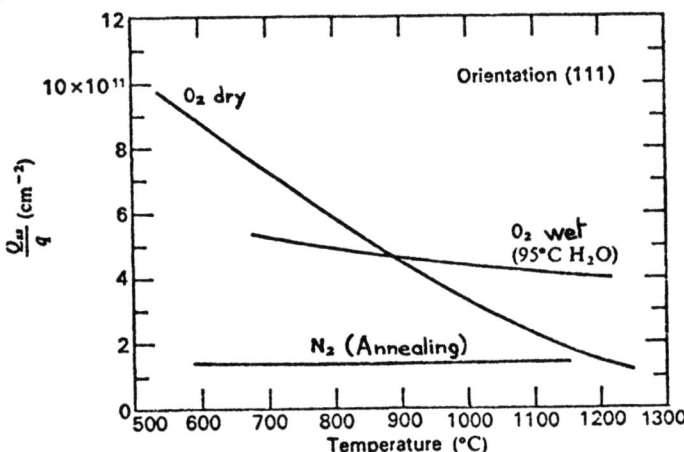

Fig. 9. Differentiation method. Effect of technological parameters on Q_{SS}. (After GROVE [2])

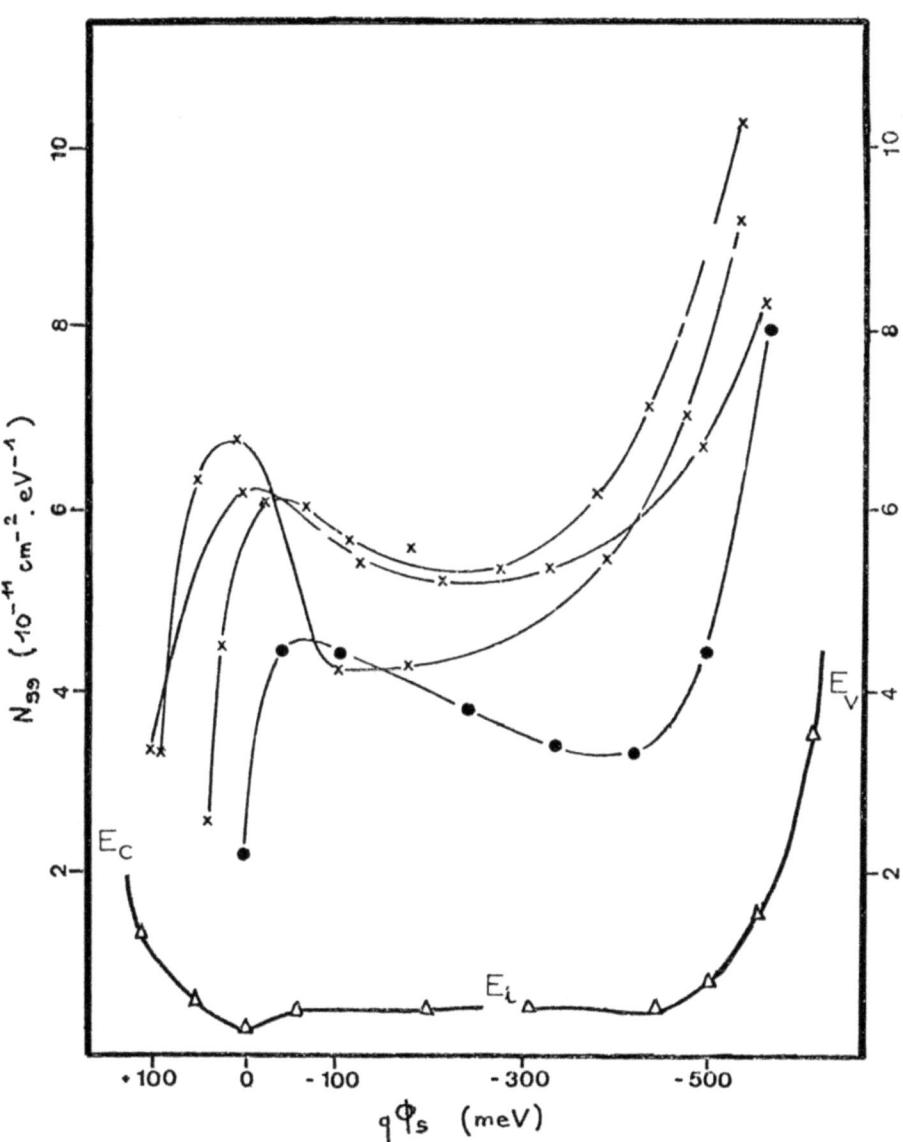

Fig. 10. Integration method - QSM.
  x In_2O_3 - SiO_2 - Si   o Al - SiO_2 - Si
  silicon <111> , Annealing : 430°C, 60 min. H_2

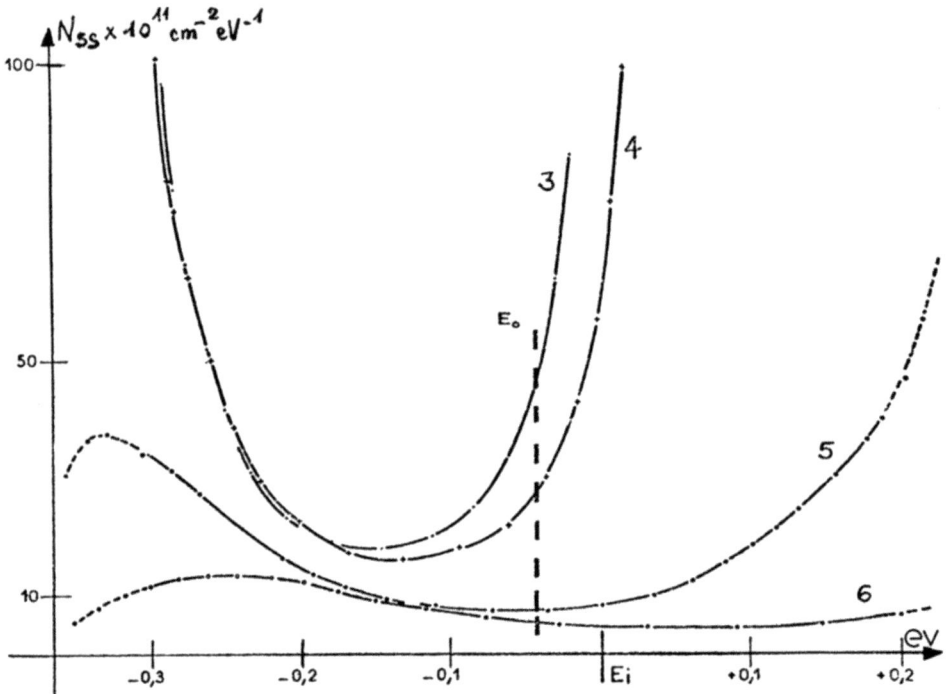

Fig. 11. Conductance and LF capacitance methods.
Annealing time 0 min. 3
(550°C) 30 min. 4
120 min. 5
420 min. 6

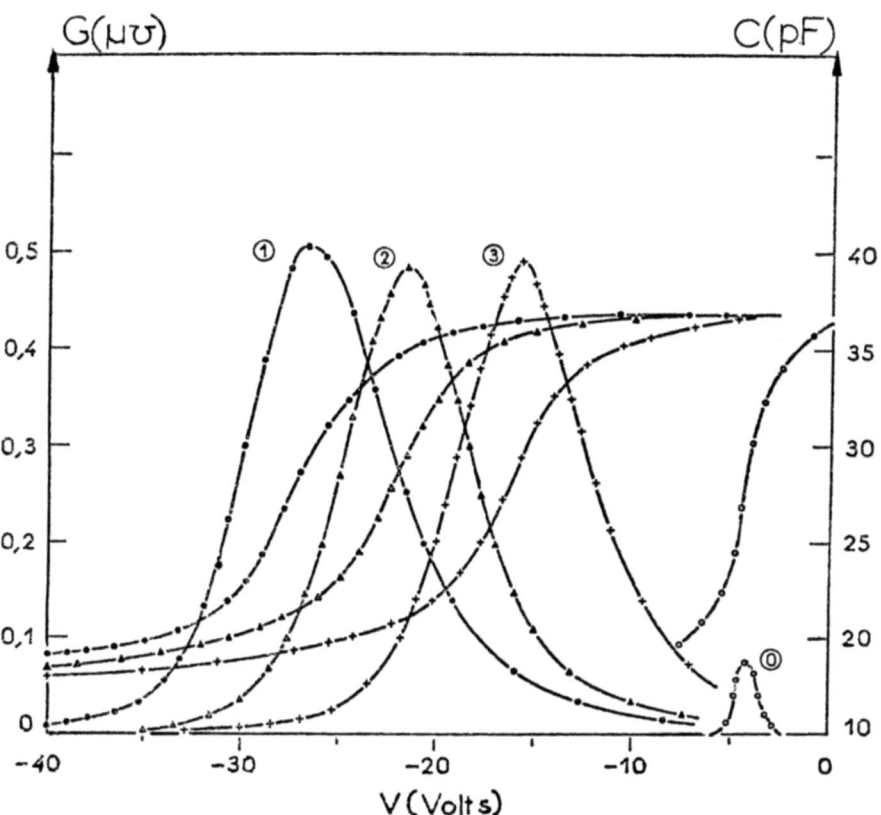

Fig. 12. G(V) and C(V)
- 0 before irradiation
- 1 after irradiation (10^{15} e/cm^2, 0V)
- 2 after 100 hours ⎫
- 3 after 300 hours ⎭ 300°K, 0V

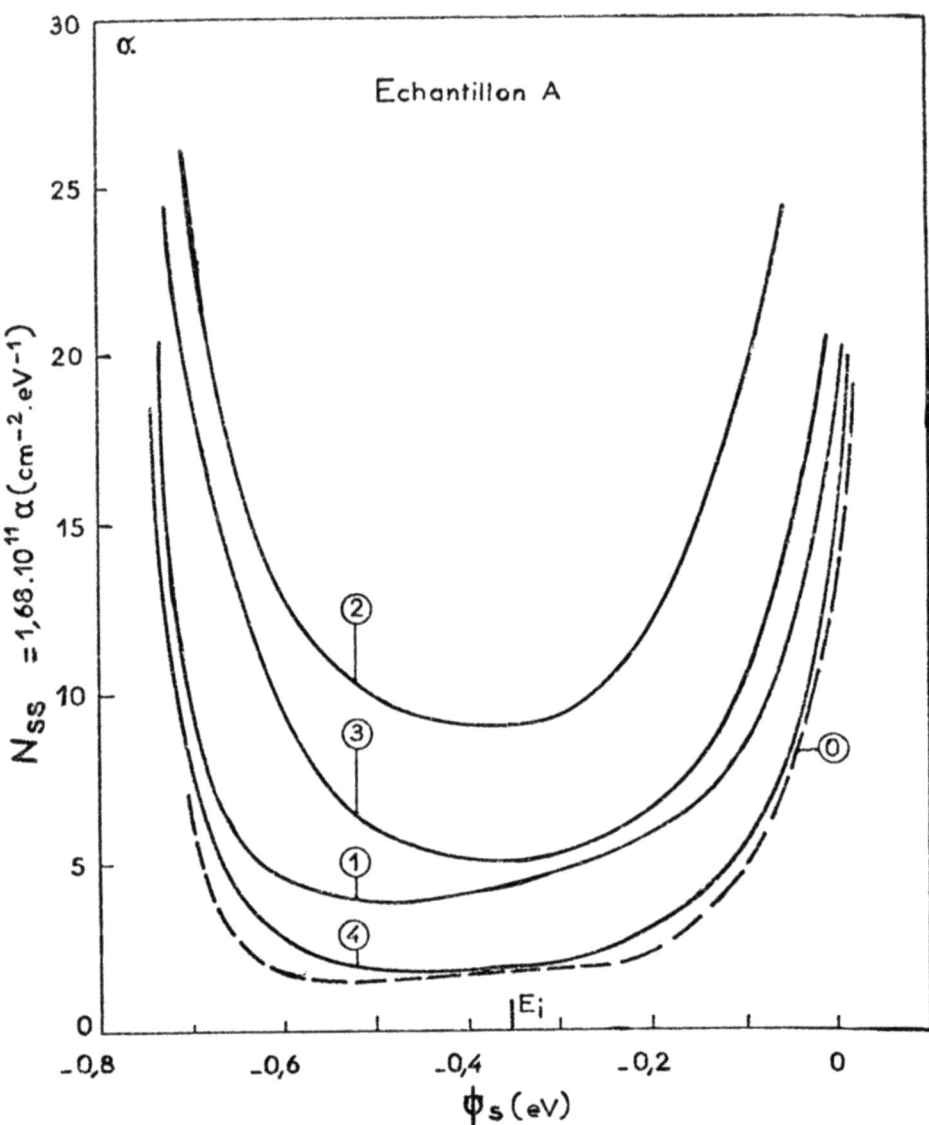

Fig. 13. N_{SS} after irradiation and annealing (G(V)).
0 before irradiation
1 after irradiation (10^{14} e/cm^2)
2 annealing 150 h. 80°C
3 annealing 150 h. 150°C
4 annealing 150 h. 250°C

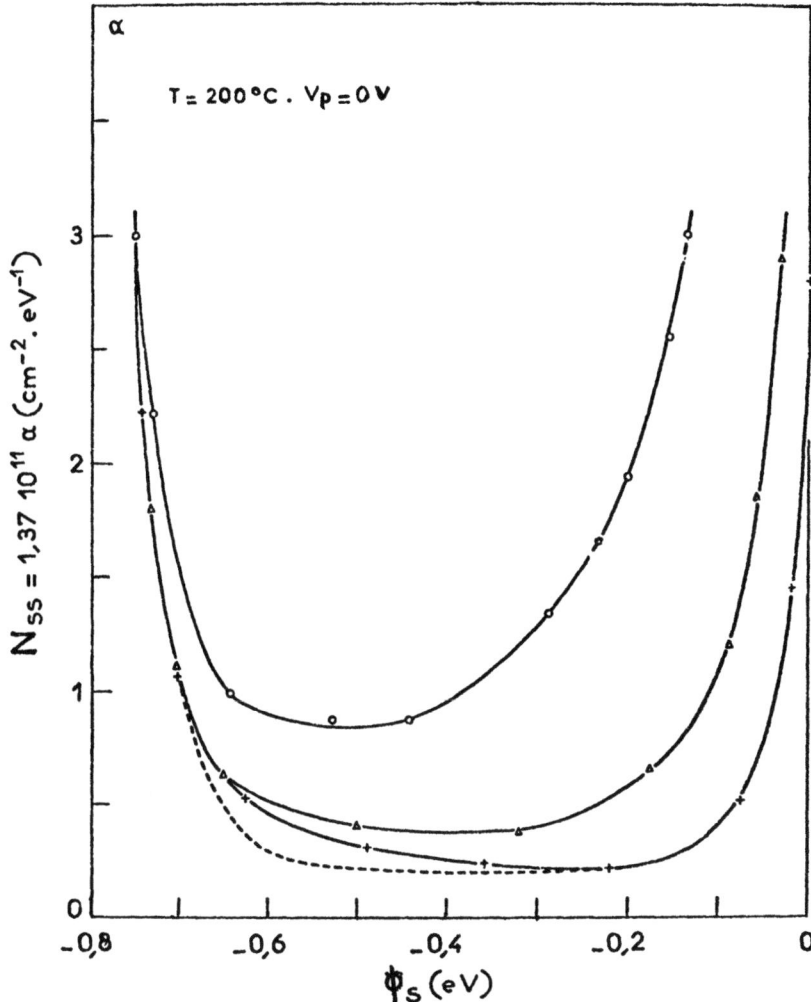

Fig. 14. N_{SS} after irradiation and isothermal annealing (200°C, 0V) (G(V))
- - - - - before irradiation
o   after irradiation (10^{14} e/cm^2)
Δ   after 1h. annealing
+   after 2,5h. annealing

REFERENCES

[1] S.M. Sze, Physics of Semiconductor Devices, Wiley, N.Y.
     Figures reprinted by permission of J.Wiley & Sons,Inc.
[2] A.S. Grove, Physique et technologie des dispositifs à semiconducteur, Dunod, Paris.
[3] C.G.B. Garret, W.H. Brattain, Phys. Rev. 99,376 (1955).
[4] R.H. Kingston, S.F. Neustadter, J. Appl. Phys. 26, 718 (1955).
[5] K.H. Zaininger, F.P. Heiman, Solid State Techn. May-June 1970, 49, 46.
[6] L.M. Terman, Sol. State Elect. 5, 285 (1962).
[7] C.T. Sah, A.B. Tole, R.F. Pierret, S.SE 12,689 (1969).
[8] C.N. Berglund, IEEE Trans. on E-D 13, 701 (1966).
[9] R. Castagne, Thesis, University of Paris (1970).
[10] K. Saminadayar, Thesis, University of Grenoble (1975).
[11] E.H. Nicollian, A. Goetzberger, Bell Syst. Techn. J. 46, 1055 (1967).
[12] P.V. Gray, D.M. Brown, Appl. Phys. Lett. 8, 31 (1966).

SURFACE CHARACTERIZATION - WEAK INVERSION

G. Merckel

Centre d'Etudes Nucléaires de GRENOBLE - LETI/MEA -
85 X 38041 Grenoble Cedex, France

ABSTRACT. In weak inversion regime, the slope of the $I_D(V_G)$ characteristics of the MOS transistor is a function of the density of surface states. After an analysis of the behavior of the transistor, the technique for measurement of N_{SS} is presented.

1. ELECTRICAL CHARACTERISTICS OF THE TRANSISTOR - WEAK INVERSION REGIME

1.1 Q_n. The charge in the inversion layer

One considers an N channel transistor. Fig. 1 shows the energy band diagram of a semiconductor in weak inversion. The potential $\emptyset$ is defined to be zero in the bulk (for $x \to \infty$) and is measured with respect ot the intrinsic level E_i. The electron and hole concentrations are given by :

$$n = \frac{n_i^2}{N_A} \exp\left(\frac{\emptyset - \emptyset_C}{v_T}\right) \qquad (1)$$

$$p = N_A \exp\left(\frac{-\emptyset}{v_T}\right) \qquad (2)$$

With n_i, the intrinsic carrier concentration in Si (1.45×10^{10} at/cm^3 at 300°K), N_A the doping density of the substrate and $v_T = kT/q$, the thermal voltage (26mV at 300°K).
Integrating Poisson's equation, the total charge Q_S per unit area in the semiconductor is given by [1] :

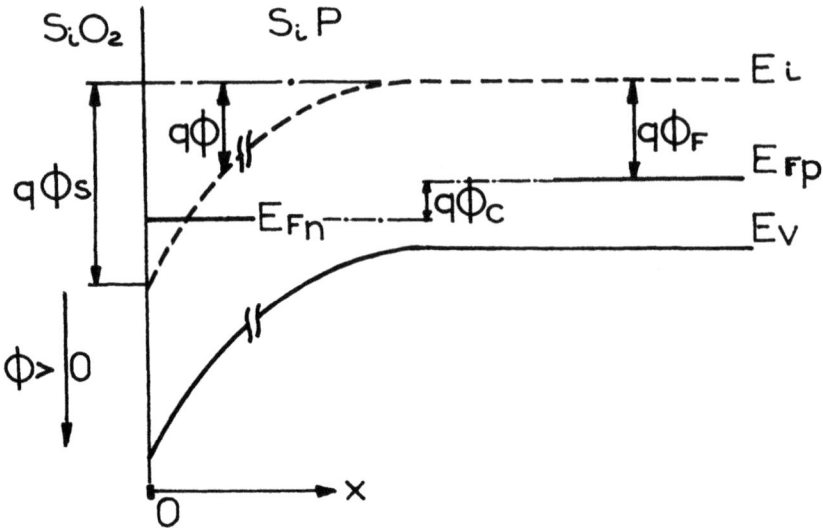

Fig. 1. Energy band diagram at the surface of a P-type semiconductor.

$$Q_S = - (2q\varepsilon_{Si}N_A)^{1/2} \left\{ \phi_S + v_T \left[-1 + \exp\left(\frac{\phi_S - 2\phi_F - \phi_C}{v_T}\right) \right. \right.$$
$$\left. \left. + \exp\left(\frac{-\phi_S}{v_T}\right) - \exp\left(\frac{-2\phi_F - \phi_C}{v_T}\right) \right] \right\}^{1/2} \quad (3)$$

with ε_{Si} the permittivity of Si and $\phi_F = v_T \log(N_A/n_i)$.
Neglecting the last two terms in (3) gives :

$$Q_S = - (2q\varepsilon_{Si}N_A)^{1/2} \left[\phi_S - v_T + v_T \exp\left(\frac{\phi_S - 2\phi_F - \phi_C}{v_T}\right) \right]^{1/2} \quad (4)$$

The charge Q_S is the sum of the charge Q_n in the inversion layer and the charge Q_B in the substrate. Therefore, $Q_n = Q_S - Q_B$ (5)
Q_B is given by [1] :

$$Q_B = - (2q\varepsilon_{Si}N_A)^{1/2} (\phi_S - v_T)^{1/2} \quad (6)$$

The charge Q_n is obtained by combining (4), (5) and (6).
In weak inversion, Q_n is small compared to Q_B and so, referring

to (4) one has,

$$\exp\left(\frac{\emptyset_S - 2\emptyset_F - \emptyset_C}{v_T}\right) < \frac{\emptyset_S}{v_T} - 1$$

and so

$$\emptyset_S < 2\emptyset_F + \emptyset_C + v_T \log\left(\frac{\emptyset_S}{v_T} - 1\right)$$

Under this condition, Q_S becomes (by a series expansion to the first order) [2]

$$Q_S = -\left[2q\varepsilon_{Si} N_A (\emptyset_S - v_T)\right]^{1/2} \left[1 + \frac{v_T}{2(\emptyset_S - v_T)} \exp\left(\frac{\emptyset_S - 2\emptyset_F - \emptyset_C}{v_T}\right)\right] \quad (7)$$

combining (5), (6) and (7) now gives Q_n

$$Q_n = - C_D (\emptyset_S) v_T \exp\left(\frac{\emptyset_S - 2\emptyset_F - \emptyset_C}{v_T}\right) \quad (8)$$

with $C_D(\emptyset_S) = \left[\dfrac{q\varepsilon_{Si} N_A}{2(\emptyset_S - v_T)}\right]^{1/2}$ the depletion capacitance (9)

For calculating the current I_D, one has to establish the relationship between $\emptyset_S$, $\emptyset_C$ and the applied voltages V_G, V_D.

1.2 Current voltage characteristics
   Treatment of Van Overstraeten and al [2]

All voltages are referred to the source and it is assumed that $V_B = 0$ (V_B = source-substrate polarization).

The relation between the gate voltage V_G and the surface potential $\emptyset_S$ is * :

$$V_G = V_{FB} + \emptyset_S - \frac{Q_{SSO}}{C_{ox}} + \frac{qN_{SS}}{C_{ox}}(\emptyset_S - 2\emptyset_F - \emptyset_C) - \frac{Q_S}{C_{ox}} \quad (10)$$

---

* assuming a constant density of surface states.

With $V_{FB} = \phi_{MS} - Q_{ox}/C_{ox}$ Flat-band voltage

ϕ_{MS} — Metal semiconductor work function

Q_{ox} — Fixed charge in the oxide per unit area

C_{ox} — Oxide capacitance per unit area

Q_{SSO} — Charge of filled fast surface states per unit area when $\phi_S = 2\phi_F + \phi_C$

N_{SS} — Fast surface state density per electron volt at $\phi_S = 2\phi_F + \phi_C$

Q_S — Total charge in the semiconductor per unit area.

In weak inversion, one can assume that $Q_S \simeq Q_B$.
The charge Q_B is a function of the surface potential ϕ_S and may be expressed (by a series expansion around $\phi_S = 2\phi_F$) as :

$$Q_S = Q_B(\phi_S) = Q_B(2\phi_F) + (\phi_S - 2\phi_F)\frac{dQ_B}{d\phi_S} \qquad (11)$$

putting

$$\frac{dQ_B}{d\phi_S} = -C_D(\phi_S) \quad (12) \quad \text{where } C_D \text{ is defined by eqn (9),}$$

$$n = 1 + \frac{C_D(\phi_S) + C_{SS}}{C_{ox}} \qquad (13)$$

and $m = 1 + \dfrac{C_D(\phi_S)}{C_{ox}}$ \hfill (14)

with $C_{SS} = q N_{SS}$ (15), the surface state capacitance and combining relations (10) to (13), the surface potential is given by :

$$\phi_S = 2\phi_F + \frac{V_G - V'_T}{n} + \frac{\phi_C}{n} \cdot \frac{C_{SS}}{C_{ox}} \qquad (15)$$

with $V'_T = V_{FB} - \dfrac{Q_{SSO}}{C_{ox}} + 2\phi_F - \dfrac{Q_B(2\phi_F)}{C_{ox}}$

The current I_D is given by [1]

$$I_D = \mu \frac{Z}{L} \int_0^{V_D} |Q_n(\phi_C)|\, d\phi_C \qquad (16)$$

with μ the electron mobility ; Z, the channel width and L, the channel length.

Finally, combining (8), (14), (15) and (16)

$$I_D = \mu \frac{Z}{L} C_D(\emptyset_S) \frac{n}{m} v_T^2 \exp\left(\frac{V_G - V'_T}{n v_T}\right) \left[1 - \exp\left(-\frac{m V_D}{n v_T}\right)\right] \quad (17)$$

For reasons of accuracy, n and m should be evaluated around
$\emptyset_S = \frac{3}{2} \emptyset_F$

It is to be noted that results similar to (17) have been published [3][4],[5].

Note : the experimental threshold voltage V_T is given by [2]

$$V_T = V'_T + \Delta\emptyset \; n \; (2\emptyset_F)$$

where $\Delta\emptyset$ is the amount of $\emptyset_S$ above the strong inversion value $(2\emptyset_F + \emptyset_C)$

2. DETERMINATION OF N_{SS}

A simple and accurate method is based on the publications of Van Overstraeten [2]. If one represents the maximum current (for $V_D \rightarrow \infty$ in (17)) by I_{Dm}, the normalized (measured) current, $i_D = I_D/I_{Dm}$, can be expressed as a function of normalized (measured) drain voltage $u_D = V_D/v_T$, as follows :

$$i_D = 1 - \exp(-u_D/\alpha), \text{with } \alpha = \frac{n}{m}$$

For $N_{SS} = 0$, $\alpha = 1$ and the normalized current i_{DI} (ideal) is :

$$i_{DI} = 1 - \exp(-u_D).$$

Fig. 2 shows the two currents schematically (for V_G constant = V_{Go})
The difference, $\Delta i = i_{DI} - i_D$ as a function of u_D, has a maximum at

$$u_D = u_D^* = \frac{\alpha \log \alpha}{\alpha - 1}$$

By measuring u_D^*, one may evaluate α

Now, n is determined from the slope of the experimental curve $\log(I_{DM}) = f(V_G)$ around V_{Go} (for $V_D \simeq 1$ V).
Knowing α and n, one may calculate $m = n/\alpha$ and then,

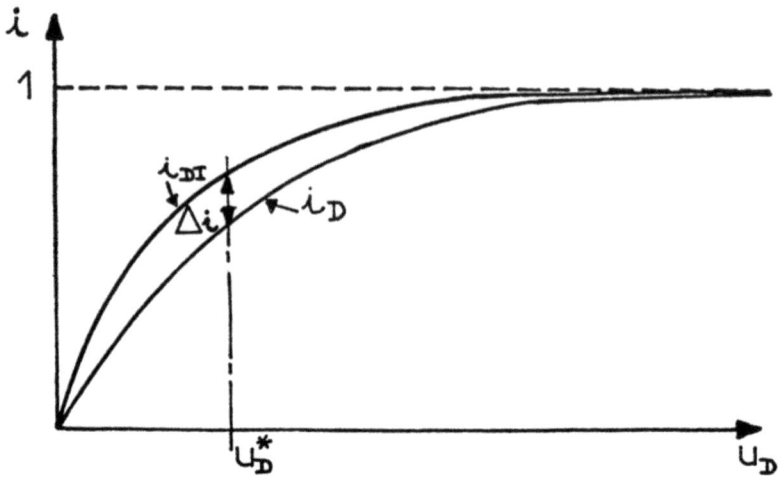

Fig. 2. $i_D(u_D)$ characteristics. i_{DI} ideal, $N_{SS}=0$; i_D measured, $N_{SS} \neq 0$

$$N_{SS} = \frac{C_{ox}}{q}(n - m)$$

3. TRANSITION ZONE BETWEEN WEAK AND STRONG INVERSION

This region can be modelled analytically, according to the following physical assumptions :

- one assumes a strong inversion in the vicinity of the source and weak inversion near the drain [3]
- the domain of validity of (7) is extended up to the onset of strong inversion [5]. One thus obtains a current voltage relation similar to (17), the detailed expressions will be given in the lecture on CAD models.

REFERENCES

[1] S.M. Sze, John Wiley and Sons 1969
[2] R. Van Overstraeten, G. Declerck, P. Muls
. IEDM, Washington, December 3-5, 1973
. IEEE Trans. on E.D., E-D 22, 5, May 1975

[3] R.M. Swanson, J. Meindl, IEEE J. of Sol. St. Circuits, vol.7, 2, April 1972
[4] M.B. Barron, Solid St. El. vol.15, 1972
[5] G.Merckel, E. Lora-Tamayo, J. Gautier, To be published.

ION IMPLANTED MOS TRANSISTORS

E. DEMOULIN and F. VAN DE WIELE

Université Catholique de Louvain
Microelectronics Laboratory,
Louvain-la-Neuve, Belgium.

ABSTRACT - *A review of the principal features of ion implanted transistors is presented. A detailed analysis of the threshold voltage shift is given, with emphasis directed towards dose and energy effects. Analytical expressions are derived for enhancement mode devices and compared to numerical solutions; the effect of the implantation on subthreshold currents is described. Some aspects of depletion mode transistors are treated.*

I - INTRODUCTION

An interesting feature of ion implantation in silicon is the fact that the impurity concentration in the surface region of the semiconductor can be modified while a thin oxide layer is left on top of the substrate. The introduction of charged atomic perticles into the substrate of MOS structures changes the electrical properties of the devices. The use of ion implantation offers, in particular, a means of changing the threshold voltage V_T of MOS devices [1,2]. This topic (for P^+P substrates) is discussed in the first part of this paper; the influence of the dose ϕ and of the energy E of the ion implantation are considered.

To describe the channel current of implanted MOS transistors one has to make a clear distinction between two types of devices. In a first case the current is flowing through a thin surface channel, which is the usual inversion layer formed at the interface between the oxide and the semiconductor. The current mechanism is the same as that for a non-implanted MOS transistor, although the current dependence on the drain and gate voltages is different. Surface conduction is always obtained when the implanted impurities are of the same type (acceptor or donor) as the substrate impurities. The second part of this paper is devoted to the description of surface channel transistors.

On the other hand, buried channel MOS transistors are obtained by incorporating within the surface region impurities of the opposite type to that of the substrate impurities. A conducting channel between the source and drain contacts exists in a direction parallel to the interface but below it. The width of the neutral channel region is now restricted by a surface depletion region and a bulk depletion region. The current mechanism for this type of device is quite similar to that of a junction field effect transistor. In the third part of this paper we describe the bulk conduction mode of implanted devices. Note that surface conduction and bulk conduction may occur, for different polarization conditions, in the same device.

Throughout this paper only n-channel MOS transistors will be considered. The substrate doping N_{AB} is taken equal to 10^{15} cm^{-3}, while the oxide thickness t_{ox} is normally equal to 1000 Å.

II - DOPING DENSITY OF THE IMPLANTED IONS

In order to illustrate, by means of numerical examples, the changes induced by ion implantation, we will consider in particular boron and phosphorus implantations performed through an oxide layer. We assume a Gaussian spatial distribution for the implanted ions; their doping density may then be expressed as [3]

$$N_{AI}(x) = \frac{\Phi}{(2\pi)^{1/2} \Delta R_p} \exp\left[-\frac{(x - R_p)^2}{2 \Delta R_p^2}\right] \qquad (1)$$

The quantity Φ is the dose (i.e. the total number of ions per unit area impinging on the surface); R_p is the effective projected range (taken from the interface), ΔR_p is the standard deviation or perpendicular straggle and x the distance with respect to the Si-SiO$_2$ interface. For the numerical values of R_p and ΔR_p we adopt the figures published by Gibbons et al. [4], since the agreement with the experimental results, in particular for boron [5], is reasonably good.

We assume that complete electrical activation is achieved by high temperature anneal [6]. Furthermore it is assumed that the high temperature steps in the fabrication process of the device do not introduce an important redistribution of the impurities; this assumption is normally valid for a standard aluminum gate technology. It is obvious that the case of redistributed profiles can be treated in a similar way.

Since a fraction of the implanted ions is located in the oxide layer, the total number of ions per unit area within the semiconductor, denoted by Φ_{eff}, is smaller than the dose Φ. The effective dose Φ_{eff} is an increasing function of the implantation energy E, if the oxide thickness t_{ox} is kept constant (see fig.1).

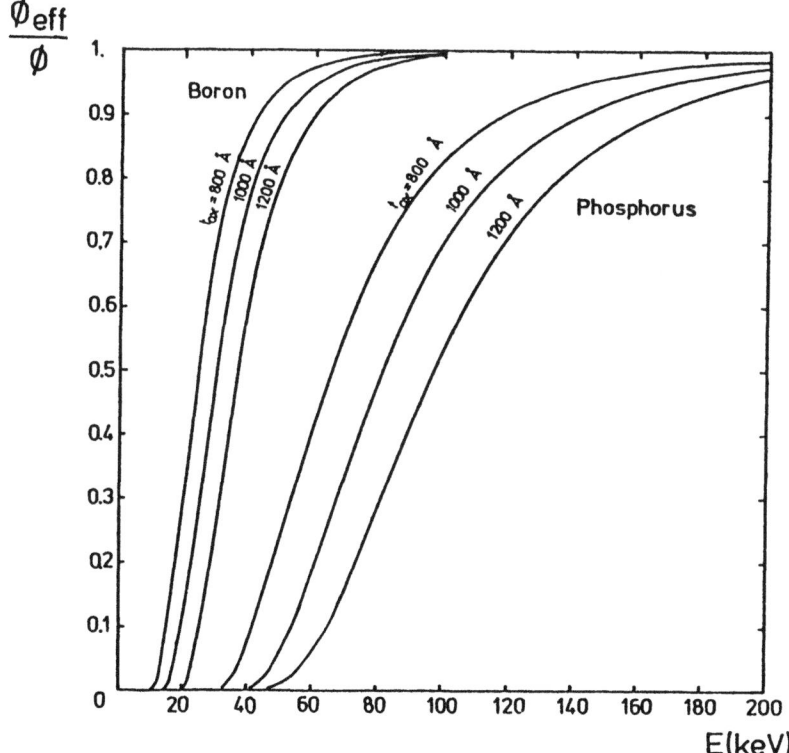

Fig. 1. Ratio Φ_{eff}/Φ, where Φ_{eff} is the effective dose in the silicon and Φ the total implanted dose.

The ratio ϕ_{eff}/ϕ becomes larger than 0.9, for the case t_{ox} = 1000 Å, when E exceeds 50 keV for boron and 140 keV for phosphorus.

In the next section we start the discussion of the threshold voltage shift resulting from an implanted profile given by (1).

III. Threshold voltage (P^+P substrates)

III-1. Introduction

The threshold voltage V_T of a MOS structure on an uniformly doped substrate is usually defined as the gate voltage V_G at the onset of strong surface inversion [7], i.e. the gate voltage for which the concentration of surface minority carriers n_S is equal to the bulk impurity concentration N_{AB}. Under non-equilibrium condition, if V_S (> 0) is the reverse bias voltage applied to an adjacent junction (fig.2), this criterion means a surface potential ψ_S given by

$$\psi_{S,inv} = 2\phi_{FB} + V_S \tag{2}$$

where ϕ_{FB} is the bulk Fermi potential :

$$\phi_{FB} = (kT/q) \ln (N_{AB}/n_i) \tag{3}$$

n_i is the intrinsic carrier concentration.

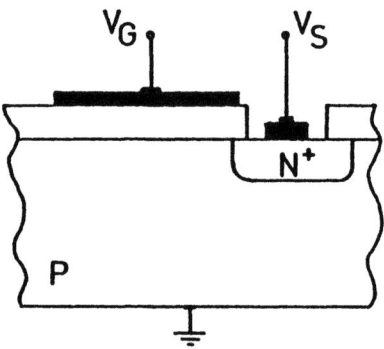

Fig.2 - Gate-controlled diode structure. V_S is positive for reverse bias.

The corresponding threshold voltage V_T^* [8] is then given by (all potentials and voltages are referred to the grounded substrate) :

$$V_T^* = V_{FB} + 2\phi_{FB} + V_S + K_B (2\phi_{FB} + V_S)^{1/2} \quad (4)$$

where

$K_B = (2q\varepsilon_S N_{AB})^{1/2}/C_{ox}$ is the body factor,
C_{ox} is the gate oxide capacitance per unit area,
ε_S is the permittivity of the semiconductor,
V_{FB} is the flat-band voltage defined by :

$$V_{FB} = \phi_{MS} - (Q_{SS}/C_{ox}) \quad (5)$$

ϕ_{MS} is the work function difference between the metal and the semiconductor, and Q_{SS} is the equivalent interface charge per unit area.

Under equilibrium condition the threshold voltage reduces to

$$V_{TO}^* = V_{FB} + 2\phi_{FB} + K_B (2\phi_{FB})^{1/2} \quad (6)$$

III-2. Threshold voltage of non-uniformly doped P^+P structures

The description of the surface phenomena becomes more difficult for MOS structures on non-uniformly doped substrates. Consider in particular the case of MOS structures for which the surface impurity concentration is larger than the bulk concentration (P^+P structures). Since the surface inversion layer is formed in a region where the local impurity concentration can be quite different from the bulk impurity concentration N_{AB}, it is evident that the relations (4) and (6) are no longer valid.

Consider the relationship between the surface potential ψ_S and the gate voltage :

$$V_G - V_{FB} = \psi_S - Q_S/C_{ox} \quad (7)$$

Q_S is the total space charge per unit area within the semiconductor

$$Q_S = q \int_0^{x_b} \left[p(x) - n(x) + N_D(x) - N_A(x) \right] dx \quad (8)$$

x_b represents the distance, with respect to the oxide-semiconductor interface, of a point located deep into the bulk of the semiconductor, beyond which the local space charge density is zero. The local donor and acceptor concentrations are $N_D(x)$ and $N_A(x)$ respectively. In this section we assume that $N_D(x) = 0$, and that

$$N_A(x) = N_{AB} + N_{AI}(x) \quad (9)$$

with $N_{AI}(x)$ defined by (1). The total space charge density Q_S depends on the shape of the local hole concentration $p(x)$ and electron concentration $n(x)$, which are functions of the surface potential ψ_S. The relationship (7) between ψ_S and V_G can be computed when the distribution functions $p(x)$ and $n(x)$ of the free carriers are known (for example, by means of a numerical analysis).

Define $\psi_{S,inv}$ as the surface potential corresponding to the onset of strong surface inversion; the threshold voltage V_T of the structure is then, according to (7),

$$V_T = V_{FB} + \psi_{S,inv} - Q_S(\psi_{S,inv})/C_{ox} \qquad (10)$$

Thus, the problem is reduced to the definition of a correct $\psi_{S,inv}$ value. Different solutions have been proposed in the literature.

III-2-1. <u>Classical criterion</u> $\psi_{S,inv} = 2\phi_{FB} + V_S$

Many authors [9,10] just consider for $\psi_{S,inv}$ the condition (2) valid for uniformly doped substrates; they assume that strong surface inversion in the implanted structure occurs when the surface concentration of minority carriers is equal to the bulk concentration of majority carriers.

This criterion is hardly acceptable from a theoretical point of view. Let us consider the current characteristics of MOS structures for small drain to source voltages. The significant parameter in this case is the total charge Q_n per unit area of the minority carriers near the semiconductor surface :

$$Q_n = -q \int_0^{x_i} n(x)\,dx \qquad (11)$$

x_i represents the width of the surface channel, i.e. the surface region where the minority carriers effectively contribute to the conduction phenomena. Figure 3 shows Q_n and the total space charge Q_S as a function of the surface potential, respectively for an uniformly doped P-type substrate and for boron implanted structures on the same substrate. These results were obtained by numerical integration of Poisson's equation using a discretization method [11]. The source voltage is taken equal to zero ($V_S = 0$). Two interesting features can be observed. The rapid increase of the total charge Q_S occurs for increasing values of the surface potential ψ_S when the dose Φ becomes larger. For a fixed surface potential equal to $2\phi_{FB}$, we see that the inversion charge Q_n varies over a wide range : from 3.25×10^{-10} Cb/cm^2 for a uniform substrate to 3.26×10^{-11} Cb/cm^2 for an implanted substrate with a dose

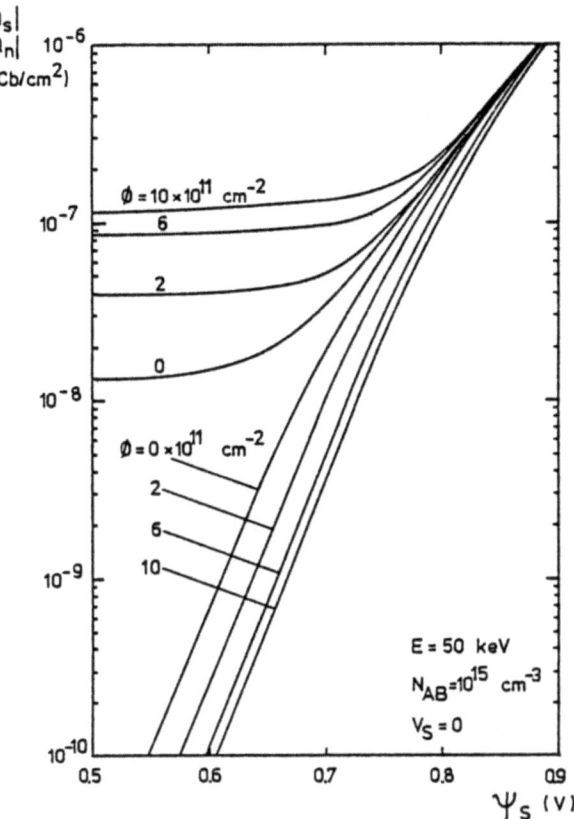

Fig. 3 - Total space charge Q_S (upper curves) and minority carrier charge Q_n (lower curves) for non-implanted and boron implanted silicon. The implantations are performed through a 1000 Å SiO_2 layer.

of 1.4×10^{12} cm^{-2}. This fact implies that the current parallel to the interface varies for the different MOS devices in a similar fashion, if the surface potential is maintained equal to $2\phi_{FB}$ by applying an appropriate gate potential. The inadequacy of the classical criterion $\psi_{S,inv} = 2\phi_{FB}$ is obvious for implanted structures under equilibrium conditions.

The same conclusion results from the study of the capacitance versus voltage curves of the MOS structures presented in figure 4; the depicted figures were obtained by numerical analysis for the

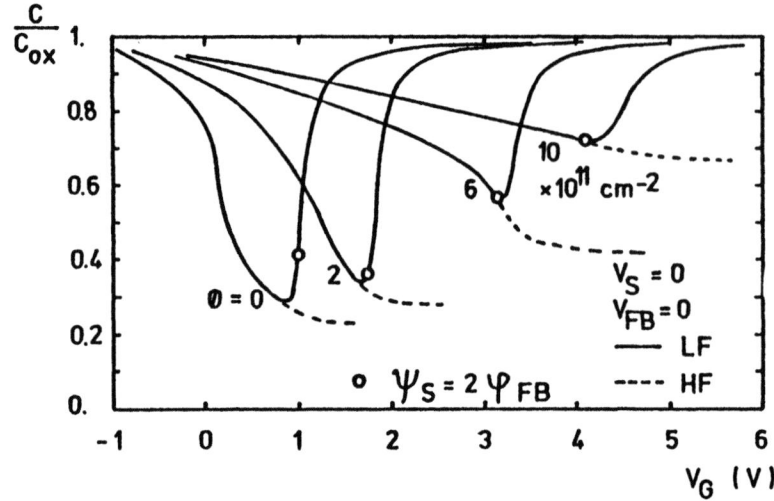

Fig. 4 - Low and high frequency capacitance of non-implanted and boron implanted MOS structures. The implantations are performed at E = 50 keV through an oxide layer of 1000 Å. Circles (o) indicate the $\psi_S = 2\phi_{FB}$ points.

case $V_S = 0$ [11]. At low frequencies the point corresponding to $\psi_S = 2\phi_{FB}$ is located at the right hand side of the minimum for the case of an uniform substrate. For increasing values of the implantation dose a gradual shift to the left of the point $\psi_S = 2\phi_{FB}$ can be observed; for a heavy implantation the representative point is even located in the depletion part of the capacitance curve. At high frequencies the agreement between the minimum of the capacitance curve and the point $\psi_S = 2\phi_{FB}$ is excellent only for the case of an uniform substrate; for implanted structures the capacitance continues to decrease beyond the point $\psi_S = 2\phi_{FB}$.

We conclude that the use for implanted structures of the criterion $\psi_{S,inv} = 2\phi_{FB} + V_S$ for the onset of strong surface inversion is questionable. Despite the drawbacks the criterion has two advantages : the simplicity of its mathematical expression, and the fact it is valid for implanted substrates with a low dose.

The relationship (7) between the surface potential ψ_S and the gate voltage V_G is depicted in figure 5; the curves are obtained from a numerical analysis. The particular gate voltage corresponding to the surface potential $\psi_S = 2\phi_{FB}$ is indicated; one obtains the threshold voltage according to the classical criterion as a function of the implantation dose.

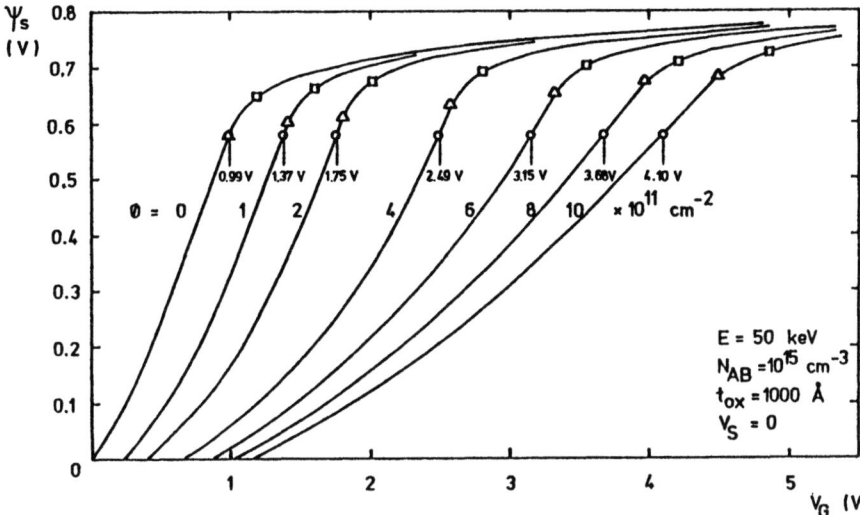

Fig. 5. Surface potential versus gate voltage for non-implanted and boron implanted MOS structures, with $V_{FB} = 0$. The circles are for $2\phi_{FB}$, the triangles for $dQ_n^{FB}/d\psi_S = dQ_B/d\psi_S$ and the squares for V_{TE}.

III.2.2. Extrapolation of the $Q_n - V_G$ curves

The total space charge Q_S of the semiconductor can be divided into two terms : the inversion charge Q_n, due to the minority carriers in the surface region, and the so called depletion charge Q_B, due to the difference between the concentration of majority carriers and the impurity concentration,

$$Q_S = Q_n + Q_B \tag{12}$$

where

$$Q_B = q \int_0^{x_b} \left[p(x) + N_D(x) - N_A(x) \right] dx \tag{13}$$

Taking (7) and (10) into account one obtains :

$$V_G - V_T = \psi_S - \psi_{S,inv} - \frac{1}{C_{ox}} \left[Q_n(\psi_S) - Q_n(\psi_{S,inv}) \right]$$

$$- \frac{1}{C_{ox}} \left[Q_B(\psi_S) - Q_B(\psi_{S,inv}) \right] \tag{14}$$

Before continuing the discussion of the criterion to be used for $\psi_{S,inv}$, it is worthwhile to comment the preceding equations.

A. If the threshold voltage is defined as the gate voltage between the conducting state and the non-conducting state (or pinch-off state) of the MOS transistor, one may introduce the following approximation for the inversion charge :

$$Q_n(\psi_{S,inv}) = 0 \tag{15}$$

B. The surface potential, when the semiconductor surface is inverted, may be considered as almost constant for gate voltages exceeding the threshold voltage :

$$\psi_S(V_G > V_T) \simeq \psi_{S,inv} \tag{16}$$

This is evident from figure 5 for ψ_S = 0.7 V (for the case V_S = 0).
The depletion charge Q_B varies very slowly as a function of the surface potential for ψ_S near $\psi_{S,inv}$ (see fig. 3), and consequently may be approximated by :

$$Q_B(V_G > V_T) \simeq Q_B(\psi_{S,inv}) \tag{17}$$

The preceding comments allow to simplify the relation (15) and to find the following approximation for the inversion charge

$$Q_n \simeq -C_{ox}(V_G - V_T) \tag{18}$$

The validity of this result is restricted for gate voltages near the threshold voltage V_T : the magnitude of the neglected terms in (14) is of the same order as the value of Q_n given by (18). However, the linear relationship (18) between the inversion charge Q_n and the gate voltage V_G holds over a large range of usual V_G values. This fact clearly appears from the figure 6, where the Q_n - V_G curves, obtained from a numerical analysis, are plotted for boron implanted structures. The intercept on the voltage axis of the extrapolated straight line yields the corresponding threshold voltage of each device. We call V_{TE} the threshold voltage deduced in that way.

It is important to note that the relationship between Q_n and V_G is identical, except for a scaling factor, to the relationship between the drain current and the gate voltage of the MOS transistor, if a small, constant voltage is applied to the drain with respect to the source. The threshold voltage can be experimentally determined by measuring the drain current under these

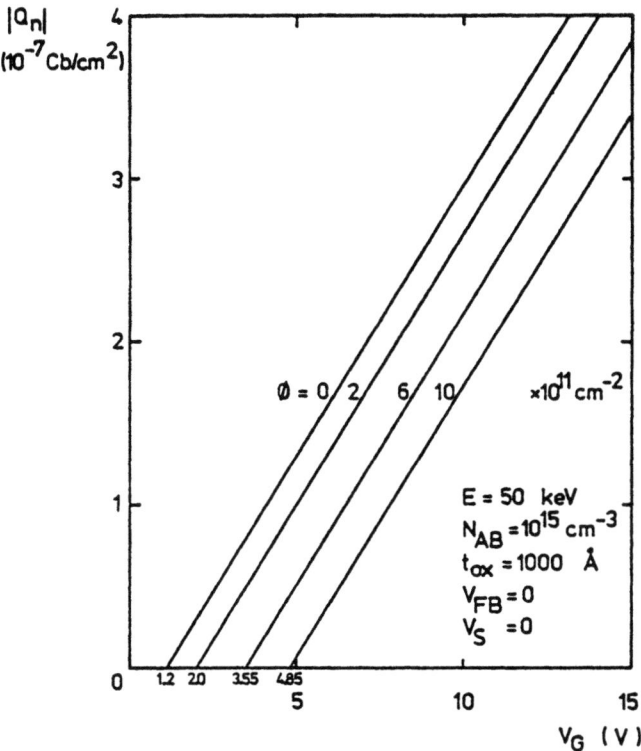

Fig. 6. Inversion charge Q_n versus gate voltage V_G for non-implanted and boron implanted n-channel structures. The implantations are performed through the oxide layer.

particular conditions. We conclude that the described extrapolation method of the $Q_n - V_G$ curves offers the advantage of a direct comparison between the theoretical V_T values and the experimental ones resulting from the current characteristics of the MOS device. Furthermore we principally need the threshold voltage to introduce it in such a relation and to compute the drain current in an MOS transistor from this relation.

The threshold voltage values obtained by the extrapolation method are indicated in figure 5. The corresponding surface potentials $\psi_{S,inv}$ are larger than $2\phi_{FB}$ and shift to higher values for increasing implanted doses. Finally a good approximation is :

$$\psi_{S,inv} = 2\phi_{FB} + \Delta\psi \qquad (19)$$

For a given structure Δψ is a constant which takes into account the substrate doping and the implant profile.

III.2.3. Criterion based on the variation of the charges Q_n and Q_B

A sharp increase of the inversion charge occurs when the surface potential ψ_S increases from the depletion to the inversion range (see fig. 3). The depletion charge Q_B, for the same interval of ψ_S values, tends to a constant value, since the space charge region within the semiconductor attains its maximum width. These facts imply that any increase of ψ_S has very little influence on Q_B, but enhances in a significant way the number of minority carriers located near the surface, even when Q_n is still smaller than Q_B.

Tobey and Gordon [12] have shown, for the case of an uniform substrate, that the following condition on the variation of Q_n and Q_B

$$dQ_n/d\psi_S = dQ_B/d\psi_S \qquad (20)$$

is mathematically equivalent to the classical criterion (2). So, strong inversion occurs when a variation of the surface potential yields an equal variation of the inversion and the depletion charges.

Feltl [13] suggested to consider condition (20) as criterion for strong surface inversion in implanted devices.

The values of $dQ_n/d\psi_S$ and $dQ_B/d\psi_S$, obtained from a numerical analysis of Poisson's equation, are plotted in figure 7 as a function of the surface potential ψ_S for an uniform substrate and for boron implanted devices. We can see that the conditions (2) and (20) are indeed equivalent for an uniform substrate, and that the surface potentials $\psi_{S,inv}$ for implanted structures are larger than $2\phi_{FB}$.
The resulting surface potentials are plotted in figure 8 as a function of the dose; also shown are the figures obtained by the extrapolating method. One can observe an increasing disagreement between $2\phi_{FB}$ and the $\psi_{S,inv}$ values resulting from (20) or those obtained by the extrapolation method. The corresponding points are also represented in figure 5 (Δ).

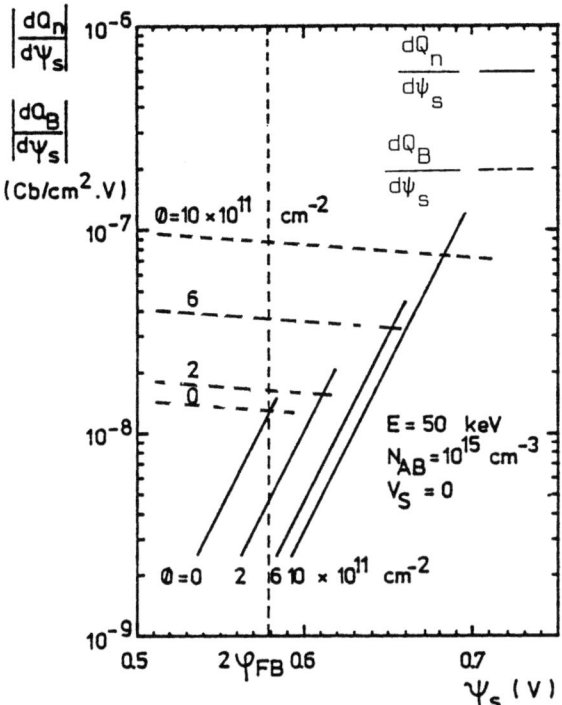

Fig. 7. $(dQ_B/d\psi_S)$ and $(dQ_n/d\psi_S)$ for non-implanted and boron implanted n-channel structures. The implantation are performed through the oxide layer.

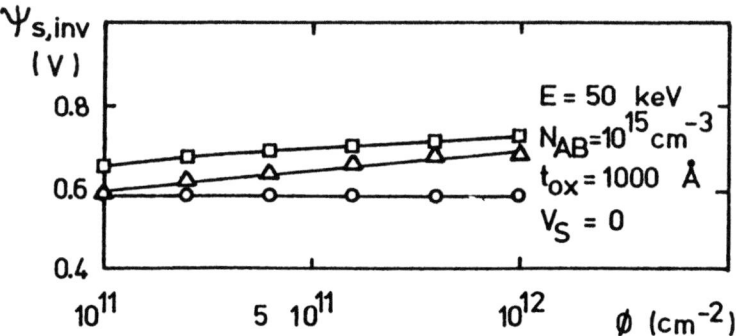

Fig. 8. Surface potential at strong inversion onset. The implantations are performed through the oxide layer. The circles are for $2\psi_{FB}$, the triangles for $dQ_n/d\psi_S = dQ_B/d\psi_S$ and the squares for V_{TE}.

III.3. Threshold voltage calculations based on the depletion approximation

The discussion presented in the preceding section makes use of numerical data obtained from a computer program [14] based on a discretization method of Poisson's equation. Such a numerical analysis is time consuming and expensive. Hence it is worthwhile to introduce appropriate approximations, in order to obtain models requiring more simple mathematical tools.

III.3.1. The depletion approximation

The surface space charge region of an uniformly doped MOS structure is usually described by means of the well known depletion approximation [15]. By applying the same approximation for implanted P^+-P structures, one obtains the following relations valid for gate voltages smaller than the threshold voltage [16]:

$$Q_B = -q \int_0^{x_d} N_A(x)\, dx - \varepsilon_S\, \xi(x_d) \quad (21)$$

$$\psi_S = \frac{q}{\varepsilon_S} \int_0^{x_d} x\, N_A(x)\, dx + x_d\, \xi(x_d) + \psi(x_d) \quad (22)$$

x_d is the surface depletion width; $\xi(x_d)$ and $\psi(x_d)$ are respectively the electric field and the potential at the boundary x_d of the depletion region.

The width of the depletion region attains its maximum value, denoted by $x_{d,max}$, at the onset of strong surface inversion [15, 16]; the surface potential becomes than constant and is given by

$$\psi_{S,inv} = \frac{q}{\varepsilon_S} \int_0^{x_{d,max}} x N_A(x)\, dx + x_{d,max}\, \xi(x_{d,max}) + \psi(x_{d,max}) \quad (23)$$

Combining (11), (13), (16) and (21) one obtains

$$V_T = V_{FB} + \psi_{S,inv} + \frac{q}{C_{ox}} \int_0^{x_{d,max}} N_A(x)\, dx + \frac{\varepsilon_S\, \xi(x_{d,max})}{C_{ox}} \quad (24)$$

The problem is again reduced to find a correct definition for the surface potential $\psi_{S,inv}$. Once this potential is known, one may determine the maximum depletion width $x_{d,max}$ from (23) and the value of the threshold voltage from (24).

Before discussing the different proposals made for $\psi_{S,inv}$ in the literature, we briefly comment on the mathematical relations.

A.- The electric field and potential at the boundary of the depletion region x_d are equal to zero for uniformly doped substrates. This is not necessarily the case for implanted structures, since x_d can be located in a region where the impurity density is not constant. For example, for a boron implanted structure with a dose of 10^{12} cm^{-2}, the value of x_d lies in the range of 1500 Å (see fig.9-a) where the majority carrier concentration is smaller than the corresponding concentration of implanted impurities; both the electric field (fig.9-b) and the potential (fig.9-c) are different from zero in the same interval. The influence of the terms $\xi(x_d)$ and $\psi(x_d)$ becomes significant for heavy ion implantations.

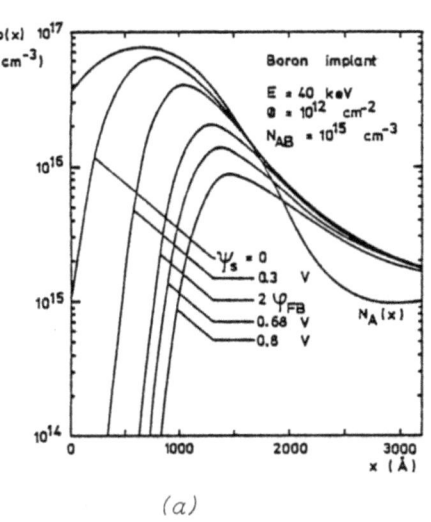

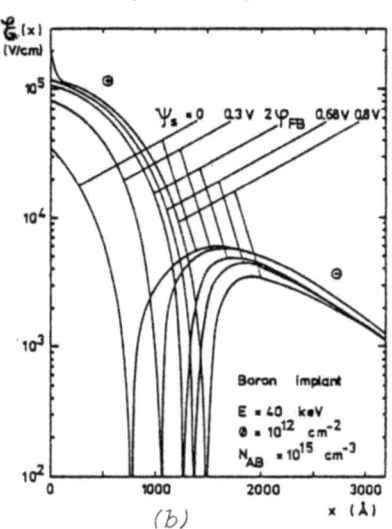

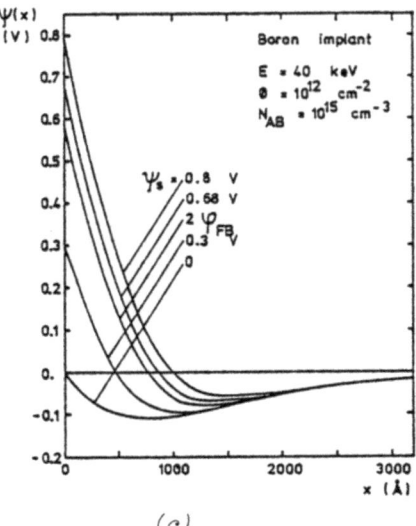

Fig. 9. Dependence of the minority carrier (a), electric field (b) and potential (c) distributions on the surface potential for boron implanted P-type silicon. The implantation ($E = 50$ keV, $\Phi = 10^{12}$ cm^{-2}) is performed through a 1000 Å SiO_2 layer. The impurity distribution is also plotted (a).

B.- The equations (23) and (24) are valid for any impurity profile. For the particular case of Gaussian impurity profiles one obtains the following analytical expressions :

$$\psi_{S,inv} = \frac{q\Phi}{2\varepsilon_S} R_p \left[erf\left(\frac{x_{d,max} - R_p}{\sqrt{2}\,\Delta R_p}\right) + erf\left(\frac{R_p}{\sqrt{2}\,\Delta R_p}\right) \right]$$

$$- \frac{q\,\Phi\,\Delta R_p}{\sqrt{2\pi}\,\varepsilon_S} \left\{ \exp\left(-\frac{(x_{d,max} - R_p)^2}{2\,\Delta R_p^2}\right) - \exp\left(-\frac{R_p^2}{2\,\Delta R_p^2}\right) \right\} \quad (25)$$

$$+ \frac{qN_{AB}}{2\varepsilon_S} x_{d,max}^2 + x_{d,max}\,\zeta(x_{d,max}) + \psi(x_{d,max})$$

and

$$V_T = V_{FB} + \psi_{S,inv} + \frac{q\Phi}{2C_{ox}} \left[erf\left(\frac{x_{d,max} - R_p}{\sqrt{2}\,\Delta R_p}\right) + erf\left(\frac{R_p}{\sqrt{2}\,\Delta R_p}\right) \right]$$

$$+ \frac{1}{C_{ox}} \left[qN_{AB}\, x_{d,max} + \varepsilon_S\,\zeta(x_{d,max}) \right] \quad (26)$$

Similar relations have been proposed by Kamoshida and Kudoh [17] for the case $\zeta(x_{d,max}) = \psi(x_{d,max}) = 0$. An iterative numerical method is required for the determination of $x_{d,max}$ from (25).

C. Let us consider the influence of the penetration depth of the implanted impurities. For a very shallow implant, and if all the charge is considered to be located at the interface (x=o), then equation (23) becomes identical to the expression valid for an uniform substrate of doping N_{AB}

$$\psi_{S,inv} = (qN_{AB}/2\varepsilon_S)\,(x_{d,max}^*)^2 \quad (27)$$

If the flat-band voltage is identical for the implanted and non-implanted cases, the resulting threshold voltage shift is simply given by :

$$\Delta V_T = V_T - V_T^* = q\,\Phi_{eff}/C_{ox} \quad (28)$$

For real implants of dose Φ_{eff} the width $x_{d,max}$ given by (23) becomes smaller than $x_{d,max}^*$; this is due to the spatial distribution of $N_A(x)$. According to (24) a similar reduction is expected for V_T^*; hence the real threshold voltage shift is smaller than

$(q\phi_{eff}/C_{ox})$.

We now start the discussion of the different conditions for $\psi_{S,inv}$ mentionned in the literature.

III.3.2. Classical condition

By analogy with the case of a uniform substrate, many authors [17,18] introduce the classical condition :

$$\psi_{S,inv} = 2\phi_{FB} + V_S \qquad (2)$$

to characterize the strong surface inversion state of implanted P⁺P devices.

III.3.3. Condition using the concentration at the boundary of the depletion region.

A criterion, including the fact that the boundary x_d of the depletion region can be located in the implanted region of the device, has been proposed by Doucet and Van de Wiele [16]. This criterion states that strong surface inversion occurs, when the surface concentration of minority carriers is equal to the concentration of impurities at the boundary of the depletion region, i.e. when

$$n_S(\psi_{S,inv}) = N_A(x_{d,max}) \qquad (29)$$

The corresponding surface potential is then given by

$$\psi_{S,inv} = \frac{kT}{q} \ln\left[N_A(x_{d,max}) N_{AB}/n_i^2\right] + V_S \qquad (30)$$

The inversion condition (29) has also been suggested by Tanaka [19] ; it has been used by many authors, for example by Troutman [20].

It is obvious that both conditions (30) and (2) become equal when the boundary of the depletion region is located in the uniform part of the substrate, i.e. when $N_A(x_{d,max})=N_{AB}$. The last case occurs for shallow implants with a low dose, or for large values of V_S.

III.3.4. Condition using the mean concentration

The inversion condition (20) based on the variation of the charges Q_n and Q_B with respect to ψ_S, can be transformed according to Feltl [13] into the following condition for the surface minority concentration by considering the depletion approximation:

$$n_S(\psi_{S,inv}) = \frac{1}{x_{d,max}} \int_0^{x_{d,max}} N_A(x)\, dx \tag{31}$$

In fact a few additional simplifying assumptions are required in order to obtain (31); however these assumptions seem to be valid for the doses usually considered for P⁺P structures. The related surface potential is given by,

$$\psi_{S,inv} = \frac{kT}{q} \ln\left[n_S(\psi_{S,inv}) \cdot N_{AB}/n_i^2\right] + V_S \tag{32}$$

For the particular case of a depletion region extending to the uniform part of the substrate, one obtains from (31) the condition

$$n_S(\psi_{S,inv}) \simeq N_{AB} + (\bar{\Phi}_{eff}/x_{d,max}) \tag{33}$$

The related surface potential may be expressed as

$$\psi_{S,inv} \simeq 2\phi_{FB} + \frac{kT}{q} \ln\left(1 + \frac{\bar{\Phi}_{eff}}{N_{AB}\, x_{d,max}}\right) \tag{34}$$

III.3.5. Comparison between the different conditions

The surface potential versus the gate voltage is plotted in figure 10 for uniform substrate and for a boron implanted structure (E = 50 keV, $\Phi = 10^{12}$ cm^{-2}). The exact numerical results are shown, together with the values obtained by the depletion approximation; a reasonable agreement exists.

The threshold voltages resulting from the described inversion conditions are indicated in figure 10, together with the threshold voltage V_{TE} obtained by the extrapolation method. One can conclude that the values given by (2) are in general too small. Better results are obtained with conditions (30) and (32), which include the influence of the non-uniform profile; the corresponding threshold voltages shift towards V_{TE}. The same conclusion holds for the threshold voltage data displayed in table I for different boron doses in a P-type substrate (E = 50 keV, t_{ox} = 1000 Å, $N_{AB} = 10^{15}$ cm^{-3}, V_S = 0, V_{FB} = 0).

We conclude that the conditions (30) and (32) are more appropriate than condition (2) for the surface inversion of implanted P⁺P devices, especially for heavy implants. Condition (2) has the advantage to be an explicit expression for $\psi_{S,inv}$, while (30) and (32) are implicit equations of the surface potential.

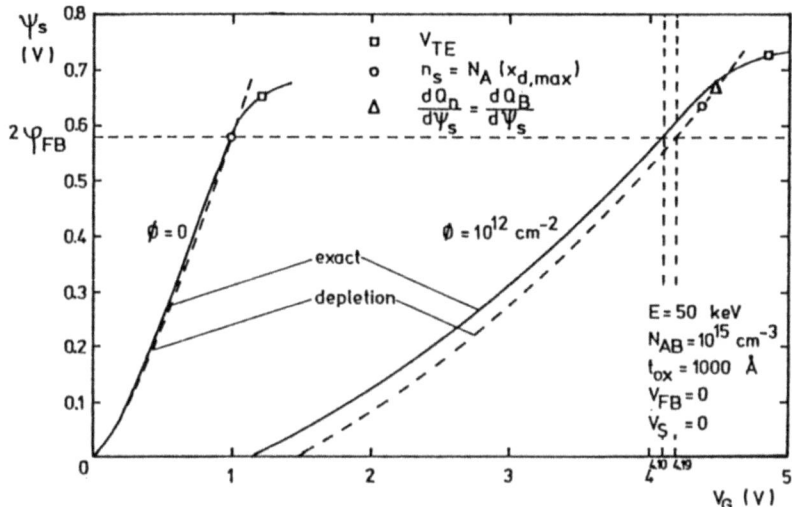

Fig. 10. Surface potential versus gate voltage for non-implanted and boron implanted structures. Squares (□) indicate the values V_{TE} obtained by extrapolation from the Q_n/V_G relations. Circles (O) are for Doucet and Van de Wiele criterion [16] while triangles (Δ) refer to the Feltl condition [13].

Table I. Threshold voltage for boron implanted MOS structures.

ϕ (cm^{-2})	V_{TE}(V) (extrapolation)	V_T(V) $\psi_{S,inv}=2\phi_{FB}$	V_T(V) $n_S^T=N_A(x_{d,max})$	V_T(V) $dQ_n/d\phi_S=dQ_B/d\phi_S$
0	1.22	0.99	0.99	0.99
2×10^{11}	2.04	1.76	1.76	1.80
6×10^{11}	3.53	3.17	3.19	3.33
10^{12}	4.78	4.19	4.43	4.53
1.4×10^{12}	5.60	4.90	5.32	5.36

III.4. Influence of energy and dose of the ion implantation

For a technologist only differences in threshold voltages are important, and in particular the shift indiced by an ion implantation if a constant V_{FB} is assumed :

$$\Delta V_T = V_T \text{ (implanted)} - V_T^* \text{ (non-implanted)} \quad (35)$$

In this section we study the influence of the energy and dose of boron implantations on the threshold voltage shift, calculated from the V_{TE} values given by the extrapolation method (a few results will also be presented for the other models). We are still referring to n-channel MOS structures.

III.4.1. Introduction

Consider an implant of ϕ ions per cm^2; if all ions are located at the semiconductor interface the threshold voltage shift is, according to (28), given by [21] :

$$\Delta V_{T,max} = q\,\phi/C_{ox} \tag{36}$$

This is the largest possible shift; for real implants a number of physical effects will reduce the actual shift.

A. A fraction of the implanted ions remain in the oxide layer and are electrically inactive for the threshold voltage shift. The number of these ions becomes larger for increasing implant energies, for thicker oxides or for an increasing mass of the implanted ions. Figure 1 illustrates these facts; for example, 11.4% of the implanted boron ions are located in the 1000 Å oxide layer for an energy of 50 keV.

The relation (36) must be corrected in the following way :

$$\Delta V_{T,max} = q\,\phi_{eff}/C_{ox} \tag{37}$$

where ϕ_{eff} represents the effective dose of ions within the semiconductor.

B. After ion implantation a high temperature treatment (for example, T > 950° during 30 minutes) is required to make the implanted impurities electrically active [6]. Equation (37) is only valid if a complete activation may be assumed; for other ions a reduced effective dose must be introduced.

C. The implanted ions are distributed within the semiconductor over a specific range; we assumed a Gaussian distribution function [3]. According to (23) and (24) a deeper penetration depth of the implanted ions at a constant dose requires a smaller gate voltage in order to maintain the surface potential constant.

D. We know that the boundary of the depletion region can be located in the non-uniform part of the impurity profile; this occurs especially in heavily implanted structures (for example, for boron doses exceeding 5.10^{11} cm^{-2}). The implanted ions lying outside the depletion region do not contribute to the space charge; this reduces the effectiveness of the implanted charge [22].

III.4.2. Influence of the implanted dose

Figure 11 shows the variation of the threshold voltage shift ΔV_T as a function of the total boron dose Φ into a P-type substrate (E = 50 keV, t_{ox} = 1000 Å, N_{AB} = 10^{15} cm^{-3}). The straight lines corresponding respectively to equation (36) and (37) are depicted; for the last case a maximum shift of 420 mV for Φ = 10^{11} cm^{-2} is expected. Included in figure 11 are the threshold voltage shifts resulting from the extrapolation (18) and from the conditions (2) and (20); these values are obtained by applying the respective condition to the exact numerical solution of the Poisson equation. The agreement between the condition (20) and

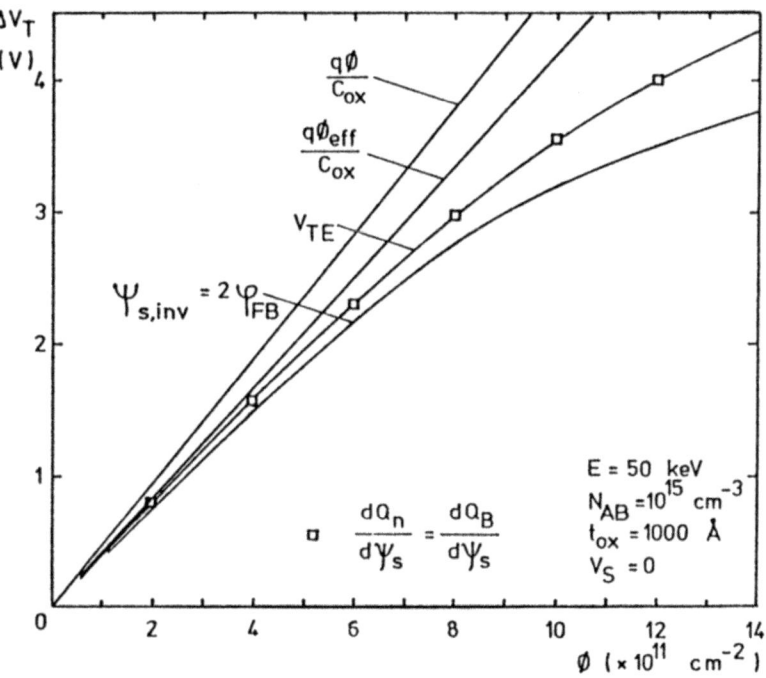

Fig. 11. ΔV_T as a function of the dose for boron implanted MOS structures. These values are obtained by applying the conditions (18), (2) or (20) to the exact numerical solution of the Poisson equation. The references are the straight lines $q\Phi/C_{ox}$ and $q\Phi_{eff}/C_{ox}$.

the extrapolation method is excellent; condition (2) is not valid for heavy implant doses.

The difference between the actual value of ΔV_T and $q\phi_{eff}/C_{ox}$ is due, for small doses ($< 5.10^{11}$ cm^{-2}), to the location of the implanted ions inside the semiconductor and not at the interface (point C of III.4.1); for heavy doses the discrepancy increases because the depletion region ends in the non-uniform part of the doping profile (point D of III.4.1).

Similar results are depicted in figure 12 for the models (2), (29) and (31) based on the depletion approximation; a Gaussian profile is assumed. By comparison with the extrapolation method (□) one concludes that the criterion (31) proposed by Feltl gives excellent results. A similar conclusion holds for the criterion (29) proposed by Doucet and Van de Wiele, although a slight deviation appears for doses in the range $4-12.10^{11}$ cm^{-2}.

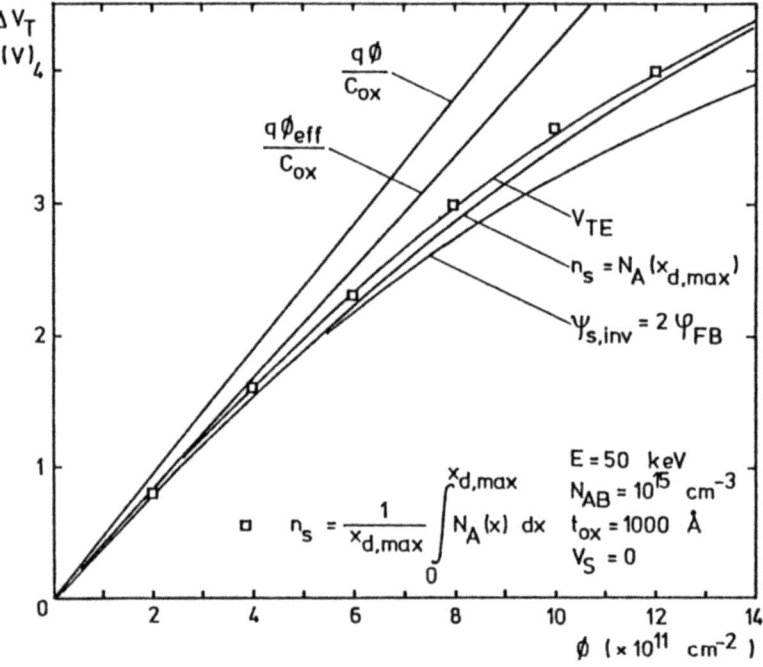

Fig. 12. ΔV_T *as a function of the dose for boron implanted MOS structures. The corresponding results are based on the depletion approximation together with the conditions (2), (29) or (31). The reference V_{TE} curve is obtained by extrapolation (18) of the Q_n/V_G relationship.*

Finally it is important to note that the efficiency η of the implanted dose, defined as

$$\eta = \Delta V_T / \Phi \tag{38}$$

decreases for increasing doses, the energy of the implantation being kept constant.

III.4.3. Influence of the energy

Since the penetration depth of the implanted ions is an increasing function of the energy, one may expect that the mechanism

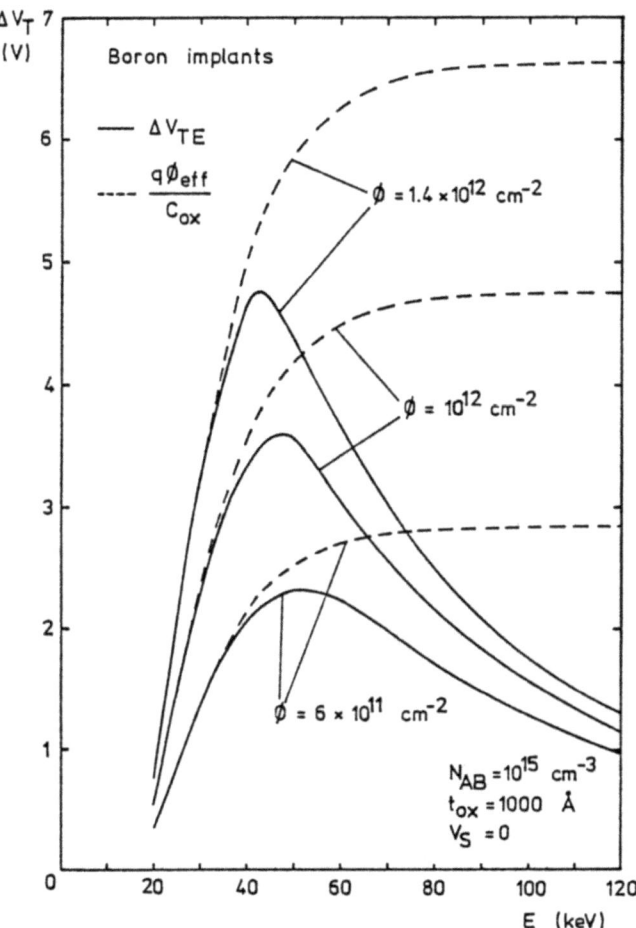

Fig. 13. Dependence of the threshold voltage shift on the implantation energy: —— ΔV_{TE}; --- $q\Phi_{eff}/C_{ox}$.

A described in section III.4.1 is less important at high energies and that mechanisms C and D become simultaneously dominant. The data of figure 13 illustrate the influence of the energy. The ideal threshold voltage shift given by (37) increases as a function of the energy and attains its maximum possible value, given by (36), for energies larger than 100 keV.

The actual ΔV_T data, obtained by the extrapolation method, coincide with the $q\Phi_{eff}/C_{ox}$ values in the low energy range; they attain a maximum near 50 keV and decrease for larger energies. This decrease can be explained by the mechanisms C and D of section III.4.1. Experimental results in general agreement with the data of figure 13 have been reported by Kudoh et al.[23].

An important conclusion from these results is that the optimum implant efficiency, for the conditions under consideration, is obtained for energies in the range of 40 to 50 keV. A particular threshold voltage shift ΔV_T can then be realized with a minimal dose; this is desirable in order to maintain a high carrier mobility (cfr. section IV. 4).

An other point of view, concerning the choice of an optimal implantation energy, is to consider the influence on the threshold

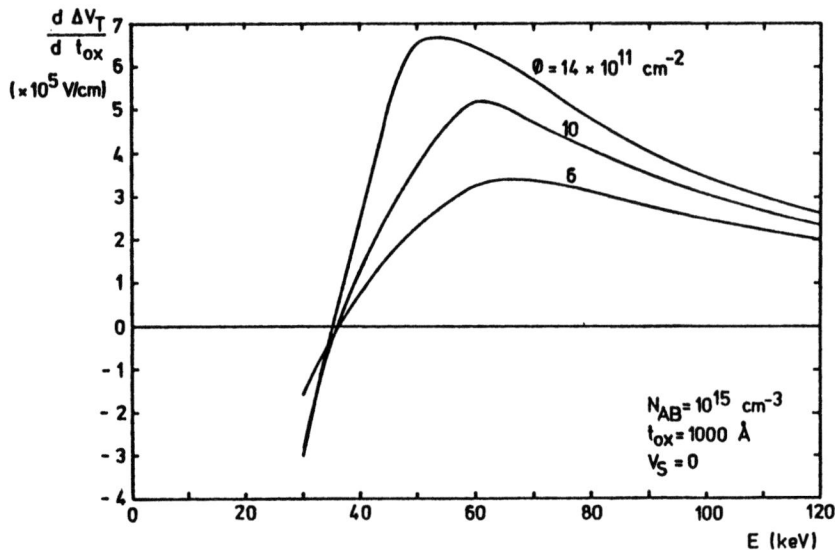

Fig. 14. $d(\Delta V_T)/dt_{ox}$ for boron implanted structures as a function of the energy. Threshold voltages are found by the extrapolation method.

voltage shift of fluctuations in the technological parameters. It is well known, for example, that processing fluctuations of ± 5 % in the oxide thickness with respect to the nominal value may be expected in standard MOS technologies; it is desirable that these variations have a minimal effect. The concomitant variations of ΔV_T are illustrated in figure 14, as a function of the energy for three different doses; the threshold voltages were calculated by means of the extrapolation method. The influence on ΔV_T is minimal, i.e.

$$d(\Delta V_T)/dt_{ox} = 0 \qquad (39)$$

for energies in the range 30 to 40 keV, for an oxide layer of 1000 Å. A detailed study of this problem has been made by Schemmert and Zimmer [24]; according to their results, the variation in ΔV_T is a minimum at the energy for which

$$\Phi_{eff} = \frac{\varepsilon_{Si}}{\varepsilon_{Si} - \varepsilon_{ox}} t_{ox} N_{AI}(0) \qquad (40)$$

where $N_{AI}(0)$ is the concentration of implanted impurities at the interface.

The optimal implantation energy is thus obtained when condition (39) is satisfied and when the implant efficiency figure is high. For a nominal oxide thickness of 1000 Å , energies around 40 keV seem to be appropriate; this conclusion agrees with the results published by Ishiwara et al.[25].

III-5. Effect of substrate polarization

For circuit designers it is desirable to maintain a low substrate sensitivity; the threshold voltage should be as insensitive as possible to changes in the source-to-substrate bias. This condition is required to avoid current impairing levels in source-follower applications. We consider this problem for boron implanted MOS devices with a P-type substrate.

For the theoretical analysis we use the threshold voltages V_{TE} obtained by extrapolating the numerically computed Q_n versus V_G curves, which are shown in figure 15 for a boron dose of 10^{12} cm^{-2}. The inversion charge Q_n, for a fixed gate voltage V_G, is a decreasing function of the source voltage V_S (we recall that V_G is the gate voltage with respect to the grounded substrate). Figure 16 shows (V_T-V_S) versus V_S for an uniform substrate ($N_{AB} = 10^{15}$ cm^{-3}) and for implanted structures with different doses (E = 50 keV); a comparison is made between the extrapolation method and

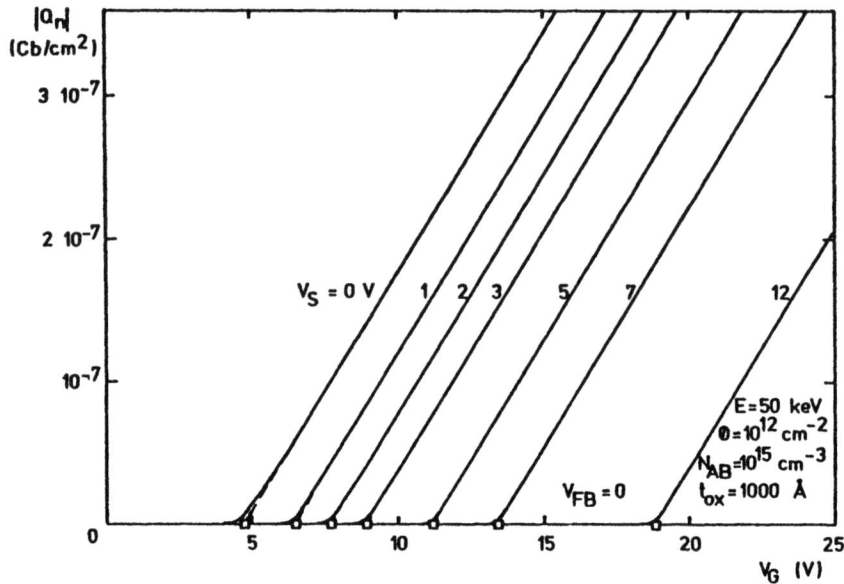

Fig. 15. Inversion charge for different source voltages V_S in a boron implanted MOS structure ($E=50$ keV, $\Phi=10^{12}$ cm^{-2}).

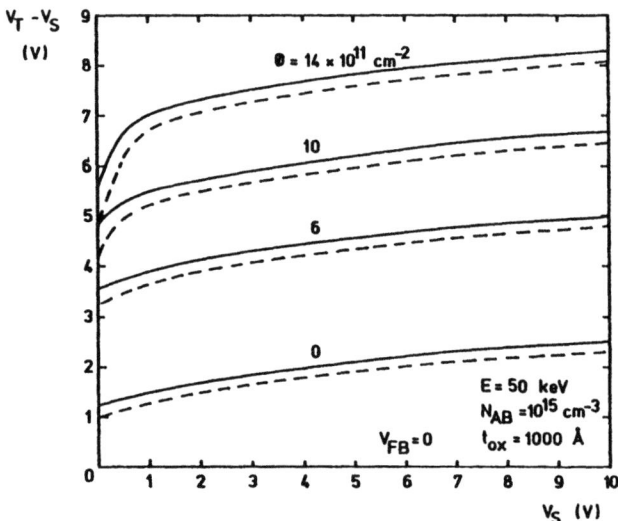

Fig. 16. V_T-V_S for non-implanted and boron implanted MOS structures ($E=50$ keV). The solid curves represent V_{TE}; the dashed ones represent the values for $\psi_{S,inv} = 2\phi_{FB} + V_S$.

Table II. Overall substrate sensitivity for non-implanted and implanted MOS structures (boron implants, $E = 50$ keV, $t_{ox} = 1000$ Å).

ϕ (cm^{-2})	$[V_T - V_S]_{V_S=10\ V} - [V_T - V_S]_{V_S=0}$
0	1.31 V
6×10^{11}	1.50 V
10^{12}	1.93 V
1.4×10^{12}	2.79 V

criterion (2). According to figure 16 and table II, one concludes that the overall sensitivity of $(V_T - V_S)$ on variations in V_S becomes larger for increasing doses. The slope of the $(V_T - V_S)$ curves is particularly large for small values of V_S and is then strongly dose dependent. For large values of V_S, on the contrary, all the curves have the slope of the uniform substrate curve. This is due to the fact that the depletion region reaches the uniform part of the profile for large V_S values; the behavior of the threshold voltage is then similar to that of the uniform substrate case.

The threshold voltage shifts are usually presented as a function of $(V_S + 2\phi_{FB})^{1/2}$, since a straight line is then obtained

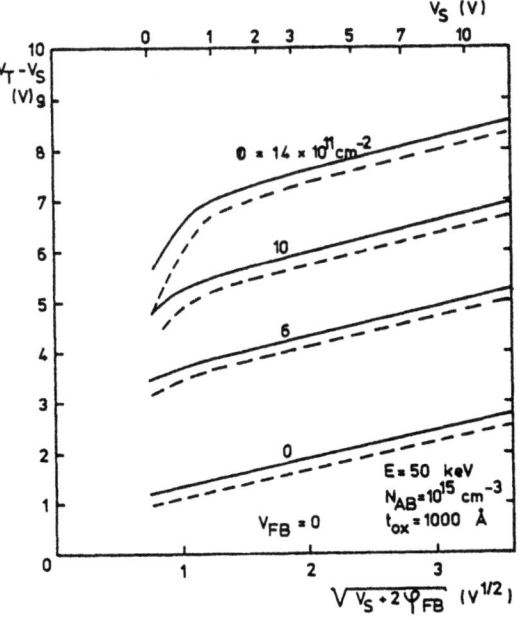

Fig. 17. $(V_T - V_S)$ vs. $(V_S + 2\phi_{FB})^{1/2}$. The solid lines represent V_{TE}; the dashed ones represent the values for $\psi_{S,inv} = 2\phi_{FB} + V_S$.

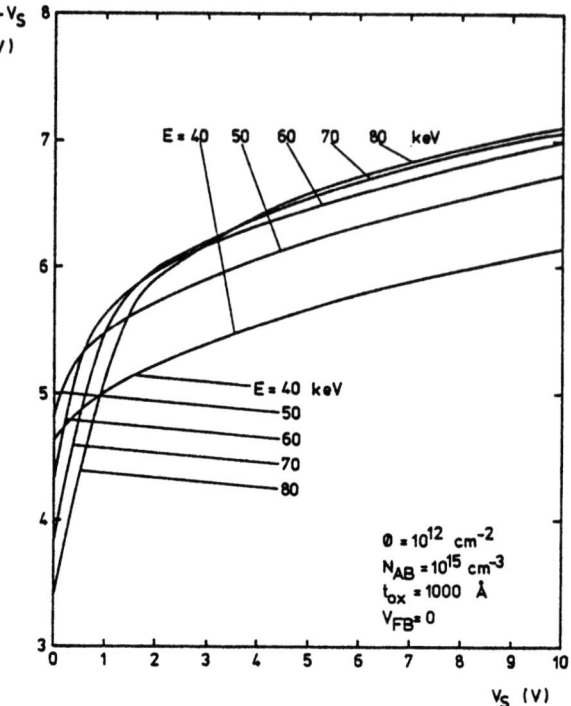

Fig.18. $(V_T - V_S)$ *for different energies but a constant boron dose* $(\Phi = 10^{12}\ cm^{-2})$. *The values are determined by extrapolation from the* Q_n/V_G *relations.*

Table III : Overall substrate sensitivity for different energies (boron implants, $\Phi = 10^{12}\ cm^{-2}$, $t_{ox} = 1000\ Å$).

E (keV)	$[V_T - V_S]_{V_S=10\ V} - [V_T - V_S]_{V_S=0}$
40	1.53 V
50	1.85 V
60	2.70 V
70	3.31 V
80	3.75 V

for the case of an uniform substrate (see fig.17). For implanted structures one obtains parallel straight lines for large V_S values, even for heavy implants.

The dependence of V_T on V_S, with the energy E as parameter, is depicted in figure 18 (for a constant dose of 10^{12} cm^{-2}). By analysing the extension of the surface space charge region, we again conclude that the body effect is especially important for deep implants, i.e. for high implantation energies; this is also evident from table III ($\Phi = 10^{12}$ cm^{-2}, t_{ox} = 1000 Å, N_{AB} = 10^{15} cm^{-3}).

The body effect imposes an additional constraint to the choice of the optimal energy. It is quite evident that energies exceeding 60 keV must be rejected, since the body effect is then too important. However we have seen in a preceding section that for energies below 40 keV the efficiency of the ion implantation becomes small, and very large doses are then necessary to attain a given ΔV_T; we already known it is desirable to keep the impurity concentration as low as possible. It is clear that the optimal choice, for the case of an oxide layer of 1000 Å, is around 40 keV. If a constant substrate (or source) polarization can be adopted, in order to maintain the boundary of the depletion region in the non-implanted part of the substrate, then one should consider a slightly larger energy as optimal choice; for a substrate polarization of 1 or 2 V, an implantation energy of 50 keV is appropriate.

III-6. Step doping profile approximation

The main disadvantage of the different theoretical approaches, presented in the previous sections, is that the mathematical treatment is cumbersome when the exact doping profile of the implanted ions is considered. For practical purposes it is desirable to derive simple models by introducing appropriate approximations.

Rideout et al.[26] suggested to approximate the actual doping profile of an implanted structure by a step doping profile of width D (see fig.19). The surface concentration N_{AS} is related to the dose of the ion implantation :

$$N_{AS} = N_{AB} + \Phi_{eff}/D \tag{41}$$

By substituting this rectangular doping profile in the relations (23) and (24), which are based on the depletion approximation, one obtains simple explicit and analytical expressions for the maximum depletion width $x_{d,max}$ and the threshold voltage V_T. They are given in table IV [26]; the quantity $K_S = (2q\epsilon_S N_{AS})^{1/2}/C_{ox}$ is the equi-

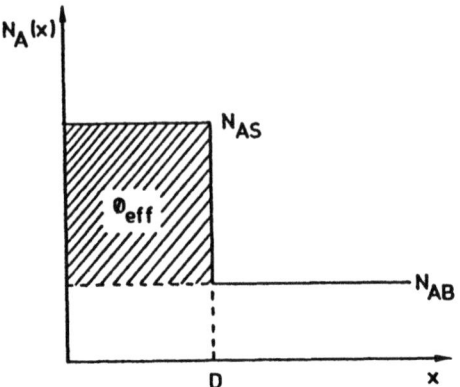

Fig. 19. Step doping profile for implanted region.

Table IV : Maximum depletion layer width and threshold voltage formulae for step doping profile.

For $x_{d,max} \leq D$:

$$x_{d,max} = \left[(2\varepsilon_S/qN_{AS})(\psi_{S,inv} + V_S) \right]^{1/2} \qquad (42)$$

$$V_T = V_{FB} + \psi_{S,inv} + K_S(\psi_{S,inv} + V_S)^{1/2} \qquad (43)$$

For $x_{d,max} \geq D$:

$$x_{d,max} = \left[\left(\frac{2\varepsilon_S}{qN_{AB}}\right)\left(\psi_{S,inv} + V_S - \frac{q\Phi_{eff}D}{2\varepsilon_S}\right) \right]^{1/2} \qquad (44)$$

$$V_T = V_{FB} + \psi_{S,inv} + \frac{q\Phi_{eff}}{C_{ox}} + K_B\left(\psi_{S,inv} + V_S - \frac{q\Phi_{eff}D}{2\varepsilon_S}\right)^{1/2} \qquad (45)$$

---

valent body factor for the surface region of impurity concentration N_{AS}.

The problem is reduced to the definition of the surface potential $\psi_{S,inv}$ and to the correct choice of the width D of the step doping profile.

It is evident that N_{AS} is not very different from the substrate concentration N_{AB} for implantations with low doses, or for smooth doping profiles resulting, for example, from a redistribution. In this case a good simulation of the V_T versus V_S curve is obtained, according to Rideout et al.[26], by considering the classical approximation $\psi_{S,inv} = 2\phi_{FB} + V_S$. This yields excellent results for boron implantations at 50 keV, without redistribution, and for doses smaller than 6×10^{11} cm^{-2}. An optimal value for the width D can be found from the theoretical or experimental V_T versus V_S curve by applying a least squares method in the range of the useful voltages. An example is given in figure 20 for an effective dose of 5.32×10^{11} cm^{-2}, i.e. for an implanted dose of 6×10^{11} cm^{-2} at 50 keV through an oxide layer of 1000 Å. The agreement between the threshold voltages V_{TE}, obtained from the extrapolation of the Q_n versus V_G curves, and the values resulting from (43) and (45) for D = 1200 Å is excellent, if however a constant correction factor $\Delta V_{FB} = 0.23$ V is introduced.

Increasing deviations are observed when the classical condition $\psi_{S,inv} = 2\phi_{FB} + V_S$ is used for larger doses. The condition no longer applies, especially for low V_S voltages, when the depletion region is completely contained within the highly doped surface region; the surface potential is then clearly larger than $2\phi_{FB} + V_S$. According to the criterion (31), one must now impose the following condition to the surface concentration of minority carriers :

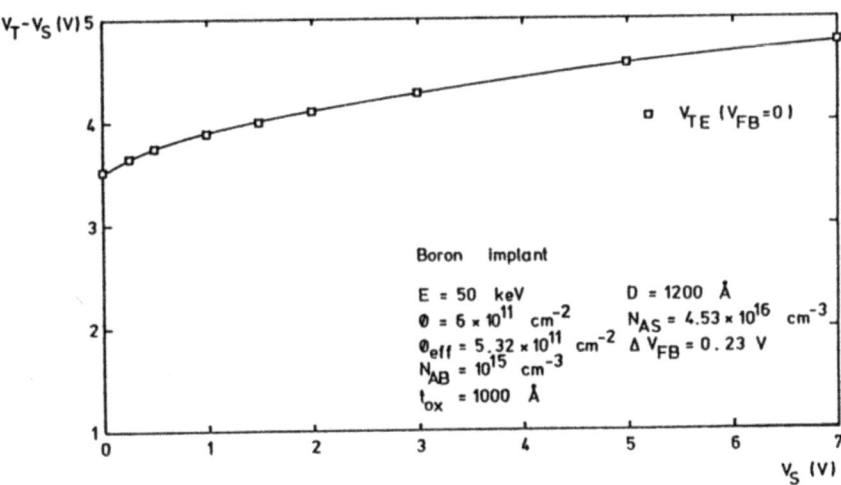

Fig. 20. $(V_T - V_S)$ vs. V_S for a boron implant (E = 50 keV, $\Phi = 10^{12}$ cm^{-2}) approximated by a step doping profile (D = 1200 Å). Squares (□) represent V_{TE} values. The flat-band voltage for the approximate case is 0.23 V larger than the one for V_{TE}.

$$n_S(\psi_{S,inv}) = N_{AS} \qquad \text{for } x_{d,max} < D \qquad (46)$$

$$n_S(\psi_{S,inv}) = N_{AB} + \Phi_{eff}/x_{d,max} \qquad \text{for } x_{d,max} > D \qquad (47)$$

The surface potential under strong surface inversion is then given by

$$\psi_{S,inv} = \phi_{FB} + \frac{kT}{q} \ln(N_{AS}/n_i) + V_S \qquad (48)$$

$$= 2\phi_{FB} + \frac{kT}{q} \ln\left(1 + \frac{\Phi_{eff}}{N_{AB}D}\right) + V_S, \quad \text{for } x_{d,max} < D \qquad (49)$$

or

$$\psi_{S,inv} = 2\phi_{FB} + \frac{kT}{q} \ln\left(1 + \frac{\Phi_{eff}}{N_{AB}x_{d,max}}\right) + V_S,$$

$$\text{for } x_{d,max} > D \qquad (50)$$

For the last case one obtains two implicit equations, (44) and (50), from which $x_{d,max}$ and $\psi_{S,inv}$ can be numerically calculated without difficulty.

Figure 21 clearly shows the necessity of an optimal choice for the parameter D in order to assure an adequate simulation of the V_T versus V_S curve, especially in the range of low V_S values. The curves of figure 21 also schematically depict the influence of

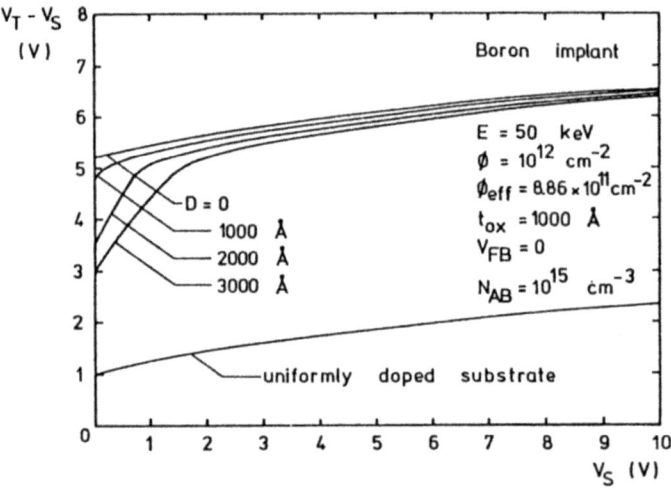

Fig. 21. $(V_T - V_S)$ vs. V_S for step doping profile approximations of a boron implant $(E = 50$ keV, $\Phi = 10^{12}$ cm$^{-2})$: the parameter is the width D of the step. Also shown is the curve for the **uniformly** doped substrate.

the implantation energy on the body effect : larger step
widths D, corresponding to deeper implants, introduce a larger
body effect, in accordance with the conclusions presented in previous sections.

Note that a redistribution of the implanted ions, as a result
of subsequent thermal treatments, increases the width of the doping
profile (the effective dose remains almost constant in the absence
of oxide growth [27]) and enhances the body effect. High temperature steps in the fabrication process of the devices should be reduced to a minimum in order to avoid this enhancement.

IV - CURRENT CHARACTERISTICS OF IMPLANTED P^+P TRANSISTORS

IV-1. Introduction

Consider an n-channel MOS transistor (see fig.22). The gradual
channel approximation is valid for not too small channel lengths L.
The current density of minority carriers within the channel is
given by the general relation

$$J_{ny}(x,y) = q\mu_n \left[n(x,y) \mathcal{E}_y(x,y) + \frac{kT}{q} \frac{\partial n(x,y)}{\partial y} \right] \quad (51)$$

where μ_n is the electron mobility, $n(x,y)$ the local electron densi-

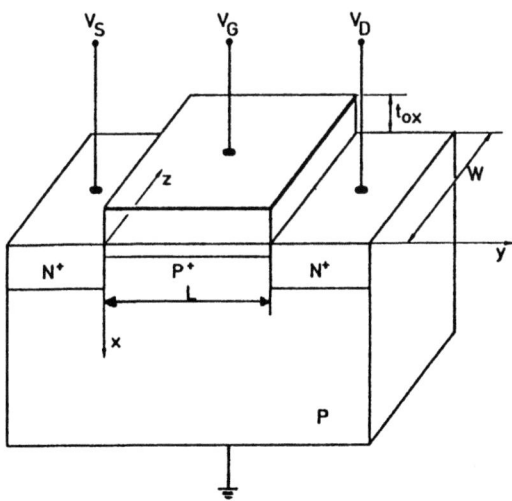

Fig. 22. An MOS n-channel transistor.

ty and $\mathcal{E}_y$ the electric field component parallel to the interface. Calling V(y) the electron quasi-Fermi level (measured with respect to the bulk Fermi level), one can transform expression (51) to [28]

$$J_{ny}(x,y) = - q\mu_n n(x,y) \frac{dV}{dy} \tag{52}$$

After integration in the z-direction, over the channel width W, and in the x-direction, over a distance x_i including the entire conducting region of minority carriers, one obtains the following expression for the drain current, assuming a constant mobility equal to μ_{no}, [29] :

$$I_D = - W \mu_{no} \frac{dV}{dy} \int_0^{x_i} \left[- q\, n(x,y)\right] dx \tag{53}$$

The integral appearing in this expression is the inversion charge Q_n, already defined in (11); Q_n is now, besides its dependence on the gate voltage V_G, a function of the distance y (with respect to the source contact) along the channel, and consequently depends on the local quasi-Fermi level V(y). Thus equation (53) can be written as :

$$I_D = - W\mu_{no} Q_n(V_G,V) \frac{dV}{dy} \tag{54}$$

A final integration over the channel length L yields

$$I_D = - \frac{W}{L} \mu_{no} \int_{V_S}^{V_D} Q_n(V_G,V)\, dV \tag{55}$$

V_S and V_D being the source and drain voltages. Note that both the diffusion current and the conduction current along the channel are taken into account in (55).

IV-2. <u>Drain current by numerical integration</u>

The drain current, for the case of an uniform substrate, was studied by Pao and Sah [29] by means of a numerical integration of (55). A discretization of the channel potential V(y) is performed between the limits V_S and V_D. The inversion charge Q_n is computed, for each V(y), by integrating Poisson's equation for a constant gate voltage V_G (the parameter V_S of section III is now replaced by V). This yields the current elements

$$\Delta I_D(V_G,V) = -\frac{W}{L} \mu_{no} Q_n(V_G,V) \Delta V \tag{56}$$

The final drain current is then given by the following numerical sum :

$$I_D = \sum_{V_S \leqslant V \leqslant V_D} \Delta I_D(V_G, V) \qquad (57)$$

The complete static current characteristics can be computed in this way. It should be noted that saturation occurs when the inversion charge decreases to zero for large channel potential V. However the described one-dimensional model does not simulate in a perfect way the saturation regime where bi-dimensional effects are dominant [30]. This is particularly true for short-channel transistors.

The same method applies for long-channel implanted MOS transistors. The charge $Q_n(V_G, V)$ must now be obtained by a numerical integration of Poisson's equation in the direction perpendicular to the interface, taking the implantation profile into account.

We present a few results obtained by this numerical integration method [31]; the computer program was developed by R. Sinon of our laboratory. Table V gives the data used for the transistors.

Table V : MOST parameters

W/L	10
μ_{no}	500 cm^2/V.sec
N_{AB}	10^{15} cm^{-3}
t_{ox}	1000 Å
V_{FB}	0

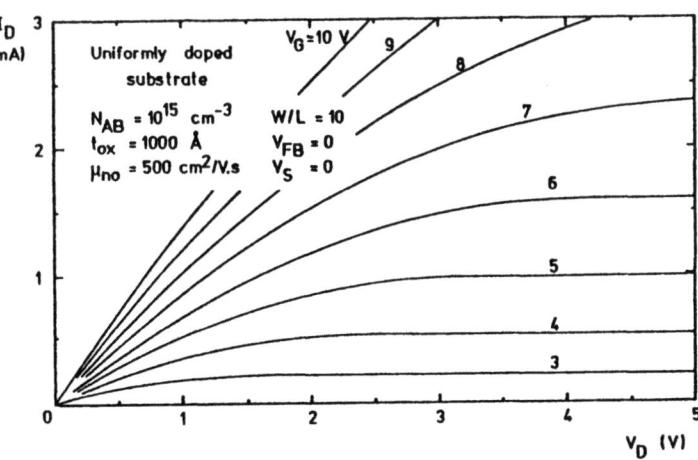

Fig.23. Theoretical current characteristics of a MOS transistor on an uniformly doped substrate.

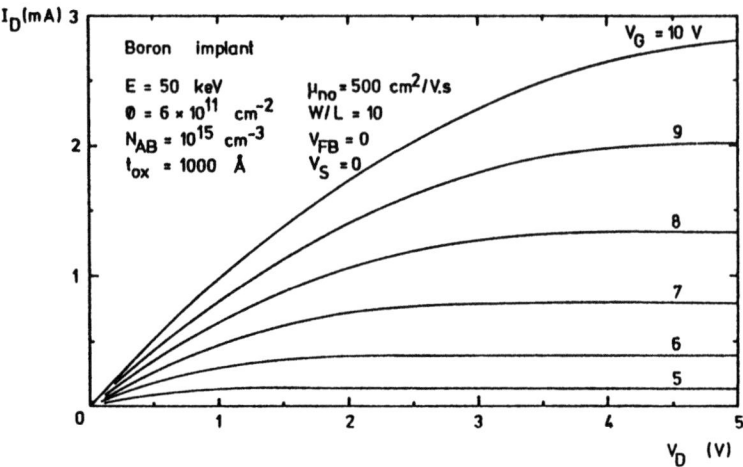

Fig. 24. Theoretical current characteristics of a boron implanted MOS transistor ($E = 50$ keV, $\Phi = 6.10^{11}$ cm^{-2}).

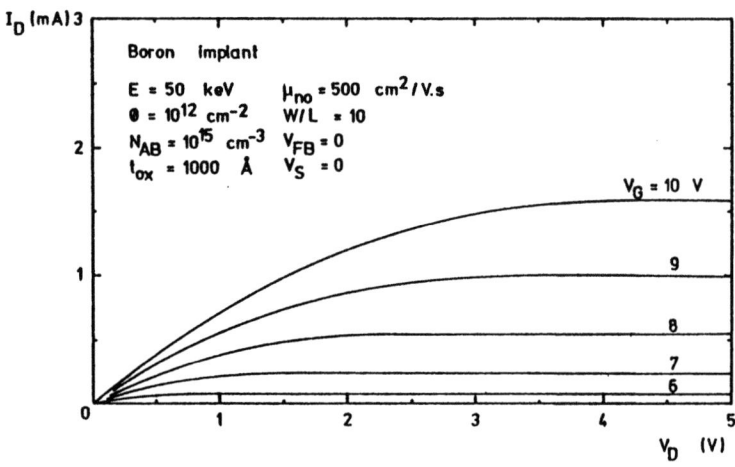

Fig. 25. Theoretical current characteristics of a boron implanted MOS transistor ($E = 50$ keV, $\Phi = 10^{12}$ cm^{-2}).

The resulting current characteristics I_D versus V_D are shown in figures 23, 24 and 25 for an uniform substrate and for boron implanted devices at 50 keV with a dose of respectively 6.10^{11} and

10^{12} ions per cm^2. The behavior of low dose implanted transistors is very similar to that of non-implanted devices. The shape of the current curves is modified for larger doses (> 6.10^{11} cm^{-2}), especially for gate voltages near the threshold voltage. Figures 26 and 27 illustrate this fact; note that in figure 27 the threshold voltage V_T has been substracted from all V_G values, in order to allow a better comparison.

IV-3. Modeling of the current characteristics I_D versus V_D

The method, we just described, requires the development and the use of an expensive and time consuming computer program, based on the numerical integration of Poisson's equation for each discretization point. Hence it is worthwhile to look for simpler numerical or analytical models.

IV-3-1. Modeling of the inversion region

In order to avoid the numerical integration of Poisson's equation, Sinon et al. [32] divided the semiconductor in a direction perpendicular to the interface into three distinct regions : a surface inversion region ($0 \leqslant x \leqslant w_i$) of finite width w_i, a depletion region ($w_i \leqslant x \leqslant x_{d,max}$) and a quasi-neutral bulk region ($x > x_{d,max}$). By considering only the dominant term in the expression of the space charge density (i.e. the minority carriers for the inversion region, the ionized impurities in the depletion region), it is possible to integrate twice Poisson's equation [33]. The integration constants of the resulting analytical expressions, together with the specific parameters (w_i, $x_{d,max}$) of the model, are given by a system of 8 simultaneous non-linear equations, which is solved by a Newton-Raphson iteration method. Current characteristics of the MOS device can then be calculated. In particular, the threshold voltage and the saturation voltage can easily be computed (they only require a subsystem of 3 equations).

This method has been applied to the study of MOS transistors with a shallow thermal diffusion profile in the gate region [34]; the theoretical results are in excellent agreement with the experimental data (see fig.28)[32]. The same method applied to implanted structures gives results with a maximum deviation of 1 % compared to the complete numerical integration method. The accuracy of the model proposed by Sinon et al. is due to the excellent simulation of the inversion charge Q_n; the surface potential considered in the model is not constant, but depends on the gate voltage V_G and on the electron quasi-Fermi level along the channel.

The model of Sinon et al., although simpler than the complete numerical model requiring a large memory space, is essentially a

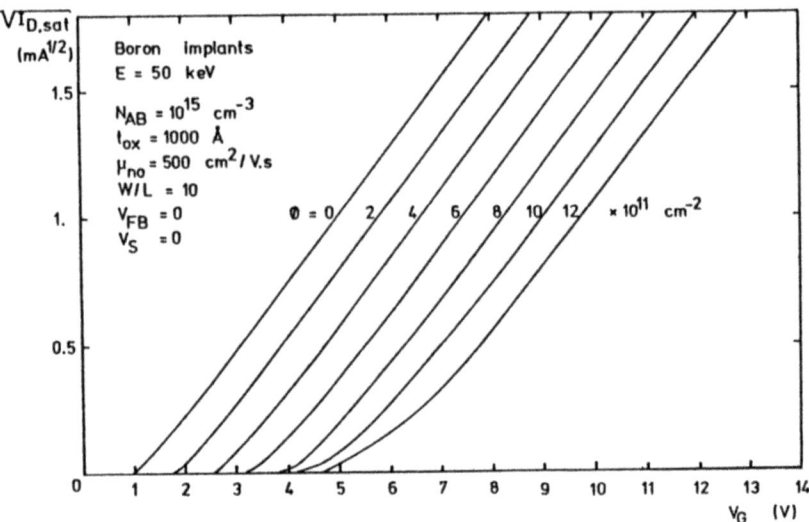

Fig. 26. $(I_{D,sat})^{1/2}$ *vs.* V_G *for non-implanted and boron implanted MOS transistors.*

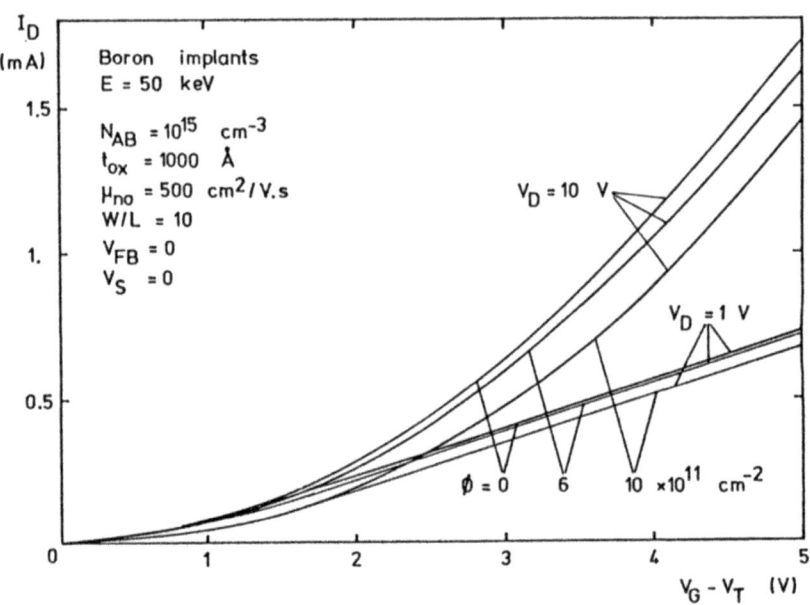

Fig. 27. Drain current vs. $(V_G - V_T)$ *for non-implanted and boron implanted MOS transistors* $(V_D = 1$ *and* 10 $V)$.

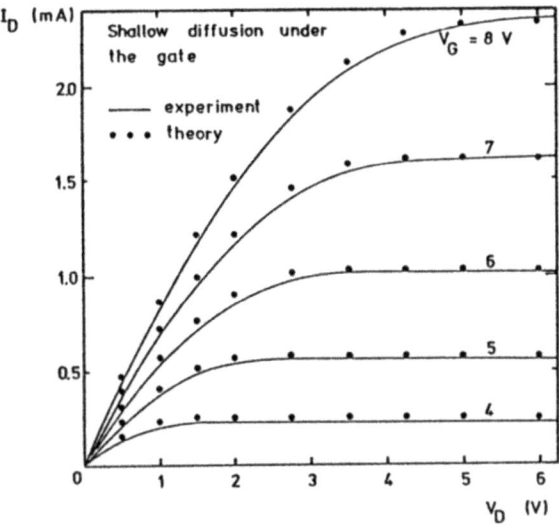

Fig.28. Theoretical and experimental current characteristics of a MOS transistor with an increased boron density under the gate [32]. Fabrication process includes a shallow boron diffusion from a doped oxide; technological details are given in ref. [34].

device oriented analysis tool, and is difficult to apply for circuit analysis purposes.

IV.3.2. Model with a constant surface potential

The inversion charge along the channel can be expressed in the following way by combining (7) and using (16) and (17) :

$$Q_n(V_G,V) = - C_{ox}\left[V_G - V_{FB} - \psi_{S,inv}(V)\right] - Q_B\left[\psi_{S,inv}(V)\right] \quad (58)$$

The bulk charge Q_B is defined by (13). To compute Q_n one has to know the surface potential which can be written as

$$\psi_{S,inv}(V) = 2\phi_{FB} + \Delta\psi_S(V) + V \quad (59)$$

The term $\Delta\psi_S(V)$ accounts for the deviation of $\psi_{S,inv}(V)$ with respect to the classical value (2) and in general depends on V, as shown in figure 29. The calculation of the drain current, from (55), (58) and (59), remains simple if $\Delta\psi_S$ is considered as a constant, denoted by $\Delta\psi$, with respect to V and V_G (in the classical approximation (2) $\Delta\psi$ is taken to be zero). The drain

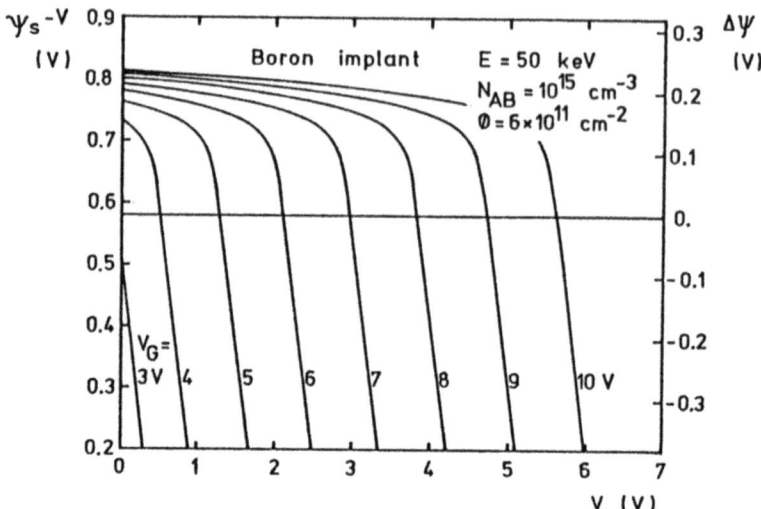

Fig. 29. Differences ψ_S-V (left scale) or $\Delta\psi_S(V) = \psi_S - 2\phi_{FB} - V$ (right scale) vs. V for a boron implanted MOS structure. The flat-band voltage equals zero. The oxide thickness is 1000 Å.

current in the non-saturation region ($V_D \leq V_{D,sat}$) is then given by :

$$I_D = \frac{W}{L} \mu_{no} C_{ox} \{(V_G - V_{FB} - 2\phi_{FB} - \Delta\psi)(V_D - V_S) - \frac{1}{2}(V_D^2 - V_S^2)$$
$$+ \frac{1}{C_{ox}} \int_{V_S}^{V_D} Q_B(2\phi_{FB} + \Delta\psi + V) \, dV \} \quad (60)$$

and the threshold voltage is

$$V_T = V_{FB} + 2\phi_{FB} + \Delta\psi + V_S - \frac{1}{C_{ox}} Q_B(2\phi_{FB} + \Delta\psi + V_S) \quad (61)$$

The saturation voltage $V_{D,sat}$ is the drain voltage for which the inversion charge Q_n, given by (58), vanishes [35].

The remaining problem is the determination of the bulk charge, Q_B given by (13). Different cases must be considered.

IV.3.3. Shallow implants with a low dose

By applying the depletion approximation and assuming the electric field at the boundary of the depletion region as negligeable, one obtains :

$$Q_B = - q \int_0^{x_{d,max}(V)} N_A(x) \, dx \qquad (62)$$

For shallow implants with a low dose $x_{d,max}(V)$ is located in the uniform part of the substrate behind the implanted region; consequently :

$$Q_B = - q \left[\Phi_{eff} + N_{AB} \, x_{d,max}(V) \right] \qquad (63)$$

where $x_{d,max}(V)$ is, according to (22) and (59), given by

$$x_{d,max}(V) = \left[(2 \, \varepsilon_S/qN_{AB}) \, (V + 2 \, \phi_{FB} + \Delta\psi - \Delta\phi) \right]^{1/2} \qquad (64)$$

$\Delta\phi$ being a constant equal to

$$\Delta\phi = \frac{q}{\varepsilon_S} \int_0^{x_{d,max}(V_S=0)} x N_{AI}(x) \, dx \qquad (65)$$

The drain current can then be calculated from (60); for $V_D \leq V_{D,sat}$ one obtains [36,34] :

$$I_D = \frac{W}{L} \mu_{no} C_{ox} \{ (V_G - V_{FB} - 2\phi_{FB} - \Delta\psi - \frac{q\Phi_{eff}}{C_{ox}})(V_D - V_S) - \frac{1}{2}(V_D^2 - V_S^2)$$

$$- \frac{2}{3} K_B \left[(V_D + 2\phi_{FB} + \Delta\psi - \Delta\phi)^{3/2} - (V_S + 2\phi_{FB} + \Delta\psi - \Delta\phi)^{3/2} \right] \} \qquad (66)$$

where K_B is the substrate body factor. The related threshold voltage and saturation voltage are given by :

$$V_T = V_{FB} + 2 \phi_{FB} + \Delta\psi + \frac{q\Phi_{eff}}{C_{ox}} + V_S + K_B (V_S + 2\phi_{FB} + \Delta\psi - \Delta\phi)^{1/2} \qquad (67)$$

and

$$V_{D,sat} = V_G - V_{FB} - 2\phi_{FB} - \Delta\psi - \frac{q\Phi_{eff}}{C_{ox}}$$

$$+ \frac{1}{2} K_B^2 \{ 1 - \left[1 + \frac{4}{K_B^2} (V_G - V_{FB} - \Delta\phi - \frac{q\Phi_{eff}}{C_{ox}}) \right]^{1/2} \} \qquad (68)$$

The expression (66) requires the knowledge of 4 constant parameters : $W \mu_{no} C_{ox}/L$, K_B, $V_1 = V_{FB} + 2 \phi_{FB} + \Delta\psi + q\Phi_{eff}/C_{ox}$ and $V_2 = 2 \phi_{FB} + \Delta\psi - \Delta\phi$.
The determination of these constants is based on the following measurements [36-38] :

- the drain conductance as a function of the gate voltage for small V_D-V_S (the results can also be used to evaluate the parameter characterizing the mobility reduction under high transverse electric field).
- the threshold voltage as a function of the source voltage.
- the variations of the drain current as a function of the source voltage for small V_D-V_S and constant gate voltage.

Figure 30 shows the current characteristics of a boron implanted MOS transistors (E = 50 keV, Φ = 6 x 10^{11} cm^{-2}), computed by using this model (circles) and compared to the numerical solution (solid lines).

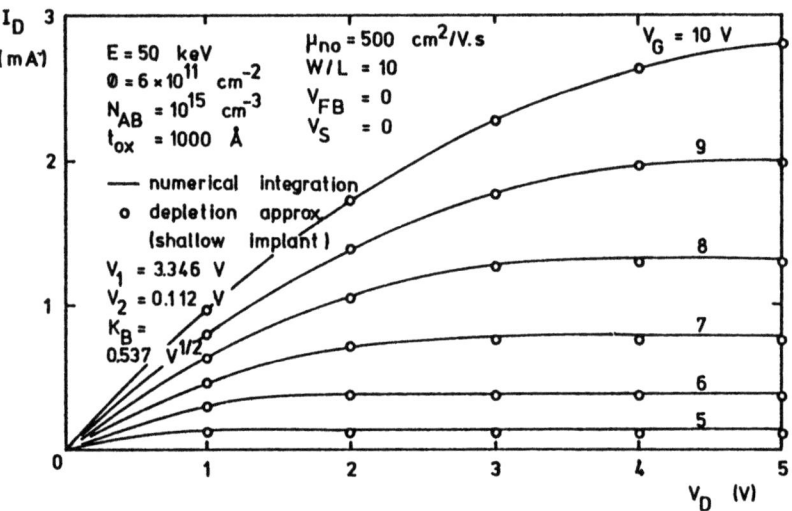

Fig. 30. *Current characteristics of a boron implanted MOS transistor (E = 50 keV, Φ = 6 x 10^{11} cm^{-2}). The solid lines represent the numerical integration; the circles (o) represent the depletion model (66). The used parameters are K_B = 0.537 $V^{1/2}$, V_1 = 3.346 V and V_2=0.112V.*

IV.3.4. Step profile approximation

By approximating the actual doping profile by a one-step profile of width D (see fig.19), and by introducing the depletion hypothesis, one obtains analytical expressions for the drain current, even for the case of strong implants.

Define V_I as the quasi-Fermi level for which the width of the depletion region is equal to the width D of the step profile :

$$V_I = \frac{qN_{AS}}{2\epsilon_S} D^2 - \psi_{S,inv} \qquad (69)$$

Three cases are to be considered, depending on the location of the maximum depletion boundary :

A. $V_S < V_D \leq V_I$ or $x_{d,max}(V_S) < x_{d,max}(V_D) \leq D$

B. $V_S \leq V_I \leq V_D$ or $x_{d,max}(V_S) \leq D \leq x_{d,max}(V_D)$

C. $V_I \leq V_S \leq V_D$ or $D \leq x_{d,max}(V_S) < x_{d,max}(V_D)$

The depletion boundary for case A is, for all applied voltages, located in the implanted surface region; this corresponds to an ion implantation with a strong dose. The electrical behavior of the transistor is similar to that of a non-implanted device with a substrate doping equal to $N_{AS} = \Phi_{eff}/D$. Case C occurs for shallow implants with a low dose; one obtains the step profile approximation of the model already discussed in the previous section. The current is thus given by (65) after substitution of the appropriate parameters depending on the step profile. Finally, the drain current for the intermediate case B is obtained by superposition of A-type and B-type current terms; the limiting voltage V_I is the drain voltage for the first component and the source voltage for the second one.

The detailed expressions for the drain current, the threshold voltage and the saturation voltage are given in table VI. Equation (59) was used for the surface potential, assuming a constant $\Delta\psi$ value.

It should be noted that the relation (76) is identical to (66). In the case of shallow and low dose implants, the current simulation resulting from the depletion approximation appears independent of the profile, provided that the parameters $\Delta\psi$ and $\Delta\phi$ are determined by curve fitting from the experimental values. The shape of the impurity profile influences the electrical characteristics of the transistor through the term $\Delta\phi$ (65).

Table VI. *Drain current of an implanted MOS transistor with the step doping profile approximation.*

A. $V_S < V_D \leq V_I$ $I_{DA} = \frac{W}{L} \mu_{no} C_{ox} \{ (V_G - V_{FB} - 2\phi_{FB} - \Delta\psi)(V_D - V_S)$

$$- \frac{1}{2}(V_D^2 - V_S^2) - \frac{2}{3} K_S \left[(V_D + 2\phi_{FB} + \Delta\psi)^{3/2} - (V_S + 2\phi_{FB} + \Delta\psi)^{3/2} \right] \}$$

for $V_D \leq V_{D,sat,A}$ (70)

$V_{TA} = V_{FB} + 2\phi_{FB} + \Delta\psi + V_S + K_S (V_S + 2\phi_{FB} + \Delta\psi)^{1/2}$ (71)

$V_{D,sat,A} = V_G - V_{FB} - 2\phi_{FB} - \Delta\psi + \frac{1}{2} K_S^2 \{ 1 - \left[1 + \frac{4}{K_S^2}(V_G - V_{FB}) \right]^{1/2} \}$ (72)

B. $V_S \leq V_I \leq V_D$ $I_{DB}(V_S, V_D) = I_{DA}(V_S, V_I) + I_{DC}(V_I, V_D)$ (73)

$V_{TB} = V_{TA}$ (74)

$V_{D,sat,B} = V_{D,sat,C}$ (75)

C. $V_I \leq V_S < V_D$ $I_{DC} = \frac{W}{L} \mu_{no} C_{ox} \{ (V_G - V_{FB} - 2\phi_{FB} - \Delta\psi - \frac{q\Phi_{eff}}{C_{ox}})(V_D - V_S)$

$$- \frac{1}{2}(V_D^2 - V_S^2) - \frac{2}{3} K_B \left[(V_D + 2\phi_{FB} + \Delta\psi - \Delta\phi)^{3/2} \right.$$

$$\left. - (V_S + 2\phi_{FB} + \Delta\psi - \Delta\phi)^{3/2} \right] \}$$ (76)

for $V_D \leq V_{D,sat,C}$

$V_{TC} = V_{FB} + 2\phi_{FB} + \Delta\psi + \frac{q\Phi_{eff}}{C_{ox}} + V_S$

$\quad\quad + K_B (V_S + 2\phi_{FB} + \Delta\psi - \Delta\phi)^{1/2}$ (77)

$V_{D,sat,C} = V_G - V_{FB} - 2\phi_{FB} - \Delta\psi - \frac{q\Phi_{eff}}{C_{ox}}$

$\quad\quad + \frac{1}{2} K_B^2 \{ 1 - \left[1 + \frac{4}{K_B^2}(V_G - V_{FB} - \Delta\phi - \frac{q\Phi_{eff}}{C_{ox}}) \right]^{1/2} \}$ (78)

where $\Delta\phi = \frac{q}{\varepsilon_S} \Phi_{eff} \frac{D}{2}$ (79)

IV.3.5. Linear variation of Q_B

We know from section IV.3.3. that the electrical behavior of a low dose implanted transistor and of a non-implanted transistor are very similar. The mathematical expressions for both types of transistors differ only by constant quantities $q\phi_{eff}/C_{ox}$ and $\Delta\phi$.
Hence we may assume, as for non-implanted transistors [39], a linear variation for the bulk charge Q_B near $V = V_S$; this yields for the non-saturation regime

$$I_D = \frac{W}{L} \mu_{no} C_{ox} \left[(V_G-V_T)(V_D-V_S) - \frac{1}{2}(1+\delta)(V_D-V_S)^2 \right] \quad (80)$$

The threshold voltage V_T is given by (67). The parameter δ can be approximated by [40]

$$\delta \simeq 0.4 \, K_B (V_S + \psi_{S,inv})^{-1/2} \quad (81)$$

or can be determined by a curve fitting technique from the slope of the curve (V_T-V_S) versus V_S. The data presented in figure 16 clearly show that a linear variation of Q_B may only be considered for implanted devices with a low dose. The saturation voltage in their case is given by :

$$V_{D,sat} = V_S + (V_G-V_T)/(1+\delta) \quad (82)$$

IV.4. <u>Effective mobility</u>

Until now the carrier mobility was assumed to be constant; this makes the integration of the relations (53) and (55) possible. However, the surface mobility depends on both the transversal and the longitudinal components of the electric field [41]. The last component is less important for transistors with a long channel length. The influence of the transversal electric field component is, for uniform substrates, usually modeled by an effective mobility of the form [39]

$$\mu_n = \mu_{no} / \left[1 + \theta(V_G-V_T) \right] \quad (83)$$

This expression must be substituted in equation (55) and in the subsequent current relations. An accurate simulation of the electrical behavior of an MOS transistor on an uniform substrate requires a variable mobility relation of that form [42].

The same relationship has succesfully been applied to MOS transistors on non-uniformly doped substrates [34,36]. The additional parameter θ can be determined, for example, from conductance measurements [34].

The effective mobility μ_{no} depends on the impurity distribution under the gate oxide; this dependance can be qualitatively understood in terms of the change of the impurity scattering effect and the surface scattering effect. A boron implantation in a P-type substrate increases the surface impurity concentration; accordingly this situation increases the impurity scattering effect, which reduces the mobility monotically with dose increase. On the other hand, the surface scattering effect depends on the surface electric field; this field increases with a dose increase [18], and consequently the resulting mobility decreases in that case. These interpretations appear consistent with experimental results [23] ; a pronounced decrease of the effective mobility results from dose increase, in the case of boron implanted n-channel transistors : μ_{no} =470 cm^2/Vsec for Φ = 0 and approximatively 250 cm^2/Vsec for $\Phi \cong$ 1.2 x 10^{12}cm^{-2} (E = 50 keV).

IV.5. Subthreshold current

For gate voltages between the flat band voltage and the threshold voltage, the drain current is essentially a diffusion current depending on the band bending near the source contact [43]; this component is included in the general expression (55) for the current [29]. For long channel transistors the subthreshold current I_{ST} shows virtually no dependance on the drain-to-source voltage [43].

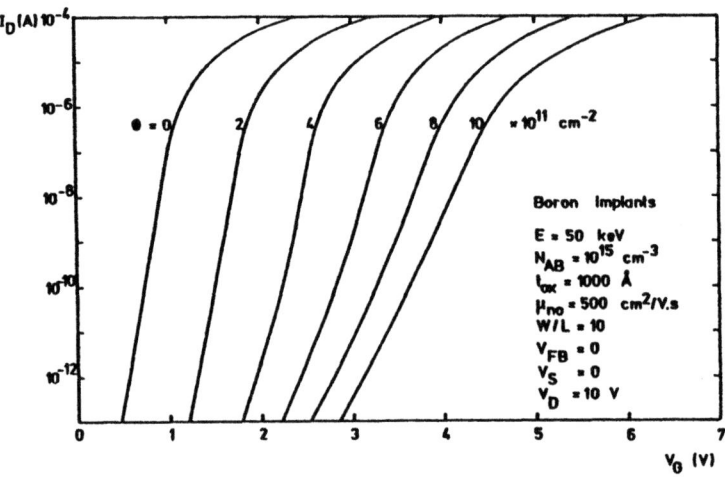

Fig. 31. *Subthreshold current as a function of the gate voltage for non-implanted and boron implanted MOS transistors (E = 50 keV).*

Figure 31 shows the results of a numerical integration of equation (55) for gate voltages near the threshold voltages for implanted devices with doses ranging from $2 \cdot 10^{11}$ cm^{-2} to 10^{12} cm^{-2} [31] ; also shown is the reference curve of an unimplanted device. These calculations do not include the influence of the surface states [44]. One observes an important variation of the subthreshold slope $S = \partial V_G / \partial (\log I_{ST})$ as the dose increases (88 mV/decade for the uniform substrate, 250 mV/decade for $\phi = 10^{12}$ cm^{-2}). This corresponds to the expected effect : the subthreshold slope increases with the substrate doping [45]. Note furthermore, that S becomes larger for smaller gate voltages in the case of doses exceeding 6.10^{11} cm^{-2}, since the depletion region is then completely located in the implanted region.

The subthreshold slope S is also a decreasing function of the source polarization [45], as shown in figure 32 for a dose of 10^{12} cm^{-2}. This effect, already present for an uniform substrate, is enhanced for this implanted structure since the edge of the depletion region, initially in the implanted region, is shifted towards the bulk when V_S increases; this gives an apparent doping reduction. S varies from 250 mV/dec for $V_S=0$ to 68 mV/dec for $V_S=5V$.

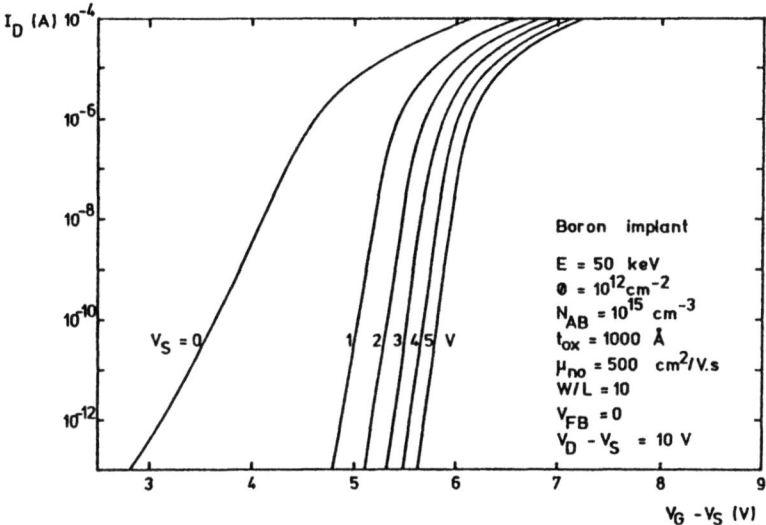

Fig. 32. Influence on the subthreshold current of the source polarization, for an implanted MOS transistor (boron, E=50keV, $\phi= 10^{12}$ cm^{-2}).

If the actual impurity profile is approximated by a step profile one observes (fig. 33) that the subthreshold slope S increases for a fixed dose as the width D of the step profile becomes larger; this is due to the fact that the depletion region is then entirely located within the implanted surface region.

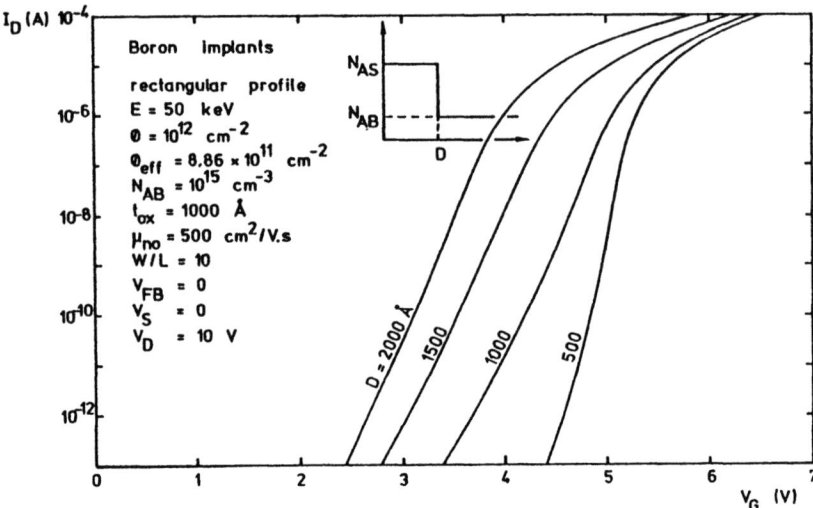

Fig. 33. Subthreshold current for the step profile approximation of an implanted MOS transistor (boron, E=50 keV, $\phi = 10^{12}$ cm^{-2}). The parameter is the width D of the step profile.

These results are important for the optimization of implanted transistors used in integrated circuits [26]. It is well known that a sharp subthreshold conduction characteristic is desirable in order to reduce the limiting gate voltage range between the cut-off and the conduction regime of the MOS transistor. The last point is particularly important for memories and dynamic digital circuits requiring, furthermore, small residual currents when the transistor is cut-off [46].

According to figure 31 it is necessary to consider the lowest possible dose; this yields an additional constraint for the optimisation of the ion implantation. The data of figure 33 for step profiles of increasing width can be considered as simulating the effect of an increasing implantation energy, or of an increasing redistribution process. Hence we again conclude that it is desirable to use low implantation energies and to reduce the subsequent high temperature steps during fabrication to a minimum. Finally, according to the data of figure 32 sharp conduction characteristics can be obtained by applying an appropriate source polarization (larger than 1 V for the case under consideration). This method offers more advantages, compared to the method of increasing the implant dose, if the purpose is to increase the threshold voltage of specific transistors especially in the subthreshold regime; however an additional voltage supply is required.

V. BURIED CHANNEL MOS TRANSISTORS

V.1. Introduction

We consider now the case of N^+P implanted MOS transistors, i.e. n-channel transistors for which the type of the implanted impurities is opposite to that of the substrate. If the doping level is sufficiently high a quasi-neutral channel region can be formed between the source and the drain. This doping procedure is used to make depletion-mode transistors, with a buried channel. The resulting bulk conduction mode is particularly attractive because the bulk electron mobility is higher than the surface mobility of enhancement MOS devices, which is reduced due to surface scattering effects.

Before starting the study of buried channel transistors, it is worthwhile to pay attention for an extreme case. Consider for example a donor implantation into an uniformly doped P-type substrate. If the implant is sufficiently shallow the ionized implanted acceptor atoms act as a thin additional surface charge layer; the bulk properties of the P-type substrate are not modified. Consequently the characteristics of an MOS transistor on such a substrate are similar to those of an unimplanted enhancement device, but the threshold voltage of the implanted device is shifted by an amount proportional to the active implanted impurities within the surface region [1,2]. Q_n/V_G curves are presented in figure 34 for phosphorus implants (E = 130 keV) in a P-type substrate (N_{AB} = 10^{15} cm^{-3}). For low doses (up to 10^{12} cm^{-2}) the electron charge Q_n can be reduced to zero by applying a negative gate voltage. For strong doses, however, this is no longer possible, as will be explained in the next section on the operation modes.

For low doses we can find the threshold voltage which is the extrapolation of the linear part of the Q_n/V_G relation. The variation of this threshold voltage as a function of the source voltage is illustrated in figure 35.

A deeper implant is used for buried channel depletion-mode MOS devices; within the P-type substrate a N-type subsurface or buried channel is created whose characteristics depend on both the N-type surface region and the P-type substrate.

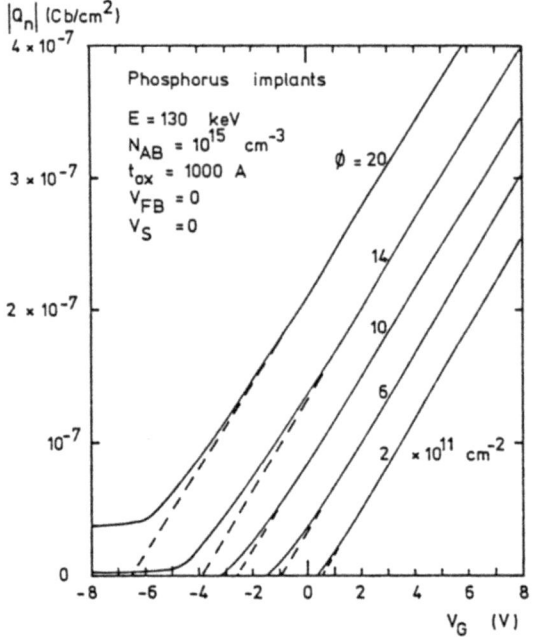

Fig. 34. Q_n versus V_G curves for phosphorus implanted n-channel MOS transistors ($E = 130$ keV, $N_{AB} = 10^{15}$ cm^{-3}, $t_{ox} = 1000$ Å, $V_{FB} = 0$).

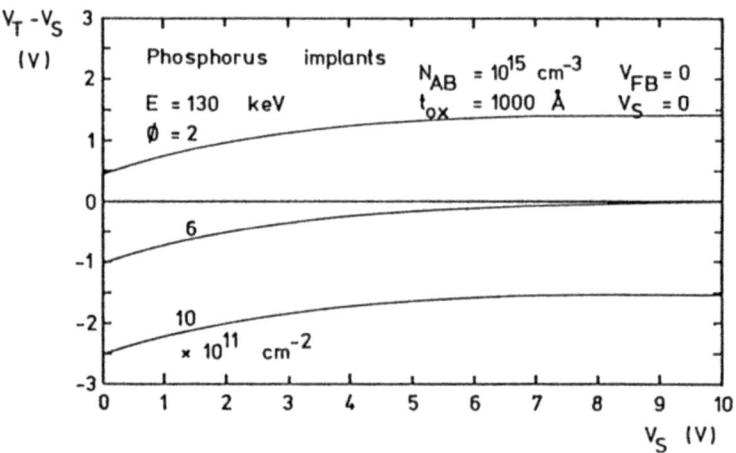

Fig. 35. Threshold voltage as a function of the source voltage for phosphorus implanted n-channel MOS transistors ($E = 130$ keV, $N_{AB} = 10^{15}$ cm^{-3}, $t_{ox} = 1000$ Å, $V_{FB} = 0$).

V.2. Operating modes

The actual profile of the acceptor impurities beneath the gate oxide depends on the particular fabrication process. In order to describe the different operating modes of a buried channel transistor, we approximate for simplicity the exact profile by a one-step profile with a junction depth x_j (fig. 36).

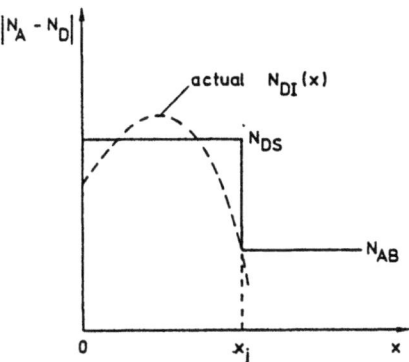

Fig. 36. Actual profile and one-step approximation.

A cross-section of the transistor is represented in figure 37. The source and drain N^+ regions are connected by the N-type surface region. A junction depletion region is formed around the metallurgical junction x_j; its local width depends on the external applied voltages. The gate electrode also modulates the local width of the buried channel from the surface side. We now consider the different operating modes.

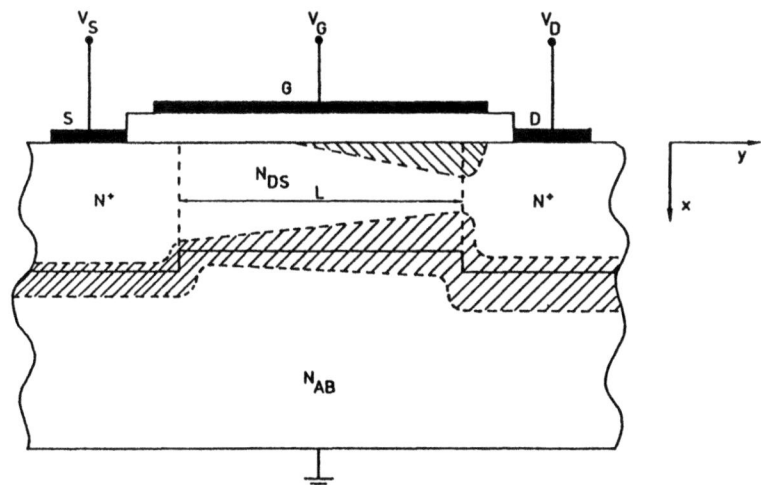

Fig. 37. Cross-section of a buried n-channel MOS transistor.

V.2.1. Complete enhancement mode

For large positive values of V_G electrons from the N-type surface region accumulate near the semiconductor surface over the entire channel length L between the source and the drain electrodes. Looking to the channel along the direction perpendicular to the surface (x-direction for fig. 37), we see that the thin surface accumulation layer (near x = 0) is followed by an almost neutral P-type buried channel (fig. 38a). The local width of the channel is only limited by the depletion region extending around the bulk PN junction ($x_1 \leq x \leq x_r$). The channel conductance increases with larger positive values of V_G. The current flowing between source and drain increases non linearly for larger positive values of $V_D - V_S$ since the depletion region of the PN junction becomes larger. A similar effect results by modifying the source (or substrate) polarization.

One may expect a lower electron mobility in the surface region than in the buried channel region, due to surface scattering effects.

V.2.2. Partial enhancement mode

For smaller values of $V_G - V_S$, compared to the previous mode, electron accumulation only occurs between the source contact (y ≠ 0) and some specific point y_a (< L) along the channel. The rest of the channel ($y_a < y < L$) is now depleted in the x direction over a local distance $x_d(y)$. An almost neutral N-type buried channel still exists if $x_d(y)$ is smaller than the N-side limit $x_1(y)$ of the PN junction depletion region. The local width of the buried channel for $y \geq y_a$ is now restricted to $x_1(y) - x_d(y)$, assuming $x_1 > x_d$ (fig. 38-b).

V.2.3. Normal depletion mode

For negative values of $V_G - V_S$ (more exactly for negative values of $V_G - V_S - V_{FB}$) the surface depletion of the N-region extends over the entire channel length (i.e. y_a = 0). The width $x_1(y) - x_d(y)$ of the buried channel decreases for increasing y-values (assuming $x_m > x_d$) and for increasing $V_D - V_S$ values. This is due to the local potential V(y) along the channel ($V_S \leq V(y) \leq V_D$).

V.2.4. Depletion mode with partial surface inversion

By still increasing the negative gate potential $V_G - V_S$ one may partially invert the surface region of the N region, i.e. a hole inversion layer is formed between the source contact (y = 0) and a specific point y_i (< L) along the channel. The rest of the channel ($y_i \leq y \leq L$) is still surface depleted.

Within the region $0 \leq y \leq y_i$ where surface inversion occurs, we now have that the depletion region in the x-direction attains its maximum value $x_{d,max}(y)$, since any change in the local semiconductor charge essentially occurs through a change in the local surface inversion charge (fig. 38-c). Note that $x_{d,max}(y)$ remains a function of the local potential $V(y)$ along the channel and, as a result, increases with increasing values of y.

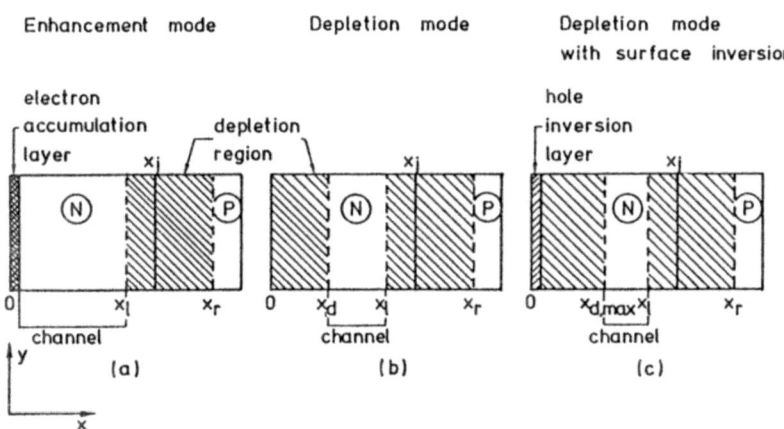

Fig. 38. *Schematic representation of the buried channel structure for the different operating modes.*

V.2.5. Depletion mode with complete surface inversion

For still more negative values of V_G the surface region of the N-region becomes inverted over the entire channel length, and the surface depletion region attains its maximum width $x_{d,max}(y)$. The width of the buried channel is no longer a function of the gate potential $V_G - V_S$.

In the previous description of the operating modes we assumed that the width $x_d(y)$ or $x_{d,max}(y)$ of the surface depletion region is smaller than the limit $x_1(y)$ of the PN junction depletion region. This is indeed the case if the junction depth x_j is large and (or) if the doping level N_{DS} of the N-type region is sufficiently high. However for a lightly doped and (or) a shallow surface region a punch-through effect occurs when the surface depletion region and the junction depletion region collapse into a single depletion region for particular values of the applied potentials. This punch-through effect may occur for all operating modes (except the complete enhancement mode of operation) and results into a saturation of the drain current.

V.3. Current characteristics

The majority carrier current flow through the buried channel from the source to the drain contact can be calculated, quite easily, by using the depletion approximation for the region near the metallurgical junction x_j, a surface depletion region and eventually a strong surface accumulation region or inversion region [47 - 49]. The drain current is then given by

$$I_D = -\frac{W}{L} \int_{V_S}^{V_D} \mu_n Q_n(V) \, dV \qquad (84)$$

L being the channel length, W the channel width, $Q_n(V)$ the local charge of free electrons along the channel and μ_n the mean mobility of these electrons. In order to obtain Q_n one must integrate the free electron concentration within the channel along the x-direction. The charge Q_n may be written in the following form :

$$Q_n = -Q_I + Q_S + Q_D + Q_B \qquad (85)$$

where

$$Q_I = \int_0^{x_j} qN_{DI}(x) \, dx = q \, N_{DS} x_j = q \, \Phi_{eff} \qquad (86)$$

is the implanted donor charge of the surface N-region,

$$Q_S = -C_{ox} [V_G - V_{FB} - V(y) - \psi_C] \qquad (87)$$

is the surface accumulation charge [existing only when $V_G > V_{FB} + V(y) + \psi_C$]; ψ_C is the contact potential equal to $(kT/q) \ln (N_{AB} N_{DS}/n_i^2)$,

$$Q_D = qN_{DS} x_d = qN_{DS} \left[-\frac{\epsilon_S}{C_{ox}} + \sqrt{(\frac{\epsilon_S}{C_{ox}})^2 - \frac{2\epsilon_S}{qN_{DS}} [V_G - V_{FB} - V(y) - \psi_C]} \right]$$

(88)

is the surface depletion charge [Q_D exists if $V_G < V_{FB} + V(y) + \psi_C$] and

$$Q_B = \sqrt{\frac{2\epsilon_S q N_{AB} N_{DS}}{N_{AB} + N_{DS}}} \, [V(y) + \psi_C] \qquad (89)$$

is the depletion charge at the N-side of the bulk junction.

Note that the surface depletion width x_d attains its maximum value $x_{d,max}$ when the surface becomes inverted. Surface inversion starts when the electron surface concentration equals or exceeds the substrate concentration N_{AB} [49] ; hence in the case of surface inversion we have to use for Q_D the following expression :

$$Q_{D,max} = qN_{DS} x_{d,max} = \sqrt{2\epsilon_S qN_{DS} \left[V(y) + \psi_C\right]} \tag{90}$$

Since, according to the previous relations, the total charge Q_n is a known function of the channel potential V(y), one obtains the drain current by a simple integration.

The electron mobility near the surface is a function of the surface scattering effects. In order to account for this influence, we consider a surface mobility μ_{nS} for the charge Q_S different from the bulk minority μ_{nB} for the charges Q_I, Q_D and Q_B.

The drain current is now given by the following relations :

a) Complete enhancement mode operation :

$$I_D = -\frac{W}{L} \int_{V_S}^{V_D} \left[\mu_{nS} Q_S + \mu_{nB}(Q_I - Q_B)\right] dV \tag{91}$$

b) Partial enhancement mode operation :

$$I_D = -\frac{W}{L} \int_{V_S}^{V(y_a)} \left[\mu_{nS} Q_S + \mu_{nB}(Q_I - Q_B)\right] dV$$

$$- \frac{W}{L} \int_{V(y_a)}^{V_D} \mu_{nB}(Q_I - Q_D - Q_B) dV \tag{92}$$

where $V(y_a) = V_G - V_{FB}$ \hfill (93)

c) Normal depletion mode

$$I_D = -\frac{W}{L} \int_{V_S}^{V_D} \mu_{nB}(Q_I - Q_D - Q_B) dV \tag{94}$$

d) Depletion mode with partial surface inversion

$$I_D = -\frac{W}{L} \int_{V_S}^{V(y_i)} \mu_{nB}(Q_I - Q_{D,max} - Q_B) dV$$

$$- \frac{W}{L} \int_{V(y_i)}^{V_D} \mu_{nB}(Q_I - Q_D - Q_B) dV \tag{95}$$

where $V(y_i)$ is given by [50] :

$$V(y_i) = -\psi_C + \frac{C_{ox}}{2\epsilon_S qN_{DS}} (V_G - V_{FB})^2 \tag{96}$$

e) Depletion mode with complete surface inversion

$$I_D = -\frac{W}{L} \int_{V_S}^{V_D} \mu_{nB}(Q_I - Q_{D,max} - Q_B) \, dV$$

For all cases saturation occurs when $Q_n(L) = 0$; however channel pinch-off may become impossible if the junction depth x_j is too large.

The current-voltage equations, obtained after integration, are rather complex as a result of the non-linear relationships for the depletion widths. Similar expressions were proposed by Edwards and Marr [51] ; they approximated the implanted channel concentration by a double-stepped profile. A general scheme for a variable donor doping $N_D(x)$ has been proposed by Hatert et al [49] .

A slightly different scheme has been introduced by Huang [48, 52] in order to simplify the mathematical relations. Instead of using the explicit equation for $Q_D = qN_{DS} x_d$, he approximates the depletion charge by :

$$Q_D = \bar{C} \left[V_G - V_{FB} - V(y) - \psi_C \right] \tag{98}$$

$\bar{C}$ being an average semiconductor capacitance. Its value is determined from the slope of the threshold voltage versus $\left[V_S + (kT/q) \ln (N_{AB}N_{DS}/n_i^2) \right]^{1/2}$ plot.

Examples of the current characteristics of implanted depletion mode transistors are treated in other papers [53,54] .

ACKNOWLEDGEMENTS

The authors want to thank R. Sinon and C. Gilles for their fruitful collaboration and their assistance in preparing the manuscripts. They also acknowledge M. Kaisin for the excellent typing.

REFERENCES

1. K.G. Aubuchon, Intern. Conf. on Properties and Use of M.I.S. Structures, Grenoble, France (June 1969).

2. M.R. Mc. Pherson, Appl. Phys. Lett., 18, 502 (1971).

3. J.F. Gibbons, Proc. IEEE, 56, 295 (1968).

4. J.F. Gibbons, W.S. Johnson and S.W. Mylroie, "Projected Range Statistics", Dowden, Hutchinson and Ross, Stroudsburg, Pa. (1975).

5. H. Ryssel, H. Kranz, K. Müller, R.A. Henkelmann and J. Biersack, Appl. Phys. Lett., 30, 399 (1977).

6. For a review of activation treatments, see : G. Dearnaley, J.H. Freeman, R.S. Nelson and J. Stephen, "Ion Implantation", chap. 5, § 4, North-Holland Publishing Co., Amsterdam (1973).

7. W.L. Brown, Phys. Rev., 91, 518 (1953).

8. A.S. Grove and D.J. Fitzgerald, Solid-State Electron. 9, 783 (1966).

9. M.R. MacPherson, Solid-State Electron., 15, 1319 (1972).

10. M. Kamoshida, Appl. Phys. Lett., 22, 404 (1973).

11. C. Gilles, private communication.

12. M.C. Tobey and N. Gordon, IEEE Trans. Electron Dev., ED-21, 649 (1974).

13. H. Feltl, IEEE Trans. Electron. Dev. ED-24, 288 (1977).

14. This program has been developped by R. Hatert, R. Sinon and C. Gilles.

15. A.S. Grove, "Physics and Technology of Semiconductor Devices", Wiley, N.Y. (1967).

16. G. Doucet and F. Van de Wiele, Solid-State Electron., 16, 417 (1973).

17. M. Kamoshida and O. Kudoh, Appl. Phys. Lett., 24, 501 (1974).

18 M.R. McPherson, Solid-State Electron., 15, 1319 (1972).

19 T. Tanaka, Japanese J. Appl. Phys. 10, 84 (1971).

20 R.R. Troutman, IEEE Trans. Electron Dev., ED-24, 182 (1977).

21 M.R. McPherson, Appl. Phys. Lett., 18, 502 (1971).

22 P.P. Peressini and W.S. Johnson, IEDM Tech. Dig., 467 (1973).

23 O. Kudoh, K. Nakamura and M. Kamoshida, J. Appl. Phys., 45, 4514 (1974).

24 W. Schemmert and G. Zimmer, Electronics Lett., 10, 151 (1974).

25 H. Ishiwara, S. Furukawa, J. Yamada and M. Kawamura, "Ion implantation in Semiconductors - Science and Technology", ed. by S. Namba, pp. 423-428, Plenum Press, N.Y. (1975).

26 V.L. Rideout, F.H. Gaensslen and A. Leblanc, IBM J. Res. Develop., 19, 50 (1975).

27 E.C. Douglas and A.G.F. Dingwall, IEEE Trans. Electron. Dev., ED-21, 324 (1974).

28 W. Shockley, "Electrons and Holes in Semiconductors", Van Nostrand, Princeton, N.J. (1950).

29 H.C. Pao and C.T. Sah, Solid-State Electron., 9, 927 (1966).

30 D. Van Dorpe, J. Borel, G. Merckel and P. Saintot, Solid-State Electron., 15, 547 (1972).

31 R. Sinon, private communication.

32 R. Sinon, R. Hatert and F. Van de Wiele, Phys. Stat. Sol. (a), 33, 661 (1976).

33 J.R. Hauser and M.A. Littlejohn, Solid-State Electron., 11, 667 (1968).

34 G. Doucet, F. Van de Wiele and P. Jespers, Solid-State Electron., 19, 191 (1976).

35   Ref. [15], p. 325.

36   J.R. Verjans and R.J. Van Overstraeten,
     IEEE Trans. Electron. Dev., ED-22, 862 (1975).

37   P. Rossel,
     "Propriétés statiques et dynamiques du transistor à
     effet de champ à grille isolée", Doctoral Thesis,
     Université Paul Sabatier, Toulouse, France (1973).

38   F.M. Klaassen, "Characterization and Measurements of MOST
     Devices",
     NATO Advanced Study Institute on "Process and Device
     Modeling for Integrated Circuit Design" (1977).

39   G. Merckel, J. Borel and N.Z. Cupcea,
     IEEE Trans. Electron. Dev., ED-19, 681 (1972).

40   F.M. Klaassen, Philips Res. Repts., 31, 71 (1976).

41   For a review of these effects, see : G. Vassilieff,
     "Modèle du transistor MOS-Influence des variations
     de la mobilité des porteurs", Doctoral Thesis, Uni-
     versité Paul Sabatier, Toulouse, France (1971).

42   P. Robaux and G. Fallon, "Système automatique de mesure
     des paramètres du transistor MOS", Thesis, Université
     Catholique de Louvain, Louvain-la-Neuve, Belgium (1977).

43   R.R. Troutman, IEEE of J. Solid-State Circ., SC-9, 55
     (1974).

44   R.J. Van Overstraeten, G.J. Declerck and P.A. Muls,
     IEEE Trans. Electron Dev., ED-22, 282 (1975).

45   R.R. Troutman, IEEE Trans. Electron. Dev., ED-22, 1049
     (1975).

46   R.M. Swanson and J.D. Meindl, IEEE J. Solid-State Circ.,
     SC-7, 146 (1972).

47   H. Hara, Electronics Communic. Japan, 55C, 99 (1972).

48   J.S.T. Huang and G.W. Taylor, IEEE Trans. Electron. Dev.,
     ED-22, 995 (1975).

49   R. Hatert, R. Sinon and F. Van de Wiele, Phys. Stat. Sol.
     (a), 36, 235 (1976).

50 R. Hatert, "Simulations des dispositifs semiconducteurs-Influence de la forte inversion", Doctoral Thesis, Université Catholique de Louvain, Louvain-la-Neuve, Belgium (1975).

51 J.R. Edwards and G. Marr, IEEE Trans. Electron. Dev., ED-20, 283 (1973).

52 J.S.T. Huang, IEEE Trans. Electron. Dev., ED-20, 513 (1973).

53 G. Merckel, "Ion implanted MOS transistors - Depletion mode devices", NATO Advanced Study Institute on "Process and Device Modeling for Integrated Circuit Design" (1977).

54 W.H. Schroen, "Physical MOS models", NATO Advanced Study Institute on "Process and Device Modeling for Integrated Circuit Design" (1977).

ION IMPLANTED MOS TRANSISTORS - DEPLETION MODE DEVICES

G. Merckel

Centre d'Etudes Nucléaires de Grenoble, LETI-MEA
85 X   38041 Grenoble Cedex, France

ABSTRACT. Depletion mode transistors are often employed as load devices in the basic invertor. After an analysis of the specific operating modes, the model for the transistor and its application as a load device in an invertor are presented.

1. STRUCTURE AND BEHAVIOUR OF THE TRANSISTOR

1.1 Structure of the transistor

A cross-section view of an N channel depletion mode transistor is shown in Fig. 1. The channel is modulated by the gate electrode and by the substrate through the extension of the space charge region of the N-P junction. The doping profile of the implanted zone (N) is Gaussian. For simplifying the analysis of the electrical behaviour, we assume a rectangular profile (Fig.2) the same junction depth, d, and with a doping level such that :

$$N_I \simeq \frac{1}{d} \int_0^d C(x) \, dx$$

1.2 The transistor behaviour

All voltages are referred to the source. V_P represents the channel pinch-off voltage. The various modes of working are represented in Fig. 3 :

- $V_G < V_P$ transistor "off" (no channel)

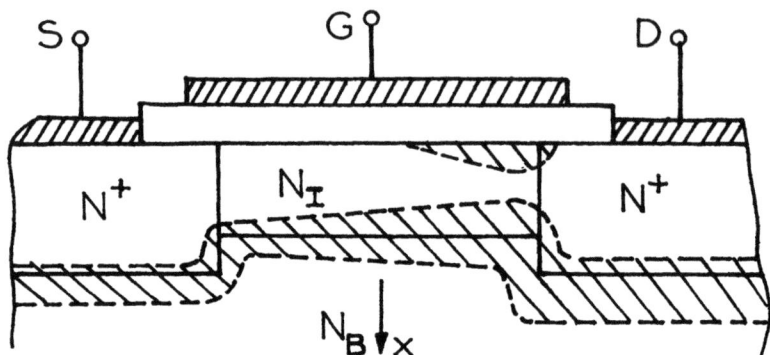

Fig. 1. Cross section of the depletion-mode device (N channel).

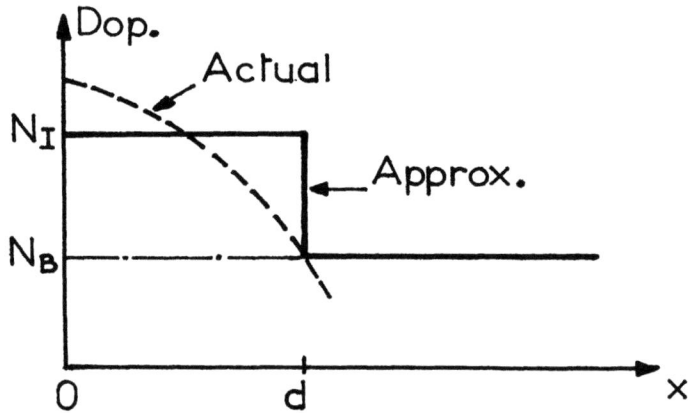

Fig. 2. Implanted profile in channel region.   ---- actual
                                                ——— approximated

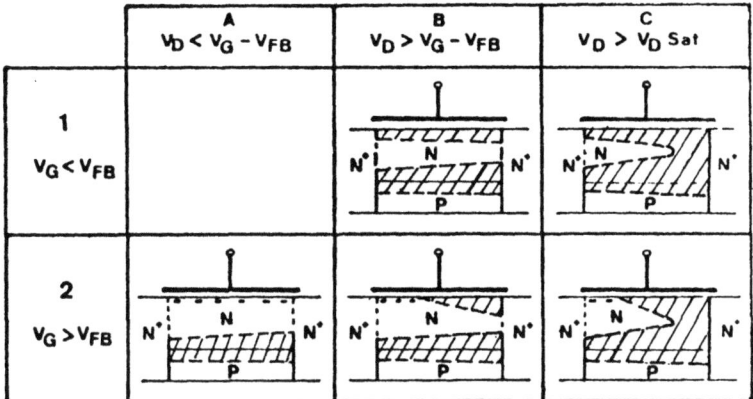

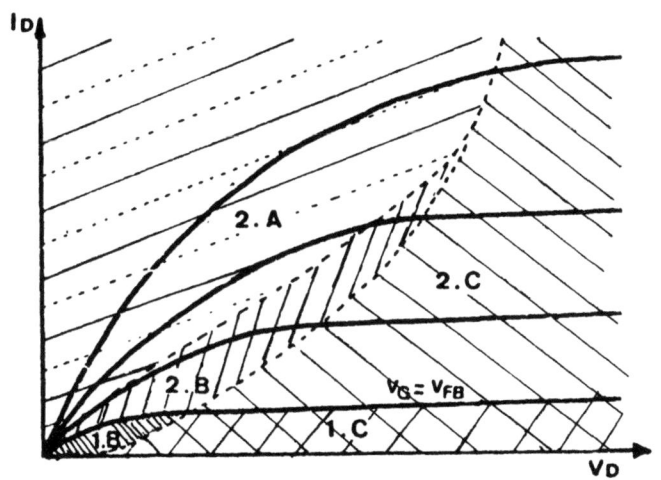

Fig. 3. Working modes.

- $V_P < V_G < V_{FB}$ depletion mode (burried channel) (B1)
- $V_G \geqslant V_{FB}$
  - if $V_G - V_{FB} \geqslant V_D \geqslant 0$ complete enhancement mode (A2)
  - if $V_D > V_G - V_{FB} \geqslant 0$ partial enhancement (B2)

C represents the behaviour in saturation ($V_D > V_{DSS}$)

2. TRANSISTOR MODEL

2.1 Channel pinch-off voltage V_P (or threshold voltage w.r.t. source)

A cross section view through the region near the source is shown in Fig. 4. The symbols used in the calculations are given in the figure. In normal operation (no residual current when the transistor is off) if $V_G \to V_P$, x_o increases and the thickness of the channel $e \to o$, hence $I_D \to o$. Under these conditions, the threshold voltage is calculated by solving Poisson's equation (in Si) and applying Gauss' law at the Si-SiO$_2$ interface. This gives [1]

$$V_P = V_{FB} - \frac{N'_B}{N_I}(V_\emptyset + V_B) - \frac{qN_I d}{C_{ox}}\left(1 + \frac{C_{ox}}{2C_I}\right) + \left(\frac{1}{C_I} + \frac{1}{C_{ox}}\right) \cdot \left[(2q\varepsilon_{Si} N'_B)(V_\emptyset + V_B)\right]^{1/2} \quad (1)$$

with V$_{FB}$ the flat band voltage (the channel pinch off being in the volume)

$N'_B = \dfrac{N_B N_I}{N_B + N_I}$ the composite concentration

$V_\emptyset = v_T \log\left(\dfrac{N_B N_I}{n_i^2}\right)$ the potential barrier of the NP junction.

$C_I = \dfrac{\varepsilon_{Si}}{d}$ capacitance of the implanted zone per unit area

C_{ox} gate oxide capacitance per unit area.

Expression (1) is valid if $x_I < d$ (see Fig. 4) ie. if

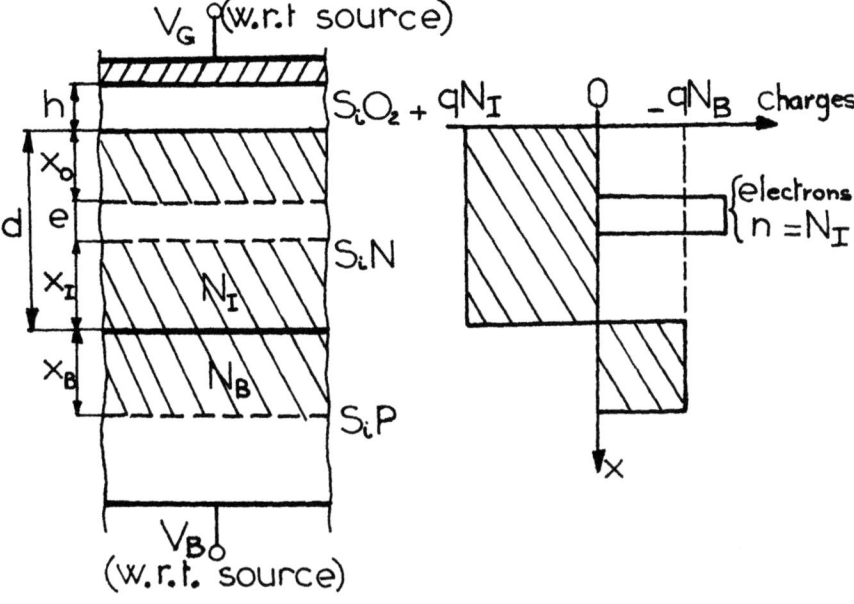

Fig. 4. Cross section and symbols.

$$d \geq \left[\frac{2\varepsilon_{Si} N'_B}{q N_I^2} (V_\emptyset + V_B) \right]^{1/2}$$

In most of the cases encountered in practice, $N_I \gg N_B$; so $N'_B \simeq N_B$ and (1) reduces to

$$V_P \simeq V_{FB} - \frac{q N_I d}{C_{ox}} \left(1 + \frac{C_{ox}}{2 C_I} \right) + K \left(1 + \frac{C_{ox}}{C_I} \right) (V_\emptyset + V_B)^{1/2} \quad (2)$$

Where $K = \left(2\varepsilon_{Si} q N_B \right)^{1/2} / C_{ox}$ is the body factor for an enhancement mode transistor on the same substrate (without ion implant).

As an example, Fig. 5 shows the variation of V_P (for $V_B=0$) as a function of the depth, d, and concentration, N_I, of the implanted zone. The curve V_{PL} is the limiting pinch off voltage defined in section 2.2.

2.2 Channel not pinched off

    2.2.1 Limiting pinch off voltage V_{PL}

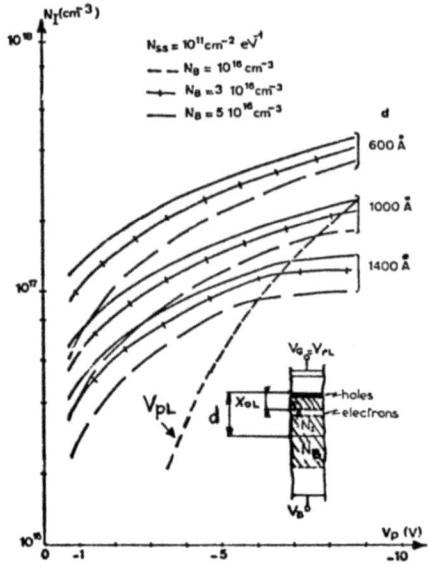

Fig. 5. Pinch off voltage V_P versus concentration N_I and channel junction depth d.

If the concentration, N_I, and/or the junction depth, d, are large and if one tries to put the transistor off, the width of the space charge layer (see Fig. 4) tends to a limited value x_{oL}. All further decrease in V_G will increase the hole concentration in the surface inversion layer. Because of a finite thickness of the conduction channel, (e ≠ o), a residual current results. Under these conditions, the conduction channel can be eventually pinched off by substrate polarization (if $x_o = x_{oL}$ and if $V_B\uparrow$, $x_I\uparrow$ and e↓). V_{PL} is given by :

$$V_{PL} = V_{FB} + 2\emptyset_{FI} + K_I (2\emptyset_{FI})^{1/2}$$

with $\emptyset_{FI} = v_T \log (N_I/n_i)$ and $K_I = (2\varepsilon_{Si} q N_I)^{1/2}/C_{ox}$

One can similarly define a limiting junction depth d_L by

$$d_L = \frac{1}{N_I} \left(\frac{2\varepsilon_{Si} N_B}{q} \right)^{1/2} \left[\left(2\emptyset_{FI} \frac{N_I}{N_B} \right)^{1/2} + (V_B + V_\emptyset)^{1/2} \right] \quad (3)$$

Such that if $d > d_L$, it is impossible to turn the transistor 'off' completely
    and if $d \leqslant d_L$, it is possible to turn it off completely.

2.2.2 Residual current I_R

If the transistor can not be put off completely, the residual current I_R is given to the first order by

$$I_R = q\, N_I\, \mu_I\, \frac{Z}{L}\, (d - d_L)\, V_D$$

with μ_I majority carrier mobility in the implanted layer

Z and L channel width and length

d_L is given by (3).

2.3 Simplified model

In this section, we shall first describe the model and then give the definitions of the parameters.
The basic model [4,5], relatively complex, is obtained by following the principle of calculation shown in Fig. 6. One thus obtains the analytical expression of the current I_D as a function of the applied voltages. The basic model has 3 sections according to the relative values V_D, V_G, V_{FB} and V_P.
The simplified model is obtained from the basic model by a series expansion limited to second order in V_D.

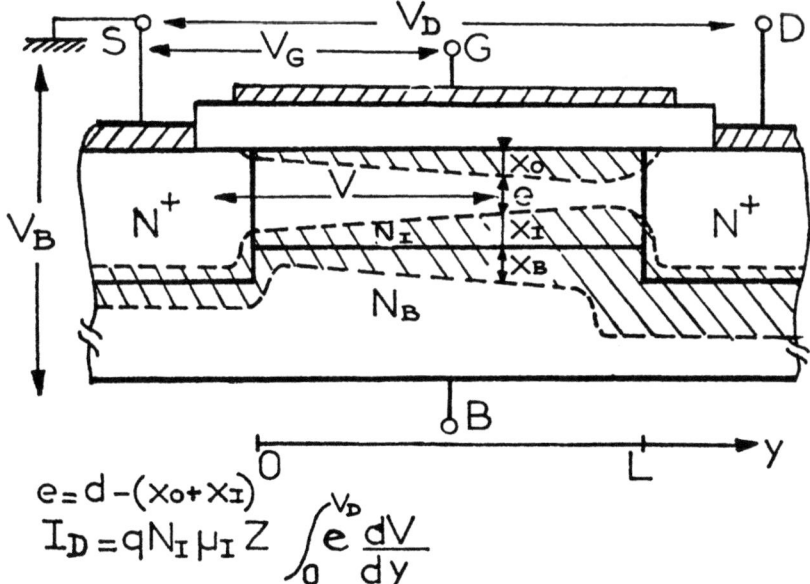

Fig. 6. Calculation of the I(V) characteristics.

2.3.1 Current voltage characteristics

The current voltage equation are (before saturation):

a) If $V_{FB} > V_G \gg V_P$ (burried channel)

$$I_D = M_D \left[V_{GP} V_D - \frac{V_D^2}{2} (1+\delta) \right] \quad (4)$$

with $V_{GP} = V_G - V_P$

$$V_P \simeq V_{PO} + K_D \left[(V_B + V_\emptyset)^{1/2} - (V_\emptyset)^{1/2} \right] \quad (5)$$

b) . If $V_G \geqslant V_{FB}$ and $V_G - V_{FB} \geqslant V_D \geqslant 0$ (surface fully enhanced)

$$I_D = M_D \left\{ V_{GP} V_D - \frac{V_D^2}{2} (1+\delta) + (r-1) \left[V_{GF} V_D - \frac{V_D^2}{2} \right] \right\} \quad (6)$$

with $V_{GF} = V_G - V_{FB}$ and $r = r_o (1 + \Theta_G V_{GF})^{-1}$

. If $V_D \geqslant V_G - V_{FB} > 0$ (surface partially enhanced)

$$I_D = M_D \left[V_{GP} V_D - \frac{V_D^2}{2} (1+\delta) + (r-1) \frac{V_{GF}^2}{2} \right] \quad (7)$$

In the saturation zone, a simple model reflecting the physical aspect of the behaviour of the transistor remains to be done. A two dimensional analysis in the vicinity of the drain would be necessary to define the assumptions and the boundary conditions required for a simple solution of Poisson's equation.

Keeping these remarks in mind, one can simply define the $I_D(V_D)$ relation under the following form :

$$I_D = I_{DSS} \left[1 + \frac{V_D - V_{DSS}}{V_E + V_{DSS}} \right] \quad (8)$$

where V_E is the equivalent of Early voltage.

Assuming (8), all the relations defining the current in saturation zone are given in the lecture "CAD Models of MOSFET'S"

2.3.2 Parameters

The model, as defined above, contains the following parameters :

- $M_D = \dfrac{\mu_{BI}}{1+\alpha} C_{ox} \dfrac{Z}{L}$

with μ_{BI} = majority carrier mobility in the implanted zone
C_{ox} = gate oxide capacitance per unit area
Z = channel width
L = channel length

$\alpha = \dfrac{C_{ox}}{C_I} \left(\dfrac{C_{ox}}{2C_I} + 1 \right)$

$C_I = \dfrac{\varepsilon_{Si}}{d}$ capacitance of the implanted zone per unit area.

- V_{PO} = pinch off voltage for $V_B = 0$ (see (2))
- $V_\emptyset = v_T \log (N_I N_B/n_i^2)$ potentiel barrier of the N P junction
- V_{FB} = flat band voltage
- $K_D = K (1+C_{ox}/C_I)$ body factor

   K is the body factor of an enhancement transistor (on the same substrate)

- $r_o = \mu_{So}(1+\alpha)/\mu_{BI}$ μ_{So} = max. surface mobility
- $\Theta_G \sim C_{ox}$ constant for dependence of mobility on the normal electric field.
- $\delta = 0.25\ K\ (1+\alpha)\ V_\emptyset^{-1/2}$
- $V_E \sim L\ (N_B)^{1/2}$

3. MEASUREMENT OF THE PARAMETERS

The parameters Z, L, C_{ox} and N_B are generally known.
- C_{ox} and N_B can be determined by C - V measurements on a test capacitor, thus K is known.
- If one has an enhancement transistor (without implantation) on the same substrate (threshold voltage V_T), one can calculate K from the slop of a V_T versus $(V_B+2\emptyset_F)^{1/2}$ plot.

- K_D is obtained from the slope of V_P (eqn.5) versus $(V_B+V_\emptyset)^{1/2}$ (in general, one can take $V_\emptyset \simeq 0.7V$)
- α is obtained from K_D (gives d)
- C_I is obtained from α and C_{ox} (gives d)
- μ_{BI} is obtained from the slope dI_D/dV_G of the $I_D(V_G)$ curve for small values of V_D (< 0.1 V) and for $V_G < V_{FB}$.
- V_{FB} is either obtained from the C-V curves of a test capacitor or from the 'knee' point of the $I_D(V_G)$ curve for small drain voltages (Fig. 7). One should speak of a "pseudo flat band voltage" since the concentration in silicon is not uniform.
- r_o and Θ_G are obtained by fitting the $I_D(V_G)$ curves, for $V_G > V_{FB}$ to eqn. (6) (V_D small, so V_D^2 is negligible).
- δ can either be calculated from parameters already known or by fitting experimental results to eqn. (4).
- Finally, if V_{FB} is accurately known, an extrapolation of the straight line V_P versus $(V_B + V_\emptyset)^{1/2}$ to $V_B + V_\emptyset = 0$ gives (eqn.(2)) :

$$V_P \simeq V_{FB} - \frac{q\,N_I\,d}{C_{ox}}\left(1 + \frac{C_{ox}}{2\,C_I}\right)$$

from where N_I may be calculated.

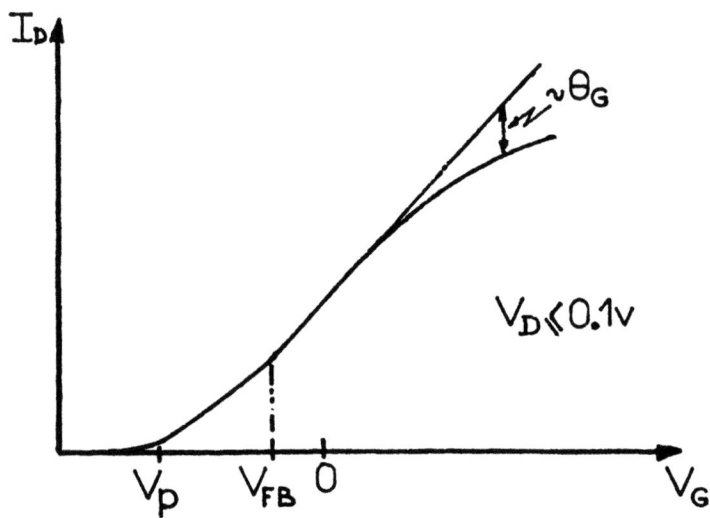

ig. 7. $I_D(V_G)$ plots in the linear region (V_D small).

4. TRANSISTOR USED AS A LOAD ELEMENT

The depletion mode transistor is very often associated with an enhancement mode transistor (Fig. 8) to constitute an invertor stage. As a load element, it acts likes a current source ($I_D \simeq V_P^2$) over a large range of voltages between 0 and V_{DD}.

One assumes that the drain current of the depletion transistor is constant in the saturation zone. For $V_G = 0$, one obtains the following relationships for the load transistor :

. If $-V_P \leqslant V_D$ ie. if $V_o \leqslant V_{DD} + V_P$

(7) gives
$$I_D = \frac{M_D}{2} \left[\frac{V_P^2}{1+\delta} + V_{FB}^2 (r-1) \right] \quad (9)$$

with $r = r_o (1 + \Theta_G V_{FB})^{-1}$

. If $-V_{FB} \leqslant V_D \leqslant -V_P$ ie. if $V_{DD} + V_P \leqslant V_o \leqslant V_{DD} + V_{FB}$

(7) gives
$$I_D = M_D \left[-V_P V_D - \frac{V_D^2}{2}(1+\delta) + (r-1)\frac{V_{FB}^2}{2} \right] \quad (10)$$

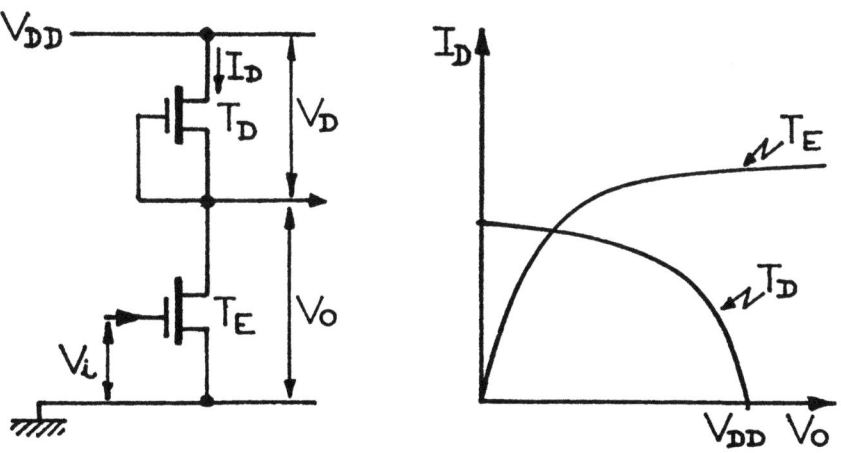

Fig. 8. NOR gate with a depletion-mode device as a load.

. If $V_D \leqslant -V_{FB}$ ie. if $V_{DD} + V_{FB} \leqslant V_O$

(6) gives
$$I_D = M_D \left\{ -V_P V_D - \frac{V_D^2}{2}(1+\delta) - (r-1)\left[V_{FB}V_D + \frac{V_D^2}{2}\right] \right\} \tag{11}$$

Eqn. (9) shows that the drain current I_D of the load transistor changes little for a range of output voltage

$V_O \leqslant V_{DD} + V_P$ (the variation is due to the dependence of V_P on V_B)

Compared to a technology using only enhancement type devices, the use of depletion mode transistors as load devices offers the following advantages :

- large voltage gain per stage
- improvement of the power-delay product
- reduced size per invertor.

REFERENCES

[1] Y. Gris, J. Gautier, G. Merckel, J.P. Suat, ESSDERC 76, Munich, 13 - 16 Sept. 1976.
[2] G. Merckel, Int. Techn. Rept.
[3] B. Baylac, G. Merckel, ESSCIRC 76, Toulouse, 21-24 Sept. 1976.
[4] J. Gautier, Int. Techn. Rept.
[5] Y. Gris, J.P. Suat, G. Merckel, Techn. Rept. LETI/MEA 1185, Nov. 1976.
[6] J.S.T. Huang, G.W. Taylor, IEEE Trans. on E.D., E-D22, 11, Nov. 1975.

PHYSICAL MOS MODELS

Walter H. Schroen

Texas Instruments Incorporated, Dallas, Texas, U.S.A.

1. INTRODUCTION

The key circuit element in MOS logic is the MOS inverter, realized most simply by using a resistor as the load for an enhancement mode driver transistor (EDT). The reduced area required when an enhancement mode transistor is used as an active load (ELT) has allowed practical MOS LSI in volume production. However, the ELT has slower switching transitions than an equivalent resistor and reduces logic swing by the value of its threshold voltage. With the advent of ion implantation, the depletion mode transistor has become a practical reality. When used as a load, the depletion load transistor (DLT) remains in the on-stage and increases logic swing to a value close to the supply voltage. In theory the DLT has current sourcing ability superior to a resistor, hence, providing improved switching speed. As a result, the use of an EDT/DLT inverter allows speed margins to be achieved with power consumption approaching that of CMOS, thus facilitating low voltage battery operation.

Sections 2 and 3 describe calculations for p-channel depletion mode transistors. The gradual channel approximation is used with a quasi-two-dimensional solution of the transport equations in which Fermi-Dirac statistics and an arbitrary quantum density of states are assumed. The implanted boron profile is calculated from process inputs. The techniques developed for p-channel transistors are also applicable to n-channel devices.

Comparison of I_D-V_G characteristics to experimental data are given for several ion energies and doses. A voltage dependent shift in the I_D-V_G curves is shown to be a direct consequence of hole degeneracy. The calculated curves corrected to account for this effect are in excellent agreement in all cases with the measured $\partial I_D/\partial V_G$, V_{FB} and leakage current.

Experimentally observed distortion in the I_D-V_G curve at low positive gate voltage is due to fast state density (N_{st}). In view of the magnitude of N_{st} on (111) material, the determination of an extrapolated pinch-off or equivalent threshold voltage, as currently practiced, is precluded.

2. MATHEMATICAL MODEL OF DEPLETION LOAD TRANSISTOR*

The following description is based on the model by Evans and co-workers[1].

The initial distribution of implanted boron is assumed to be Gaussian with the projected range (RP) and range straggling (DRP) determined by experiment[2]. Using these starting conditions, including an effective oxide thickness correction to RP, the linear diffusion equation is solved to get the redistribution of impurities after anneal[3]. The boron atoms tend to stack up at the surface due to the non-oxidizing anneal (passivation) conditions, resulting in a skewed boron distribution in the silicon.

Although charge transport is primarily in a direction parallel to the surface channel, the conduction properties of the channel are controlled by fields perpendicular to the surface to first order. Since the implanted charge density is greater than the n-type substrate, a p-n junction is formed and the conducting channel is controlled by both surface and junction voltages. A cross section of a DLT showing qualitatively the depletion and accumulation regions is given in Figure 1.

In order to calculate the surface potential ψ_S, Figure 2 represents a detailed band diagram near the source and near the drain. Valence band ψ_V is the top line, the conduction band ψ_C the lower line. The following comments should be considered for the diagram near the source:

- The negative gate potential V_G attracts holes bending the silicon valence band ψ_V towards the Fermi level ϕ_F at the surface.

- The surface potential ψ_S is the change of ψ_V from the equilibrium value.

- The p-n junction is at zero bias.

The following comments hold for the diagram near the drain:

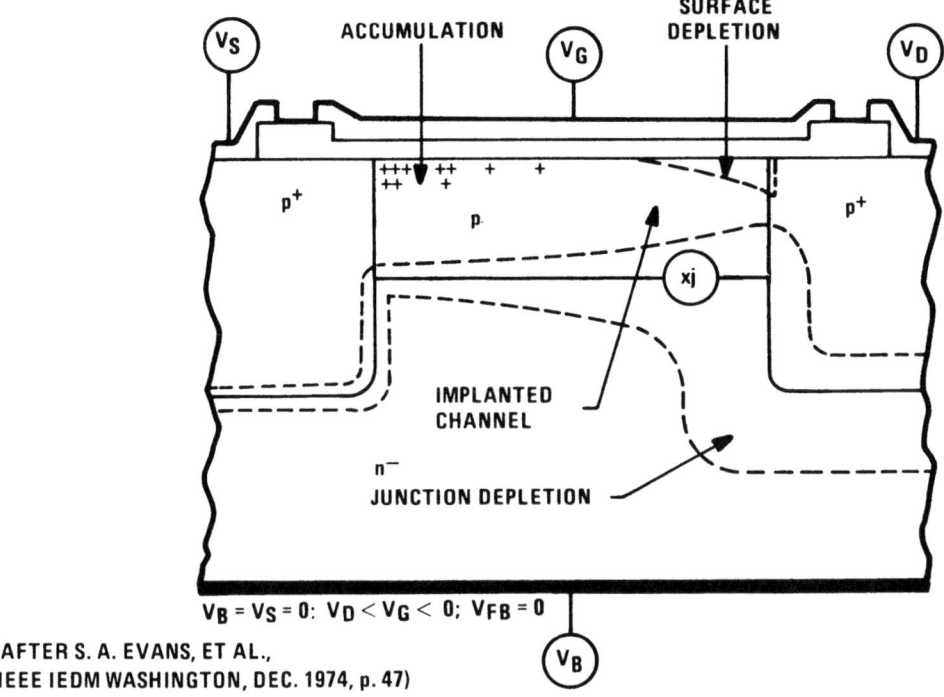

Fig. 1. MOS depletion load transistor.

- The drain voltage causes a drop in ϕ_{Fp} by V_D. Also, the p-n junction is reverse biased by V_D.

- $d\phi_{Fn}=0$ since there is no electron current.

- A depletion region is formed at the surface, but no inversion region, because the reverse biased junction blocks electrons.

- The surface potential ψ_s includes the combined effects of V_D and V_G.

The carrier distribution in the channel is found by solving one dimensional diode equations[4] subject to the following boundary conditions:

(1) At the Si-SiO$_2$ interface (x = 0), it is required that

$$p(o) = p_o(0) \exp(-q\psi_s/kT), \qquad (1)$$

BOUNDARY CONDITIONS: $V_D < V_G < 0; t_{ox} = d; V_{FB} = 0$

BAND DIAGRAM NEAR SOURCE

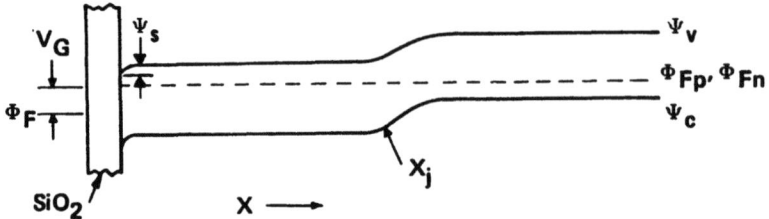

BAND DIAGRAM NEAR DRAIN

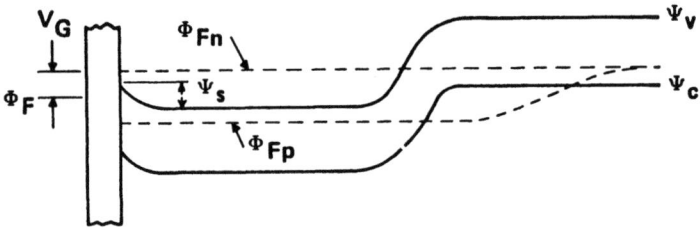

BASED ON THE BAND DIAGRAM ABOVE THE SURFACE POTENTIAL IS GIVEN BY

$$\Psi_s(y) = (V_G - V_{ox}) - \Phi_{Fp}(y),$$

OR

$$\Psi_s(y) = V_G - \frac{K_s}{K_{ox}} E_s t_{ox} - \Phi_{Fp}(y).$$

Fig. 2. Band diagram of depletion load MOS transistor.

$$n(o) = n_o(0) \exp(q\psi_s/kT), \qquad (2)$$

where p_o and n_o are the equilibrium carrier densities and ψ_s is the difference between the valence band edge and its equilibrium value. The drain voltage is assumed to be small. The surface potential is related to the gate voltage by

$$\psi_s = V_G - (K_{si} t_{ox} E_s / K_{ox}), \qquad (3)$$

where $E_s = E(0)$ is the surface field, t_{ox} is the oxide thickness (see Figure 2).

(2) At the substrate contact, the usual ohmic contact boundary conditions are used.

When the hole density exceeds ~ 10^{19} ions/cm^3 and/or the impurity density exceeds ~ 10^{17}, the Boltzmann relation in Eq. (1) is not a valid approximation. One must use a more exact expression for the hole density[5].

$$p = \int_{-\infty}^{\infty} \frac{\rho(\varepsilon)d\varepsilon}{[1+\exp(q\varepsilon/kT-\zeta_p/kT)]} \quad (4)$$

where $\rho(\varepsilon)$ is an arbitrary quantum density of states function and ζ_p is the chemical potential. Using the above definition of p, a more general form of the hole transport equation can be easily derived as

$$J_p(x) = q\mu_p[p(E+E_p) - \frac{kT}{q}\alpha_p \nabla_x p], \quad (5)$$

where α_p is the generalized Einstein relation for holes and E_p is a quasifield due to the valence band edge gradient. If a parabolic band structure is assumed, $E_p=0$ and α_p can be written as

$$\alpha_p = \frac{q}{kT}\frac{D_p}{\mu_p} = \frac{q}{kT}\frac{\partial \zeta_p}{\partial \ln p}. \quad (6)$$

Clearly the relation between mobility and diffusivity is altered when α_p exceeds unity. The net effect of Eqs. (4)-(6) is an increase in the surface field associated with a given carrier density and, therefore, at a given gate bias, the channel conductance is reduced. This is observed as a voltage dependent shift in the I_D-V_G curve relative to the conventional calculation.

The procedure for calculating the channel current has been described in some detail by Pao and Sah[6]. The expression for the total drain current is

$$I_D = -q\frac{W}{L}\int_0^{V_D} d\Phi_{F_p} \int_0^{x_j} \mu_p p \, dx, \quad (7)$$

where Φ_{F_p} is the hole quasi-Fermi level. In the case of small drain voltages, Eq. (7) simplifies to

$$I_D = -q\frac{W}{L} V_D \int_0^{x_j} \mu_p p\, dx \ .$$

The bulk mobility model was taken from Caughey and Thomas[7] using a modified reference concentration determined by Wagner[8] and verified. In accumulation the surface mobility may be degraded by surface effects similar to those observed in inverted surface channels. Since the voltage induced surface charge is actually spread over 50 to 100 Å layer, a voltage dependent mobility correction would have to be made to the integrated surface charge based on experimental surface mobility measurements. For the time being we neglect corrections due to surface mobility[9].

The procedure for solving the diffusion and transport equations is displayed in Figure 3. The diode equations are solved applying the boundary conditions in Eq. (1) and (2) with ψ_s varied such that the Fermi level scans from the conduction band to the balence band. Since only a lower limit for the dose is known for these devices, the dose was varied until calculated and measured leakage currents were in agreement. The Boltzmann approximation is used to obtain the dashed curve in Figure 4. The surface densities at large negative gate bias are used as input boundary conditions to a bipolar device calculator program (GENTRAN) which is designed to simulate the emitter base junction of a bipolar transistor (or any n-p or p-n junction) including heavy doping and high carrier density effects. The program allows the user to choose an arbitrary density of states and either Fermi or Boltzmann statistics. The calculator is extremely fast: computation time is independent of statistics and only slightly dependent on the density of states model. By examining the difference in surface field between GENTRAN and the Boltzmann calculation, the shifted dotted curve in Figure 4 is obtained. Inclusion of Fermi statistics is necessary in order to find a voltage region where the experimental and calculated I-V curves are parallel.

3. DEVICE FABRICATION, RESULTS AND DISCUSSION

Standard MOS slices were withdrawn from the process prior to passivation, exposed with photoresist leaving DLT gate areas exposed, and implanted with an ion dose suitable for the device final characteristics. Activation of the implanted species was achieved during the normal passivation process.

In the work reported here, the energy range 35-80 keV was examined for ion doses of 1 to 2×10^{12} ions/cm^2. A standard pro-

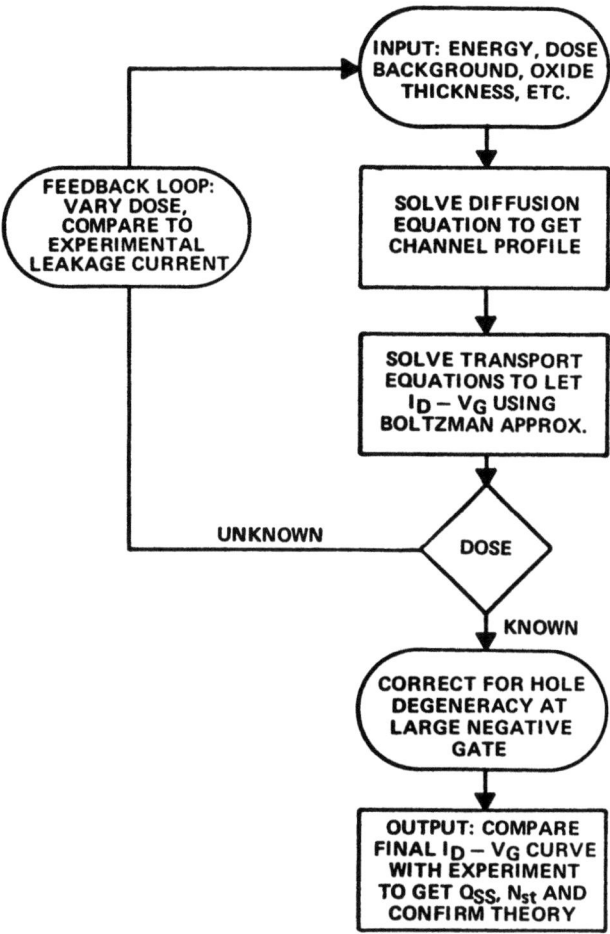

Fig. 3. Procedure.

duction implanter was used with no special precautions taken; hence, the measured doses are accurate on a relative basis only.

Large sample measurements of I-V curves were generated using a high speed parametric tester, such that characteristics matched are not from individual devices but are averaged I-V's having a known small standard deviation. These measurement statistics can be related to process parameters such as substrate doping density, oxide thickness, and flatband voltage, which were also determined on a statistical basis. Since the ion implanted capacitor is not readily conducive to analysis, flatband voltage and fast surface state density (N_{st}) close to midgap were determined using non-

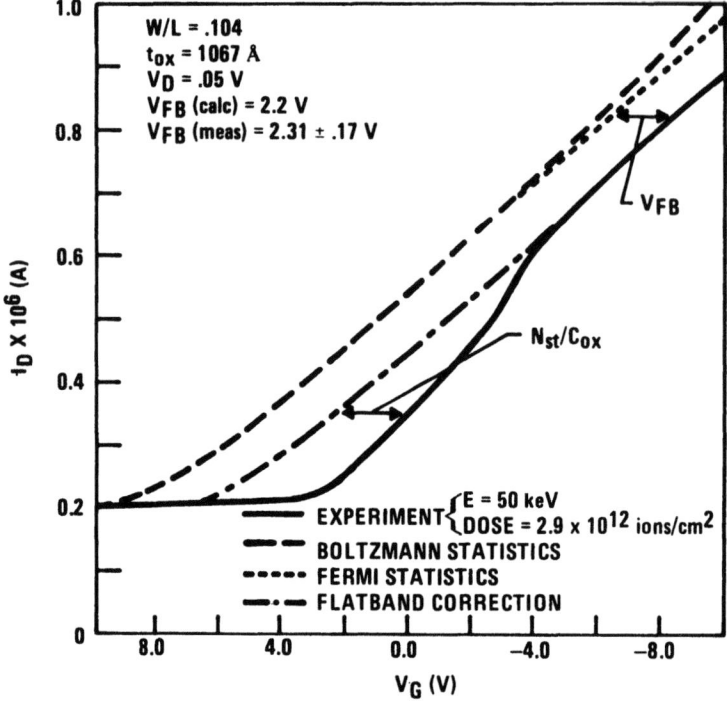

(AFTER S. A. EVANS, ET AL., IEEE IEDM, DEC. 1974, WASHINGTON, p. 47)

Fig. 4. I_D-V_G characteristics of depletion load transistor.

implanted samples. The ϕ_{ms} contributed to V_{FB} was then corrected to account for the difference in the surface doping (ϕ_{ms} = simple metal-silicon work function, V_{FB} = flatband voltage).

The calculated I_D-V_G characteristic is compared to experiment in Figure 4. V_{FB} is determined by taking the difference between calculated and measured voltages at a current level such that N_{st}=0 and ∂I_D (meas)/∂V=∂I_D (calc)/∂V. In Figure 5 other combinations of dose and energy are considered and the calculated I_D-V_G curves are corrected for V_{FB}. The agreement demonstrated in Figures 4 and 5 shows that the low drain voltage model presented here can adequately predict leakage current, V_{FB}, and $\partial I/\partial V$ if process conditions (i.e., implant dose, oxide thickness, etc.) are known. It should be noted that due to the high value of N_{st}, there is no clearly defined region in which the calculated and measured curves are parallel when degenerate or Fermi statistics are neglected. For the doping densities considered, we found that the departure from parabolic band structure was a small effect compared to the effect of degeneracy. In addition,

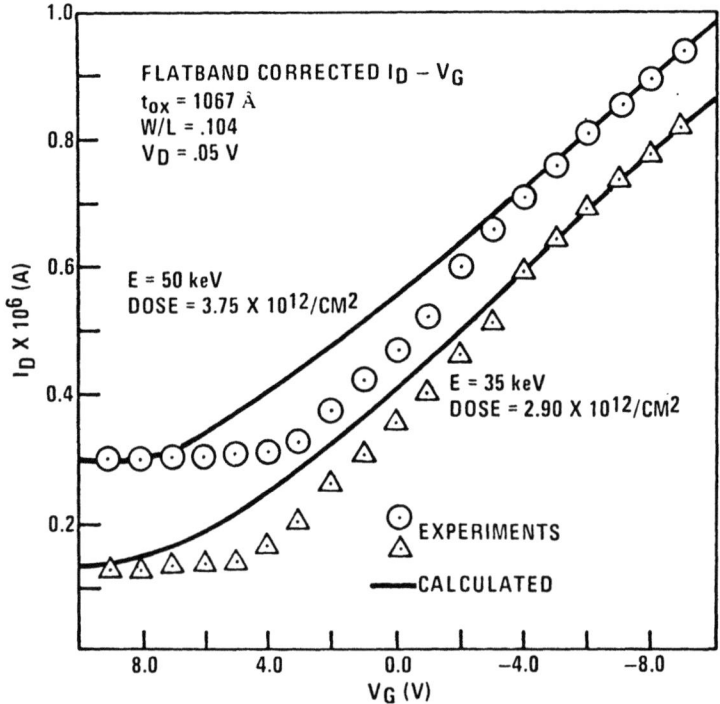

Fig. 5. I_D-V_G curves: theory and experiment.

N_{st} as a function of surface Fermi level can be found since the actual surface density is calculated along with I_D.

A larger range of drain bias conditions are presented in Figure 6. It is clear from these measured I_D-V_G characteristics that the voltage threshold V_{px}, determined by extrapolation, is a function of both V_G and V_D. If V_{px} is used in design or production testing, then care must be taken to choose a sampling point in the region of device operation. A better approach would be to define new, more meaningful design parameters based on a good understanding of the device physics.

4. PHYSICAL MODEL OF ENHANCEMENT MODE TRANSISTOR

Numerous approaches have been undertaken to formulate models for enhancement mode MOS transistors[10]. For the case of a one-dimensional enhancement mode transistor, i.e., a transistor with channel length > 5 μm, an expression has been derived[10] which contains only physically meaningful parameters. In this equation,

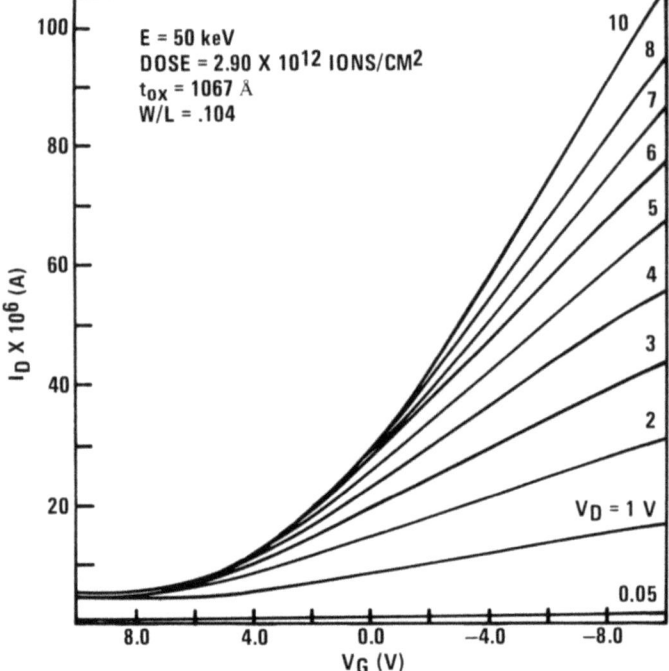

Fig. 6. I_D-V_G curves of depletion load transistor.

the current I_D from source to drain is expressed as an analytical function of independently measurable and controllable process parameters.

The equations are shown in Figure 7 both for the linear regime and the saturation regime. The various symbols are explained in Figure 8. It should be noted, however, that the equations in Figure 7 have to be modified when short channel transistors are considered; for long channels, they are in excellent agreement with measurements[10].

The schematic of the verification of existing physical MOS models is depicted in Figure 9. The process used to fabricate MOS devices is characterized by test structures or plug bars[11] The measurement of these test bars delivers one set of process parameters. On the other hand, the I-V measurements performed using the MOS devices deliver another set of process parameters from a model fitter which is based on the physical MOS model. It is the comparison between the process parameters obtained by the process characterization and the device I-V measurements which is used to verify the physical MOS model incorporated in the model

ENHANCEMENT MODE – LINEAR REGIME:

$$I_D = \frac{\mu \epsilon_{ox} \epsilon_s W}{t_{ox} L} \left\{ \left(V_G - \phi_{ms} - \phi_B + \frac{Q_{ox} t_{ox}}{\epsilon_{ox} \epsilon_s} \right) (V_D - V_S) \right.$$

$$-1/2 \left(V_D^2 - V_S^2 \right) - \frac{2 t_{ox} (2 \epsilon_s q N)^{1/2}}{3 \epsilon_{ox} \epsilon_s} \left([\mp V_D \mp \phi_B]^{3/2} \right.$$

$$\left. \left. - [\mp V_S \mp \phi_B]^{3/2} \right) \right\}$$

ENHANCEMENT MODE – SATURATION REGIME:
SUBSTITUTE V_{DP} BELOW FOR V_D IN ABOVE EQUATION

$$V_{DP} = V_G - \phi_{ms} - \phi_B + \frac{Q_{ox} t_{ox}}{\epsilon_{ox} \epsilon_s}$$

$$\pm \frac{\epsilon_s q N t_{ox}^2}{\epsilon_{ox}^2 \epsilon_s^2} \left\{ \left[1 \mp \frac{2 \epsilon_{ox}^2 \epsilon_s^2}{\epsilon_s q N t_{ox}^2} \left(V_G - \phi_{ms} + \frac{Q_{ox} t_{ox}}{\epsilon_s \epsilon_{ox}} \right) \right]^{1/2} - 1 \right\}$$

(UPPER SIGNS FOR P-CHANNEL, LOWER SIGNS FOR N-CHANNEL

Fig. 7. Physically meaningful MOS equations.

INDEPENDENT PROCESS PARAMETERS
N = BULK DOPING DENSITY = f (STARTING MATERIAL DOPING)
Ns = SURFACE DOPING DENSITY = f (N, Ox & II PROCESSES)
t_{ox} = INSULATOR THICKNESS = f (INSULATOR GROWTH PROCESS)
Q_{ox} = TOTAL INSULATOR CHARGE = f (CLEANING, GROWTH, METAL, ANNEALS)

DEPENDENT PROCESS PARAMETERS
ϕ_{MS} = GATE-SUBSTRATE WORK FN. DIFF. = f (GATE MAT. & PASSIVATION)
ϕ_B = STRONG INVERSION POTENTIAL = f (N)
ϵ_{ox} = DIELECTRIC CONSTANT OF INS. = f (INSULATOR CHOICE)
μ = EFFECTIVE SURFACE MOBILITY = f (Ns, OXIDATION, AVERAGE GATE POTENTIAL)

Fig. 8. Physically meaningful MOS process parameters.

fitter.

The process parameters, which are independently measurable and controllable, serve as inputs to a computer program written for the physical model to calculate I-V curves as illustrated in Figure 10. The threshold voltage V_T can then be found at the

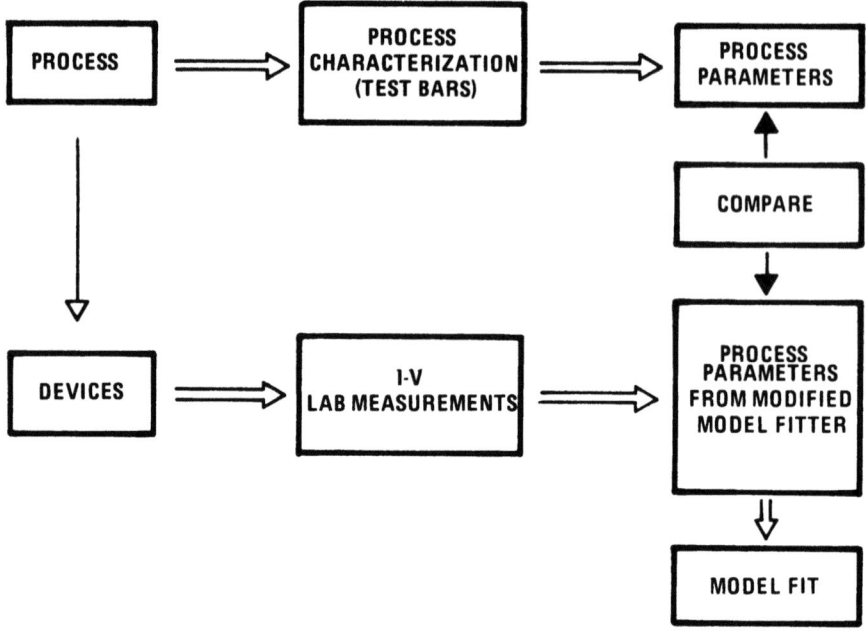

Fig. 9. Verification of existing physical MOS models.

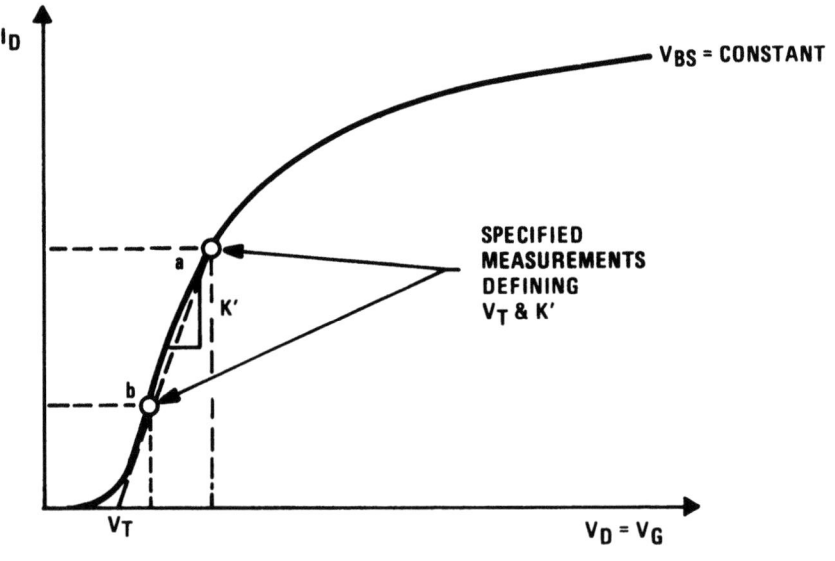

V_T = INTERCEPT OF LINE ab
K' = SLOPE OF LINE ab

Fig. 10. Determination of V_T and K' from calculated or measured I-V curves.

intercept with the voltage axis of the line through two predefined points a and b on the I-V curve. The slope of the line ab defines the conduction factor K'.

5. PROCESS SENSITIVITY ANALYSIS

. When physically meaningful process and device models are available, they can be used for a computerized process sensitivity analysis[11]. This analysis can identify those process steps which affect the device characteristics particularly strongly and therefore require special attention for effective process control.

Figure 11 describes this philosophy by the example of physically meaningful MOS equations. In this equation, the current I_D from source to drain is expressed as an analytical function of independently measurable and controllable process parameters. Figure 11 lists four parameters: bulk doping density, surface doping density, insulator thickness, and insulator charge. Furthermore, the length and width of the channel enter the equation as design geometry parameters. The expression also contains four terminal voltages.

For the process sensitivity analysis, the partial derivative for one of the variables is formed, while all the other variables are kept constant. Figure 12 demonstrates this approach by first selecting the oxide thickness t_{ox} as the variable to form the partial derivative, in the next step the fixed oxide charge Q_{ss}, etc. For each parameter, the impact of its change (= partial

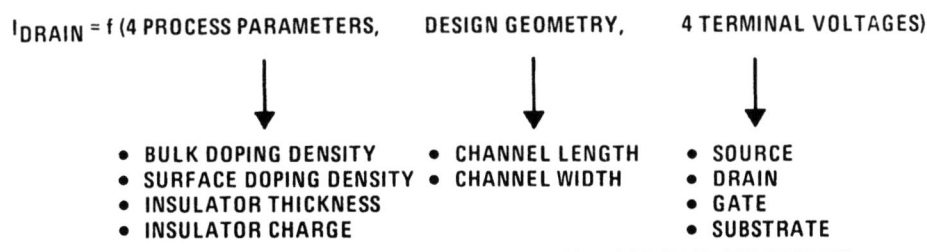

- PHYSICALLY MEANINGFUL MOS EQUATIONS PRESENT MOS DEVICE CHARACTERISTIC AS AN EXPLICIT ANALYTICAL FUNCTION OF INDEPENDENTLY MEASURABLE AND CONTROLLABLE PROCESS PARAMETERS:

  I_{DRAIN} = f (4 PROCESS PARAMETERS,   DESIGN GEOMETRY,   4 TERMINAL VOLTAGES)

  - BULK DOPING DENSITY
  - SURFACE DOPING DENSITY
  - INSULATOR THICKNESS
  - INSULATOR CHARGE

  - CHANNEL LENGTH
  - CHANNEL WIDTH

  - SOURCE
  - DRAIN
  - GATE
  - SUBSTRATE

- FOR ANY PRODUCTION PROCESS, PARTIAL DERIVATIVES IDENTIFY THOSE PROCESS PARAMETERS WHICH MOST SENSITIVELY INFLUENCE DEVICE CHARACTERISTICS.
- ACCEPTABLE RANGES OF PARAMETER VARIATION DEFINE PROCESS PARAMETER BOXES FOR HIGH QUALITY.

Fig. 11. MOS process sensitivity analysis.

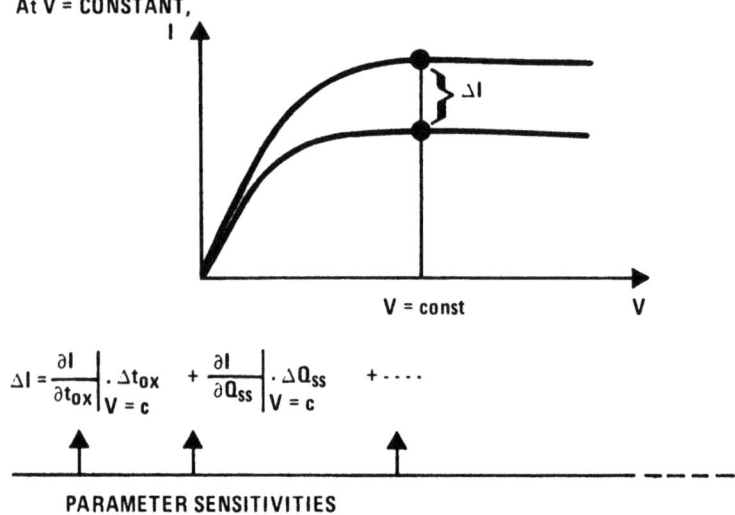

Fig. 12. MOS device sensitivity to process parameters.

derivative) on the current I_D is studied. This way it can be determined which of the parameters in an equation like the one of Figure 7 exerts the greatest influence on the I-V characteristic of the MOS transistor.

Process windows can be determined which represent the functional dependence of one parameter on another one. As Figure 13

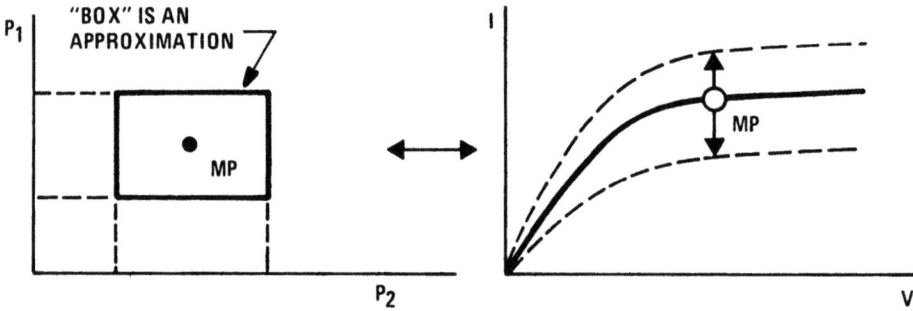

- GIVEN n PROCESS PARAMETERS P_i
- ASSUME PARAMETERS ARE INDEPENDENT (BOX RATHER THAN SPACE)
  - SET EACH P_i AT SOME ACCEPTABLE VALUE (MIDPOINT)
  - AT EACH VALUE OF V, VARY P_i UNTIL I > OR < ACCEPTABLE RANGE (GIVES ACCEPTABLE RANGE ON "BOX" FOR PROCESS PARAMETER P_i)
  - REPEAT FOR $P_2 \cdots P_n$ (HOLDING ALL OTHER P's AT MP)

Fig. 13. MOS process parameter window determination.

illustrates, the process window can also be looked at as a variation in the I-V characteristic of the MOS transistor. Taking into account the process sensitivity analysis of Figure 12, which resulted in a variation of the current voltage characteristic, one can superimpose this sensitivity analysis for the two parameters considered over the I-V characteristic of Figure 13. The resulting I-V characteristic is shown in Figure 14 where two small variations are shown for each of the limiting curves of Figure 13. Using the process sensitivity analysis in this way, the process control engineer can judge whether this variation of the boundaries of the I-V characteristics is tolerable for the circuit characteristics being fabricated.

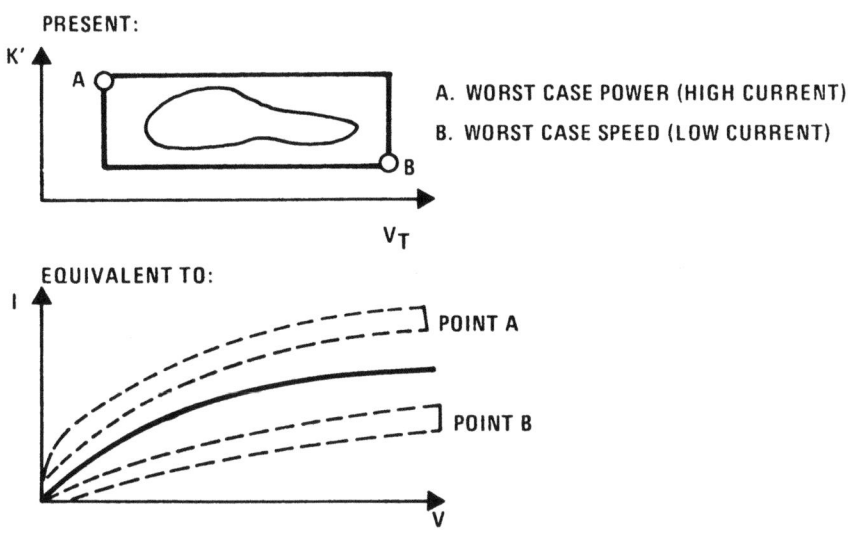

Fig. 14. Definition of process window.

REFERENCES

1. S.A. Evans, I.H. Morgan, and T.N. Jernigan, "A model for the accurate prediction of the current-voltage behavior of ion-implanted MOS transistors at low drain voltages," IEEE IEDM Meeting, Washington, D.C., 1974, p. 47.
2. J.F. Ziegler, B.L. Crowder, G.W. Cole, J.E.E. Baglin, B.J. Masters, "Boron atom distributions in ion-implanted silicon by the (n,^{4}He) nuclear reaction," Appl. Phys. Letters 21, no. 1, 16 (1972).
3. J.L. Prince and F.N. Schwettmann, "Diffusion of boron from implanted sources and oxidizing conditions," J. Electrochem. Soc. 121, 705 (1974).

4. G.D. Hachtel, R.C. Joy and J.W. Cooley, "A new efficient one-dimensional analysis program for junction device modeling," Proceedings of the I.E.E.E. $\underline{60}$, 86 (1972).
5. W. Shockley, Electron and holes in semiconductors, D. Van Nostrand Company, Inc., Princeton, N.J., 1950.
6. H.C. Pao and C.T. Sah, "Effects of diffusion current on characteristics of metal-oxide (insulators)-semiconductor transistors," Solid State Electronics $\underline{9}$, 927 (1966).
7. D.M. Caughey and R.E. Thomas, "Carrier mobilities in silicon empirically related to doping and field," Proceedings of the I.E.E.E. $\underline{55}$, 2192 (1967).
8. S. Wagner, "Diffusion of boron from shallow ion implants in silicon," J. Electrochem. Soc. $\underline{119}$, 1570 (1972).
9. V.G.K. Reddi, "Majority carrier surface mobilities in thermally oxidized silicon," I.E.E.E. Trans. Electron Devices, ED-$\underline{15}$, 151 (1968).
10. The Engineering Staff of American Micro-systems, Inc., MOS Integrated Circuits, ed. by W.M. Penney and L. Lan, Van Nostrand Reinhold Comp., New York, 1972.
11. W.H. Schroen, "Materials quality and process control in integrated circuits manufacture,", in Festkoerperprobleme XVII, Friedrich Vieweg Verlag, Wiesbaden, Germany, 1977.

* NOTE: MATERIAL EXCERPTED FROM S. A. EVANS, I. H. MORGAN, AND T. N. JERNIGAN, "A model for the accurate prediction of the current-voltage behavior of ion-implanted MOS transistors at low drain voltages," IEEE IEDM Meeting, Washington, D.C., 1974, p. 47-49.
© 1974 by THE INSTITUTE OF ELECTRICAL AND ELECTRONICS ENGINEERS, INC.
REPRODUCTION AUTHORIZED BY IEEE.

SHORT CHANNELS - SCALED DOWN MOSFET's

G. Merckel

Centre d'Etudes Nucléaires de Grenoble -LETI-MEA-
85X 38041 Grenoble Cedex, France

ABSTRACT. The main electrical and technological constraints related to reduction of geometrical parameters of the MOSFET are presented. The modifications of the electrical parameters are analyzed.

1. WHY S MOST[*]s?

With the present state of technology, it is possible to fabricate MOS transistors of very small geometries, and this evolution permits to improve the performances of integrated circuits [1-2]. The main advantages of scaling down the device geometries are the following :
. high density of integration (Very Large Scale Integration)
. cost reduction (more transistors on the chip)
. low transit time in the channel
. lower voltage and power supply
. lower delay time . power product.
The advantages are very attractive, but special attention has to be given to specific constraints which are associated with such devices.

---

[*] Scaled down (or Small) MOST.

2. INTRODUCING S MOST's

One can assume that classical transistors, have a channel length, L, between 6 to 10μm, and a channel width, Z, of a few 10μm.
If one chooses a reduction factor of 5 to 10, one obtains typical values for S MOST : 1 to 2μm for the channel length, and less than 10μm for the channel width.
To ensure good functionning of the transistor and a reasonably low threshold voltage for a small channel length, the other parameters of the device have to be modified. For instance, one has to compensate the relative increase of the space charge layer width (compared with the channel length) to avoid punch through. The space charge layer width W is dependent on the substrate doping level through :

$$W \sim N_B^{-1/2}$$

This leads to an increase of N_B by a factor 20 to 100 (compared with classical structures) and so, doping levels of the order of 10^{16} to 10^{17} at/cm^3 have to be used. As a consequence of the increase of N_B, the threshold voltage V_T is increased because of the body factor K ($\sim N_B^{1/2}/C_{ox}$). To compensate the increase of K for keeping a low threshold voltage, one has to decrease the gate oxide thickness (200 to 300 Å). In these conditions, the available current in S MOS devices is of the same order as in classical ones. Depending on these considerations, the S MOST structure is shown in Fig. 1.

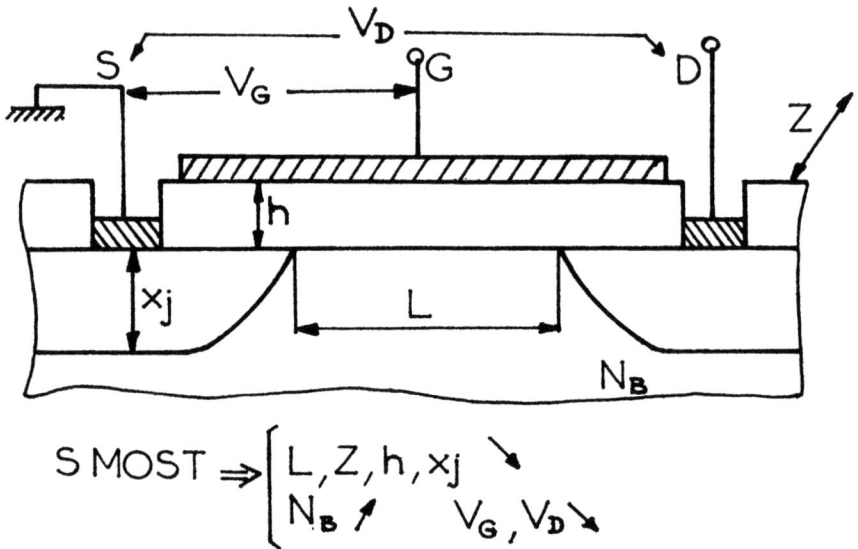

Fig. 1. S MOST structure.

The typical parameters are :
- channel length $2 \geqslant L(\mu m) \geqslant 1$
- thin oxide thickness $400 \geqslant h(\text{Å}) \geqslant 200$
- substrate doping $10^{17} \geqslant N_B \text{ (at/cm}^3) \geqslant 10^{16}$
- junction depth $x_j < 1\mu m$
- gate and drain voltages V_G and $V_D \leqslant 5V$.

Since the performance of the transistor is improved by scaling down, one would like to see to what extent to do it, and what are the associated constraints. The limiting factors are of two orders :
- the factors limiting the useful working range of the device
- the factors influencing the performances of the transistor in the useful working range.

3. FACTORS LIMITING THE USEFUL WORKING RANGE

The main factors are the following [3-4] : 1) Thin oxide break down ; 2) Junction avalanche break down ; 3) Punch-through ; 4) Field turn on voltage ; and 5) Electro-migration in small metallizations.

3.1 Thin oxide break down

The dielectric strength of SiO_2 is of the order of 700 V/μm, which limits the destructive gate break down voltage to about 7 volts/100 Å of gate oxide thickness. Thus, gate oxide thicknesses of the order of 200 Å are compatible with moderate power supply voltages. However, care has to be taken that the technological steps (in particuliar, the high temperature treatments) do not affect the Si - SiO_2 interface (migration of the ions of the gate metal across the thin oxide layer).

3.2 Junction avalanche break down

The break down voltage V_{BR} of a diode, depends on the substrate doping level N_B (first order) and on the junction radius of curvature R_o :

$$V_{BR} \sim N_B^{-1}$$

However, since the drain juction is located under the gate, the normal electric field induced by the gate voltage V_G, has a direct effect on the break down voltage [5-6].
V_{BR} corresponds to a lateral electric field, E_{BR}, of the order of

60V/µm [6]. Fig. 2 shows the variation of V_{BR} as a function of doping for different gate oxide thicknesses h and for $V_G - V_{FB} = V_{GF} = 0$ (flat band condition). The relation between V_{BR} and the other parameters is given by [6] :

$$E_{BR} = Xo^{-1/2} \left\{ \left[V_{BR} - V_{GF} + XoB \right] \coth{(W_G Xo^{-1/2})} - XoB \cdot \left[\sinh(W_G Xo^{-1/2}) \right]^{-1} \right\} \quad (1)$$

with E_{BR} the lateral critical field ($\simeq$ 60 V/µm)

$Xo = \dfrac{\varepsilon_{Si}}{\varepsilon_{ox}} h\, k'$ h gate oxide thickness

                 $k' \simeq 2\mu m$

$B = q\, N_B / \varepsilon_{Si}$

$$W_G = \left(\dfrac{2V\emptyset}{B}\right)^{1/2} \left[(1 + V_{BR}/V\emptyset)^{1/2} - (1 + V_{GF}/V\emptyset)^{1/2} \right]$$

V∅ junction potential barrier height.
It is to be noticed that for :

$N_B \ll 5 \cdot 10^{15} h^{-1/2}$ (with N_B in at/cm^3 and h in 10^3 Å)

expression (1) reduces to

$V_{BR} \simeq V_{GF} + 47 h^{1/2}$ (with V_{BR}, V_{GF} in Volts and h in 10^3 Å)

3.3 Punch-through

Punch-through occurs when the drain space charge layer width, W, reaches the space charge layer of the source.
W is related to punch-through voltage, V_{PT}, by :

$$V_{PT} = \dfrac{B}{2} \left[(Ro + W)^2 \left(\log{\dfrac{Ro + W}{Ro}} - \dfrac{1}{2} \right) + \dfrac{Ro^2}{2} \right] - V\emptyset \quad (2)$$

with $W \simeq L - (2V\emptyset/B)^{1/2}$

Ro is the radius of curvature of the junction and is assumed to be Ro $\simeq$ 0.8 xj near the Si - SiO$_2$ interface (xj = junction depth). Results for channel lengths of 1µm and 2µm are shown in Fig. 3. To increase V_{PT}, one has to increase N_B and decrease xj.

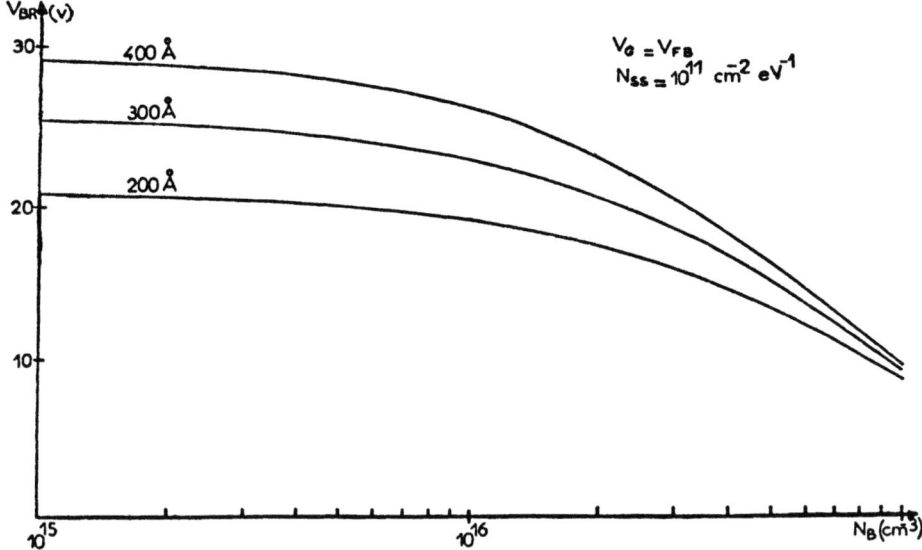

Fig. 2. Drain break-down voltage V_{BR} versus doping N_B.

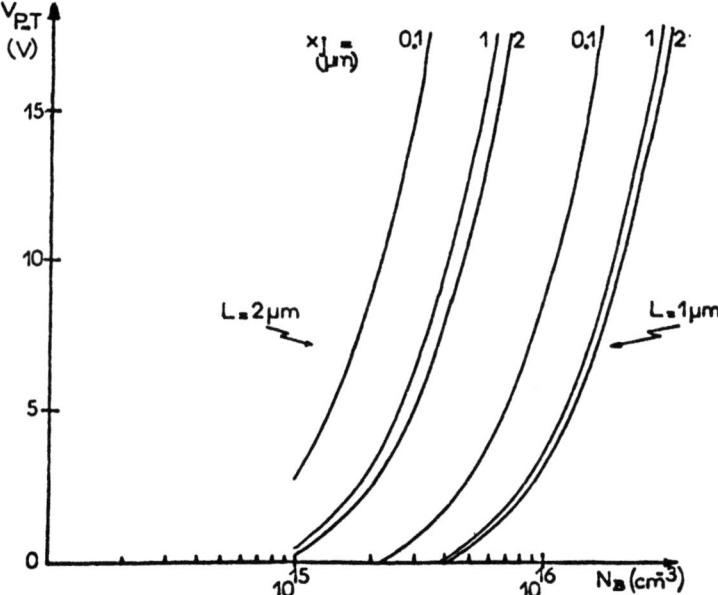

Fig. 3. Punch-through voltage V_{PT} versus doping N_B.

3.4 Field turn on voltage

For reproducibly defining small channel lengths ($\sim$ 1 to 2μm), it is necessary to have a thin layer of gate metal (to avoid deviations caused by lateral etching). Therefore, the maximum oxide step, between gate oxide and field oxide which can be traversed by the metal layer reliably, is limited. This limits the thickness of the field oxide which can be used.

Fig. 4 shows the values of required field oxide thickness versus N_B and field turn on voltage. The field turn on voltage V_{TF} is given by (N channel)

$$V_{TF} = V_{FB} + 2\phi_F + K(2\phi_F)^{1/2}$$

V_{FB} and K depend on the field oxide thickness through C_{Fox} (field oxide capacitance/unit area):

$$V_{FB} = \phi_{MS} - q N_{SS}/C_{Fox} \quad \text{and} \quad K = (2\epsilon_{Si} q N_B)^{1/2}/C_{Fox}$$

ϕ_{MS} is the work function difference.

3.5 The last parameter which can limit the working range of the device is the degradation of the gate metal due to electro migration. For avoiding this phenomenon, the cross-section area of

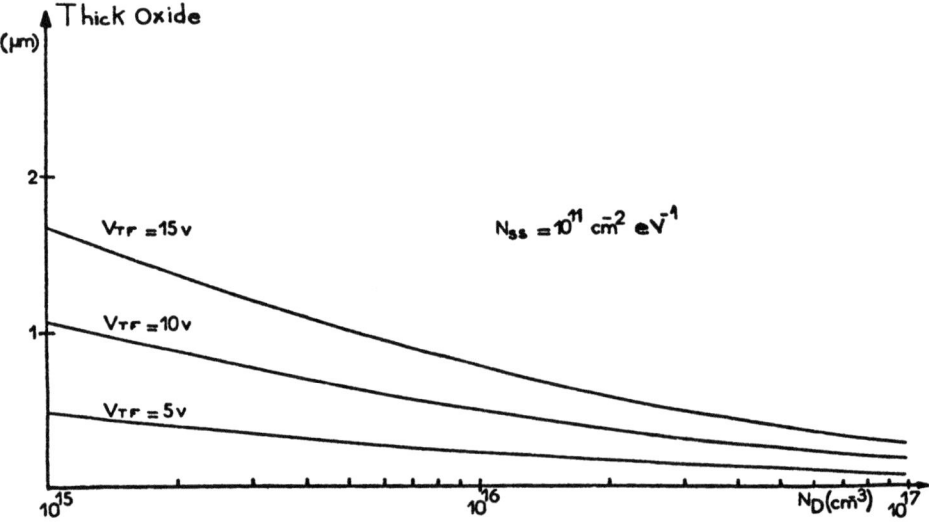

Fig. 4. Limits on field turn on voltage V_{TF}.

the metal and its nature has to be choosen according to the current density Jo in the metal [7-8]. For Aluminium, this fixes the limit at

$$J_o \leqslant 1 \text{ma } (\mu m)^{-2}$$

In conclusion to section 3, and taking into account the various limiting factors, one can say that for channel lengths of 1 to 2μm, gate oxide thicknesses of 200 to 300 Å and if one keeps a margin of 5V (over the supply voltage of 5V) for the voltages, the substrate doping levels are :

$$N_B \simeq 10^{16} \text{at/cm}^3 \text{ (for a P channel, Al gate device)}$$
$$\simeq 10^{17} \text{at/cm}^3 \text{ (for an N channel, Al gate device)}$$

These values indicate the orders of magnitude and may be reduced according to the technology used (Al gate or Si gate), the threshold voltage expected (and its fluctuations), the quality of Si - SiO$_2$ interface, etc...

4. FACTORS INFLUENCING THE PERFORMANCES OF THE TRANSISTOR WITHIN ITS USEFUL WORKING RANGE

For defining these factors, one should take the static as well as the dynamic behavior of the transistor into account.

4.1 Dynamic aspect

The main parameters to be minimized are : the drain diffusion parasitic capacitance C_p and the capacitance of the gate and its interconnections which constitute the capacitive load of an invertor (Fig. 5). Scaling down the geometry of the transistor by a factor F_r (with $F_r > 1$), leads to a reduction of C_G by the same factor F_r, since :
  . the area is reduced by a factor F_r^2
  . capacitance per unit area of the gate oxide is increased by a factor F_r.

The capacitance C_p is related to the area of the drain S_D (which is divided F_r^2) and the doping level N_B (increased by $\simeq F_r^2$) by the relation :

$$C_p \sim S_D N_B^{1/2}$$

C_P will be smaller for lower levels of doping. Fig. 6 shows the variation of the drain capacitance ($V_B = 0$) C_{TO} per unit area as a function of N_B.
  C_p is given by $C_P \simeq C_{TO} S_D$

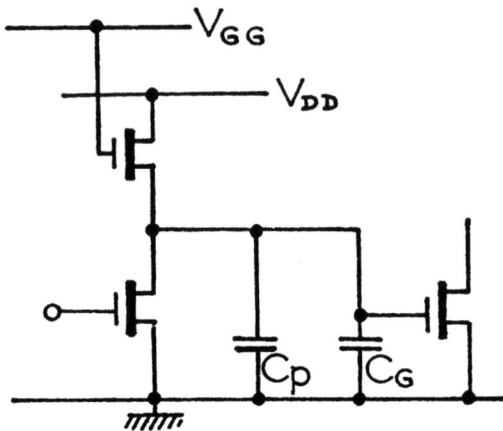

Fig. 5. Parasitic load capacitances.

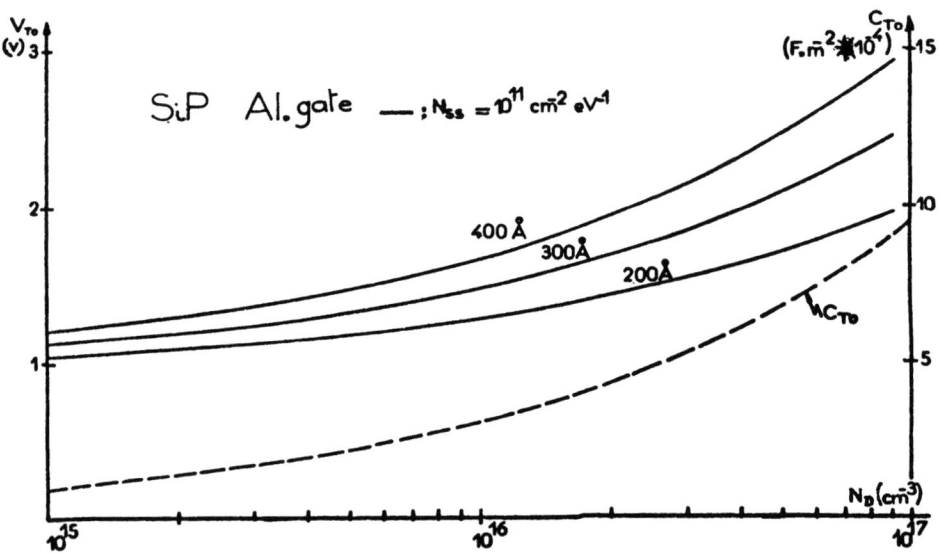

Fig. 6. Drain junction capacitance C_{TO} and threshold voltage V_{TO} versus N_B.

It is also to be noted that the transit time in the channel ($\tau_t \sim \frac{L^2}{\mu}$) is reduced by a factor between F_r and F_r^2. (L is reduced, but also the mobility since N_B is increased).

4.2 Static aspect - Comparison of parameters with respect to classical structures

The analysis of the variation of the static parameters is done using simplified current voltage relationships [9-10]. The characteristics are derived from classical expressions [11-12] by an expansion to second order [10] and are given by : (all voltages referred to the source).

. Before saturation ($V_D \leqslant V_{DSS}$)

$$I_D = \frac{\mu_o}{(1+\Theta_G V_{GTO})(1+\Theta_D V_D)} C_{ox} \frac{Z}{L} \left[V_{GT} V_D - \frac{V_D^2}{2}(1+\delta) \right] \quad (3)$$

with $V_{GTO} = V_G - V_{TO}$ and $V_{GT} = V_G - V_T$

V_T = threshold voltage, V_{TO} = threshold voltage for zero bulk-source voltage.

. In saturation region ($V_D \geqslant V_{DSS}$)

$$I_D = I_{DSS} \left[1 + \frac{V_D - V_{DSS}}{V_E + V_D} \right] \quad (4)$$

. At pinch off point ($V_D = V_{DSS}$)

Putting $V_{ED} = V_E (1 + \Theta_D V_E)^{-1}$ and $u_{ED} = V_{GT} \left[V_{ED}(1+\delta) \right]^{-1}$

one obtains

$$V_{DSS} \simeq V_{ED} \left[(1 + 2 u_{ED})^{1/2} - 1 \right] \quad (5)$$

and I_{DSS} is given by combining (5) and (3).

The expression (4) describes in a simple manner the behaviour of the transistor in saturation region and assures the continuity of current. The output impedance has been assumed to be constant for $V_D \geqslant V_{DSS}$ (a limitation for some analogic applications), and the number of carriers in transit across the drain space charge layer has been neglected compared to the doping level N_B.

The last assumption is valid, considering the high values of N_B for S MOST structures. All the parameters of the model are affected by the reduction of geometrical dimensions and, at the same time, by high values of N_B. We shall successively analyse the threshold voltage V_T ; the zero field mobility μ_o ; the factors governing the dependence of mobility on the applied electric field : Θ_G (for normal field) and Θ_D (for lateral field) ; the corrective term δ and (by analogy with the bipolar transistor) the Early voltage V_E.

4.2.1 Threshold voltage
a) Effect of the doping N_B, and of the gate oxide thickness, h. Neglecting the geometrical effects (other than that of h), the threshold voltage is given by

$$V_T = V_{FB} + 2\phi_F + K(V_B + 2\phi_F)^{1/2}$$

The body factor K ($\sim N_B^{1/2}/C_{ox}$) should be small to get reasonable threshold voltages. This is assured if the increase in N_B (factor F_r^2) is (at least) compensated by an increase in C_{ox} (by a factor F_r). Under these conditions, the value of K is not modified with respect to classical structures.
Fig. 6 shows the variation of V_{TO} (V_T for $V_B = 0$) as a function of doping level (P channel). For an S MOS structure, the threshold value is around -1.4 V. (Al·gate).

b) Effect of channel length L
Reduction of channel length L causes a reduction in the contribution of the substrate to the threshold voltage [14, 15].
The total charge in the substrate in the channel region, Q_{BT}, consists of two components (see Fig. 7).
. the charge Q_{BG} controlled by the gate
. the charge Q_{BJ} controlled by the junctions.
Referring to Fig. 7, the ratio Q_{BG}/Q_{BT} is given by

$$\frac{Q_{BG}}{Q_{BT}} = \frac{1}{2}\left(1 + \frac{L'}{L}\right) \qquad (6)$$

For very long channels, $L' \approx L$ and $Q_{BG} \simeq Q_{BT}$.
For short channels and for $L'/L \ll 1$, the relation (6) gives

$$Q_{BG} \simeq \frac{Q_{BT}}{2}$$

This shows that the body effect is reduced by a factor of 2. Taking this geometrical effect into account, the threshold voltage is given by [15] :

$$V_T = V_{FB} + 2\phi_F + K(V_B+2\phi_F)^{1/2}\left\{1-\frac{x_j}{L}\left[\left(1+\frac{2W}{x_j}\right)^{1/2} - 1\right]\right\} \qquad (7)$$

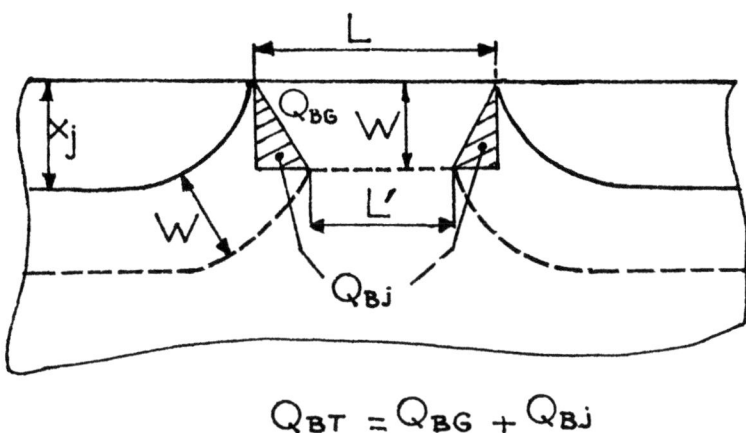

$$Q_{BT} = Q_{BG} + Q_{Bj}$$

Fig. 7. Effect of channel length L on substrate charge.

with $W \simeq \left(\dfrac{2\varepsilon_{Si}}{q\, N_B}\right)^{1/2} (V_B + 2\phi_F)^{1/2}$ \hspace{1em} (7a)

and $K = (2\varepsilon_{Si}\, q\, N_B)^{1/2}/C_{ox}$

For practical needs of computer Aided Design, the expression (7) can be approximated by an expansion to first order of the form [10] :

$$V_T \simeq V_{FB} + 2\phi_F + K\,(V_B+2\phi_F)^{1/2}\left[1+\Theta_B\,(V_B+2\phi_F)^{1/2}\right] \quad (8)$$

with :
$\Theta_B = \dfrac{1}{L}\left(\dfrac{2\varepsilon_{Si}}{q\, N_B}\right)^{1/2}$

As an example, Fig. 8 shows a comparison between experiment (circles), "exact" theory (plain line, eqn.(7)), and the approximate theory (dotted, eqn.(8)).

For example, for $N_B \simeq 10^{16}$ at/cm³, h ≃ 300 Å and V_B = 5V, one obtains a reduction of threshold voltage $\Delta V_T \leqslant -0.2V$ for $L \leqslant 6\mu m$.

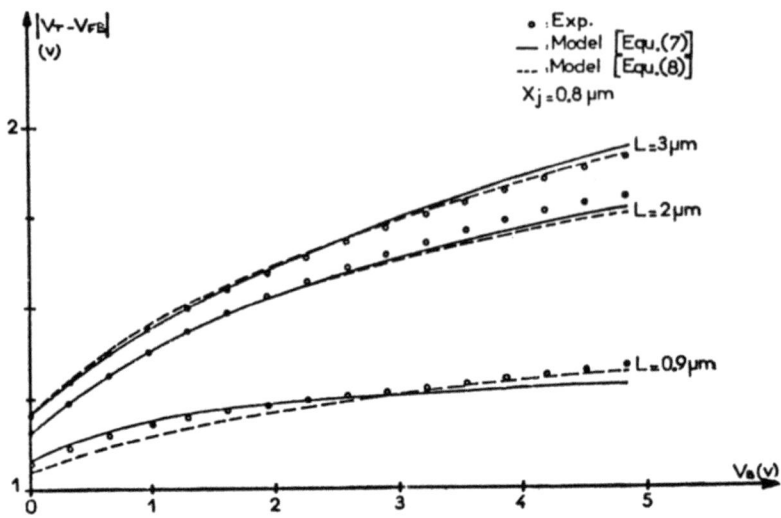

Fig. 8. Threshold voltage versus source-substrat voltage. Comparison between theory and exp.
P channel, $N_B = 10^{17} \text{at/cm}^3$, h = 200 Å.

c) Effect of channel width Z

It has been established recently that a reduction in Z contributes to an increase in the threshold voltage [16-17].
This is due to the fact that the space charge region spreads laterally in the substrate along the channel-width. Fig. 9 illustrates this effect which can be analysed in a simple way [18]. Assuming that the lateral extension of the space charge region (of a thickness W) is cylindrical, the total charge in the substrate is given by

$$Q_{BT} = q N_B Z L W \left(1 + \frac{\Pi}{2} \frac{W}{Z}\right) \quad (9)$$

This relation shows that the contribution of Z in the body effect is increased by a factor $1 + \Pi/2 \, W/Z$ (10)

Thus, one obtains for the threshold voltage,

$$V_T = V_{FB} + 2\phi_F + K (V_B + 2\phi_F)^{1/2} \left(1 + \frac{\Pi}{2} \frac{W}{Z}\right) \quad (11)$$

with W given by eqn. (7a).
(The effect of L has been ignored in this case).
For example, for $N_B \simeq 10^{16} \text{at/cm}^3$, h ≃ 300 Å and $V_B = 5V$, one obtains an increase in the threshold voltage $\Delta V_T \geqslant 0.2V$ for

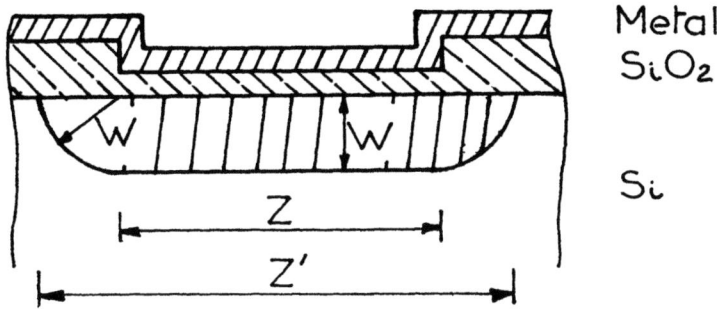

Fig. 9. Effect of channel width Z on substrate charge.

$Z \leqslant 8\mu m$.

4.2.2 Carrier mobility μ

In the expression for current (eqn.(3)) the carrier mobility in the inversion layer is given by [9] [19]

$$\mu = \frac{\mu_o}{(1 + \Theta_G V_{GTo})(1 + \Theta_D V_D)} \quad (12)$$

μ_o represents the mobility under zero field.

Θ_G the coefficient for normal field dependence.

Θ_D the coefficient for longitudinal dependence.

These three factors are appreciably modified for S MOS structures.

a) Carrier mobility under zero field (μ_o)
Based on the numerous publications on the experimental values of μ_o, one can write an empirical formula :

$$\mu_o \simeq \frac{\mu_B}{2} \quad (13)$$

with μ_B : minority carrier mobility in the substrate.

In the case of S MOST, the substrate doping being higher than for a classical transistor, μ_o will be lower. For example, Fig. 10 shows the results obtained for P channel transistors [10]. The theoretical values are calculated according to (13), μ_B being given by the ref. [20].

$$\mu_B = \frac{\mu_{max} - \mu_{min}}{1 + \left(\frac{N_B}{Nref}\right)^\alpha} + \mu_{min}$$

b) The coefficient Θ_G for dependence of mobility on normal field. This parameter is a function of the gate oxide thickness, h. One can note that Θ_G is correlated to the mobility through temperature [21] [22]. Θ_G depends on the oxide thickness by a relation of the form

$$\Theta_G = a_1 C_{ox}$$

with C_{ox} : gate capacitance per unit area. The parameter a_1 is a constant and for P channel devices, $a_1 \simeq 10^2 m^2 C^{-1}$
for N channel devices, $10^3 \gtrsim a_1 (m^2 C^{-1}) \gtrsim 10^2$

Θ_G is higher by a factor between 3 to 6 for S MOS structures because of a thinner gate oxide.

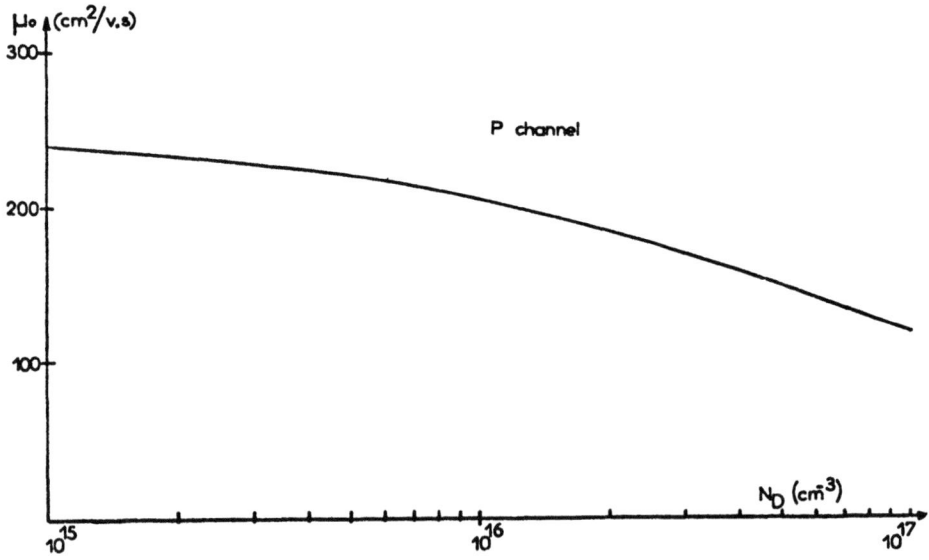

Fig. 10. Mobility μ_o versus doping N_B.

c) The coefficient Θ_D for dependence of mobility on the longitudinal field. This parameter is a function of the channel length [19] [22] and is given by the relation

$$\Theta_D \simeq (LE_o)^{-1} \qquad (14)$$

with E_o : longitudinal critical field.
It is observed experimentally that

for P channel, $E_o \geqslant 10V/\mu m$
for N channel, $E_o \leqslant 3V/\mu m$

As a result, Θ_D is often negligible for P channel devices, but should be taken into account for N channel devices.

Note : assuming $\Theta_D \simeq 0$, the expression (5) giving the saturation voltage V_{DSS} becomes :

$$V_{DSS} \simeq V_E \left[(1 + 2u_E)^{1/2} - 1 \right]$$

(if $\Theta_D \to o$, $V_{ED} \to V_E$ and $u_{ED} \to u_E = V_{GT} \left[V_E (1+\delta) \right]^{-1}$

4.2.3 Corrective term δ

The value of this term is given approximately be the relation [9-10]

$$\delta \simeq 0.25 \; K \; (2\phi_F)^{-1/2}$$

The body factor K being constant, (see 4.2.a) the parameter δ is also nearly constant.

4.2.4 Early voltage V_E

Matching the mean value of current given by (4) with the simplified theory [23] permits one to write

$$V_E \simeq a_2 L \; (N_B)^{1/2} \qquad (15)$$

a_2 is a constant, evaluated experimentally.
The results obtained for the parameter V_E are shown in Fig. 11 for $L \geqslant 1\mu m$ and $10^{15} \leqslant N_B \; (at/cm^3) \leqslant 10^{17}$.
The circles represent the measured values and the plain lines represent the mean value of V_E according to eqn.(15). From these curves, one may deduce that

$$a_2 \simeq 1.6 \times 10^7 \quad \text{(with } V_E \text{ in volts, L in } \mu m \text{ and } N_B \text{ in at/cm}^3)$$

In conclusion, one can say that all the parameters of the transistor are affected by a reduction in geometrical dimensions. In fact, for the S MOS structures, the substrate doping has to be optimized.

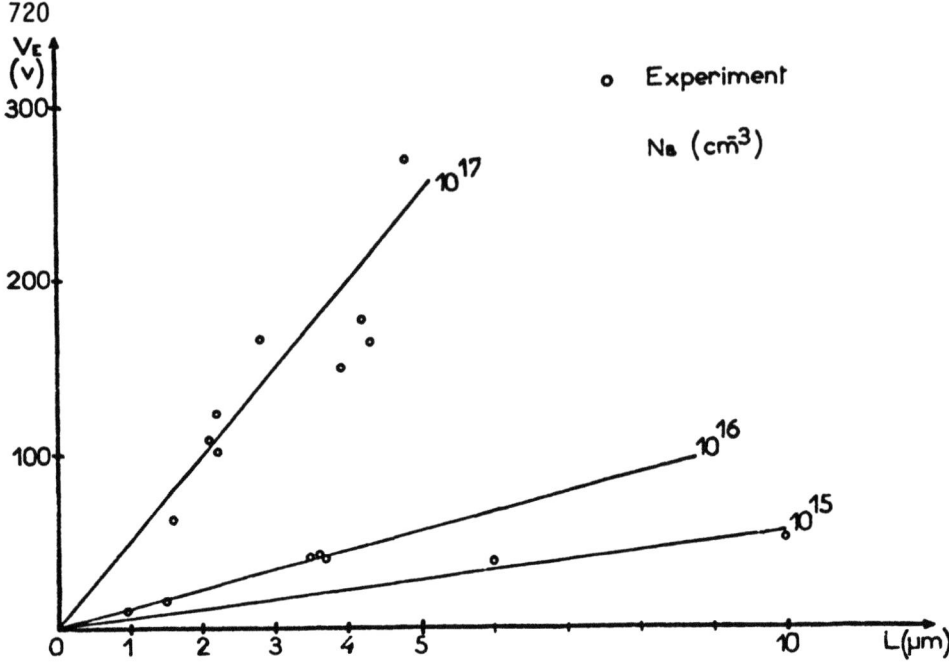

Fig. 11. Saturation parameter V_E versus channel length L.

N_B should be as low as possible, compatible with reasonable values for punch through voltage and field oxide turn on voltage. A technological solution, which permits one to take care of the above mentioned criteria, is to start with a relatively high substrate resistivity ($N_B \simeq 10^{15}$ at/cm^3) and to dope the silicon surface subsequently.
This permits one to adjust the field turn on voltage, (keeping the body effect to a minimum) while guaranteeing a reasonable punch through voltage. Such device is shown in Fig. 12.

5. OTHER TYPES OF SHORT CHANNEL DEVICES

In the previous sections we have reviewed the principal constraints and their consequences on reducing all the geometrical dimensions of a conventional MOS transistor. However, modifying the structure of the transistor itself, it is possible to realize devices of small channel lengths. This is the case for DMOS and VMOS transistors.

5.1 Double-diffused MOS transistor

In contrast with a conventional transistor, the D MOST necessitates a double diffusion near the source (Fig. 13).

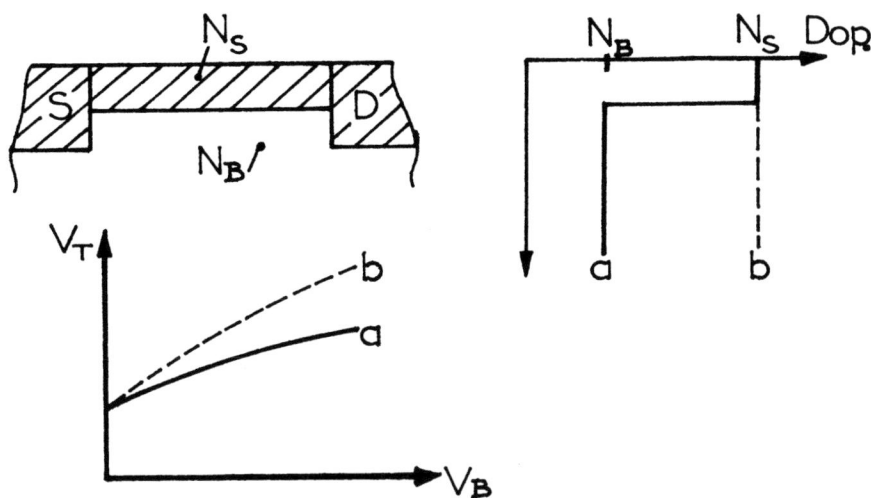

Fig. 12. S MOST - Cross section - Threshold voltage sensitivity.

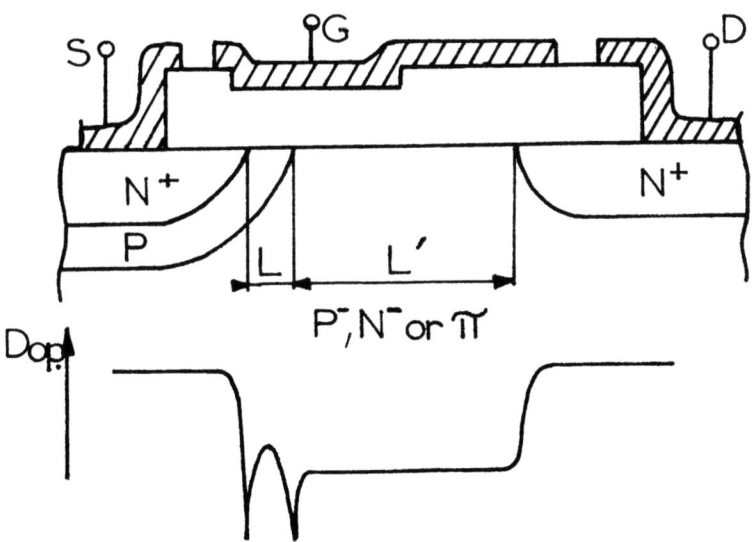

Fig. 13. D MOS - Cross section - Doping profile.

The P zone, which defines the channel length is diffused in a non oxidizing atmosphere and the following technological operations are conventional. One thus obtains in a simple way, a small channel length (L ⩽ 1µm). This does not require any sophisticated photo masking techniques. The D MOS has a better high frequency response because :

- The transit time of carriers in the channel is small ($\tau_t \sim L^2$).
- The drain parasitic capacitance is small (substrate Π).
- The gate-drain overlap capacitance is small (Miller effect small).

In addition to this, the drain diode breakdown voltage is high because :
- The substrate is lightly doped.
- The oxide thickness between the gate metal and the drain is high.

As far as its electrical characteristics are concerned, the transistor may be represented by the juxtaposition of two N channel transistors in series : a short channel enhancement transistor, T_E, and a depletion mode transistor, T_D, with a longer channel [26] (Fig. 14). Like the conventional transistor, the charge of the carriers in the inversion layer is controlled by the gate voltage and their velocity by the drain voltage. When the transistor is in saturation, due to the short channel, the

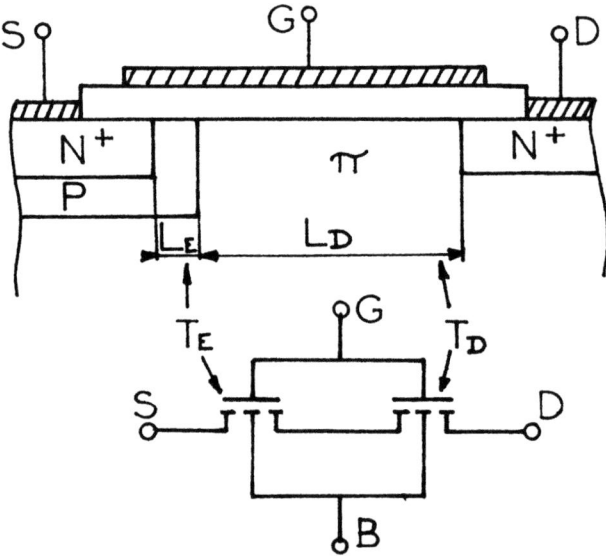

Fig. 14. Equivalent schem of D MOS.

electric field in the pinch off region becomes very high and the carrier velocity is very near the limiting velocity v_L. Under these conditions, the maximum transconductance g_m is given by [25] :

$$g_m = C_{ox} Z v_L$$

from where $\dfrac{g_m}{Z} \simeq C_{ox} v_L$ which is $\simeq 24\mu\text{mhos}/\mu\text{m}$

for $v_L = 6.5 \times 10^4 \text{m/s}$

and $h = 1000 \text{ Å}$

The transit time under limiting velocity is given by

$$\tau_t = \frac{L_T}{v_L}$$

where $L_T = L + L'$, the total length.

for example, for $L_T = 6\mu\text{m}$, $\tau_t \simeq 0.1\text{ns}$

However, it is to be noted that one should add the distributed time constant, channel resistance x gate-channel capacitance, to the transit time.

5.2 V groove MOS transistor [27][28]

A section of the V MOS is shown in Fig. 15. This device needs a V groove etch and because of its vertical structure, permits to obtain a very high density of integration and short channel lengths. The starting material is <100> and the sloping planes are <111>. So, the surface charge density, Q_{SS}, is about ten times higher for V MOS, compared with D MOS (or Scaled down MOS) on the same chip.

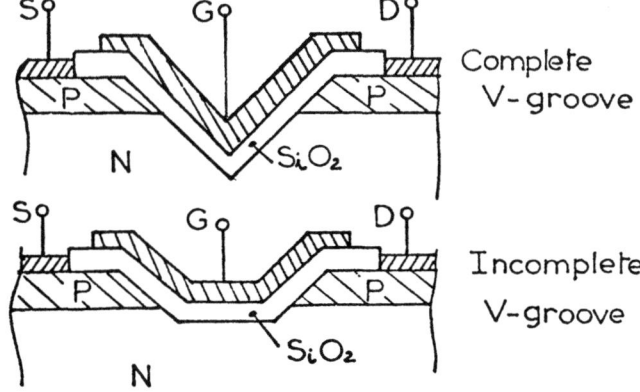

Fig. 15. V MOS - Cross section.

REFERENCES

[1] V.L. Rideout, F.H. Gaenslen, A. Le Blanc, IBM J. RES. Develop. January 1975.
[2] J. Borel, ESSCIRC, Toulouse, 21-24 Sept. 1976.
[3-4] B. Hoeneisen, C.A. Mead
   . Sol. St. El., vol 15, 1972.
   . IEEE Trans. on E.D.
[5] A.S. Grove, Ed. J. Wiley and Sons, New York.
[6] H.C. De Graaff, Philips Res. Repts 25, 1970.
[7] I.A. Blech, E.S. Meieran, Appl. Phys. Lett.,11, 1967.
[8] P. Mortini, Private communication.
[9] G. Merckel, J. Borel, N. Cupcea, IEEE Trans. on E.D., ED-19, 5, May 1972.
[10] G. Merckel, E. Lora-Tamayo, J. Gautier, To be published.
[11] H.K. Ihantola, J.L. Moll, Sol. St. El. vol. 7, 1964.
[12] J.A. Van Nielen, O.W. Memelink, Philips Res. Repts 22, 1967.
[13] A. Popa, IEEE Trans. on E.D., ED-19, 6, June 1972.
[14] G. Merckel, J. Borel, Journées d'études sur les Modèles de composants, Toulouse, March 1970.
[15] H.C. Poon, L.D. Yau, R.L. Johnston, D. Beecham, IEDM Washington, December 1973.
[16] K.Kroell, C.A. Ackerman, Sol. St. El. vol.19, 1975.
[17] W.P. Noble, P.E. Cottrell, IEDM Washington, December 1976.
[18] G. MERCKEL, Intern. Rept.
[19] P. Rossel, H. Martinot, G. Vassilieff, Sol. St. El. vol.19, 1976.
[20] R.F. Thomas, D.M. Caughey, Prog. IEEE, 65, December 1967.
[21] G. Vassilieff, Thesis, Toulouse, 1971.
[22] P. Rossel, Thesis, Toulouse, 1973.
[23] V.G. Reddi, C.T. Sah, IEEE Trans. on E.D., March 1965.
[24] V.L. Rideout, F.H. Gaensslen, A. Le Blanc, IBM I. Res. Develop. January 1975.
[25] T.P. Cauge, J. Kocsis, H.J. Sigg, G.D. Vendelin, Electronics, February 15, 1971.
[26] . T.J. Rodgers, S. Asai, M.D. Pocha, R.W. Dutton, J.D. Meindl ISSCC, Philadelphia, February 1975.
   . M.D. Pocha, R.W. Dutton, IEEE J. of Sol. St. Circts., vol.11, October 1976.
[27] T.J. Rodgers, J.D. Meindl, IEEE J. of Sol. St. Cirts. vol.9, October 1974.
[28] S.R. Combs, D.C. D'Avanzo, R.W. Dutton, IEDM Washingtion, December 1976.

SOS MOSFET's

G. Merckel

Centre d'Etudes Nucléaires de GRENOBLE - LETI/MEA -
85X 38041 Grenoble Cedex, France

ABSTRACT. Silicon on sapphire MOSFET's are very attractive in the fields of low power and high speed integrated circuits. After a brief recall of the technology, the behavior of the device, its characterization and the performance improvements are described.

1. TECHNOLOGY

To realize an N channel transistor (for ex.), the starting material is an epitaxial P type silicon film on insulating substrate, especially spinel or sapphire. The film thickness is relatively small ($\sim 0.8\mu m$) and the crystal is of <100> orientation.
The main technological steps applied to SOS MOSFET's are given in Fig. 1 and are the following : 1) Etching of Silicon ; 2) Silox deposition ; 3) Etching of SiO_2 ; 4) Phosphorus predeposition ; 5) Gate oxidation and phosphorus redistribution ; 6) Contact windows ; 7) Al. evaporation and etching.
The transistors are located on separate silicon islands and so, it is easy to achive complementary structures. These structures lead to a very low power consumption per gate. Compared to standard bulk devices, the parasitic capacitances of the drain diffusion and of the interconnections are drastically reduced. As a result, the delay power product is very low.
Fig. 2 shows the technological steps applied to complementary Silicon on sapphire MOSFET's. The doping levels can be achieved by ion implantation.

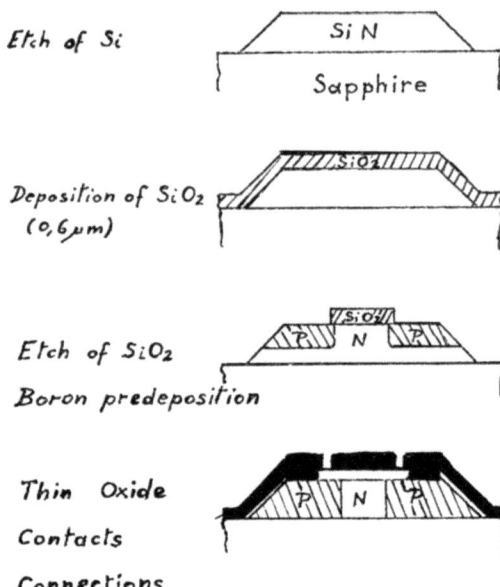

Fig. 1. Single device technology. Main steps.

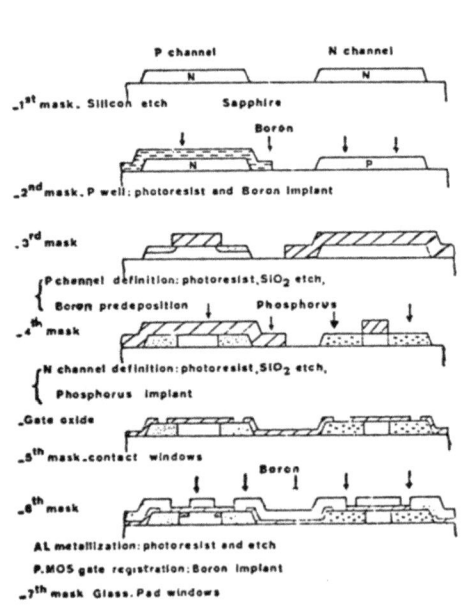

Fig. 2. CMOS/SOS Technology.

2. BEHAVIOR OF THE DEVICE

The main differences between SOS and bulk MOSFET's can be summarized in the following way :
- the silicon substrate is finite and floating (electrically)
- the leakage currents are increased due to the silicon-sapphire interface and lateral effects
- no need of field oxide
- the parasitic capacitances due to the interconnections are very low
- the junction capacitances are very low.

2.1 The floating substrate

In the normal mode of operation, the silicon substrate has a floating potential which depends on the substrate currents. Due to the small thickness of the film, the volume of silicon and therefore the charge below the gate is small. Typical $I_D(V_D)$ curves are shown in Fig. 3 and one observes that the characteristics are affected in a specific way and one can distinguish two kinks. Compared to bulk devices, the drain current I_D and the transconductance are higher ; the output conductance is higher in the current saturation zone.
The mechanisms leading to these effects are shown in Fig. 4.
The transistor is considered with a channel current $(V_G > V_T)$.

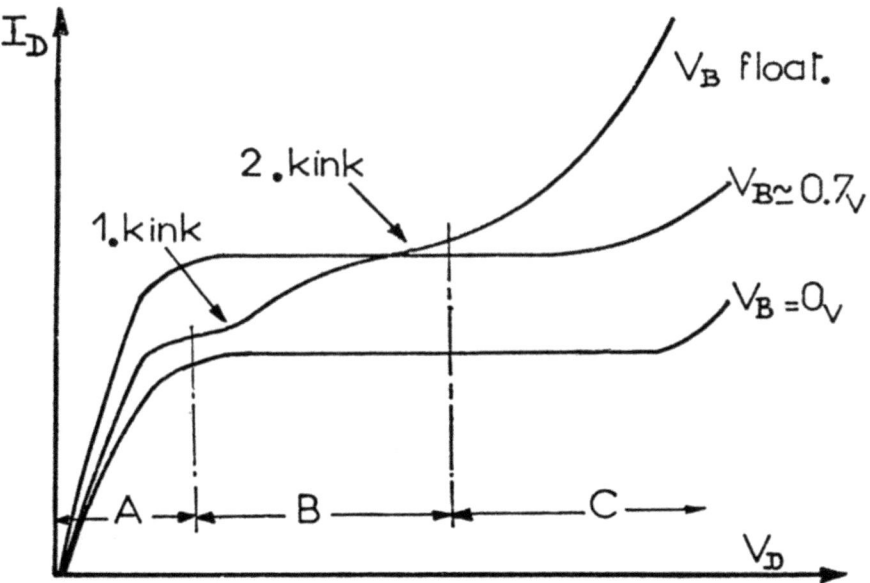

Fig. 3. $I_D(V_D)$ characteristics.

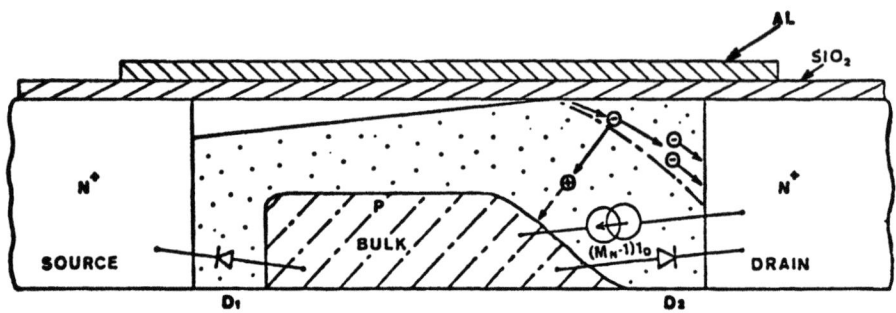

Fig. 4. Currents in the device.

For small drain to source voltages, V_D, the source-substrate diode is forward biased (voltage V_B) through the reverse current, I_R, of the reverse biased drain-substrate diode. If V_D increases, I_R and V_B increase and the threshold voltage, V_T, decreases following the well known relation :

$$V_T = V_{FB} + 2 \phi_F + K (- V_B + 2 \phi_F)^{1/2} \qquad (1)$$

with V_{FB} the flat band voltage, ϕ_F the Fermi level referred to the intrinsic level and K the body factor. This leads to a moderate increase of the drain current, at constant gate-source voltage V_G (zone A).

At higher values of V_D, the transistor is working in saturated regime and the high electric field in the drain region gives rise to carrier multiplication due to impact ionization [1-2]. The eletron flux flows into the drain while the holes flow into the bulk. At this stage, the multiplied electron current is small in comparison with the channel current. The hole flux flows out of the bulk into the source and as a result, the bulk-source diode becomes more forward biased, the threshold voltage is lowered, the drain current rises and an increase in the transistor output conductance is observed (zone B).

Finally, since the multiplication is still increasing as the drain voltage is increased (above the first kink), the hole current flowing into the substrate increases but the change in threshold

voltage is quite small (V_B nearly constant). The electron current increases and thus causes a further increase in the drain current (zone C).
The deviations of the current can be avoided by the application of a constant potential to the substrate (for ex. through a low ohmic contact between the source and the substrate).

2.2. Parasitic capacitances

If we consider the parasitic capacitances involved in the floating silicon substrate, we can distinguish 3 capacitors as shown in Fig. 5 :

- C_B is the capacitance of the depletion layer between the channel and the neutral substrate.
- C_S is the storage capacitance of the forward biased source-substrate diode.
- C_D is the capacitance of the depletion layer near the drain.

These capacitances are related to the channel length ,L, the channel width ,Z, the silicon doping level ,N_B, the life time ,τ, (a few ns near the Si-sapphire interface), the substrate current, I_B ,and the voltages :

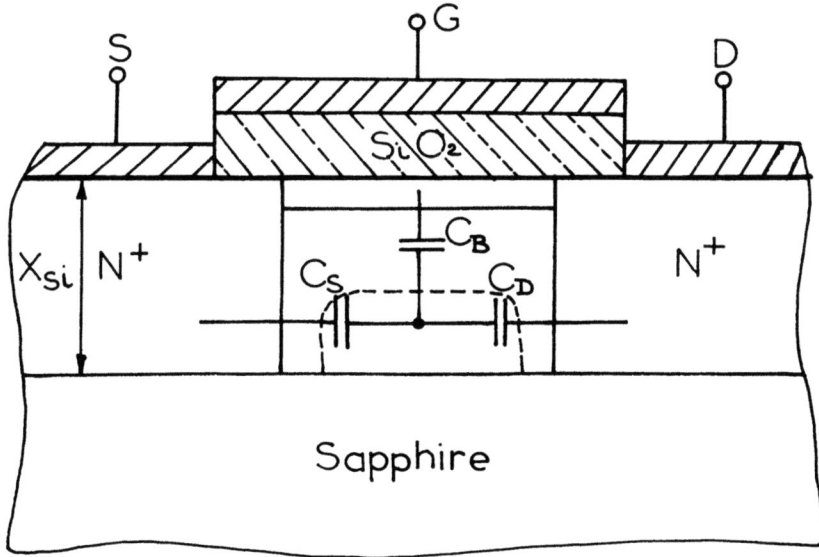

Fig. 5. Capacitances in the silicon film.

$$C_B \sim Z\, L \left[\frac{N_B}{V_\emptyset + V_D - V_B} \right]^{1/2}$$

$$C_S \sim \tau I_B$$

$$C_D \sim Z\, X_{Si} \left[\frac{N_B}{V_\emptyset + V_D - V_B} \right]^{1/2}$$

These capacitances are generally small, but they have to be considered for high frequency operation of the device. One can also notice that for supply voltages near transistor threshold, the propagation delay for digital CMOS integrated circuits, is a function of the operating frequency [3], due to the floating substrate and the associated capacitors.

3. DEVICE MODELLING - EXPERIMENTAL RESULTS

One assumes that the theoretical current voltage relations, except for the effects mentionned in section 2, are the same as those for bulk devices (see previous lectures).

3.1 Leakage currents

Treatment of Denis J.Mc GREIVY [4].
If one considers a cross section through the channel (parallel to channel width Z), the total drain to source current is the sum of currents through 4 regions, as shown in Fig. 6.
The contribution of each region is :

- $I_{L(I)}$ from the top Si <100> - SiO_2 surface (negligible)
- $I_{L(II)}$ from the top of volume space charge (and neutral) layer (negligible)
- $I_{L(III)}$ from the Si-sapphire surface (dominant)
- $I_{L(IV)}$ from the parasitic MOS transistor at the edge <111> surface (not always negligible).

The $I_D(V_G)$ characteristic, is represented schematically on the lower part of Fig. 6. The symbols g_{100} and g_{111} refer to conductances in an inversion layer at the top <100> surface and at the edge <111> surface. For both N and P channel devices, the contribution $I_{L(III)}$ is dominant and is proportionnal to So (surface

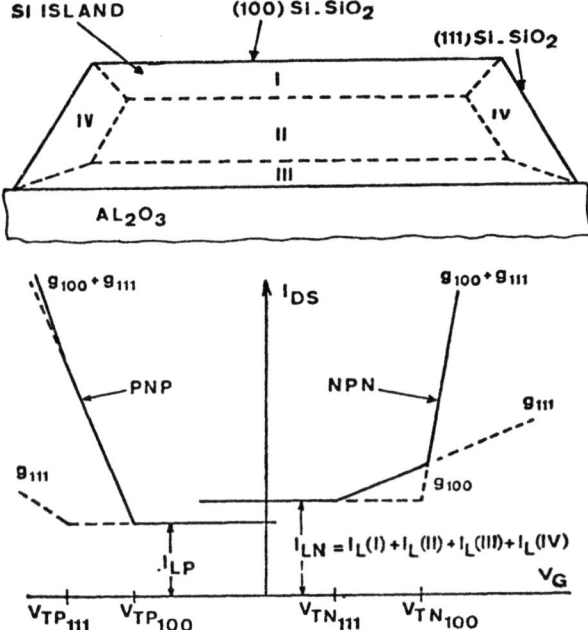

Fig. 6. Leakage current components. (After D.J.Mc GREIVY [4])

recombination velocity at the Si-sapphire interface :
So $\simeq 10^2$ to 10^3 m/s) and to τ (τ is the lifetime near the Si-sapphire interface : $\tau \simeq$ a few ns.)
$I_L(IV)$ can be neglected for P channel devices, but has to be taken into account in the case of short N channel devices.
In this case one can write :

$$I_{L(IV)} \simeq \frac{1}{2} \mu_o C_{ox} \frac{Z(IV)}{L} (V_G - V_{TN111})^2$$

$Z(IV)$ is the channel width of the edge parasitic transistor.
In conclusion, the effects of the dominant leakage current $I(III)$ can be represented by the forward biased source-substrate diode (current I_B) and the reverse biased drain-substrate diode (current I_r) :

$$I_B \simeq I_o \left[\exp(V_B/2v_T) - 1 \right] \qquad (2)$$

$$I_r \simeq I'_o \left[1 + \frac{V_D - V_B}{V_\phi} \right]^{1/2} \qquad (3)$$

V_ϕ is the barrier height of the abrupt (assumed) drain-substrate junction.

3.1 Carrier multiplication

Assuming a weak avalanche multiplication, the substrate current I_{SM} is given by :

$$I_{SM} \simeq (M_n - 1) I_D \qquad (4)$$

I_D is the channel current and M_n-1 the avalanche multiplication factor.
The factor M_n-1 is evaluated over the drain region (high field E) and can be expressed as [5] :

$$M_n - 1 \simeq \int_{E_P}^{E_D} \alpha_n(E) \frac{dy}{dE} dE \qquad (5)$$

E_D and E_P are the electric fields at the drain and at the channel pinch off point, and $\alpha_n(E)$ is given by [6-7] :

$$\alpha_n = A_n \exp(-B_n/E) \qquad (6)$$

A_n and B_n are the ionization coefficients.
Combining (4), (5) and (6), one obtains [8] :

$$\frac{I_{SM}}{I_D} \simeq A_n B_n \frac{2}{a} \left[\frac{E_i^2/B_n^2}{1 + 2 E_i/B_n} \exp(-B_n/E_i) \right]_{E_i = E_P}^{E_i = E_D} \qquad (7)$$

with $a = \dfrac{2qN_B}{\varepsilon_{Si}}$ and $E_D = \left[a(V_D - V_{DSS}) + E_P^2 \right]^{1/2}$

If the electric field at the drain E_D, is higher than the field at the channel pinch off point E_P, than eqn.(7) reduces to :

$$\frac{I_{SM}}{I_D} \simeq \frac{2}{a} \frac{A_n}{B_n} \frac{a(V_D - V_{DSS}) + E_P^2}{1 + 2\left[a(V_D - V_{DSS}) + E_P^2\right]^{1/2}/B_n} \qquad (8)$$

At this stage, one has to calculate the channel current I_D related to the gate and drain voltages in the saturated regime. The conditions of channel pinch off are given by [9-10]

$$I_{DSS} = \frac{\mu_o C_{ox} Z/L}{1 + \Theta_G V_{GT}} \left[V_{GT} V_{DSS} - \frac{V_{DSS}^2}{2}(1+\delta) \right] \qquad (9)$$

$$E_P = \frac{1}{L}\left[\frac{AI_{DSS}(1+\Theta_G V_{GT})}{\mu_o C_{ox}\frac{Z}{L}(1+\delta)}\right]^{1/3} \quad (10)$$

$$V_{DSS} = \frac{1}{1+\delta}\left[V_{GT} - \frac{I_{DSS}(1+\Theta_G V_{GT})}{\mu_o C_{ox} Z E_P}\right] \quad (11)$$

The current I_D in the saturated regime is :

$$I_D = I_{DSS}\left\{1 - \frac{\left[(LE_P)^2 + 2A(V_D-V_{DSS})\right]^{1/2} - LE_P}{A}\right\}^{-1} \quad (12)$$

with $A = qN_B L^2/\varepsilon_{Si}$ and $\delta \simeq 0{,}25K\,(2\phi_F)^{-1/2}$.

The scheme of the calculations to obtain the theoretical current voltage characteristics is shown in Fig. 7.
Typical measured values of the ionization coefficients A_n and B_n are : $A_n \simeq 10^8$ to $10^{11} m^{-1}$ and $B_n \simeq 2.10^8$ V/m. These values are higher than those measured on bulk silicon. This can be attributed to the doping profile in the silicon film (N_B decreases from the upper interface to the lower interface).
Fig. 8 shows an example of calculated and measured I(V) characteristics.

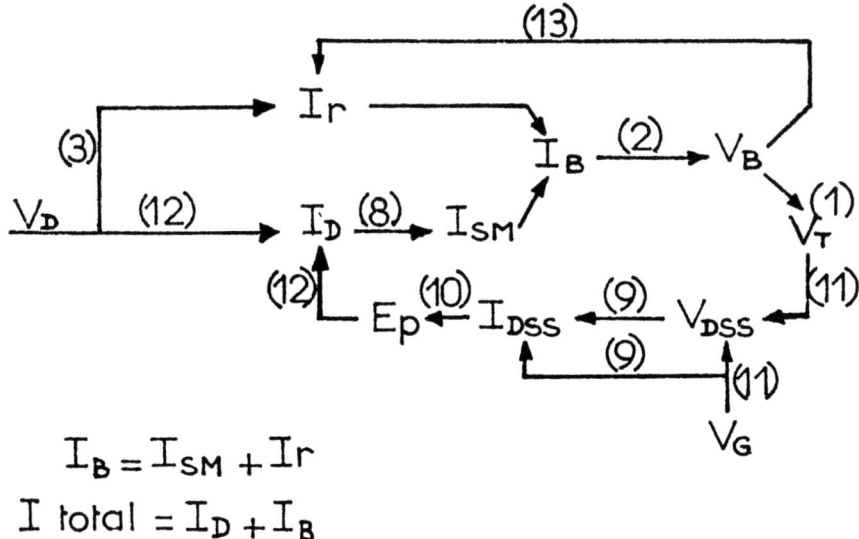

$$I_B = I_{SM} + I_r$$
$$I\text{ total} = I_D + I_B$$

Fig. 7 Iterative calculation of the current.

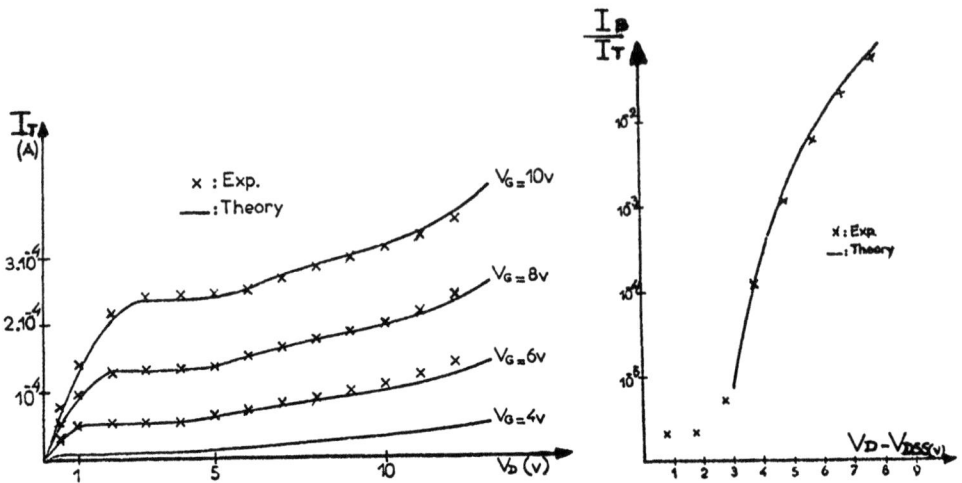

Fig. 8. Substrate current I_B and total drain current I_T versus V_D. ($I_T = I_D + I_B$)

4. SOME TYPICAL PARAMETER MEASUREMENTS. COMPARISON WITH BULK DEVICES

. Mobility μ_0. The carrier mobility is generally deduced through the $I_D(V_G)$ characteristics. The values observed on SOS devices are about 10% lower than those measured on bulk devices [10].

. Threshold voltage V_T. The values are very close to those obtained on bulk silicon. Due to surface charges at the silicon-sapphire interface (positive in Al_2O_3), the modification in threshold voltage becomes noticable for low substrate doping levels ($<2.10^{15}$ at/cm^3) and/or for small silicon film thicknesses ($\leqslant 0,5\mu m$). This is due to the finite charge in the whole depleeted volume of the substrate.

. Leakage currents. The main component $I_{L(III)}$ of the leakage current is much more important for N channel devices ($\simeq 10^{-9}$ A/μm), than for P channel devices ($\simeq 10^{-12}$ A/μm). Moreover, it has been observed that the leakage current I_{LN} of N channel devices is strongly dependent on channel length [4], as shown in Fig. 9.
These results suggest that the drain space charge region spreads along the Si-sapphire interface, possibly due to the presence of a high density of donor states at this interface (N type silicon at the interface). I_{LN} can be drastically reduced (2 to 3 orders

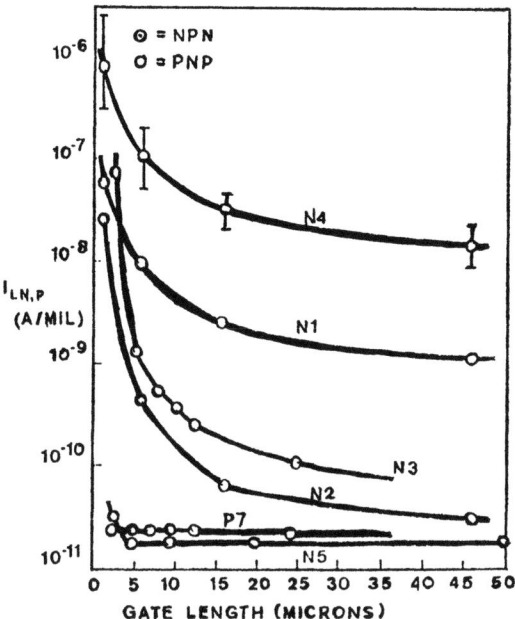

Fig. 9. Leakage current versus channel length.
(After D.J. Mc GREIVY [4])

of magnitude) by doping the N channel device with Boron atoms located at the Si-sapphire interface [4].

. Techniques for investigation of the sapphire-silicon interface. The interface can be investigated by several techniques : transient capacitance [11], transient current analysis [12], and backside C(V) measurements [13]. In the case of the transient measurements, the test structure is either an N^+N or a P^+P structure.
Fig. 10 and Fig. 11 show the principle of the transient current analysis as given in Ref. [12].

. Ionization coefficients A_n and B_n. The ionization rate being much higher for electrons than for holes, one can assume that carrier multiplication can be ignored in standard P channel devices. The coefficients A_n and B_n (for N channel) are generally obtained by curve fitting of the $I_D(V_D)$ characteristics (arround the first kink). Compared to bulk silicon, the values on SOS are much higher ($4.10^8 \lesssim A_{n(m^{-1})} \lesssim 2.10^{11}$; and $B_n \simeq 2.10^8$ V/m). This can be explained by the non uniformity of the silicon doping profile near the Si - SiO_2 interface (channel pinch off region) [14].

. Speed of SOS MOSFET's. It was seen in section 2., that SOS MOSFET's present the advantage of high speed (in comparison with bulk devices), and that CMOS/SOS technology results

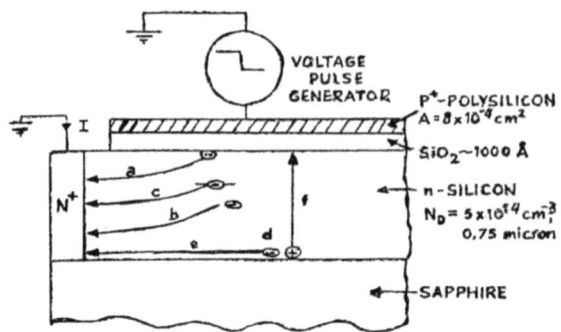

Fig. 10. Charge flow after bias voltage step from accumulation to inversion : a) Expulsion of accumulation electrons ; b) Expulsion of bulk electrons (expansion of depletion layer) ; c) Emission of electrons from traps in the depletion layer ; d) Generation of electron-hole pairs mainly at Si - Al_2O_3 interface ; e) Lateral flow of generated electrons ; f) Vertical flow of generated holes. (After K. Lehovec and R. Miller [12]).

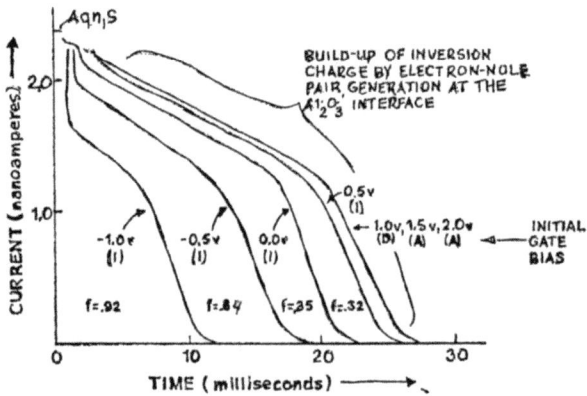

Fig. 11. Transient current of a final gate bias of - 1,5V(I) and various initial gate bias voltages. The surface recombination velocity is obtained by extrapolation to t = 0 of the current at accumulation bias voltages. (After K.Lehovec and R.Miller [12]).

in low power consumption per gate. The figure of merit F (power x delay time/gate) versus frequency, is shown in Fig. 12. One can notice the improvements obtained by scaling down the size of the device.

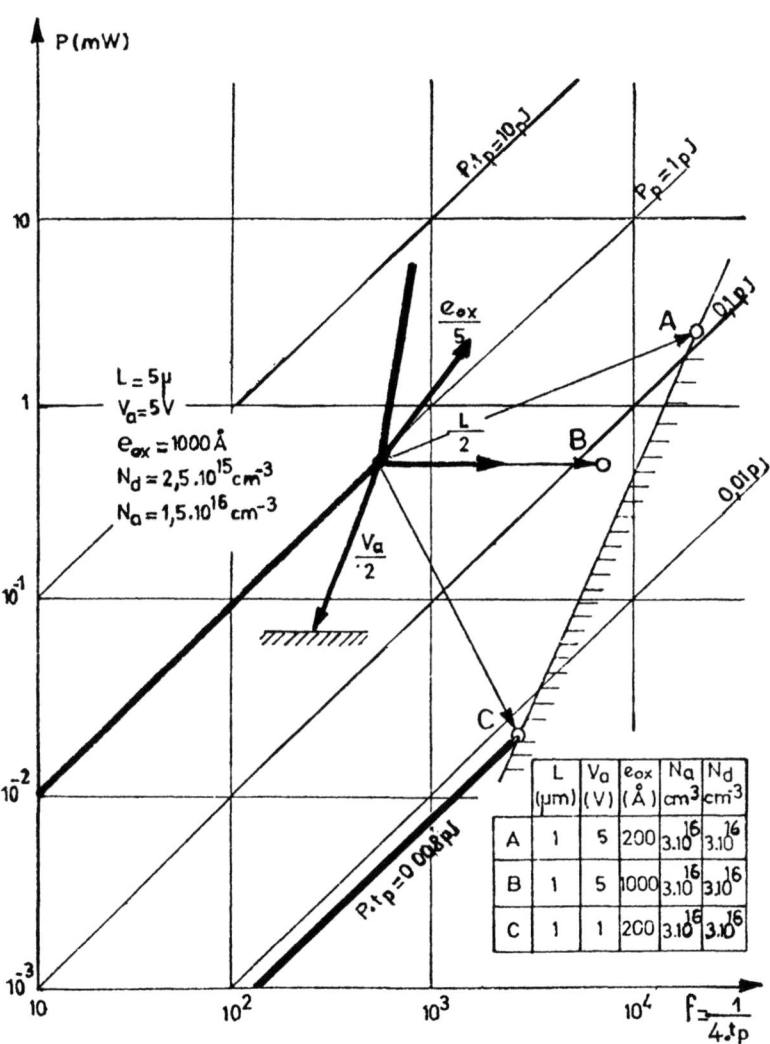

Fig. 12.   P versus frequency for CMOS/SOS.

REFRENCES

[1] H. Martinot, P. Rossel, Multiplication de porteurs dans la zone de pincement des transistors MOS, Electron. Letters, vol.7 N°5/6, 1971, pp. 118-120.
[2] W.W. Lattin, J.L. Rutledge,
Solid State Electron., Vol. 16, 1973, pp. 1043.
[3] S. Sheffield, B. Lalevic, The effects of operating frequency on propagation delay in SOS digital integrated circuits, IEDM, Washington, 1976, paper 8.6.
[4] D.J. Mc Greivy (Hughes), Factors affecting leakage currents in SOS MOS transistors, SOS Workshop, 15-26 September 1975, Lake Tahoe.
[5] J.L. Moll, R. Van Overstraeten, Charge multiplication in silicon P-N junctions, Solid State Electron., vol.6, pp. 147-157.
[6] A.G. Chynoweth, Ionisation rates for electrons and holes in silicon, Phys. Rev., vol.109, 1958, pp.1537-1540.
[7] S.M. Szee, Physics of semiconductor devices, Ed. John Wiley.
[8] P. Rossel, Propriétés statiques et dynamiques du transistor MOS, Thesis, Toulouse, 1973.
[9] G. Merckel, J. Borel, N. Cupcea, An accurate large signal MOST model for CAD, IEEE Trans. On E.D., vol.ED 19 N°5 may 1972, pp. 681-690.
[10] G. Merckel, Y. Gris, Physical modeling of SOS P channel MOSFET and comparison with bulk devices, ESSDERC, Grenoble, 1975.
[11] J. Tihanyi, Measurements of profiles of electrical properties in epitaxial silicon films on insulating substrates, European Solid State Device Research Conf. Munich, Sept.1973 paper A4.2.
[12] K. Lehovec, R. Miller, Investigation of the sapphire-silicon interface by transient current analysis, IEDM, Washington, 1976, paper 12.6.
[13] A. Choujaa and al, Etat actuel de la réalisation des dépôts de silicium en épitaxie sur substrat de corindon, Colloque Matériaux et Technologies pour la Microélectronique Tendances Actuelles, Université des Sciences et Techniques du Languedoc Montpellier, 16-19 Novembre 1976.
[14] Y. Gris, G. Merckel, J.P. Suat, Tech. Rep., LETI/MEA, 1975.

A MOST MODEL FOR CAD WITH AUTOMATED PARAMETER DETERMINATION

F.M. Klaassen

Philips Research Laboratories,
Eindhoven - The Netherlands

1. INTRODUCTION

At the present time computer-aided design techniques greatly facilitate the design of large and complicated MOS integrated circuits. For this purpose many transistor models have been made available [1,2,3]. Although these models have been employed successfully by circuit designers, several effects appear to be modelled unsatisfactorily, for example the saturation effects of the drain current, in particular for devices with shorter channel length, and the dependence of the intrinsic gate capacitances on bias conditions. In designs with large clearances model errors will no doubt be masked by parasitic and stray capacitances, but are unacceptable in self-aligned densely packed circuits.

In the model presented here [4,5,6] the shortcomings mentioned above have been eliminated by improved description of several physical effects. Generally the model equations are functions of the terminal voltages, geometrical factors, material constants and process parameters. To a large extent the success of such a model depends on the way these parameters are determined. Although graphic methods based on measured characteristics are widely used (see chapter " Characterization and measurements of MOS transistors [8]", an automatic determination is desirable for two reasons. Most graphic methods only apply near the $V_{DS}=0$ - region, whereas a close fit to the complete characteristics is more important, and, moreover for factory process evaluation these methods are too slow.

By splitting the d.c. current parameters into two groups, the model equations can be modified so that equations are obtained which are linear in the parameters to be determined.

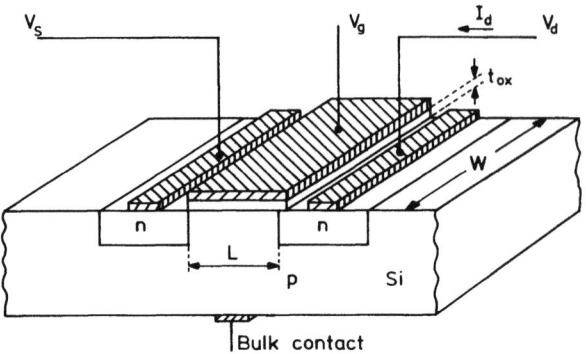

Fig. 1. A MOST structure.

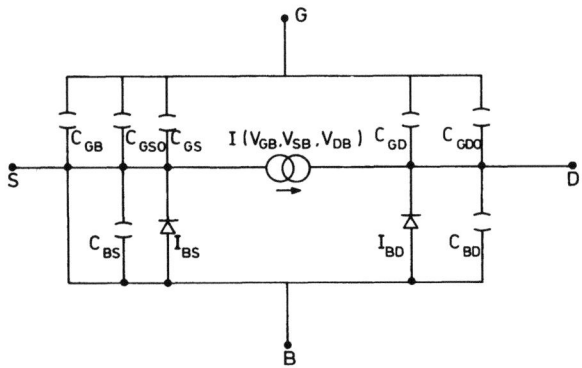

Fig. 2. Dynamic MOST model.

Actually this approach fits the standard measuring practice
quite well. Due to the linear character of the equations,
a very fast routine for parameter deternmination can be designed
[9] and will also be described here.

2. MODEL EQUATIONS

A fully dynamic model for the MOS transistor whose structure is shown in fig. 1, is given in fig.2.
In this scheme the current $I_D(V_{GB}, V_{SB}, V_{DB})$ represents the drain current, and the three gate capicitors C_{GB}, C_{GS} and C_{GD} represent the variation of depletion and inversion charge with gate voltage. The above components form the intrinsic device properties; the remaining ones represent parasitic effects Successively, we have the overlap capacitors C_{GSO}, C_{GDO} and the capacitors C_{BS} and C_{BD} of the source and drain area. In circuit simulation the last two components are often treated separately.

Relating the voltages of source (V_{SB}), gate (V_{GB}) and drain (V_{DB}) to the substrate, the drain current (I_D) is given by

$$I_D = \frac{W\beta_0 \left[V_{SB}^2 - V_R^2 - 2(V_{GB} - V_{TO} + k(2\phi_F)^{\frac{1}{2}})(V_{SB} - V_R)\right]}{2L_0\{1+\Theta_1(V_{GB}-V_{TS})+\Theta_2(V_R-V_{SB})\}(1-\alpha V_{DS}^{\frac{1}{2}}/2L_0)}$$
$$+ \frac{\left[4k/3\{(V_{SB}+2\phi_F)^{3/2} - (V_R+2\phi_F)^{3/2}\}\right]}{2L_0\{1+\Theta_1(V_{GB}-V_{TS})+\Theta_2(V_R-V_{SB})\}(1-\alpha V_{DS}^{\frac{1}{2}}/2L_0)} \quad (1)$$

As the origin of several parts of the above expression has been
fully discussed previously (see chapter " Review of physical
models for the MOS transistor [7]") we shall only briefly explain
several terms here. The functional relation between terminal
voltages in the numerator of eq. (1) is due to the simple gradual
channel approximation. In the denominator the first term represents a correction for the mobility dependence on normal and
lateral field, while the second term represents the channel
shortening effect. Compared to the expressions introduced in
section 2.2. of ref. 7, the latter term is a very simple one.
We have chosen this simplification because saturation of drain
current does not require very accurate modelling for digital
circuit simulation. We shall come back to this later on.

In the above expression the following terms may be considered
as process parameters:

V_{TO} = threshold voltage at zero substrate bias (V_{SB}=0),
k = body factor by which the threshold voltage V_{TS} increases with substrate bias,
$\beta = \dfrac{\mu \epsilon_0 W}{L\, tox}$ = gain factor, expressed in A/V^2,
Θ_1 = mobility degradation factor due to the normal field,
Θ_2 = reduction factor related to velocity saturation occurring at high lateral fields,
α = channel length modulation constant.

It should be remarked that Θ_1 also may represent current reduction due to a resistance in series with the source [8].
The above expression only applies for the voltage range

$$V_{DB} > V_{SB} ,$$

which characterises the normal operation, and

$$V_{GB} \geq V_{TS} = V_{SB} + V_{TO} + K\left[(V_{SB} + 2\phi_F)^{\frac{1}{2}} - (2\phi_F)^{\frac{1}{2}}\right] , \quad (2)$$

where the threshold voltage V_{TS} refers to the substrate and $2\phi_F(N)$ is the bulk Fermi potential. For lower gate voltages the transistor is in the so-called sub-threshold region, which we do not discuss here.
When the drain voltage satisfies the relation $V_{DB} \leq V_{DP}$, where

$$\left. \begin{array}{l} V_{DP} = V_0 - \Delta \\ V_0 = -2\phi_F + \{-\dfrac{k}{2} + (\dfrac{k^2}{4} + V_{GB} - V_{TO} + k\sqrt{2\phi_F} + 2\phi_F)^{\frac{1}{2}}\}^2 \\ \Delta = \dfrac{\Theta_2(V_0 - V_{SB})^2}{2[1 + \Theta_1(V_{GB} - V_{TS}) + \Theta_2(V_0 - V_{SB})]} \end{array} \right\} \quad (3)$$

the transistor operates in the linear region. In this case

$$V_R = V_{DB}$$

in relation (1).
For higher drain voltages the device is in the saturation regime. Then the following substitution applies,

$$V_R = V_{DP}$$

For the inverse operation ($V_{DB} \leq V_{SB}$), the symbols V_{SB} and V_{DB} have to be interchanged in eqs.(2) and (3).

Of course, the equations described here are mainly valid for relatively long channels and uniform substrate doping. Effects due to depletion layers of source and drain [10] or implanted layers [11] are not properly accounted for. However, in practice the model has been shown to cover also these effects to some extent.

The intrinsic gate capacitances have been modelled by the expressions:

$$C_{GB} = C_{ox}\left[1 + \frac{4}{k^2}(V_G - V_{FB})\right]^{-\frac{1}{2}} , \qquad (4)$$

$$C_{GS} = \frac{2}{3} C_{ox}\left[1 - \frac{(V_{GB} - V_{TD})^2}{(2V_{GB} - V_{TS} - V_{TD})^2}\right] \qquad (5)$$

$$C_{GD} = \frac{2}{3} C_{ox}\left[1 - \frac{(V_{GB} - V_{TS})^2}{(2V_{GB} - V_{TS} - V_{TD})^2}\right] \qquad (6)$$

where $C_{ox} = \varepsilon_{ox} WL/t_{ox}$ and ε_{ox}/t_{ox} is the oxide capacitance per unit area which has to be considered as a parameter,
V_{FB} is called the flatband voltage and
V_{TD} is the equivalent of the threshold voltage with respect to the drain

$$V_{TD} = V_{DB} + V_{TO} + k\left[(V_{DB} + 2\phi_F)^{\frac{1}{2}} - (2\phi_F)^{\frac{1}{2}}\right] \qquad (7)$$

In a previous chapter [8] the value of the above functions has been compared to carefully measured results. Generally it was found that the accuracy is rather low. However, since the largest capacitor C_{GS} does not vary much with bias condition and the circuit delay times are also determined by rather constant parasitic capacitors, the expressions (4) through (6) adequately satisfy practical needs.

The parasitic overlap capacitors C_{GSO} and C_{GDO}, which to the first order may be considered as constants, have to be measured on the basis of special test figures.

Finally the source and drain capacitors remain to be specified. Often these components are considered as separate elements in a circuit model. In general, it has been found sufficient to model them by the abrupt-junction capacitance expression

$$C = \frac{SC(0)}{(1 + V/2\phi_F)^{\frac{1}{2}}}, \qquad (8)$$

in which S is the area of source and drain and the parameter C(0) is the unit area capacitance at zero voltage. The last equation completes the model of fig.2. In the next section we discuss the determination of the parameters which have been introduced with the model specification.

3. PARAMETER DETERMINATION

As has been mentioned in the introduction we have decided on a fully automated procedure for obtaining the parameters associated with the drain current generator. However, when inspecting equations (1) through (3) a complication arises due to the fact that slightly different formula for I_D apply in the linear and saturated modes. Unfortunately the value of the drain voltage which separates both modes is determined by several parameters (compare eq. (3)). Hence no split in appropriate regions can be made in advance.

Fortunately, by splitting the parameters into two groups, the drain current expression can be modified so that equations arise which are linear in the parameters to be determined. Actually this approach fits the standard measuring practice quite well.

3.1. Determination of V_{TO} and k

As mentioned above, the value of the saturation voltage has to be known before a split can be made of the characteristics equation into a linear and a saturated region. For this purpose we have to determine the parameters V_{TO} and k. This is the main purpose of part 1 of our procedure. Usually both parameters are obtained from measured values of the threshold voltage V_T of the MOST as a function of the source-substrate bias V_{SB}. In practice, V_T is determined from an automated measurement of the drain current I_D as a function of the gate-source bias V_{GS} at a very low drain source voltage V_{DS}, with V_{SB} as a parameter (compare fig.3). When V_{DS} is sufficiently low, the MOST operates in the linear mode and the drain current as given by eq.(1) can be closely approximated by

$$I_D = \frac{\beta(V_{GS} - V_T)V_{DS}}{1 + \Theta_1(V_{GS} - V_T)} \qquad (9)$$

where the threshold voltage V_T now refers to the source.

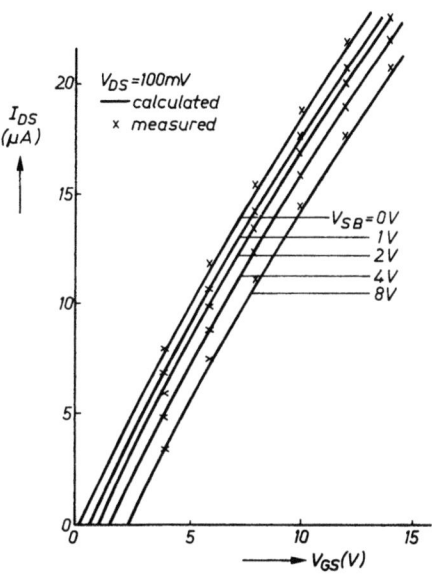

Fig.3. Measured and simulated drain current vs gate voltage.
Parameter values obtained from least-squares fit are
$V_T = .21V, k = 1.05 \ V^{\frac{1}{2}}$

Denoting I_D/V_{DS} as the drain conductance g_D, it is possible to rearrange eqs. (2) and (9) in a form more suitable for parameter determination, that is

$$V_{GS} = V_{TO} + k\left[(V_{SB} + 2\phi_F)^{\frac{1}{2}}\right] + (1-\Theta_1 g_D/\beta)^{-1} \cdot \frac{g_D}{\beta} \qquad (10)$$

For most devices a plot of g_D vs V_{GS}, with V_{SB} held constant is close to a straight line, as shown in fig.3. The slight curvature is caused by the mobility reduction factor Θ_1. As the factor $(g_D\Theta_1/\beta)$ may be considered as a correction term, it is useful, before determining the threshold voltage accurately, to obtain a mean value for Θ_1/β from the measurements When the correction term Θ_1/β is known, eq. (10) is suitable in form to determine the parameters V_{TO}, k and β from a least squares fit, with experimentally measured values of V_{GS} as a function of $g_D/(\beta - \Theta_1 g_D)$ and taking V_{SB} as a measuring parameter. During the first run of this procedure the value of the bulk Fermi potential $2\phi_F(N)$ is taken as 0.7, which value is afterwards corrected from the result obtained for the body factor k. Making use of the corrected value for $2\phi_F$, an improved value of V_{TO} and k will result from a second run. If necessary, the above loop may be repeated for higher accuracy.

The result achieved with this procedure is outlined in fig.3, which gives experimental values together with simulated results obtained after the last squares fit. The optimum value of the parameters as obtained from the fit is also given in the same figure. As can be observed, each of the 30 measured points is characterized by values: I_D, V_{DS} = 100 mV, V_{GS} and V_{SB}. The average error which results from the program amounts to 1%. Generally, no values of I_D close to the horizontal axis should be taken for two reasons. Current tails which have not been included in the MOS model are often observed near threshold. For low values of the effective gate voltage ($V_{GS} - V_T$) and a fixed drain source voltage of 100 mV, the approximation of eq.(1) will give rise to a number of errors.

Although 4 parameters can be found from this part of the program, we now take only the values obtained for two of them as definite. The values of V_{TO} and k have to be used for the computation of V_D to permit analysis of the saturated characteristics. However, we prefer to obtain the parameters β, Θ_1, Θ_2 and α from a global fit of the overall I-V characteristics.

3.2 Determination of β, Θ_1, Θ_2, and α

In this part we use an $I_D - V_{DS}$ characteristic as given in fig. 4 for the determination of the parameters β, Θ_1, Θ_2 and α. Since these characteristics are taken for a fixed gate-source and a fixed source-substrate voltage each measurement is characterized by 4 values: I_D, V_{DS}, V_{GS}, and V_{SB}.
Now the areas separated by the value of the saturation voltage have to be considered for V_{DS}. When considering Δ as a correction factor and taking the preliminary value equal to zero, V_{DP} is only determined by the bias condition and the parameters V_{TO} and k. Obviously, since k and V_{TO} have been determined in part 1, both V_{TS} and V_{DP} are known values for each measured point of the characteristics.
With $V_{DB} < V_{DP}$, the transistor is in the linear mode and the factor V_R in eq.(1) has a value $V_R = V_{DB}$.
With $V_{DB} > V_{DP}$ the transistor is in saturation and for V_R a new value V_{DP} in eq. (1) has to be substituted.
Inspection of the numerator of eq.(1) reveals that it has a known value for each measured point. We denote this term

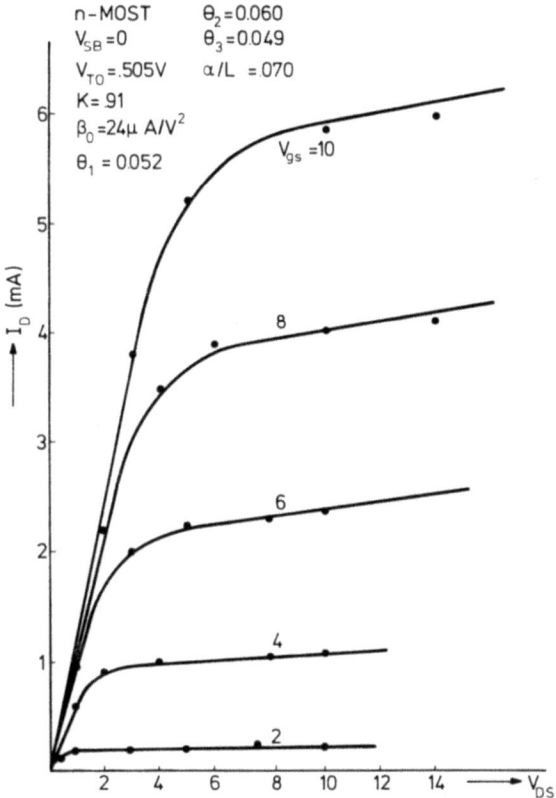

Fig. 4. Measured and simulated (fully drawn curve) I-V characteristic of n-channel MOST. (W = 100 µm, L = 10 µm)

by $F(V_{GS}, V_{SB}, V_R, k, V_{TO})$.

A form more suitable for parameter determination can be obtained by reversing eq. (1)

$$I_D^{-1} = \frac{2}{F\beta} + \frac{2\Theta_1(V_{GB} - V_{TS})}{F\beta} + \frac{2\Theta_2(V_R - V_{SB})}{F\beta} - \frac{\alpha V_{DS}^{\frac{1}{2}}}{LF\beta} + R \quad (11)$$

in which the term R may be considered as a small residual term

$$R = -\left[\Theta_1(V_{GB} - V_{TS}) + \Theta_2(V_R - V_{SB})\right] \frac{\alpha V_{DS}^{\frac{1}{2}}}{\beta LF}. \quad (12)$$

Apart from the residual term, eq. (11) is linear in the parameters β, Θ_1, Θ_2 and α Therefore a procedure similar to part 1 can be followed to evaluate the parameters. A first-order value of the parameters is obtained from a least-squares fit between measured values of I_D^{-1} and equation (11) taking R=0. Then a new value for R and V_{DP} is computed from eq. (12) and (3). Finally, by using several successive loops, accurate values for β, Θ_1, Θ_2 and α are obtained. Obviously care has to be taken that the program avoids optimizing the ratio Θ_1/β and Θ_2/β instead of the separate parameters (compare (11)). This can generally be achieved by taking an equal amount of measuring points from both the linear region and the saturated region.

In fig.4 a comparison is made between measured values of the characteristics and simulated ones after a current-fitting procedure. However, this figure covers only part of all measuring points. For the global fit we actually took fifty measured points, characterized by four values provided by I_D, V_{DS}, V_{GS} and V_{SB}, since we combined fig.4 with three comparable characteristics at V_{SB}=1, V_{SB}=4 and V_{SB}=8 volt. The value of the drain current parameters which result from the fitting procedure are also given in the figure. Again the agreement is fair; the average deviation amounts to 3%. Computation times are in the order of seconds on a mini-computer.

3.3. Determination of capacitance parameters

In order to minimize measurement errors, values for the gate oxide capacitance per unit area (C_{ox}), the overlap capacitances per unit length (C_{GSO} and G_{GDO}) and the source and drain capacitance per unit area (C_{SB} and C_{DB}) have to be obtained from special, relatively large area devices. For the first measurement the substrate is biased so that the device is in the accumulation mode. The overlap capacitances have to be measured from a bias condition, in which either the source or the drain is pinched-off with respect to the gate.

4. SOME REMARKS ON ACCURACY AND SENSITIVITY OF THE MODEL

As already mentioned, the average deviation between a simulated characteristic and a measured one amounts 3%. This is generally the case for devices with a channel length L > 4 µm. However, the deviations become larger for shorter channels and also, some parameters become dependent on L. As is well known, this is caused by two dimensional effects. Empirically it has been found that V_{TO} and k decrease linearly, but Θ_1, Θ_2 and α increase linearly with decrease of L. The effects upon V_{TO}, k and α are caused by the relative increase in importance of the depleted areas around source and drain [10]. The value of Θ_1 increases due to the influence of series resistances and Θ_2 has to be inversely proportional to L (compare eq. (20) of ref. 7).

As fluctuations in the value of the parameters have to be expected from spreads in mask dimensions and process variations, it is interesting to investigate the sensitivity of a model for these fluctuations. Therefore a check has been made to ascertain how the simulated results deviate from the measured results when one parameter is given an intentional error of 10%. For this error in the parameter values of β, V_{TO}, k, Θ_1, Θ_2 and α, the average deviation between calculated and measured characteristic amounts 10.2%, 3.5%, 3.2%, 3.8%, 2.9% and 3.1%, respectively.

With the procedures already outlined it is also very easy to investigate whether a higher accuracy may be obtained from an alternative model description. As examples we mention here the simplified equations for the drain current given in relation (14) of ref. 7 or those used in other models [1], a more complicated formulation of channel shrinkage (eq. (28) of ref. 7) and the question whether the parameter Θ_1 represents a series resistance to the source.

The usefulness of an approximation for I_D can be easily checked by replacing the F-function in (11) by the equivalent function, which follows from result (14) of ref. 7 or from other model equations [1]. Generally it has been found that the deviations between the simulated and measured characteristics become larger for simple I_D equations, in particular at higher values of the body factor. A model like the one given in ref.1 will even fail when k > 1.

Surprisingly, the above mentioned relation for channel shrinkage, which is more exact from the viewpoint of physics than the simple representation used in (1), did not show any improvement in accuracy. This result has been obtained by replacing the $\alpha V_{DS}^{\frac{1}{2}}$ - term in (11) by an equivalent term according to eq. (28) of ref. 7.

For some transistors, whose Θ_1 values were found from the above parameter determination procedure as being much higher than expected from mobility reduction effects, it has been checked to find out whether eq.(1), with a normal value for Θ_1 of about 0.02, could provide higher accuracy when combined with a resistance in series with the source. The best values for this resistance were obtained from similar fitting procedures for transconductance. In general no improvement in accuracy has in fact been found.

Summarising, we may conclude that the model given in section 2 in conjunction with the parameter determination procedure in section 3 satisfies the demands of simplicity and accuracy in describing MOST characteristics of present commercial processes and may therefore be considered as a useful tool for CAD techniques.

The author is indebted to Mr. W. de Groot for evaluation of the parameter determination procedures.

References

1. Model SPICE - D.A. Hodges and H. Schichman, IEEE J. SSC-3 (1968) 285.
2. Model FETSIM - J.E. Meyer, RCA Rev. 32 (1971) 42.
3. Model SLIC - D. Frohman-Bentchkowski, IEEE J. SSC-4 (1969).
4. H. Sibbert and B. Höfflinger - Nachrichten technische Fachberichte Bd 49 (1974).
5. F.M. Klaassen - Philips Res. Rep. 31 (1976) 71.
6. Model Mini-Msinc - T.K. Young and R.W. Dutton, Techn. Rep. Stanford University (1976).
7. Chapter "Review of physical models for MOS transistors".
8. Chapter "Characterization and measurements of MOST devices".
9. F.M. Klaassen - Philips Res. Rep. 31 (1976) 84.
10. L.D. Yau - Solid St. Electr. 17 (1974) 1059.
11. J.S.T. Huang and G.W. Taylor - IEEE Trans. ED-22 (1975) 995.

CAD MODELS OF MOSFET's

G. Merckel

Centre d'Etudes Nucléaires de GRENOBLE - LETI/MEA -
85 X 38041 Grenoble Cedex, France

ABSTRACT. CAD models of MOS transistors, a system of automatic acquisition of device parameters and the main results obtained in this field are presented.

INTRODUCTION. Computer Aided Design of circuits using MOS transistors requires simple models of devices. These models should satisfy the following two criteria : 1) A good compromise between computer time (to be minimized) and accuracy of simulated performance as compared to experimental results (good precision of the model) ; and 2) A direct correlation between the electrical parameters of the model and the physical, technological and geometrical parameters of the transistors. The purpose of this lecture is to present the results obtained in the field of automatic acquisition of the parameters for the CAD models developed at LETI. The lecture includes the following sections : 1) CAD Models and electrical parameters ; 2) System of automatic measurements and acquisition of parameters ; 3) Results obtained ; 4) Other possible applications of the system.

1. CAD MODELS AND ELECTRICAL PARAMETERS

In this section, N channel MOS transistors are considered. The results are applicable to P channel transistors with an appropriate rearrangement of signs for the applied voltages. All voltages are referred to the source.

1.1 Enhancement mode transistor on bulk silicon [1] [2]

1.1.1 Strong inversion regime

The simplified equations of working are obtained in a classical way and take into account the main physical effects ; in particular, those which are related to small channel lengths. Following these considerations, particular attention has been given to :
1) The effect of electrical field on carrier mobility ; 2) The effect of substrate polarization and channel length on threshold voltage ; 3) The effect of substrate doping and channel length on non saturation of $I_D(V_D)$ characteristics.

Before saturation $(V_D \leqslant V_{DSS})$, the simplified equation for current is given by :

$$I_D = \frac{M_o}{1+\Theta_G V_{GTo}} \left[V_{GT} V_D - \frac{V_D^2}{2}(1+\delta) \right] \quad (1)$$

Representing carrier mobility under zero field by μ_o, thin oxide capacitance per unit area by C_{ox}, channel length by L, channel width by Z and substrate doping by N_B, the relations defining the parameters used in eqn.(1) are given by :

$$V_{GT} = V_G - V_T$$

$$V_{GTo} = V_G - V_{To}$$

$$V_T = V_{To} + K \left[(V_B + 2\phi_F)^{1/2} - (2\phi_F)^{1/2} \right] \left[1 + \Theta_B (V_B + 2\phi_F)^{1/2} \right]^{-1}$$

$$K = (2\varepsilon_{Si} \, q \, N_B)^{1/2} / C_{ox}$$

$$\phi_F = v_T \log(N_B/n_i) \quad \text{with } v_T = kT/q$$

$$\Theta_B = \frac{1}{L} \left(\frac{2\varepsilon_{Si}}{q \, N_B} \right)^{1/2}$$

$$M_o = \mu_o C_{ox} Z/L$$

$$\Theta_G \sim C_{ox}$$

$$\delta = 0.25 \, K(2\phi_F)^{-1/2}$$

In saturation zone $(V_D \geqslant V_{DSS})$, a simple relation, in analogy with the bipolar transistor, is used to introduce the effect of non-saturation of drain current. The relation is of the form :

$$I_D = I_{DSS} \left[1 + \frac{V_D - V_{DSS}}{V_E + V_{DSS}} \right] \quad (2)$$

where I_{DSS} and V_{DSS} designate the drain current and voltage at the point of saturation. V_E is the equivalent of the Early voltage. Fitting the simplified theory with the experimental results leads to $V_E \simeq 5L\,(10^{-15}N_B)^{1/2}$ with V_E in volts, L in microns and N_B in at/cm³.

The saturation voltage V_{DSS} is given by the following relation

$$V_{DSS} = V_E \left[\left(1 + \frac{2V_{GT}}{V_E(1+\delta)}\right)^{1/2} - 1\right] \qquad (3)$$

The saturation current I_{DSS} is obtained by replacing V_D in eqn.(1) by V_{DSS} eqn.(3).

The main advantage of relation (2) is its simplicity. It leads to a constant output impedance for a given V_G. This could, in fact, be a limitation for some analogic applications. As an example, Fig. 1 illustrates the experimental results obtained for V_E, for $L \geqslant 1\mu$ and N_B between 10^{15} and 10^{17} at/cm³.

1.1.2 Weak inversion regime

The simplified equations are obtained according to the principle of analysis of R. VAN OVERSTRAETEN and al [3]. The procedure for obtaining the model has been presented by the author in the lecture on weak inversion. One obtains, after simplification of

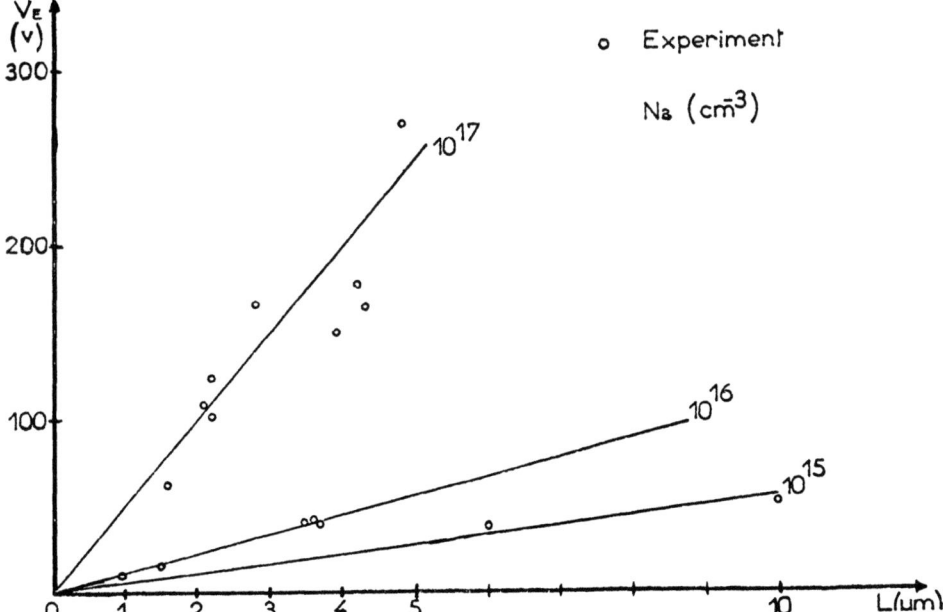

Fig. 1. Early voltage V_E (L, N_B)

the expression I(V) :

$$I_D = M_o \, a \, (nv_T)^2 \frac{\exp(V_{GT}/(nv_T)) \left[1-\exp(-mV_D/(nv_T))\right]}{1 + m_R \exp(n_R V_{GT}/(nv_T))} \quad (3)$$

The parameters used in eqn.(3), in addition to those already defined in section 1.1.1, are defined as follows :

$a = \frac{(m-1)}{mn} \exp(\Delta\phi/v_T)$ $\quad \Delta\phi$: strong inversion "excess" surface potential

$m = 1 + \frac{C_D}{C_{ox}}$ $\quad C_D$: depletion capacitance per unit area

$n = m + \frac{qN_{SS}}{C_{ox}}$ $\quad N_{SS}$: density of surface states per eV.

$m_R \sim \frac{v_T}{2\phi_F} \exp(\Delta\phi/v_T)$

$n_R = \frac{1}{u_R} \log\left[\frac{1}{m_R}\left(\frac{ma}{u_R}\right) \exp(u_R) - 1\right]$

$u_R = 2(1 + \delta)/m$ $\quad$ reduced gate voltage at the point of matching between weak and strong inversion regimes

$\left(u_R = \frac{V_{GR} - V_T}{n \, v_T}\right)$

The CAD model so defined, has the following 14 parameters : a, m, m_R, n, n_R, u_R, M_o, V_{TO}, ϕ_F, K, Θ_B, Θ_G, δ and V_E. However, 3 points are to be noted : 1) u_R and n_R are calculated from other parameters ; 2) ϕ_F may be taken to be constant and nearly equal to 0.35 volts; 3) K and Θ_B are not necessary for $V_B = 0$.

From these remarks, it follows that the characterization of the transistor necessitates the measurement of 11 parameters (of which 7 are for the region of strong inversion) in the general case ($V_B \neq 0$) and of 9 parameters in the case when $V_B = 0$.

1.2 Depletion mode transistor on bulk silicon [4]

The simplified equations of working in the strong inversion regime are obtained by an expansion, limited to second order, of the expressions I(V) presented by the author in the lecture on the ion implanted depletion mode transistor. Considering the characteristics $I_D(V_D)$ presented schematically in Fig. 2, one can

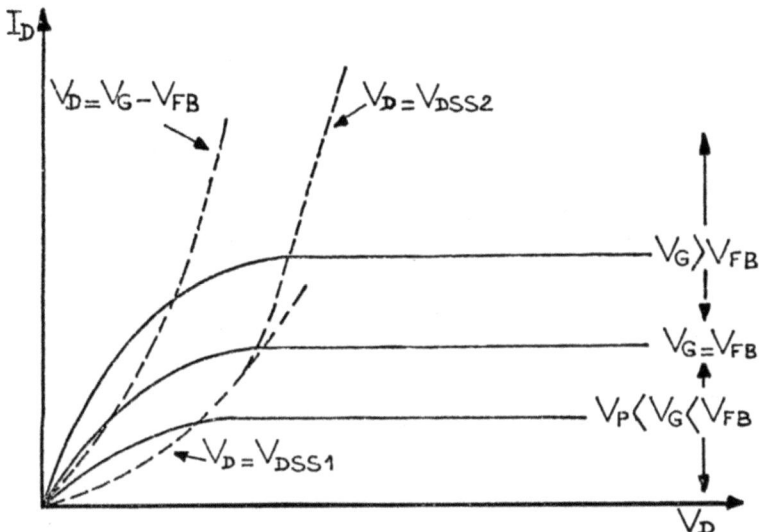

Fig. 2. Depletion mode device $I_D(V_D)$ characteristics.

distinguish 3 typical modes of functioning (according to V_G):
1) $V_G \leqslant V_P$, $I_D = 0$ (pinched channel, no residual current);
2) $V_{FB} > V_G \geqslant V_P$ (burried channel); 3) $V_G \geqslant V_{FB}$ (burried and surface channel). For these different modes of working, three relations I(V) are defined as follows.

a) Burried channel $V_{FB} > V_G \geqslant V_P$

. If $V_{DSS1} \geqslant V_D$

$$I_D = M_D \left[V_{GP} V_D - \frac{V_D^2}{2} (1+\delta) \right] \quad (4)$$

with $V_{GP} = V_G - V_P$

$$V_P \simeq V_{Po} + K_D \left[(V_B + V_\emptyset)^{1/2} - (V_\emptyset)^{1/2} \right]$$

. If $V_D \geqslant V_{DSS1}$

$$I_D = I_{DSS1} \left[1 + \frac{V_D - V_{DSS1}}{V_E + V_{DSS1}} \right] \quad (5)$$

The saturation voltage V_{DSS1} is given by a relation similar to (3)

$$V_{DSS1} = V_E \left[\left(1 + \frac{2 V_{GP}}{V_E(1+\delta)}\right)^{1/2} - 1 \right] \quad (6)$$

The saturation current I_{DSS1} is obtained by substituting the value of V_{DSS1} (equ.6) for V_D in eqn.(4).

b) Surface fully enhanced $V_G \geqslant V_{FB}$ and $V_G - V_{FB} \geqslant V_D \geqslant 0$

$$I_D = M_D \left\{ V_{GP} V_D - \frac{V_D^2}{2}(1+\delta) + (r-1)\left[V_{GF} V_D - \frac{V_D^2}{2}\right] \right\} \quad (7)$$

with $V_{GF} = V_G - V_{FB}$ and $r = r_o(1 + \Theta_G V_{GF})^{-1}$

c) Surface partially enhanced $V_D \geqslant V_G - V_{FB} \geqslant 0$

• If $V_{DSS2} \geqslant V_D$

$$I_D = M_D \left[V_{GP} V_D - \frac{V_D^2}{2}(1+\delta) + (r-1)\frac{V_{GF}^2}{2} \right] \quad (8)$$

• If $V_D \geqslant V_{DSS2}$

$$I_D = I_{DSS2} \left[1 + \frac{V_D - V_{DSS2}}{V_E + V_{DSS2}} \right] \quad (9)$$

The saturation voltage V_{DSS2} is given by the relation :

$$V_{DSS2} = V_E \left\{ \left(1 + \frac{2}{1+\delta}\left[\frac{V_{GP}}{V_E} - \frac{(r-1)}{2}\left(\frac{V_{GF}}{V_E}\right)^2\right]\right)^{1/2} - 1 \right\} \quad (10)$$

The saturation current I_{DSS2} is obtained by substituting the value of V_{DSS2} from equ.(10) for V_D in eqn.(8).

The model so defined, contains the following 9 parameters :
M_D, V_{Po}, $V_\emptyset$, V_{FB}, K_D, r_o, Θ_G, δ and V_E. All these electrical parameters are related to physical, technological and geometrical parameters of the transistor [4]. One can assume that $V_\emptyset$ (the potential barrier between the N type implanted zone and the

P type substrate) is constant and nearly equal to 0.7 V. Thus, the characterization of the transistor necessitates the measurement of 8 parameters in the general case and of 7 parameters in the particular case when $V_B = 0$.

2. SYSTEM OF AUTOMATIC ACQUISITION OF DATA AND PARAMETERS [5]

2.1 Acquisition of data

The system has been set up for being applied to the problems of research and development and has to satisfy the following criteria : 1) Accuracy of measurements ; 2) Simplicity of use ; 3) Versatility in application to various semiconductor devices ; 4) On line or off line treatment of measured data ; 5) Compatibility with a mini computer.

Fig. 3 shows the block diagram of the system (SIAM). One can distinguish 5 parts : 1) The central processor is a desk top calculator ; 2) The storage and output of data is accomplished by a magnetic cassette, a teletype, an X - Y plotter and a CRT display ; 3) 4 voltage sources ; 4) Testing of devices on wafers or in packages; 5) Measurements of current, voltage and capacitance. The main specifications of the system are given in Fig. 4 (full scale values).

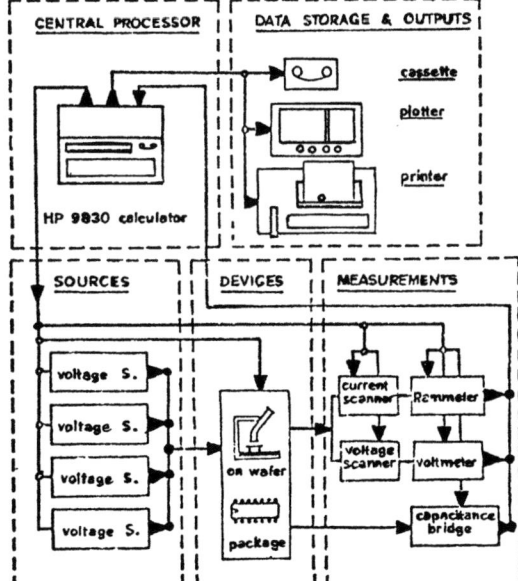

Fig. 3. Block diagram of SIAM.

SPECIFICATIONS

- Voltage sources. : up to 100 V
- Measurement
  - voltage : 0.1 V — 1000 V
  - current : 10 pA — 100 mA
  - capacitance : 10 pF — 1 μF
  - voltage scanner : 10 channels
  - current scanner : 10 channels
- Processor
  - calculator : HP 9830
  - basic language
  - memory : 12 k bytes
  - data storage : 60 k bytes (cassette)
- Outputs
  - teletype : printer and tape reader/puncher
  - plotter
  - display

Fig. 4. SIAM - specifications.

2.2 Acquisition of parameters

The acquisition of parameters and the comparison of theory with experiments takes place in 4 phases. For example, for the strong inversion regime of a MOSFET, one has :

a) Data input and options

b) Direct acquisition of M_o, V_{To} (10 points of measurement)
K, Θ_B (10 points of measurement)
Θ_G, δ (10 points of measurement)
V_E (2 points of measurement)

c) Because of possible small errors in direct acquisition, the parameters M_o, Θ_G and δ are, in general, obtained by an iterative calculation (Newton - Raphson)

d) Print out and plotting of results, comparison of theory with experiments. The total time necessary for the various operations is of the order of 6 min. and is divided as follows : 1 min. for measurements, 1 min. for calculations, 4 min. for print out and plotting of results. The size of the program is about 8 K bytes.

3. EXAMPLE OF RESULTS

Fig. 5 shows the comparison between model and experiment $I_D(V_G)$ for a P channel transistor (L=2.2µm) working in weak and strong inversion regimes. Fig. 6 shows the $I_D(V_D)$ characteristics for the same transistor in strong inversion regime. Fig. 7 shows the results obtained for the transconductance dI_D/dV_G. Finally, Fig. 8 illustrates the results obtained for a P channel depletion mode transistor.

One can see that the agreement between theory and experiment is better than 10%. At this level, one can claim that the validity of the model has been proved. For a more physical analysis of each parameter a large number of devices has been tested over the temperature range : $77°K \leqslant T \leqslant 300°K$ [2].
For example, Fig. 9 and Fig. 10 illustrate the theoretical and experimental results obtained for 2 parameters.
Fig. 11 and Fig. 12 illustrate the comparison between model and experiment at 300°K and at 77°K.

4. OTHER POSSIBLE APPLICATIONS OF THE SYSTEM

In the previous chapters we have detailed some results obtained for the parameters of CAD models (in static regime). However, other possibilities have been developed and are working.

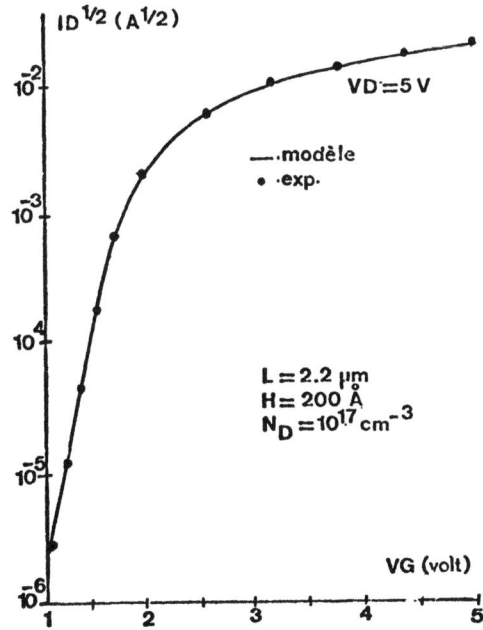

Fig. 5. $I_D(V_G)$ characteristic.

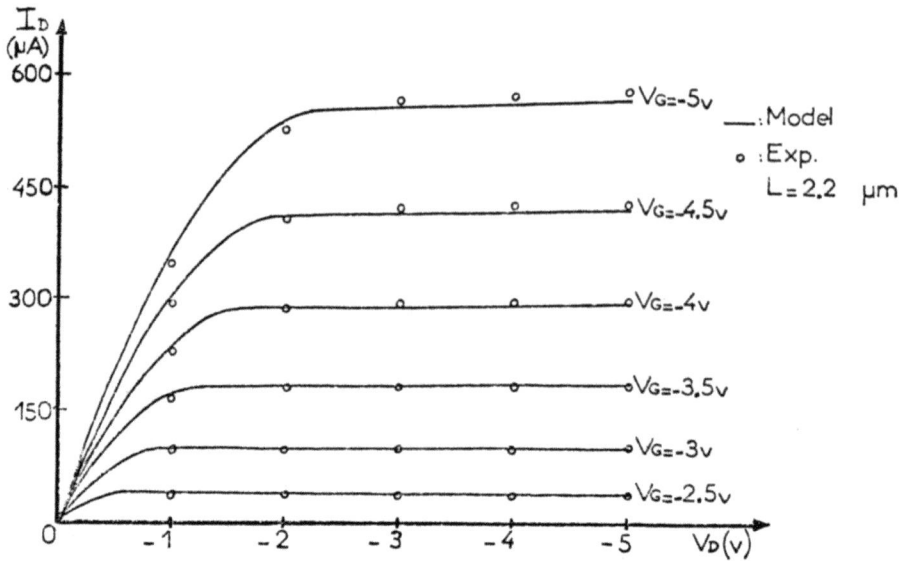

Fig. 6. $I_D(V_D)$ characteristics.

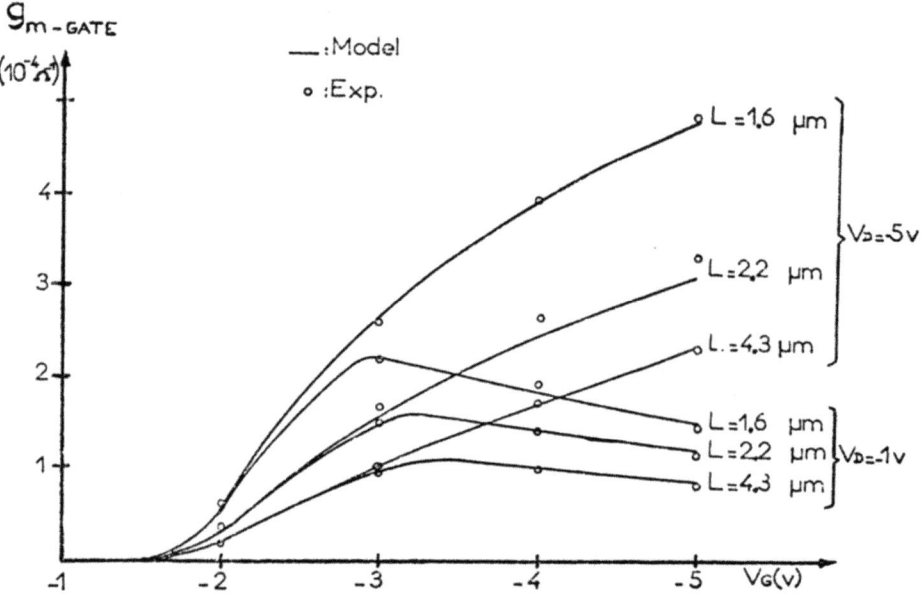

Fig. 7. Transconductance g_{m-GATE} (V_G, L).

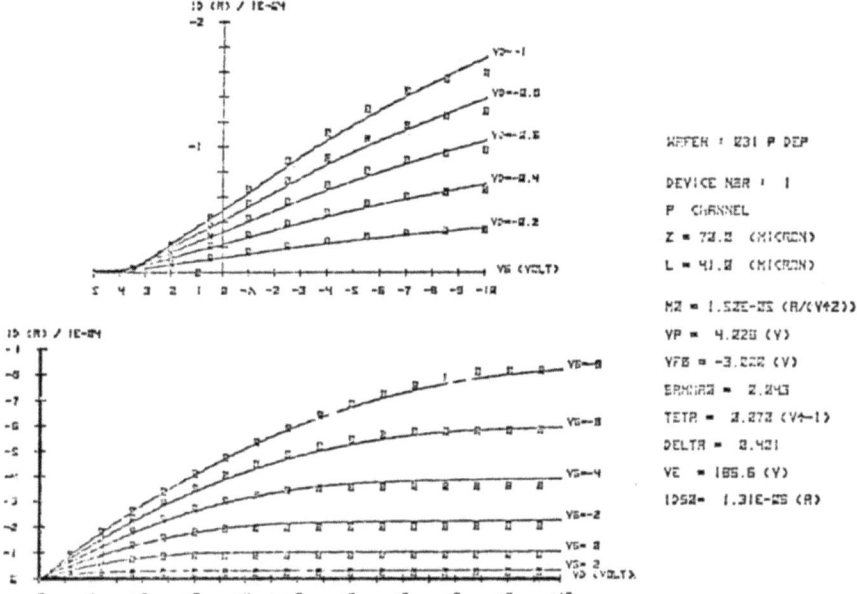

Fig. 8. Depletion mode device. $I_D(V_G)$ and $I_D(V_D)$ characteristics.

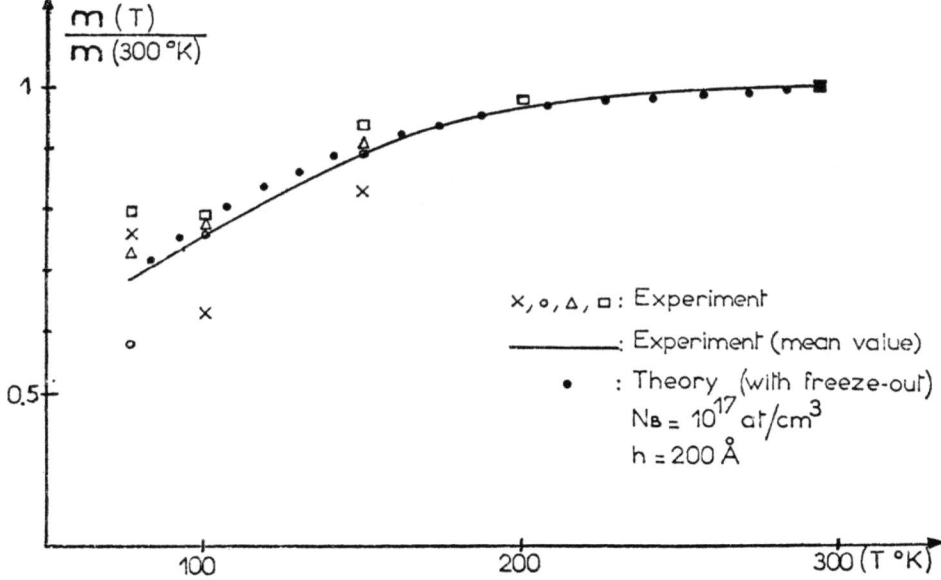

Fig. 9. Weak inversion parameter m(T).

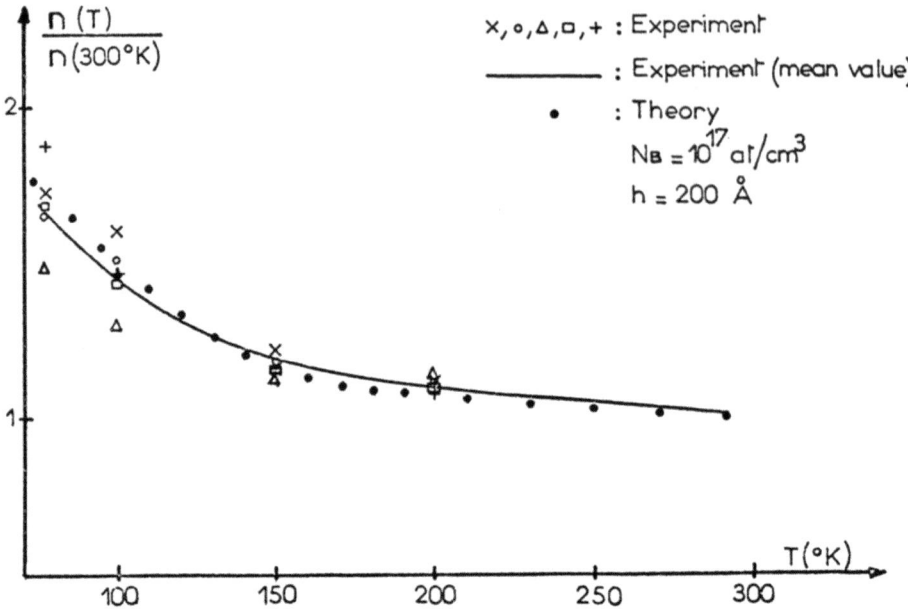

Fig. 10. Weak inversion parameter n(T).

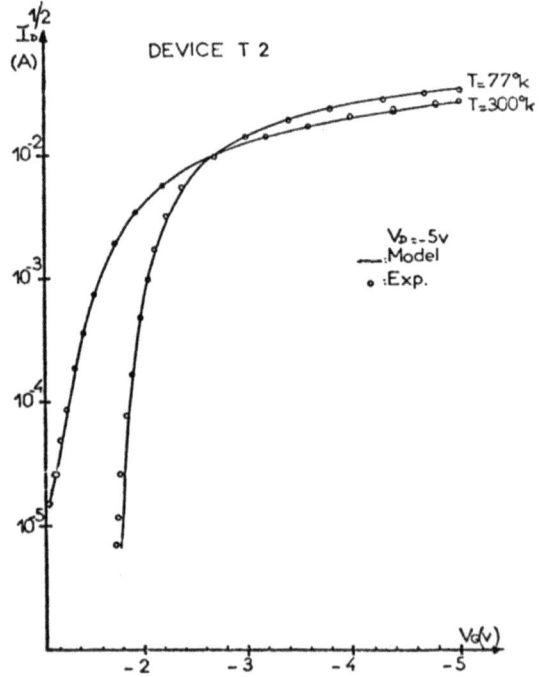

Fig. 11. $I_D(V_G)$ characteristics. Effects of temperature.

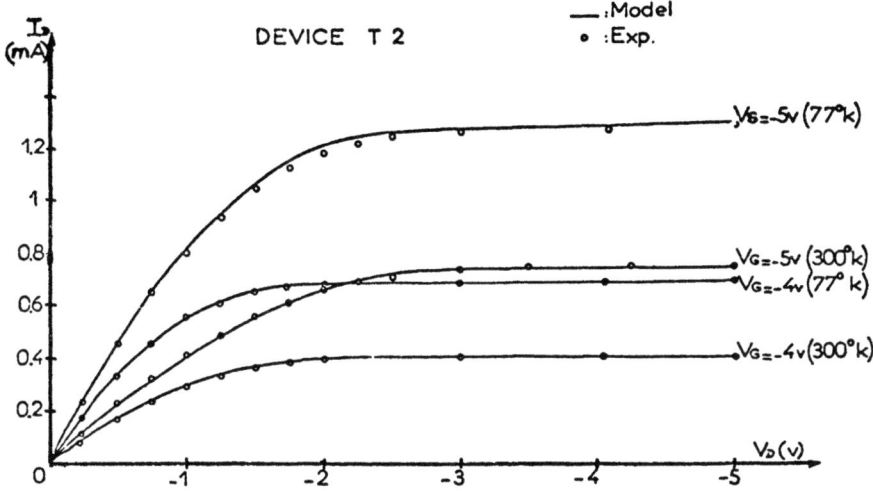

Fig. 12. $I_D(V_D)$ characteristics. Effects of temperature.

In particular :

- characterization of SOS devices
- cartography of DC parameters on a wafer
- cartography of dynamic parameters on a wafer
  (measurement of power consumption P and delay time τ_p for ring oscillators)
- measurement and parameters of C(V) curves (N_{SS}, N_B)
- measurement of low level currents (for example, of N_{SS} and method of temperature → N_{SS} in the band gap).

In addition, the system is used for functional tests of circuits.

CONCLUSION. The CAD models developed in the laboratory assure a good compromise between the complexity of formulations and accuracy of the model. The electrical parameters of the model are in direct correlation with the physical, technological and geometrical parameters. The system of automatic acquisition of data and parameters of models, due to its versatility, allows it to be used for varied problems of research and development.
In particular, this system permits one to test the validity of the model very quickly and so, to confirm or reject the initial physical assumptions which lead to the model.

REFERENCES

[1] E. Lora-Tamayo, G. Merckel, J. Gautier, Le transistor MOS de faibles dimensions géométriques, Tech. Rep. 1160, LETI/MEA, Grenoble, France, July 1976.
[2] G. Merckel, E. Lora-Tamayo, J. Gautier, Submitted for public., IEEE Trans. on E-D.
[3] R. Van Overstraeten, G. Declerk, P. Muls, Theory of the MOS transistor in weak inversion. New method to determine the number of surface states, IEEE Trans. on E-D, 22, n°5, May 1975.
[4] G. Merckel, Internal paper - ESSDERC, Munich, Germany, September 13-16, 1976.
[5] B. Baylac, G. Merckel, CAD Models of MOSFET's and their parameter acquisition on an automatic system, ESSCIRC, Toulouse, France, September 21-24, 1976.

Section V

PROCESS MODELING

PROCESS MODELING

Walter H. Schroen

Texas Instruments Incorporated, Dallas, Texas, U.S.A.

1. INTRODUCTION*

Numerous methods of characterization of impurity depth distribution in semiconductors exist[1]. These can be broadly divided into two categories, one that determines those impurities that contribute carriers (electron or holes) to the conduction and another that determines total impurity concentration. Various methods available for measuring electrically active impurity profiles are spreading resistance technique[2,3,4], incremental sheet resistance[5] involving successive electrical measurement combined with stripping procedures, and differential capacitance technique[6,7], which depends on the change in dopant distribution and corresponding change in the capacity of a formed or an induced junction. These methods suffer from accuracy due to the nature of steps involved such as accuracy of stripping, requirement of accurate knowledge of the carrier mobilities, and lack of measuring capability for sharp profile gradients. The techniques for determining the distribution of total concentration dopant are even more involved; for example, nuclear methods[8] such as radiochemical profiling when one utilizes a nuclear reaction initiated by neutron radiation. More widely used techniques are Secondary Ion Mass Spectroscopy (SIMS)[9] and Auger Electron Spectroscopy[10]. Each of these have been used for specific impurities and require rather elaborate instrumentation. A newer technique is reported which uses glow discharge[11] from impurity ions as they are sputtered. All of these techniques are extensively used in the characterization of the processes. But the destructive, cumbersome, and time consuming nature of these techniques limit their use for special purposes such as research applications and to verification of theories and models. These are only of limited use for on line process characterization and analysis as required by a semiconductor device or process engineer.

A numerical scheme has been developed[12,13,14] to fill the need for predicting and characterizing semiconductor processes in a manner such that a process and device engineer can use it for accurate, rapid and economical process and device design. The scheme requires prediction of the impurity profiles with an accuracy that will impact the device performance through process-device parameter correlation and provide data base for process control. The process simulator also was designed so that it would be flexible enough yet simple and efficient to simulate a wide range of process conditions and be applicable for most common donor and acceptor impurities normally used in LSI circuit technology.

The flexible, efficient, and fast mathematical scheme allows simulation of a wide range of processes. The capabilities of the scheme include analysis of the doping species arsenic, boron, phosphorus, and antimony, with diffused or ion-implanted initial conditions, for both unidirectional and bidirectional diffusions. The numerical scheme considers the impurity diffusion for sequential heat treatment cycles, at different temperatures, times, and ambients, with or without thermal transients. The analysis, furthermore, includes concentration and orientation dependent diffusion and temperature dependent segregation. The partial electrical activation of the species is determined as a function of concentration, based on complex formation. The following description is based on research by Shah[13,14].

2. NUMERICAL SCHEME *

An early diffusion analysis and profile modeling for device analysis impurity distribution is assumed to be of the Gaussian or complementary error function form. These solutions for linear diffusion equations are adequate as approximations in some applications; however, lead to incorrect estimation of the surface concentration or anomalous background concentration effect[15] when correlated with experimental characterization data such as sheet resistance and junction depth. Experimental doping distribution shows characteristic non-Gaussian distributions with flattened or square shaped distribution curves[16] with low concentration tails[17] or kinks[18] in the distribution. A more detailed analysis is required with consideration of phenomena such as concentration dependent diffusion, partial activation and segregation.

The universal process simulator described is based on a mathematical scheme that simulates redistribution of the impurities either implanted or deposited through diffusion. The redistribution of the dopant species arsenic, boron, phosphorus, and antimony is considered under conventional unidirectional diffusion or special cases of bidirectional diffusion encountered in cases such as up diffusion of an under layer during the epitaxial growth. An accurate prediction requires that scheme include effects such as concentration and orientation, temperature dependence of diffusion coefficient and coefficients of segregation at Si-SiO$_2$ boundary, partial electrical activation, oxidation and diffusion in various ambients such as O$_2$, steam or nitrogen. These are described in more detail in following paragraphs. The basis of the scheme is a well known diffusion equation

$$\frac{\partial C_T}{\partial t} = \frac{\partial}{\partial x} \left[D(C_A) \frac{\partial C_T}{\partial x} + \frac{dx}{dt} C_T \right] \quad (1)$$

where
- C_T = total dopant concentration (electrically active + inactive)
- C_A = active dopant concentration = $f(C_T, T)$
- $D(C)$ = concentration dependent diffusion coefficient
- dx/dt = rate of a propagation of Si-SiO$_2$ interface, silicon consumption during oxidation = f(ambient, crystalline orientation, temperature)

This nonlinear partial differential equation with a moving boundary condition due to propagation of Si-SiO$_2$ interface during the oxidation is further complicated if one considers effect of vacancies on diffusion. This equation then transforms into coupled multi-stream diffusion equations. Such an approach is required for species such as phosphorus which show two different species each diffusing with different diffusion coefficients[19,20]. The above diffusion equation is solved with a time dependent boundary condition that depends on the nature of the process. For example, a bidirectional diffusion, as in the case of deep implanted layers or up-diffusion of under-layers, requires conservation of total impurities:

$$\int_0^\infty C_T(x,t)\, dx = \text{Constant}. \quad (2)$$

However, for the normal process condition where one grows oxide at the time of diffusion (uni-directional diffusion), the boundary condition is

$$D(C) \frac{\partial C_T}{\partial x}\bigg|_{x=0} = (S-1)\, C(o,t)\, \frac{dx}{dt}, \quad (3)$$

where

$$S / X_m = C(o^-,t) / C(o^+,t) \qquad (4)$$

is the segregation coefficient of the doping species as a function of temperature and crystalline orientation and X_m is the ratio of the oxide thickness to thickness of silicon consumed. In addition, the following applies to both diffusion conditions:

$$C(\infty,t) = 0 \qquad (5)$$

3. INITIAL CONDITION *

In addition to the spatial boundary conditions, the above described diffusion equation requires a redistribution that is used as an initial condition. In the case of the implanted layers this corresponds to a distribution of dopants as they come to rest after the implantation. The unique feature of a precise control of the total change introduced and its relative depth is reflected in an accurate specification of this initial condition. For a given ion-substrate combination three major factors influence this profile; namely, implanted dose, incident ion energy, and crystallographic orientation relative to the ion beam trajectory. Impurities implanted have Gaussian distribution with characteristic range (RP) and straggling factor (DRP). Each of these are functions of ion energy[21,22]. In addition to the Gaussian portion the profile shows a low concentration exponential tail region which has been attributed to channeling[23] and diffusion effects during implantation[17]. The initial distribution is modeled as

$$C_T(X,0) = C_{max} \exp\left(-\frac{X-RP+X_o}{DRP}\right)^2 + A \exp(-B(X-X_o)) \qquad (6)$$

The first term corresponds to the Gaussian distribution with an effective range RP and a range straggling DRP. These are calculated from the function of implant energy fitted to experimental data. The second term representing the exponential tail is defined with factors A, B which are empirical functions of dose and implant energy. X_o is the equivalent thickness of the oxide through which the implantation is carried out. X_o is calculated using the effective stopping power of the oxide relative to bulk silicon.

For diffused layers distributions can be determined with appropriate diffusion conditions. However, in actual manufacturing environment these have to be adjusted for a given process and equipment condition.

4. DIFFUSION

The basis of the numerical scheme is the solution of the well-known diffusion equation for a set of process-dependent boundary conditions. The partial differential equation becomes nonlinear for most species due to the concentration dependence of the diffusion coefficient. For a given dopant species, the diffusion coefficient is also a function of process temperature and crystallographic orientation. The equation is further complicated by adding a term for the moving silicon/silicon-dioxide interface during oxide growth, which also contains a factor to account for the segregation of doping atoms into the growing oxide.

4.1 Boron diffusion model.**

The following discussion is based on the model by Prince and Schwettmann[24].

The coordinate system used in formulating the boron diffusion equation and boundary conditions is shown in Figure 1. Origin of coordinates is taken at the Si-SiO$_2$ interface. In this coordinate system, diffusion of the boron atoms is governed by

$$\frac{\partial N}{\partial t} = \frac{\partial}{\partial y}\left(D \frac{\partial N}{\partial y}\right) + m \frac{dX_o}{dt} \frac{\partial N}{\partial y} \quad (7)$$

where

$N(y,t)$ = boron concentration
D = boron diffusion coefficient in silicon
m = thickness of silicon consumes in growing unit thickness of SiO$_2$, assumed to be 0.44
$X_o(t)$ = SiO$_2$ thickness.

The initial condition is

$$N(y,o) = f(y,o) \quad (8)$$

where $f(y,o)$ is presumed known through LSS theory[21] or experiment. Boundary conditions are

$$N(\infty, t) = 0 \quad (9)$$

$$D \frac{\partial N}{\partial y} = (k-m) N(0,t) \frac{dX_o}{dt} \quad (10)$$

where k is the segregation coefficient of boron in the Si-SiO$_2$

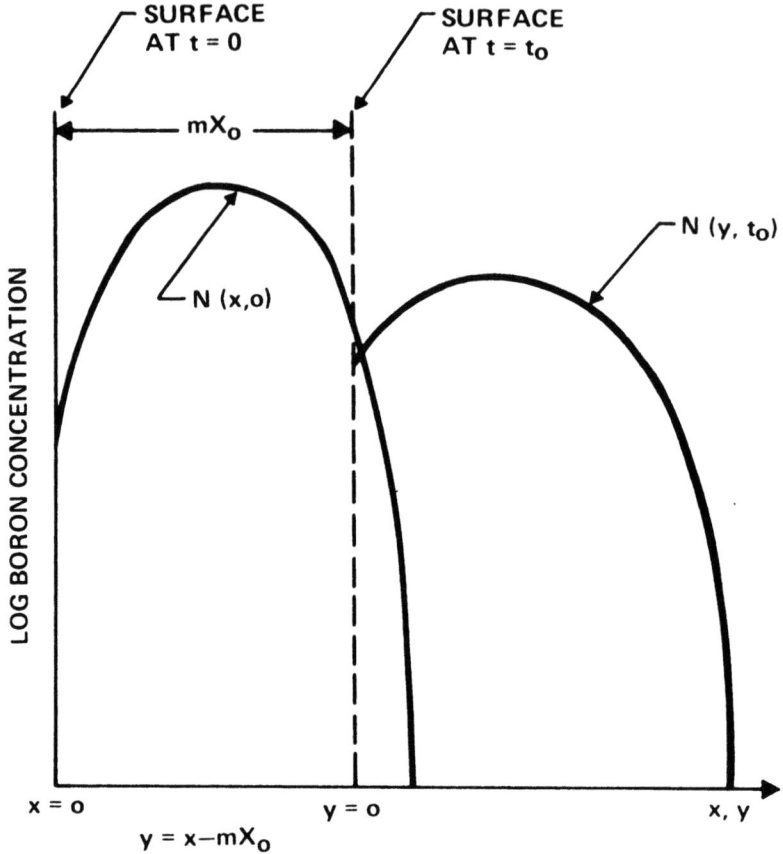

Fig. 1. Coordinate system for solution of diffusion equation.

system, $k=[N(0^-,t)]/[N(0^+,t)]$. This last boundary condition is derived from the requirement of conservation of boron atoms and under the assumption that boron diffusion in the oxide is slow relative to the oxidation rate $(dX_o)/(dt)$. This last condition is met for steam oxidation.

The boron diffusion coefficient was assumed to be the low-concentration value D_i enhanced by the local electric field effect[25]. In addition, possible increase of the diffusion coefficient due to vacancy solubility enhancement[26] caused by heavy doping was accounted for by including a multiplicative factor $[1+(AN)/n_i)]$ where A is a constant between 0 and 1 and n_i is the intrinsic carrier concentration at the diffusion temperature. Uncertainties in energy band and defect parameters at diffusion temperatures prevent accurate calculation of A. However, estimates[26,27] range from 10^{-1} to 2×10^{-2} depending on temperature. The total expression used for the diffusion constant was thus

$$D = D_i \left(1 + \frac{AN}{n_i}\right) \left[1 + \frac{1}{\sqrt{1 + 4\left(\frac{n_i}{N}\right)^2}}\right] \quad (11)$$

Oxide thickness was taken to be[28]

$$X_o = \frac{k_p}{2k_1} \left\{ \left[\frac{4(t+t^*)k_1^2}{k_p} + 1\right]^{1/2} - 1 \right\} \quad (12)$$

where k_p and k_1 are respectively the parabolic and linear oxidation rate coefficients, t is the oxidation time, and t* is a parameter related to the oxide thickness which exists prior to the beginning of an oxidation cycle.

The system of Equations (7) through (10) was solved on a computer by finite difference techniques using a Crank-Nicolson type approach[29]. The nonlinearity engendered by concentration dependence of D required use of a two-step (iterative) solution algorithm. Truncation error of the solution algorithm was found to be on the order of 0.01% for the concentration levels examined. The output of the computer solution was boron concentration versus depth into the silicon, and sheet resistance of the boron layer. Sheet resistance was calculated using the recent hole mobility data[30,31].

Boron diffusion models by other researchers[32-34] followed similar approaches and delivered comparable results. Various anomalous effects such as concentration, orientation dependent diffusion, and segregation have been discussed[32,33]. More recently more involved multistream diffusion mechanism have been suggested for implanted boron layers[34].

For verification of the model, the predicted dependences are compared with measured values. Referring to the work by Prince and Schwettmann[24], Figure 2 shows experimental values of sheet resistance as a function of steam oxidation time for three boron doses at 1200°C diffusion temperature. The maximum error is ±10% in sheet resistance for this data. Figure 3 shows boron profiles obtained by incremental sheet resistance profiling. Accuracy of this data is estimated to be ±5% for the distance scale and ±20% for concentration.

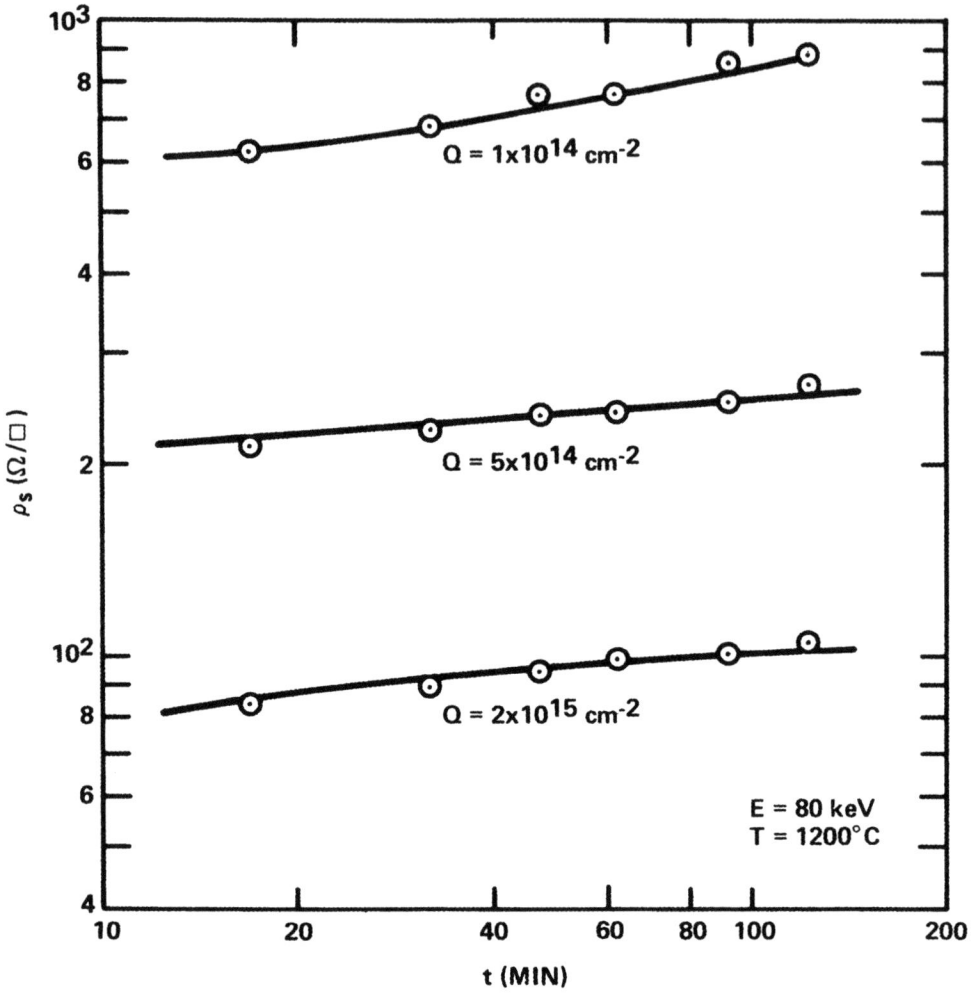

Fig. 2. Sheet resistance of boron layer versus steam oxidation time, T = 1200°C.

Boron diffusion coefficient at 1000°C, 1100°C, and 1200°C was determined by iterating the computer calculation and comparing the part of the calculated boron profile past the peak concentration point to the experimental boron profiles (analogous to Figure 3). The initial boron profile was taken to be of gaussian form with range 0.285 μm and range straggling 0.075 μm for these calculations. These values for range and range straggling are different from the theoretical values[35]. However, it has been determined through compilation of results in the literature and

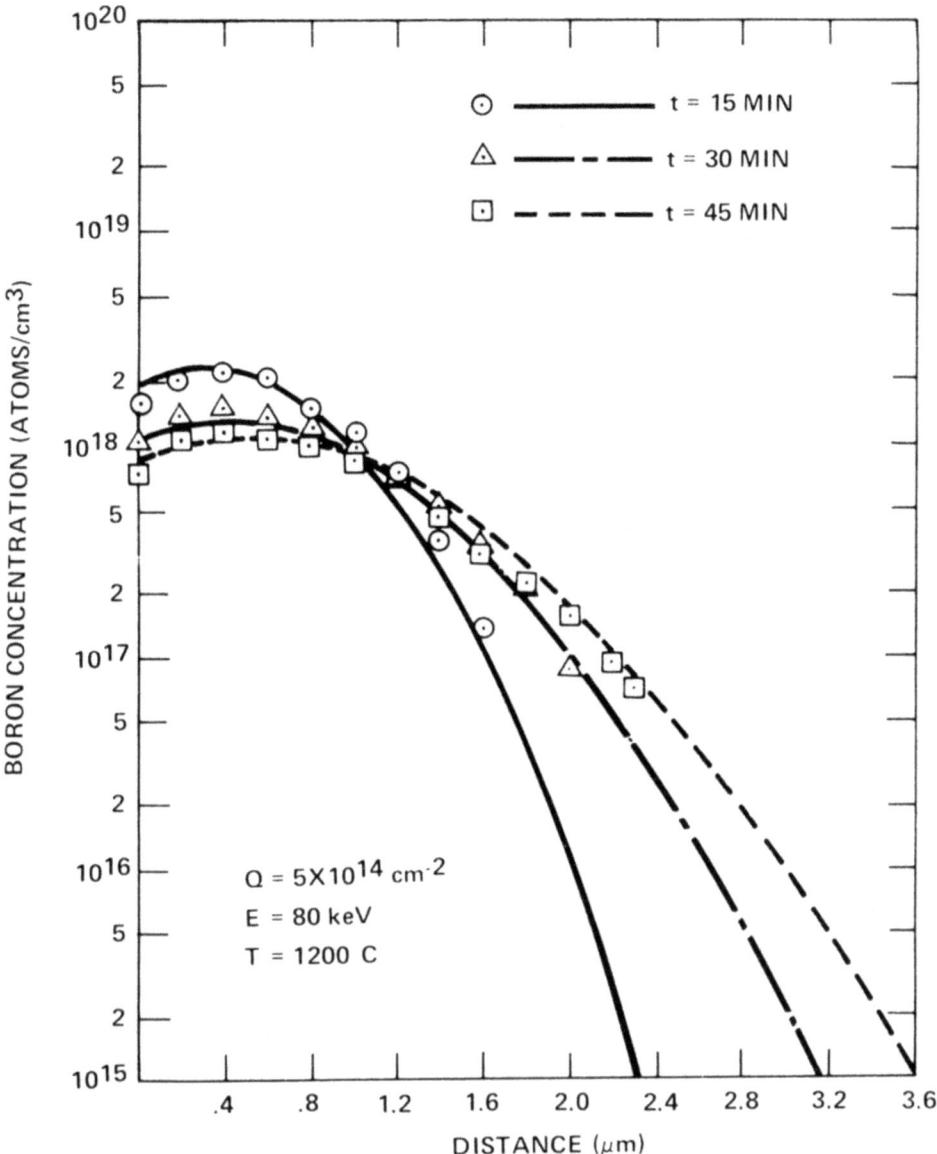

Fig. 3. Calculated and actual boron concentration profiles for oxidation at 1200°C.

independent experiments that these values best characterize an 80 keV boron profile, which is in fact not gaussian[36]. No explicit account was taken in the numerical solution of annealing of implantation damage. This annealing was assumed to occur in times short in comparison with the diffusion times. Within wide bounds, the value of segregation coefficient used in this comparison does not affect the calculated profile past the peak concentration ratio point.

Various values for the constant A were used in making the above comparison. No values for A within the limits of 10^{-1} to 2×10^{-2} gave profiles significantly different from those obtained for A=0 for the dose levels used in the experiment. Figure 4

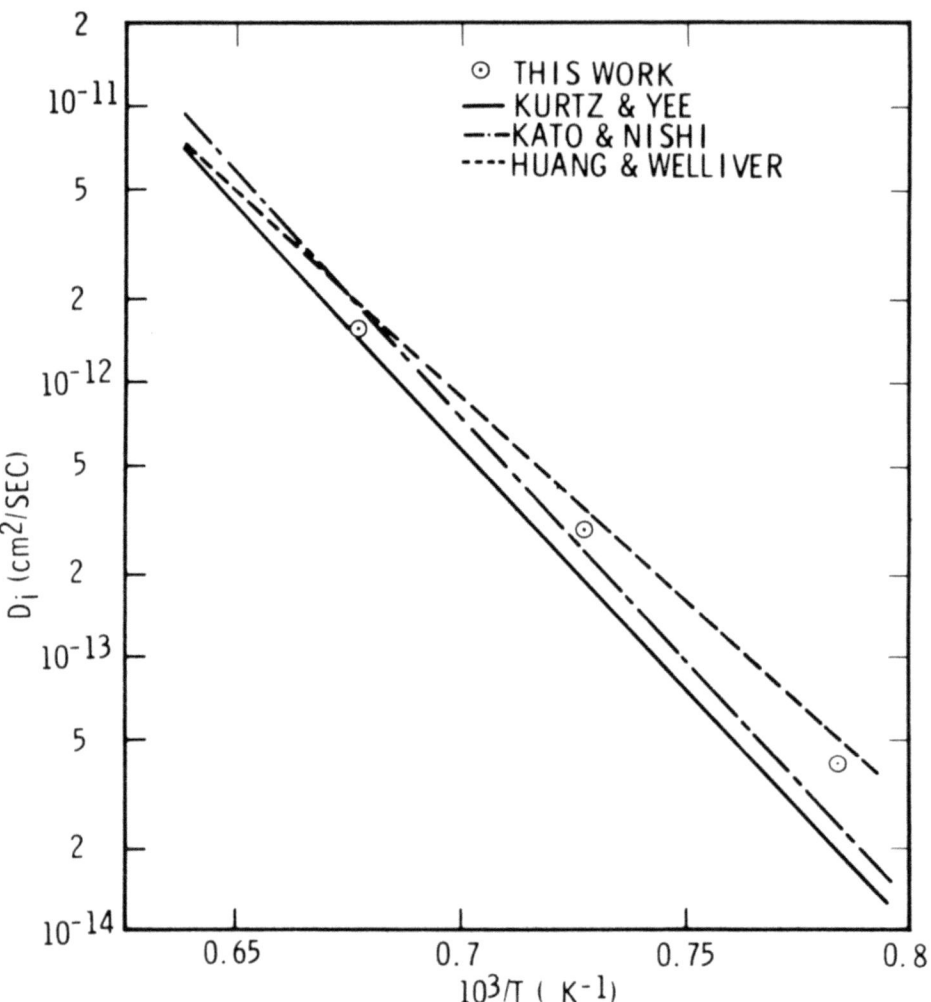

Fig. 4. Intrinsic boron diffusion coefficient versus inverse temperature (after Prince and Schwettmann[24]).

shows boron diffusion coefficient as a function of inverse temperature, as determined by the comparison. For diffusion at 1000°C, heavier weight was given to matching experimental profiles at longer times. For this temperature, experimental profiles exhibited apparent enhanced diffusion for short (e.g., 15 minutes) diffusion times. Also shown in Figure 4 are examples of boron diffusion coefficient as determined by other workers. It is seen that the results of this work are in general agreement with previous results over the range of common diffusion temperatures. The diffusion coefficient shown in Figure 4 is approximated by

$$D_i = 0.0322 \exp(-3.02/kT). \tag{13}$$

Segregation coefficient k was determined by iteration of the computer solution, using values for the diffusion coefficient calculated from Equation (13). Boron concentration profiles and values of sheet resistance calculated in the iteration process were compared to experimental profiles and experimental values of sheet resistance. Sensitivity of the calculated results to variations in value of the segregation coefficient is illustrated in Figures 5 and 6. This comparison was made for diffusion at 1200°C and for boron dose equal to 2×10^{15} cm^{-2}. Results calculated for temperatures lower than 1200°C show reduced sensitivity to k. This is due to increase in the ratio of parabolic oxidation rate coefficient to boron diffusion coefficient, with decreasing temperature.

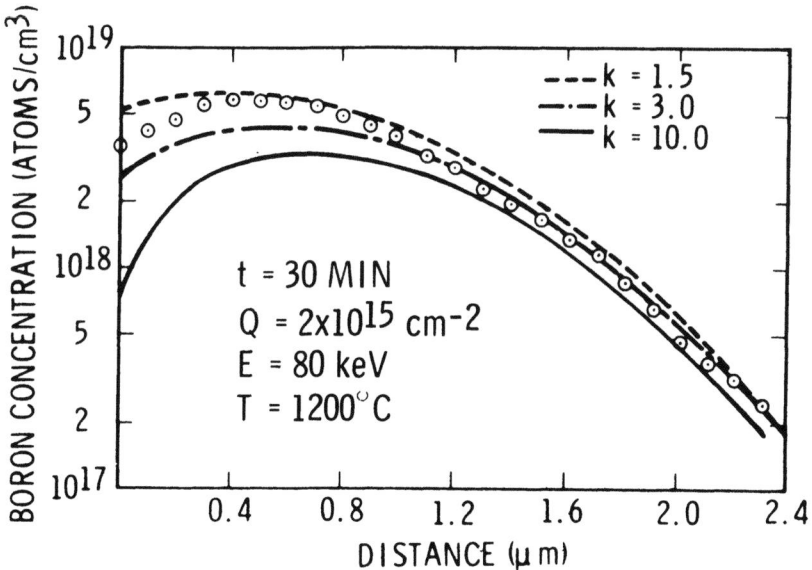

Fig. 5. Actual boron concentration profile and calculated profile for three values of segregation coefficients.

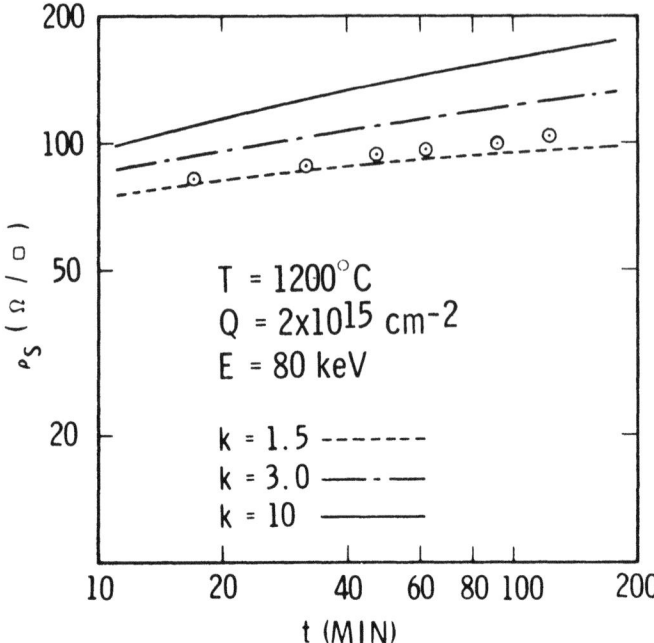

Fig. 6. Actual sheet resistance of boron layer and calculated behavior for three values of segregation coefficient.

Figure 7 shows segregation coefficient determined in this work, as a function of inverse temperature. The "error bars" in Figure 7 indicate the range of values of k for which satisfactory fit to experimental data was obtained. The value of k was found to range from 1.8 to approximately 10 over the range 1200°C to 1000°C. The relative sizes of boron diffusion coefficient and the parabolic oxidation rate coefficient at 1000°C are such that the comparison of experiment and calculation is insensitive to the value of segregation coefficient. Thus, the value k = 10 at 1000°C is an estimation and is not shown in Figure 7. It could be in error by a factor of two in either direction. The segregation coefficient of Figure 7 is best fit by the expression

$$k = 2.33 \times 10^{-4} \exp(1.135/kT). \tag{14}$$

The best-fit calculated boron concentration profiles and sheet resistance of the boron layers are superimposed on the experimental data of Figures 2 and 3. These calculated quantities were obtained assuming A = 0, using the diffusion coefficient of Figure 4, and using values of segregation coefficient shown as experimental points in Figure 7. Note that in the iteration

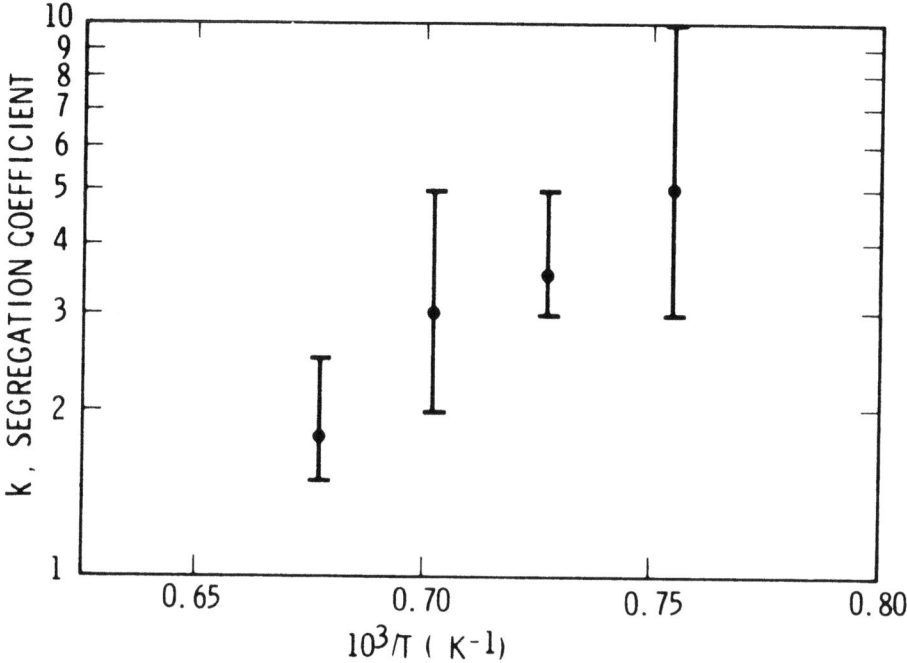

Fig. 7. Segregation coefficient versus inverse temperature.

process and in the calculated curves of Figures 2 and 3, segregation coefficient was allowed to vary with temperature but was assumed to be independent of surface concentration.

In comparing the best fit calculated boron profiles and experimental profiles, it is seen that agreement is good for all cases but diffusion for short times at low temperatures (e.g., 15 minutes at 100°C). The discrepancy may be due to the annealing of implantation-related damage, which at 1000°C could take an amount of time appreciable compared to 15 minutes. Investigations by the authors of 8×10^{14} cm^{-2} B^{11} implantations annealed without oxidation at 900°C indicate a value for boron diffusion coefficient at 900°C in agreement with Equation (13).

4.2 Arsenic and phosphorus diffusion model.*

For general diffusion features, impurity diffusion in semiconductors have been very extensively studied[37-40]. Concentration dependent diffusion and complex formation play an important role in accurate prediction of impurity distributions. The diffusion coefficient is modeled as

$$D(C) = D_o \exp\left(\frac{-E_A}{KT}\right)\left[1 + A\frac{C_A^2}{n_i C_T}\right] \qquad (15)$$

where the multiplier term outside the bracket is a temperature dependent intrinsic diffusion coefficient. The term in the bracket has the familiar linear concentration dependence. The diffusion coefficient peaks due to the complex formation described in more detail below.

Donor impurities show strong dependence on the concentration [16,41,42]. The diffusion coefficient is shown to increase linearly depending on the ratio of dopant and intrinsic carrier concentrations, change in the Fermilevel[43] and corresponding change in vacancy solubility[26]. A monotonic increase in the diffusion coefficient, however, peaks at concentration determined by the complex formation[44]. The diffusion coefficient for arsenic is modeled as

$$D(C) = D_i \left[1 + \frac{A\, C_A/n_i}{1 + \eta_1 C_A^3 + \eta_2 C_A^7}\right], \qquad (16)$$

where A is the enhancement factor and η_1 and η_2 are the factors determined from the complex formation analysis and n_i is the intrinsic carrier concentration. Intrinsic diffusion coefficient D_i has a functional form of

$$D_i = D_o \exp(-E_A/kT) \qquad (17)$$

The diffusion coefficient as a function of concentration for arsenic is shown in Figure 8 for two temperatures.

The impurities such as arsenic, phosphorus, and boron show partial electrical activation in silicon especially when the concentration is high in the range of $>10^{20}/cm^3$ and approach solid solubility. Accuracy of the profile determination and characterization of the layer in terms of the sheet resistances critically depends on the accurate modeling of the partial activation of the impurities. No universal theory or model exists for explaining the nature or mechanisms for the partial activation of the dopants. Discrepancies between concentration dependent diffusion and actual diffusivity at concentrations beyond 10^{20} were observed for arsenic. These have been explained in terms of cluster formations[38,45] and arsenic vacancy complexes[44]. It was shown by Fair that a fraction of the dopants that are activated in a well annealed implanted layers is considerably higher as compared to the diffused layers.

Based on the theory of arsenic complex formation, the model

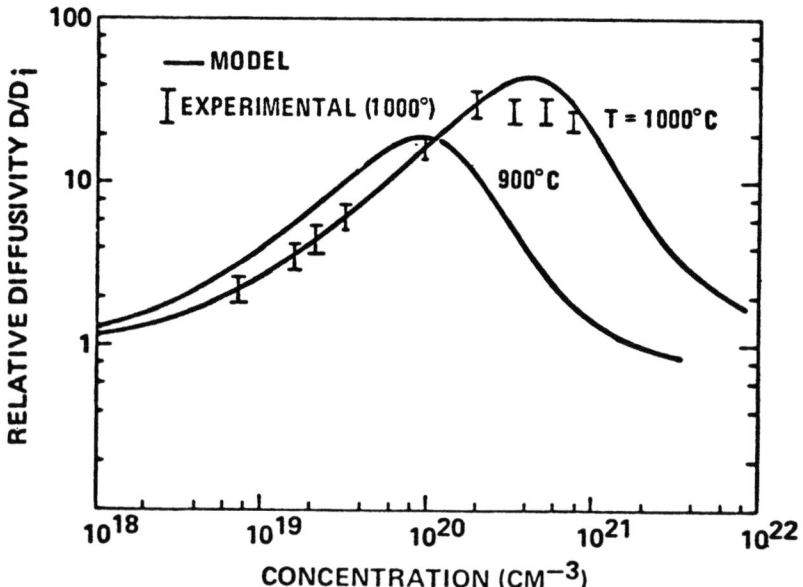

Fig. 8. Relative diffusivity of arsenic in silicon.

by Shah[13,14] yields a relation between total and electrically active concentration that is a function of concentration and temperature. The model can be easily extended to boron and phosphorus by fitting experimental activation data. This model assumes formation of vacancy-arsenic complexes that make two or four arsenic ions immobile and electrically inactive.

$$C_T = C_A + \sum_m C_A [V]^m \qquad (18)$$

where [V] is vacancy concentration.

Each term in Eq. 18 is determined by a rate reaction

$$mC_A^+ + m\,e^- + [V]^m \rightleftarrows [V^m\,mC_A] \qquad (19)$$

The rate reactions can be used under condition that $C_A >> n_i$ to arrive at an equation relating active ion concentration to the total impurity concentration.

$$C_T = C_A + n_1 C_A^4 + n_2 C_A^8 \qquad (20)$$

for $C_A < C_{Amax}$, where C_{Amax} is experimentally determined maximum active concentration satisfying a relation

$$C_{Amax} = C_{Amax_0} \exp(-E_{cmax}/kT) \tag{21}$$

Effective diffusivity can be related to the concentration to arrive at Eq. 16 using an analysis similar to the one described by Fair[44]. Constants η_1, η_2 depend on the activity coefficients of the vacancies, electrons and the active arsenic ions as described by Hu[38]. These coefficients determine the activation as a function of the temperature. Functional relation of activation and total concentration for different temperatures modeled for arsenic are shown in Figure 9.

Figure 10 depicts the sheet resistance as a function of ion implant at anneal time and arsenic implant dose. The sheet resistance is a very important parameter that is used for rapid "on line" characterization of the implanted or diffused layers and for process control. Variation of the sheet resistance with time and temperature under different experimental conditions is also very often used as a check on the process models. Knowledge of the carrier mobility in an implanted layer is required to determine the resistivity of the layers. An experimental technique of incremental sheet resistance for determining the electrically active concentration depends on the functional dependence of the mobility and concentration. Irvin's curves[46] are generally

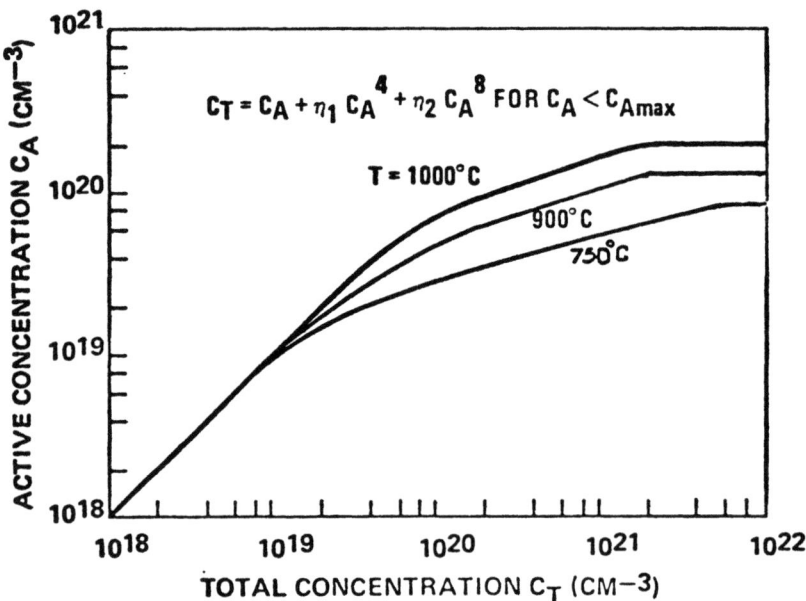

Fig. 9. Function dependence of active and total arsenic concentration in silicon.

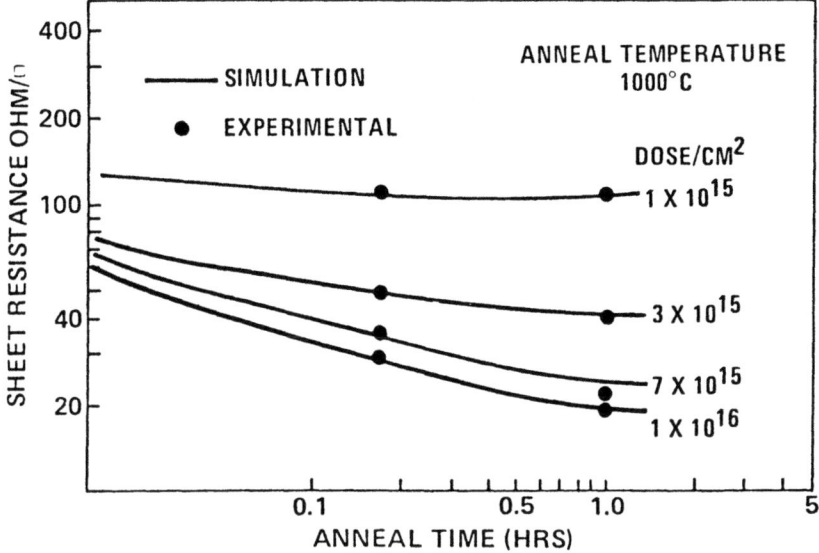

Fig. 10. Sheet resistance of implanted arsenic layer as a function of anneal time.

used for profiling. A numerical model was used which has a functional form shown in the following equation:

$$\mu(n,p) = \frac{\mu^o_{(n,p)1}}{1+(C_A/C^o_{(n,p)1})^{\alpha(n,p)1}}$$

$$+ \frac{\mu^o_{(n,p)2}}{1+(C_A/C^o_{(n,p)2})^{\alpha(n,p)2}} , \qquad (22)$$

where $\mu_{np1,2}$, and $\alpha_{np1,2}$ are the fitting factors for n and p type mobilities $\mu_{n,p}$ respectively. The sheet resistance of the layer is calculated from a conventional integral relation:

$$R_s = 1 \bigg/ \left[\int_0^x C_A(x) \cdot \mu(c) dx \right] . \qquad (23)$$

Equation 23 can be fitted to the experimental mobility data at high concentrations which has been shown to be different from the Irvin's curves[47].

Figure 11 shows an arsenic implanted layer with various doses at 60 keV all annealed at 1000°C for 60 minutes. The predicted active and total concentration distributions are superposed with

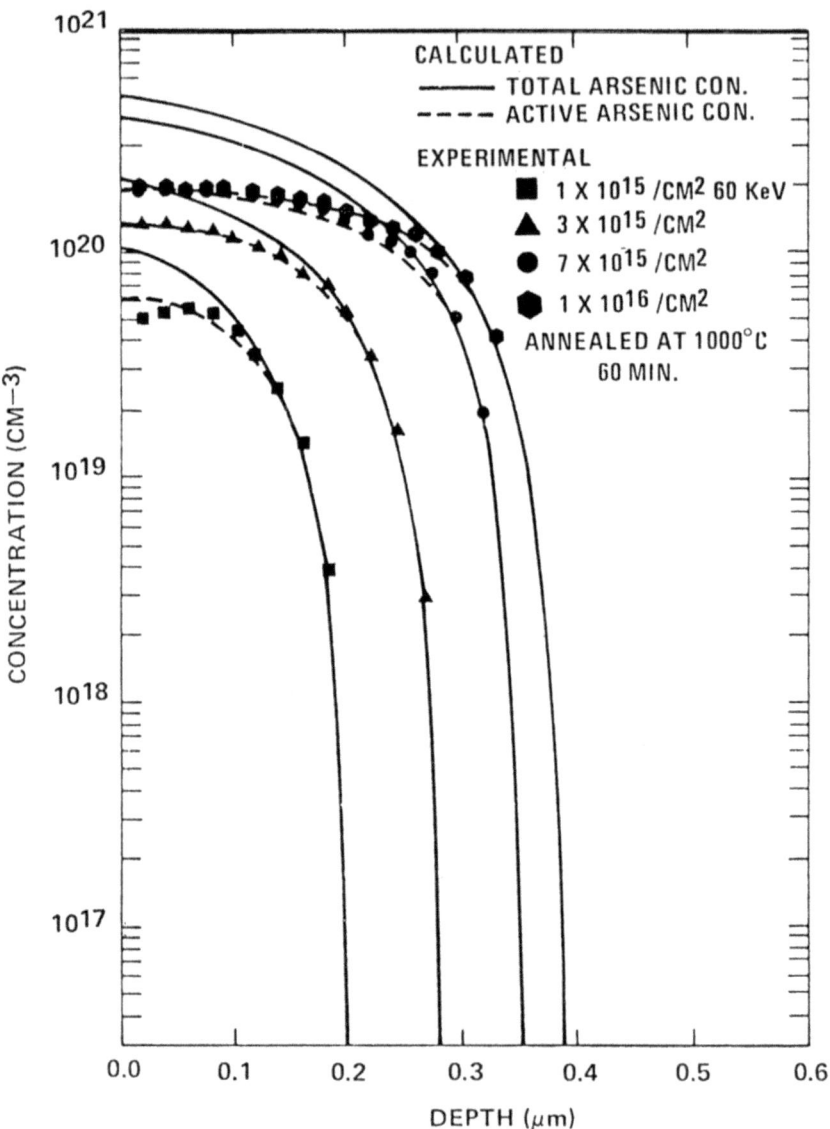

Fig. 11. Arsenic diffusion for different implant conditions.

incremental sheet data. The agreement is excellent. The effect of the sequential diffusions and annealing is typified in Figure 12. The arsenic implanted layer is annealed in an air-stream - N_2 cycle for 5-4-25 minutes respectively at 1000°C. In the oxidizing ambient, especially during steam oxidation, the pileup of arsenic at the interface is quite obvious. The tail present at the beginning is no longer present after 25 minutes N_2 cycle. This is due to enhanced diffusion at high concentrations.

A modeling of phosphorus diffusion is more complex due to the fact that profiles, especially after low temperature anneals, show pronounced low concentration tails which diffuse with anomalously

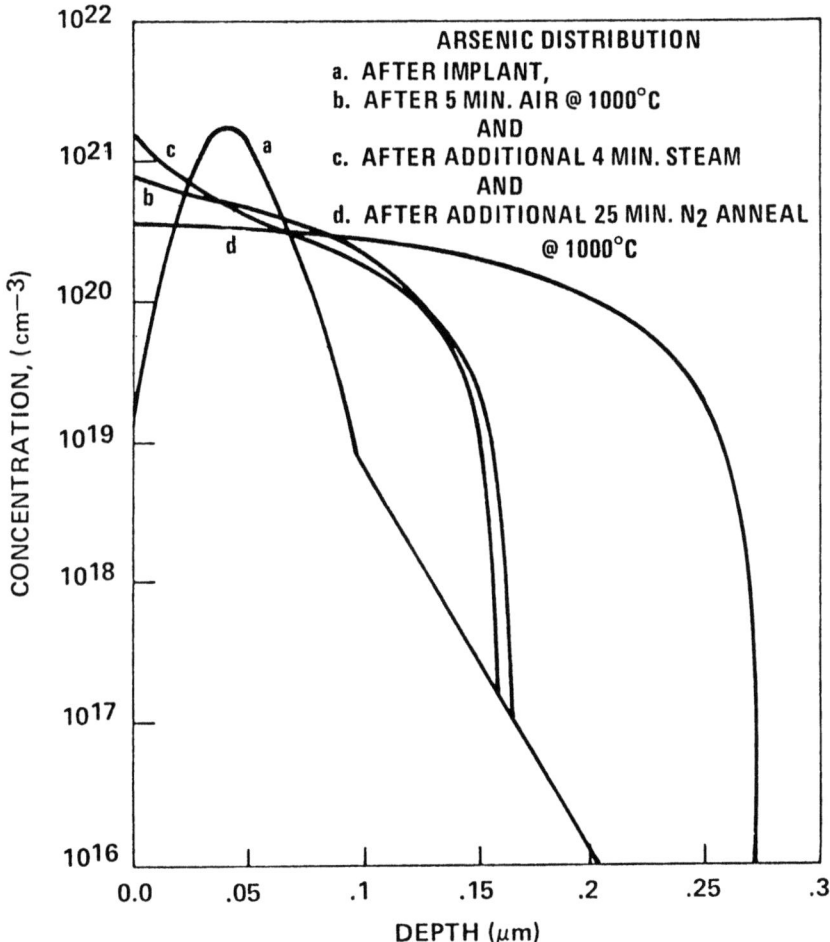

Fig. 12. Sequential diffusion of implanted arsenic layer.

high diffusion coefficients. An empirical diffusion model with two streams and a surface layer was suggested by Tsai[19]. The diffusion coefficients for slow and fast species were calculated by fitting the experimental data to a model. Although no specific diffusion mechanism has been associated for each of these species, various diffusion mechanisms such as through excess vacancy generation[37,48] and through interstitial processes[49] have been suggested. The work of Shah[13,14] uses an interactive two-stream model with partial activation, each stream diffusing with different characteristic diffusivity. Shah proposed[50] that the interaction between the species is charge as well as vacancy assisted. Typical diffusion profiles, their relation to the previous model, and experimental data are shown in Figure 13.

4.3 Heat treatment, oxidation and segregation.*

The redistribution of the dopants is determined by solving the partial differential equation. Since various aspects of the distribution such as activation and segregation between oxide Si interface are dependent on temperature and also on the time, the differential equation is solved with time dependent boundary conditions. A unique feature of the modeling scheme is that it computes redistribution due to sequential anneal cycles in different ambients and at different temperatures. Any number of steps can be simulated in a sequence as is required in a complex semiconductor process technology.

The rates of oxidation are functions of temperature, ambient and orientation. The oxidation is modeled using the thermal oxidation model by Deal and Grove[51] based on the diffusion of oxidant from an ambient through the existing oxide at Si-SiO_2 interface. The oxide thickness is governed by an equation

$$X_o^2 + AX_o = B(t + \tau) \tag{24}$$

where X_o is oxide thickness.

A, (B/A) are parabolic and linear rate constants respectively, each of these are functions of temperature, orientation and ambient and are modeled accordingly. τ is the characteristic time determined by initial oxide thickness. Impurity segregation defined by a factor S in Eq. 3 determines the redistribution of the impurity at Si-SiO_2 interface. Donor ions tend to segregate into Si causing the impurity pile up at interface when oxide is grown. However, acceptor ions such as boron tend to segregate into oxide leading to loss of boron to oxide during oxidation. This has been modeled using the temperature and orientation dependent segregation coefficients.

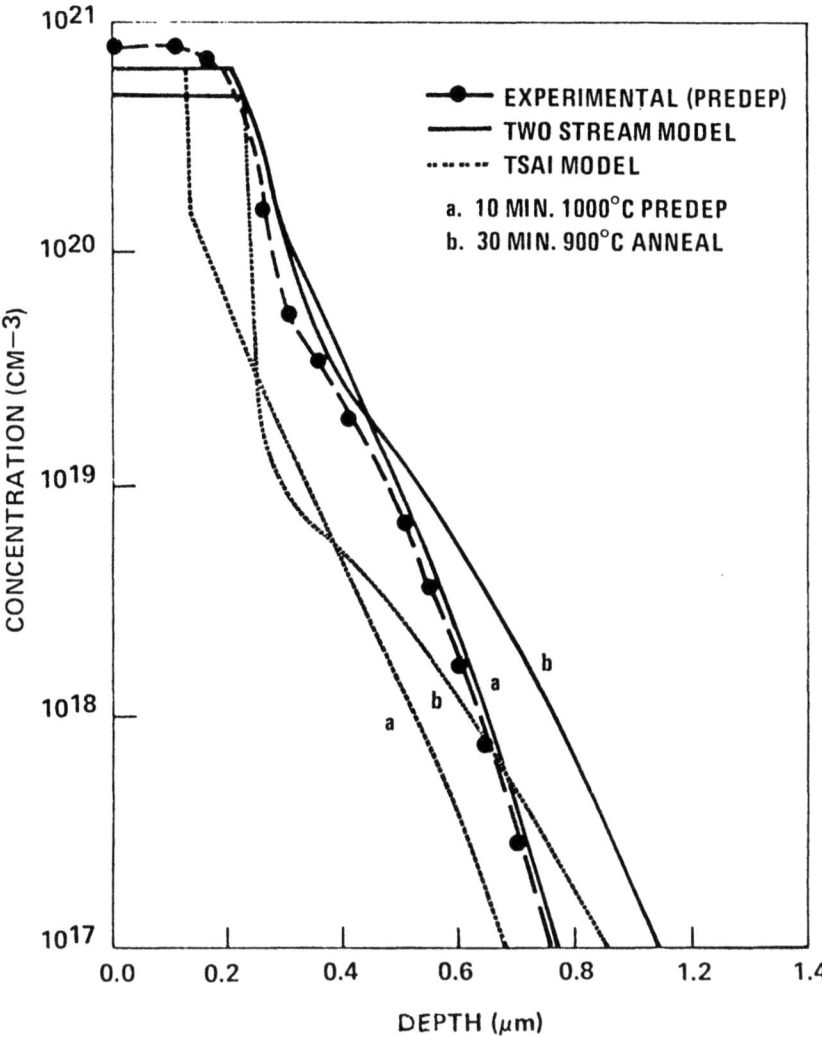

Fig. 13. Two-stream model for phosphorus deposition and diffusion.

Use of time dependent boundary condition described above allows simulation of thermal transients and their effects on the impurity distribution by appropriately modeling heat, cool down temperature-time variation.

Usefulness of the solution scheme as an effective design aid depends not only on simulation accuracy but also on its numerical attributes such as flexibility, efficiency and operational economy for repeated use in a design optimization cycle. The program uses a universal set of equations for diffusion, implantation, activation and mobility variation with a different set of parameters for a given dopant allowing flexibility in terms of dopants and process conditions analyzed. Important numerical parameters such as spatial and time steps are automatically determined from the boundary condition as the solution evolves. The result being a numerical scheme that is user oriented requiring only process inputs normally available to process engineers.

4.4 Process simulation as design aid.*

Shah has pointed out[13,14] that a wide range of processes have been simulated using the simulator. The entire LSI process can be simulated for optimization of the process[52-55] considering one dopant at a time. Typical dopant profiles through an NPN transistor determined numerically are shown in Figures 14 and 15. Figure 14 refers to a standard bipolar transistor with phosphorus emitter at the surface. Figure 15 describes an inverse vertical transistor of an I^{2}L circuit with surface collector, high-doped and low-doped bases, epitaxial layer, buried layer, and substrate. Direct verification of the profiles was done using a number of

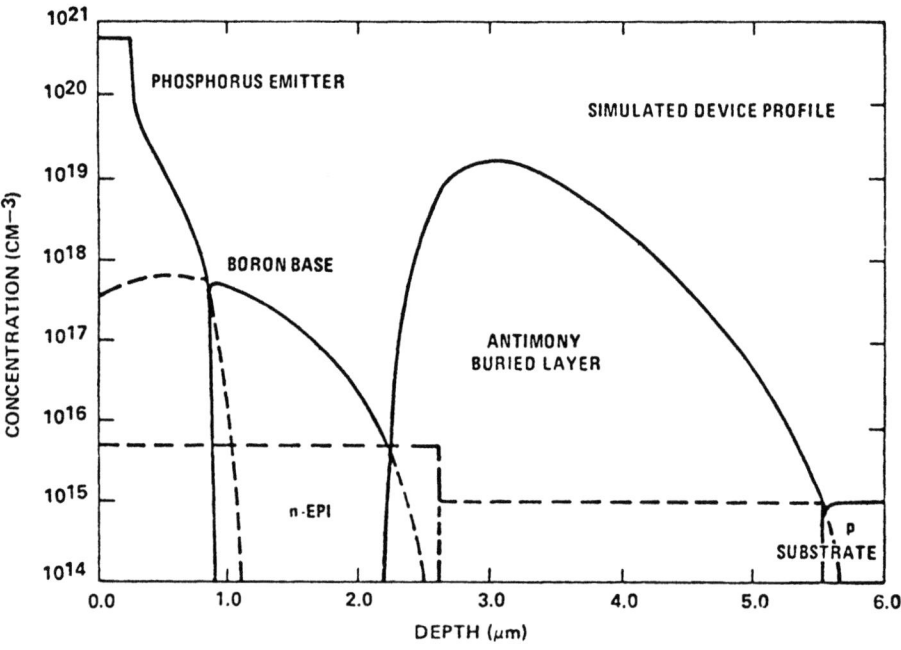

Fig. 14. Composite impurity profile for a device generated using process simulator.

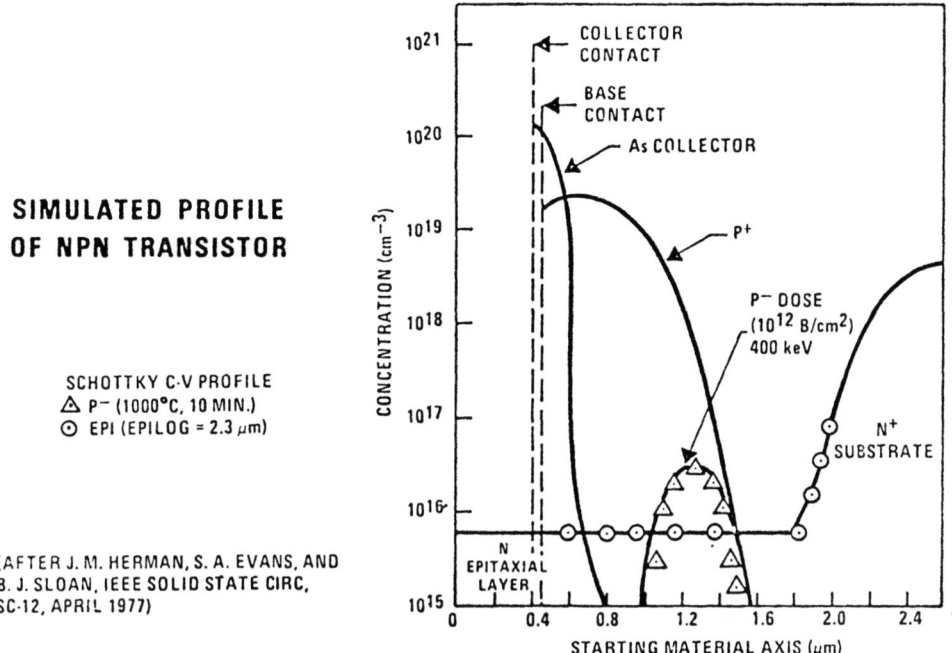

Fig. 15. Simulated profile of NPN transistor.

the characterization schemes such as incremental sheet, spreading resistance and CV measurements. Indirect verification can be done through data such as sheet resistance and junction depth under different implant and anneal conditions to indicate good model fit for diffusion, activation and segregation.

In Figure 16, boron was diffused into the substrate, before antimony was diffused at high concentrations. The question deciding the process design was, under which conditions of time, temperature, and concentration the boron dopant would be prevented from diffusing ahead of the antimony dopant into the growing epitaxial layer. A thin p-layer as depicted in Figure 16, forming junctions between the n-epi and the n-antimony, was to be avoided. Process decisions and definitions like these make the computer process simulator a valuable design aid.

Process Design Analysis and Optimization. Detailed device profiles can be generated for a given process. The process conditions such as temperature, duration in a given ambient can be optimized to give required sheet resistance, junction depth or to control the total charge in a bipolar transistor base or the surface concentration for a MOS device threshold control. Process

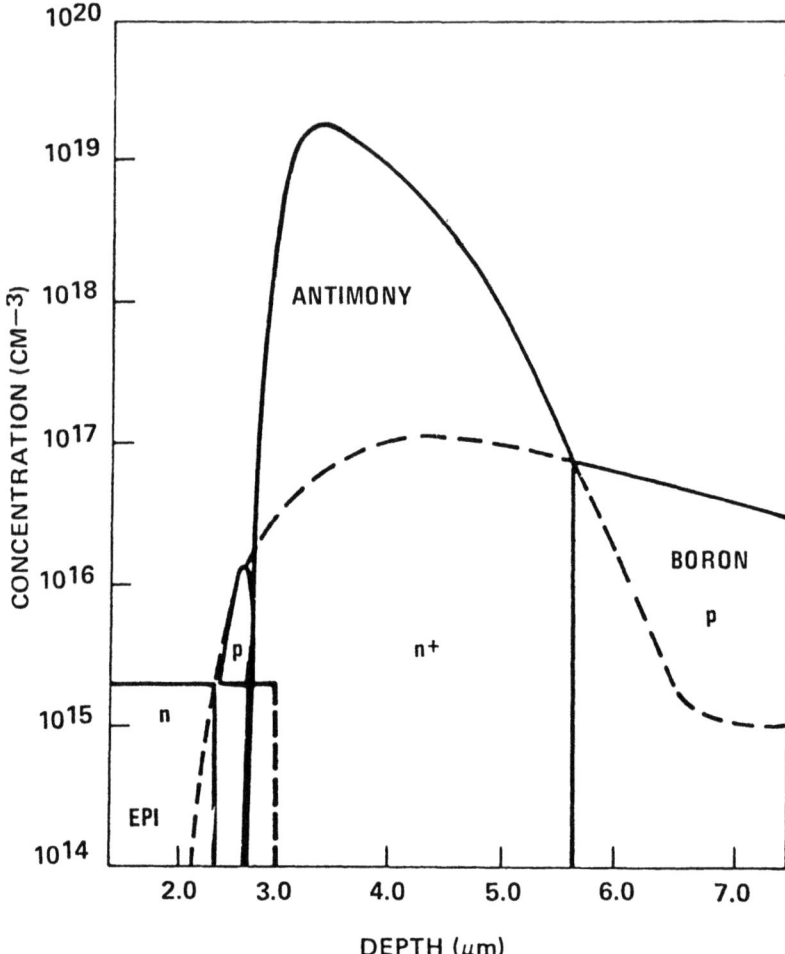

Fig. 16. Distribution of antimony and boron during epi growth and subsequent process.

parameter optimization is possible through selection of appropriate initial conditions such as dose and/or energy of implant to obtain required final distribution. Using the detailed profiles device parameters such as gain, junction capacitances, resistance, threshold voltage, and leakage current can be optimized.

Sensitivity Analysis. The process simulators can be useful in understanding a process dependence on various process parameters. This is a very important application in a manufacturing environment where it is necessary to design process with optimum yield and widest possible design window for various process parameters.

The sensitivity analysis also plays an important role in process control and optimization.

Thermal Transient Analysis. Interprocess thermal transient can play an important role in determining the device parameter such as gain of transistor. In such cases estimation of equivalent heat treatment cycle will be cumbersome and inaccurate. Numerical characterization schemes can very effectively simulate such thermal transients and their effect or redistributions.

REFERENCES

1. Various depth profiling techniques are recently reviewed in Proc. on Silicon Device Processing, ed. by C.P. Marsden, National Bureau of Standards, Spec. Publ. 337, Gaithersburg, Md., pp. 99-155, 1972, and also in J. Vac. Sci. Tech. 12, p. 356-405, 1975.
2. W.H. Schroen, G.A. Lee, and F.W. Voltmer, "Comparison of the spreading resistance probe with other silicon characterization techniques," Proc. Spreading Resistance Symp., ed. by J.R. Ehrstein, NSB-SP 400-10, Gaithersburg, Md., 1974, pp. 155-168.
3. W.H. Schroen, "Application of the spreading resistance technique to silicon characterization for process and device modeling," Proc. Spreading Resistance Symp., ed. by J.R. Ehrstein, NSB SP 400-10, Gaithersburg, Md., 1974, p. 235-247.
4. Y.T. Yeh, "Current status of the spreading resistance probe and its application," in Silicon Device Processing, ed. by C.P. Marsden, NBS Spec. Publ. 337, Gaithersburg, Md., 1972, pp. 111-122.
5. R.P. Donovan and R.A. Evans, "Incremental sheet resistivity technique for determining diffusion profiles," in Silicon Device Processing, ed. by C.P. Marsden, NSB Spec. Publ. 337, Gaithersburg, Md., 1972, pp. 123-131.
6. J. Hilderbrand and R.D. Gold, "Determination of Impurity Distribution in Junction Diodes from Capacitance-Voltage Measurements," RCA Review 21, pp. 245-252, 1960.
7. T.E. Seidel, "Ion implantation in semiconductors," ed. by I. Ruge and J. Graul. Springer-Verlag, 1971.
8. R.J. Masters, "Nuclear methods for the determination of diffusion profiles," in Silicon Device Processing, ed. by C.P. Marsden, NBS Spec. Publ. 337, Gaithersburg, Md., 1972, pp. 132-140.
9. H. Liebel, "Secondary ion mass spectrocopy," J. Phys. Sci. Instruments, 8, 787, 1975.
10. J.M. Morabito, "Auger electron spectrocopy," Thin Solid Films, 19, 21, 1973.
11. J.E. Greene and J.M. Wheelan, "Glow discharge optical spectroscopy for the analysis of thin films," J. Appl. Phys., 44,

pp. 2509-2513, 1973.
12. P.L. Shah and W.H. Schroen, "A process model for sequential diffusions and redistributions in silicon LSI technology," Electrochem. Soc. Meeting May 1975, Extended Abstr. No. 170.
13. P.L. Shah, "Analysis of implanted layers in silicon for semiconductor device processing," IEEE Conf. Sci. Indust. Applic. Small Accelerators, 1976, pp. 411-419.
14. P.L. Shah, "Computer-aided design and optimization for semiconductor device fabrication," Semiconductor Silicon/1977 ed. by H.F. Huff and E. Sirtl.
15. S.M. Hu, "A fictitious background doping effect," J. Appl. Phys., 42, 4102-4105, 1971.
16. E. Tannenbaum, "Detailed analysis of thin phosphorus diffused layers in p-type silicon," Solid State Electronics, 2, 14, 1961.
17. F.N. Schwettmann,"Enhanced diffusion during the implantation of arsenic in silicon," Appl. Phys. Lett., 22, pp 570-572, 1973.
18. F.N. Schwettmann and D.L. Kendall, "On the nature of the kink in the carrier profile for phosphorus diffused layers in silicon," Appl. Phys. Lett., 21, pp 2-4, 1972.
19. J.C.C. Tsai,"Shallow phosphorus diffusion profiles in silicon," Proc. IEEE, 57, pp 1499-1506, 1969.
20. R.B. Fair and J.C.C. Tsai, "A quantitative model for the diffusion of phosphorus in silicon and the emitter dip effect," to be published.
21. J. Linhard, M. Scharff and H.E. Schiott, "Range concepts and heary ion ranges," Kgl. Danske Videnskap Selskab, Mat.-Fys. Model, 33, pp 1-42, 1963.
22. W.S. Johnson and J.F. Gibbons, Projected Range Statistics in Semiconductors, Stanford, California, Stanford University Press, 1969.
23. J.W. Mayer, L. Eriksson, and J.A. Davies, Ion Implantation in Semiconductors. New York: Academic Press, 1970.
24. J.L. Prince, F.N. Schwettmann, "Diffusion of boron from implanted sources under oxidizing conditions," J. Electrochem Soc., 121, pp 705-710, 1974.
25. K. Lehovec and A. Slobodskoy, "Diffusion of charged particles into a semiconductor under consideration of the built-in field," Solid-State Electron. 3, 45, 1961.
26. S.M. Hu and S. Schmidt, "Interactions in sequential diffusion processes in semiconductors," J. Appl. Phys., 39, pp 4272-4283, 1968.
27. J. Borel, G. Merckel, J. Monnier, P. Saintot, D. VanDorpe, M. Stern and M. Maffei, "Computer aids to device modeling and design," 1973 IEEE International Solid State Circuits Conf., Philadelphia, Pa., February 14-16, 1973.
28. A. Grove, "Physics and technology of semiconductor devices," John Wiley and Sons, New York, 1967.
29. D.U. Von Rosenberg, "Methods for the numerical solution of

partial differential equations," American Elsevier Publishing Co., Inc., New York 1969.
30. S. Wagner, "Diffusion of boron from shallow ion implants in silicon," J. Electrochem. Soc. 119, 1570, 1972.
31. B.L. Crowder, "The influence of the amorphous phase on ion distributions and annealing behavior of Group III and Group V ions implanted into silicon," J. Electrochem. Soc. 118, 943, 1971.
32. R.B. Fair, "Boron diffusion in silicon concentration and orientation dependence, background effects and profile estimation," J. Electrochem Soc., 122, pp 800-805, 1976.
33. J.W. Kolby and L.E. Katz, "Boron segregation at Si-SiO$_2$ interface," J. Electrochem Soc., 123, pp 409-412, 1976.
34. J.R. Anderson and J.F. Gibbons, "New model for boron diffusion in silicon," Appl. Phys. Lett., 28, pp 184-186, 1976.
35. W.S. Johnson and J.F. Gibbons, "Projected range statistics in semiconductors," Distributed by Stanford University Bookstore, 1969.
36. J.F. Gibbons and S. Mylroie, "Estimation of impurity profiles in ion-implanted amorphous targets using joined half-gaussian distributions," Appl. Phys. Lett. 22, 568, 1973.
37. D.L. Kendall and D.B. DeVries, "Diffusion in silicon," in Semiconductor Silicon/1969, ed. by R.R. Haberecht and E.L. Kern, Electrochem. Soc., New York, pp. 358-421.
38. S.M. Hu, "Diffusion in silicon and germanium," in Atomic Diffusion in Semiconductors, ed. by D. Shaw, Plenum Press 1973, Chapter 5.
39. A. Seeger and K.P. Chik, "Diffusion mechanism and point defects in silicon and germanium," Phys. Stat. Solidi. 89 pp. 455-542, 1968.
40. R.B. Fair, "Recent advances in implantation and diffusion modeling for the design and process control of bipolar ICs," in Semiconductor Silicon/1977, ed. by. H.R. Huff and E. Sirtl, Electrochem. Soc., Princeton, N.J., pp. 968-987.
41. S. Maekawa, "Diffusion of phosphorus into silicon," J. Phys. Soc. Japan, 17, pp. 592-597, 1962.
42. P.L. Shah and F.N. Schwettmann, "Redistribution of implanted arsenic in silicon," Electrochem. Soc. Meeting May 1975, Extend. Abstr. p. 437.
43. R.K. Jain and J. Van Overstraten, "Theoretical calculations of the Fermilevel and other parameter in phosphorus doped silicon at diffusion temperatures," IEEE Trans. on Electron Devices, ED-21, pp. 155-165, 1974.
44. R.B. Fair and G.R. Weber, "Effect of complex formation on diffusion of arsenic in silicon," J. Appl. Phys., 44, pp. 273-279, 1973.
45. R.O. Schwenker, E.S. Pan and R.F. Lever, "Arsenic clustering in silicon," J. Appl. Phys., 42, pp. 3195-3200, 1971.
46. J.C. Irvin, "Resistivity of bulk silicon and of diffused layers in silicon," Bell Sys. Tech. J. 41, pp. 387-410, 1962.

47. R.B. Fair and J.C.C. Tsai, "Diffusion of ion implanted arsenic in silicon," J. Electrochem Soc., 122, pp. 1689-1696, 1975.
48. M. Yoshida, E. Arai, H. Nakamura and Y. Terunuma, "Excess vacancy generation mechanism at phosphorus diffusion in silicon," J. Appl. Phys. 45, pp. 1498-1506, 1974.
49. S.M. Hu, "Formation of stacking faults and enhanced diffusion in the oxidation of silicon," J. Appl. Phys. 45, pp. 1567-1573, 1974.
50. P.L. Shah, Private communication, 1974.
51. A.E. Deal and A.S. Grove, "General relationship for the thermal oxidation of silicon," J. Appl. Phys., 36, pp. 3770-3778, 1965.
52. J.M. Herman, III, S.A. Evans, and B.J. Sloan, Jr., "Second generation I^2L/MTL: A 20 nsec process/structure," IEEE Solid State Circuits SC-12, April 1977.
53. S.A. Evans, "An analytic model for the design and optimization of ion implanted I^2L devices," IEEE Solid-State Circuits SC-12, April 1977.
54. S.A. Evans, J.M. Herman, III, and B.J. Sloan, Jr., "On the electrical properties of the I^2L n-p-n transistor," IEEE Trans. Electron Devices ED-23, 1192, 1976.
55. S.A. Evans and J.S. Fu, "Application of a new, accurate bipolar simulator to correlate device performance to process selection," Electrochem. Soc. Meeting May 1975, Extended Abstract No. 168.

* NOTE: MATERIAL EXCERPTED FROM P.L. SHAH, "Analysis of implanted layers in silicon for semiconductor device processing," in SCIENTIFIC AND INDUSTRIAL APPLICATIONS OF SMALL ACCELERATORS, Fourth Conf. 1976, pp. 411-419
© 1976 INSTITUTE OF ELECTRICAL AND ELECTRONICS ENGINEERS, INC. REPRODUCTION AUTHORIZED BY IEEE.

** NOTE: MATERIAL EXCERPTED FROM J.L. PRINCE AND F.N. SCHWETTMANN, "Diffusion of boron from implanted sources under oxidizing conditions," J. Electrochem. Soc. 121, pp. 705-710, 1974.
© 1974 THE ELECTROCHEMICAL SOCIETY, INC., REPRODUCTION AUTHORIZED BY ECS.
This information reprinted by permission of the publisher, the Electrochemical Society Inc.

PROCESS MODELING

H. S. Rupprecht

IBM System Products Division--East Fishkill
Hopewell Junction, New York 12533 USA

ABSTRACT. A mathematical model that permits predictions on the
impurity redistribution during interactive diffusion processes
is described. The discussion is based upon work by
F. F. Morehead and A. Michel, and comparisons will be made
between theoretical and experimental data.

1. INTRODUCTION

The complexity that is so frequently found in today's electronic
systems necessitates extensive reliance upon computer-aided
design (CAD) concepts. Computer modeling has been successfully
used to simulate the operation of intricate data processing
systems, and to establish data flow patterns and data rates on
communication lines. In integrated silicon-technology layout
patterns, interconnection schemes as well as basic circuit
designs are based upon electronic design systems.

In this lecture series you are getting acquainted with the
application of CAD to device modeling. It seems, therefore,
just a natural extension to use similar approaches for process
modeling.

In the case of circuit design, the underlying theoretical
concepts are relatively simple, once the necessary parameters of
the individual devices are established. The controlling physical
laws are summarized by the well-known Maxwell equations of
electrodynamics. The situation becomes more complex for device
modeling. Here we find that many of the materials parameters

are not known to the degree needed to model high-performance devices, particularly bipolar devices, from first principles.

Compared with the two preceding examples, namely circuit and device modeling, the complexity of the underlying theoretical concepts becomes rather formidable as one progresses to process modeling. All of a sudden one is confronted with practically every discipline of silicon technology--physics, chemistry, crystallography, metallurgy, etc.

It is therefore not surprising that the state of the art in process modeling is trailing that of the other two areas, i.e., circuit and device design.

In this lecture, we shall concentrate particularly on the discussions of impurity redistribution during hot processing. For this purpose we shall make extensive use of simulation programs that have been worked out by Morehead [1] and Michel [2].

2. PHYSICAL MODEL

In the lecture on the theory of diffusion, we showed that the dominant physical phenomena governing the diffusion of As, P, and B in silicon could be summarized as follows:

(1) Field Effect $N_d^+ \leftrightarrow e^-$ (1)

$$N_a^- \leftrightarrow h^+ \quad (2)$$

(2) Vacancy Enhancement $V_o + e^- \rightarrow V^-$ (3)

$$V^- + e^- \rightarrow V^= \quad (4)$$

(3) Clustering $mN_d \leftrightarrow N_d^m$ (5)

(4) Pairing $N_a^- + N_d^+ \leftrightarrow (N_a N_d)^o$ (6)

It seems advisable to rewrite the expressions for the impurity flux, which were derived in the lecture on diffusion phenomena, in a form more amenable to computer modeling.

The total impurity flux will be partitioned into a diffusion current and a field current. For donors we obtain

$$j_d = -D_d \frac{dN_d}{dx} + \mu_d N_d E, \quad (7)$$

and a similar relation holds for acceptors,

$$j_a = -D_a \frac{dN_a}{dx} - \mu_a N_a E. \tag{8}$$

The electric field, E, is obtained by making the simplifying assumption that the carrier concentration is non-degenerate, and, therefore, Boltzmann statistics apply.

We can then write

$$\frac{n}{n_i} = \exp\left(\frac{E_F - E_i}{kT}\right) = \exp\left\{\frac{q}{kT}(\phi - \phi_i)\right\}, \tag{9}$$

where ϕ, ϕ_i is the Fermi potential for doped and intrinsic material, respectively.

Hence,

$$E = -\frac{d\phi}{dx} = -\frac{kT}{q} \frac{d}{dx} \ln\left(\frac{n}{n_i}\right). \tag{10}$$

The ratio n/n_i can be derived from the impurity concentration under the assumptions that (1) all impurities are ionized and (2) one has local charge neutrality.

$$p - n + N_d - N_a = 0 \quad \text{(charge neutrality)}. \tag{11}$$

With $n_i^2/n = p$ and $N = N_d + mN_d^m - N_a$ (implying that an m-fold cluster is also m-fold ionized), one gets from Eq. (11)

$$\frac{n}{n_i} = \frac{N}{2n_i} + \left(1 + \left(\frac{N}{2n_i}\right)^2\right)^{1/2}. \tag{12}$$

One can now rewrite Eqs. (7) and (8) in the following way:

$$j_d = -D_d \frac{dN_d}{dx} - D_d N_d \frac{d}{dx} \ln\left(\frac{n}{n_i}\right), \tag{13}$$

$$j_a = -D_a \frac{dN_a}{dx} + D_a N_a \frac{d}{dx} \ln\left(\frac{n}{n_i}\right). \tag{14}$$

The next task is to derive an expression for the diffusion coefficient, D, including (1) Vacancy Enhancement, (2) Clustering, and (3) Pairing.

We use the example of donor diffusion; analogous expressions are obtained for the diffusion of acceptors. We include, in addition to the interaction of donors with neutral and singly charged vacancies, the interaction with doubly ionized vacancies.

The effective diffusion coefficient, D_{eff}, is written as the sum of the three contributing components:

$$D_{eff} = A^o[V_o] + A^-[V_-] + A^=[V_=], \qquad (15)$$

$$\begin{array}{ccc} \text{neutral} & \text{singly} & \text{doubly} \\ \text{vacancy} & \text{charged} & \text{charged} \\ & \text{vacancy} & \text{vacancy} \end{array}$$

where A^o, A^-, and $A^=$ are appropriately chosen constants. Dividing Eq. (15) by $A^o[V_o]$ results in

$$D_{eff} = A^o[V_o]\left\{1 + \frac{A^-[V_-]}{A^o[V_o]} + \frac{A^=[V_=]}{A^o[V_o]}\right\}. \qquad (16)$$

From the relation $V_o + e^- = V_-$ one finds

$$[V_o]n = K^-[V_-], \qquad (17)$$

and, for the intrinsic condition,

$$[V_o]n_i = K^-[V_{i-}], \qquad (18)$$

where K^- is the equilibrium constant for the reaction. Hence,

$$n/n_i = [V_-]/[V_{i-}]. \qquad (19)$$

In a similar way one obtains from the relation $V_- + e^- = V_=$

$$(n/n_i)^2 = [V_=]/[V_{i=}]. \qquad (20)$$

The intrinsic effective diffusion coefficient can be written by use of Eq. (16) in the following way:

$$D_{eff,i} = A^o[V_o](1 + \beta + \gamma), \qquad (21)$$

where $\beta = A^-[V_{i-}]/A^o[V_o]$

and $\gamma = A^=[V_{i=}]/A^o[V_o]$.

Hence we find for the effective diffusion coefficients by use of Eqs. (16), (19), (20), and (21)

$$D_{eff} = D_{eff,i} \frac{\{1 + \beta \frac{n}{n_i} + \gamma (\frac{n}{n_i})^2\}}{1 + \beta + \gamma} . \qquad (22)$$

In order to incorporate clustering and pairing effects, one has to multiply the effective diffusion coefficient as given in Eq. (22) by the terms derived in the lecture on diffusion [Eqs. (53) and (56)]:

$$(1 + m\, K_c N_d^{m-1})^{-1} \quad \text{(clustering)} \qquad (23)$$

and

$$(1 + K_p N_d N_a)^{-1} \quad \text{(pairing)}. \qquad (24)$$

3. MOREHEAD'S MODEL

In 1974, Morehead [1] reported on a redistribution model that permits the determination of impurity distributions under predeposition schemes followed by a drive-in diffusion. Those conditions are usually encountered in subcollector and emitter process steps involving ion implantation. His model takes into account interactive diffusion phenomena, described earlier in the lecture on diffusion. Morehead used, in his original concept, a clustering factor m=2 for arsenic, based on Schwenker's work [3] on arsenic precipitation. This is in contrast to Hu's results [4], which showed that the activity coefficient could be better fitted to the vapor pressure data by the use of a clustering factor of 4. This apparent discrepancy will be clarified in Section 4, when we discuss Michel's approach.

Morehead neglects the interaction with doubly charged vacancies, by setting $\gamma = 0$ in Eq. (22), as well as pairing effects. For β he chooses the following set of data at temperatures of $1000^\circ C$:

$\beta = 100$ (arsenic)

$\beta = 3$ (phosphorus)

$\beta = 3$ (boron)

Morehead partitions his sample, in which the diffusion takes place, into a series of bins normal to the direction of the diffusion. The total particle flux between those bins is separated into a current due to the concentration gradient and a current due to the electric field. [Compare Eqs. (7) and (8).]

In order to explain Morehead's model in more detail, we shall make the following definitions:

$N(y)$ is number of atoms in bin y
Δx is bin width
A is cross section of bin
Δt is time element.

The total change, ΔN_{tot}, of impurity atoms in a given bin is then the sum of a gradient-induced component, ΔN_{grad}, and a field-induced component, ΔN_{field}:

$$\Delta N_{tot} = \Delta N_{grad} + \Delta N_{field}. \qquad (25)$$

The diffusion term can be expressed by means of the previous definitions as

$$\Delta N(1)_{grad} = D_{1/2} \frac{(\frac{N(2)}{\Delta x} - \frac{N(1)}{\Delta x})}{\Delta x} \Delta t$$

$$= j_{grad} \, A \, \Delta t. \qquad (26)$$

$D_{1/2}$ is an average diffusion coefficient corresponding to the average carrier concentration $<N(1,2)>$ in bins 1 and 2.

In a similar way we can define the field term:

$$\Delta N(1)_{field} = -D_{1/2} \frac{\Delta \ln(\frac{n}{n_i})}{\Delta x} \frac{<N(1,2)>}{\Delta x} \Delta t. \qquad (27)$$

The net change of impurity atoms in bin 1 at time $t+\Delta t$ is therefore given by

$$\Delta N^{\{(1), \, t+\Delta t\}}_{total} = \Delta N^{\{(1), \, t+\Delta t\}}_{grad} + \Delta N^{\{(1), \, t+\Delta t\}}_{field} \qquad (28)$$

This situation is schematically depicted in Fig. 1. It is relatively easy from a conceptual point of view to extend this one-dimensional model to a two-dimensional configuration. Figures 2 through 4 show a base-emitter redistribution obtained by Morehead [5] with a two-dimensional model. Figure 2 represents the boron distribution after a double energy implant (E = 50 keV, dose = 10^{14} cm^{-2}, E_2 = 80 keV, dose = 10^{13} cm^{-2}). Figure 3 describes the arsenic redistribution after a predeposition drive-in step (implant energy E = 40 keV, dose 3 · 10^{15} cm^2) at 1000°C for 30 minutes. The final boron distribution is given in Fig. 4.

4. MICHEL'S MODIFICATION

Michel [2] has rewritten Morehead's original program, maintaining, however, its essential features. Michel puts an upper limit of 3 x 10^{20} atoms/cm^3 on the maximum concentration of monomer As donors. The excess arsenic is converted into dimer clusters according to the rate equation

$$dN_m/dt = k_{c\ ass.} N_d^m - k_{c\ diss} N_m, \qquad (29)$$

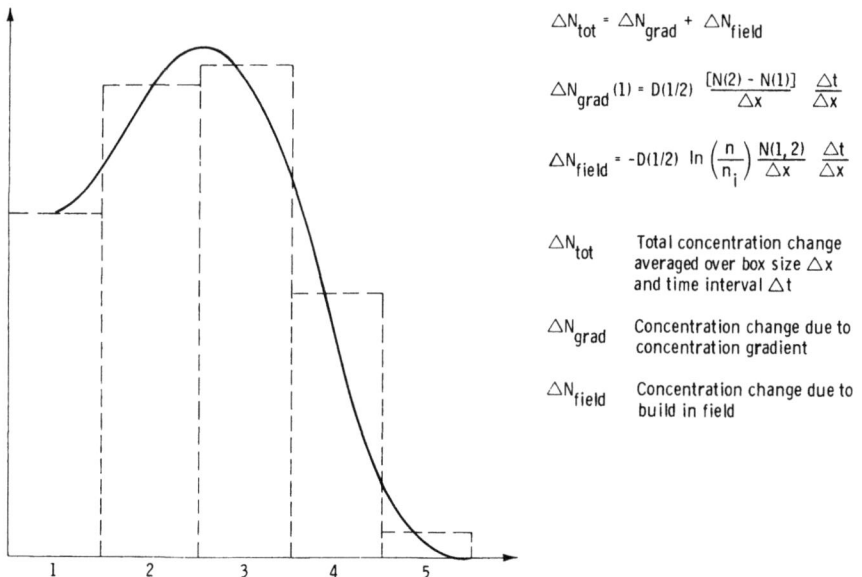

$\triangle N_{tot} = \triangle N_{grad} + \triangle N_{field}$

$\triangle N_{grad} (1) = D(1/2) \dfrac{[N(2) - N(1)]}{\triangle x} \dfrac{\triangle t}{\triangle x}$

$\triangle N_{field} = -D(1/2) \ln\left(\dfrac{n}{n_i}\right) \dfrac{N(1,2)}{\triangle x} \dfrac{\triangle t}{\triangle x}$

$\triangle N_{tot}$ Total concentration change averaged over box size $\triangle x$ and time interval $\triangle t$

$\triangle N_{grad}$ Concentration change due to concentration gradient

$\triangle N_{field}$ Concentration change due to build in field

Fig. 1. Schematic diagram depicting the subdivision of the sample into individual bins.

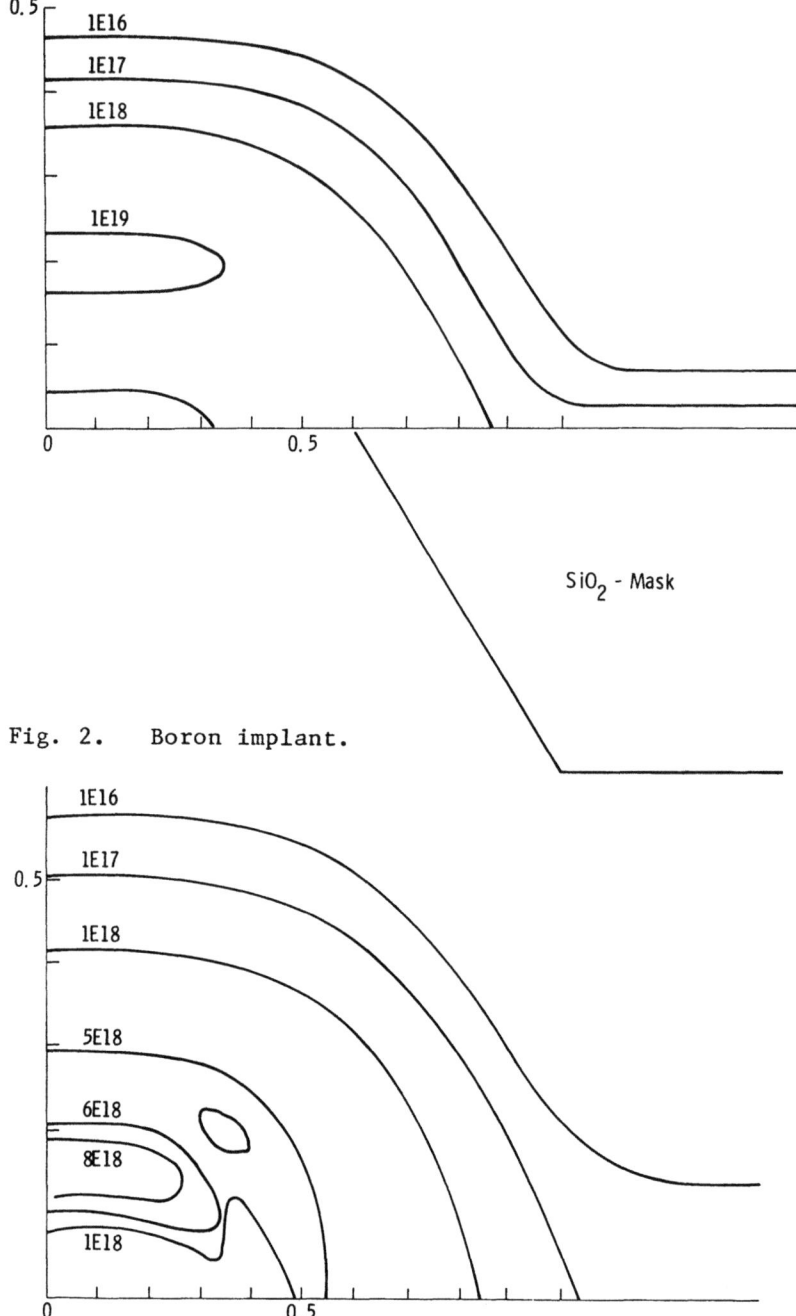

Fig. 2. Boron implant.

Fig. 3. Boron redistribution after 1000°C and 30-minute arsenic drive-in cycle.

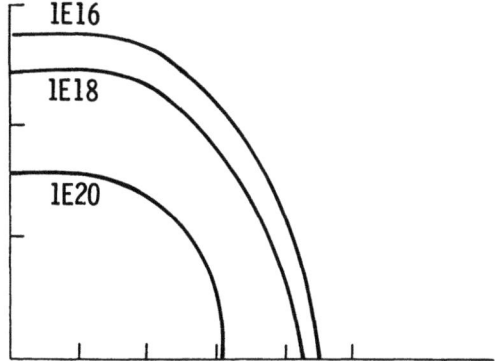

Fig. 4. Arsenic distribution after 1000°C and 30-minute drive-in cycle.

where m=2, in accordance with Morehead,

$k_{c\ ass.}$ is the rate constant of cluster association, and

$k_{c\ diss}$ is the rate constant of cluster dissocation

$k_{c\ ass.}$, $k_{c\ diss}$ are related to K_c in Eq. (51) of the lecture on diffusion) in the conventional way: $k_c = k_{c\ ass.}/k_{c\ diss}$.

The clusters are assumed to be (1) immobile and (2) doubly ionized. They therefore contribute to the field term according to Eqs. (11) and (12). The intrinsic carrier concentration n_i in those equations is computed using the procedure of Morin and Maita [6]. The activation energies of the intrinsic diffusion coefficients are determined for arsenic from data published by Chiu and Gosh [7] and, in the case of boron, from data by Kendall and DeVries [8].

It seems appropriate at this point to discuss the question of the proper choice of the clustering factor for arsenic in more detail. Should it be m=4, as Hu used in fitting the vapor pressure data, or is m=2 the better value, as derived by Schwenker et al. from electrical conductivity changes?

Michel explains his choice of m=2 as follows.

In the high-concentration range the free-carrier concentration (as determined by Hall effect measurements at room temperature) can be best fitted to the total arsenic concentration (as determined either by SIMS or by NAA) by a clustering factor m=4.

One can, however, obtain an excellent first approximation to this relation by using an upper concentration limit of 3×10^{20} cm^{-3} for the monomer arsenic concentration N_d and a clustering factor of $m=2$ for monomer concentrations less than 3×10^{20} cm^{-3}. The advantage of this procedure is a simplified mathematical formalism and reduction of the programming complexity.

On the other hand, the particular choice of m is of no consequence to the electric-field expression, since in deriving Eq. (12) one has assumed that every arsenic atom, clustered or unclustered, is fully ionized at the diffusion temperature.

Figure 5 gives a comparison between theoretical As redistributions and experimentally measured ones from an initial As implant of 150 keV and a dose of 1×10^{16} cm^{2} for three different drive-in times. The experimental data were derived from spreading-resistance measurements and neutron activation analysis.

Table I lists the theoretical and experimental sheet-resistance values for the same set of distributions.

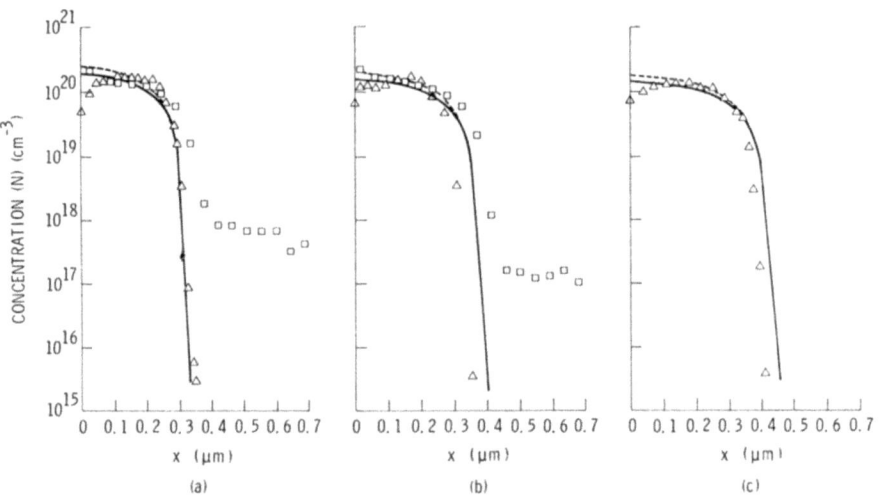

Fig. 5. Profiles for 150-keV, 5×10^{15} cm^{-2} implants after 1000°C drive-in cycle at (a) 30 minutes, (b) 60 minutes, and (c) 90 minutes. The solid curves are the calculated carrier concentrations, the dashed curves the total arsenic concentration, the triangles the spreading-resistance data, and the squares the neutron activation data.

Table I. Comparison of calculated and measured sheet resistance values

Energy (keV)	Dose (10^{16} cm^{-2})	Drive-in Time at 1000°C (min)	ρs (Calc) ($\Omega/\square$)	ρs (S.R.) ($\Omega/\square$)	ρs 4-pt probe ($\Omega/\square$)
150	0.5	30	25.8	25.2	26.8
		60	24.1	26.3	25.5
		90	23.0	24.1	24.9
150	1.0	30	16.7	18.2	17.7
		60	14.9	15.8	16.0
		90	14.1	14.1	14.8
150	2.0	30	13.4	13.7	14.0
		60	11.1	11.1	11.9
		90	9.9	10.8	10.7

5. FUTURE DIRECTIONS

It is well known that crystalline defects can lead to enhanced diffusion and hence influence final impurity distributions markedly. In the lecture on diffusion, we discussed radiation-enhanced diffusion under steady-state conditions. In the redistribution of ion-implanted impurities, one frequently finds modifications of the standard diffusion due to transient phenomena. For instance, we have found that the diffusion of ion-implanted boron is enhanced in the initial redistribution phase because of the radiation damage created by the implant processes. These phenomena play a major role in device fabrication and can lead to discrepancies between predicted and actual impurity profiles. The accurate simulation of narrow base structures is just one example. An exact knowledge of the kind of defects and a quantitative understanding of the annealing kinetics are needed before one can successfully incorporate such phenomena into a process model. Much more work has to be done in this area.

The final reward will be a process simulation model that permits complete process optimization and yield prediction.

ACKNOWLEDGMENT

I should like to thank my colleagues F. F. Morehead and A. Michel for many enlightening discussions and for their helpful critique of the manuscript. I should also like to thank them for making some of their latest results available to me prior to publication.

REFERENCES

1. F. F. Morehead, Abstract 194, p. 474, The Electrochemical Society Extended Abstracts, Fall Meeting, New York, N.Y., Oct. 13-17, 1974.
2. A. Michel, Private communication, (to be published).
3. R. O. Schwenker, E. S. Pan, and R. F. Lever, J. Appl. Phys. $\underline{42}$, 3195 (1971).
4. S. M. Hu, in Atomic Diffusion in Semiconductors (Plenum Press, London and New York, 1973), pp. 306-310.
5. F. F. Morehead, Private communication (to be published).
6. F. J. Morin and J. P. Maita, Phys. Rev. $\underline{96}$, 28 (1954).
7. T. L. Chiu and H. M Ghosh, IBM J. Res. Dev. $\underline{15}$, 472 (1971).
8. D. L. Kendall and D. B. DeVries, in Semiconductor Silicon 1969, edited by Rolf R. Haberecht and Edward L. Kern, The Electrochemical Society, Inc. (New York, N.Y.,1969), p. 358.

Process Oriented IC Design

D. L. Scharfetter
J. G. Ruch

Bell Laboratories
Murray Hill, New Jersey 07974

Introduction

The design process, for large integrated circuits, requires the determination of numerous models parameters, before use can be made of computer-aided circuit analysis programs. This paper presents an approach which allows the automatic generation of parameters as determined by the mask layout dimensions and IC processing variables. Furthermore, if process-tolerant designs are to be obtained, the parameter values must be obtained over the expected range of processing variations and the corresponding circuit simulations performed. The IC designer must produce a layout, containing from hundreds to perhaps thousands of active devices, which must function over a fairly broad range of wafer processing variations. This paper reports on an approach to process-oriented circuit design by the combined use of circuit, device, and process simulators.[1]

Computer-aided design is employed when it is cost effective. Accurate results must be obtainable at reasonable computer cost and with minimal input preparation time. With regard to process-oriented circuit design this requires a) a device model which strikes a good balance between accuracy and complexity, and b) a "front end" for automatic model parameter prediction, as determined by wafer processing variables. The expanded Gummel-Poon model* takes account of basewidth modulation, current-dependence of forward current gain and high level injection. Parameters for the model are automatically calculated by a *PAR*ameter *CA*lculator program PARCA. The user supplied inputs to PARCA are device mask dimensions. PARCA has access to device characteristics, produced by a *TRAN*sistor *SIM*ulator program TRANSIM[2] or to experimentally derive device characteristics.

---

* See, "Bipolar Model for IC Design," D. L. Scharfetter, this digest.

Parameter Calculation

Bipolar integrated circuits generally contain a wide variety of transistors of differing shapes and sizes. Hence experimental characterization of all devices employed in a design is generally not performed. For the same reason full two-dimensional analysis of all devices would be prohibitively expensive. In this section a procedure is described for the efficient and accurate calculation of bipolar transistor models parameters, based upon the existence of one-dimensional solutions for active and parasitic portions of the structure.

The active and parasitic elements are shown schematically in Fig. 1. Transistor T_E represents the active transistor of given emitter geometry. Diodes D_E and D_C represent the emitter sidewall and inactive collector junction diodes, respectively. A conductively modulated base resistance, R'_B, is also included. For a given emitter geometry, multiple one-dimensional solutions are arranged in parallel to account for emitter crowding and result in the value for R'_B and the Gummel-Poon transported (collector) parameters.

The effects of the diodes D_E and D_C are incorporated into the base current and into the dynamic charge storage components of the model. The I-V and Q-V characteristics of the diode D_C are scaled in proportion to the area of the inactive collector junction. The delay time of D_C is modified according to the factor dI_{D_C}/dI_{CC} to reflect the component of diode stored charge controlled by the emitter current (inverse mode assumed). In a similar fashion forward delay time and emitter capacitive components are calculated for the emitter sidewall diode D_E, except that dc parameters are scaled in proportion to the emitter perimeter, and this delay time is modified by the factor dI_{D_E}/dI_{CC}. Specifically, effective delay time values are obtained from the relationship

$$\tau = \tau_{T_E} + \tau_D \frac{dI_D}{dI_{CC}},$$

where τ_{T_E} is the active transistor value, and τ_D the diode value.

The recombination parameters are obtained from the curve for the total base current of the composite structure. The diode current I_{D_E} is added to the base current of T_E, taking account of the voltage drop accross R_B. In a similar fashion the inverse gain is calculated by adding the diode current I_{D_C} to the base current of T_E in inverse operation.

Summary

The I-V and Q-V curves for the three devices T_E, D_E, and D_C are accessed and Gummel-Poon parameters automatically generated by the PARCA program. The curves, on a "per-unit-area" basis, result from either process characterization, by use of specialized test patterns, or by device and process simulation. The device simulation[2] proceeds from a doping profile and silicon materials parameters. The doping profile is produced by a process simulation program, for example, SUPREM.[3]

The "per-unit-area" characteristics of the three devices must be determined for nominal and worst case processing conditions. These characteristics then represent "process files" which are accessed by the IC designer. For a given technology these files should accurately reflect the known photolithographic and doping related process variations.

The parameter calculator program, PARCA, takes as input the nominal description of the transistor structure (mask dimensions) and calculates the worst case properly correlated set of model parameters. Process Oriented IC Design results when the PARCA program is incorporated as a "front-end" to the IC designer's circuit simulation program, and the "process files" are properly maintained.

References

1. J. G. Ruch, H. C. Poon, W. J. McCalla, D. L. Scharfetter, "Automated Transistor Characterization and Parameter Generation for Linear Integrated Circuit Design," 1974 ISSCC, Philadelphia, Pa., Feb. 13-15, 1974.
2. J. G. Ruch, D. L. Scharfetter, "Characterization of Bipolar Devices," 1973 IEDM, Washington, D.C., December, 1973.
3. R. W. Dutton, A. G. Gonzalez, R. D. Rung, D. A. Antoniadis, "IC Process Engineering Models and Applications," Electrochemical Society Meeting, Philadelphia, Pa., May, 1977.

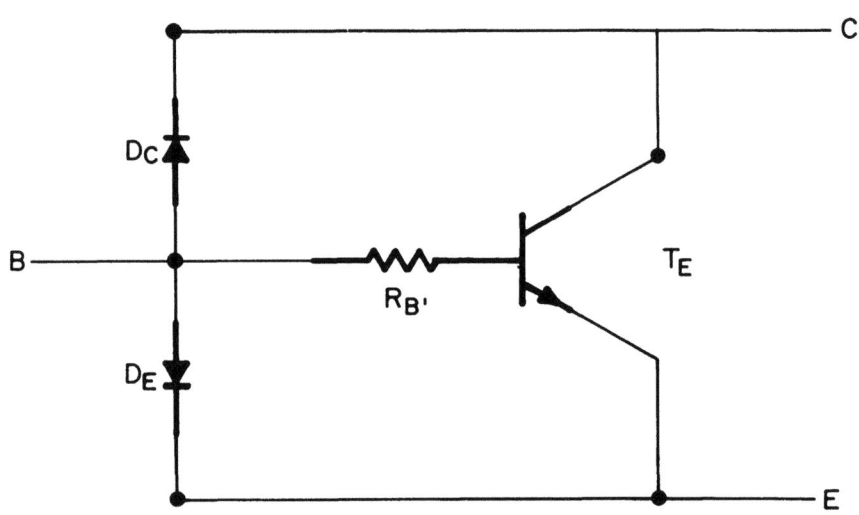

FIG. 1 ACTIVE TRANSISTOR AND PARASITIC DIODE ELEMENTS

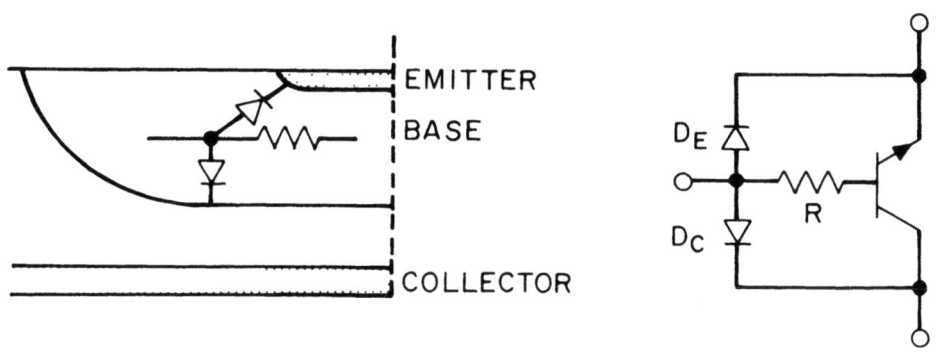

a) PARASITICS

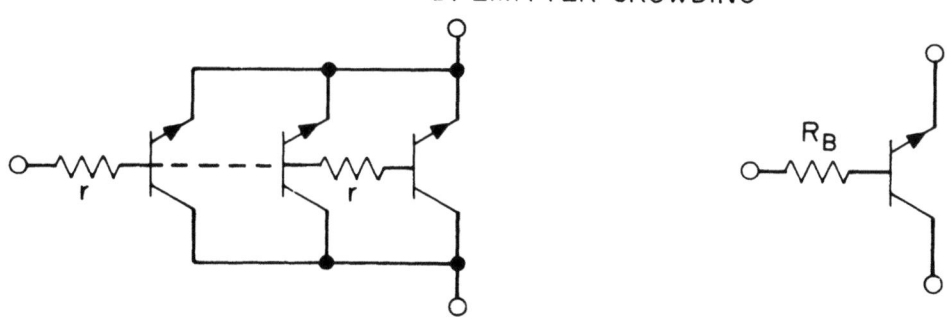

b) EMITTER CROWDING

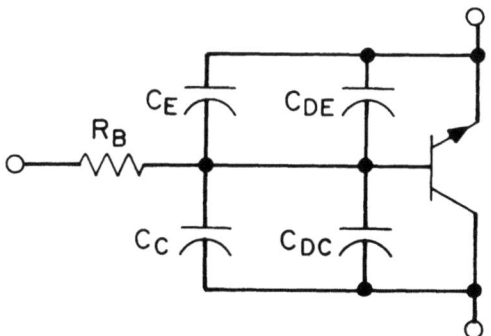

c) COMPOSITE MODEL

FIG. 2   TRANSISTOR MODEL (DC)

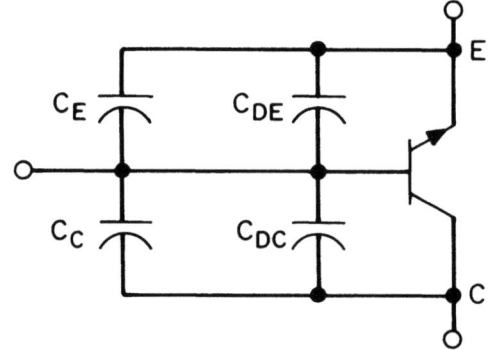

$$I_B = \left(I_1 e^{\frac{V_{BE}}{V_T}} + I_2 e^{\frac{V_{BE}}{n_E V_T}}\right) + I_{BC}$$

$$I_C = \frac{I_S Q_0 \left(e^{\frac{V_{BE}}{V_T}} - e^{\frac{V_{BC}}{V_T}}\right)}{(Q_0 + Q_E + Q_C + Q_D)} - I_{BC}$$

$$C_E = \left.\frac{\partial Q_E}{\partial V_{BE}}\right|_{V_{BC}} \qquad C_C = \left.\frac{\partial Q_C}{\partial V_{BC}}\right|_{V_{BE}}$$

$$C_{DE} = \left.\frac{\partial Q_{DE}}{\partial V_{BE}}\right|_{V_{BC}} \qquad C_{DC} = \left.\frac{\partial Q_{DC}}{\partial V_{BC}}\right|_{V_{BE}}$$

FIG. 3 TRANSISTOR MODEL (AC)

MODELING OF I^{2}L AND PROCESS SELECTION

Walter H. Schroen

Texas Instruments Incorporated, Dallas, Texas, U.S.A.

1. INTRODUCTION: MODELING AND PROCESS "WINDOWS"

The trend in semiconductor manufacturing pushes for smaller geometries and higher complexity of the integrated circuits, both bipolar and MOS, combined with faster product cycle time. In order to perform effective control of the more and more interacting processes, models are needed as standards or ideals so that the actual results of the processes can be compared against the ideal results of the models. Decisions then can be made early in the process sequence, whether the processes have performed "right" and delivered the expected results, or whether something in the process has gone "wrong," and what corrective actions, if any, are available.

This effective process control can best be performed using the same analytical and computerized models which have been developed as guidelines and trade-off analysis tool to optimize bipolar and MOS fabrication technologies with minimum process iterations. This "analytical" process control is based on the availability of modeling schemes which allow process simulation, device parameter calculation, and circuit analysis. Using this insight in the functioning of bipolar and MOS devices and their process correlation, a process sensitivity analysis can be performed using computer programs. This process sensitivity analysis identifies those process steps which most sensitively affect device performance, and also suggests how tight the process controls have to be to assure device parameters in the right "windows." This interactive control of process and device builds performance and reliability into the circuit and thus avoids excessive testing.

Figure 1 summarizes, for the example of a bipolar integrated injection logic (I²L) circuit, the principal simulators and their mutual coupling. The three main parts are the process simulator, the device calculator, and the circuit analysis program.

2. PROCESS SIMULATION

A flexible, efficient, and fast mathematical scheme has been developed[1,2,3] that allows simulation of a wide range of processes. The capabilities of this process simulator are described in detail in the paper entitled "Process Modeling," to be discussed at this NATO Advanced Study Institute.

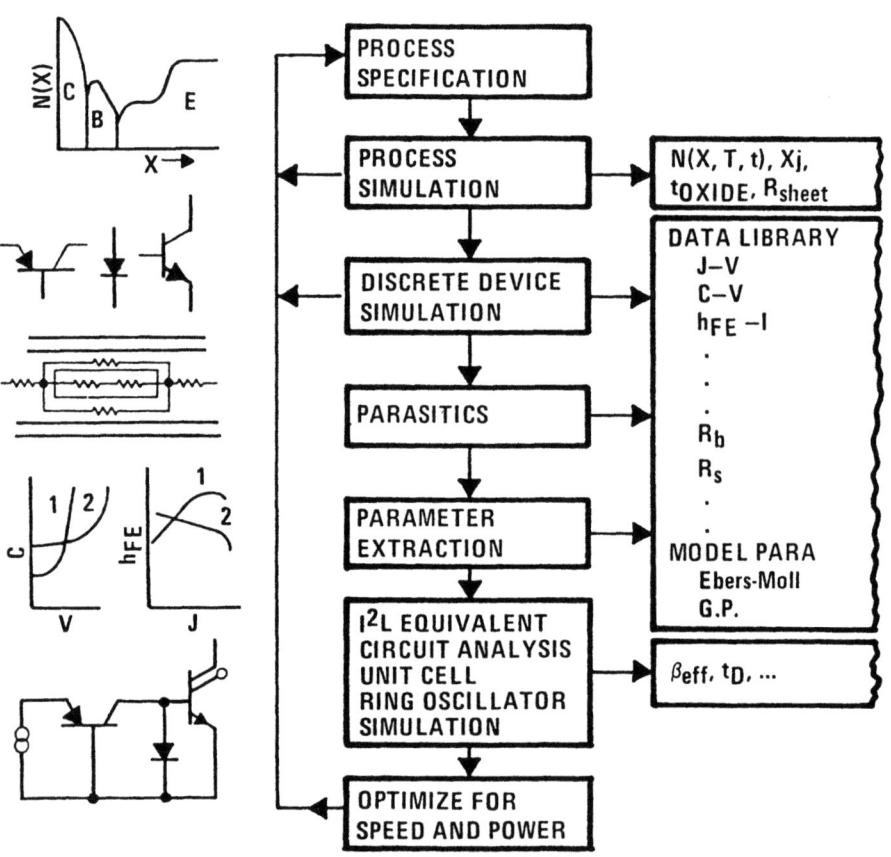

Fig. 1. I²L device design and optimization.

3. BIPOLAR DEVICE SIMULATION

To bridge the gap between process and circuit simulators, a fast efficient one-dimensional diode and transistor simulator (GENTRAN) has been developed[4]. The primary input for the simulator is measured or calculated impurity profiles. The output is electrical parameters such as current gain, delay time and capacitance. GENTRAN, while being relatively easy to use and cheap to run, includes all the important physical mechanisms that control device performance.

The transport of carriers in a bipolar device in steady state is completely determined by the current, continuity and Poisson's equations for holes and electrons. An algorithm has been developed and programmed for solving a general form of these equations. Included in the solution are the following effects:

1. Fermi statistics
2. Arbitrary quantum density of states
3. Concentration and field dependent mobility
4. Recombination-generation (bulk)
5. Concentration dependent lifetime
6. Surface recombination

The models for mobility and generation-recombination can be arbitrarily varied to simulate exactly the degrading effects of external radiation.

The computation time is independent of the statistics used and only moderately dependent on the quantum density of states model. To simplify the algorithm, only a two-terminal device was considered. Thus, the program will calculate device characteristics for transistors in low injection or for diodes at any injection level.

The device characteristics calculated are h_{FE} (I_c), t_d (I_c), $C(V_{eb})$, $\alpha(I_c)$, I_c - V_{eb}, I_b - V_{eb}. The capacitance is broken into diffusion and depletion components. The delay time includes contributions from the emitter, base, and junction regions. Also given is a composite of the emitter and base profiles, the emitter sheet resistance, the base sheet resistance under the emitter, and the junction depth.

Examples for application of the device simulator have been published by Evans and co-workers[5,6,7] for I^2L designs and processes. They are discussed in the Sections 5, 6 and 7, following closely the approach by Evans and co-workers.

4. CIRCUIT SIMULATION

A general purpose circuit simulation program (SPICE 2) is available to make nonlinear dc and nonlinear transient analyses of I^2L circuits. Circuits may contain resistors, capacitors, inductors, mutual inductors, independent voltage and current sources, four types of dependent sources, transmission lines, and the four most common semiconductor devices: diodes, Bipolar Junction Transistors (BJTS), JFETS, and MOSFETs.

SPICE 2 has built-in models for the semiconductor devices, and the user specifies only the pertinent model parameter values. The model for the BJT is based on the integral charge model of Gummel and Poon. However, if the Gummel-Poon parameters are not specified, the model reduces to the simpler Ebers-Moll model. In either case, charge storage effects, ohmic resistances, and a current-dependent output conductance may be included. The diode model can be used for either junction diodes or Schottky Barrier diodes.

5. CORRELATION OF I^2L PERFORMANCE TO PROCESS SELECTION*

A simple merged transistor logic (MTL) structure is chosen to show the interrelation of the lateral pnp and the inverse npn device characteristics for several process options. A schematic of an MTL structure is shown in Figure 2. The current densities are defined as follows: J_{nv} = injected electron current from buried layer under pnp emitter and collector contacts; J_{pv} = injected hole current into buried layer; J_{inv} = injected electron current under npn collector contact; J_{pl} = laterally injected

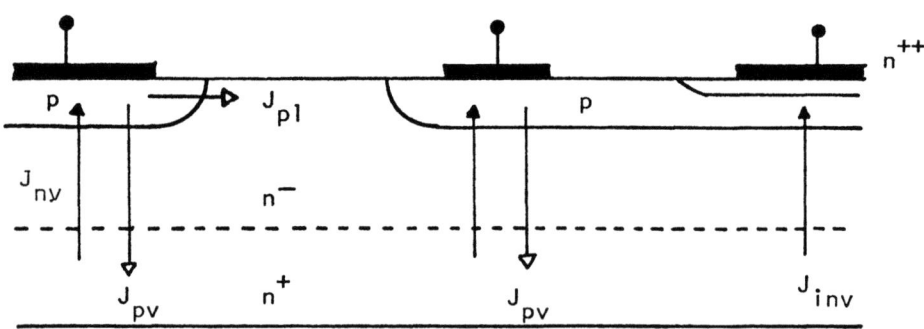

Fig. 2. Schematic of I^2L structure. Definition of currents: J_{nv} = injected electron current under pnp emitter and collector contacts; J_{pv} = injected hole current into buried layer; J_{inv} = injected electron current under npn collector contact; J_{pl} = laterally injected hole current.

hole current. The vertical injection currents (J_{nv}, J_{pv}, J_{inv}) are used to characterize the process options. A knowledge of the two dimensional current flow would be necessary if geometry were included as a variable.

Four hypothetical process options which are representative of diffusion and ion implant technology are described in Table 1. The associated profiles, given in Figure 3, were calculated using assumed process conditions (heat cycles, predeposition conditions, etch back, etc.) as inputs. The profile for process option D is the same as A except for the substitution of a higher buried layer sheet resistance requirement (shorter predeposition time). The models used for boron, arsenic and antimony diffusion have been verified[1]; the phosphorus model is adequate only in a uniformly doped substrate.

Below, the effects of the process conditions are discussed on the injection properties of the inverse npn transistor in low

Case		A	B	C	D
Description		Shallow Diff. Base	Deep Diff. Base	Implanted Base	Shallow Base Short Pre-dep
Si consumed by oxidation (μm)		.18	.61	~.09	.18
Epi thickness (μm)		3.0	3.0	1.0	3.0
Up diffusion (μm)		.3	.9	minimal	.3
R_s, Ω/□	Buried layer	14.1	15.4	15.7	19.9
	Base	178	104	505	178
	Base under emitter	1704	1466	1335	1704
n$^+$pn$^+$	$J_{inv}(\frac{A}{cm^2}) V_{EB}=.8V$	154	333	259	124
	h_{FE} - max	82	173	237	69
	$f_T(MH_z) V_{EB}=.8V$	37	142	158	34
pn$^-$n$^+$	$J_{pv}(\frac{A}{cm^2}) V_{EB}=.8V$	.97	.23	.49	1.17
	$J_{nv}(\frac{A}{cm^2}) V_{EB}=.8V$	13.3	5.6	27.8	15.6

Table 1. Process/device characterization table.

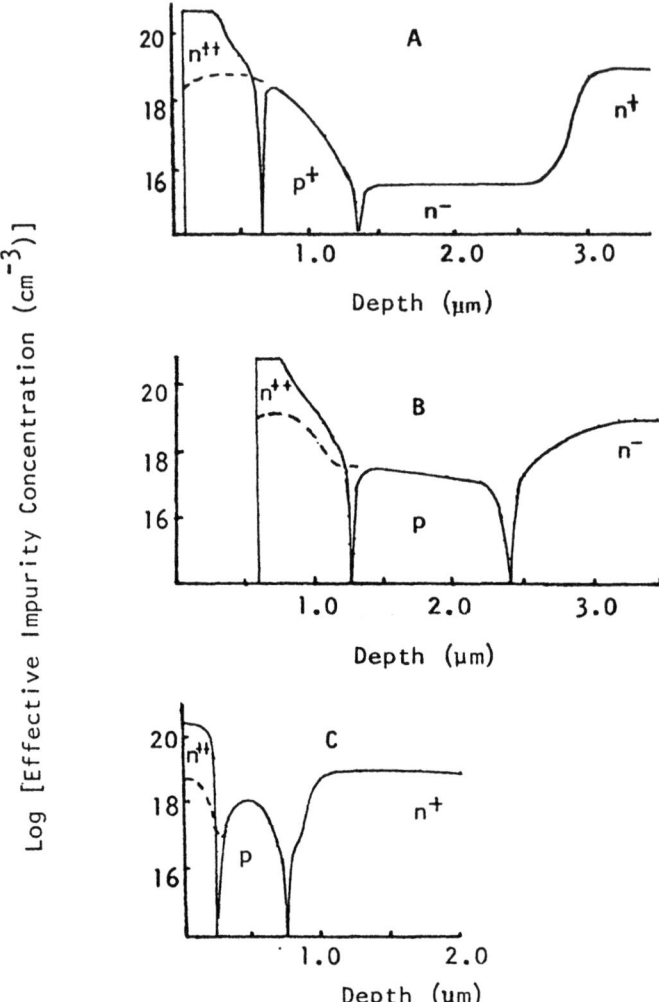

Fig. 3. ——Inverse npn profiles for process options A, B, and C
- - -Parasitic $p^+n^-n^+$ diode.
The zero of the depth scale is taken with reference to the buried layer interface directly after epitaxial process step.

injection (collector acts as a sink for electrons) and on the parasitic pn^-n^+ diode using the data tabulated in Table 1.

5.1 Deep base and steam oxidation.

A deep base may be used to reduce the sensitivity of the base sheet to the emitter diffusion and to improve inverse beta; however, the silicon removed in the long steam oxidation should be considered in the selection of the emitter and base processes. For case B shown in Fig. 3, the silicon consumed in the oxidation cycles is .61μm.

5.2 Epi-buried layer interface.

The heavy doping in the n^+ buried layer gives rise to quasi-fields[8] which are proportional to the impurity concentration and its gradient. The quasi-fields enhance the injection currents J_{pv}, J_{nv} across the n^+ barrier while the normal electrostatic field retards these currents. For example, the hole current loss to the substrate is reduced by a factor of three by using process B instead of A because the antimony up-diffusion reduces the gradient of the n^+ barrier. The base current J_{nv} of the pnp emitter is reduced by a lower base sheet R_s and a reduction in the electron quasi-field at the n^- - n^+ interface.

5.3 Buried layer up-diffusion and predep condition.

Consideration should be given to the antimony up-diffusion from the n^- - n^+ interface that occurs when long high temperature cycles (case B) are used. If this up-diffusion is combined with the Si consumed by oxidation, the thickness of the uniform epi layer left for the emitter and base diffusion is only 1.49μm. The predeposition time and temperature for the buried layer will determine the amount of charge in the emitter of the inverse npn. The effect of the change in predeposition time is seen directly by comparing case A and D. In case A, the injected current J_{pv} is less, and h_{FE} is greater due to the larger emitter charge. It should be noted that the high h_{FE}'s shown here are due to a reduced importance of heavy doping at these relatively low emitter concentrations[9,10].

5.4 Phosphorus vs. arsenic emitter.

For thin epi or shallow base processes As will reduce the parameter sensitivity to slight process variations. If a deep base is desired, the faster diffusing phosphorus can be used.

5.5 Frequency response.

Shorter delay times are attainable when junctions are more heavily doped and thinner epi layers are used (compare cases B and C to A and D); however, a tradeoff will be necessary if high breakdown voltages are required.

6. ANALYTIC MODEL OF EFFECTIVE BETA FOR ION-IMPLANTED I^2L DEVICES**

The betas of the up-transistor and the I^2L unit cell were investigated recently by Evans and co-workers[7]

A profile of the intrinsic n-p-n transistor is given in Figure 4. The current-voltage characteristics are given in Figure 5 for base implant doses of 1 and 4×10^{12} ions/cm^2. The base current of the up or inverted transistor was obtained by subtracting the parasitic diode current from the total measured base current. The calculated values of intrinsic base current were obtained using a general bipolar simulator[4] which includes the effects of bandgap shrinkage[10], Fermi-Dirac statistics, and SRH and Auger recombination[11]. By omitting these physical effects one at a time in the simulation, it was found that recombination mechanisms, bandgap shrinkage, and the substrate concentration determine the base current of the up transistor, while bandgap shrinkage, degeneracy, and the emitter charge determine the base current of the down (normal) transistor. The down transistor is

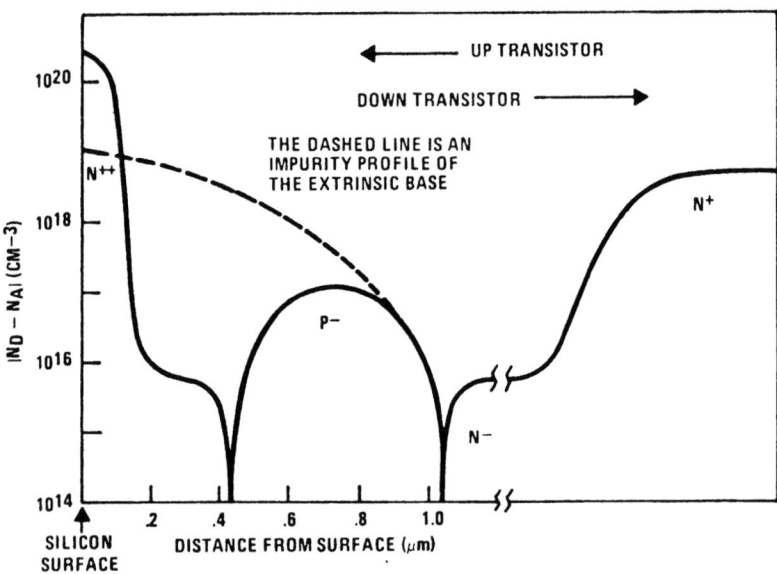

Fig. 4. Impurity profile of the npn transistor.

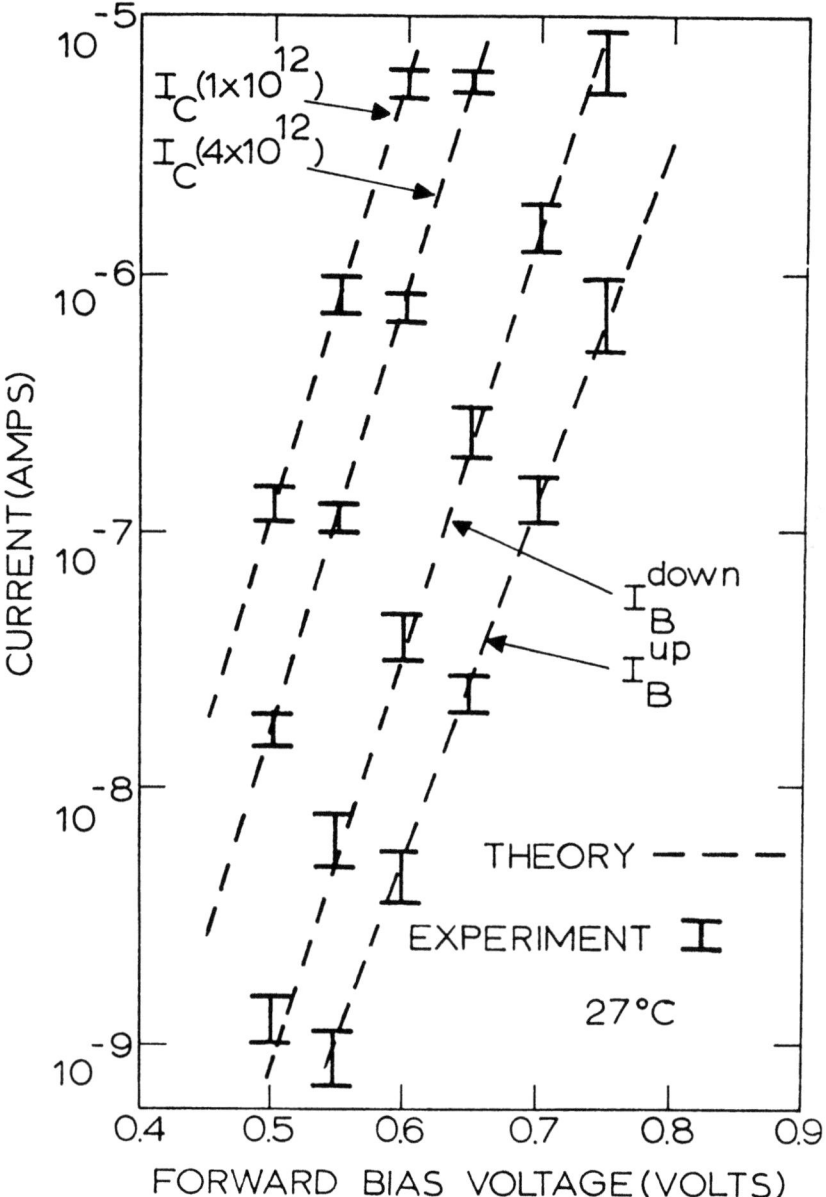

Fig. 5. Current-voltage characteristics of the I²L intrinsic vertical n-p-n. The collector current is the same for both up and down transistors. The error bars represent the maximum variation of good devices on many slices.

insensitive to lifetime at medium and high currents due to the shallow junction depth and heavy doping effects. A small controlled base charge combined with the large effective emitter charge, which is characteristic of moderately doped substrates, gives rise to the high n-p-n up beta demonstrated in Figure 6. It has been shown that the effective emitter concentration decreases above ~5×10^{18} Ref.[12,13]. High betas for the down transistor are difficult to obtain for this structure, since the emitter charge is defined by a thin highly concentrated n^{++} layer. The explanation given here for high n-p-n up beta also applies to the LEC transistor since the essential structure and profile are identical.

The I^2L gate has one or more collectors isolated from each other by the extrinsic p^+ base which also serves as the collector of the lateral p-n-p. An effective beta β_{eff} of this structure is measured by forcing I_B at the input, grounding the injector,

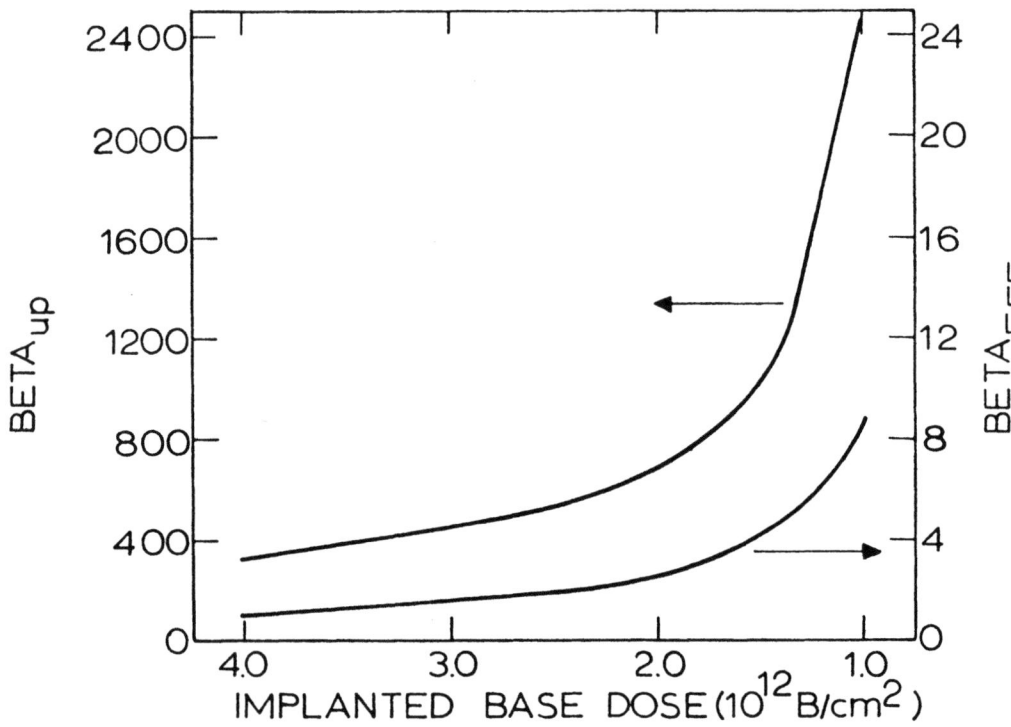

Fig. 6. n-p-n beta versus implanted dose. Top curve corresponds to intrinsic beta while the bottom is β_{eff} as defined in Fig. 7.

and monitoring I_C at one of the output collectors. The equivalent circuit for a simple five collector unit cell is shown in Figure 7. The base current is distributed at low and medium injection levels among the elements in this equivalent circuit according to

$$I_B = \{[I_s^{pnp} + I_s^{diode} + M(1-\alpha_u)I_s^u]e^{qV_1/kT} \\ + (M-1)(1-\alpha_d)I_s^d e^{qV_2/kT}\} \qquad (1)$$

where I_s^{pnp}, I_s^{diode}, I_s^u, and I_s^d are the ideal saturation currents for the base-collector diode of the inverse lateral p-n-p, the parasitic diode, the base-emitter diode of the up transistor, and the base-emitter diode of the down transistor, respectively. V_1 and V_2 are the base-emitter and base-collector voltages of the up transistor, M is the number of collectors, and α is the common base current gain. Typically, the intrinsic n-p-n base current represented by $(1-\alpha_u)I_s^u e^{(qV_1/kT)}$ is less than 1 percent of I_B.

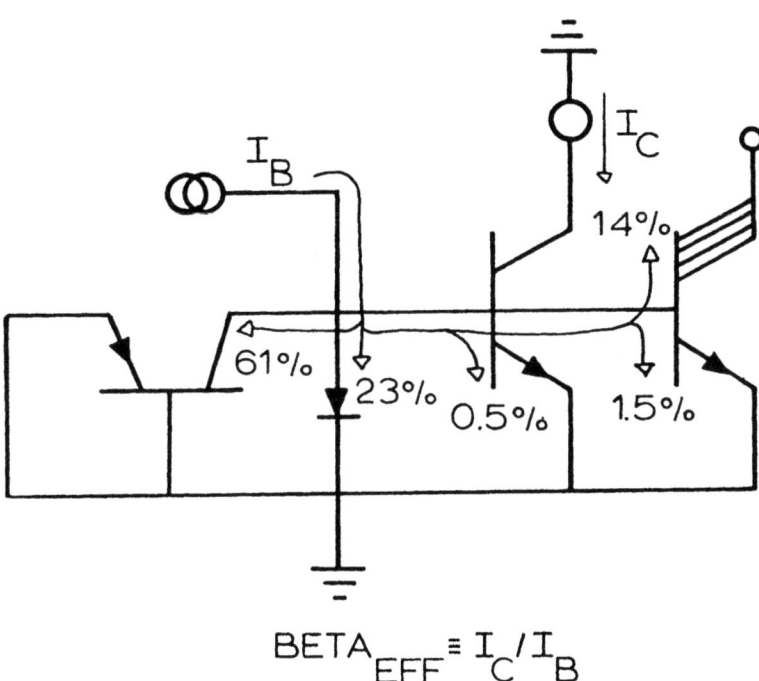

Fig. 7. Equivalent circuit of a five collector I^2L unit cell. Effective beta is defined as the input current I_B divided the current into one collector with the other collectors open. The distribution of base current within the cell is given by the percentages.

Therefore, the gain of the up transistor must be in the hundreds in order for the I²L gate to have enough effective beta to perform its switching function. β_{eff} as a function of implant dose is given in Fig. 6. The use of ion implantation for the intrinsic base dose allows a very precise control of beta, with β_{max} limited primarily by BV_{CEO} requirements on the up n-p-n transistor.

The propagation delay time τ_d for low currents at a given injector voltage of an I²L gate depends on the sum of the capacitances at the input, the voltage swing, p-n-p alpha and n-p-n β_{eff}. This time is approximately defined by[14]

$$\tau_d \approx \frac{1}{2} \frac{\Delta V}{\alpha_{pnp} I_{inj}} \left[C + \frac{C}{(\beta_{eff}-1)} + \frac{C_{cb}\beta_{eff}}{(\beta_{eff}-1)} \right] \quad (2)$$

where C and C_{cb} are the total input node capacitance and n-p-n collector-base capacitance, respectively, averaged over the voltage swing ΔV. The first term in (2) represents t_{on} and the second and third terms represent t_{off}. To make a significant reduction in t_{on} would require technology innovations impacting the geometry and process (i.e., oxide isolation, Schottky diodes on input or output, vertical p-n-p logic, e-beam lithography, etc.). However, t_{off} can be minimized by reducing the intrinsic n-p-n base dose. Referring to Fig. 6, the second term in (2) can be reduced to 12 percent of t_{on} by selecting a base dose such that $\beta_u \gtrsim 2400$ ($\beta_{eff} \gtrsim 8.6$).

7. ANALYTIC MODEL FOR OPTIMIZATION OF I²L DEVICES***

The model described in the following sections was developed by Evans[6,15] to aid in the design and optimization of implanted I²L structures like the one pictured in Figures 8 and 9. The standard double diffused process is replaced by a single p⁺ diffusion or implant for the extrinsic base and a separate deep p⁻ implant for the intrinsic base. The decoupling of the extrinsic and intrinsic base elements and the precise control of base charge afforded by implant technology allows devices to be routinely built on thin epitaxial material with a heavily doped p region to minimize bulk resistive drops. Typical profiles of the two base regions are shown in Figure 4. The substrate is the up transistor emitter while an arsenic implant serves as the down transistor emitter. The p⁺ extrinsic base combined with epi-substrate constitutes a parasitic diode in parallel with the emitter-base of the up transistor.

This structure described here is particularly amendable to one dimensional injection current modeling[9] because of the heavily

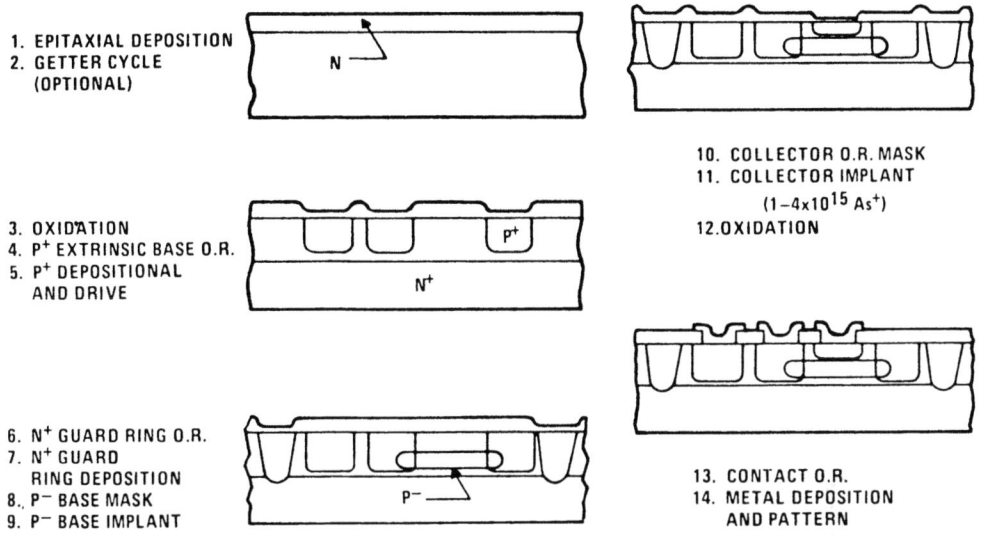

Fig. 8. Material process flow of advanced non-isolated I²L.

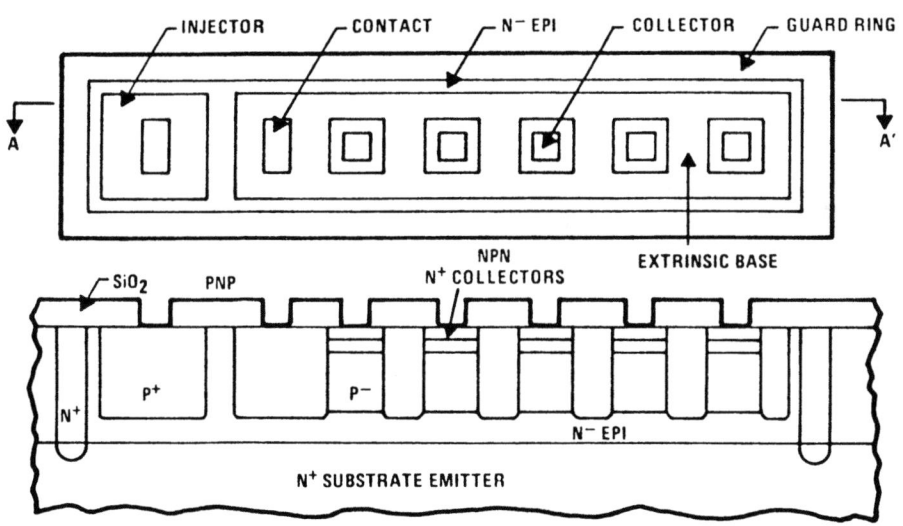

Fig. 9. Second generation I²L/MTL gate.

doped p base reduces the importance of surface recombination under oxide layers and the resistive drop from the injector side of the base to the far collector.

Some variations of this structure have already been discussed in literature: double diffused pnp[16], Schottky contact on output[17], oxide isolation[18,19], Schottky contact on input (p-epi)[20], and up-diffused intrinsic base[18].

7.1 Model formulation.

The ring oscillator or inverter chain is used to characterize the transient performance of an I^2L process/device design. An equivalent circuit for two adjacent inverters in the chain is shown in Fig. 10, where V_1 is the base-emitter and V_2 is the base-collector voltage of the npn up transistor in units of kT/q. The unit cell effective beta, β_{eff}, is measured by applying a base current I_B at the input, and observing collector current I_C at the output with the injector grounded. t=0 is the time at which transistor Q_{n-1} has completely turned off and transistor Q_n begins to turn on. The capacitances at node n are charged until $V_1=V_{BE}$ [i.e., the voltage at which $I_C(t)=I_p$]. The period of time required to reach V_{BE} is t_1 or t_{on}. At t_1, node n+1 begins to discharge due to $I_C(t)$. When the charge at node n+1 is exhausted, the voltage at node n is V_{BEI} where $V_{BE} \leq V_{BEI} \leq V_{BESAT}$. The time required for node n to go from 0 to V_{BEI} is the total delay, t_T, and $t_{off} = t_T - t_{on}$. The average propagation delay is defined as

$$t_D = (t_{on} + t_{off}) / 2.0. \tag{3}$$

The time dependence of the node voltages is shown in Figure 11. Node n goes from V=0 to V=V_{BEI} during t_T while node n+1 goes from V=V_{BESAT} to V=0. Since node n+1 is discharged before $V_1=V_{BESAT}$ at

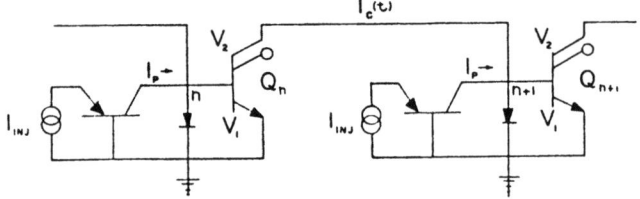

Fig. 10. I^2L unit cell equivalent circuit for two consecutive inverter elements of a ring oscillator. V_1 and V_2 are the base-emitter and base-collector voltages, respectively, of the npn.

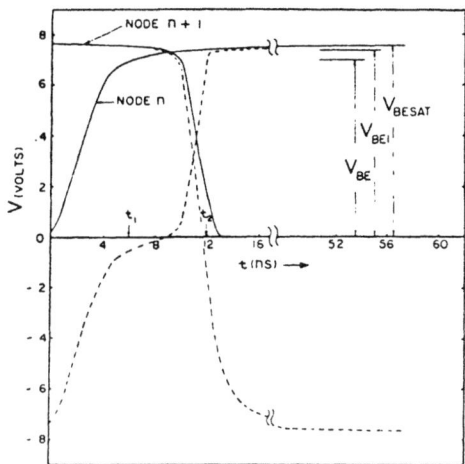

Fig. 11. The time dependence of voltages at node n and n+1. Solid lines are the input or base-emitter voltage of the npn transistor and the dashed lines are the npn base-collector voltage.

node n, $\beta_{eff} I_p$ is never available as a switching current.

The current at node n is distributed at times $<t_{off}$ according to

$$I_p = [MC_{ne} + (M-1)C_{nc} + C_p + C_{pn}] \frac{dV_1}{dt} + \frac{I_c}{\beta_{eff}} \quad (4)$$

where M is the number of collectors, C is the total capacitance (diffusion + depletion) and the subscripts ne, nc, p, and pn label the emitter and collector of the npn, the pnp and the parasitic diode, respectively. It is assumed that $V_1 \approx V_2$ for all open collectors due to the symmetry of ion-implanted npn and that for normal multicollector stick geometries used in logic design

$$C_{nc}(V_2) << MC_{ne}(V_1) + (M-1)C_{nc}(V_1). \quad (5)$$

Making similar approximations, the current at node n+1 is described by

$$I_p = [M(C_{ne} = C_{nc}) + C_p + C_{pn}] dV_1/dt + I_c \quad (6)$$

where the capacitance discharge current is assumed to be large compared to the diode currents of the various elements. An expression for t_{on} can be easily found from Eq. (4) by neglecting I_c/β_{eff} during the turn-on phase, substituting an Ebers-Moll Model (21) for the capacitances and integrating from V-0 to V_{BE}. Two expressions for t_{off} as a function of V_{BEI} can be found from Eq. (4) and (6). The first expression, derived by consideration of Eq. (4) alone is given by:

$$t_{off} = [(M\tau_{ne}I_{sn} + (M-1)\tau_{nc} + \tau_p I_{sp} + \tau_{pn} I_{spn}) \times \beta_{eff} F_1 / I_{sn}] + \bar{C}_T^{dpl}(V_{BEI} - V_{BE} + F_1) / I_p, \quad (7)$$

where I_s is the saturation current, τ is the transit time, $\bar{C}_T^{dpl}$ is the total depletion component of capacitance evaluated at $V=(V_{BEI}+V_{BE})/2$ [22] and

$$F_1 = \ln\left[\frac{\beta_{eff} - 1}{\beta_{eff} - (I_{sn}\exp(V_{BEI}) / I_p)}\right]. \quad (8)$$

A second expression for t_{off} is found by simultaneously solving Eq. (4) and (6). The resulting equation is given by

$$t_{off} = \frac{1}{(\beta_{eff}-1)} \frac{1}{I_p} \times$$

$$\left\{\beta_{eff}\int_{V_{BE}}^{V_{BEI}}\left[MC_{ne} + (M-1)C_{nc} + C_p + C_{pn}\right]dV_1 \right. \quad (9)$$

$$\left. + \int_0^{V_{BESAT}}\left[M(C_{ne} + C_{nc}) + C_p + C_{pn}\right]dV_1\right\}.$$

Combining Eq. (7) and (9) to eliminate t_{off} yields a single transcendental equation in V_{BEI}. This equation can be easily solved either graphically or with a bisection technique requiring only a few iterations. The expression for t_{on} along with Eq. (7) and (9) completely determine the switching characteristic of a multicollector I²L gate as a function of physically meaningful process and device parameters at low and medium currents. The parameters can be calculated using process/device simulators[1,2,3] or measured from carefully designed test structures[5]. Of primary importance is the validity of the equations at current levels in which both diffusion and depletion capacitance are important.

a) Low current operation. At the extremes of current operation the equations discussed above reduce to forms similar to those already given in the literature[23]. At low currents, V_{BEI} approaches V_{BESAT} asymtotically. By letting $V_{BES} \sim V_{BEI}$ and assuming that $M >> 1$, Eq. (9) reduces to

$$t_{off} \cong \left[\frac{t_{on}}{(\beta_{eff}-1)}\right] + \left[\frac{(\beta_{eff}+1)}{I_p(\beta_{eff}-1)}\right] \int_{V_{BE}}^{V_{BESAT}} C_T dV \qquad (10)$$

In the low current range the capacitance terms are dominated by their slowly varying depletion components. As $\beta_{eff} \to 1.0$, the first term on the right in Eq. (10) dominates giving rise to extremely long gate delay times.

b) High current operation. The high current regime is dominated by the diffusion component of capacitance. Using the Ebers-Moll model for capacitance and again assuming $M >> 1$, Eq. (9) may be simplified to

$$t_{off} \cong \tau \frac{\beta_{eff}}{(\beta_{eff}-1)} e^{V_{BEI}-V_{BE}} \qquad (11)$$

where τ is an effective time constant defined as

$$\tau = \left[\tau_p I_{sp} + \tau_{pn} I_{spn} + M(\tau_{ne}+\tau_{nc})I_{sn}\right]/I_{sn} \qquad (12)$$

As β_{eff} becomes small, V_{BEI} approaches V_{BESAT} and Eq. (11) may be approximated by

$$t_{off} \sim \tau \frac{\beta_{eff}^2}{(\beta_{eff}-1)} . \qquad (13)$$

Just as in the low current case, long delay times result if β_{eff} is very close to 1.0. For large betas Eq. (11) may be approximated by

$$t_{off} \cong \tau \, \beta_{eff}^{\frac{1}{n}}, \tag{14}$$

where $n > 1$. Typically, for a β_{eff} range of $10 < \beta_{eff} < 50$, n will be $1.4 \lesssim n \lesssim 1.8$.

Due to the integrated nature of the device under discussion, it is not meaningful as a general rule to make arbitrary changes in a device parameter such as beta to examine the impact of that parameter on device performance. For example, process or geometry changes which increase beta will usually decrease τ and vice versa.

Because of the prominence of β_{eff} in all the equations derived above, one process variation has been chosen for detailed analysis. Beta of the down transistor, β_d, can be arbitrarily scaled with little impact on τ by changing the active emitter charge of the down npn transistor. The minimum delay is plotted versus β_{eff} in Figure 12. β_d is labeled at specific points along the curve. The results presented in Figure 12 are based on an exact solution of Eq. (4) and (6) which includes base conductivity modulation in all device elements. Therefore, low current β_{eff} is used in this plot since the gain at t_{off}^{min} is somewhat lower. As β_d increases the percentage of base current required to drive the down npn transistor diminishes, making β_{eff} almost independent of the down gain as shown in Fig. 12. One might conclude from this plot and from Eq. (13) and (14) that extremely high or low beta is undesirable. However, other variations such as scaling

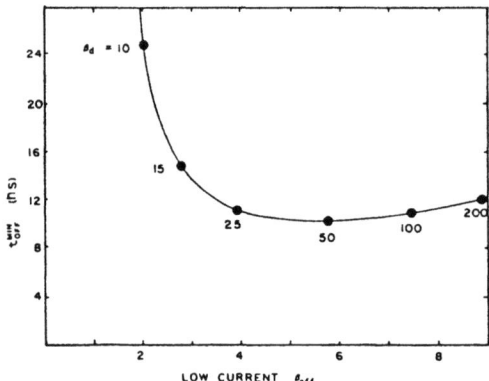

Fig. 12. Minimum t_{off} versus low current β_{eff}. The corresponding β_d is labeled on the graph at several points. The variation in $\beta_d(\beta_{eff})$ is realized by arbitrarily changing the effective emitter charge of the down npn transistor.

the collector area or changing the p⁻ dose have significant impact on both β_{eff} and τ. It is even possible to improve t_{off} by increasing β_{eff} ($5 < \beta_{eff} < 10$) for particular combinations of p⁻ dose and collector area.

Modulation effects. Collector fall-off in both the pnp and npn transistors due to base conductivity modulation is a factor at medium and high currents. Fall-off in the pnp I_c, smears the transition from depletion capacitance dominated to diffusion dominated regions. However, the same fall-off mechanism reduces the pnp back injected current (i.e. inverse pnp I_c) and, thus, improves β_{eff} since more base current is available to drive the intrinsic npn. At higher currents the npn base modulates and β_{eff} falls off. The delay time versus injector current is plotted for three levels of model complexity and compared to experiment in Figure 13. The most important factor is the fall-off in the source current which degrades the speed-power product at medium and high currents. The current dependence of β_{eff} is also important when this parameter is large.

7.2 Application to I²L device design

An I²L gate simulation program based on Eq. (7) and (9) was

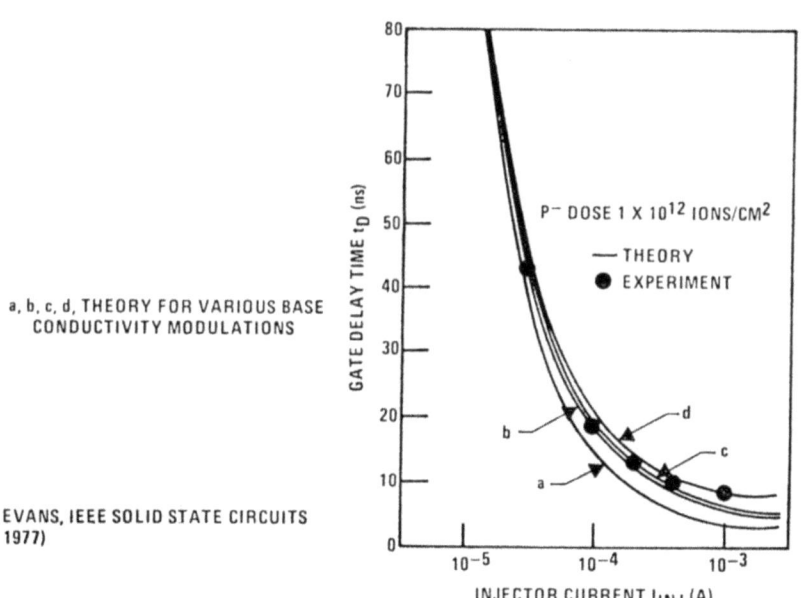

Fig. 13. Ring oscillator delay time.

written to analyze the large signal switching properties of numerous process and geometry variations. Input parameters are determined for a standard process and geometry (i.e. devices already built and characterized) and then scaled or modified as required to model changes in structure. The independent variables are $\underline{x}$, $\underline{y}$, $\underline{z}$, pnp base width, β_u, β_d, collector area, extrinsic base sheet resistance, and parasitic diode capacitance. The basic structure on which variations are made is shown in Fig. 9. $\underline{x}$ is the vertical dimension while $\underline{y}$ and $\underline{z}$ are lateral dimensions. Several assumptions made in allowing variations are the following: (1) The extrinsic base is doped in a range, 10^{18} to 10^{19} cm^{-3}, in which the parasitic diode current is not a strong function of the base sheet resistance[13], (2) The width of the lightly doped n⁻ regions in Figure 4 are adjusted to keep the time constants of the vertical elements constant.

The gate simulator has been checked by comparing calculated t_D and V_{BEI} with the results of a large circuit analysis program (SPICE) using the same set of Ebers-Moll parameters as inputs. Agreement is good when the chain of inverters is long enough to allow a complete voltage swing of the input. The advantages of an I²L gate simulator based on simple analytic equations are two fold: (1) the model requires 100 times less computer time than SPICE, making it more suitable as an optimizer, (2) the model provides a tractable mathematical formalism to explain performance.

a) <u>Intrinsic base process.</u> A high energy p- implant supplies the base of the intrinsic npn. The depth of the implant is chosen to make the base charge independent of epi thickness and collector process variations. An example of the relationship of p- dose to β_{eff} and β_u is given in Figure 6 for the structure described in Figure 9. It is interesting to note the extremely high intrinsic npn up beta required to obtain $\beta_{eff} \geq 1$. At low currents it is easy to see from Figure 6 and Eq. (10) that the lowest p⁻ dose compatible with breakdown requirements should yield the best performance. At medium and high currents the optimal p⁻ dose is less obvious. Decreasing the dose results in both a larger β_{eff} and a smaller intrinsic npn time constant. According to Eq. (14), t_{off} is directly proportional to both of these parameters.

The influence of parasitic diode capacitance is best observed on devices with inherently high betas ($\beta_{eff} \gtrsim 7.5$). The total unit cell parasitic diode capacitance, measured at $V = 0$, is plotted versus delay time at 100µA injector current in Figure 14. The scatter in the experimental data is due to variations in beta of the intrinsic npn. The solid curve represents model calculations including both pnp base width and p⁺ junction depth variations due to changes in diffusion drive time. The dashed curve is obtained by making simultaneous mask changes to keep the

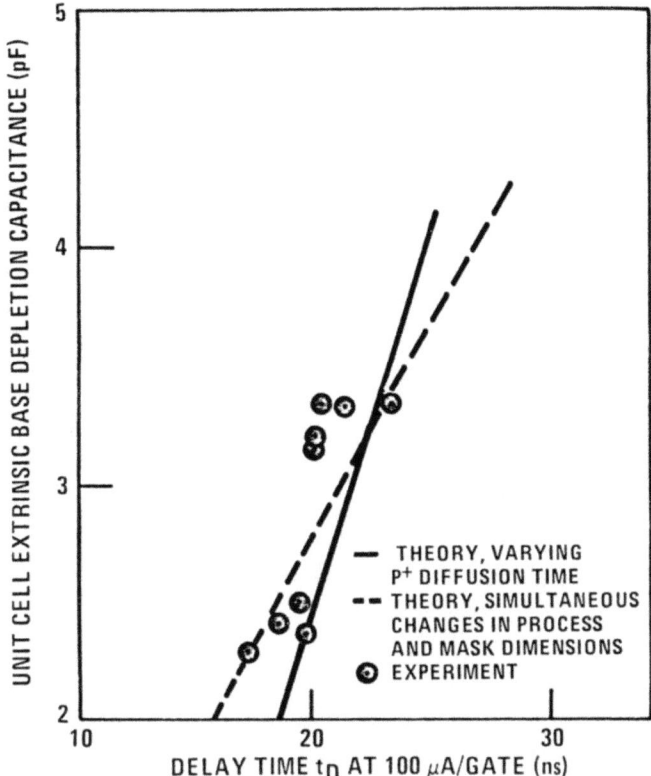

(AFTER S. A. EVANS, IEEE SOLID STATE CIRCUITS SC 12, APRIL 1977)

Fig. 14. Capacitance and delay time in I²L unit cell.

pnp base width constant. One can conclude that improving speed at medium currents by reducing parasitic diode capacitance is most effective when both process and mask changes are allowed.

b) <u>Device scaling</u>. The performance of I²L devices has been greatly enhanced by replacing the heavily doped phosphorus guard ring with oxide and by adding Schottkys to the output. Another method of obtaining substantial improvements in speed is to shrink device dimensions using advanced processing techniques such as e-beam and X-ray lithography. Given that one can arbitrarily scale device designs in three dimensions using one of the newer process technologies, how could current designs be changed for ultimate device performance? One possibility is to scale down x, y, and z dimensions while holding the pnp base width constant. As expected it was found that the speed-power product is reduced by the scale factor squared; however, the maximum speed is almost independent of scale factor. In this current range, the delay time is determined almost entirely by β_{eff} and τ. One can easily show that these parameters are approximately independent of the scale for the conditions stated above.

The results of scaling the pnp and extrinsic base areas while holding the npn collector area constant is shown in Figure 15. In the operating region where the depletion capacitance of the parasitic elements controls the speed, the speed-power product is again proportional to the scale factor squared. However, there is a definite improvement in the maximum speed as the scale factor gets smaller. This is primarily due to the increasing importance of the intrinsic npn time constant with decreasing extrinsic base area. Figure 15 also shows approximately the effect of bulk resistance when the far collector is hooked up in the circuit. When scaling $\underline{x}$, $\underline{y}$, and $\underline{z}$, the IR drop at a given delay goes as the scale factor.

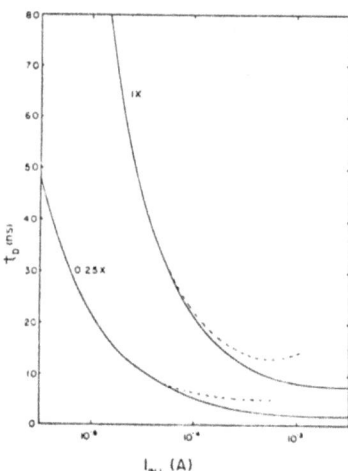

Fig. 15. Changes in gate delay time characteristic due to scaling $\underline{x}$, $\underline{y}$, and $\underline{z}$ while holding the npn collector area constant. Device area changes as the square of the scale factor (1x, .25x, etc.).

REFERENCES

1. P.L. Shah and W.H. Schroen, "A process model for sequential diffusions and redistributions in silicon LSI technology," Electrochem. Soc. Meeting May 1975, Extended Abstr. No. 170.
2. P.L. Shah, "Analysis of implanted layers in silicon for semiconductor device processing," IEEE Conf. Sci. Indust. Applic. Small Accelerators, 1976, pp. 411-419.
3. P.L. Shah, "Computer-aided design and optimization for semiconductor device fabrication," Semiconductor Silicon/1977, ed. by H.R. Huff and E. Sirtl, Electrochem. Soc., Princeton, N.J., p. 923.
4. S.A. Evans and J.S. Fu., "Application of a new, accurate bipolar simulator to correlate device performance to process selection," Electrochem. Soc. Meeting May 1975, Extended Abstract No. 168.
5. J.M. Herman, III, S.A. Evans, and B.J. Sloan, Jr., "Second generation I^2L/MTL: A 20 nsec process/structure," IEEE Solid State Circuits SC-12, April 1977.
6. S.A. Evans, "An analytic model for the design and optimization of ion implanted I^2L devices," IEEE Solid-State Circuits SC-12, April 1977.
7. S.A. Evans, J.M. Herman, III, and B.J. Sloan, Jr., "On the electrical properties of the I^2L n-p-n transistor," IEEE Trans, Electron Devices ED-23, 1192, 1976.
8. S.A. Evans, I.H. Morgan, and T.N. Jernigan, "A model for the accurate prediction of the current-voltage behavior of ion-implanted MOS transistors at low drain voltages," IEEE IEDM Meeting, Washington, D.C., 1974, p. 47.
9. H.H. Berger, "The injection model--A structure-oriented model for merged transistor logic (MTL)," IEEE Journal of Solid State Circuits SC-9, 218, 1974.
10. M.S. Mock, "Transport equation in heavily doped silicon, and the current gain of a bipolar transistor," Solid State Electronics 16, 1251 (1973).
11. J. Krausse, "Auger-Rekombination im Mittelgebiet durchlassbelasteter Silicium-Gleichrichter und Thyristoren," Solid-State Electronics, 17, 427-429, 1974.
12. H.J. deMan, "The influence of heavy doping on the emitter efficiency of a bipolar transistor," IEEE Trans. Electron Devices, ED-18, 833-835, 1971.
13. M.S. Mock, "On heavy doping effects and the injection efficiency of silicon transistors," Solid-St. El., 17, 819-824, 1974.
14. F.M. Klaassen, "Device physics of integrated injection logic," IEEE Trans. Electron Devices, ED-22, 145-152, 1975.
15. S.A. Evans, "An analytic model for the design and optimization of ion-implanted I^2L devices," Semiconductor Silicon/1977, ed. by H.R. Huff and E. Sirtl, Electrochem. Soc., Princeton, N.J., p. 944.
16. B. Sander and J.M. Early, "A4096X1(I^3L) bipolar dynamic RAM,"

ISSCC Dig. Tech. Papers, 182-183, 1976.
17. F.W. Hewlett, "Schottky I^{2}L," IEEE Solid State Circuits, SC-10, 343-348, 1975.
18. C. Mulder and H.E.J. Wulms, "High-speed I^{2}L," IEEE Solid State Circuits, SC-11, 379-385, 1976.
19. H.H. Berger and S.K. Wiedmann, "Schottky transistor logic," ISSCC Dig. Tech. Papers, pp 42-43, Feb. 1975.
20. V. Blatt, S. Walsh, and L.W. Kennedy, "Substrate fed logic," IEEE Solid State Circuits, SC-10, 336-342, 1975.
21. J.J. Ebers and J.L. Moll, "Large-signal behavior of junction transistors," I.R.E. 42, 1761-1772, 1954.
22. M. Spivak, Calculus, New York: W.A. Benjamin, Inc., 1967, p. 237.
23. F.M. Klaassen, "Device physics of integrated injection logic," IEEE Trans. Electron Devices, ED-22, 145-152, 1975.

* NOTE: MATERIAL EXCERPTED FROM S.A. EVANS AND J.S. FU, "Application of a new, accurate bipolar simulator to correlate device performance to process selection," Electrochem. Soc. Meeting May 1975, Extended Abstracts 75-1, pp. 394-396 (Abstract No. 168).

© 1975 by THE ELECTROCHEMICAL SOCIETY INC. REPRODUCTION AUTHORIZED BY ECS.
This information was originally presented at the 147th Spring Meeting, Toronto, Canada, 1975, of the Electrochemical Society Inc. This information reprinted by permission of the publisher, the Electrochemical Society Inc.

** NOTE: MATERIAL EXCERPTED FROM S.A. EVANS, J.M. HERMAN, III, AND B.J. SLOAN, "On the electrical properties of the I^{2}L n-p-n transistor," IEEE Trans. Electron. Devices ED-23, pp. 1192-1194, 1976.

© 1976 by THE INSTITUTE OF ELECTRICAL AND ELECTRONICS ENGINEERS, INC. REPRODUCTION AUTHORIZED BY IEEE.

*** NOTE: MATERIAL EXCERPTED FROM S.A. EVANS, "An analytic model for the design and optimization of ion implanted I^{2}L devices," SEMICONDUCTOR SILICON/1977, ed. by H. R. Huff and E. Sirtl, Electrochem. Soc., Princeton, N.J., pp. 944-954.

© 1977 by THE ELECTROCHEMICAL SOCIETY INC. REPRODUCTION AUTHORIZED BY ECS.
This information was originally presented at the 151st Spring Meeting, Philadelphia, Pennsylvania, 1977, of the Electrochemical Society Inc. This information reprinted by permission of the publisher, the Electrochemical Society Inc.

SIMULATION OF INTEGRATED CIRCUIT FABRICATION PROCESSES *

D.A. Antoniadis and R.W. Dutton

Stanford University Integrated Circuits Laboratory
Stanford, California 94305

ABSTRACT. The Stanford University IC process simulator program is used as a vehicle for the discussion of the various aspects of modern process simulation. The development of numerical models for the one-dimensional redistribution of impurities in the presence of moving boundaries as in oxidation and epitaxy is outlined. The dilemma posed by the complete physical vs engineering type of diffusion simulation is discussed.

1. INTRODUCTION

The Stanford University PRocess Engineering Models (SUPREM) program is a computer simulator capable of simulating most typical IC fabrication steps. The program is designed so that these steps can be simulated either individually or sequentially, just as they would occur during the actual fabrication process. The output of the program, available at the end of each step, consists of the one-dimensional profiles of all the dopants present in the silicon and silicon-dioxide materials. These profiles are displayed in various formats including line-printer printout, line-printer plots, and high-resolution (Calcomp type) plot. It is understood that, in sequential step simulation, the output of a processing step constitutes the initial conditions for the following one. The junction depths and sheet resistances of all n or p layers formed during the process are also calculated.

The fabrication-step simulation is based on several process models. The models implemented in SUPREM are the following:

---

* This work has been supported through ARO Contract DAAG-G29-77-C-006.

(a) oxidation/drive-in
(b) Predeposition through the surface (gaseous or solid)
(c) epitaxial growth
(d) ion implantation
(e) etching
(f) oxide deposition

Solid-state diffusion is fully accounted for in any of the above models involving high temperature.

Communication between the user and SUPREM has been made as simple as possible. The input file consists of free format statements involving key words and numbers. Typically, a single processing step can be simulated with an input specification involving less than 60 alphanumeric characters. The details of input specification are available [1].

Several physical parameters have been stored in the program and constitute the default values used by the models. These values are described in Ref. 1 with the appropriate references. They may be overridden by user-specified parameters.

2. PROGRAM DESCRIPTION

The various process models have been implemented in SUPREM as subprograms, each consisting of a number of subroutines and special functions. Figure 1 is a schematic of this arrangement. Input parameter specifications have been designed to resemble actual process runsheet data and documentation. A run may consist of a series of process steps. A typical input file is illustrated in Figure 2. The sequence of steps and the specification of the correct model parameters are controlled by a supervisor subprogram that evokes the appropriate step subprogram. All communication between the various subprograms is directed through the common variable area of the computer memory.

One essential part of the common area contains the impurity concentration arrays. In the present version of SUPREM, there is a capability for handling up to four different impurity species. The impurity concentration is stored in terms of a discrete profile with a maximum number of 500 points. Each concentration value corresponds to a point in the discrete space (spatial grid) defined along a vertical axis, with its origin at the surface of the material -- silicon (Si) or silicon dioxide (SiO_2). Typically, the spatial grid is divided into the SiO_2 grid points, high-resolution-depth grid points, and low-resolution grid points, as illustrated in Figure 3. Within each of the above three areas, the grid-point separation is uniform. The spatial steps for each area and the three interface point indexes are stored at the end

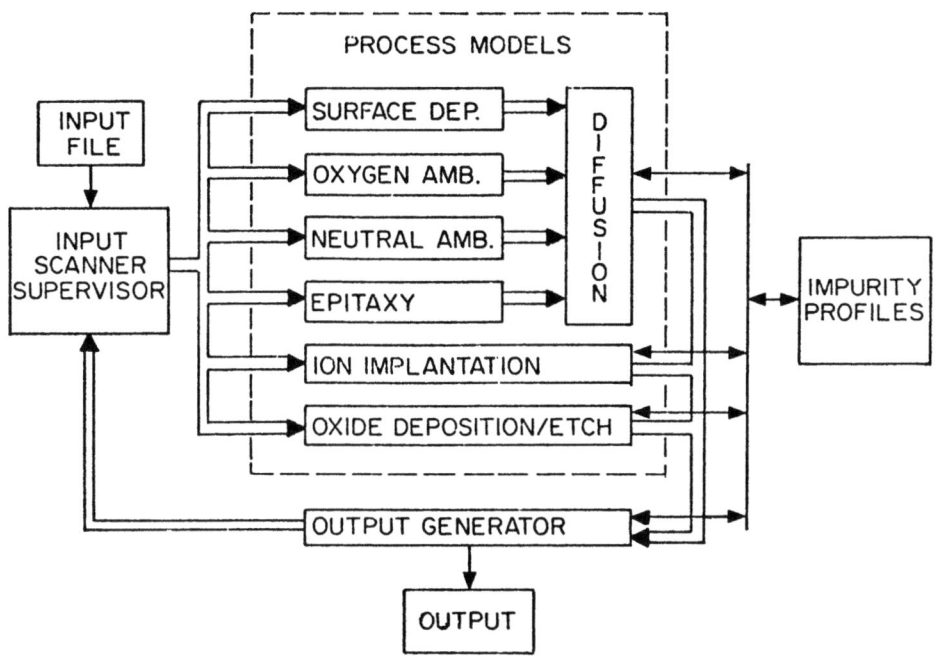

Fig. 1. General block diagram of the SUPREM process simulator.

of the concentration array, together with additional information pertaining to the various physical characteristics (such as diffusivities, segregation coefficients, etc.) of the impurity species represented in this array. Storage (and recovery) of intermediate or final results in nonvolatile records is therefore greatly facilitated.

Very often, during a processing step, the physical dimensions of the simulated discrete space may change, as happens for example during oxidation, etch, deposition, and epitaxy. The distance between spatial grid points may not be uniform during any of the above processes. For this reason, a cubic spline interpolation routine has been incorporated into the program and is used at the end of each of the processing steps to restore the uniformity of the spatial grid.

```
TITLE  SUPREM TEST EXAMPLE #1
GRID   DXS1=0.005, DPTH=0.5, XMAX=1.0                    ⎫  INITIALIZATION
SUBS   ELEM=B, CONC=2.0E15, ORNT=111                     ⎭

PLOT   IDIV=Y, TOTL=Y, YMIN=14, NDEC=7, WIND=0.8         ⎫  OUTPUT CONTROL
PRINT  HEAD=Y, IDIV=Y, TOTL=Y                            ⎭

MODEL  NAME=DIF1, ELEM=B, SIDC=1.56E9, SIEA=3.3          ⎫  EXTERNAL
MODEL  NAME=DIF2, ELEM=AS, SIDC=1.44E11, SIEA=4.08       ⎬  MODEL
+      FILE=ASDF2                                        ⎭  SPECIFICATION

COMM   ION IMPLANT (BORON)                               ⎫
STEP   TYPE=IMPL, ELEM=B, DOSE=1.0E15, AKEV=50           ⎬  STEP 1

COMM   OXIDATION (DRY/WET/DRY)                           ⎫
STEP   TYPE=OXID, TEMP=1000, TIME=10, MODE=DRYO          ⎬  STEP 2
STEP   TYPE=OXID, TEMP=1000, TIME=20, MODE=WETO          ⎬  STEP 3
STEP   TYPE=OXID, TEMP=1000, TIME=5, MODE=DRYO           ⎭  STEP 4

COMM   REMOVE OXIDE                                      ⎫
STEP   TYPE=ETCH, TEMP=25                                ⎬  STEP 5

COMM   GASEOUS PREDEPOSITION (ARSENIC)                   ⎫
STEP   TYPE=PDEP, TEMP=1050, TIME=20, ELEM=AS            ⎬  STEP 6
+      CONC=5.0E20, MODL=DIF1, MODL=DIF2                 ⎭

END
```

Fig. 2. Typical input file with identification of the various control sections.

3. PROCESS MODELS

3.1 Diffusion models

3.1.1 <u>Introduction</u>. Solid-state diffusion is the mechanism responsible for impurity migration within the silicon body during high-temperature process steps. The diffusive flux, F, of impurities is related to their concentration gradient. In one dimension, the relation is

$$F = -D \frac{\partial C}{\partial x} \qquad (1)$$

where D is the diffusion coefficient of the impurity and C is its concentration. Under the assumption of single-species migration, there is no generation or loss rates within the material. The impurity conservation equation, therefore, becomes

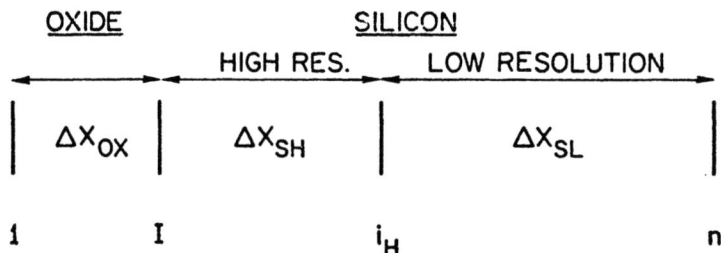

Fig. 3. COMPOSITE SILICON/SILICON DIOXIDE DISCRETE SPACE.
Indices identify the nodes at the various interfaces, and the ΔX refer to the nodal spacing in each region.

$$\frac{\partial C}{\partial t} = D \frac{\partial^2 C}{\partial x^2} \qquad (2)$$

which is well known as Fick's Law.

Accurate knowledge of diffusivity is of great importance in process simulation, and a great amount of work has been spent on the experimental determination of D; however, various published data have rarely been consistent with each other. The reason for this is that Eq. (2) is a phenomenological law with a limited range of applicability and does not reflect the physical mechanisms of the diffusion process. Although Eq. (2) appears to apply under relatively low impurity concentration, it does not apply for concentrations at or above the intrinsic carrier concentration, n_i, in the semiconductor at the process temperature. Figure 4 is a plot of n_i vs temperature [2].

Historically, one of the first attempts to extend the applicability of Fick's Law to higher concentrations was to include the effect of the introduced free carriers on the impurity ion migration [3] in the same way as for ambipolar diffusion in plasmas. The diffusion coefficient therefore becomes a function of impurity concentration given by

$$D = D_o f_e = D_o \left\{ 1 + \left[1 + 4 \left(\frac{n_i}{C} \right)^2 \right]^{-1/2} \right\} \qquad (3)$$

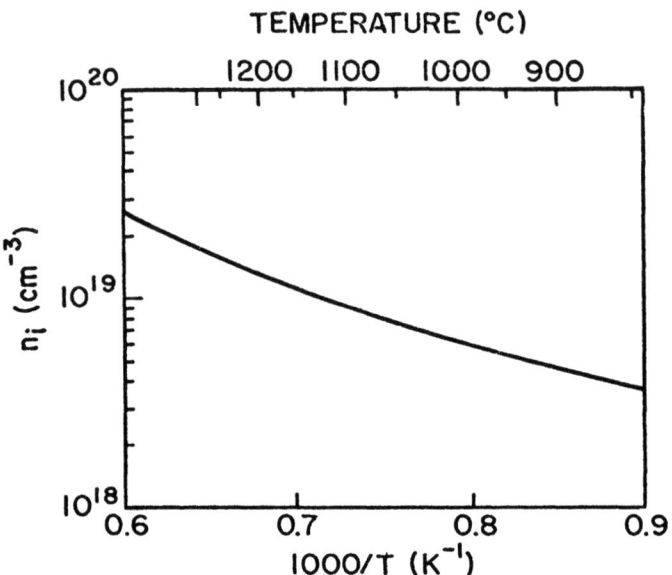

Fig. 4. Intrinsic silicon carrier concentration vs temperature.

where D_o is a "constant" diffusion coefficient and f_e is the enhancement factor caused by the E-field between ionized impurities and free carriers. The range of applicability of Eq. (3), however, was still limited. It became apparent [4] that D_o is also a function of concentration at high concentrations. Recently Fair [5] summarized some of the latest works on diffusion and proposed that the actual diffusion coefficient is the sum of various diffusivities, each accounting for impurity interactions with different charge states of lattice vacancies. The diffusion coefficient is then expressed as

$$D_o = \left[D^x + D^- \left[V^- \right] + D^= \left[V^= \right] + D^+ \left[V^+ \right] \right] \quad (4)$$

where $[V^r]$ is the normalized excess charged vacancy concentrations for each charge state, $r = -, =, +$. These concentrations depend on the position of the Fermi level and may be related to the impurity concentrations by

$$\left[V^-\right] = \frac{n}{n_i}, \quad \left[V^=\right] = \left(\frac{n}{n_i}\right)^2 \text{ and } \left[V^+\right] = \frac{n_i}{n} \tag{5}$$

Although the above formulation appears to extend the validity of Fick's Law, the law still remains a phenomenological approach to the diffusion problem with applicability only near thermal equilibrium conditions. Phenomena such as the anomalous annealing behavior of ion-implanted boron [6] or proton-enhanced diffusivity [7] are not possible to explain with such a simple model.

Recently, Gibbons et al. [8] proposed a model that can describe both the nonequilibrium and near-equilibrium behavior of impurity migration. The model assumes that the impurities migrate as two or more identifiable species mutually interacting by means of first-order kinetic reactions. This model can be formulated as

$$\frac{\partial C_1}{\partial t} = F_1 + G - R_1 \tag{6}$$

$$\frac{\partial C_2}{\partial t} = F_2 + G - R_2, \quad \cdots\cdots\cdots$$

where $C_{1,2...}$ are the various species concentrations, $(G-R)_{1,2...}$ the generation recombination terms, and $F_{1,2}$ the diffusive fluxes for each species. For a two-species model, for example, C_1 could be the electrically active impurity concentration and C_2 might be the concentration of an impurity vacancy complex. This last model, which deviates significantly from the simple Fick's Law formulation, has a broad range of applicability but requires the experimental determination of several physical constants.

3.1.2 <u>Engineering models of diffusivity</u>. As the generality of diffusion models increases so does their complexity and the number of physical parameters necessary for model implementation. For engineering applications, of prime importance is the consideration of model complexity vs expected result accuracy. In addition, the completeness of the physical approximation achieved by a model must be weighted against the reliability of the necessary physical parameters. The first consideration relates to the cost-effectiveness of the process simulator as an engineering tool, and the second relates to its expected reliability. The type of diffusion model currently implemented in SUPREM is expected to perform well in several situations of practical interest such as under heavy doping and in the presence of impurity coupling. No attempt has been made, however, to include anomalous phosphorus diffusion effects and

and anomalous annealing behavior of boron. The inclusion of such effects [9] is under study.

The diffusion coefficient for any impurity simulated by SUPREM is

$$D = D_i (1 + \beta f_v)/(1 + \beta) \qquad (7)$$

where D_i is the "intrinsic" or low impurity concentration diffusivity and $f_v = n/n_i$ for donors and p/n_i for acceptors. The factor f_v is thus related to the relative concentration of charged vacancies and while β is related to the relative impurity diffusion effectiveness of charged vacancies as compared to neutral ones. The above form is essentially identical to that proposed by Fair, given the fact that typically only one type of charged vacancies is responsible for diffusive migration [5]. In addition, the formulation for the diffusive flux has been modified to

$$F = -\frac{\partial}{\partial x}(DC) \qquad (8)$$

This form has been suggested by Morehead [10]. The advantage is that it predicts impurity migration even in the absence of impurity concentration gradients provided there exists a non-uniform diffusivity D; therefore, the effect of impurity interaction, as is the case for the diffused arsenic on top of uniformly distributed boron [11], shown in Figure 5, is immediately taken into account. As a result, using Eq. (7) and (8), it is not necessary to evoke direct E-field coupling between impurities or ambipolar diffusion effects for single impurities. These phenomena are essentially accounted for by the position of the Fermi level. It may be shown that the traditional forms used for predicting E-field effects [3, 12] can be derived from Eq. (7) and (8).

3.2 Oxidation - Diffusion

3.2.1 <u>Introduction</u>. Oxidation-diffusion is probably the most heavily used step during IC fabrication. In SUPREM, this step is simulated by means of a moving boundary diffusion model. This model has been implemented in the subprogram OXIDI which also handles the predeposition step (no surface oxide) and the anneal step (fixed surface-oxide thickness).

Nonuniform diffusivities, coupling between diffusing species, time- and concentration-dependent segregation coefficients, and oxide growths are within the capabilities of the model. In the current version of SUPREM, however, the first two features only have been implemented. Both experimental and developmental work is in progress to further extend the range of effects modeled in SUPREM.

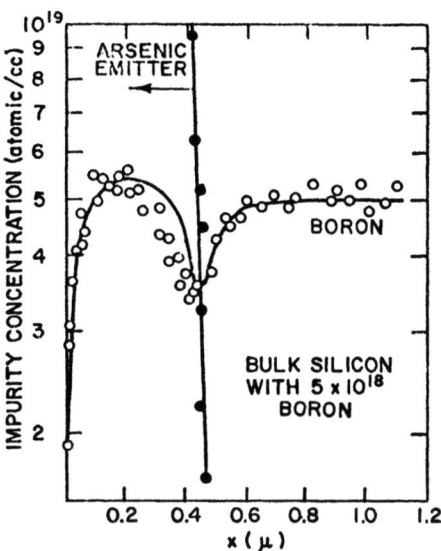

Fig. 5. Effect of diffused arsenic on uniformly distributed Boron. Solid line is simulation and the data are from Ziegler et al. [11].

The oxidation model as now implemented assumes a time-dependent oxide thickness determined by the Deal-Grove quadratic law [13] with linear and parabolic growth rates as functions of temperature and ambient conditions (dry O_2, partial O_2 pressure, wet O_2, etc.). Other ambient and silicon surface doping effects have been described earlier [14]. These effects are soon to be modeled as modifications of the B and B/A oxide-growth terms. For each impurity species the silicon-silicon dioxide segregation coefficient is assumed to be a function of temperature. The impurity concentrations are determined from the (impurity) continuity equation subject to the above moving boundary conditions. The development of the appropriate algorithm is described in the following sections. Unlike other published numerical treatments of the oxidation-diffusion problem [15], no coordinate transformation has been used; instead, the algorithm was based on the integral form of the continuity equation in the space defined by the time-dependent silicon-dioxide-silicon system with correct accounting of the flux resulting from the moving interface separating the two materials.

3.2.2 <u>The conservation equation</u>. The general equation describing the conservation of particles in the absence of chemical generation and loss can be written as

$$\oint_{S(t)} \vec{F} \cdot \vec{n} \, da = - \frac{d}{dt} \left[\int_{V(t)} C \, dv \right] \tag{9}$$

where

$\vec{F}$ = flux vector
$\vec{n}$ = outward unit normal
$S(t)$ = closed surface (function of time)
$V(t)$ = volume enclosed by S
C = concentration

This equation states that the net number of particles leaving through S is equal to the net rate of change of the number of particles in the enclosed volume. Flux F is related functionally to concentration, C, via specific physical approximations.

The formulation of numerical methods for the solution of Eq. (9) requires both spatial and time descretization. These discretizations are considered separately.

3.2.3 <u>Discretization in time</u>. The numerical solution of Eq. (9) for oxidation (and also for epitaxy, as described later), has been implemented by using implicit methods. Implicit means that the solution for the concentrations from one time step to the next proceeds simultaneously over all of the simulated space. Given N grid nodes, over which Eq. (9) applies, a system of N equations with N unknowns is thus solved for every time step. This is different from explicit methods where the solution at each node, for the next time step, is based only on the previous time-step concentrations and proceeds independently. Although implicit methods require substantially more computational overhead than do explicit ones, they are advantageous because their stability does not depend on the space and time step size [16], and they ultimately result in less computer time usage to achieve the same solution accuracy.

The discretization of Eq. (9) in time is straightforward. It is of the general form

$$H(t) = \frac{d}{dt} G(t) \tag{10}$$

where

$$H(t) = \oint_{S(t)} \vec{F}\,[S(t),\,t] \cdot \vec{n}\, da$$

$$G(t) = \oint_{V(t)} C\,[x,\,t]\, dV$$

There are several discrete time approximations to Eq. (10) [16]. For the oxidation-diffusion algorithm, we have implemented a first-order approximation given by

$$H(t^n) \doteq \frac{G(t^n) - G(t^{n-1})}{t^n - t^{n-1}} \tag{11}$$

where t^n is the time at which the equation is to be solved and t^{n-1} is the previous time increment. Although the order of the discretization error in this method is larger than that in all second-order methods, it has been found to be acceptable. Second-order approximations like the Crank-Nicolson method where

$$1/2 \left[H(t^n) + H(t^{n-1}) \right] = \frac{G(t^n) - G(t^{n-1})}{t^n - t^{n-1}} \tag{12}$$

could be relatively easy to implement, but the resulting operational overhead did not appear to be justified.

3.2.4 <u>Spatial discretization</u>. To perform spatial discretization, the region over which Eq. (10) is to be solved is divided into sub-regions (subvolumes) and a set of grid points (nodes) is defined. The functions H and G in Eq. (10) are then approximated in terms of these node quantities. The time variation of the subvolumes and subsurfaces is taken into account when calculating the time-derivative approximations. As a result, time and space functions are related.

The development of methods to approximate the spatial dependence of the solution is not as straightforward as the development of the time methods described above. The difficulty in approximating the spatial dependence is caused by changes over time in the physical structure near the Si/SiO_2 interface. It is necessary, therefore, to consider the discrete space subvolumes separately, according to their relationship to the moving interface.

Because, in the continuity equation, volumes and surfaces appear explicitly, there is one specific advantage of the method developed in that it can be adapted to the solution of the two-dimensional oxidation problem. It should be noted that the total time derivative in Eq. (4) cannot be easily moved under the volume integral for problems of more than one dimension because of the volume time dependence. As a result, moving boundary problems in two or three dimensions cannot be expressed in a partial differential form.

Unlike time discretization, no general equations can be written for spatial discretizations; instead, each case must be taken separately. The remaining sections describe the development of the numerical formulation of the impurity conservation equation for silicon oxidation in one dimension.

3.2.5 <u>The method in one dimension</u>. Defining a one-dimensional frame, the conservation equation can be written as

$$F(x_2) - F(x_1) = -\frac{1}{dt} \int_{x_1}^{x_2} C \, dx \qquad (13)$$

where x_1 and x_2 are two points along the x-axis. Figure 6 illustrates the discretization of the space determined by the silicon and the silicon-dioxide materials. In the algorithm developed here, the SiO_2/Si interface always lies on a node, indexed I. Because the interface may be moving (under oxidation conditions) the value of I may be a function of time. It is also noteworthy that the nodal spacing in either material may be other than uniform.

The basic algorithmic constraint on the motion of the interface in our discrete space and time is that the interface may not move within one time step, Δt, by more than the distance Δx_I to the next node. Given that the amount of consumed Si within Δt, $\Delta Y(t, \Delta t)$ is a function of both t and Δt [Eq. (14)], the following two cases concerning the time evolution of the interface index can be distinguished.

Case 1.

If

$$\frac{\Delta x_I}{2} < \Delta Y(t \, \Delta t) \leq \Delta x_I$$

for a Δt such that

$$0 < \Delta t \leq \Delta t_{max}$$

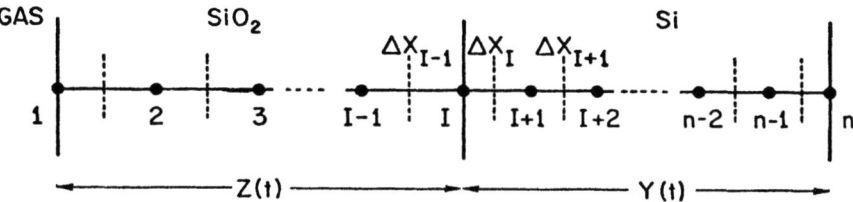

Fig. 6. THE DISCRETIZED SiO_2/Si REGION.

Numbers refer to nodes where the concentration is calculated. The dashed lines indicate the boundaries between the subvolumes associated with each node. The interface between the two media lies on node I. The flux exchanged between adjacent subvolumes is calculated at these boundaries. The node spacing Δx_i does not need to be uniform.

where Δt_{max} is a program/user maximum specified time step, the interface index is incremented by one $[I(t + \Delta t) = I(t) + 1]$.

Case 2.
　　If
$$\Delta Y(t, \Delta t_{max}) \leq \frac{\Delta x_I}{2}$$
the interface index is not incremented.

Note that, in both cases, the interface does move, but the interface node index is incremented only in Case 1. It can also be seen that, although the distance between nodes in Si may be generally uniform, this will often not be true for nodes in SiO_2.

In the models that describe the oxide growth and the resulting impurity redistribution, the oxide thickness, $Z(t)$, is determined by the standard equation [13,14]

$$Z(t)^2 + A\, Z(t) = B[t + \tau] \qquad (14)$$

where t is time, τ is related to the initial oxide thickness by

$$\tau = \frac{Z^2(0) + A\, Z(0)}{B} \qquad (15)$$

and A and B are related to the linear and parabolic growth coefficients K_L and K_P by

$$\begin{aligned} A &= K_P/K_L \\ B &= K_P \end{aligned} \qquad (16)$$

These coefficients are measurable quantities and depend on the crystal orientation and ambient conditions such as temperature, pressure and the presence or absence of H_2O [13].

The thickness of the consumed silicon, Y(t), is related to the thickness of the oxide grown by

$$Y(t) = \alpha\, Z(t) \qquad (17)$$

The factor α is fairly constant and equal to 0.44. Figure 7 illustrates the relationship between Z(t) and Y(t) at three equally spaced instances of time.

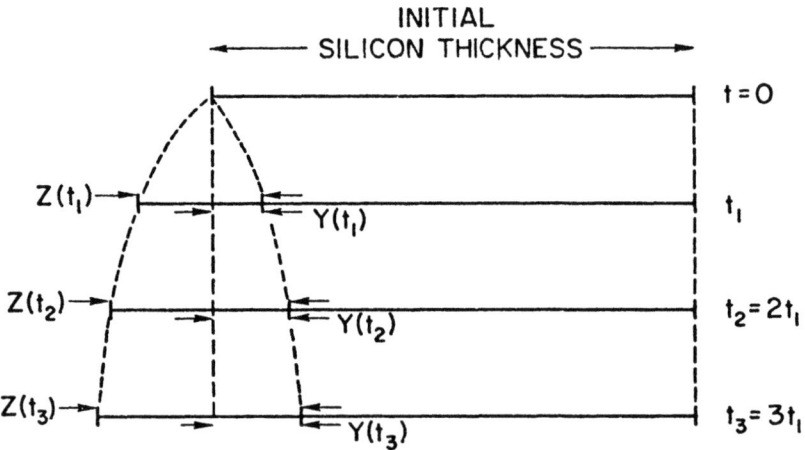

Fig. 7. Relationships between initial silicon thickness, oxide thickness, and oxidized silicon thickness, at three equispaced instances of time.

At the SiO_2/Si interface, the ratio of the impurity concentrations in the two materials is

$$m = \frac{C_{Si}}{C_{ox}}\bigg|_{boundary} \qquad (18)$$

where m is referred to as the thermodynamic segregation coefficient for the particular impurity species and is exponentially dependent on temperature. Segregation occurs because the chemical potential must be continuous (in thermal equilibrium) from one material (SiO_2) to the other (Si). Because of the nonunity values of α and m, a flux exists at the interface between SiO_2 and Si caused by the motion of the interface. Its relationship to α, m, and the velocity of the interface is derived below. Figure 8 illustrates a laboratory frame (in which the silicon is fixed) and a moving SiO_2 frame at two instances of time.

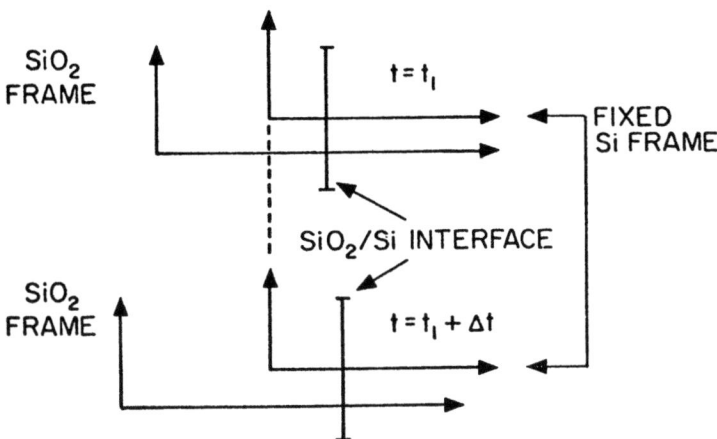

Fig. 8. MOVING SiO_2/Si INTERFACE AT TWO INSTANCES IN TIME.
Showing relationship of interface to a fixed (Si) frame and a moving (SiO_2) frame.

Because the SiO_2 frame recedes from the moving interface, the velocity of the moving boundary is V_{ox} for an observer on the SiO_2 frame and V_{Si} for an observer on the stationary laboratory frame,

$$V_{ox} = \frac{dZ(t)}{dt}$$

$$V_{Si} = \frac{dY(t)}{dt} \tag{19}$$

therefore, $V_{ox} = \alpha V_{Si}$ [using Eq. (17)]. As a result, for the SiO_2 observer, there is a flux F_{ox} emerging from the moving boundary, given by

$$F_{ox} = C_{ox} V_{ox} \tag{20}$$

and, for the observer in the Si, there is a flux lost into the moving boundary given by

$$F_{Si} = C_{Si} V_{Si} \tag{21}$$

Generally, these two fluxes are not equal; their difference is a net flux flowing into SiO_2 or Si and is defined as

$$F(b) = F_{ox} - F_{Si} = C_{ox} V_{ox} - C_{Si} V_{Si} \tag{22}$$

and, from Eqs. (17) and (18),

$$F(b) = \left(\frac{1}{m} - \alpha\right) V_{ox} C_{Si} \tag{23}$$

Clearly, F(b) must be supplied by a diffusive flow of impurities in the Si, because there is no generation of impurities at the interface. This is illustrated in Figure 9, where it can be observed that the impurities in the Si near the interface have a positive slope. The flux, F(b), is included in the total flux terms of Eq. (13).

In the simplest form (first order in time), Eq. (13) can be rewritten as

$$F_{i+1/2}^n - F_{i-1/2}^n = \frac{1}{\Delta t}\left(Q_i^n - Q_i^{n-1}\right) \tag{24}$$

where

$$Q_i^n = \int_{-1/2(\Delta x_{i-1}^n)}^{1/2(\Delta x_i^n)} C_i^n \, dx$$

F is the flux across the cell boundary and the upper indices identify the time increment and the lower ones identify the nodes in space. The exact form of F and Q in terms of C^n, C^{n-1} depends on the particular numerical approximation chosen (for example, first order in time for F and trapezoidal integration in space for Q). For cells lying either entirely in Si or in SiO_2, where flux is purely diffusive, Eq. (9) thus becomes

$$D_{i+1/2} \frac{C_{i+1}^n - C_i^n}{\Delta x_i} - D_{i-1/2} \frac{C_i^n - C_{i-1}^n}{\Delta x_{i-1}} = \frac{1}{\Delta t}\left(Q_i^n - Q_i^{n-1}\right) \tag{25}$$

In this equation, the moving boundary does not appear. The effect of the moving interface is included explicitly only in the flux at the boundary between the cell that contains the interface and its neighboring cell in the SiO_2.

Denoting the index of the interface node by I, the flux exchanged by the two cells is

$$F_{I-1/2} = D_{I-1/2} \frac{\frac{1}{m} C_I^n - C_{I-1}^n}{\Delta x_{I-1}} + \frac{1}{2}\left(\frac{1}{m} - \alpha\right) v_{ox}^n C_I^n \tag{26}$$

where

$$v_{ox}^n = \frac{\Delta x_{I-1}}{\Delta t}$$

The reason for including only a portion of the net flux resulting from the moving interface [see Eq. (23)] is that this flux flows across the cell boundary only after the interface has moved beyond it. For the order of the numerical approximations used here, this portion is exactly 1/2.

When the oxide growth is such that the interface index I does not need to be incremented, the moving interface flux does not appear. In other words, the nondiffusive flux term is present in Eq. (26) only if the moving interface crosses a flux boundary during the course of the time increment for which the problem is solved.

The concentration stored at the interface node belongs to the silicon side of the interface. When the concentration in the SiO_2 side is required, as in the diffusive term in Eq. (26), use is made of Eq. (18) (the silicon-side concentration is divided by the segregation coefficient, m).

The SUPREM simulated results have been compared to available analytic solutions of the moving boundary oxidation problem; Figure 15 is one such example. Here, 1000 Å of oxide were grown on a uniformly doped silicon wafer. It was assumed that the growth rate was purely parabolic. The SUPREM results were compared to the analytic solutions obtained by Av Ron et al. [17], and agreement was excellent after the initial oxidation phase and after the subsequent annealing phase.

3.2.6 <u>Other boundary conditions</u>. In addition to the SiO_2/Si interface boundary, there exist two more boundaries of the simulated space. In SUPREM, the boundary conditions (at these two boundaries) are determined as fluxes. The outer surface condition is either zero flux (reflecting) if the surface material is SiO_2 or, if the surface is Si, a flux determined by the following equation

$$F_{surface} = h(C^* - C_{surface}) \tag{27}$$

where h is a surface velocity or evaporation coefficient, and C* is either the solid solubility of the impurity or the gas-phase impurity concentration, depending on the type of the process step. Because of the lack of experimental knowledge of h, no attempt has been made to assure accurate default values; "h" has been internally set to 1 µm/sec to ensure fast relaxation of the surface concentration to C*.

The other boundary lies inside the silicon substrate at the point where the simulated space terminates. The condition at this point is always zero flux. Because the depth of simulation is user-determined, care must be taken to ensure that the presence of a reflecting boundary at that point does not affect the simulation results.

3.2.7 <u>Particle conservation and numerical errors</u>. The presence of reflecting boundary conditions at the two ends of the simulated space allows easy verification of total charge conservation

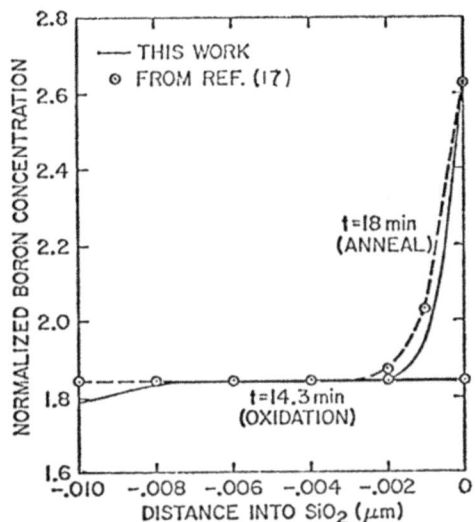
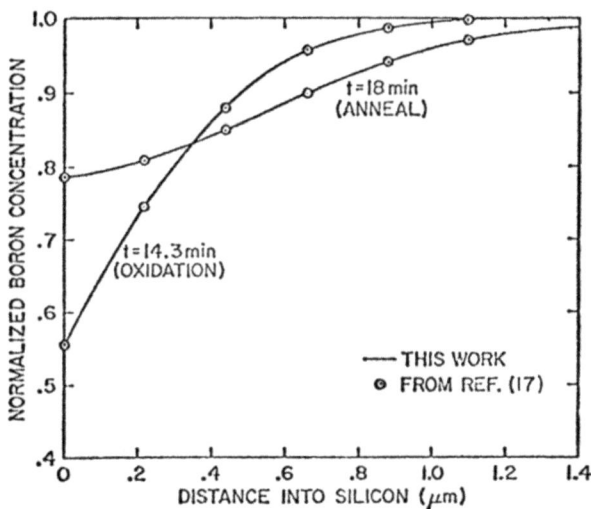

Fig. 9. IMPURITY DISTRIBUTION IN OXIDE AND SILICON FOLLOWING OXIDATION OF A UNIFORMLY BORON DOPED SAMPLE AND FOLLOWING AN 18 MIN POSTOXIDATION ANNEAL.

Oxide growth rate was assumed purely parabolic.

by the algorithm under many conditions. This conservation has been found to be accurate to within machine precision over a wide range of impurity profiles, provided the following condition holds:

$$\theta = \frac{D \Delta t}{\Delta x^2} \lesssim 1 \qquad (28)$$

where D is the highest diffusivity, Δt is the time step, and Δx is the smallest spatial step. The simulation results are acceptable, however, even when $\theta \simeq 100$. It is clear that the higher θ is, the smaller the required computation time is for a process step of a given simulated time and space length.

As θ becomes large, the numerical error introduced is a round-off error and depends on the number of bits used to represent real numbers in the computer. Nevertheless, it must be noted that perfect particle conservation does not imply a perfect solution because there is no guarantee on the magnitude of the discretization errors (both spatial and time) introduced by the discrete representation of continuous phenomena. These errors typically increase as both Δt and Δx increase.

3.3 Epitaxy

3.3.1 Introduction. Epitaxy is a processing step commonly used in modern semiconductor fabrication technologies. As with oxide growht, the epitaxial silicon growth step defines a moving boundary numerical problem. The objective of SUPREM is the simulation of impurity redistribution occurring during this step. The subprogram that performs this simulation is called EPTAX.

The model for impurity redistribution is similar to that reported by Langer and Goldstein [18]. Although redistribution caused by diffusion and evaporation is well simulated, front-side autodoping is omitted. Several plausible models have been proposed [19,20] that can reproduce this autodoping phenomenon, but knowledge of the quantitative physical processes involved is very limited. Further investigative work is necessary before a reliable front autodoping model can be incorporated in the process simulator.

Backside autodoping has also been neglected to conform with the one-dimensional constraint imposed on SUPREM. This, however, should not be a severe limitation because backside autodoping is often avoided either by backside sealing or by the use of ion-implantation on lightly doped substrates. On the other hand, frontside autodoping is a nonnegligible effect, particularly when lightly doped layers are grown on heavily doped ones.

3.3.2 **The model.** Figure 10 illustrates the epitaxial growth model used in SUPREM. It is assumed that the bulk gas phase has a uniform concentration of dopant, C_m. The dopant concentration, C_{gI}, in the gas phase at the surface of the silicon is simply related to C_m by

$$C_{gI} = k_g C_m \tag{29}$$

where k_g is an arbitrary constant that depends on the gas velocity distribution in the boundary layer [18]. The actual physical details of the boundary layer are not considered.

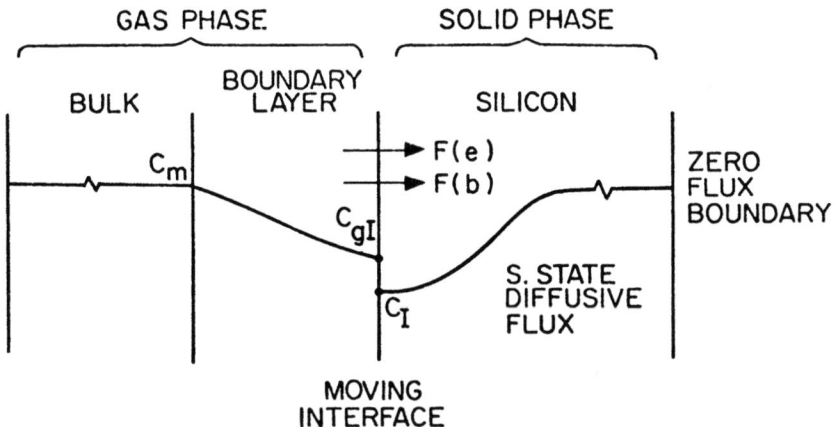

$F(e)$ = Impurity "evaporation" flux

$F(b)$ = Moving interface-induced flux

Fig. 10. Model for the redistribution of impurities during epitaxial silicon growth.

At the solid-gas interface, there exist two flux components, $F(e)$, and $F(b)$, as indicated in Figure 10. $F(e)$ is an impurity "evaporation" flux,

$$F(e) = h(kC_{gI} - C_I) \tag{30}$$

where h is a restriction coefficient with units of velocity commonly referred to as an "evaporation coefficient." As evaporation flux, we define any impurity flux exchange between the gas and the solid, other than direct moving boundary incorporation; C_I is the impurity concentration at the solid surface, and k is the equilibrium segregation coefficient defined as

$$k = \frac{C_I}{C_{gI}}\bigg|_{equilibrium} \qquad (31)$$

The term F(b) is a flux induced by the interface motion and can be calculated as in the oxidation case (Section 3.2.5). It is given by

$$F(b) = V(C_{gI} - C_I) \qquad (32)$$

where V is the interface velocity (growth rate).

Within the solid (silicon) body, diffusive flow of impurities is accounted for as has been discussed in previous sections. The simulated solid-phase space is terminated at a point where the diffusive flux can be assumed negligible.

3.3.3 **Numerical model implementation.** As with oxidation-diffusion, discussed in Section 3.2, the epitaxial growth-diffusion algorithm has been derived from the impurity conservation equation [Eq. (13)]. Special treatment of fluxes is necessary only at the moving gas/solid interface. Again the algorithmic constrain is imposed -- the interface must always lie on a spatial grid node and, during any simulated time increment t, no more than one new node of solid can be generated. This is illustrated in Figure 11. The solid vertical lines identify the gas/solid interface, and the broken lines delineate the discrete subcell boundaries across which the fluxes flow. The concentration in each subcell is considered uniform with value equal to the concentration at the grid node.

The numerical method used for the solution of the conservation equation is implicit, second order in time (Crank-Nicolson), and uses midpoint integration in space. The procedure for the solution of the resulting system of equations from one time increment to the next can be best understood by referring to Figure 11.

(a) At time t^{n-1}, the concentrations in the solid and the gas (near the interface) are known. These concentrations are either the initial conditions or the results of the solution up to the simulated time t^{n-1} (Figure 11a).

(b) Still at time t^{n-1}, an intermediate step (Figure 11b) is taken at which the contents of the two cells near the interface are rearranged. Specifically, the concentration at the new node, i-1, to be added during the next time increment Δt, is fixed at the gas concentration C_{gI}, and the concentration at the interface node, i, is modified to account for particle conservation in the ith subcell. The second operation is necessary because one nodal concentration value can be used to account for the content in each subcell. The results of this intermediate step serve now as initial conditions for the next time increment.

(c) Time is incremented to $t^{n-1}+ \Delta t$, and the system of coupled impurity conservation equations for each subcell is solved. The interface has now advanced by one grid node. This is shown in Figure 11c. From this point on, the cycle of operations is repeated as just described.

In setting up the system of equations, step (c) above, the two interface fluxes, described in Section 3.3.2, are also included; however, although the evaporation flux, (e), flows out of or into the composite system shown in Figure 11c, the moving boundary flux F(b) does not; F(b) only results in rearrangement of the contents of the (i-1)th and i^{th} cells in going from the situation shown in Figure 11b to that in 11c. This is a key point if particles are to be fully conserved by the algorithm,

The equations for the impurity conservation in the first two cells from time t^{n-1} to t^n becomes

$$F(D)_{i-1/2} - F(e)_{i-1} + \frac{1}{2}F(b)_{i-1/2} = \frac{1}{\Delta t}\left(Q_{i-1}^n - Q_{i-1}^{n-1}\right) \quad (33)$$

$$F(D)_{i+1/2} - F(D)_{i-1/2} - \frac{1}{2}F(b)_{i-1/2} = \frac{1}{\Delta t}\left(Q_i^n - Q_i^{n-1}\right) \quad (34)$$

where F(D) identifies the diffusive flux and the remaining symbols and indexing have been defined earlier in this section and in Section 3.2.5.

3.4 Ion-implantation model

Experimental distributions for many ions, like boron and arsenic, are found to be asymmetrical. The gaussian approximation of

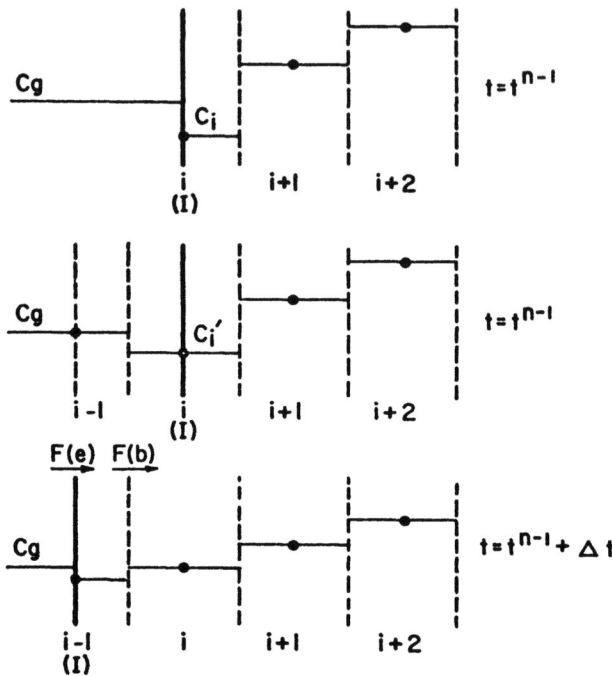

Fig. 11. DISCRETIZATION OF SPACE NEAR THE SOLID/GAS INTERFACE DURING EPITAXIAL GROWTH.

Solid vertical line is the interface and the broken lines are the discrete cell boundaries. Indices refer to nodes; I is the interface index. Horizontal levels indicate concentrations.

implanted impurity profiles is inadequate so that odd-order moments must be used to construct range distributions. Gibbons and Mylroie [21] have shown that the third central moment alone provides sufficient information to construct accurate distributions when the asymmetry is not too extreme. In these cases, the distribution can be represented by two half-gaussian profiles, each with a different standard deviation, σ_1 and σ_2, joined together at a modal range R_m as shown in Figure 12.

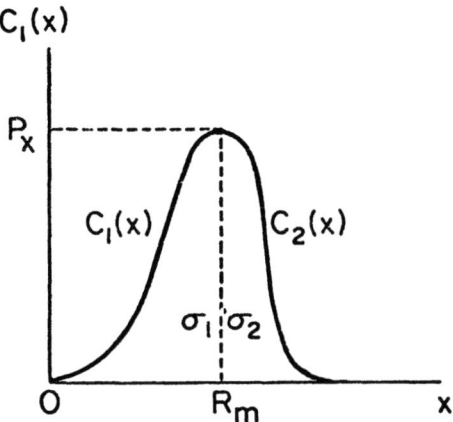

Fig. 12. The joint half-gaussian representation of the implanted ion profile.

The two sides of the distrubution are given by

$$C_1(x) = P_x \exp\left[-(x - R_m)^2/2\sigma_1^2\right] \qquad 0 \leq x \leq R_m \qquad (35)$$

$$C_2(x) = P_x \exp\left[-(x - R_m)^2/2\sigma_2^2\right] \qquad R_m < x \leq \infty \qquad (36)$$

In SUPREM, the values for σ_1, σ_2, and R_m are obtained from the Gibbons-Mylroie algorithm which makes use of the projected range, projected standard deviation, and third-moment ratio. These last three values are obtained by interpolation from a look-up table for each element in silicon or silicon dioxide as a function of implant energy [22]. In addition to the acceleration energy, the normal way of specifying an implant is by selecting the ion dose. As a result, a relationship between the dose and peak concentration must be used. Defining the normalized ion dose by

$$Q(\sigma,R,a,b) = \int_a^b \exp\left[-(x - R)^2/2\sigma^2\right] dx \qquad (37)$$

the peak concentration, P_x, and the total dose, Q_D, are related by

$$P_x = Q_D / Q(\sigma_1, R_m, 0, R_m) + Q(\sigma_2, R_m, R_m, \infty) \qquad (38)$$

In implantation through silicon dioxide, two sets of σ_1, σ_2, and R_m must be used in the two regions. Becuase the stopping power of silicon dioxide is different from that of silicon, the modal range in silicon must be modified to account for this difference in the following approximate way,

$$R'_{mSi} = R_{mSi} + (1 - R_{PSi}/R_{Pox}) Z_{ox}$$

where Z_{ox} is the silicon dioxide thickness, R'_{mSi} and R_{mSi} are the corrected and the uncorrected values of the silicon modal range, and R_{PSi} and R_{Pox} are the silicon-dioxide projected ranges.

4. CONCLUSION

A program structure for simulating multi-step IC fabrication processes has been described. Details of algorithms for moving-boundary oxidation and epitaxy process models have been discussed. The program presently models one-dimensional ion-implantation, diffusion, oxidation and epitaxy. The diffusion models use first-order approximations for Fermi-level dependent effects. Thus, enhanced arsenic and boron diffusivity, as well as coupling of the two species is reasonably simulated. Model improvements are being investigated to provide for species coupling and anomalous diffusion of phosphorus. In addition, research is in progress at Stanford, on fundamental IC process models which will provide needed data concerning details of the process physics. A portion of this work has been described earlier. Accurate engineering models are definitely needed for epitaxial autodoping and second-order oxidation effects including two-dimensional simulation. The problems associated with multi-species diffusion will also require further work to minimize the cost and complexity of simulation and yet retain the essential physics.

REFERENCES

1. D.A. Antoniadis, S.E. Hansen, R.W. Dutton, A.G. Gonzalez, and M. Rodoni, "SUPREM I -- A Program for IC Process Modeling and Simulation," SEL-77-006, Stanford Electronics Laboratories, Stanford University, Stanford, California, May 1977.

2. R. B. Fair, J. Electrochem. Soc. (122), 6, Jun 1975, p. 800.

3. K. Lehovec and A. Slobodskoy, Solid-State Electron. (3), 1961, p. 45.

4. J.S. Makris and B. J. Masters, J. Appl. Phys. (42), 1971, p. 3750.

5. R.B. Fair, Proc. of Third International Symp. on Silicon Materials Science and Technology (77-2), The Electrochem. Soc., May 1977, p. 968.

6. W.K. Hofker, H.W. Werner, D.P. Oosthoek, and H.A.M. de Grefte, J. Appl. Phys. (2), Springer-Verlag, 1974, p. 125.

7. R.L. Minear, D.G. Nelson, and J.F. Gibbons, J. Appl.Phys. (43), 1972, p. 3468.

8. A. Chu and J.F. Gibbons, Proc. of Fifth International Conference on Ion Implantation in Semiconductors and Other Materials, Plenum Press, New York, Aug 1976.

9. R.B. Fair and J.C.C. Tsai, paper presented at meeting of Electrochem. Soc., Las Vegas, 1976 (to be published).

10. F. Morehead, private communication.

11. J.F. Ziegler, G.W. Cole, and J.E.E. Baglin, Appl. Phys. Lett. (21), 1972, p. 177.

12. R.B. Fair, J. Electrochem. Soc. (44), 1, Jan 1973, p. 283

13. B.E. Deal and A.S. Grove, J. Appl. Phys. (36), 1965, p. 3770.

14. R.W. Dutton and D.A. Antoniadis, Oxidation, in this report.

15. J.L. Prince and F.N. Schwettmann, J. Electrochem. Soc.(121), 5, May 1974, p. 705.

16. R.D. Richtmyer "Difference Methods for Initial Value Problems", Interscience, New York and London, 1957.

17. M. Av-Ron, M. Shatzker and P.J. Burkhardt, J. Appl. Phys. (47), 7, Jul 1976, p. 3159.

18. P.H. Langer and J.I. Goldstein, J. Electrochem. Soc. (121), 4, Apr 1974, p. 563.

19. D. Kahng, C.O. Thomas and R.C. Manz, J. Electrochem. Soc. (110), 5, May 1963, p. 394.

20. R.W. Dutton and D.A. Antoniadis, Epitaxy, in this report.

21. J. Gibbons and S. Mylroie, Appl. Phys. Letts. (22), Jun 1973, p. 568.

22. Mayer, Erihasen and Davies, Ion Implantation in Semiconductors, Academic Press, New York, 1974.

PARTICIPANTS

ABELARD C., Motorola, Toulouse, France
ALLMAN P., University College Swansea, Swansea, England
AMON S., University of Ljubljana, Ljubljana, Yugoslavia
ANTOGNETTI P., Universita di Genova, Genova, Italy
ANTONIADIS D., Stanford University, Stanford, U.S.A.
AUBERT A., Thomson-CSF, Gennevilliers, France
BAETEN R.L., Carborandum, Overijse, Belgium
BENDEKOVIC Z., RIZ, Zagreb, Yugoslavia
BOCKEL C., CIT-Alcatel, Vélizy, France
BORGES TEIXEIRA P., Instituto Superior Tecnico, Lisboa, Portugal
BOUSSE L., Vrije Universiteit Brussel, Brussel, Belgium
BRAND T., Philips N.V., Nijmegen, The Netherlands
CASCO L.M., Fabrica Espanola Magnetos S.A., Madrid, Spain
CAVALIER C., LTT, Conflans-Ste-Honorine, France
CHARIL J.P., AOIP, Paris, France
CHAUDHARI P., IBM Corporation, Poughkeepsie, U.S.A.
CONTE G., Politecnico di Torino, Italy
DARWISH M., Centre Electronique Horloger, Neuchâtel, Switzerland
DESTINE J., Université de Liège, Liège, Belgium
DRECKMANN U., IBM, Boeblingen, West-Germany
ENTENMANN W., Technische Universität München, München, West-Germany
ERIKSEN K., Norwegian Defence Research Establishment, Kjeller, Norway
FAGG S., Mullard Research Laboratories, Redhill Surrey, England
FICHTNER W., Technical University Wien, Wien, Austria
GIEBEL B., Siemens A.G., München, West-Germany
GLASL A., Siemens A.G., München, West-Germany
GOKHALE B., IBM, Hopewell Junction, U.S.A.
GUECKEL H., Siemens A.G., München, West-Germany
GUERIN F., IBM, Corbeil Essonnes, France
GULDBERG J., Technical University of Denmark, Lyngby, Denmark
HAMMAR C., Microwave Institute Foundation, Stockholm, Sweden
HEBER K., Fraunhofer-Gesellschaft, Freiburg-i-Br., West-Germany
HELWIG K., IBM, Boeblingen, West-Germany
HILLEN M., State University, Groningen, The Netherlands
HUGEN M., Philips N.V., Nijmegen, The Netherlands
JAKOBSSON L., Telefonaktiebolaget L.M., Ericsson, Tyresö, Sweden
KARLSEN D., Aksjeselskapet Mikro-Elektronikk, Horten, Norway

KAYIHAN I., Technical University, Istanbul, Turkey
KOSTKA A., AEG-Telefunken, Heilbronn, West-Germany
KUB F., University of Maryland, Baltimore, U.S.A.
KUMESAWA T., Sony Corporation, Hodogaya-ku, Japan
LAGOS A., Ecole Polytechnique Fédérale de Lausanne, Lausanne, Switzerland
LEBESNERAIS G., IBM France, Corbeil Essonnes, France
LEDUC P., RTC-La Radiotechnique-Compelec, Caen, France
LINDEMAN H., Philips N.V., Nijmegen, The Netherlands
MAHR von STASZEWSKI G., Rijksuniversiteir Centrum Antwerpen, Antwerpen, Belgium
MALDONADO C., Rockwell International Corporation, San Juan Capistrano, U.S.A.
MALMROS D., Topsil, Fr. Sund, Denmark
MANCK O., Technische Hochschule Aachen, Aachen, West-Germany
MANGAS J.J., C.I.A.Telefonica Nacional de Espana, Madrid, Spain
MARTIN P.Y., R.T.C.-La Radiotechnique-Compelec, Caen, France
MASETTI G., LAMEL Laboratory, Bologna, Italy
MATTHEUS W., Barco-Cobar Electronic, Kortrijk, Belgium
MEIJER G., Technische Hogeschool Delft, Delft, The Netherlands
MOLLERBERG R., ASEA-HAFO AB, Vällingby , Sweden
MORANDI C., Universita di Bologna, Bologna, Italy
NAHAB F.F., University of Kent, Canterbury, England
OKTER I., Technical University, Istanbul, Turkey
PATTERSON D.O., U.S. Naval Research Laboratory, Washington, U.S.A.
PEYKOV P., Institute of Solid State Physics, Sofia, Bulgaria
PIQUERAS J., Universidad Autonoma de Madrid, Madrid, Spain
POPP G., Brown, Boveri & Cie, Lampertheim, West-Germany
PROFUMO E., SGS-ATES, Agrate Brianza, Italy
RICHOU F., CNET, Lannion, France
RIDEOUT V.L., IBM, Yorktown Heights, U.S.A.
ROCCHI M., LEP, Limeil-Brévannes, France
RYSSEL H., Fraunhofer-Gesellschaft, München, West-Germany
SALEH M.N., Ain Shams University, Cairo, Egypt
SALSANO A., Universita di Roma, Roma, Italy
SCHMIDT P.E., Instituto Venezolano de Investigaciones Cientificas, Caracas, Venezuela
SCHRADER L., Siemens A.G., München, West-Germany
SCHRAS P., Philips N.V., Nijmegen, The Netherlands
SEEGEBRECHT P., Technische Hochschule Aachen, Aachen, West-Germany
SIDERIS M., Nuclear Research Center "Democritos", Athens, Greece
SKARELVEN H., Central Institute of Industrial Research, Oslo, Norway
SKARLATOS Y., Bogazici Universitesi, Istanbul, Turkey
SOMMER E., Universität Stuttgart, Stuttgart, West-Germany
SPAANENBURG L., Twente University of Technology, Twente, the Netherlands
SPEIGHT J., Post Office Research Centre, Ipswich, England
STEENHAUT O., Vrije Universiteit Brussel, Brussel, Belgium

STROEHLE D., SEL-AG, Stuttgart, West-Germany
SZABO A., University of Zagreb, Zagreb, Yugoslavia
TREMAIN R., Xerox Corporation, Palo Alto, U.S.A.
TRULLEMANS C., Bell Telephone Mfg. Co., Antwerpen, Belgium
URSIC S., RIZ, Zagreb, Yugoslavia
VAN HORN J.H., ITT Europe Inc., Brussels, Belgium
VAN RUMSTE M., Bell Telephone Mfg. Co., Antwerpen, Belgium
VANZANTEN A., Philips N.V., Eindhoven, The Netherlands
VIDIMARI F., Centro Informazioni Studi Esperienze, Milano, Italy
VIDKJAER J., Technical University of Denmark, Lyngby, Denmark
VON DEWITZ H., Siemens A.G., München, West-Germany
VOUNCKX R., Vrije Universiteit Brussel, Brussel, Belgium
WALDVOGEL U., Intermetall, Freiburg-im-Breisgau, West-Germany
WEISE E., FAVAG S.A., Bevaix, Switzerland
WILLMANN M., Technische Hochschule Aachen, Aachen, West-Germany
WRIGHT A.J., University of Manchester, Manchester, England
ZIMMER G., Universität Dortmund, Dortmund, West-Germany

LECTURERS

DE GRAAFF H.C., Philips N.V., Eindhoven, The Netherlands
DEMOULIN E., Université Catholique de Louvain, Louvain-la-Neuve, Belgium
DUTTON R., Stanford University, Stanford, U.S.A.
ENGL W., Technische Hochschule Aachen, Aachen, West-Germany
JESPERS P., Université Catholique de Louvain, Louvain-la-Neuve, Belgium
KLAASSEN F., Philips N.V., Eindhoven, The Netherlands
MERCKEL G., LETI, Grenoble, France
RUPPRECHT H., IBM, East Fishkill, N.Y., U.S.A.
SCHARFETTER D., Bell Labs, Murray Hill, N.J., U.S.A.
SCHROEN W., Texas Instruments Inc., Dallas, Texas, U.S.A.
WIDMANN D., Siemens A.G., München, West-Germany
WIEDER A., Technische Hochschule Aachen, Aachen, West-Germany

SCIENTIFIC ORGANIZING COMMITTEE

Prof. F. Van de Wiele, ASI Director, Université Catholique de
   Louvain, Louvain-la-Neuve, Belgium
Prof. W.L. Engl, Technische Hochschule Aachen, West-Germany
Prof. P. Jespers, Université Catholique de Louvain, Louvain-la-Neuve, Belgium

MIX
Papier aus verantwortungsvollen Quellen
Paper from responsible sources
FSC® C105338

If you have any concerns about our products,
you can contact us on
ProductSafety@springernature.com

In case Publisher is established outside the EU,
the EU authorized representative is:
**Springer Nature Customer Service Center GmbH
Europaplatz 3, 69115 Heidelberg, Germany**

Printed by Libri Plureos GmbH
in Hamburg, Germany